PLANE-WAVE THEORY OF TIME-DOMAIN FIELDS

IEEE PRESS SERIES ON ELECTROMAGNETIC WAVE THEORY

The IEEE Press Series on Electromagnetic Wave Theory consists of new titles as well as reprintings and revisions of recognized classics that maintain long-term archival significance in electromagnetic waves and applications.

BOOKS IN THE IEEE PRESS SERIES ON ELECTROMAGNETIC WAVE THEORY

Chew, W. C., *Waves and Fields in Inhomogeneous Media*
Christopoulos, C., *The Transmission-Line Modeling Methods: TLM*
Clemmow, P. C., *The Plane Wave Spectrum Representation of Electromagnetic Fields*
Collin, R. E., *Field Theory of Guided Waves,* Second Edition
Dudley, D. G., *Mathematical Foundations for Electromagnetic Theory*
Elliott, R. S., *Electromagnetics: History, Theory, and Applications*
Felsen, L. B., and Marcuvitz, N., *Radiation and Scattering of Waves*
Harrington, R. F., *Field Computation by Moment Methods*
Hansen, T. B., and Yaghjian, A. D., *Plane-Wave Theory of Time-Domain Fields: Near-Field Scanning Applications*
Jones, D. S., *Methods in Electromagnetic Wave Propagation*, Second Edition
Lindell, I. V., *Methods for Electromagnetic Field Analysis*
Peterson et al., *Computational Methods for Electromagnetics*
Tai, C. T., *Generalized Vector and Dyadic Analysis: Applied Mathematics in Field Theory*
Tai, C. T., *Dyadic Green Functions in Electromagnetic Theory*, Second Edition
Van Bladel, J., *Singular Electromagnetic Fields and Sources*
Volakis et al., *Finite Element Method for Electromagnetics: Antennas, Microwave Circuits, and Scattering Applications*
Wait, J., *Electromagnetic Waves in Stratified Media*

PLANE-WAVE THEORY OF TIME-DOMAIN FIELDS

Near-Field Scanning Applications

IEEE Press Series on
Electromagnetic Wave Theory

Thorkild B. Hansen
Schlumberger-Doll Research
Ridgefield, Connecticut

Arthur D. Yaghjian
Visiting Scientist
Air Force Research Laboratory
Hanscom AFB, Massachusetts

IEEE Antennas & Propagation Society, *Sponsor*

IEEE
PRESS

The Institute of Electrical and Electronics Engineers, Inc., New York

This book and other books may be purchased at a discount
from the publisher when ordered in bulk quantities. Contact:

IEEE Press Marketing
Attn: Special Sales
Piscataway, NJ 08855-1331
Fax: (732) 981-9334

For more information about IEEE PRESS products,
visit the IEEE Press Home Page: http://www.ieee.org/organizations/pubs/press

Printed in the United States of America

10 9 8 7 6 5 4 3 2 1

ISBN 0-7803-3428-0
IEEE Order Number: PC5702

Library of Congress Cataloging-in-Publication Data

Hansen, Thorkild, 1965–
Plane-wave theory of time-domain fields: near-field scanning applications / Thorkild B. Hansen, Arthur D. Yaghjian.
p. cm. -- (IEEE Press series on electromagnetic wave theory)
"IEEE Antennas & Propagation Society, sponsor."
Includes bibliographical references and index.
ISBN 0-7803-3428-0
1. Antennas (Electronics)--Mathematical models. 2. Electromagnetic fields--Mathematical models. 3. Time-domain analysis. 4. Green's functions. 5. Electro-acoustic transducers--Mathematical models. I. Yaghjian, Arthur D. II. IEEE Antennas and Propagation Society. III. Title. IV. Series.
TK7871.6.H352 1999
621.382'4--dc21 99-17759
CIP

A. D. Yaghjian dedicates this book to the memory of his father,

Nubar Yaghjian

IEEE Press
445 Hoes Lane, P.O. Box 1331
Piscataway, NJ 08855-1331

Contents

Preface

We began the research that led to this book with the sole intent of formulating probe-corrected planar near-field measurement techniques for electromagnetic antennas (and electroacoustic transducers) that are excited by wide-band pulses rather than by a single-frequency (or narrow-band) signal. As the title of the book indicates, however, the scope of the research expanded beyond its initial intent and finally encompassed a general plane-wave theory for time-domain fields. In working toward the initial goal of formulating time-domain measurement techniques, we found it necessary to systematically derive Green's function formulas and plane-wave spectrum representations in the frequency domain, and to keep the frequency dependence of all the expressions explicit. This systematic development of frequency-domain representations for electromagnetic and acoustic fields is contained in Chapter 3 and the corresponding representations for static electric and magnetic fields are given in Chapter 4. Chapter 5 develops the Green's function and plane-wave representations for time-domain fields, and investigates various wave phenomena, such as "electromagnetic missiles," encountered only in the time domain. Within Chapters 3 and 5 are the formulas needed for planar near-field measurements without probe correction in the frequency domain and time domain, respectively.

With the help of the plane-wave spectrum representations of Chapter 3, the formulas required for the probe-corrected planar near-field measurement of antennas and electroacoustic transducers in the frequency domain are derived in Chapter 6. In Chapter 7 we derive the analogous time-domain, probe-corrected planar near-field measurement formulas that apply to pulsed electromagnetic and acoustic radiators. Chapter 8 contains the sampling theorems, reconstruction formulas, and computation schemes that determine the space-time sampling increments required to accurately and efficiently evaluate far fields from near-field data.

The plane-wave theory of time-domain and frequency-domain fields in Chapters 3 through 8 rests on the fundamental classical equations of electromagnetics and linear acoustics. Thus, Chapter 2 introduces these fundamental equations for time-domain and frequency-domain fields. We have made a special effort to develop linear acoustics along the same lines as electromagnetics, since linear acoustics is usually treated as an addendum to fluid mechanics or general acoustics. Chapter 2 also includes a great deal of background material that is either new or omitted in the classic textbooks. With Chapter 2 especially, we found that the writing fostered more research, which in turn led to a longer chapter than we had originally anticipated.

The final product, however, may be used as the basis for a graduate course in electromagnetics and linear acoustics.

We have not included a chapter that discusses either the practical implementation of planar near-field measurements in the time domain or the results obtained from time-domain near-field measurements in the laboratory. To date, there simply have not been enough near-field measurements taken in the time domain to warrant such a chapter. The implementation of time-domain near-field measurements in the laboratory has been hindered in the past by the need for a rigorous probe-corrected formulation in the time domain. We hope this book will fill this need, and, in so doing, not only develop the general plane-wave theory of frequency-domain and time-domain fields, but also help advance the science and utility of near-field measurements.

Thorkild B. Hansen
Arthur D. Yaghjian

Acknowledgments

We are indebted to the staff of the Research Laboratory and Research Library at Hanscom Air Force Base, MA, for providing the facilities, interactions, and expertise conducive to the completion of the book. We are grateful to Arje Nachman of the U.S. Air Force Office of Scientific Research, Arlington, VA, for supporting and encouraging the research that forms the basis of the book. We would also like to thank (in alphabetical order) the people who reviewed parts of the manuscript and made thoughtful suggestions during various stages of its preparation: A.J. Devaney, A.K. Dominek, E. Heyman, A.T. de Hoop, P.M. Johansen, J.M. Meyers, E.B. Miller, T.M. Roberts, and R.C. Wittmann. Thorkild B. Hansen is grateful for the financial support he received from the Danish Technical Research Council during the initial stages of the work. Arthur D. Yaghjian is deeply indebted to Lucretia B. Yaghjian, who so generously accepted the extra responsibilities that her husband neglected while coauthoring this book.

Thorkild B. Hansen
Arthur D. Yaghjian

1

Introduction

Plane-wave expansions have proven useful for solving numerous problems involving the radiation, reception, propagation, and scattering of electromagnetic and acoustic fields. Several textbooks and monographs have dealt with the subject of expanding fields in a spectrum of plane waves [1]–[10]. However, most of the previous work with plane-wave representations has been confined to the frequency domain, and relatively little work has been devoted to deriving general plane-wave representations for time-domain fields. In geophysics, Schultz and Claerbout [11], Chapman [12], [13], Phinney et al. [14], and Stoffa et al. [15] have used time-domain Radon transforms and "slant stacks" to analyze seismic reflection and refraction data. Tygel and Hubral [16] developed a plane-wave theory for transient acoustic waves in layered media. Heyman et al. [17]–[19] presented plane-wave expressions for scalar transient fields, and Marengo and Devaney [20] derived time-domain angular spectrum expressions for electromagnetic fields. Our own previous contributions [21]–[26] to the theory and application of time-domain plane-wave representations form the nucleus of this book in which we develop a general plane-wave theory of electromagnetic and acoustic fields produced by sources that have arbitrary time dependence and are located in homogeneous, isotropic media. The resulting time-domain plane-wave spectrum representations have a variety of applications, ranging from the general decomposition and analysis of transient fields to the formulation of time-domain near-field measurement techniques.

A large body of literature exists that documents the development of near-field measurements in the frequency domain. The discovery by Kerns [27] in the early 1960s of a simple, rigorous method to correct for the effects of the measurement probe used in planar near-field scanning initiated the development of practical, accurate near-field measurement techniques. Since that time, the theory for probe-corrected near-field measurements has been successfully formulated and applied to the measurement of electromagnetic and acoustic radiators on planar [28]–[31], cylindrical [32], [33], and spherical [34]–[37] surfaces. Accurate experimental and computational techniques have been developed and implemented for near-field scanning [32], [38]–[51], and a number of books [8], [40], [52] and overviews [42], [53]–[55] have been published that document the modern development of near-field measurements and attest to its status as a mature science.

Because all this development of probe-corrected near-field measurement techniques during the past 35 years has been limited to the frequency domain, an important objective of this book is to formulate planar near-field measurement techniques for radiators that are excited by wide-band short pulses and must be measured in the time domain. However, the time-domain formulation of planar near-field scanning may also be applied to radiators that are designed to operate at many discrete frequencies. If such a radiator is excited by a pulse with a frequency bandwidth that covers all the discrete operating frequencies, the radiator can be measured in the time domain, and by means of the Fourier transform, its near and far fields can be determined in the frequency domain at each of the discrete operating frequencies. Alternatively, the radiator could be measured in the frequency domain at each of the discrete operating frequencies. Since the frequency response of a pulsed radiator can be determined from a single scan in the time-domain near field, time-domain near-field measurements can be used to determine the wide-band or out-of-band frequency response of radiators at a reduced measurement time [56], [57]. We find in Chapter 8 that time gating can be used with time-domain measurements to remove the errors caused by the finite size of the scan plane. These finite-scan errors can be prohibitively large for the frequency-domain near-field measurement of broadbeam radiators. Therefore, for broadbeam radiators it may be advantageous to excite the radiator by a relatively wide-band, short pulse and take near-field measurements in the time domain, even if the broadbeam radiator is to be used at a single frequency.

Although this book does not treat time-domain measurements on cylindrical and spherical surfaces, we mention the recent work in these areas. The theory, sampling theorems, and computation schemes for probe-corrected spherical near-field scanning of time-domain acoustic and electromagnetic fields are developed in references [58]–[61]. Preliminary results on the formulation of cylindrical scanning for scalar time-domain fields were presented in [62].

An outline of the material contained in each of the succeeding chapters of the book is given as follows.

Chapter 2 presents the basic electromagnetic and acoustic field equations that are used in Chapters 3 through 8 of the book. We also include as background material in Chapter 2, a number of derivations, theorems, and expressions that are new or not found in the classic textbooks on electromagnetics and acoustics. The time-domain and frequency-domain fields are treated in separate sections. For each domain, the electromagnetic field equations are given first, and the acoustic field equations are then developed along parallel lines. The basic time-independent (static) field equations are included as part of the time-domain sections.

In Chapter 2 we prove, with the help of causality and far-field conditions, the equivalence of the first- and second-order differential equations for the electromagnetic and acoustic fields. Rigorous uniqueness and existence theorems are also proven, and sufficient conditions on the fields and sources are found for the validity of surface and volume integral representations of the frequency-domain and time-domain fields. Expressions are derived for cavity-defined and mathematically defined fields in the source regions. The electromagnetic volume integral representations are applied to obtain the Liénard-Wiechert fields of accelerating point charges used in Chapter 5 for analyzing "electromagnetic missiles." Care is taken to derive unambiguous expressions for the force, power, momentum, and energy supplied by the electromagnetic and acoustic fields. In the last section of Chapter 2, the basic electromagnetic and acoustic field equations are extended to lossy dispersive media, and a direct method is used to derive the dispersion relations.

In Chapter 3, we derive the frequency-domain Green's function and plane-wave spectrum representations for the fields in the half space in front of an electromagnetic or acoustic radiator in terms of the fields in a scan plane. (In Chapter 3, the measurement probe is assumed to be ideal in the sense that it can be used to directly measure the required near fields in the scan plane.) The equations are given in a form that is convenient for their translation into the time domain in Chapter 5. The far fields and plane-wave spectra are expressed in terms of volume sources, Huygens' sources on a closed surface, and the fields in the scan plane. Relationships between the low-frequency behavior of the far fields and their sources are found from the volume-source expressions.

The frequency-domain plane-wave spectra of electromagnetic and acoustic sources are first derived in Chapter 3 without regard to mathematical rigor and are then verified rigorously in Appendix D. The plane-wave spectrum divides into propagating and evanescent parts, and explicit expressions are found for the propagating and evanescent contributions to the fields of an acoustic point source and to the far fields of arbitrary electromagnetic and acoustic sources. We determine the singularities and asymptotic behavior of the spectra, and we find that the spectra, the far-field patterns, and the fields outside the source region are analytic functions in certain complex domains of their arguments. Many of the plane-wave representations are rewritten in terms of the far-field patterns with complex angles of observation. Homogeneous fields and inverse source formulas are found that can be expressed throughout all space in terms of far-field patterns evaluated at real angles of observation only. The asymptotic behavior of the plane-wave spectrum is used to derive simple expressions for the smallest possible convex region that can contain the sources of a given far-field pattern. The power and radiation force transmitted through an infinite plane in front of the electromagnetic or acoustic radiator are expressed in terms of the plane-wave spectrum and far-field pattern. In the last section of Chapter 3, we determine the changes required in the preceding equations if the electromagnetic and acoustic sources radiate into lossy regions. A major difference between the equations in the lossless and lossy media is that the entire spectrum of plane waves, not just the evanescent part of the spectrum, is attenuated in lossy media.

In Chapters 3 and 6, which derive the frequency-domain formulas for planar near-field scanning with and without probe correction, respectively, the frequency is chosen positive ($\omega > 0$) because the fields, sources, and equations for negative frequencies can be found from those for positive frequencies simply through use of the reality condition ($f_{-\omega} = f_{\omega}^{*}$). However, many of the formulas and relationships in Chapters 3 and 6 do not hold for $\omega = 0$, that is, for time-independent fields. Moreover, many of the time-domain formulas and relationships in Chapters 5 and 7 dealing with time-domain planar near-field scanning with and without probe correction, respectively, do not hold for time-independent fields. Therefore, Chapter 4 is devoted to deriving Green's function and plane-wave spectrum representations for static electric and magnetic fields. Emphasis is placed on formulas that determine the total charge and dipole moments of the static sources, as well as on formulas that determine the static fields in a half space in front of the sources in terms of the static fields measured in a scan plane. Many of the electrostatic results in Chapter 4 can also be applied to the measurement of Newtonian gravitational fields.

In Chapter 5 the frequency-domain representations of Chapter 3 are translated into the time domain. The time-domain plane-wave representations are expressed in terms of real time-domain plane-wave spectra and fields, and, alternatively, in terms of complex analytic-signal spectra and fields. The analytic-signal formulas are simpler in form than the real time-domain formulas, and thus they prove useful for analysis. However, they involve complex functions of complex times, and, at least for planar near-field measurements, are ultimately more cumbersome to use for computations than the real time-domain formulas. Both the analytic-signal and real time-domain spectra are written as space-time Radon transforms of the fields in the scan plane.

The time-domain Green's function and plane-wave spectrum representations are derived in Chapter 5 for the acoustic and electromagnetic fields in the half space in front of the sources in terms of the time-domain fields measured in a scan plane. The time-domain far fields and plane-wave spectra are expressed in terms of the volume sources, the Huygens' sources on a closed surface, and the fields in the scan plane. Closed-form expressions are found for the separate contributions of the propagating and evanescent parts of the time-domain spectrum to the fields of an acoustic point source and a rectangular-pulse point source. Time-domain far fields are written in terms of their spectra, and the analytic properties of the far-field patterns and the spectra are determined. At times after all sources have been turned off, the time-domain fields everywhere can be expressed in terms of their far-field patterns integrated over the far-field sphere. Expressions are obtained in terms of the time-domain plane-wave spectra for the total acoustic and electromagnetic energy and momentum radiated into a half space. In addition to presenting the time-domain field expansions, Chapter 5 also derives a number of general theorems on the behavior of time-domain fields and investigates the phenomena of electromagnetic "bullets" and "missiles."

In Chapter 6, we derive the frequency-domain probe-corrected formulas that determine the acoustic and electromagnetic fields in the half space in front of the sources (test antenna) in terms of the output on a scan plane of an arbitrary measurement probe (with known receiving characteristic). A straightforward approach based on linearity is taken in the frequency-domain derivations that proves useful in the corresponding derivations in Chapter 7 of the time-domain probe-corrected formulas. With this approach it is not necessary to introduce the full scattering-matrix description of antennas and electroacoustic transducers [8], [29], or the Lorentz reciprocity theorem [30], to obtain the probe-corrected transmission formula first derived by Kerns.

The frequency-domain probe-corrected formulas for planar near-field measurements are also derived in Chapter 6 by a novel method based on the probe output satisfying the homogeneous Helmholtz equation. This latter result and its region of validity are rigorously proven for both acoustic and electromagnetic probes. The receiving characteristic for a probe is defined, and reciprocity relations are given for the plane-wave transmitting spectra and receiving characteristics of reciprocal electromagnetic antennas and electroacoustic transducers. General "system 2-port" reciprocity relations between the outputs of two reciprocal antennas are also found. For a reciprocal acoustic monopole or a reciprocal electric dipole, it is proven that the output of the probe becomes proportional to the field, and the probe-corrected formulas of Chapter 6 reduce to the non-probe-corrected formulas of Chapter 3.

The probe-corrected formulas for planar near-field scanning in the time-domain are obtained in Chapter 7 by first taking the inverse Fourier transform of the corresponding frequency-domain probe-corrected formulas in Chapter 6, then by deriving the formulas directly in the time domain, and finally by using the result that the time-dependent output of the measurement probe satisfies the time-dependent homogeneous wave equation. These general probe-corrected formulas are then shown to greatly simplify through the use of "time-derivative" probes. Reciprocity relations in the time domain are derived between the plane-wave transmitting spectrum and the receiving characteristic of a single reciprocal antenna, as well as between the outputs of any two reciprocal antennas excited by identical input pulses. As in the frequency domain, the time-domain probe-corrected formulas are shown to reduce to the non-probe-corrected formulas if the probe is a reciprocal acoustic monopole or a reciprocal electric dipole. Lastly, in Chapter 7, we consider the time-domain near-field scanning of antennas with a reciprocal probe whose far field is frequency independent over the measurement bandwidth.

To determine the time-domain near and far fields of an antenna from the time-domain output data of the near-field measurement probe, one can compute the required integrals over the scan plane directly in the time domain using the formulas of Chapters 5 and 7. Alternatively, one can compute the near and far fields of the antenna by Fourier transforming the time-domain near-field data to the frequency domain, evaluating the frequency-domain integrals over the scan plane found in Chapters 3 and 6, and finally Fourier transforming back into the time domain. Chapter 8 derives the sampling theorems and reconstruction formulas needed to adequately sample the near field and accurately evaluate the time-domain far fields using either of the two computation schemes. The advantages and disadvantages of both computation schemes are illustrated by applying the two schemes to compute the fields of simple time-domain acoustic and electromagnetic radiators. We compare the number of operations required by each computation scheme and the errors in the time-domain caused by the finite size of the scan plane. With either computation scheme, it is found that one of the advantages of time-domain near-field measurements is that time gating the output of the probe can be used to remove the errors caused by the finite size of the scan plane as well as by the multiple interactions between the probe and test antenna.

Throughout the book, we have used the International System of Units (mksA). We have consistently defined the Fourier transform of a time-domain function $f(t)$ as

$$f_\omega = \frac{1}{2\pi}\int_{-\infty}^{+\infty} f(t)e^{i\omega t}\,dt \tag{1.1}$$

so that $f(t)$ is given in terms of its frequency-domain function f_ω by the inverse Fourier transform

$$f(t) = \int_{-\infty}^{+\infty} f_\omega e^{-i\omega t}\,dt. \tag{1.2}$$

The choice of $e^{i\omega t}$ and the $1/(2\pi)$ factor in the Fourier transform of $f(t)$ given by (1.1) produces a time-harmonic function $f_\omega e^{-i\omega t}$ in (1.2) that conforms to the time-harmonic functions used in previous frequency-domain, plane-wave formulations of near-field measurement techniques [8]. However, f_ω has the dimensions of $f(t)$ multiplied by time t, and this fact must be taken into account when checking and comparing the dimensions of the frequency-domain and time-domain expressions.

2

Electromagnetic and Acoustic Field Equations

This chapter serves as a reasonably self-contained introduction to the electromagnetic and acoustic field equations that are used throughout the rest of the book. The definitions of electromagnetic and acoustic fields, as well as the basic equations they satisfy, are discussed in detail in many classic textbooks on electromagnetics and acoustics [1], [63]–[70]. Here we review only briefly the basics and concentrate on the less familiar derivations, theorems, and expressions that are useful directly or as background material in the analyses in the subsequent chapters of the book.

Except when Dirac delta functions ($\delta(t)$) are used, all the functions, differentiations, and integrations throughout the book are assumed to be defined in the ordinary classical sense, and, in particular, integral signs refer to ordinary Riemann integration. Although this is a mathematical limitation, it is not a practical limitation when integrating measured quantities, since measured quantities can presumably be approximated by bounded functions that are continuous or at worst have a denumerable set of discontinuities on closed finite intervals, and thus they are Riemann integrable [71, vol. 1, p. 468]. Delta functions are used nonrigorously in some of the derivations, but only if the result of the derivation, or the expressions in which a result containing delta functions is to be used, can be proven valid by classical means as well. This caution against the nonrigorous use of delta functions is especially observed here in Chapter 2 when deriving integral expressions for the electromagnetic fields and the acoustic velocity field in source regions. For these derivations, the cavalier application of nonrigorous delta-function methods fails to produce the correct integral expressions that are derivable from either classical analysis [72]–[74] or rigorous delta-distribution theory [75]. Also, in this chapter dealing with the basic equations, we have made an effort to state explicitly the conditions on the fields and sources that rigorously ensure the validity of the primary expressions and theorems.

All curves, surfaces, and volumetric regions (which we shall simply call "volumes") of integration in this chapter are assumed "regular" [76, ch. IV]. Regular curves and surfaces are bounded (i.e., can be enclosed by a sphere of finite radius). A regular volume is either a bounded closed region of space whose boundary is a closed regular surface [76, ch. IV] or an infinite region of space outside a closed regular surface [76, ch. VIII, sec. 3]. The precise definition of a regular curve and surface can be found in Kellogg [76, ch. IV]. It suffices here to note that the Riemann integration of piecewise continuous functions can be defined for regular curves, surfaces, and volumes. Regular surfaces include ordinary open and closed surfaces such as circles, polygons, polyhedrons, spheres, cylinders, cones, and toroids.

Throughout the rest of the book, all curves, surfaces and volumes of integration are regular except for the infinite plane, which cannot strictly be called a regular surface according to Kellogg's definition because it is an unbounded surface. However, the definition of a regular surface can be extended to include the infinite plane in the same way that Kellogg extended the definition of regular volumes to include infinite volumes [76, ch. VIII, sec. 3].

Here in Chapter 2, unlike Chapters 3, 5, 6, 7, and 8, we begin with the electromagnetic field equations rather than the acoustic field equations. Although we develop the basic electromagnetic and acoustic equations along parallel lines, the lines of development we use are more commonly found in electromagnetics than acoustics, and thus the electromagnetic equations are presented first.

Also, here in Chapter 2, the time-domain electromagnetic and acoustic equations are introduced before the frequency-domain equations to emphasize the fundamental nature of the time-domain equations. Afterward, the frequency-domain equations are found, often by taking the Fourier transform of the time-domain equations. Even though many of the derivations in the frequency domain are streamlined or omitted altogether, sufficient conditions are still given for the validity of the primary resulting frequency-domain expressions and theorems.

Throughout the book, the term *frequency domain* refers to complex time-harmonic quantities with the $e^{-i\omega t}$ factor suppressed. We use the term *time domain* to refer to time-dependent quantities that are zero for time $t \leq t_0$. However, time-independent equations and fields are also included here in the time-domain sections of Chapter 2. Time-independent equations and fields are not included in the frequency-domain sections of Chapter 2, where frequency-domain fields are considered for all frequencies ω except $\omega = 0$. Moreover, as explained in Sections 2.3.1 and 2.3.2, the reality condition allows us to consider only positive frequencies ($\omega > 0$) throughout most of the frequency-domain sections of the rest of the book.

Green's function and plane-wave representations for static electric and magnetic fields are derived in Chapter 4. The term *statics* is reserved for the time-independent electromagnetic equations and is not used to refer to the time-independent acoustic equations, because "statics" in fluid mechanics usually implies zero fluid velocity. In the acoustics equations of this chapter, we allow time-independent fields to have nonzero fluid velocity.

Most of the plane-wave theory and applications contained in the book are formulated for electromagnetic sources that radiate into lossless free space and for acoustic sources that radiate into lossless fluids. However, many of the results are readily extendible to lossy media, as exemplified in Section 3.3. Thus, in the last section of this chapter, basic electromagnetic and acoustic equations are given for sources located in lossy media.

2.1 TIME-DOMAIN ELECTROMAGNETIC FIELD EQUATIONS

The basic time-domain Maxwell equations are initially presented in their traditional first-order differential form and then in a form involving integrals over time rather than derivatives with respect to time. We use this latter form of Maxwell's equations to prove more general uniqueness and existence theorems that can be applied to rapidly varying fields, such as those of electromagnetic "missiles" discussed in Section 5.4. Second-order differential equations for the electric and magnetic fields are derived from the first-order differential equations and are shown to be equivalent to the first-order equations. Weak and strong causality are defined, and the second-order differential equations are used to derive the strong causality condition from the weak causality condition. Sufficient conditions on the fields and their sources are given to ensure the validity of general volume and surface integral expressions for the time-domain electric and magnetic fields. The volume integral expressions are immediately used to determine cavity-defined electric and magnetic fields. The volume integral expressions are also applied to derive the fields of a moving point charge used in Section 5.4.1 to determine the necessary conditions for the existence of electromagnetic missiles. In the last subsection, 2.1.10, convenient expressions are derived and interpreted for the force-power and momentum-energy associated with electromagnetic fields and their sources.

We devote a number of pages to deriving the time-domain electromagnetic force and power expressions for point dipoles as well as for a continuum of electric and magnetic polarization densities, because of the many different expressions that have been published for the instantaneous electromagnetic force and power. We confirm the results of Penfield and Haus [77] and de Groot and Suttorp [78] that the force-power supplied by electromagnetic fields to their polarization densities is independent of the model (electric charge-current, which we call the Amperian model, or magnetic charge-current model) chosen to represent the polarization densities. The force-power supplied to point dipoles outside the source region of an externally applied field is also independent of the model chosen for the point dipoles. However, the force-power supplied to point dipoles that enter the source region of an externally applied field can differ depending on the model chosen for the point dipoles (and the model chosen for the polarization densities of the external field). Apparently, all experimental evidence to date supports the conclusion that the intrinsic magnetic dipole moments of elementary particles behave according to the electric-current (Amperian) model of the magnetic dipole [63, p. 191].

2.1.1 First-Order Differential Equations

The electric and magnetic fields, $\mathbf{E}(\mathbf{r}, t)$ and $\mathbf{H}(\mathbf{r}, t)$, produced by the volume charge and current densities, $\rho(\mathbf{r}, t)$ and $\mathbf{J}(\mathbf{r}, t)$, located in free space (free of all electromagnetic sources except ρ and $\mathbf{J}$), are subject to Maxwell's equations, which can be written in differential form as [1, ch. 1]

$$\nabla \times \mathbf{E}(\mathbf{r}, t) + \mu \frac{\partial}{\partial t}\mathbf{H}(\mathbf{r}, t) = 0 \tag{2.1}$$

$$\nabla \times \mathbf{H}(\mathbf{r}, t) - \epsilon \frac{\partial}{\partial t}\mathbf{E}(\mathbf{r}, t) = \mathbf{J}(\mathbf{r}, t) \tag{2.2}$$

$$\nabla \cdot \mathbf{H}(\mathbf{r}, t) = 0 \tag{2.3}$$

$$\nabla \cdot \mathbf{E}(\mathbf{r}, t) = \rho(\mathbf{r}, t)/\epsilon \tag{2.4}$$

where μ and ϵ are the real, scalar, constant permeability and permittivity of homogeneous, isotropic, lossless free space, and the International System of Units (mksA) is used throughout. The equations (2.1)–(2.4) are assumed to hold for all time t at observation points $\mathbf{r} = r\hat{\mathbf{r}} = x\hat{\mathbf{x}} + y\hat{\mathbf{y}} + z\hat{\mathbf{z}}$ throughout all space. (The unit vector $\hat{\mathbf{r}}$ is defined as $\mathbf{r}/r$ where $r = |\mathbf{r}| = \sqrt{x^2 + y^2 + z^2}$.) This means that all sources, whether applied or induced (for example, antennas or scatterers), are accounted for entirely by the charge and current densities, $\rho(\mathbf{r}, t)$ and $\mathbf{J}(\mathbf{r}, t)$. Taking the divergence of (2.2), interchanging the space and time derivatives, and substituting $\nabla \cdot \mathbf{E}$ from (2.4) produce the equation of continuity between ρ and $\mathbf{J}$

$$\nabla \cdot \mathbf{J}(\mathbf{r}, t) + \frac{\partial}{\partial t}\rho(\mathbf{r}, t) = 0. \tag{2.5}$$

Throughout most of the book, we have chosen to work with the electromagnetic field vectors, $\mathbf{E}$ and $\mathbf{H}$, and in free space we limit the source distributions to ρ and $\mathbf{J}$. In free space, there is little loss in generality by not including sources of electric and magnetic polarization densities, $\mathbf{P}$ and $\mathbf{M}$, because equations (2.1)–(2.4) can be made to include $\mathbf{P}$ and $\mathbf{M}$ by merely replacing $\mathbf{H}$ with $\mathbf{B}/\mu$ (the magnetic induction divided by the permeability of free space), $\mathbf{J}$

with the modified current density $\mathbf{J} + \partial\mathbf{P}/\partial t + \nabla \times \mathbf{M}$, and ρ with the modified charge density $\rho - \nabla \cdot \mathbf{P}$ [1, p. 12]. Alternatively, we can retain the vector $\mathbf{H}$, as well as $\mathbf{E}$, by replacing (2.1)–(2.4) with the following equations that include $\mathbf{P}$ and $\mathbf{M}$[1]

$$\nabla \times \mathbf{E} + \mu \frac{\partial \mathbf{H}}{\partial t} = -\mu \frac{\partial \mathbf{M}}{\partial t}, \quad \nabla \cdot \mathbf{H} = -\nabla \cdot \mathbf{M} \tag{2.6}$$

$$\nabla \times \mathbf{H} - \epsilon \frac{\partial \mathbf{E}}{\partial t} = \mathbf{J} + \frac{\partial \mathbf{P}}{\partial t}, \quad \nabla \cdot \mathbf{E} = (\rho - \nabla \cdot \mathbf{P})/\epsilon. \tag{2.7}$$

These equations with $\mathbf{P}$ and $\mathbf{M}$ show that the equations throughout the book for $\mathbf{E}$ and $\mathbf{H}$ in terms of $\mathbf{J}$ and ρ can be generalized to include $\mathbf{P}$ and $\mathbf{M}$ by replacing $\mathbf{J}$ and ρ in the equations for $\mathbf{E}$ and $\mathbf{H}$ with $\mathbf{J}^e = \mathbf{J} + \partial\mathbf{P}/\partial t$ and $\rho^e = \rho - \nabla \cdot \mathbf{P}$, and adding the $\mathbf{E}$ and $\mathbf{H}$ of the dual equations formed by replacing $\mathbf{E}$ with $\mathbf{H}$, $\mathbf{H}$ with $-\mathbf{E}$, μ with ϵ, ϵ with μ, $\mathbf{J}^e$ with $\mathbf{J}^m = \mu\partial\mathbf{M}/\partial t$, and ρ^e with $\rho^m = -\mu\nabla \cdot \mathbf{M}$. The equivalent electric and magnetic charges, ρ^e and ρ^m, are related to the equivalent electric and magnetic currents, $\mathbf{J}^e$ and $\mathbf{J}^m$, through the same continuity equation (2.5) that relates ρ and $\mathbf{J}$. Of course, Maxwell's equations can be further generalized to include quadrupole and higher order multipole densities [63, sec. I.4].[2]

Electromagnetic equations that apply to homogeneous, isotropic, lossy media and that include induced and applied sources of current, electric polarization, and magnetic polarization are given in Section 2.5.1. Also, polarization densities are included in some of the force and power expressions of Sections 2.1.10 and 2.3.9.

For time-dependent sources and fields that are zero at some initial time, (2.1) implies (2.3), and (2.2) with (2.5) implies (2.4). To prove this, take the divergence of (2.1) and (2.2),

[1] The electric and magnetic dipole moments per unit volume ($\mathbf{P}$ and $\mathbf{M}$) can be defined as follows in terms of "bound" charge and current densities (ρ_b and $\mathbf{J}_b$) [79, pp. 436–437]

$$\mathbf{P}(\mathbf{r}, t) = \lim_{D \to 0} \frac{1}{\Delta V} \int_{V_D} \rho_b(\mathbf{r}', t)\mathbf{r}' dV' \tag{2.8}$$

$$\mathbf{M}(\mathbf{r}, t) = \lim_{D \to 0} \frac{1}{2\Delta V} \int_{V_D} \mathbf{r}' \times \mathbf{J}_b(\mathbf{r}', t) dV' - \frac{1}{2}\mathbf{r} \times \frac{\partial}{\partial t}\mathbf{P}(\mathbf{r}, t) \tag{2.9}$$

where V_D is a region of material that contains $\mathbf{r}$, has maximum dimension D, and has volume ΔV. For example, V_D could be a sphere with diameter D and volume $\Delta V = \pi D^3/6$. If the origin of the coordinate system is chosen within V_D, that is, at $\mathbf{r} = 0$, the $\mathbf{r} \times \partial\mathbf{P}/\partial t$ term in (2.9) vanishes. It can be shown that if ρ_b and $\mathbf{J}_b$ are set equal to $-\nabla \cdot \mathbf{P}$ and $\nabla \times \mathbf{M} + \partial\mathbf{P}/\partial t$, respectively, (2.8) and (2.9) become identities, provided V_D includes the equivalent surface charge density $\hat{\mathbf{n}} \cdot \mathbf{P}$ and surface current density $\mathbf{M} \times \hat{\mathbf{n}}$ on the surface of V_D. For a point-charge molecular model of matter, ΔV in (2.8) and (2.9) should not approach zero but remain large enough to contain many molecules and yet have linear dimensions much smaller than the shortest effective wavelength [63, sec. 6.6]. The electric charge-current definitions of $\mathbf{P}(\mathbf{r}, t)$ and $\mathbf{M}(\mathbf{r}, t)$ in (2.8) and (2.9) are also called the Amperian definitions of polarization densities [80].

[2] In Maxwell's "Treatise on Electricity and Magnetism" [81, vol. 2, art. 619], the four primary equations of the electromagnetic fields were given as

$$\mathbf{B} = \nabla \times \mathbf{A} \tag{2.10}$$

$$\mathbf{E} = -\frac{\partial \mathbf{A}}{\partial t} - \nabla\Psi \tag{2.11}$$

$$\nabla \times \mathbf{H} = \mathbf{J} + \epsilon \frac{\partial \mathbf{E}}{\partial t} \tag{2.12}$$

$$\nabla \cdot \mathbf{D} = \rho \tag{2.13}$$

where $\mathbf{A}$ and Ψ are vector and scalar potentials, and the electric displacement $\mathbf{D}$ and magnetic induction $\mathbf{B}$ are related to $\mathbf{E}$, $\mathbf{P}$, $\mathbf{H}$, and $\mathbf{M}$ by $\mathbf{B} = \mu(\mathbf{H} + \mathbf{M})$ and $\mathbf{D} = \epsilon\mathbf{E} + \mathbf{P}$. (Actually, Maxwell did not introduce the electric polarization vector $\mathbf{P}$ in his Treatise, but wrote the linear relationship $\mathbf{D} = K\mathbf{E}$, with K called the "dielectric inductive capacity." He stated in [81, vol. 1, art. 60] that electric polarization "may be studied to the greatest advantage as exemplified in permanent magnets," which he studied extensively in terms of the three vectors $\mathbf{B}$, $\mathbf{H}$, and $\mathbf{M}$. However, it was apparently H. A. Lorentz who first introduced the electric polarization vector $\mathbf{P}$ [82, vol. 1, p. 397].) Taking the divergence of (2.10) gives $\nabla \cdot \mathbf{B} = 0$, and taking the curl of (2.11) with $\nabla \times \mathbf{A}$ inserted from (2.10) gives $\nabla \times \mathbf{E} = -\partial\mathbf{B}/\partial t$.

interchange the time derivatives and divergence operators, and insert $\nabla \cdot \mathbf{J}$ from (2.5) to get

$$\frac{\partial}{\partial t}(\nabla \cdot \mathbf{H}) = 0 \tag{2.14}$$

and

$$\frac{\partial}{\partial t}(\nabla \cdot \mathbf{E} - \rho/\epsilon) = 0 \tag{2.15}$$

which imply that $\nabla \cdot \mathbf{H}$ and $\nabla \cdot \mathbf{E} - \rho/\epsilon$ do not change with time. Thus, if the sources and fields are zero everywhere at some point in time, (2.14) and (2.15) imply (2.3) and (2.4), respectively. We want to retain (2.3) and (2.4) explicitly, however, in order to allow for static fields (which are never zero everywhere) and to show that the traditional Maxwell differential equations require the existence of the divergence of **E** and **H**. The existence of the divergence of **E** and **H** is not guaranteed by (2.1), (2.2), and (2.5), but had to be assumed in the above derivation of (2.3) and (2.4) from (2.1), (2.2), and (2.5).

Note that if $\rho(\mathbf{r}, t)$ is zero at some initial time t_0, the continuity equation (2.5) can be integrated over time to find $\rho(\mathbf{r}, t)$ in terms of $\mathbf{J}(\mathbf{r}, t)$, that is

$$\rho(\mathbf{r}, t) = -\int_{t_0}^{t} \nabla \cdot \mathbf{J}(\mathbf{r}, t')dt'. \tag{2.16}$$

Thus, $\rho(\mathbf{r}, t)$ can be eliminated from the time-dependent Maxwell equations (2.1)–(2.4) for sources that are zero at some initial time t_0. Also, if the current lies inside a bounded volume V, so that the surface S_V of the volume lies in free space outside the source region, the continuity equation (2.5) can be integrated over V to prove

$$\frac{\partial}{\partial t}\int_{V} \rho(\mathbf{r}, t)dV = -\int_{V} \nabla \cdot \mathbf{J}(\mathbf{r}, t)dV = -\int_{S_V} \mathbf{J}(\mathbf{r}, t) \cdot \hat{\mathbf{n}}dS = 0 \tag{2.17}$$

where $\hat{\mathbf{n}}$ is the unit normal to S_V pointing out of V. That is, the total charge remains constant for all time (charge is conserved), and, in particular, the total charge remains zero for all time, if it is zero at some initial time.

All of the time derivatives can be removed from the time-domain electromagnetic field equations by writing Maxwell's equations in the form

$$\nabla \times \int_{t_0}^{t} \mathbf{E}(\mathbf{r}, t')dt' + \mu\mathbf{H}(\mathbf{r}, t) = 0 \tag{2.18}$$

$$\nabla \times \int_{t_0}^{t} \mathbf{H}(\mathbf{r}, t')dt' - \epsilon\mathbf{E}(\mathbf{r}, t) = \int_{t_0}^{t} \mathbf{J}(\mathbf{r}, t')dt' \tag{2.19}$$

$$\nabla \cdot \int_{t_0}^{t} \mathbf{H}(\mathbf{r}, t')dt' = 0 \tag{2.20}$$

$$\nabla \cdot \int_{t_0}^{t} \mathbf{E}(\mathbf{r}, t')dt' = -\frac{1}{\epsilon}\nabla \cdot \int_{t_0}^{t}\int_{t_0}^{t'} \mathbf{J}(\mathbf{r}, t'')dt''dt' \tag{2.21}$$

when the sources and fields are zero for $t \le t_0$, where t_0 is a finite initial time. Thus, (2.18)–(2.21) assume the static fields are zero, and (2.18)–(2.21) cannot be used to find nonzero static electric and magnetic fields, which are determined by the following equations obtained from

(2.1)–(2.4)

$$\nabla \times \mathbf{E}(\mathbf{r}) = 0, \quad \nabla \cdot \mathbf{E}(\mathbf{r}) = \rho(\mathbf{r})/\epsilon \tag{2.22}$$

$$\nabla \times \mathbf{H}(\mathbf{r}) = \mathbf{J}(\mathbf{r}), \quad \nabla \cdot \mathbf{H}(\mathbf{r}) = 0. \tag{2.23}$$

The equations (2.18)–(2.21) are mathematically less restrictive than the original equations (2.1)–(2.4) in that, unlike (2.1)–(2.4), they do not require the existence of the time derivatives of **E** and **H**. Thus, they allow uniqueness and existence of solution to be proven in the following sections for a larger class of electromagnetic fields.[3] For most engineering applications and for most applications in this book, even discontinuous fields can be adequately handled by equations (2.1)–(2.4) through the use of delta functions. Nonetheless, for the sake of obtaining rigorous, more general uniqueness and existence theorems within the context of classical analysis and Riemann integration, equations (2.18)–(2.21) become a useful, alternative form of Maxwell's time-domain equations, which retain their traditional differential form with respect to the spatial coordinates.

2.1.2 Second-Order Differential Equations

Separate second-order differential equations for $\mathbf{E}(\mathbf{r}, t)$ and $\mathbf{H}(\mathbf{r}, t)$ can be found by differentiating and combining the first-order equations (2.1)–(2.4). Taking the time derivative of (2.2) and (2.1), interchanging the space and time derivatives, and then substituting $\partial\mathbf{E}/\partial t$ and $\partial\mathbf{H}/\partial t$ from (2.1) and (2.2), respectively, produce the time-domain, inhomogeneous, "curlcurl" vector wave equations for $\mathbf{E}(\mathbf{r}, t)$ and $\mathbf{H}(\mathbf{r}, t)$

$$\nabla \times \nabla \times \mathbf{E} + \frac{1}{c^2}\frac{\partial^2 \mathbf{E}}{\partial t^2} = -\mu \frac{\partial \mathbf{J}}{\partial t} \tag{2.24}$$

$$\nabla \times \nabla \times \mathbf{H} + \frac{1}{c^2}\frac{\partial^2 \mathbf{H}}{\partial t^2} = \nabla \times \mathbf{J}. \tag{2.25}$$

The speed of light in free space is denoted by c and is equal to $1/\sqrt{\mu\epsilon}$.

Expressing $\nabla \times \nabla\times$ as the $-\nabla^2 + \nabla\nabla\cdot$ operator [83, app. 1, eq. 18] and substituting $\nabla \cdot \mathbf{E}$ and $\nabla \cdot \mathbf{H}$ from (2.4) and (2.3) into (2.24) and (2.25) produce the alternative time-domain, inhomogeneous vector wave equations that involve the Laplacian (∇^2) instead of the curlcurl ($\nabla \times \nabla\times$) spatial operator

$$\nabla^2 \mathbf{E} - \frac{1}{c^2}\frac{\partial^2 \mathbf{E}}{\partial t^2} = \mu \frac{\partial \mathbf{J}}{\partial t} + \frac{\nabla \rho}{\epsilon} \tag{2.26}$$

$$\nabla^2 \mathbf{H} - \frac{1}{c^2}\frac{\partial^2 \mathbf{H}}{\partial t^2} = -\nabla \times \mathbf{J}. \tag{2.27}$$

Like the first-order differential equations (2.1)–(2.4), the second-order differential equations (2.24)–(2.27) hold for all time t and at all points in space $\mathbf{r}$.

2.1.3 Causality

Equations (2.26) and (2.27) imply that each rectangular component of the electric and magnetic fields satisfies the inhomogeneous scalar wave equation

$$\nabla^2 \psi(\mathbf{r}, t) - \frac{1}{c^2}\frac{\partial^2}{\partial t^2}\psi(\mathbf{r}, t) = -f(\mathbf{r}, t). \tag{2.28}$$

[3] The time derivatives in (2.1) and (2.2) imply that $\mathbf{E}(\mathbf{r}, t)$ and $\mathbf{H}(\mathbf{r}, t)$ are continuous functions of t for each $\mathbf{r}$. The curls and divergences in (2.1)–(2.4) imply that $\mathbf{E}(\mathbf{r}, t)$ and $\mathbf{H}(\mathbf{r}, t)$ are continuous functions of $\mathbf{r}$ for each t. These continuities are not implied by (2.18)–(2.21).

Using the divergence theorem, Phillips [84, sec. 91] has proven that a function $\psi(\mathbf{r}, t)$, that has continuous second derivatives with respect to $\mathbf{r}$ and t, and that satisfies the scalar wave equation (2.28) with $f(\mathbf{r}, t)$ zero outside a bounded volume V can be expressed as

$$\begin{aligned}\psi(\mathbf{r}, t) = &\lim_{\delta\to 0} \int_{V-V_\delta} \frac{f(\mathbf{r}', t - |\mathbf{r} - \mathbf{r}'|/c)}{4\pi|\mathbf{r} - \mathbf{r}'|} dV' \\ &+ \int_{S_o} \left[\frac{\hat{\mathbf{n}}' \cdot \{\nabla'\psi\}(\mathbf{r}', t - |\mathbf{r} - \mathbf{r}'|/c)}{4\pi|\mathbf{r} - \mathbf{r}'|} \right. \\ &\left. + \nabla \cdot \left(\frac{\hat{\mathbf{n}}'\psi(\mathbf{r}', t - |\mathbf{r} - \mathbf{r}'|/c)}{4\pi|\mathbf{r} - \mathbf{r}'|} \right) \right] dS'\end{aligned} \tag{2.29}$$

where S_o is a surface (with outward normal $\hat{\mathbf{n}}$) that surrounds the bounded source region V and the observation point $\mathbf{r}$. The "principal volume" V_δ and the $\lim_{\delta\to 0}$ on the volume integral are explained after equation (2.55). They are unessential to the discussion of causality in this section. The curly brackets on $\{\nabla'\psi\}$ indicate differentiation with respect to just the first $\mathbf{r}'$ variable in $\psi(\mathbf{r}', t - |\mathbf{r} - \mathbf{r}'|/c)$. (Phillips also proves in [84, sec. 92] that sufficient conditions for $\psi(\mathbf{r}, t)$ to have continuous second derivatives are that the source function $f(\mathbf{r}, t)$ has continuous first spatial derivatives and continuous second time derivatives.)

If one invokes the weak causality condition, which states that all fields are zero for $t \le t_0$ if the sources $f(\mathbf{r}, t)$ are zero for $t \le t_0$, then for any fixed $(\mathbf{r}, t)$ the integrand of the surface integral in (2.29) is zero provided the surface S_o is chosen large enough to make $t - |\mathbf{r} - \mathbf{r}'|/c < t_0$ for all $\mathbf{r}'$ on S_o. Therefore, (2.29) reduces simply to

$$\psi(\mathbf{r}, t) = \lim_{\delta\to 0} \int_{V-V_\delta} \frac{f(\mathbf{r}', t - |\mathbf{r} - \mathbf{r}'|/c)}{4\pi|\mathbf{r} - \mathbf{r}'|} dV' \tag{2.30}$$

for sources in a bounded volume that are zero for $t \le t_0$ and that obey the weak causality condition. In addition, (2.30) implies that the surface integral in (2.29) is zero for any S_o that encloses the source region and the observation point. This result can also be obtained from a time-domain "extinction theorem" [85, p. 500], [86], which can be proven using weak causality by applying Green's theorem to the source-free volume outside S_o when the observation point is inside S_o.

Now (2.30) shows that the field radiated by the source located at $\mathbf{r}'$ is zero for $|\mathbf{r} - \mathbf{r}'| > c(t - t_0)$, that is, for all observation distances larger than $c(t - t_0)$. In other words, Maxwell's equations, from which (2.26) and (2.27) are derived, imply the strong causality condition (that the fields are zero beyond a distance $c(t - t_0)$ from a point source that turns on at $t = t_0$) from the weak causality condition (that the fields, produced by sources that turn on at $t = t_0$, are zero everywhere for $t \le t_0$). In short, weak causality and Maxwell's equations imply that electromagnetic fields propagate away from sources with constant speed c.

Since the electromagnetic and, later in this chapter, the acoustic field equations imply strong causality from weak causality, strong causality can be assumed whenever we refer to mere causality throughout the rest of the chapter and book.

2.1.4 Equivalence of First- and Second-Order Differential Equations

The second-order differential equations (2.24)–(2.25) and (2.26)–(2.27) were derived from Maxwell's first-order differential equations (2.1)–(2.4), assuming the existence of the second-order derivatives and the validity of interchanging space and time derivatives. If, in addition, it is assumed that:

1. equations (2.24)–(2.27) can be differentiated and the derivatives interchanged to get (2.31)–(2.34) and (2.36)–(2.38) below,
2. the functions in parentheses in (2.33)–(2.34) and (2.36)–(2.38) have continuous second derivatives with respect to $\mathbf{r}$ and t, and
3. the sources lie in a bounded region of space and are zero for $t \le t_0$,

then (2.24)–(2.25) or (2.26)–(2.27), along with the continuity equation (2.5) and the causality condition, imply (2.1)–(2.4). In other words, under quite general conditions, the second-order equations (2.24)–(2.25) or (2.26)–(2.27) are equivalent to the first-order Maxwell equations (2.1)–(2.4). We also prove the equivalence of the second-order equations (2.26)–(2.27) and the first-order Maxwell equations for static fields that approach zero as $r \to \infty$.

To prove this equivalence, we begin by deriving the divergence equations (2.3)–(2.4) from (2.24)–(2.25) and from (2.26)–(2.27). Take the divergence of (2.24) and (2.25), bring the divergence inside the time derivatives, and insert $\nabla \cdot \mathbf{J}$ from the continuity equation (2.5) to get

$$\frac{\partial^2}{\partial t^2}(\nabla \cdot \mathbf{E} - \rho/\epsilon) = 0 \tag{2.31}$$

and

$$\frac{\partial^2}{\partial t^2}(\nabla \cdot \mathbf{H}) = 0. \tag{2.32}$$

If all the sources and fields are zero everywhere before some initial time, (2.31) and (2.32) imply (2.4) and (2.3), respectively.

Similarly, taking the divergence of (2.26) and (2.27), bringing the divergence inside the second derivatives, and substituting $\nabla \cdot \mathbf{J}$ from (2.5) will produce the homogeneous scalar wave equations

$$\left(\nabla^2 - \frac{1}{c^2}\frac{\partial^2}{\partial t^2}\right)(\nabla \cdot \mathbf{E} - \rho/\epsilon) = 0 \tag{2.33}$$

and

$$\left(\nabla^2 - \frac{1}{c^2}\frac{\partial^2}{\partial t^2}\right)(\nabla \cdot \mathbf{H}) = 0. \tag{2.34}$$

Equation (2.29) states that a function $\psi(\mathbf{r}, t)$ that has continuous second derivatives with respect to $\mathbf{r}$ and t, and satisfies the homogeneous scalar wave equation, can be expressed as the following integral over a surface S_o surrounding the observation point at $\mathbf{r}$

$$\begin{aligned}\psi(\mathbf{r}, t) = \frac{1}{4\pi}\int_{S_o}\Bigg[&\frac{\hat{\mathbf{n}}' \cdot \{\nabla'\psi\}(\mathbf{r}', t - |\mathbf{r} - \mathbf{r}'|/c)}{|\mathbf{r} - \mathbf{r}'|} \\ &+ \nabla \cdot \left(\frac{\hat{\mathbf{n}}'\psi(\mathbf{r}', t - |\mathbf{r} - \mathbf{r}'|/c)}{|\mathbf{r} - \mathbf{r}'|}\right)\Bigg]\, dS'.\end{aligned} \tag{2.35}$$

If $\psi(\mathbf{r}, t)$ is zero for all $\mathbf{r}$ and $t \le t_0$, then for any fixed $(\mathbf{r}, t)$, the integrand of the surface integral in (2.35) equals zero provided the surface S_o is chosen large enough to make $t - |\mathbf{r} - \mathbf{r}'|/c < t_0$ for all $\mathbf{r}'$ on S_o. Therefore, the trivial solution, $\psi(\mathbf{r}, t) = 0$ for all $(\mathbf{r}, t)$, is the only solution to the homogeneous scalar wave equation with continuous second derivatives, given that $\psi(\mathbf{r}, t)$ is equal to zero everywhere before a finite initial time t_0. (Incidentally, (2.35) also holds for the solution to the homogeneous scalar wave equation if the advanced time $(t + |\mathbf{r} - \mathbf{r}'|/c)$ replaces the retarded time $(t - |\mathbf{r} - \mathbf{r}'|/c)$. This advanced-field surface integral could then be used to prove that $\psi(\mathbf{r}, t) = 0$ for all $(\mathbf{r}, t)$ if $\psi(\mathbf{r}, t)$ is zero everywhere for $t \ge t_0$. In Sections 5.2.6 and 5.3.5, the advanced-field surface integral is used to find time-domain near fields, after all sources turn off, in terms of an integration of the far-field pattern.)

Assume the electric and magnetic fields satisfy causality (they cannot travel away from the sources faster than c), the sources are located in a bounded volume, and the sources are zero everywhere for $t \le t_0$. Then $(\nabla \cdot \mathbf{E} - \rho/\epsilon)$ and $\nabla \cdot \mathbf{H}$ in (2.33) and (2.34) are zero everywhere for $t \le t_0$, and it follows that the only solution to (2.33) or (2.34) is the trivial solution. Thus, (2.33) and (2.34) imply (2.4) and (2.3), respectively.

For static fields (which are never zero everywhere) the time derivatives in (2.24) and (2.25) vanish; therefore, taking the divergence of (2.24) and (2.25) no longer leads to (2.4) and (2.3). However, taking the divergence of (2.26) and (2.27) for static fields produces Laplace's equation in (2.33) and (2.34). In Appendix A, Green's identities are used to prove that the only twice continuously differentiable function that satisfies Laplace's equation throughout all space and approaches zero as $r \to \infty$ is the trivial one (zero). Thus, for static fields, (2.33) and (2.34) still imply (2.4) and (2.3), respectively, if $(\nabla \cdot \mathbf{E} - \rho/\epsilon)$ and $\nabla \cdot \mathbf{H}$ have continuous second spatial derivatives and approach zero as $r \to \infty$.

Now that the divergence equations (2.3) and (2.4) have been derived from (2.24)–(2.25) and (2.5) (except for static fields), they can be inserted into (2.24)–(2.25) to derive (2.26) and (2.27), respectively. Consequently, we can concentrate on (2.26) and (2.27) to complete the proof that the second-order differential equations for $\mathbf{E}$ and $\mathbf{H}$ are equivalent to Maxwell's equations (2.1)–(2.4). Specifically, take the curl of (2.26) and substitute $\nabla \times \mathbf{J}$ from (2.27) to get

$$\left(\nabla^2 - \frac{1}{c^2}\frac{\partial^2}{\partial t^2}\right)\left(\nabla \times \mathbf{E} + \mu\frac{\partial \mathbf{H}}{\partial t}\right) = 0. \tag{2.36}$$

This equation, like (2.33) and (2.34), has only the trivial solution, provided the causality condition is invoked for time-dependent sources in a bounded region that are zero for $t \le t_0$. Even for static fields, (2.36) reduces to Laplace's equation, $\nabla^2(\nabla \times \mathbf{E}) = 0$, which has only the trivial solution if the curl of the static electric field is twice continuously differentiable and approaches zero as $r \to \infty$. Thus, (2.36) implies equation (2.1).

Having derived equation (2.1) from (2.26)–(2.27), we can take the curl of (2.27) and substitute $\partial \mathbf{H}/\partial t$ from (2.1) to obtain after some manipulation

$$\nabla^2\left(\nabla \times \mathbf{H} - \epsilon\frac{\partial \mathbf{E}}{\partial t} - \mathbf{J}\right) + \nabla\nabla \cdot \left(\epsilon\frac{\partial \mathbf{E}}{\partial t} + \mathbf{J}\right) = 0. \tag{2.37}$$

The second term on the left-hand side of (2.37) is zero because of (2.4) and (2.5). (Recall that (2.4) has been derived from (2.26) and (2.5).) Thus, (2.37) reduces to

$$\nabla^2\left(\nabla \times \mathbf{H} - \epsilon\frac{\partial \mathbf{E}}{\partial t} - \mathbf{J}\right) = 0. \tag{2.38}$$

Again, Laplace's equation (2.38) has only the solution $(\nabla \times \mathbf{H} - \epsilon\partial\mathbf{E}/\partial t - \mathbf{J}) = 0$ throughout space if $(\nabla \times \mathbf{H} - \epsilon\partial\mathbf{E}/\partial t - \mathbf{J})$ is twice continuously differentiable and approaches zero as $r \to \infty$. For time-dependent fields produced by sources in a bounded region of space that are zero for $t \le t_0$, causality demands that the fields be precisely zero at infinity. For static fields, (2.38) reduces to $\nabla^2(\nabla \times \mathbf{H} - \mathbf{J}) = 0$, which has only the trivial solution if $\nabla \times \mathbf{H} - \mathbf{J}$ approaches zero as $r \to \infty$. Thus, (2.38) implies the Maxwell equation (2.2).

To recapitulate, the first-order differential Maxwell equations (2.1)–(2.4) were used to derive the continuity equation (2.5) and the second-order differential equations (2.24)–(2.27). Moreover, for causal time-domain fields and for static fields that approach zero as $r \to \infty$, equations (2.26)–(2.27) along with (2.5) were used to derive (2.1)–(2.4). Except for static fields, (2.24)–(2.25) along with (2.5) were also used to derive (2.1)–(2.4). Thus, for causal fields produced by sources in a bounded region of space, the second-order differential equations (2.26)–(2.27) along with the continuity equation (2.5) are equivalent to Maxwell's first-order differential equations (2.1)–(2.4). And except for static fields, the second-order differential equations (2.24)–(2.25) along with the continuity equation (2.5) are also equivalent to Maxwell's first-order equations (2.1)–(2.4). (For static fields, (2.24)–(2.25) are also equivalent to (2.1)–(2.4) if the divergence conditions (2.3) and (2.4) are imposed separately on (2.24)–(2.25).) We have assumed that the differentiations and interchanges of derivatives required to obtain (2.24)–(2.27) from (2.1)–(2.4), and to obtain (2.31)–(2.34) and (2.36)–(2.38) from (2.24)–(2.27) are valid, and that the second derivatives of the fields and sources within the parentheses of (2.33)–(2.34) and (2.36)–(2.38) are continuous functions of $\mathbf{r}$ and t. Continuity of these derivatives allowed their insertion into Green's first and second identities, which were required in the above proof of this somewhat surprising, though well-used, and generally accepted equivalence of the first- and second-order electromagnetic field equations.

Another way to prove the equivalence of the first- and second-order equations would be to prove the uniqueness of solution to each of the four equations (2.24)–(2.27) for fields that have the proper far-field behavior. (For static fields, uniqueness of solution to (2.24) and (2.25) would also require extra conditions, namely, the divergence equations (2.4) and (2.3), respectively.) If, then, a solution to (2.1)–(2.4) exists, a unique solution exists to (2.24)–(2.25), (2.26)–(2.27), and (2.1)–(2.4), because (2.24)–(2.27) are derivable from (2.1)–(2.4), and thus the equations (2.24)–(2.25) or (2.26)–(2.27) are equivalent to (2.1)–(2.4).

By either method, one proves the equivalence of the first- and second-order electromagnetic field equations with all sources of the electromagnetic field given by the charge and current densities, $\rho(\mathbf{r},t)$ and $\mathbf{J}(\mathbf{r},t)$. This means that we have assumed there are no boundaries, such as the surfaces of perfect conductors, for which boundary values of the fields are given instead of the charge-current densities. If we relax this assumption, for example, by introducing perfectly conducting electric or magnetic surfaces, the second-order differential equations (2.26)–(2.27) would need the first-order divergence equations to derive the first-order Maxwell equations. Thus, the second-order time-domain equations (2.26)–(2.27) and the first-order Maxwell equations would no longer be equivalent [87].

2.1.5 Uniqueness of Solution

Uniqueness theorems are of general interest because they can tell us whether a particular solution to Maxwell's equations is the only allowable solution. We also have a special interest in proving uniqueness of solution to Maxwell's equations in both the time domain and (as explained later in Section 2.3.5) the frequency domain.

For time-domain fields, the traditional uniqueness proof [1, sec. 9.2], [67, sec. 10.12] uses Poynting's theorem, which combines equations (2.1)–(2.2) with the divergence theorem in a way that requires the integration of the first time derivatives of the electric and magnetic fields. Here, we give a slightly modified, less restrictive proof of uniqueness, based on the time-domain equations (2.18)–(2.21), that requires the electric and magnetic fields to be continuous functions of $\mathbf{r}$ and t,[4] yet does not require the existence of the first time derivatives of the electric and magnetic fields. The fields of electromagnetic "missiles" (pulses, discussed in Section 5.4, with energy decay slower than $1/r^2$ as $r \to \infty$) are examples of time-domain fields with first time derivatives that are either infinite or do not exist, and yet the fields themselves can be continuous functions of $\mathbf{r}$ and t.

The proof of uniqueness for time-domain fields begins by assuming there are two solutions to equations (2.18)–(2.21) with the same source distributions $\mathbf{J}$ and ρ. The difference of the fields of the two solutions ($\mathbf{E}_d(\mathbf{r},t)$ and $\mathbf{H}_d(\mathbf{r},t)$) satisfies the homogeneous equations

[4] When we state that a function is continuous in $\mathbf{r}$ and t, we mean it is continuous in both variables, or more precisely, in all four variables x, y, z, and t. This continuity with respect to $(\mathbf{r},t)$ is slightly stronger than continuity with respect to $\mathbf{r}$ for each t, and with respect to t for each $\mathbf{r}$ [88, p. 67].

corresponding to (2.18)–(2.21)

$$\nabla \times \int_{t_0}^{t} \mathbf{E}_d(\mathbf{r}, t')dt' + \mu \mathbf{H}_d(\mathbf{r}, t) = 0 \tag{2.39}$$

$$\nabla \times \int_{t_0}^{t} \mathbf{H}_d(\mathbf{r}, t')dt' - \epsilon \mathbf{E}_d(\mathbf{r}, t) = 0 \tag{2.40}$$

$$\nabla \cdot \int_{t_0}^{t} \mathbf{H}_d(\mathbf{r}, t')dt' = 0 \tag{2.41}$$

$$\nabla \cdot \int_{t_0}^{t} \mathbf{E}_d(\mathbf{r}, t')dt' = 0. \tag{2.42}$$

These equations hold for all time t and for all observation points $\mathbf{r}$ throughout all space.

To prove that the only solution to (2.39)–(2.42) is the trivial solution (thereby proving uniqueness of solution for (2.18)–(2.21)), we form the divergence

$$\begin{aligned}\nabla \cdot \left(\int_{t_0}^{t} \mathbf{E}_d dt' \times \int_{t_0}^{t} \mathbf{H}_d dt' \right) &= \int_{t_0}^{t} \mathbf{H}_d dt' \cdot \nabla \times \int_{t_0}^{t} \mathbf{E}_d dt' \\ &\quad - \int_{t_0}^{t} \mathbf{E}_d dt' \cdot \nabla \times \int_{t_0}^{t} \mathbf{H}_d dt'\end{aligned} \tag{2.43}$$

and then substitute the curls from (2.39) and (2.40) into (2.43) to get

$$\begin{aligned}\nabla \cdot \left(\int_{t_0}^{t} \mathbf{E}_d dt' \times \int_{t_0}^{t} \mathbf{H}_d dt' \right) &= -\left(\epsilon \mathbf{E}_d \cdot \int_{t_0}^{t} \mathbf{E}_d dt' + \mu \mathbf{H}_d \cdot \int_{t_0}^{t} \mathbf{H}_d dt' \right) \\ &= -\frac{1}{2}\frac{\partial}{\partial t}\left(\epsilon \left| \int_{t_0}^{t} \mathbf{E}_d dt' \right|^2 + \mu \left| \int_{t_0}^{t} \mathbf{H}_d dt' \right|^2 \right).\end{aligned} \tag{2.44}$$

The fundamental theorem of calculus [89, th. 6.18] ensures that the last equality in (2.44) holds if $\mathbf{E}_d(\mathbf{r}, t)$ and $\mathbf{H}_d(\mathbf{r}, t)$ are continuous functions of t for each $\mathbf{r}$. The electromagnetic field equations (2.39)–(2.42) imply that the curl and divergence of $\int_{t_0}^{t} \mathbf{E}_d dt'$ and $\int_{t_0}^{t} \mathbf{H}_d dt'$ exist. Therefore, $\int_{t_0}^{t} \mathbf{E}_d dt'$ and $\int_{t_0}^{t} \mathbf{H}_d dt'$ are continuous functions of $\mathbf{r}$ for each t (as well as continuous functions of t for each $\mathbf{r}$). If, in addition, $\mathbf{E}_d(\mathbf{r}, t)$ and $\mathbf{H}_d(\mathbf{r}, t)$ are continuous functions of $\mathbf{r}$ (for each t), (2.39)–(2.42) imply that the first spatial partial derivatives of $\int_{t_0}^{t} \mathbf{E}_d dt'$ and $\int_{t_0}^{t} \mathbf{H}_d dt'$ are continuous functions of $\mathbf{r}$, and thus the divergence theorem can be applied rigorously to (2.44) to show that

$$\begin{aligned}\frac{1}{2}\frac{d}{dt}\int_{V_a} \left(\epsilon \left| \int_{t_0}^{t} \mathbf{E}_d dt' \right|^2 + \mu \left| \int_{t_0}^{t} \mathbf{H}_d dt' \right|^2 \right) dV \\ = -\int_{S_a} \hat{\mathbf{n}} \cdot \left(\int_{t_0}^{t} \mathbf{E}_d dt' \times \int_{t_0}^{t} \mathbf{H}_d dt' \right) dS\end{aligned} \tag{2.45}$$

where S_a is the closed surface bounding any volume V_a, and $\hat{\mathbf{n}}$ is the unit normal to S_a pointing out of V_a. The time derivative has been brought outside the volume integral in (2.45). This interchange is allowed by Leibnitz's rule [90, sec. 4.9] if $\mathbf{E}_d(\mathbf{r}, t)$ and $\mathbf{H}_d(\mathbf{r}, t)$ are continuous functions of $(\mathbf{r}, t)$, since then the integrand of the volume integral and the time derivative of the integrand are continuous functions of $(\mathbf{r}, t)$. Note that (2.45) is simply Poynting's theorem [1, sec. 9.2] with the fields replaced by the time integrals of the fields.

Suppose the sources $\mathbf{J}$ and ρ in the electromagnetic time-domain equations (2.18)–(2.21) are located in a bounded volume. Because the sources are zero for $t \le t_0$ where t_0 is a finite initial time, the causality condition, which requires fields to propagate away from sources no faster than the speed of light, demands that the surface integral on the right-hand side of (2.45) becomes zero as the volume V_a becomes infinitely large (V_∞) and the surface S_a recedes to an infinite distance from the sources. Therefore, (2.45) becomes

$$\frac{d}{dt}\int\limits_{V_\infty}\left(\epsilon\left|\int\limits_{t_0}^{t}\mathbf{E}_d dt'\right|^2 + \mu\left|\int\limits_{t_0}^{t}\mathbf{H}_d dt'\right|^2\right) dV = 0 \tag{2.46}$$

which implies

$$\int\limits_{V_\infty}\left(\epsilon\left|\int\limits_{t_0}^{t}\mathbf{E}_d dt'\right|^2 + \mu\left|\int\limits_{t_0}^{t}\mathbf{H}_d dt'\right|^2\right) dV = \beta \tag{2.47}$$

where β is a constant independent of time. Since (2.47) holds for all time, it holds for $t = t_0$. This implies that β in (2.47) is zero, and thus

$$\int\limits_{t_0}^{t}\mathbf{E}_d dt' = 0, \quad \int\limits_{t_0}^{t}\mathbf{H}_d dt' = 0 \tag{2.48}$$

for all time t (since $\int_{t_0}^{t}\mathbf{E}_d dt'$ and $\int_{t_0}^{t}\mathbf{H}_d dt'$ are continuous functions of $\mathbf{r}$ at each t). Then (2.39) and (2.40) demand that $\mathbf{H}_d(\mathbf{r}, t)$ and $\mathbf{E}_d(\mathbf{r}, t)$ are zero.

In other words, the uniqueness of solution to (2.18)–(2.21) at all $(\mathbf{r}, t)$ is assured for causal electric and magnetic fields that are continuous functions of $\mathbf{r}$ and t, provided the sources are located in a bounded volume, and the sources and fields are zero for $t \le t_0$, where t_0 is a finite initial time. Also, the proof of uniqueness remains valid inside or outside perfect electric or magnetic conductors on which the tangential electric or magnetic field is zero. This can be seen from (2.45) by letting the volume of integration V_a be the region inside or outside the perfect conductors, and noting that the surface integration on the right-hand side of (2.45) over the perfect conductors is zero, because $\hat{\mathbf{n}} \times \mathbf{E}_d$ or $\hat{\mathbf{n}} \times \mathbf{H}_d$ is zero.

The homogeneous equations (2.39)–(2.42) for the difference fields hold not only if the fields and sources are zero for $t \le t_0$, but also if the fields and sources are simply specified at $t = t_0$. Consequently, the proof of uniqueness we have given in this section applies also to the following generalization of (2.18)–(2.21) for sources and fields that are specified at $t = t_0$ and were zero everywhere for $t < t_{00} \le t_0$

$$\nabla \times \int\limits_{t_0}^{t}\mathbf{E}(\mathbf{r}, t')dt' + \mu(\mathbf{H}(\mathbf{r}, t) - \mathbf{H}(\mathbf{r}, t_0)) = 0 \tag{2.49}$$

$$\nabla \times \int\limits_{t_0}^{t}\mathbf{H}(\mathbf{r}, t')dt' - \epsilon(\mathbf{E}(\mathbf{r}, t) - \mathbf{E}(\mathbf{r}, t_0)) = \int\limits_{t_0}^{t}\mathbf{J}(\mathbf{r}, t')dt' \tag{2.50}$$

$$\nabla \cdot \int_{t_0}^{t} \mathbf{H}(\mathbf{r}, t')dt' = 0 \tag{2.51}$$

$$\nabla \cdot \int_{t_0}^{t} \mathbf{E}(\mathbf{r}, t')dt' = \frac{1}{\epsilon} \int_{t_0}^{t} \rho(\mathbf{r}, t')dt'. \tag{2.52}$$

If bounded surfaces are present on which the tangential electric or magnetic field is specified for all time, the proof of uniqueness also remains valid.

These generalized equations (2.49)–(2.52) hold formally for all fields, even static fields. However, the above uniqueness proof assumed causal fields equal to zero beyond a finite distance from the sources, and static fields do not satisfy this condition. Fortunately, uniqueness of solution for the static electric- and magnetic-field equations, (2.22) and (2.23), follows from the uniqueness of solution to Laplace's equation for twice continuously differentiable static potentials that approach zero as $r \to \infty$ (see Appendix A).

2.1.6 Existence of Solution

The uniqueness theorems of Section 2.1.5 neither imply that a solution exists nor provide sufficient conditions for a solution to exist to the time-domain Maxwell equations (2.18)–(2.21). To prove that a causal solution to (2.18)–(2.21) exists, we can first construct causal vector-potential expressions for $\mathbf{E}(\mathbf{r}, t)$ and $\mathbf{H}(\mathbf{r}, t)$ and then determine sufficient conditions on the current $\mathbf{J}(\mathbf{r}, t)$ that ensure the vector-potential expressions obey the time-domain Maxwell equations (2.18)–(2.21).

The vector-potential expressions for $\mathbf{H}(\mathbf{r}, t)$ and $\mathbf{E}(\mathbf{r}, t)$ can be written as

$$\mathbf{H}(\mathbf{r}, t) = \frac{1}{\mu} \nabla \times \mathbf{A}(\mathbf{r}, t) \tag{2.53}$$

$$\mathbf{E}(\mathbf{r}, t) = c^2 \nabla \times \nabla \times \int_{t_0}^{t} \mathbf{A}(\mathbf{r}, t')dt' - \frac{1}{\epsilon} \int_{t_0}^{t} \mathbf{J}(\mathbf{r}, t')dt' \tag{2.54}$$

where the vector potential is given by

$$\mathbf{A}(\mathbf{r}, t) = \frac{\mu}{4\pi} \lim_{\delta \to 0} \int_{V - V_\delta} \frac{\mathbf{J}(\mathbf{r}', t - |\mathbf{r} - \mathbf{r}'|/c)}{|\mathbf{r} - \mathbf{r}'|} dV'. \tag{2.55}$$

The source current $\mathbf{J}(\mathbf{r}, t)$ lies within the volume V of integration. The "principal volume" V_δ, which excludes the singularity at $\mathbf{r}' = \mathbf{r}$ from the volume V of integration in (2.55) and approaches zero as its maximum chord length δ approaches zero, is usually omitted from (2.55) because the value of the integral is independent of the shape of V_δ and the position of the singularity within V_δ. However, to differentiate the vector potential correctly in the source region, the excluding principal volume must be retained because it defines the integration near the singularity and reveals that the region of integration, and thus the limits of integration, depend on the differentiation variable $\mathbf{r}$ [76, ch. 6], [84, secs. 55–57], [73], [74].

The electric field in (2.54) is found by substituting $\mathbf{H}$ from (2.53) into (2.2), then integrating over time for sources and fields that are zero for $t \le t_0$. Thus, (2.54) is not valid for electrostatic fields, whereas (2.53) is a valid expression for static as well as time-dependent magnetic fields. (A static current, $\mathbf{J}(\mathbf{r}, t) = \mathbf{J}(\mathbf{r})$, $\nabla \cdot \mathbf{J}(\mathbf{r}) = 0$, inserted into (2.54) gives a zero electric field, because $\nabla \times \nabla \times \mathbf{A} = \mu \nabla \times \mathbf{H} = \mu \mathbf{J}$ for magnetostatic fields.)

We want to determine sufficient conditions on the current distribution $\mathbf{J}(\mathbf{r}, t)$ for (2.53)–(2.55) to be a solution to the time-domain form of Maxwell's equations given in (2.18)–(2.21).

Müller [91] has given sufficient conditions on the frequency-domain current for the frequency-domain vector-potential solution to satisfy the frequency-domain Maxwell differential equations; see Section 2.3.6. However, we have not found in the published literature an analogous determination of sufficient conditions on the time-domain current for (2.53)–(2.55) to satisfy Maxwell's time-domain equations (2.18)–(2.21).

In most applications of physics and engineering, one is concerned only with the fields outside the source region. Thus, as long as the current integrates to produce fields that obey Maxwell's homogeneous equations outside the source region, one is usually not concerned with the detailed behavior of the fields within the source region. Even if the fields do not obey Maxwell's differential equations at every point within the source region, it is usually sufficient for most purposes to assume the fields in the source region will satisfy Maxwell's integral equations or a locally integrated form of Maxwell's equations [83, app. 6], [92]. For example, the electric and magnetic fields of the electromagnetic "missiles" in Section 5.4.3, produced by a circular current disk with step-function time dependence, satisfy the Maxwell equations (2.18)–(2.21) everywhere outside the circular current disk. Whether or not the differential Maxwell equations (2.18)–(2.21) are also satisfied within the current disk is a moot concern.

Nonetheless, applications that are especially concerned with fields in the source region may require more rigorous analyses that state explicitly the sufficient conditions on sources to ensure the existence of a well-defined solution. For example, the expressions in Section 2.1.7 for the electric and magnetic fields in terms of the volume current density $\mathbf{J}(\mathbf{r}, t)$ involve "principal volume" integrations and "source dyadic" terms that are ill-defined within the source region without sufficient conditions on the current such as those determined in this section.

As we shall prove shortly, sufficient conditions on the current $\mathbf{J}(\mathbf{r}, t)$ for (2.53)–(2.55) to satisfy (2.18)–(2.21) are:

(i) $\mathbf{J}(\mathbf{r}, t)$ is located within a bounded volume V,

(ii) $\mathbf{J}(\mathbf{r}, t)$ and its first partial derivatives (with respect to x, y, z, and t) are continuous functions of $\mathbf{r}$ and t, and

(iii) $\nabla \cdot \mathbf{J}(\mathbf{r}, t)$ is Hölder continuous in $\mathbf{r}$ for all t.

Then $\mathbf{E}(\mathbf{r}, t)$ and $\mathbf{H}(\mathbf{r}, t)$ given by (2.53)–(2.55) are continuous functions of $\mathbf{r}$ and t that satisfy the time-domain electromagnetic field equations (2.18)–(2.21). Moreover, because this solution satisfies causality, the uniqueness theorem of Section 2.1.5 implies that (2.53)–(2.55) is the unique causal solution to the Maxwellian equations (2.18)–(2.21).

To derive these existence conditions, we use the following two subsidiary theorems which are proven in Appendix B.

Theorem 2-I If $f(\mathbf{r}, t)$ and its time derivative $\partial f(\mathbf{r}, t)/\partial t$ are continuous functions of $\mathbf{r}$ and t, then

$$\frac{\partial}{\partial x_i} \lim_{\delta \to 0} \int\limits_{V-V_\delta} \frac{f(\mathbf{r}', t - |\mathbf{r} - \mathbf{r}'|/c)}{|\mathbf{r} - \mathbf{r}'|} dV' = \lim_{\delta \to 0} \int\limits_{V-V_\delta} \frac{\partial}{\partial x_i} \left[\frac{f(\mathbf{r}', t - |\mathbf{r} - \mathbf{r}'|/c)}{|\mathbf{r} - \mathbf{r}'|} \right] dV' \tag{2.56}$$

where the left- and right-hand sides of (2.56) exist and are continuous functions of $\mathbf{r}$ and t. The x_i is equal to x, y, or z for $i = 1, 2$, or 3, respectively; that is, $\mathbf{r} = x\hat{\mathbf{x}} + y\hat{\mathbf{y}} + z\hat{\mathbf{z}} = x_1\hat{\mathbf{x}} + x_2\hat{\mathbf{y}} + x_3\hat{\mathbf{z}}$.

Theorem 2-II If $f(\mathbf{r}, t)$, $\partial f(\mathbf{r}, t)/\partial t$, and $\partial^2 f(\mathbf{r}, t)/\partial t^2$ are continuous functions of $\mathbf{r}$ and t, and $f(\mathbf{r}, t)$ is Hölder continuous in $\mathbf{r}$ for all t, then

$$\begin{aligned}\frac{\partial^2}{\partial x_i \partial x_j} \lim_{\delta \to 0} \int\limits_{V-V_\delta} \frac{f(\mathbf{r}', t - |\mathbf{r} - \mathbf{r}'|/c)}{|\mathbf{r} - \mathbf{r}'|} dV' \\ = -\frac{4}{3}\pi f(\mathbf{r}, t)\delta_{ij} + \lim_{\delta \to 0} \int\limits_{V-V_\delta^s} \frac{\partial^2}{\partial x_i \partial x_j}\left[\frac{f(\mathbf{r}', t - |\mathbf{r} - \mathbf{r}'|/c)}{|\mathbf{r} - \mathbf{r}'|}\right] dV'\end{aligned} \tag{2.57}$$

where $i, j = 1, 2, 3$ and the left- and right-hand sides of (2.57) exist.

It is assumed that $f(\mathbf{r}, t)$ is zero for $\mathbf{r}$ outside a bounded volume V. For functions $f(\mathbf{r}, t)$ that are also zero for $t \le t_0$, continuity of $\partial f(\mathbf{r}, t)/\partial t$ with respect to $(\mathbf{r}, t)$ implies continuity of $f(\mathbf{r}, t)$ with respect to $(\mathbf{r}, t)$ as well [93, pp. 324–325, ex. 8]. Thus, for these functions that begin at $t = t_0$, it is redundant to include $f(\mathbf{r}, t)$ in the first line of Theorem 2-I, or $f(\mathbf{r}, t)$ and $\partial f(\mathbf{r}, t)/\partial t$ in the first line of Theorem 2-II. Also, $\mathbf{J}(\mathbf{r}, t)$ need not be included in addition to its partial derivatives in condition (ii) above.

A function is Hölder continuous in a closed domain if it satisfies a Hölder condition at every point of that domain. The Hölder condition is defined in a number of textbooks, for example, [76, p. 152], and in Appendix B. If a function satisfies the Hölder condition at a point, it is smoother in a neighborhood of that point than ordinary continuity demands, yet the function need not be differentiable at that point. If a function is differentiable at a point, it is Hölder continuous at that point. Thus, the existence of the first partial derivatives of $\mathbf{J}(\mathbf{r}, t)$ in condition (ii) ensures that $\mathbf{J}(\mathbf{r}, t)$ is Hölder continuous in $\mathbf{r}$ and t.

The symbol δ_{ij} in (2.57) is the Kronecker delta defined by

$$\delta_{ij} = \begin{cases} 1 & \text{for} \quad i = j \\ 0 & \text{for} \quad i \ne j \end{cases} \tag{2.58}$$

and V_δ^s refers to a spherical principal volume with the singularity at the center of the sphere. For another shaped principal volume, the integral on the right-hand side of (2.57) can have a different value. Then the 4/3 factor on the right-hand side of (2.57) will also change so as to keep the combined value of the two terms on the right-hand side of (2.57) equal to the unique value of the left-hand side of (2.57), which is independent of the shape and orientation of V_δ. For the purpose of deriving our existence conditions, we need only consider a spherical principal volume on the right-hand side of (2.57).

To prove that (2.53)–(2.55) satisfy Maxwell's equations under the conditions (i), (ii), and (iii), we begin by showing that (2.54) can also be expressed as

$$\mathbf{E}(\mathbf{r}, t) = -\frac{\partial}{\partial t}\mathbf{A}(\mathbf{r}, t) - \nabla\Psi(\mathbf{r}, t) \tag{2.59}$$

where the scalar potential $\Psi(\mathbf{r}, t)$ is given by

$$\Psi(\mathbf{r}, t) = \frac{1}{4\pi\epsilon} \lim_{\delta \to 0} \int\limits_{V-V_\delta} \frac{\rho(\mathbf{r}', t - |\mathbf{r} - \mathbf{r}'|/c)}{|\mathbf{r} - \mathbf{r}'|} dV'. \tag{2.60}$$

For sources and fields that are zero for $t \le t_0$, the charge density ρ can be expressed in terms of $\mathbf{J}$ through (2.16), so that

$$\rho(\mathbf{r}', t - |\mathbf{r} - \mathbf{r}'|/c) = -\int_{t_0}^{t} \{\nabla' \cdot \mathbf{J}\}(\mathbf{r}', t' - |\mathbf{r} - \mathbf{r}'|/c)dt' \tag{2.61}$$

where the curly brackets on $\{\nabla' \cdot \mathbf{J}\}$ in (2.61) indicate that $\nabla' \cdot$ operates only on the first $\mathbf{r}'$ variable in $\mathbf{J}(\mathbf{r}', t' - |\mathbf{r}-\mathbf{r}'|/c)$, that is

$$\{\nabla' \cdot \mathbf{J}\}(\mathbf{r}', t' - |\mathbf{r}-\mathbf{r}'|/c) = \nabla \cdot \mathbf{J}(\mathbf{r}, t)|_{\mathbf{r}=\mathbf{r}', t=t'-|\mathbf{r}-\mathbf{r}'|/c}. \tag{2.62}$$

To derive (2.59) from (2.54), we write (2.54) as

$$\begin{aligned}\mathbf{E}(\mathbf{r}, t) = {} & \frac{1}{4\pi\epsilon}(-\nabla^2 + \nabla\nabla\cdot) \lim_{\delta\to 0} \int_{V-V_\delta} \left[\frac{1}{|\mathbf{r}-\mathbf{r}'|} \int_{t_0}^{t} \mathbf{J}(\mathbf{r}', t' - |\mathbf{r}-\mathbf{r}'|/c)dt'\right] dV' \\ & - \frac{1}{\epsilon} \int_{t_0}^{t} \mathbf{J}(\mathbf{r}, t')dt'.\end{aligned} \tag{2.63}$$

The interchange of the time and volume integrations required to obtain the first term on the right-hand side of (2.63) is justified by writing the volume integration in spherical coordinates centered at $\mathbf{r}$ (to obtain a continuous integrand and eliminate the limiting process), and applying the theorem of integral calculus [90, p. 238] that allows the order of integrations to be interchanged when the integrand is continuous in all the integration variables and the limits of integration are finite.

With the help of Theorems 2-I and 2-II, (2.63) becomes

$$\begin{aligned}\mathbf{E}(\mathbf{r}, t) = {} & -\frac{1}{4\pi\epsilon} \lim_{\delta\to 0} \int_{V-V_\delta^s} \nabla^2 \left[\frac{1}{|\mathbf{r}-\mathbf{r}'|} \int_{t_0}^{t} \mathbf{J}(\mathbf{r}', t' - |\mathbf{r}-\mathbf{r}'|/c)dt'\right] dV' \\ & + \frac{1}{4\pi\epsilon} \nabla \lim_{\delta\to 0} \int_{V-V_\delta} \nabla \cdot \left[\frac{1}{|\mathbf{r}-\mathbf{r}'|} \int_{t_0}^{t} \mathbf{J}(\mathbf{r}', t' - |\mathbf{r}-\mathbf{r}'|/c)dt'\right] dV'\end{aligned} \tag{2.64}$$

if $\mathbf{J}(\mathbf{r}, t)$ satisfies the above stated conditions (i), (ii), and (iii). We have used the fact that the integral of the current over time can be rewritten as

$$\int_{t_0}^{t} \mathbf{J}(\mathbf{r}', t' - |\mathbf{r}-\mathbf{r}'|/c)dt' = \int_{t_0}^{t-|\mathbf{r}-\mathbf{r}'|/c} \mathbf{J}(\mathbf{r}', t')dt' \tag{2.65}$$

and, therefore, it is a function of $\mathbf{r}'$ and $t - |\mathbf{r}-\mathbf{r}'|/c$. The integral $\int_{t_0}^{t} \mathbf{J}(\mathbf{r}, t')dt'$ is Hölder continuous in $\mathbf{r}$ for all t because $\mathbf{J}(\mathbf{r}, t)$ is Hölder continuous in $\mathbf{r}$ for all t. We can now apply the identities [83, pp. 497–498]

$$\begin{aligned}\nabla \cdot & \left[\frac{1}{|\mathbf{r}-\mathbf{r}'|} \int_{t_0}^{t} \mathbf{J}(\mathbf{r}', t' - |\mathbf{r}-\mathbf{r}'|/c)dt'\right] \\ & = -\int_{t_0}^{t} \mathbf{J}(\mathbf{r}', t' - |\mathbf{r}-\mathbf{r}'|/c)dt' \cdot \nabla'\left(\frac{1}{|\mathbf{r}-\mathbf{r}'|}\right) \\ & \quad - \frac{\mathbf{J}(\mathbf{r}', t - |\mathbf{r}-\mathbf{r}'|/c) \cdot (\mathbf{r}-\mathbf{r}')}{c|\mathbf{r}-\mathbf{r}'|^2}\end{aligned} \tag{2.66}$$

and

$$\nabla^2 \left[\frac{1}{|\mathbf{r}-\mathbf{r}'|} \int_{t_0}^{t} \mathbf{J}(\mathbf{r}', t' - |\mathbf{r}-\mathbf{r}'|/c)dt'\right] = \frac{\frac{\partial}{\partial t}\mathbf{J}(\mathbf{r}', t - |\mathbf{r}-\mathbf{r}'|/c)}{c^2|\mathbf{r}-\mathbf{r}'|}, \quad \mathbf{r}' \neq \mathbf{r} \tag{2.67}$$

to convert (2.64) to

$$\mathbf{E}(\mathbf{r},t) = -\frac{\partial}{\partial t}\mathbf{A}(\mathbf{r},t) + \frac{1}{4\pi\epsilon}\nabla \lim_{\delta\to 0}\int\limits_{V-V_\delta}\left[\frac{1}{|\mathbf{r}-\mathbf{r}'|}\nabla'\cdot\int\limits_{t_0}^{t}\mathbf{J}(\mathbf{r}',t'-|\mathbf{r}-\mathbf{r}'|/c)dt' - \frac{\mathbf{J}(\mathbf{r}',t-|\mathbf{r}-\mathbf{r}'|/c)\cdot(\mathbf{r}-\mathbf{r}')}{c|\mathbf{r}-\mathbf{r}'|^2}\right]dV' \tag{2.68}$$

where the divergence theorem has been used to obtain the first term in the integrand of the volume integral in (2.68), and $\mathbf{A}(\mathbf{r},t)$ is given in (2.55). Use of the divergence theorem is allowed because $\mathbf{J}(\mathbf{r},t)$ and its first partial derivatives are continuous functions of $\mathbf{r}$ and t, in accordance with the condition (ii). The time derivative on $\mathbf{J}$ has been brought outside the volume integral for $\mathbf{A}(\mathbf{r},t)$ defined in (2.55) in order to get the $\partial\mathbf{A}/\partial t$ term on the right-hand side of (2.68). This is permissible because the volume integration can be written in spherical coordinates centered at $\mathbf{r}$ to eliminate the singularity and the limiting process. Then Leibnitz's rule [90, p. 266] can be applied, given that $\mathbf{J}(\mathbf{r}',t-|\mathbf{r}-\mathbf{r}'|/c)$ and $\partial\mathbf{J}(\mathbf{r}',t-|\mathbf{r}-\mathbf{r}'|/c)/\partial t$ are continuous functions of $\mathbf{r}'$ and t for each $\mathbf{r}$, to bring the time derivative outside the integral sign. Moreover, $\partial\mathbf{A}(\mathbf{r},t)/\partial t$ is a continuous function of $(\mathbf{r},t)$ [90, p. 273].

By means of Leibnitz's rule, the continuity of $\mathbf{J}$ and $\nabla\cdot\mathbf{J}$ also allows the ∇' operator in (2.68) to be brought under the integral sign, so that

$$\nabla'\cdot\int\limits_{t_0}^{t}\mathbf{J}(\mathbf{r}',t'-|\mathbf{r}-\mathbf{r}'|/c)dt' = \int\limits_{t_0}^{t}\{\nabla'\cdot\mathbf{J}\}(\mathbf{r}',t'-|\mathbf{r}-\mathbf{r}'|/c)dt' + \frac{\mathbf{J}(\mathbf{r}',t-|\mathbf{r}-\mathbf{r}'|/c)\cdot(\mathbf{r}-\mathbf{r}')}{c|\mathbf{r}-\mathbf{r}'|} \tag{2.69}$$

and (2.68) becomes identical to the desired result (2.59).

We have just proven that the expression (2.54) for the electric field implies (2.59) if the current satisfies the conditions (i), (ii), and (iii) stated above. Since all the steps in the proof are reversible, (2.59) also implies (2.54). Thus, (2.54) and (2.59) are equivalent expressions for the electric field, under the stated conditions on the behavior of the current, when sources and fields are zero for $t \le t_0$.

The conditions (i), (ii), and (iii) are also sufficient to prove from (2.55) and (2.60), using Theorem 2-II, that the vector and scalar potentials satisfy the equations[5]

$$\nabla^2\int\limits_{t_0}^{t}\mathbf{A}(\mathbf{r},t')dt' - \frac{1}{c^2}\frac{\partial}{\partial t}\mathbf{A}(\mathbf{r},t) = -\mu\int\limits_{t_0}^{t}\mathbf{J}(\mathbf{r},t')dt' \tag{2.72}$$

---

[5]Taking the time derivatives of (2.72) and (2.73) produces the familiar wave equations for the vector and scalar potentials

$$\nabla^2\mathbf{A} - \frac{1}{c^2}\frac{\partial^2\mathbf{A}}{\partial t^2} = -\mu\mathbf{J} \tag{2.70}$$

and

$$\nabla^2\Psi - \frac{1}{c^2}\frac{\partial^2\Psi}{\partial t^2} = -\rho/\epsilon \tag{2.71}$$

if the second derivatives of $\mathbf{A}$ and Ψ exist, and the time derivatives can be interchanged with the ∇^2 operators. We keep the form of the equations (2.72) and (2.73), however, because the conditions (i), (ii), and (iii) are not sufficient to prove that the second derivatives of $\mathbf{A}$ and Ψ exist or that the interchanges of the time derivatives and the ∇^2 operators are valid. Theorem 2-II tells us that sufficient conditions for the vector and scalar potentials in (2.55) and (2.60) to satisfy the familiar wave equations (2.70) and (2.71) everywhere are that $\mathbf{J}(\mathbf{r},t)$ and $\rho(\mathbf{r},t)$ are located within a bounded volume V, that $\mathbf{J}(\mathbf{r},t)$ and $\rho(\mathbf{r},t)$ along with their first and second time derivatives are continuous functions of $(\mathbf{r},t)$, and that $\mathbf{J}(\mathbf{r},t)$ and $\rho(\mathbf{r},t)$ are Hölder continuous in $\mathbf{r}$ for all t.

and

$$\nabla^2 \int_{t_0}^{t} \Psi(\mathbf{r}, t')dt' - \frac{1}{c^2}\frac{\partial}{\partial t}\Psi(\mathbf{r}, t) = -\frac{1}{\epsilon}\int_{t_0}^{t} \rho(\mathbf{r}, t')dt'. \tag{2.73}$$

We are now in a position to prove that the fields in (2.53) and (2.59), which are equivalent to (2.53) and (2.54), satisfy Maxwell's equations (2.18)–(2.21). Integrate (2.59) over time to get

$$\int_{t_0}^{t} \mathbf{E}(\mathbf{r}, t')dt' = -\mathbf{A}(\mathbf{r}, t) - \nabla \int_{t_0}^{t} \Psi(\mathbf{r}, t')dt' \tag{2.74}$$

The interchange of the time integral and the ∇ operator in (2.74) is allowed by Leibnitz's rule because $\Psi(\mathbf{r}, t)$ and $\nabla\Psi(\mathbf{r}, t)$ are continuous functions of $(\mathbf{r}, t)$. The continuity of Ψ and $\nabla\Psi$ is assured by Theorem 2-I for current $\mathbf{J}(\mathbf{r}, t)$ that satisfies the conditions (i) and (ii). Next, take the curl of (2.74) and use (2.53) to get

$$\nabla \times \int_{t_0}^{t} \mathbf{E}(\mathbf{r}, t')dt' = -\nabla \times \mathbf{A}(\mathbf{r}, t) = -\mu\mathbf{H}(\mathbf{r}, t) \tag{2.75}$$

which is identical to the first Maxwell equation (2.18). In obtaining (2.75) from (2.74), the curl of the gradient on the right-hand side of (2.74) was set equal to zero. This is valid provided the second spatial derivatives of $\int_{t_0}^{t} \Psi(\mathbf{r}, t')dt'$ exist. The existence of these derivatives is assured by Theorem 2-II for $\mathbf{J}(\mathbf{r}, t)$ that obeys conditions (i), (ii), and (iii).

To obtain Maxwell's second equation (2.19), integrate (2.53) over time so that (2.53) becomes

$$\int_{t_0}^{t} \mathbf{H}(\mathbf{r}, t')dt' = \frac{1}{\mu}\nabla \times \int_{t_0}^{t} \mathbf{A}(\mathbf{r}, t')dt'. \tag{2.76}$$

Then take the curl of (2.76) to get

$$\nabla \times \int_{t_0}^{t} \mathbf{H}(\mathbf{r}, t')dt' = \frac{1}{\mu}\nabla \times \nabla \times \int_{t_0}^{t} \mathbf{A}(\mathbf{r}, t')dt' \tag{2.77}$$

or from (2.54)

$$\nabla \times \int_{t_0}^{t} \mathbf{H}(\mathbf{r}, t')dt' = \epsilon\mathbf{E}(\mathbf{r}, t) + \int_{t_0}^{t} \mathbf{J}(\mathbf{r}, t')dt' \tag{2.78}$$

which is identical to (2.19). The interchange of the time integration and the curl operation used to obtain the right-hand side of (2.76) is again allowed by Leibnitz's rule because $\mathbf{A}(\mathbf{r}, t)$ and $\nabla \times \mathbf{A}(\mathbf{r}, t)$ are continuous functions of $(\mathbf{r}, t)$ when $\mathbf{J}(\mathbf{r}, t)$ satisfies the conditions (i) and (ii).

In a similar fashion, the Maxwell divergence equations (2.20) and (2.21) result from taking the divergence of (2.76) and (2.74), respectively, and invoking Theorems 2-I and 2-II along with the current conditions (i), (ii), and (iii). Under these current conditions, Theorem 2-I also shows that $\nabla \times \mathbf{A}(\mathbf{r}, t)$ in (2.53) and $\nabla\Psi$ in (2.59) are continuous functions of $(\mathbf{r}, t)$. And, as explained after (2.68), $\partial\mathbf{A}(\mathbf{r}, t)/\partial t$ is a continuous function of $(\mathbf{r}, t)$. Thus, $\mathbf{H}(\mathbf{r}, t)$ and $\mathbf{E}(\mathbf{r}, t)$ are continuous functions of $\mathbf{r}$ and t.

In summary, $\mathbf{E}(\mathbf{r}, t)$ and $\mathbf{H}(\mathbf{r}, t)$ given in (2.53)–(2.55) are unique, causal, continuous fields that satisfy Maxwell's time-domain equations (2.18)–(2.21) for current $\mathbf{J}(\mathbf{r}, t)$ that satisfies the above stated conditions (i), (ii), and (iii). At the end of the next section, where the static solutions to Maxwell's equations are shown explicitly, similar conditions on the current $\mathbf{J}(\mathbf{r})$ and charge $\rho(\mathbf{r})$ are given for the existence of solution to Maxwell's time-independent equations, that is, to the magnetostatic and electrostatic equations, (2.23) and (2.22).

2.1.7 Volume Integral Expressions for the Electric and Magnetic Fields

The theorems and current conditions used in Section 2.1.6 to prove that the vector-potential expressions satisfy Maxwell's equations also allow the differential operators in (2.53), and (2.54) or (2.59) to be brought inside the integrals so that $\mathbf{H}(\mathbf{r}, t)$ and $\mathbf{E}(\mathbf{r}, t)$ can be written as

$$\mathbf{H}(\mathbf{r}, t) = \frac{1}{4\pi} \lim_{\delta \to 0} \int\limits_{V-V_\delta} \left[\frac{1}{c} \frac{\partial}{\partial t} \mathbf{J}(\mathbf{r}', t - R/c) + \frac{\mathbf{J}(\mathbf{r}', t - R/c)}{R} \right] \times \frac{\mathbf{R}}{R^2} dV' \tag{2.79}$$

and

$$\begin{aligned} \mathbf{E}(\mathbf{r}, t) &= \frac{1}{4\pi\epsilon} \nabla \times \lim_{\delta \to 0} \int\limits_{V-V_\delta} \left[\frac{1}{c} \mathbf{J}(\mathbf{r}', t - R/c) + \frac{1}{R} \int_{t_0}^{t} \mathbf{J}(\mathbf{r}', t' - R/c) dt' \right] \times \frac{\mathbf{R}}{R^2} dV' \\ &\quad - \frac{1}{\epsilon} \int_{t_0}^{t} \mathbf{J}(\mathbf{r}, t') dt' \\ &= \frac{1}{4\pi\epsilon} \lim_{\delta \to 0} \int\limits_{V-V_\delta^s} \Bigg[\frac{1}{c^2 R} \frac{\partial \mathbf{J}(\mathbf{r}', t - R/c)}{\partial t} \cdot \left(\frac{\mathbf{RR}}{R^2} - \bar{\mathbf{I}} \right) \\ &\qquad + \frac{1}{R^2} \left(\frac{1}{c} \mathbf{J}(\mathbf{r}', t - R/c) + \frac{1}{R} \int_{t_0}^{t} \mathbf{J}(\mathbf{r}', t' - R/c) dt' \right) \\ &\qquad \cdot \left(\frac{3\mathbf{RR}}{R^2} - \bar{\mathbf{I}} \right) \Bigg] dV' - \frac{1}{3\epsilon} \int_{t_0}^{t} \mathbf{J}(\mathbf{r}, t') dt' \end{aligned} \tag{2.80}$$

or from (2.59) and the dyadic divergence theorem in [83, app. 3, eq. 53]

$$\begin{aligned} \mathbf{E}(\mathbf{r}, t) = \frac{1}{4\pi\epsilon} \lim_{\delta \to 0} \int\limits_{V-V_\delta} \Bigg[&\frac{1}{c^2 R} \frac{\partial \mathbf{J}(\mathbf{r}', t - R/c)}{\partial t} \cdot \left(\frac{\mathbf{RR}}{R^2} - \bar{\mathbf{I}} \right) \\ &+ \frac{1}{cR^2} \mathbf{J}(\mathbf{r}', t - R/c) \cdot \left(\frac{2\mathbf{RR}}{R^2} - \bar{\mathbf{I}} \right) \\ &+ \rho(\mathbf{r}', t - R/c) \frac{\mathbf{R}}{R^3} \Bigg] dV' \end{aligned} \tag{2.81}$$

where $\mathbf{R}$ has been substituted for $\mathbf{r} - \mathbf{r}'$, $\bar{\mathbf{I}} = \hat{\mathbf{x}}\hat{\mathbf{x}} + \hat{\mathbf{y}}\hat{\mathbf{y}} + \hat{\mathbf{z}}\hat{\mathbf{z}}$ is the identity or unit dyadic, and use has been made of the fundamental relation

$$\nabla f(t - R/c) = -\frac{1}{c} \frac{\partial}{\partial t} f(t - R/c) \frac{\mathbf{R}}{R}. \tag{2.82}$$

Recall that (2.54), and thus (2.80), assume the sources and fields are zero for $t \leq t_0$ and do not hold for electrostatic fields. However, (2.79) and (2.81) remain valid expressions for the magnetostatic and electrostatic fields, respectively, produced by static current and charge distributions.

Equation (2.80) has been derived using Theorem 2-II of Section 2.1.6, and thus contains the spherical principal volume V_δ^s. For an arbitrarily shaped limiting principal volume V_δ, equation (2.80) becomes [72]–[75]

$$\mathbf{E}(\mathbf{r},t) = \frac{1}{4\pi\epsilon}\lim_{\delta\to 0}\int\limits_{V-V_\delta}\left[\frac{1}{c^2R}\frac{\partial\mathbf{J}(\mathbf{r}',t-R/c)}{\partial t}\cdot\left(\frac{\mathbf{RR}}{R^2}-\bar{\mathbf{I}}\right)\right.$$
$$+\frac{1}{R^2}\left(\frac{1}{c}\mathbf{J}(\mathbf{r}',t-R/c)+\frac{1}{R}\int\limits_{t_0}^{t}\mathbf{J}(\mathbf{r}',t'-R/c)dt'\right) \tag{2.83}$$
$$\left.\cdot\left(\frac{3\mathbf{RR}}{R^2}-\bar{\mathbf{I}}\right)\right]dV' - \frac{1}{\epsilon}\bar{\mathbf{L}}\cdot\int\limits_{t_0}^{t}\mathbf{J}(\mathbf{r},t')dt'$$

where the source or depolarization dyadic $\bar{\mathbf{L}}$ is defined as

$$\bar{\mathbf{L}} \equiv \frac{1}{4\pi}\int\limits_{S_\delta}\frac{\hat{\mathbf{n}}\hat{\mathbf{r}}}{r^2}dS \tag{2.84}$$

with $\hat{\mathbf{n}}$ being the outward unit normal to the surface S_δ of the principal volume V_δ. ($\bar{\mathbf{L}}$ is a real symmetric dyadic with unity trace [72].) The result (2.83) can be proven directly from (2.80) by dividing the volume $V - V_\delta^s$ into $V - V_\delta$ and $V_\delta - V_\delta^s$, and then converting the volume integral over $V_\delta - V_\delta^s$ to surface integrals over S_δ and S_δ^s as $\delta \to \infty$. The value of both the volume integral and the $\bar{\mathbf{L}}$ term on the right-hand side of (2.83) depends on the shape and position with respect to the singularity of the principal volume V_δ, but add to produce the unique electric field on the left-hand side of (2.83). For observation points $\mathbf{r}$ outside the source current $\mathbf{J}$, the integrand of the volume integral is no longer singular, the limiting principal volume can be removed, and the $\bar{\mathbf{L}}$ dyadic term vanishes.

Equations (2.79), (2.80) or (2.83), and (2.81) can also be derived by taking the inverse Fourier transform of the corresponding frequency-domain formulas, which can be found in Section 2.3.7. However, the advantage of having rigorously derived (2.79), (2.80) or (2.83), and (2.81) in the time domain from the theorems of Section 2.1.6 is that one is assured that the integrals in (2.79), (2.80) or (2.83), and (2.81) converge to give continuous $\mathbf{E}(\mathbf{r},t)$ and $\mathbf{H}(\mathbf{r},t)$ fields, and that (2.79) and (2.80), or (2.79) and (2.83), or (2.79) and (2.81) represent the unique causal solution to Maxwell's equations (2.18)–(2.21) for all $\mathbf{r}$ and t if the current satisfies the conditions (i), (ii), and (iii) of Section 2.1.6.

Equations (2.79) and (2.83) determine the time-dependent "Maxwellian" fields, that is, the magnetic and electric fields defined mathematically by Maxwell's equations, in terms of the time-dependent volume current density. These equations are also in a form that shows that "cavity-defined" [1, sec. 3.28] magnetic and electric fields within a time-dependent current distribution are identical to the Maxwellian magnetic and electric fields. To show this, consider the definition of cavity-defined fields within a time-dependent current distribution. They are defined as the fields measured at a point $\mathbf{r}$ at time t within that current distribution when an infinitesimal volume V_D of current that contains the point $\mathbf{r}$ and has maximum dimension D is removed for an infinitesimal amount of time Δt about the time t [73]. Since (2.79) and (2.83) already are written in terms of a principal volume V_δ that excludes the observation point $\mathbf{r}$ from the integration, we can replace V_δ by V_D, set $\mathbf{J}(\mathbf{r},t)$ equal to zero inside V_D during the time Δt, take the limit as $D \to 0$, and (2.79) shows immediately that there is no difference between the Maxwellian and cavity-defined magnetic field in the time-dependent current distribution. Similarly, (2.83) shows that the difference between the Maxwellian and cavity-defined electric field approaches $-\Delta t \bar{\mathbf{L}} \cdot \mathbf{J}(\mathbf{r}, t)/\epsilon$, that is, zero in the limit as $\Delta t \to 0$. (Note that cavity-defined fields are determined by letting D approach zero before Δt approaches zero.)[6]

As mentioned above, equations (2.79) and (2.81) are valid expressions for static as well as time-dependent electromagnetic fields. Specifically, (2.79) and (2.81) give the magnetostatic and electrostatic fields, respectively, when $\mathbf{J}$ and ρ are independent of time. (The contribution from the current terms in (2.81) is zero when $\mathbf{J}$ is independent of time and $\nabla \cdot \mathbf{J} = 0$.) The magnetostatic field can therefore be written from (2.79), and the electrostatic field from (2.81) as

$$\begin{aligned}\mathbf{H}(\mathbf{r}) &= \frac{1}{4\pi} \lim_{\delta \to 0} \int\limits_{V-V_\delta} \mathbf{J}(\mathbf{r}') \times \nabla' \left(\frac{1}{|\mathbf{r} - \mathbf{r}'|} \right) dV' \\ &= \nabla \times \lim_{\delta \to 0} \int\limits_{V-V_\delta} \frac{\mathbf{J}(\mathbf{r}')}{4\pi |\mathbf{r} - \mathbf{r}'|} dV'\end{aligned} \tag{2.85}$$

and

$$\begin{aligned}\mathbf{E}(\mathbf{r}) &= \frac{1}{4\pi\epsilon} \lim_{\delta \to 0} \int\limits_{V-V_\delta} \rho(\mathbf{r}') \nabla' \left(\frac{1}{|\mathbf{r} - \mathbf{r}'|} \right) dV' \\ &= -\nabla \lim_{\delta \to 0} \int\limits_{V-V_\delta} \frac{\rho(\mathbf{r}')}{4\pi\epsilon |\mathbf{r} - \mathbf{r}'|} dV'.\end{aligned} \tag{2.86}$$

As explained at the end of Section 2.1.5, if a solution to the magnetostatic equations (2.23) or the electrostatic equations (2.22) exists, it is unique for twice continuously differentiable static potentials that approach zero as $r \to \infty$. Sufficient conditions for the existence of a magnetostatic solution, that is, for the magnetostatic field in (2.85) to obey Maxwell's equations (2.23) are that $\mathbf{J}(\mathbf{r})$ lies in a bounded volume, that $\mathbf{J}(\mathbf{r})$ is Hölder continuous, and that $\nabla \cdot \mathbf{J}(\mathbf{r})$ exists and equals zero. Sufficient conditions for the existence of an electrostatic solution, that is, for the electrostatic field in (2.86) to obey Maxwell's equations (2.22), are that $\rho(\mathbf{r})$ lies in a bounded volume and is Hölder continuous. (These sufficient conditions follow from Theorems 2-I and 2-II.) Since both the divergence and curl of $\mathbf{E}(\mathbf{r})$ and $\mathbf{H}(\mathbf{r})$ exist under these conditions, $\mathbf{E}(\mathbf{r})$ and $\mathbf{H}(\mathbf{r})$ are continuous functions of $\mathbf{r}$. Equations (2.85) and (2.86) also show that the cavity-defined magnetostatic and electrostatic fields in a current and charge distribution are identical to the Maxwellian magnetostatic and electrostatic fields, respectively.

2.1.8 Electric and Magnetic Fields of a Moving Point Charge

In the analysis of electromagnetic "missiles" in Section 5.4.1, we use expressions for the electric and magnetic fields of a moving point charge. These expressions can be found in a relatively simple manner from (2.79) and (2.81). Begin by rewriting (2.79) outside the source

---

[6] In electric polarization density $\mathbf{P}(\mathbf{r}, t)$ and magnetic polarization density $\mathbf{M}(\mathbf{r}, t)$, the cavity-defined electric and magnetic fields ($\mathbf{E}_c$ and $\mathbf{H}_c$) differ from the Maxwellian electric and magnetic fields ($\mathbf{E}$ and $\mathbf{H}$) by the amounts $\mathbf{E}(\mathbf{r}, t) - \mathbf{E}_c(\mathbf{r}, t) = -\bar{\mathbf{L}} \cdot \mathbf{P}(\mathbf{r}, t)/\epsilon$ and $\mathbf{H}(\mathbf{r}, t) - \mathbf{H}_c(\mathbf{r}, t) = (\mathbf{B}(\mathbf{r}, t) - \mathbf{B}_c(\mathbf{r}, t))/\mu - \mathbf{M} = \bar{\mathbf{L}} \cdot \mathbf{M}(\mathbf{r}, t)$, respectively, because the integral over time of $\mathbf{J}(\mathbf{r}, t)$ in the last term of (2.83) is absent from the corresponding term for $\mathbf{P}(\mathbf{r}, t)$ and $\mathbf{M}(\mathbf{r}, t)$ [73]. This is true for frequency-domain fields in electric and magnetic polarization as well. Moreover, for frequency-domain fields, the cavity-defined electric field in a current distribution also differs from the Maxwellian electric field because the frequency-domain current is removed from the cavity for all time [72], [73] (see Section 2.3.7). Nonetheless, for the real time-harmonic fields found by taking the real part of the frequency-domain fields multiplied by $\exp(-i\omega t)$, the cavity-defined and Maxwellian electric fields are identical within a distribution of real time-harmonic current density, when the current in the infinitesimal cavity is removed only for an infinitesimal amount of time Δt.

region in the form

$$\begin{aligned}\mathbf{H}(\mathbf{r},t) &= \frac{1}{4\pi}\int_{-\infty}^{+\infty}\int_V \left[\frac{1}{c}\frac{\partial}{\partial t}\mathbf{J}(\mathbf{r}',t') + \frac{\mathbf{J}(\mathbf{r}',t')}{R}\right] \times \frac{\mathbf{R}}{R^2}\delta(t'-t+R/c)dV'dt' \\ &= \frac{1}{4\pi}\int_{-\infty}^{+\infty}\int_V \mathbf{J}(\mathbf{r}',t') \times \frac{\mathbf{R}}{cR^2}\frac{\partial}{\partial t}\delta(t'-t+R/c)dV'dt' \\ &+ \frac{1}{4\pi}\int_{-\infty}^{+\infty}\int_V \mathbf{J}(\mathbf{r}',t') \times \frac{\mathbf{R}}{R^3}\delta(t'-t+R/c)dV'dt'.\end{aligned} \tag{2.87}$$

The charge and current densities for an electric point charge q_e at the position $\mathbf{r}_q(t)$ and moving with velocity $\mathbf{u}(t) = d\mathbf{r}_q/dt = \dot{\mathbf{r}}_q(t)$ are given by

$$\rho(\mathbf{r},t) = q_e\delta\left(\mathbf{r} - \mathbf{r}_q(t)\right) \tag{2.88}$$

and

$$\mathbf{J}(\mathbf{r},t) = q_e\delta\left(\mathbf{r} - \mathbf{r}_q(t)\right)\mathbf{u}(t). \tag{2.89}$$

(Here we allow delta functions in (2.79) and (2.81) because we are interested in the fields outside the source region only.) Substituting (2.89) into (2.87) and performing the volume integrations over the delta functions give

$$\begin{aligned}\mathbf{H}(\mathbf{r},t) &= \frac{q_e}{4\pi}\int_{-\infty}^{+\infty} \frac{\mathbf{u}(t') \times \left(\mathbf{r} - \mathbf{r}_q(t')\right)}{c|\mathbf{r} - \mathbf{r}_q(t')|^2}\frac{\partial}{\partial t}\delta\left(t' - t + |\mathbf{r} - \mathbf{r}_q(t')|/c\right)dt' \\ &+ \frac{q_e}{4\pi}\int_{-\infty}^{+\infty} \frac{\mathbf{u}(t') \times \left(\mathbf{r} - \mathbf{r}_q(t')\right)}{|\mathbf{r} - \mathbf{r}_q(t')|^3}\delta\left(t' - t + |\mathbf{r} - \mathbf{r}_q(t')|/c\right)dt', \quad \mathbf{r} \neq \mathbf{r}_q(t).\end{aligned} \tag{2.90}$$

With the relationship

$$\begin{aligned}\frac{\partial}{\partial t}\delta\left(t' - t + |\mathbf{r} - \mathbf{r}_q(t')|/c\right) &= \frac{-\frac{\partial}{\partial t'}\delta(t' - t + |\mathbf{r} - \mathbf{r}_q(t')|/c)}{\frac{\partial}{\partial t'}(t' + |\mathbf{r} - \mathbf{r}_q(t')|/c)} \\ &= \frac{-\frac{\partial}{\partial t'}\delta(t' - t + |\mathbf{r} - \mathbf{r}_q(t')|/c)}{\left(1 - \frac{\mathbf{u}(t')\cdot(\mathbf{r} - \mathbf{r}_q(t'))}{c|\mathbf{r} - \mathbf{r}_q(t')|}\right)}\end{aligned} \tag{2.91}$$

(2.90) becomes

$$\begin{aligned}\mathbf{H}(\mathbf{r},t) = \frac{q_e}{4\pi}\int_{-\infty}^{+\infty} &\left\{\frac{\partial}{\partial t'}\left[\frac{\mathbf{u}(t') \times (\mathbf{r} - \mathbf{r}_q(t'))}{c\left(1 - \frac{\mathbf{u}(t')\cdot(\mathbf{r} - \mathbf{r}_q(t'))}{c|\mathbf{r} - \mathbf{r}_q(t')|}\right)|\mathbf{r} - \mathbf{r}_q(t')|^2}\right]\right. \\ &\left. + \frac{\mathbf{u}(t') \times (\mathbf{r} - \mathbf{r}_q(t'))}{|\mathbf{r} - \mathbf{r}_q(t')|^3}\right\} \\ &\cdot\delta\left(t' - t + |\mathbf{r} - \mathbf{r}_q(t')|/c\right)dt', \quad \mathbf{r} \neq \mathbf{r}_q(t)\end{aligned} \tag{2.92}$$

or, after taking the t' derivative

$$\mathbf{H}(\mathbf{r},t)=\frac{q_e}{4\pi}\int_{-\infty}^{+\infty}\left[\frac{\dot{\mathbf{u}}(t')\times(\mathbf{r}-\mathbf{r}_q(t'))}{c\left(1-\frac{\mathbf{u}(t')\cdot(\mathbf{r}-\mathbf{r}_q(t'))}{c|\mathbf{r}-\mathbf{r}_q(t')|}\right)|\mathbf{r}-\mathbf{r}_q(t')|^2}+\frac{\mathbf{u}(t')\times(\mathbf{r}-\mathbf{r}_q(t'))}{\left(1-\frac{\mathbf{u}(t')\cdot(\mathbf{r}-\mathbf{r}_q(t'))}{c|\mathbf{r}-\mathbf{r}_q(t')|}\right)^2|\mathbf{r}-\mathbf{r}_q(t')|^3}\cdot\left(1-\frac{u^2}{c^2}+\frac{\dot{\mathbf{u}}\cdot(\mathbf{r}-\mathbf{r}_q)}{c^2}\right)\right]\delta\left(t'-t+|\mathbf{r}-\mathbf{r}_q|/c\right)dt',\quad \mathbf{r}\neq\mathbf{r}_q(t). \tag{2.93}$$

Even though the $\mathbf{r}_q(t')$ in the delta function in (2.93) is a function of t', we can apply the identity

$$\delta\left[f(t')\right]=\frac{\delta(t'-T_0)}{[df(t')/dt']_{t'=T_0}},\quad f(T_0)=0 \tag{2.94}$$

to get

$$\delta\left(t'-t+\frac{|\mathbf{r}-\mathbf{r}_q(t')|}{c}\right)=\frac{\delta\left(t'-t+\frac{|\mathbf{r}-\mathbf{r}_q(t-|\mathbf{r}-\mathbf{r}_q(t-\cdots)|/c)|}{c}\right)}{1-\frac{\mathbf{u}(t')\cdot(\mathbf{r}-\mathbf{r}_q(t'))}{c|\mathbf{r}-\mathbf{r}_q(t')|}} \tag{2.95}$$

which when substituted into (2.93) gives the final result for the magnetic field of a moving point charge

$$\mathbf{H}(\mathbf{r},t)=\frac{q_e}{4\pi\left(1-\frac{\mathbf{u}\cdot(\mathbf{r}-\mathbf{r}_q)}{c|\mathbf{r}-\mathbf{r}_q|}\right)^2|\mathbf{r}-\mathbf{r}_q|^2}\cdot\left[\frac{\dot{\mathbf{u}}\times(\mathbf{r}-\mathbf{r}_q)}{c}+\frac{\mathbf{u}\times(\mathbf{r}-\mathbf{r}_q)}{\left(1-\frac{\mathbf{u}\cdot(\mathbf{r}-\mathbf{r}_q)}{c|\mathbf{r}-\mathbf{r}_q|}\right)|\mathbf{r}-\mathbf{r}_q|}\cdot\left(1-\frac{u^2}{c^2}+\frac{\dot{\mathbf{u}}\cdot(\mathbf{r}-\mathbf{r}_q)}{c^2}\right)\right],\quad \mathbf{r}\neq\mathbf{r}_q(t). \tag{2.96}$$

The position ($\mathbf{r}_q$), velocity ($\mathbf{u}$), and the acceleration ($\dot{\mathbf{u}}$) in (2.96) are all evaluated at the retarded time τ

$$\tau=t-\frac{|\mathbf{r}-\mathbf{r}_q(\tau)|}{c}=t-\frac{\left|\mathbf{r}-\mathbf{r}_q\left(t-|\mathbf{r}-\mathbf{r}_q(t-\cdots)|/c\right)\right|}{c}. \tag{2.97}$$

The electric field of a moving point charge can be found with the same procedure by substituting ρ and $\mathbf{J}$ from (2.88) and (2.89) into (2.81) to get the expression corresponding to (2.92)

$$\mathbf{E}(\mathbf{r},t)=\frac{q_e}{4\pi\epsilon}\int_{-\infty}^{+\infty}\left\{\frac{\partial}{\partial t'}\left[\frac{\mathbf{u}\cdot\left(\frac{(\mathbf{r}-\mathbf{r}_q)(\mathbf{r}-\mathbf{r}_q)}{|\mathbf{r}-\mathbf{r}_q|^2}-\bar{\mathbf{I}}\right)}{c^2\left(1-\frac{\mathbf{u}\cdot(\mathbf{r}-\mathbf{r}_q)}{c|\mathbf{r}-\mathbf{r}_q|}\right)|\mathbf{r}-\mathbf{r}_q|}\right]+\frac{\mathbf{u}\cdot\left(\frac{2(\mathbf{r}-\mathbf{r}_q)(\mathbf{r}-\mathbf{r}_q)}{|\mathbf{r}-\mathbf{r}_q|^2}-\bar{\mathbf{I}}\right)}{c|\mathbf{r}-\mathbf{r}_q|^2}+\frac{(\mathbf{r}-\mathbf{r}_q)}{|\mathbf{r}-\mathbf{r}_q|^3}\right\}\cdot\delta\left(t'-t+\frac{|\mathbf{r}-\mathbf{r}_q|}{c}\right)dt',\quad \mathbf{r}\neq\mathbf{r}_q(t). \tag{2.98}$$

Taking the time derivative in the integrand of (2.98) and using the relation (2.95) converts (2.98) to

$$\mathbf{E}(\mathbf{r},t) = \frac{q_e}{4\pi\epsilon\left(1-\frac{\mathbf{u}\cdot\mathbf{R}_q}{cR_q}\right)^2 R_q^2}\left[-\frac{\dot{\mathbf{u}}R_q}{c^2} + \frac{\left(\mathbf{R}_q - \frac{\mathbf{u}R_q}{c}\right)}{\left(1-\frac{\mathbf{u}\cdot\mathbf{R}_q}{cR_q}\right)R_q}\left(1-\frac{u^2}{c^2}+\frac{\dot{\mathbf{u}}\cdot\mathbf{R}_q}{c^2}\right)\right], \quad \mathbf{R}_q(t)\neq 0 \tag{2.99}$$

where $\mathbf{R}_q(t) = \mathbf{r} - \mathbf{r}_q(\tau)$, and the retarded time is defined in (2.97). Comparing (2.96) and (2.99) and letting $\hat{\mathbf{R}}_q = \mathbf{R}_q/R_q$ reveals that

$$\mathbf{H}(\mathbf{r},t) = \sqrt{\frac{\epsilon}{\mu}}\hat{\mathbf{R}}_q \times \mathbf{E}(\mathbf{r},t), \quad \mathbf{R}_q(t)\neq 0 \tag{2.100}$$

whereas

$$\mathbf{E}(\mathbf{r},t) = -\sqrt{\frac{\mu}{\epsilon}}\hat{\mathbf{R}}_q \times \mathbf{H}(\mathbf{r},t) + \frac{q_e\left(1-\frac{u^2}{c^2}\right)\hat{\mathbf{R}}_q}{4\pi\epsilon\left(1-\frac{\mathbf{u}\cdot\hat{\mathbf{R}}_q}{c}\right)^2 R_q^2}, \quad \mathbf{R}_q(t)\neq 0. \tag{2.101}$$

The expressions (2.96) and (2.99) are the classical results for the fields of a moving point charge obtained traditionally from the Liénard-Wiechert potentials [1, p. 475], [64, sec. 20–1]. Direct differentiation of the magnetic and electric fields in (2.96) and (2.99) verifies that these fields satisfy Maxwell's homogeneous equations outside the moving point charge. Moreover, integrating Maxwell's equations (2.1)–(2.4) over a volume containing the sources $\mathbf{J}$ and ρ, then applying the divergence and Stoke's volume-surface integral theorems [1, p. 604, eqs. (19) and (20)], confirm that the fields in (2.96) and (2.99) with sources given in (2.88)–(2.89) satisfy Maxwell's volume-surface integral equations. Also, as $R_q \to 0$, the electric and magnetic fields approach those of a uniformly moving point charge. Obtaining these results from (2.79) and (2.81) confirms the validity of the volume integral expressions (2.79) and (2.81) for determining the time-domain electric and magnetic fields.

2.1.9 Surface Integral Expressions for the Electric and Magnetic Fields (Huygens' Sources)

The electric and magnetic fields outside the source region can also be written in terms of integrals of the tangential electric and magnetic fields (Huygens' sources) on a surface S in free space that surrounds the sources located within a bounded volume V (see Figure 2.1). In the frequency domain, one version of these formulas is commonly referred to as the Stratton-Chu formulas or sometimes as the Larmor-Tedone formulas [1, sec. 8.14], and a second version as the Kottler-Franz formulas [86], [94], [95]. In this section, we derive time-domain surface integral expressions corresponding to the Stratton-Chu and Kottler-Franz formulas. The derivations are performed directly in the time domain rather than by Fourier transforming the corresponding frequency-domain formulas in Section 2.3.8.

To perform these derivations, begin with the vector integral identity [83, app. 1, eq. 40]

$$\int_{V_0}[a(\mathbf{r}')\nabla'\times\nabla'\times\mathbf{b}(\mathbf{r}') + \mathbf{b}(\mathbf{r}')\nabla'^2 a(\mathbf{r}') + (\nabla'\cdot\mathbf{b}(\mathbf{r}'))\nabla' a(\mathbf{r}')]dV'$$
$$= -\int_S [a(\mathbf{r}')\hat{\mathbf{n}}'\times\nabla'\times\mathbf{b}(\mathbf{r}') + (\hat{\mathbf{n}}'\times\mathbf{b}(\mathbf{r}'))\times\nabla' a(\mathbf{r}') + (\hat{\mathbf{n}}'\cdot\mathbf{b}(\mathbf{r}'))\nabla' a(\mathbf{r}')]dS' \tag{2.102}$$

where S is the surface enclosing the sources, V_0 is the infinite volume outside S, and $\hat{\mathbf{n}}'$ is the unit normal to S pointing into V_0. (The contribution from the surface integration at an

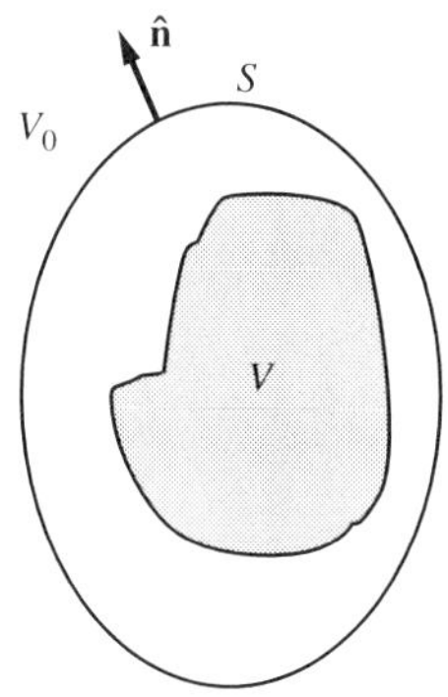

Figure 2.1 The source region V, the surrounding surface S in free space, and the infinite volume V_0 outside S.

infinite radius is zero for causal fields excited by sources in a bounded region of space that are zero before some initial time t_0. The identity (2.102) is also used below to derive the surface integral expressions, (2.124) and (2.125), for static fields that approach zero as $r \to \infty$.) Let the scalar function $a(\mathbf{r}')$ equal the retarded (with respect to t) time-domain free-space Green's function [63, p. 245]

$$a(\mathbf{r}') = G(\mathbf{r}, \mathbf{r}', t, t') = \frac{\delta(t' - t + |\mathbf{r} - \mathbf{r}'|/c)}{4\pi|\mathbf{r} - \mathbf{r}'|} \tag{2.103}$$

and the vector function $\mathbf{b}(\mathbf{r}')$ equal the electric field $\mathbf{E}(\mathbf{r}', t')$. Then (2.102) becomes

$$\begin{aligned} \mathbf{E}(\mathbf{r}, t')\delta(t' - t) = &-\frac{1}{c^2}\int\limits_{V_0}\left(G\frac{\partial^2 \mathbf{E}}{\partial t'^2} - \mathbf{E}\frac{\partial^2 G}{\partial t'^2}\right)dV' \\ &+ \int\limits_{S}\left[-\mu G\hat{\mathbf{n}}' \times \frac{\partial \mathbf{H}}{\partial t'} + (\hat{\mathbf{n}}' \times \mathbf{E}) \times \nabla' G + (\hat{\mathbf{n}}' \cdot \mathbf{E})\nabla' G\right]dS' \end{aligned} \tag{2.104}$$

where $\mathbf{r}$ is in V_0 (outside S), and use has been made of the relations

$$\nabla \times \mathbf{E}(\mathbf{r}, t) = -\mu\frac{\partial}{\partial t}\mathbf{H}(\mathbf{r}, t) \tag{2.105}$$

$$\nabla \times \nabla \times \mathbf{E}(\mathbf{r}, t) + \frac{1}{c^2}\frac{\partial^2}{\partial t^2}\mathbf{E}(\mathbf{r}, t) = 0, \quad \mathbf{r} \in V_0 \tag{2.106}$$

$$\nabla \cdot \mathbf{E}(\mathbf{r}, t) = 0, \quad \mathbf{r} \in V_0 \tag{2.107}$$

$$\nabla^2 G - \frac{1}{c^2}\frac{\partial^2 G}{\partial t^2} = -\delta(\mathbf{r} - \mathbf{r}')\delta(t - t'). \tag{2.108}$$

The delta function on the right-hand side of (2.108) has extracted the electric field on the left-hand side of (2.104) from the volume integration in (2.102). The application of the vector integral identity (2.102) is rigorously valid if the second spatial derivatives of $a(\mathbf{r}')$ and $\mathbf{b}(\mathbf{r}')$ are continuous functions of $\mathbf{r}'$. Thus, it may not be valid to apply (2.102) to G, which contains a delta function and a singularity at $\mathbf{r}' = \mathbf{r}$. Fortunately, however, the derivation of (2.104) from (2.102) can be made rigorous by isolating the singularity of G by a small principal volume V_δ with surface S_δ, as was done in Section 2.1.6, and by representing the delta function as a sequence of well-defined classical functions [75]. Then the surface integration over S_δ as $\delta \to 0$ produces the electric field on the left-hand side of (2.104). Here, we avoid this lengthier, rigorous derivation because the volume V_0 lies outside the source region, and thus the final expressions for the fields will not involve volume integrals or the principal volume V_δ. For scalar time-domain fields, the rigorous derivation involving V_δ and S_δ can be found in Stratton [1, sec. 8.1] or Phillips [84, sec. 91–93]. Sufficient conditions for the validity of the final vector expressions obtained here are given below.

Rewriting the volume integrand in (2.104) as a time derivative and noting that $\nabla' G = -\nabla G$ allows (2.104) to be recast in the form

$$\begin{aligned}\mathbf{E}(\mathbf{r}, t')\delta(t' - t) = &-\frac{1}{c^2}\frac{\partial}{\partial t'}\int\limits_{V_0}\left(G\frac{\partial \mathbf{E}}{\partial t'} - \mathbf{E}\frac{\partial G}{\partial t'}\right)dV' \\ &-\mu\int\limits_S \hat{\mathbf{n}}' \times \frac{\partial \mathbf{H}}{\partial t'} G dS' + \nabla \times \int\limits_S (\hat{\mathbf{n}}' \times \mathbf{E}) G dS' \\ &-\nabla\int\limits_S \hat{\mathbf{n}}' \cdot \mathbf{E} G dS'.\end{aligned} \tag{2.109}$$

Then integrating (2.109) over t' from $-\infty$ to $+\infty$ eliminates the volume integral (because $\mathbf{E}(\mathbf{r}', t') = 0$ for $t' < t_0$ and $G(\mathbf{r}, \mathbf{r}', t, t') = 0$ for $t' > t$) and reduces (2.109) to

$$\begin{aligned}\mathbf{E}(\mathbf{r}, t) = &-\frac{\partial}{\partial t}\int\limits_S \frac{\mu\hat{\mathbf{n}}' \times \mathbf{H}(\mathbf{r}', t - \frac{|\mathbf{r}-\mathbf{r}'|}{c})}{4\pi|\mathbf{r} - \mathbf{r}'|}dS' \\ &+\nabla \times \int\limits_S \frac{\hat{\mathbf{n}}' \times \mathbf{E}(\mathbf{r}', t - \frac{|\mathbf{r}-\mathbf{r}'|}{c})}{4\pi|\mathbf{r} - \mathbf{r}'|}dS' - \nabla\int\limits_S \frac{\hat{\mathbf{n}}' \cdot \mathbf{E}(\mathbf{r}', t - \frac{|\mathbf{r}-\mathbf{r}'|}{c})}{4\pi|\mathbf{r} - \mathbf{r}'|}dS'.\end{aligned} \tag{2.110}$$

A similar derivation with $\mathbf{b}(\mathbf{r}')$ equal to $\mathbf{H}(\mathbf{r}', t')$ in (2.102) leads to

$$\begin{aligned}\mathbf{H}(\mathbf{r}, t) = &\frac{\partial}{\partial t}\int\limits_S \frac{\epsilon\hat{\mathbf{n}}' \times \mathbf{E}(\mathbf{r}', t - \frac{|\mathbf{r}-\mathbf{r}'|}{c})}{4\pi|\mathbf{r} - \mathbf{r}'|}dS' \\ &+\nabla \times \int\limits_S \frac{\hat{\mathbf{n}}' \times \mathbf{H}(\mathbf{r}', t - \frac{|\mathbf{r}-\mathbf{r}'|}{c})}{4\pi|\mathbf{r} - \mathbf{r}'|}dS' - \nabla\int\limits_S \frac{\hat{\mathbf{n}}' \cdot \mathbf{H}(\mathbf{r}', t - \frac{|\mathbf{r}-\mathbf{r}'|}{c})}{4\pi|\mathbf{r} - \mathbf{r}'|}dS'.\end{aligned} \tag{2.111}$$

Equations (2.110) and (2.111) are the time-domain analogs of the Stratton-Chu frequency-domain formulas [1, p. 466] that express the fields outside a surface S surrounding the source region in terms of the electric and magnetic fields on S. They require the normal components of the electric and magnetic fields on S as well as the tangential components. However, for time-dependent fields, the normal components of the fields on S are determined from the tangential components through the Maxwell equations (2.18)–(2.19), and the relations between differential operators on a surface [83, app. 2, eqs. 21 and 22], specifically

$$\hat{\mathbf{n}} \cdot \mathbf{E}(\mathbf{r}, t) = -\frac{1}{\epsilon}\nabla_s \cdot \int\limits_{t_0}^{t} \hat{\mathbf{n}} \times \mathbf{H}(\mathbf{r}, t')dt', \quad \mathbf{r} \in S \tag{2.112}$$

and

$$\hat{\mathbf{n}} \cdot \mathbf{H}(\mathbf{r}, t) = \frac{1}{\mu}\nabla_s \cdot \int\limits_{t_0}^{t} \hat{\mathbf{n}} \times \mathbf{E}(\mathbf{r}, t')dt', \quad \mathbf{r} \in S \tag{2.113}$$

where $\nabla_s \cdot$ is the surface divergence operator and the fields are assumed zero for $t \le t_0$. Thus, (2.112) and (2.113) can be used to express (2.110) and (2.111) in terms of the tangential components alone of the electric and magnetic fields on the surface S surrounding the source region. If the normal components, $\epsilon\hat{\mathbf{n}} \cdot \mathbf{E}$ and $\mu\hat{\mathbf{n}} \cdot \mathbf{H}$, are interpreted as equivalent electric and magnetic surface charge densities, and the tangential components, $\hat{\mathbf{n}} \times \mathbf{H}$ and $-\hat{\mathbf{n}} \times \mathbf{E}$, are interpreted as equivalent electric and magnetic surface current densities, (2.112) and (2.113) are merely restatements of the continuity equations which require that the negative time rate of change of the electric and magnetic surface charge densities equal the divergence of the electric and magnetic surface current densities, respectively.

The time-dependent electric and magnetic fields given by the expressions (2.110) and (2.111), combined with the surface relationships (2.112) and (2.113), can be shown to obey the Maxwellian equations (2.18)–(2.21) (in the free-space region outside the surface S where $\mathbf{J}=0$) under the sufficient conditions that $\partial[\hat{\mathbf{n}}\times\mathbf{E}(\mathbf{r},t)]/\partial t$, $\partial[\hat{\mathbf{n}}\times\mathbf{H}(\mathbf{r},t)]\partial t$, $\nabla_s\cdot[\hat{\mathbf{n}}\times\mathbf{E}(\mathbf{r},t)]$ and $\nabla_s\cdot[\hat{\mathbf{n}}\times\mathbf{H}(\mathbf{r},t)]$ are continuous functions of $(\mathbf{r},t)$ for $\mathbf{r}$ on S. (These conditions imply that $\hat{\mathbf{n}}\times\mathbf{E}(\mathbf{r},t)$ and $\hat{\mathbf{n}}\times\mathbf{H}(\mathbf{r},t)$, and through (2.112)–(2.113), that $\partial(\hat{\mathbf{n}}\cdot\mathbf{E})/\partial t$, $\partial(\hat{\mathbf{n}}\cdot\mathbf{H})/\partial t$, $\hat{\mathbf{n}}\cdot\mathbf{E}$ and $\hat{\mathbf{n}}\cdot\mathbf{H}$ are also continuous functions of $(\mathbf{r},t)$ on S, since the fields are assumed to be zero for $t\le t_0$.) This is proven straightforwardly using Leibnitz's rule in a way similar to the way Theorems 2-I and 2-II of Section 2.1.6 were used to find sufficient conditions on the volume current for (2.53)–(2.55) to satisfy Maxwell's equations. Unlike the volume integrals involved in (2.53)–(2.55), the surface integrals in (2.110)–(2.111) do not require either principal value limits or Hölder continuous functions because the observation points $\mathbf{r}$ in (2.110) and (2.111) lie outside the source region. It is emphasized that these conditions on the tangential fields are sufficient conditions and not necessary conditions for the fields in (2.110) and (2.111) to obey the homogeneous Maxwell equations corresponding to (2.18)–(2.21). For example, these sufficient conditions imply that $\mathbf{E}(\mathbf{r},t)$ and $\mathbf{H}(\mathbf{r},t)$ given in (2.110) and (2.111) are continuous functions of $\mathbf{r}$ and t for all $\mathbf{r}$ outside S. The electromagnetic missile in Section 5.4.3 of a circular current disk with step-function time dependence has fields that are discontinuous functions of $\mathbf{r}$ and t and yet satisfy the Maxwell equations (2.18)–(2.21) outside the current disk, if across the discontinuity the spatial derivatives are defined "from the left" and "from the right." (Other electromagnetic missiles have sources and fields that are continuous functions of $\mathbf{r}$ and t.)

If, in addition to these continuity conditions, we have that $\mathbf{E}(\mathbf{r},t)$ and $\mathbf{H}(\mathbf{r},t)$ are Hölder continuous on S at each t, and S is a continuously curved closed surface, then (2.110) and (2.111) recover the electric and magnetic fields on S, respectively, as $\mathbf{r}$ approaches the surface S from outside S. To prove this, repeat the derivation of (2.110) and (2.111) with $\mathbf{r}$ inside S to obtain the time-domain "extinction theorems" [85, p. 500], [86]

$$\begin{aligned}0=-\frac{\partial}{\partial t}\int\limits_S \frac{\mu\hat{\mathbf{n}}'\times\mathbf{H}(\mathbf{r}',t-\frac{|\mathbf{r}-\mathbf{r}'|}{c})}{4\pi|\mathbf{r}-\mathbf{r}'|}dS'\\+\nabla\times\int\limits_S\frac{\hat{\mathbf{n}}'\times\mathbf{E}(\mathbf{r}',t-\frac{|\mathbf{r}-\mathbf{r}'|}{c})}{4\pi|\mathbf{r}-\mathbf{r}'|}dS'-\nabla\int\limits_S\frac{\hat{\mathbf{n}}'\cdot\mathbf{E}(\mathbf{r}',t-\frac{|\mathbf{r}-\mathbf{r}'|}{c})}{4\pi|\mathbf{r}-\mathbf{r}'|}dS'\end{aligned}\tag{2.114}$$

$$\begin{aligned}0=\frac{\partial}{\partial t}\int\limits_S \frac{\epsilon\hat{\mathbf{n}}'\times\mathbf{E}(\mathbf{r}',t-\frac{|\mathbf{r}-\mathbf{r}'|}{c})}{4\pi|\mathbf{r}-\mathbf{r}'|}dS'\\+\nabla\times\int\limits_S\frac{\hat{\mathbf{n}}'\times\mathbf{H}(\mathbf{r}',t-\frac{|\mathbf{r}-\mathbf{r}'|}{c})}{4\pi|\mathbf{r}-\mathbf{r}'|}dS'-\nabla\int\limits_S\frac{\hat{\mathbf{n}}'\cdot\mathbf{H}(\mathbf{r}',t-\frac{|\mathbf{r}-\mathbf{r}'|}{c})}{4\pi|\mathbf{r}-\mathbf{r}'|}dS'.\end{aligned}\tag{2.115}$$

As $\mathbf{r}$ approaches S from outside S ($\mathbf{r}\to S^+$) in (2.110), and from inside S in (2.114), the analysis in [84, sec. 63] shows that the first terms on the right-hand sides of (2.110) and (2.114) are continuous across S, whereas the second and third terms have jump discontinuities across S equal to $(\hat{\mathbf{n}}\times\mathbf{E}(\mathbf{r},t))\times\hat{\mathbf{n}}$ and $(\hat{\mathbf{n}}\cdot\mathbf{E}(\mathbf{r},t))\hat{\mathbf{n}}$, respectively, where $\mathbf{r}\in S$. Thus, we have

$$\mathbf{E}(\mathbf{r},t)_{\mathbf{r}\to S^+}=[(\hat{\mathbf{n}}\times\mathbf{E}(\mathbf{r},t))\times\hat{\mathbf{n}}+(\hat{\mathbf{n}}\cdot\mathbf{E}(\mathbf{r},t))\hat{\mathbf{n}}]_{\mathbf{r}\in S}=\mathbf{E}(\mathbf{r},t)_{\mathbf{r}\in S}.\tag{2.116}$$

Similarly, we find from (2.111) and (2.115) that

$$\mathbf{H}(\mathbf{r},t)_{\mathbf{r}\to S^+} = [(\hat{\mathbf{n}} \times \mathbf{H}(\mathbf{r},t)) \times \hat{\mathbf{n}} + (\hat{\mathbf{n}} \cdot \mathbf{H}(\mathbf{r},t))\hat{\mathbf{n}}]_{\mathbf{r}\in S} = \mathbf{H}(\mathbf{r},t)_{\mathbf{r}\in S}. \tag{2.117}$$

That is, as the observation point $\mathbf{r}$ approaches the surface S from outside S, the values of the fields at $\mathbf{r}$, found by integrating the fields specified on S, approach the values of these specified fields.

The gradient terms in (2.110) and (2.111) involving the normal components of the fields can be eliminated altogether by integrating these equations over time from t_0 to t, taking the curl of the resulting equations, and substituting from Maxwell's equations (2.18) and (2.19) to get

$$\begin{aligned}\mathbf{H}(\mathbf{r},t) &= \nabla \times \int_S \frac{\hat{\mathbf{n}}' \times \mathbf{H}(\mathbf{r}', t - \frac{|\mathbf{r}-\mathbf{r}'|}{c})}{4\pi|\mathbf{r}-\mathbf{r}'|} dS' \\ &\quad - \nabla \times \nabla \times \int_S \frac{\hat{\mathbf{n}}' \times \int_{t_0}^{t} \mathbf{E}(\mathbf{r}', t' - \frac{|\mathbf{r}-\mathbf{r}'|}{c}) dt'}{4\pi\mu|\mathbf{r}-\mathbf{r}'|} dS'\end{aligned} \tag{2.118}$$

$$\begin{aligned}\mathbf{E}(\mathbf{r},t) &= \nabla \times \int_S \frac{\hat{\mathbf{n}}' \times \mathbf{E}(\mathbf{r}', t - \frac{|\mathbf{r}-\mathbf{r}'|}{c})}{4\pi|\mathbf{r}-\mathbf{r}'|} dS' \\ &\quad + \nabla \times \nabla \times \int_S \frac{\hat{\mathbf{n}}' \times \int_{t_0}^{t} \mathbf{H}(\mathbf{r}', t' - \frac{|\mathbf{r}-\mathbf{r}'|}{c}) dt'}{4\pi\epsilon|\mathbf{r}-\mathbf{r}'|} dS'.\end{aligned} \tag{2.119}$$

These last two compact equations are the time-domain analogs to the Kottler-Franz formulas [86], [94], [95]. They explicitly involve only the tangential electric and magnetic fields on the surface S outside the source region. These equivalent electric and magnetic surface currents ($\hat{\mathbf{n}} \times \mathbf{H}(\mathbf{r},t)$ and $-\hat{\mathbf{n}} \times \mathbf{E}(\mathbf{r},t)$) may well be referred to as the time-domain Huygens' sources of the electromagnetic fields outside S. Thus, (2.118) and (2.119) represent a Hertz potential [1, sec. 1.11] formulation of Maxwell's equations in terms of equivalent surface currents.

With the help of the vector identities

$$\nabla \times \left[\frac{\mathbf{F}\left(t - \frac{|\mathbf{r}-\mathbf{r}'|}{c}\right)}{|\mathbf{r}-\mathbf{r}'|}\right] = -\frac{\mathbf{R}}{R^2} \times \left(\frac{1}{c}\frac{\partial \mathbf{F}}{\partial t} + \frac{\mathbf{F}}{R}\right), \quad \mathbf{r} \neq \mathbf{r}' \tag{2.120}$$

and

$$\begin{aligned}\nabla \times \nabla \times \left[\frac{\mathbf{F}\left(t - \frac{|\mathbf{r}-\mathbf{r}'|}{c}\right)}{|\mathbf{r}-\mathbf{r}'|}\right] &= \frac{1}{c^2 R}\frac{\partial^2 \mathbf{F}}{\partial t^2} \cdot \left(\frac{\mathbf{RR}}{R^2} - \bar{\mathbf{I}}\right) \\ &\quad + \frac{1}{R^2}\left(\frac{1}{c}\frac{\partial \mathbf{F}}{\partial t} + \frac{\mathbf{F}}{R}\right) \cdot \left(\frac{3\mathbf{RR}}{R^2} - \bar{\mathbf{I}}\right), \quad \mathbf{r} \neq \mathbf{r}'\end{aligned} \tag{2.121}$$

in which the $\mathbf{F}$'s on the right-hand sides of (2.120) and (2.121) are also evaluated at the retarded time $t - |\mathbf{r}-\mathbf{r}'|/c$, the operators can be brought inside the surface integrals to convert (2.118) and (2.119) to

$$\mathbf{H}(\mathbf{r},t) = \frac{1}{4\pi}\int_S \Bigg[\left(\frac{1}{c}\hat{\mathbf{n}}' \times \frac{\partial}{\partial t}\mathbf{H}(\mathbf{r}',t-R/c) + \frac{1}{R}\hat{\mathbf{n}}' \times \mathbf{H}(\mathbf{r}',t-R/c)\right) \times \frac{\mathbf{R}}{R^2}$$
$$-\frac{\epsilon}{R}\left(\hat{\mathbf{n}}' \times \frac{\partial}{\partial t}\mathbf{E}(\mathbf{r}',t-R/c)\right)\cdot\left(\frac{\mathbf{R}\mathbf{R}}{R^2} - \bar{\mathbf{I}}\right) \tag{2.122}$$
$$-\frac{1}{\mu R^2}\left(\frac{1}{c}\hat{\mathbf{n}}' \times \mathbf{E}(\mathbf{r}',t-R/c) + \frac{1}{R}\hat{\mathbf{n}}' \times \int_{t_0}^{t-R/c} \mathbf{E}(\mathbf{r}',t')dt'\right)$$
$$\cdot\left(\frac{3\mathbf{R}\mathbf{R}}{R^2} - \bar{\mathbf{I}}\right)\Bigg]dS'$$

$$\mathbf{E}(\mathbf{r},t) = \frac{1}{4\pi}\int_S \Bigg[\left(\frac{1}{c}\hat{\mathbf{n}}' \times \frac{\partial}{\partial t}\mathbf{E}(\mathbf{r}',t-R/c) + \frac{1}{R}\hat{\mathbf{n}}' \times \mathbf{E}(\mathbf{r}',t-R/c)\right) \times \frac{\mathbf{R}}{R^2}$$
$$+\frac{\mu}{R}\left(\hat{\mathbf{n}}' \times \frac{\partial}{\partial t}\mathbf{H}(\mathbf{r}',t-R/c)\right)\cdot\left(\frac{\mathbf{R}\mathbf{R}}{R^2} - \bar{\mathbf{I}}\right) \tag{2.123}$$
$$+\frac{1}{\epsilon R^2}\left(\frac{1}{c}\hat{\mathbf{n}}' \times \mathbf{H}(\mathbf{r}',t-R/c) + \frac{1}{R}\hat{\mathbf{n}}' \times \int_{t_0}^{t-R/c} \mathbf{H}(\mathbf{r}',t')dt'\right)$$
$$\cdot\left(\frac{3\mathbf{R}\mathbf{R}}{R^2} - \bar{\mathbf{I}}\right)\Bigg]dS'$$

where, as usual, $\mathbf{R} = \mathbf{r} - \mathbf{r}'$.

If $\hat{\mathbf{n}} \times \mathbf{H}$ is interpreted as equivalent electric surface current, one sees that the surface integrals over electric current in (2.122) and (2.123) agree with the volume integrals over electric current in (2.79) and (2.80), respectively. Similarly, interpreting $-\hat{\mathbf{n}} \times \mathbf{E}$ as equivalent magnetic surface current reveals that the surface integrals in (2.122) and (2.123) over magnetic current are the duals of the integrals over the electric current.

The time-dependent electric and magnetic fields given by (2.118)–(2.119) or (2.122)–(2.123) are continuous functions of $(\mathbf{r}, t)$ that satisfy the Maxwell equations (2.18)–(2.21) outside S if $\partial(\hat{\mathbf{n}} \times \mathbf{E})/\partial t$, $\partial(\hat{\mathbf{n}} \times \mathbf{H})/\partial t$, $\nabla_s \cdot (\hat{\mathbf{n}} \times \mathbf{E})$ and $\nabla_s \cdot (\hat{\mathbf{n}} \times \mathbf{H})$ are continuous functions of $(\mathbf{r}, t)$ for $\mathbf{r}$ on S. If, in addition to these continuity conditions, we have that $\mathbf{E}(\mathbf{r}, t)$ and $\mathbf{H}(\mathbf{r}, t)$ are Hölder continuous on S at each t, and S is a continuously curved closed surface, then (2.118)–(2.119) or (2.122)–(2.123) recover the electric and magnetic fields on S as $\mathbf{r}$ approaches the surface S from outside S (because of the surface relationships (2.112)–(2.113)). These results can be proven much like the corresponding results were proven for (2.110)–(2.111).

Even if the tangential electric and magnetic fields on S, that is, the Huygens' sources, in (2.118)–(2.119) or (2.110)–(2.113) are chosen independently, the fields on the left-hand sides of these equations will obey Maxwell's homogeneous equations for $\mathbf{r}$ outside S. However, only if the tangential electric and magnetic fields on S are compatible with Maxwell's equations in free space will (2.114)–(2.117) hold, and the fields be recovered from (2.118)–(2.119) or (2.110)–(2.113) as $\mathbf{r}$ approaches S from outside S. In other words, if designated tangential fields (called Huygens' sources or equivalent electric and magnetic surface currents) on S are incompatible with Maxwell's equations in free space, the surface integral expressions determine electromagnetic fields outside S that can be excited only if electric and magnetic surface currents reside on S (or an infinitesimal distance in front of S) in addition to the designated equivalent electric and magnetic surface currents on S [96, ch. II, sec. 4].

For static fields, the vector integral identity (2.102) from [83, app. 1, eq. 40] can also be used to show that the time-domain Stratton-Chu formulas (2.110) and (2.111) remain valid expressions for the fields in terms of equivalent surface charges and currents. Specifically, for the electrostatic field, let $a(\mathbf{r}') = 1/(4\pi|\mathbf{r}-\mathbf{r}'|)$ and $\mathbf{b}(\mathbf{r}') = \mathbf{E}(\mathbf{r}')$ in (2.102), apply $\nabla\times\mathbf{E} = 0$ and $\nabla\cdot\mathbf{E} = 0$ outside the source region, use the far-field expression (1.6) in Chapter 1 to prove that there is no contribution from the surface at infinity, and thereby obtain the result

$$\mathbf{E}(\mathbf{r}) = \nabla\times\int_S \frac{\hat{\mathbf{n}}'\times\mathbf{E}(\mathbf{r}')}{4\pi|\mathbf{r}-\mathbf{r}'|}dS' - \nabla\int_S \frac{\hat{\mathbf{n}}'\cdot\mathbf{E}(\mathbf{r}')}{4\pi|\mathbf{r}-\mathbf{r}'|}dS'. \tag{2.124}$$

Similarly, for the magnetostatic field, we have

$$\mathbf{H}(\mathbf{r}) = \nabla\times\int_S \frac{\hat{\mathbf{n}}'\times\mathbf{H}(\mathbf{r}')}{4\pi|\mathbf{r}-\mathbf{r}'|}dS' - \nabla\int_S \frac{\hat{\mathbf{n}}'\cdot\mathbf{H}(\mathbf{r}')}{4\pi|\mathbf{r}-\mathbf{r}'|}dS'. \tag{2.125}$$

However, the time-domain Kottler-Franz formulas (2.118)–(2.119) or (2.122)–(2.123) do not apply to static fields because their derivation requires (2.18) and (2.19), which are not valid for static fields. Equations (2.124) and (2.125) are mathematical statements of Helmholtz's theorem [84, secs. 83–86] when the curl and divergence of the field are both zero outside S.

Continuity of $\hat{\mathbf{n}}\times\mathbf{E}(\mathbf{r})$ and $\hat{\mathbf{n}}\cdot\mathbf{E}(\mathbf{r})$ on S assures that the electric field in (2.124) is continuous and satisfies the electrostatic divergence equation, $\nabla\cdot\mathbf{E} = 0$, for all $\mathbf{r}$ outside S. In addition, it follows from the static electric field satisfying the relationship $\nabla_s\cdot(\hat{\mathbf{n}}\times\mathbf{E}(\mathbf{r})) = 0$ on S, that the electric field in (2.124) also satisfies the electrostatic curl equation, $\nabla\times\mathbf{E} = 0$, for all $\mathbf{r}$ outside S. If $\mathbf{E}(\mathbf{r})$ is Hölder continuous on S, and S is a continuously curved surface, the electric field on the surface S is recovered by evaluating the right-hand side of (2.124) as $\mathbf{r}$ approaches S from outside S. The same conclusions hold for the magnetostatic field in (2.125).

2.1.10 Force-Power and Momentum-Energy Relations

The electromagnetic force per unit volume $\mathbf{f}_{el}(\mathbf{r},t)$ exerted on a distribution of electric charge and current density (ρ and $\mathbf{J}$) by the electromagnetic fields ($\mathbf{E}$ and $\mathbf{H}$), which are produced by the entire distribution of charge and current, is given by the Lorentz force expression

$$\begin{aligned}\mathbf{f}_{el}(\mathbf{r},t) &= \rho(\mathbf{r},t)\mathbf{E}(\mathbf{r},t) + \mathbf{J}(\mathbf{r},t)\times\mathbf{B}(\mathbf{r},t)\\ &= \rho(\mathbf{r},t)\mathbf{E}(\mathbf{r},t) + \mu\mathbf{J}(\mathbf{r},t)\times\mathbf{H}(\mathbf{r},t).\end{aligned} \tag{2.126}$$

The corresponding electromagnetic power per unit volume $p_{el}(\mathbf{r},t)$ supplied to this charge-current distribution is given by

$$p_{el}(\mathbf{r},t) = \mathbf{J}(\mathbf{r},t)\cdot\mathbf{E}(\mathbf{r},t). \tag{2.127}$$

The force-power density $[\mathbf{f}_{el}, p_{el}/c]$ transforms as a relativistic four-vector [64, sec. 18–3]. The integration of $\mathbf{f}_{el}$ and p_{el} over a volume produces the electromagnetic force and power delivered by the electromagnetic fields to the charge and current within that volume.

Substituting ρ and $\mathbf{J}$ from Maxwell's differential equations (2.2) and (2.4) into the right-hand sides of (2.126) and (2.127), and making use of the remaining two Maxwell equations

(2.1) and (2.3), lead to the identities

$$\mathbf{f}_{el}(\mathbf{r},t) = \rho\mathbf{E} + \mu\mathbf{J}\times\mathbf{H} = \nabla\cdot\bar{\mathbf{T}}_{el} - \frac{1}{c^2}\frac{\partial}{\partial t}(\mathbf{E}\times\mathbf{H}) \tag{2.128}$$

and

$$p_{el}(\mathbf{r},t) = \mathbf{J}\cdot\mathbf{E} = -\nabla\cdot(\mathbf{E}\times\mathbf{H}) - \frac{1}{2}\frac{\partial}{\partial t}(\epsilon|\mathbf{E}|^2 + \mu|\mathbf{H}|^2) \tag{2.129}$$

where the electromagnetic stress dyadic (tensor) is defined as

$$\bar{\mathbf{T}}_{el}(\mathbf{r},t) = \epsilon(\mathbf{EE} - \frac{1}{2}\bar{\mathbf{I}}|\mathbf{E}|^2) + \mu(\mathbf{HH} - \frac{1}{2}\bar{\mathbf{I}}|\mathbf{H}|^2). \tag{2.130}$$

When (2.128) and (2.129) are integrated over an arbitrary volume V_a with surface S_a, and the divergence theorem is applied, we obtain the equations for the electromagnetic force $\mathbf{F}_{el}(t)$ and power $P_{el}(t)$ exerted on the charge and current in V_a

$$\begin{aligned}\mathbf{F}_{el}(t) &= \int_{V_a}(\rho\mathbf{E} + \mu\mathbf{J}\times\mathbf{H})dV = \int_{V_a}\left[\nabla\cdot\bar{\mathbf{T}}_{el} - \frac{1}{c^2}\frac{\partial}{\partial t}(\mathbf{E}\times\mathbf{H})\right]dV \\ &= \int_{S_a}\hat{\mathbf{n}}\cdot\bar{\mathbf{T}}_{el}dS - \frac{1}{c^2}\frac{d}{dt}\int_{V_a}\mathbf{E}\times\mathbf{H}dV\end{aligned} \tag{2.131}$$

$$\begin{aligned}P_{el}(t) &= \int_{V_a}\mathbf{J}\cdot\mathbf{E}dV = -\int_{V_a}\left[\nabla\cdot(\mathbf{E}\times\mathbf{H}) + \frac{1}{2}\frac{\partial}{\partial t}(\epsilon|\mathbf{E}|^2 + \mu|\mathbf{H}|^2)\right]dV \\ &= -\int_{S_a}\hat{\mathbf{n}}\cdot(\mathbf{E}\times\mathbf{H})dS - \frac{1}{2}\frac{d}{dt}\int_{V_a}(\epsilon|\mathbf{E}|^2 + \mu|\mathbf{H}|^2)dV\end{aligned} \tag{2.132}$$

where $\hat{\mathbf{n}}$ is the unit normal to S_a pointing out of V_a.

One may question whether the first volume integral in (2.131) truly gives the force on the charge and current in V_a because the force exerted on $\rho(\mathbf{r},t)dV$ and $\mathbf{J}(\mathbf{r},t)dV$ should be $\rho(\mathbf{r},t)\mathbf{E}_c(\mathbf{r},t)dV$ and $\mathbf{J}(\mathbf{r},t)\times\mathbf{B}_c(\mathbf{r},t)dV$ where $\mathbf{E}_c(\mathbf{r},t)$ and $\mathbf{B}_c(\mathbf{r},t) = \mu\mathbf{H}_c(\mathbf{r},t)$ are the "cavity-defined" electric field and magnetic induction at $(\mathbf{r},t)$ produced by all charge and current other than $\rho(\mathbf{r},t)dV$ and $\mathbf{J}(\mathbf{r},t)dV$, that is, with $\rho(\mathbf{r},t)dV$ and $\mathbf{J}(\mathbf{r},t)dV$ removed from $\mathbf{r}$ for an infinitesimal time interval Δt about the time t. However, in Section 2.1.7 it was shown that the cavity-defined fields in a charge and current distribution are identical to the Maxwellian fields, that is, $\mathbf{E}_c(r,t) = \mathbf{E}(\mathbf{r},t)$ and $\mathbf{H}_c(\mathbf{r},t) = \mathbf{H}(\mathbf{r},t) = \mathbf{B}(\mathbf{r},t)/\mu$. Thus, (2.131) and (2.132) are indeed the force and power supplied by the electromagnetic fields to the charge and current in V_a.

Force and Power with Polarization Densities Included. If electric or magnetic polarization densities $\mathbf{P}(\mathbf{r},t)$ and $\mathbf{M}(\mathbf{r},t)$ are present, the cavity-defined fields differ from the Maxwellian fields, as explained in Footnote 6 of Section 2.1.7. Nonetheless, since $\nabla\cdot\mathbf{P}$ and $\partial\mathbf{P}/\partial t + \nabla\times\mathbf{M}$ are equivalent to macroscopic electric charge and current densities [1, p. 12], respectively, the integral

$$\int_{V_a}\left[(\rho - \nabla\cdot\mathbf{P})\mathbf{E} + \left(\mathbf{J} + \frac{\partial\mathbf{P}}{\partial t} + \nabla\times\mathbf{M}\right)\times\mathbf{B}\right]dV \tag{2.133}$$

unambiguously gives the electromagnetic force on all the sources within V_a, under two conditions. First, an additional force that is produced by the induced charge-current on the microscopic electric-current (Amperian) model of the magnetic dipoles that comprise $\mathbf{M}$ is ignored.

This force is not taken into account by the equivalent macroscopic current $\nabla \times \mathbf{M}$ in (2.133) but will be included later in this section. And second, the surface S_a of the volume V_a does not intersect the polarization densities **P** and **M**. Ambiguity arises in defining force on all the sources in V_a when its surface S_a intersects **P** and **M** because the force on **P** and **M** in V_a differs depending on whether or not the equivalent surface charge density $\hat{\mathbf{n}} \cdot \mathbf{P}$ and equivalent surface current density $\mathbf{M} \times \hat{\mathbf{n}}$ on S_a are included as part of the sources in V_a. With the stipulation that S_a lies in free space, or at least does not intersect polarization densities **P** and **M**, equivalent surface charge and current densities at the surfaces of polarized material are included in the volume integral (2.133) through delta functions in the derivatives $\nabla \cdot \mathbf{P}$ and $\nabla \times \mathbf{M}$ at the material surfaces.[7] Therefore, no ambiguity, in regard to equivalent surface charge and current densities, arises in the total force on the sources in V_a, and (2.131) generalizes to[8]

$$\mathbf{F}_{el}^0(t) = \int_{V_a} \left[(\rho - \nabla \cdot \mathbf{P})\mathbf{E} + \left(\mathbf{J} + \frac{\partial \mathbf{P}}{\partial t} + \nabla \times \mathbf{M} \right) \times \mathbf{B} \right] dV$$
$$= \int_{V_a} \left[\nabla \cdot \bar{\mathbf{T}}_{el} - \epsilon \frac{\partial}{\partial t}(\mathbf{E} \times \mathbf{B}) \right] dV = \int_{S_a} \hat{\mathbf{n}} \cdot \bar{\mathbf{T}}_{el} dS - \epsilon \frac{d}{dt} \int_{V_a} \mathbf{E} \times \mathbf{B} dV \tag{2.134}$$

where $\mathbf{B} = \mu(\mathbf{H} + \mathbf{M})$. Similarly, with S_a not intersecting polarization densities, (2.132) generalizes to

$$P_{el}^0(t) = \int_{V_a} \left(\mathbf{J} + \frac{\partial \mathbf{P}}{\partial t} + \nabla \times \mathbf{M} \right) \cdot \mathbf{E} dV$$
$$= -\int_{V_a} \left[\nabla \cdot (\mathbf{E} \times \mathbf{H}) + \frac{1}{2} \frac{\partial}{\partial t} \left(\epsilon |\mathbf{E}|^2 + \frac{1}{\mu} |\mathbf{B}|^2 \right) \right] dV \tag{2.135}$$
$$= -\int_{S_a} \hat{\mathbf{n}} \cdot (\mathbf{E} \times \mathbf{H}) dS - \frac{1}{2} \frac{d}{dt} \int_{V_a} \left(\epsilon |\mathbf{E}|^2 + \frac{1}{\mu} |\mathbf{B}|^2 \right) dV.$$

The superscript "0" on $\mathbf{F}_{el}^0$ and P_{el}^0 in (2.134) and (2.135) indicates that they do not yet include the additional force and power, mentioned above, on the electric-current (Amperian) magnetic polarization.

With S_a not intersecting **P** and **M**, the first volume integrals in (2.134) and (2.135) for electromagnetic force and power can be converted, respectively, to

$$\int_{V_a} \left[(\rho - \nabla \cdot \mathbf{P})\mathbf{E} + \mu \left(\mathbf{J} + \frac{\partial \mathbf{P}}{\partial t} \right) \times \mathbf{H} - \mu(\nabla \cdot \mathbf{M})\mathbf{H} + \epsilon\mu \mathbf{M} \times \frac{\partial \mathbf{E}}{\partial t} \right] dV$$
$$= \int_{V_a} \left[\rho \mathbf{E} + \mathbf{P} \cdot \nabla \mathbf{E} + \mu \left(\mathbf{J} + \frac{\partial \mathbf{P}}{\partial t} \right) \times \mathbf{H} \right. \tag{2.136}$$
$$\left. + \mu \mathbf{M} \cdot \nabla \mathbf{H} - \epsilon\mu \frac{\partial \mathbf{M}}{\partial t} \times \mathbf{E} + \epsilon\mu \frac{\partial}{\partial t}(\mathbf{M} \times \mathbf{E}) \right] dV$$

---

[7]The Riemann integration of delta functions in (2.133) can be justified rigorously by replacing the abrupt change in polarization densities at each surface of the polarized material by polarization densities that vary smoothly from zero to a finite value in a decreasingly thinner surface layer.

[8]If S_a intersects the polarization densities, not only do the integrals in (2.134) and (2.135) fail to unambiguously define the force and power delivered to all the polarization sources in V_a, but also the equalities between the integrals in (2.134) and (2.135) no longer hold unless **H** in $\bar{\mathbf{T}}_{el}$ of (2.134) and in $\mathbf{E} \times \mathbf{H}$ of (2.135) is replaced by $\mathbf{B}/\mu$. This replacement still does not remove the ambiguity in the definition of force and power on the polarization in V_a when its surface S_a intersects the polarization densities.

and

$$\int_{V_a}\left[\left(\mathbf{J}+\frac{\partial \mathbf{P}}{\partial t}\right)\cdot\mathbf{E}-\mathbf{M}\cdot\frac{\partial \mathbf{B}}{\partial t}\right]dV$$

$$=\int_{V_a}\left[\left(\mathbf{J}+\frac{\partial \mathbf{P}}{\partial t}\right)\cdot\mathbf{E}+\frac{\partial \mathbf{M}}{\partial t}\cdot\mathbf{B}-\frac{\partial}{\partial t}(\mathbf{M}\cdot\mathbf{B})\right]dV \tag{2.137}$$

$$=\int_{V_a}\left[\left(\mathbf{J}+\frac{\partial \mathbf{P}}{\partial t}\right)\cdot\mathbf{E}+\mu\frac{\partial \mathbf{M}}{\partial t}\cdot\mathbf{H}+\frac{1}{2}\frac{\partial}{\partial t}\left(\mu|\mathbf{H}|^2-\frac{1}{\mu}|\mathbf{B}|^2\right)\right]dV$$

so that (2.134) and (2.135) become

$$\mathbf{F}_{el}^0(t)=\int_{V_a}\left[(\rho-\nabla\cdot\mathbf{P})\mathbf{E}+\mu\left(\mathbf{J}+\frac{\partial \mathbf{P}}{\partial t}\right)\times\mathbf{H}\right.$$

$$\left.-\mu(\nabla\cdot\mathbf{M})\mathbf{H}+\epsilon\mu\mathbf{M}\times\frac{\partial \mathbf{E}}{\partial t}\right]dV$$

$$=\int_{V_a}\left[\rho\mathbf{E}+\mathbf{P}\cdot\nabla\mathbf{E}+\mu\left(\mathbf{J}+\frac{\partial \mathbf{P}}{\partial t}\right)\times\mathbf{H}\right. \tag{2.138}$$

$$\left.+\mu\mathbf{M}\cdot\nabla\mathbf{H}-\epsilon\mu\frac{\partial \mathbf{M}}{\partial t}\times\mathbf{E}+\epsilon\mu\frac{\partial}{\partial t}(\mathbf{M}\times\mathbf{E})\right]dV$$

$$=\int_{V_a}\left[\nabla\cdot\bar{\mathbf{T}}_{el}-\epsilon\frac{\partial}{\partial t}(\mathbf{E}\times\mathbf{B})\right]dV=\int_{S_a}\hat{\mathbf{n}}\cdot\bar{\mathbf{T}}_{el}dS-\epsilon\frac{d}{dt}\int_{V_a}\mathbf{E}\times\mathbf{B}dV$$

and

$$P_{el}^0(t)=\int_{V_a}\left[\left(\mathbf{J}+\frac{\partial \mathbf{P}}{\partial t}\right)\cdot\mathbf{E}-\mathbf{M}\cdot\frac{\partial \mathbf{B}}{\partial t}\right]dV$$

$$=\int_{V_a}\left[\left(\mathbf{J}+\frac{\partial \mathbf{P}}{\partial t}\right)\cdot\mathbf{E}+\frac{\partial \mathbf{M}}{\partial t}\cdot\mathbf{B}-\frac{\partial}{\partial t}(\mathbf{M}\cdot\mathbf{B})\right]dV$$

$$=\int_{V_a}\left[\left(\mathbf{J}+\frac{\partial \mathbf{P}}{\partial t}\right)\cdot\mathbf{E}+\mu\frac{\partial \mathbf{M}}{\partial t}\cdot\mathbf{H}+\frac{1}{2}\frac{\partial}{\partial t}\left(\mu|\mathbf{H}|^2-\frac{1}{\mu}|\mathbf{B}|^2\right)\right]dV \tag{2.139}$$

$$=-\int_{V_a}\left[\nabla\cdot(\mathbf{E}\times\mathbf{H})+\frac{1}{2}\frac{\partial}{\partial t}\left(\epsilon|\mathbf{E}|^2+\frac{1}{\mu}|\mathbf{B}|^2\right)\right]dV$$

$$=-\int_{S_a}\hat{\mathbf{n}}\cdot(\mathbf{E}\times\mathbf{H})dS-\frac{1}{2}\frac{d}{dt}\int_{V_a}\left(\epsilon|\mathbf{E}|^2+\frac{1}{\mu}|\mathbf{B}|^2\right)dV.$$

With S_a not intersecting **P** and **M**, the dyadic identity $\nabla\cdot(\mathbf{ab})=(\nabla\cdot\mathbf{a})\mathbf{b}+(\mathbf{a}\cdot\nabla)\mathbf{b}$ and the dyadic divergence theorem [83, app. 3, eq. 53] has been used in (2.136) to convert the $-(\nabla\cdot\mathbf{P})\mathbf{E}$ and $-\mu(\nabla\cdot\mathbf{M})\mathbf{H}$ terms to $\mathbf{P}\cdot\nabla\mathbf{E}$ and $\mu\mathbf{M}\cdot\nabla\mathbf{H}$, which are in the form of Kelvin electrostatic and magnetostatic polarization force densities, respectively [97, pp. 514–516].

To illustrate the importance of keeping the integration surface S_a outside **P** and **M**, consider the problem of determining the force on a stationary piece of permanently polarized material (for example, an electret or magnet) placed in a nonuniform external static field. The equivalent surface charges and currents on the surface of the electret or magnet must be included in the first volume integrals of (2.134) and (2.138) to correctly determine the total force on the electret or magnet. These surface contributions appear as delta functions in $\nabla \cdot \mathbf{P}$ and $\nabla \times \mathbf{M}$ in the first volume integral of (2.134), as mentioned above, and as delta functions in $\nabla \cdot \mathbf{P}$ and $\nabla \cdot \mathbf{M}$ in the first volume integral of (2.138). If, for example, the polarization $\mathbf{P}$ or $\mathbf{M}$ did not vary over the electret or magnet, one would obtain the erroneous result from the first volume integrals in (2.134) and (2.138) that the total electric force on the electret or the total magnetic force on the magnet in the nonuniform external field were zero, unless the volume integration included the delta functions at the surface of the electret or magnet. That is, the surface S_a of the volume of integration V_a must lie outside the electret or magnet. Incidentally, the self force on the electret or magnet, that is, the total force exerted on itself by its own field, is zero. An easy way to prove this is to observe that in the absence of the external field $\int_{S_a} \hat{\mathbf{n}} \cdot \bar{\mathbf{T}}_{el} dS = 0$ for S_a equal to a sphere of indefinitely large radius r surrounding the electret or magnet (since the static field of the electret or magnet decreases with increasing radius as $O(1/r^3)$ and thus $\hat{\mathbf{n}} \cdot \bar{\mathbf{T}}_{el} = O(1/r^6)$ as $r \to \infty$).

The force and power expressions in (2.134) and (2.135) assume an electric-current (Amperian) model for the magnetic polarization (and an electric-charge (Amperian) model for electric polarization). (The electric-charge and current models of electric and magnetic polarization are herein both referred to as "Amperian" models; see Footnote 1.) With a magnetic-charge model for the magnetic polarization (and an electric-charge (Amperian) model for electric polarization), the Lorentz force and power in (2.134) and (2.135) must be replaced by [80, eqs. (10.7) and (10.23)]

$$
\begin{aligned}
\mathbf{F}_{el}(t) &= \int_{V_a} \Big[(\rho - \nabla \cdot \mathbf{P})\mathbf{E} + \mu\left(\mathbf{J} + \frac{\partial \mathbf{P}}{\partial t}\right) \times \mathbf{H} \\
&\qquad - \mu(\nabla \cdot \mathbf{M})\mathbf{H} - \epsilon\mu \frac{\partial \mathbf{M}}{\partial t} \times \mathbf{E}\Big] dV \\
&= \int_{V_a} \Big[\rho\mathbf{E} + \mathbf{P} \cdot \nabla\mathbf{E} + \mu\left(\mathbf{J} + \frac{\partial \mathbf{P}}{\partial t}\right) \times \mathbf{H} \\
&\qquad + \mu\mathbf{M} \cdot \nabla\mathbf{H} - \epsilon\mu \frac{\partial \mathbf{M}}{\partial t} \times \mathbf{E}\Big] dV \\
&= \int_{V_a} \Big[\nabla \cdot \bar{\mathbf{T}}_{el} - \epsilon\mu \frac{\partial}{\partial t}(\mathbf{E} \times \mathbf{H})\Big] dV = \int_{S_a} \hat{\mathbf{n}} \cdot \bar{\mathbf{T}}_{el} dS - \epsilon\mu \frac{d}{dt} \int_{V_a} \mathbf{E} \times \mathbf{H} dV
\end{aligned}
\tag{2.140}
$$

$$
\begin{aligned}
P_{el}(t) &= \int_{V_a} \Big[\left(\mathbf{J} + \frac{\partial \mathbf{P}}{\partial t}\right) \cdot \mathbf{E} + \mu \frac{\partial \mathbf{M}}{\partial t} \cdot \mathbf{H}\Big] dV \\
&= -\int_{V_a} \Big[\nabla \cdot (\mathbf{E} \times \mathbf{H}) + \frac{1}{2}\frac{\partial}{\partial t}\left(\epsilon|\mathbf{E}|^2 + \mu|\mathbf{H}|^2\right)\Big] dV \\
&= -\int_{S_a} \hat{\mathbf{n}} \cdot (\mathbf{E} \times \mathbf{H}) dS - \frac{1}{2}\frac{d}{dt} \int_{V_a} \left(\epsilon|\mathbf{E}|^2 + \mu|\mathbf{H}|^2\right) dV.
\end{aligned}
\tag{2.141}
$$

This force and power supplied to all the sources in V_a, with $\mathbf{M}$ generated by magnetic charge, are given by the force and power in (2.138) and (2.139) for $\mathbf{M}$ generated by electric current minus the volume integrals of $\epsilon\mu\partial(\mathbf{M} \times \mathbf{E})/\partial t$ and $\partial[(\mu|\mathbf{H}|^2 - |\mathbf{B}|^2/\mu)/2]/\partial t$, respectively. We shall now show, in accordance with Penfield and Haus [77, pp. 214–216 and 244], that these terms should also be subtracted for the electric-current (Amperian) model of magnetic polarization, because of an additional force and power, mentioned above, on the Amperian magnetic dipoles that are not taken into account by the equivalent current $\nabla \times \mathbf{M}$ in (2.134) and (2.135). Consequently, the macroscopic force and power supplied by the fields to their own source densities are independent of the microscopic model used for the magnetic dipoles. (As we shall see shortly, this is not true for point dipoles placed within the source region of an external field. That is, the forces and powers on an electric-current (Amperian) model and a magnetic-charge model of a point magnetic dipole can differ if the point dipole enters the source region of an externally applied field.)

Force and Power for Amperian Dipoles. Let us determine from first principles the force and power caused by external fields, $\mathbf{E}_{ext}(\mathbf{r}, t)$ and $\mathbf{B}_{ext}(\mathbf{r}, t)$, applied to an infinitesimal region V_b of "bound" charge and current characterized by electric and magnetic dipole moments $\mathbf{p}(t)$ and $\mathbf{m}(t)$. The total Lorentz force on the bound electric charge and current distribution, that is, on the electric-charge electric dipole and electric-current magnetic dipole (Amperian electric and magnetic dipoles), is given by

$$\mathbf{F}_{el}^{d}(t) = \int_{V_b} \{\rho_b(\mathbf{r}, t)[\mathbf{E}_{ext}(\mathbf{r}, t) + \mathbf{E}_b(\mathbf{r}, t)] + \mathbf{J}_b(\mathbf{r}, t) \times [\mathbf{B}_{ext}(\mathbf{r}, t) + \mathbf{B}_b(\mathbf{r}, t)]\} dV \tag{2.142}$$

where the integration over V_b covers the bound charge and current, ρ_b and $\mathbf{J}_b$. The internal fields $\mathbf{E}_b$ and $\mathbf{B}_b$ are produced by the bound charge and current. The bound charge and current may be induced in part or entirely by the externally applied fields. The applied fields $\mathbf{E}_{ext}$ and $\mathbf{B}_{ext}$ are produced by sources (ρ_{ext}, $\mathbf{J}_{ext}$, $\mathbf{P}_{ext}$, $\mathbf{M}_{ext}$) other than ρ_b and $\mathbf{J}_b$.

First, consider the "external part" of the force

$$\mathbf{F}_{el}^{ext}(t) = \int_{V_b} [\rho_b(\mathbf{r}, t)\mathbf{E}_{ext}(\mathbf{r}, t) + \mathbf{J}_b(\mathbf{r}, t) \times \mathbf{B}_{ext}(\mathbf{r}, t)]\, dV. \tag{2.143}$$

Let the bound charge-current be distributed about the point $\mathbf{r}_0$ and expand $\mathbf{E}_{ext}(\mathbf{r}, t)$ and $\mathbf{B}_{ext}(\mathbf{r}, t)$ in a power series about the point $\mathbf{r}_0$ to get

$$\mathbf{E}_{ext}(\mathbf{r}, t) = \mathbf{E}_{ext}(\mathbf{r}_0, t) + (\mathbf{r} - \mathbf{r}_0) \cdot \nabla_0 \mathbf{E}_{ext}(\mathbf{r}_0, t) + O(|\mathbf{r} - \mathbf{r}_0|^2) \tag{2.144}$$

$$\mathbf{B}_{ext}(\mathbf{r}, t) = \mathbf{B}_{ext}(\mathbf{r}_0, t) + (\mathbf{r} - \mathbf{r}_0) \cdot \nabla_0 \mathbf{B}_{ext}(\mathbf{r}_0, t) + O(|\mathbf{r} - \mathbf{r}_0|^2) \tag{2.145}$$

where ∇_0 denotes the del operator with respect to the $\mathbf{r}_0$ coordinates. To simplify the following derivation, we shall initially choose $\mathbf{r}_0 = 0$; that is, the origin of the coordinate system is chosen within V_b.

Substituting (2.144) and (2.145) into the integrand of (2.143), then using the results that the total bound charge is zero

$$\int_{V_b} \rho_b(\mathbf{r}, t) dV = 0 \tag{2.146}$$

and that

$$\int_{V_b} \mathbf{J}_b(\mathbf{r}, t) dV = -\int_{V_b} [\nabla \cdot \mathbf{J}_b(\mathbf{r}, t)]\mathbf{r} dV = \int_{V_b} \frac{\partial \rho_b}{\partial t}(\mathbf{r}, t)\mathbf{r} dV = \frac{d\mathbf{p}(t)}{dt} \tag{2.147}$$

which follows from the identity $\nabla \cdot (\mathbf{J}_b \mathbf{r}) = (\nabla \cdot \mathbf{J}_b)\mathbf{r} + \mathbf{J}_b$ and the definition of the electric dipole moment

$$\mathbf{p}(t) = \int_{V_b} \rho_b(\mathbf{r}, t)\mathbf{r}dV \tag{2.148}$$

we obtain for the direct external part of the force

$$\begin{aligned}\mathbf{F}_{el}^{ext}(t) &= \mathbf{p}(t) \cdot \nabla \mathbf{E}_{ext}(0, t) + \frac{d\mathbf{p}(t)}{dt} \times \mathbf{B}_{ext}(0, t) \\ &+ \int_{V_b} \mathbf{J}_b(\mathbf{r}, t) \times [\mathbf{r} \cdot \nabla_0 \mathbf{B}_{ext}(0, t)]dV.\end{aligned} \tag{2.149}$$

All higher order terms in (2.149) have been omitted because they vanish as the size of the region V_b in which the bound charge-current distribution is located approaches zero (assuming the multipole moments of higher order than dipole moments are zero). With the identity

$$\mathbf{J}_b \times (\mathbf{r} \cdot \nabla_0 \mathbf{B}_{ext} \cdot \hat{\mathbf{a}})\hat{\mathbf{a}} = -\hat{\mathbf{a}} \times (\mathbf{J}_b \mathbf{r}) \cdot \nabla_0 \mathbf{B}_{ext} \cdot \hat{\mathbf{a}} \tag{2.150}$$

(where $\hat{\mathbf{a}}$ can be $\hat{\mathbf{x}}$, $\hat{\mathbf{y}}$, or $\hat{\mathbf{z}}$) and the identities

$$\nabla \cdot (\mathbf{J}_b \mathbf{r}\mathbf{r}) = (\nabla \cdot \mathbf{J}_b)\mathbf{r}\mathbf{r} + \mathbf{J}_b \mathbf{r} + \mathbf{r}\mathbf{J}_b \tag{2.151}$$

$$\mathbf{J}_b(\mathbf{r} \cdot \nabla_0 \mathbf{B}_{ext}) - \mathbf{r}(\mathbf{J}_b \cdot \nabla_0 \mathbf{B}_{ext}) = (\mathbf{r} \times \mathbf{J}_b) \times \nabla_0 \mathbf{B}_{ext} \tag{2.152}$$

the integral in (2.149) can be written as

$$\begin{aligned}\int_{V_b} \mathbf{J}_b(\mathbf{r}, t) \times [\mathbf{r} \cdot \nabla_0 \mathbf{B}_{ext}(0, t)]dV &= \nabla_0 \mathbf{B}_{ext}(0, t) \cdot \mathbf{m}(t) \\ &= \mathbf{m}(t) \cdot \nabla_0 \mathbf{B}_{ext}(0, t) + \mathbf{m}(t) \times [\nabla_0 \times \mathbf{B}_{ext}(0, t)]\end{aligned} \tag{2.153}$$

in which the electric-current (Amperian) magnetic dipole moment is defined as[9]

$$\mathbf{m}(t) = \frac{1}{2} \int_{V_b} \mathbf{r} \times \mathbf{J}_b(\mathbf{r}, t)dV. \tag{2.155}$$

Inserting (2.153) into (2.149) gives the final expression for the direct external part of the force exerted on the point Amperian electric and magnetic dipoles

$$\begin{aligned}\mathbf{F}_{el}^{ext}(t) &= \mathbf{p}(t) \cdot \nabla_0 \mathbf{E}_{ext}(\mathbf{r}_0, t) + \frac{d\mathbf{p}(t)}{dt} \times \mathbf{B}_{ext}(\mathbf{r}_0, t) + \mathbf{m}(t) \cdot \nabla_0 \mathbf{B}_{ext}(\mathbf{r}_0, t) \\ &+ \mathbf{m}(t) \times [\nabla_0 \times \mathbf{B}_{ext}(\mathbf{r}_0, t)]\end{aligned} \tag{2.156}$$

where $\mathbf{r}_0$, which is the position of the point electric and magnetic Amperian dipoles with dipole moments $\mathbf{p}(t)$ and $\mathbf{m}(t)$, can be an arbitrary point in space.

---

[9] If $\mathbf{r}_0 \neq 0$, that is, the origin of the coordinate system is chosen outside V_b, the magnetic dipole moment is defined as (see Footnote 1)

$$\mathbf{m}(t) = \frac{1}{2} \int_{V_b} \mathbf{r} \times \mathbf{J}_b(\mathbf{r}, t)dV - \frac{1}{2}\mathbf{r}_0 \times \frac{d\mathbf{p}(t)}{dt}. \tag{2.154}$$

Repeating the derivation of $\mathbf{F}_{el}^{ext}(t)$ with $\mathbf{r}_0 \neq 0$ and with (2.154) produces the same final expression (2.156) for $\mathbf{F}_{el}^{ext}(t)$.

Applying a similar procedure to the total Lorentz power

$$P_{el}^{d}(t) = \int_{V_b} \mathbf{J}_b(\mathbf{r},t) \cdot [\mathbf{E}_{ext}(\mathbf{r},t) + \mathbf{E}_b(\mathbf{r},t)]dV \tag{2.157}$$

supplied to the same bound charge-current distribution forming the Amperian electric and magnetic dipoles, produces the result

$$\begin{aligned} P_{el}^{ext}(t) = \int_{V_b} \mathbf{J}_b(\mathbf{r},t) \cdot \mathbf{E}_{ext}(\mathbf{r},t)dV &= \frac{d\mathbf{p}(t)}{dt} \cdot \mathbf{E}_{ext}(\mathbf{r}_0,t) \\ &+ \mathbf{m}(t) \cdot [\nabla \times \mathbf{E}_{ext}(\mathbf{r}_0,t)] \\ &= \frac{d\mathbf{p}(t)}{dt} \cdot \mathbf{E}_{ext}(\mathbf{r}_0,t) - \mathbf{m}(t) \cdot \frac{\partial \mathbf{B}_{ext}}{\partial t}(\mathbf{r}_0,t) \end{aligned} \tag{2.158}$$

for the "external" part of the power.

The foregoing derivations allowed the bound electric charge and current densities, ρ_b and $\mathbf{J}_b$, to vary arbitrarily in space and time within their infinitesimal region as long as they could be characterized by Amperian electric and magnetic dipole moments $\mathbf{p}(t)$ and $\mathbf{m}(t)$. To the force and power in (2.156) and (2.158) must be added the internal force and power found from (2.142) and (2.157) as

$$\mathbf{F}_{el}^{int}(t) = \int_{V_b} [\rho_b(\mathbf{r},t)\mathbf{E}_b(\mathbf{r},t) + \mathbf{J}_b(\mathbf{r},t) \times \mathbf{B}_b(\mathbf{r},t)]\,dV \tag{2.159}$$

and

$$P_{el}^{int}(t) = \int_{V_b} \mathbf{J}_b(\mathbf{r},t) \cdot \mathbf{E}_b(\mathbf{r},t)dV. \tag{2.160}$$

The fields $\mathbf{E}_b(\mathbf{r},t)$ and $\mathbf{B}_b(\mathbf{r},t)$ produced by the bound charge and current $\rho_b(\mathbf{r},t)$ and $\mathbf{J}_b(\mathbf{r},t)$ vary rapidly (like ρ_b and $\mathbf{J}_b$) about the point $\mathbf{r}_0$ where the infinitesimal region V_b of bound charge and current is located. Therefore, $\mathbf{E}_b(\mathbf{r},t)$ and $\mathbf{B}_b(\mathbf{r},t)$ cannot be represented near $\mathbf{r}_0$ by two terms of a power series, as were $\mathbf{E}_{ext}(\mathbf{r},t)$ and $\mathbf{B}_{ext}(\mathbf{r},t)$ in (2.144) and (2.145).

In order to simplify the evaluation of (2.159) and (2.160), assume the Amperian electric dipole is formed by equal and opposite electric charges at the ends of a fixed infinitesimal nonconducting rod. The charges are stationary, and the equal and opposite electrostatic forces exerted by these charges on each other cancel. Thus this static Amperian electric dipole will not contribute to the electromagnetic force and power in (2.159) and (2.160).

Assume the Amperian magnetic dipole consists of magnetostatic current on a perfectly conducting sphere placed in the external field $(\mathbf{E}_{ext}(\mathbf{r},t), \mathbf{B}_{ext}(\mathbf{r},t))$, and take the Fourier transform of this time-domain external field to obtain the frequency-domain external field. At each frequency, the frequency-domain external field will induce surface charge and current densities on the sphere that can be determined exactly, along with their fields, from the Mie series solution [98, ch. 10]. An asymptotic expansion of the Mie series solution [99], as the radius of the sphere approaches zero about $\mathbf{r}_0$ reveals (after a considerable amount of algebra followed by transforming back to the time domain) that $\mathbf{F}_{el}^{int}$ and P_{el}^{int} for this sphere with static magnetic dipole moment $\mathbf{m}$ are given by

$$\mathbf{F}_{el}^{int}(t) = -\epsilon\mu\frac{\partial}{\partial t}[\mathbf{m} \times \mathbf{E}_{ext}(\mathbf{r}_0,t)] \tag{2.161}$$

and

$$P_{el}^{int} = \frac{\partial}{\partial t}[\mathbf{m} \cdot \mathbf{B}_{ext}(\mathbf{r}_0, t)]. \tag{2.162}$$

Moreover, as the radius of the sphere approaches zero, the radius to which the induced fields extend approaches zero. In other words, the induced fields of the sphere produce the nonzero internal force and power (on the current of the sphere) given in (2.161) and (2.162), but these induced fields are undetectable externally as the radius of the sphere approaches zero. Furthermore, the electric and magnetic dipole moments of the induced charge-current become zero as the radius of the sphere approaches zero.

Although (2.161)–(2.162) can be derived, as just outlined, from a perfectly conducting spherical model of a static Amperian magnetic dipole, Penfield and Haus also derive these terms for a perfectly conducting current-loop model of an arbitrarily time varying Amperian magnetic dipole [77, pp. 214–216 and 244], [100]. Also, De Groot and Suttorp [78, pp. 195–196] derive the terms in (2.161) and (2.162) for an Amperian model consisting of charged point particles (electrons and nuclei) grouped into stable "atoms." If one abandons the idea of a passive physically realizable model (such as current on a perfect conductor or electrons orbiting a nucleus) for the Amperian magnetic dipole, and simply postulates that the electric current forming the magnetic dipole is constrained to flow in free space without the external field generating extra charge-current (for example, a battery connected to a highly resistive wire loop), then the terms in (2.161) and (2.162) would not arise [101].

Adding the internal force and power in (2.161) and (2.162) to the direct external force and power in (2.156) and (2.158) produces the total force and power exerted by the fields ($\mathbf{E}_{ext}$, $\mathbf{B}_{ext}$) (directly as well as indirectly through the induced current on the perfectly conducting model of the Amperian magnetic dipole)

$$\begin{aligned} \mathbf{F}_{el}^{d}(t) = \mathbf{p} \cdot \nabla_0 \mathbf{E}_{ext} + \frac{d\mathbf{p}}{dt} \times \mathbf{B}_{ext} + \mathbf{m} \cdot \nabla_0 \mathbf{B}_{ext} \\ + \mathbf{m} \times (\nabla_0 \times \mathbf{B}_{ext}) - \epsilon\mu \frac{\partial}{\partial t}(\mathbf{m} \times \mathbf{E}_{ext}) \end{aligned} \tag{2.163}$$

$$P_{el}^{d} = \frac{d\mathbf{p}}{dt} \cdot \mathbf{E}_{ext} + \frac{d\mathbf{m}}{dt} \cdot \mathbf{B}_{ext}. \tag{2.164}$$

Force and Power for Magnetic-Charge Model of Magnetic Dipole. If the magnetic dipole moment is produced by magnetic charges rather than electric currents, the internal force and power in (2.161) and (2.162) vanish (for the magnetic dipole formed by equal and opposite magnetic charges at the ends of a a fixed rod), and the total Lorentz force is given by just the external part [80, sec. 10.1]

$$\mathbf{F}_{el}^{md}(t) = \int_{V_b} (\rho_b \mathbf{E}_{ext} + \mu \mathbf{J}_b \times \mathbf{H}_{ext} + \rho_{mb} \mathbf{H}_{ext} - \epsilon \mathbf{J}_{mb} \times \mathbf{E}_{ext}) dV \tag{2.165}$$

in which the magnetic dipole moment of the electric bound charge-current (ρ_b, $\mathbf{J}_b$) is zero, and the electric dipole moment of the magnetic bound charge-current (ρ_{mb}, $\mathbf{J}_{mb}$) is zero. (Note that the force in (2.165) is given in terms of $\mathbf{H}_{ext}$ and $\mathbf{E}_{ext}$ instead of $\mathbf{B}_{ext}$ and $\mathbf{D}_{ext}$, respectively, because it is assumed that $\mathbf{M}_{ext}$ and $\mathbf{P}_{ext}$ are also created by magnetic charge and electric charge, respectively. Of course, $\mu \mathbf{H}_{ext} = \mathbf{B}_{ext}$ and $\epsilon \mathbf{E}_{ext} = \mathbf{D}_{ext}$ if the bound charges and currents lie outside the sources of the external fields.)

Evaluating the integral in (2.165), like we evaluated the integral in (2.143), yields

$$\mathbf{F}_{el}^{md}(t) = \mathbf{p}\cdot\nabla_0\mathbf{E}_{ext} + \mu\frac{d\mathbf{p}}{dt}\times\mathbf{H}_{ext} + \mu\mathbf{m}\cdot\nabla_0\mathbf{H}_{ext} - \epsilon\mu\frac{d\mathbf{m}}{dt}\times\mathbf{E}_{ext}. \tag{2.166}$$

Similarly, the total Lorentz power supplied by the external fields to the electric-charge (Amperian) electric dipole and the magnetic-charge magnetic dipole is given by

$$P_{el}^{md}(t) = \frac{d\mathbf{p}}{dt}\cdot\mathbf{E}_{ext} + \mu\frac{d\mathbf{m}}{dt}\cdot\mathbf{H}_{ext}. \tag{2.167}$$

(If $\mathbf{M}_{ext}$ is created by circulating electric currents rather than magnetic charge separation, $\mathbf{B}_{ext}$ replaces $\mu\mathbf{H}_{ext}$ in (2.166) and (2.167).)

Comparing (2.166) with (2.163) and (2.167) with (2.164) reveals that the force and power on point dipoles for Amperian and magnetic-charge models of magnetic polarization (and electric-charge or Amperian electric polarization) are the same if the sources of the external field lie outside the point dipoles. However, the force-power, on point dipoles located within the sources of the external field, predicted by (2.163)–(2.164) for an Amperian model of magnetic polarization differs, in general, from the force-power predicted by (2.166)–(2.167) for a magnetic-charge model of magnetic polarization. (In both cases, the electric polarization is created by electric charge.) In other words, the force needed to keep the dipoles from moving, when they lie within the sources of an external field, depends on the model of the dipoles. The difference between the forces on point Amperian and magnetic-charge magnetic dipoles within the sources of an external field has been used to demonstrate experimentally that neutrons scattered by the fields in ferromagnetic materials act like Amperian rather than magnetic-charge magnetic dipoles [102], [63, p. 191]. (For these experiments, the magnetization of the ferromagnetic materials is caused by circulating electric currents; therefore, $\mu\mathbf{H}_{ext}$ in (2.166) and (2.167) is replaced by $\mathbf{B}_{ext}$.)

We have just determined from first principles the force and power supplied by external fields to an infinitesimal source region characterized by electric and magnetic dipole moments. In particular, to correctly derive the force and power in (2.163) and (2.164) on Amperian dipoles, we had to take into account the internal force and power in (2.161)–(2.162) (induced by the external fields) on the Amperian model of the magnetic dipole, in addition to the direct force and power in (2.156) and (2.158) applied by the external fields to the electric and magnetic dipoles. In general, a self reaction force and power may also arise from the evaluation of the integrals in (2.159) and (2.160) (and in the corresponding expressions for a magnetic-charge model of the magnetic dipole) that did not arise for the simple models of the dipoles that led to (2.163)–(2.164) and (2.166)–(2.167). This self reaction force-power is analogous to the self reaction force-power on the moving charge distribution of a Lorentz electron [103], and its evaluation need not be considered in the present context.

Force and Power on Polarization Densities Derived from Point-Dipole Force and Power. The general force and power equations for a distribution of electric and magnetic polarization densities $\mathbf{P}(\mathbf{r}, t)$ and $\mathbf{M}(\mathbf{r}, t)$ can be found from (2.163) and (2.164) by replacing $\mathbf{p}(t)$ and $\mathbf{m}(t)$ in (2.163) and (2.164) with $\mathbf{P}(\mathbf{r}, t)\Delta V$ and $\mathbf{M}(\mathbf{r}, t)\Delta V$ (see Footnote 1), where ΔV is the volume of a small region V_D of the polarized material at the point $\mathbf{r}$. The maximum dimension of V_D is D. In particular, the force equation (2.163) becomes

$$\Delta\mathbf{F}_{el}(\mathbf{r}, t) = \Big\{\mathbf{P}(\mathbf{r}, t)\cdot\nabla\mathbf{E}_c(\mathbf{r}, t) + \frac{\partial\mathbf{P}}{\partial t}(\mathbf{r}, t)\times\mathbf{B}(\mathbf{r}, t) + \mathbf{M}(\mathbf{r}, t)\cdot\nabla\mathbf{B}_c(\mathbf{r}, t)$$
$$+\mathbf{M}(\mathbf{r}, t)\times[\nabla\times\mathbf{B}_c(\mathbf{r}, t)] - \epsilon\mu\frac{\partial}{\partial t}[\mathbf{M}(\mathbf{r}, t)\times\mathbf{E}(\mathbf{r}, t)]\Big\}\Delta V \tag{2.168}$$

where $\mathbf{E}_c$ and $\mathbf{B}_c$ are the fields in V_D produced by all sources except $\mathbf{P}\Delta V$ and $\mathbf{M}\Delta V$; that is, inside V_D they are the cavity-defined fields discussed in Footnote 6. In the $(\partial\mathbf{P}/\partial t)\times\mathbf{B}$ term or the $-\epsilon\mu\partial(\mathbf{M}\times\mathbf{E})/\partial t$ term of (2.168), we use the total $\mathbf{B}$ field or $\mathbf{E}$ field because $\mathbf{B}$ or $\mathbf{E}$ is not changed by the removal of $\mathbf{P}\Delta V$ or $\mathbf{M}\Delta V$, respectively, from the polarized material as $D\to 0$. Also, the self reaction force and power, mentioned in the previous paragraph in regard to $\mathbf{p}$ and $\mathbf{m}$, do not contribute to the force and power on $\mathbf{P}\Delta V$ and $\mathbf{M}\Delta V$ as $D\to 0$ because the self reaction force and power on $\mathbf{P}\Delta V$ and $\mathbf{M}\Delta V$ equal $o(\Delta V)$.

The force equation (2.163) was derived assuming that the external fields varied slowly across the dipoles as the size of the dipoles approached zero. This assumption is satisfied by the total fields in (2.168) for sufficiently smoothly varying $\mathbf{P}(\mathbf{r},t)$ and $\mathbf{M}(\mathbf{r},t)$, but not in general for the cavity-defined fields. For a generally shaped volume element V_D, the difference between the cavity-defined and total fields within V_D depends on the depolarization dyadic $\bar{\mathbf{L}}$ (see Footnote 6), and $\bar{\mathbf{L}}$ varies in general for different points within V_D [72]. Moreover, the $\bar{\mathbf{L}}\cdot\mathbf{P}/\epsilon$ and $\bar{\mathbf{L}}\cdot\mathbf{M}$ terms in Footnote 6 do not determine the cavity-defined fields across the surface S_D of the volume V_D. Thus, the terms with $\mathbf{E}_c$ and $\mathbf{B}_c$ in (2.168) should be replaced by an integration over V_D. Fortunately, we can avoid these complications by expressing the cavity-defined fields in terms of the total fields as

$$\mathbf{E}_c(\mathbf{r},t)=\mathbf{E}(\mathbf{r},t)-\mathbf{E}_s(\mathbf{r},t) \tag{2.169}$$

$$\mathbf{B}_c(\mathbf{r},t)=\mu\mathbf{H}_c(\mathbf{r},t)=\mu[\mathbf{H}(\mathbf{r},t)-\mathbf{H}_s(\mathbf{r},t)] \tag{2.170}$$

where $\mathbf{E}_s$ and $\mathbf{H}_s$ are equal to $-\bar{\mathbf{L}}\cdot\mathbf{P}/\epsilon$ and $-\bar{\mathbf{L}}\cdot\mathbf{M}$ inside V_D as $D\to 0$. These self fields of the volume element V_D of polarization densities $\mathbf{P}$ and $\mathbf{M}$ are produced by the equivalent surface charge and current densities $\hat{\mathbf{n}}\cdot\mathbf{P}$ and $\mathbf{M}\times\hat{\mathbf{n}}$, respectively, at the surface S_D of V_D. Since these surface charge and current densities at S_D are canceled by equal and opposite surface charge and current densities on the adjacent volume elements within the polarized material, their fields $\mathbf{E}_s$ and $\mathbf{H}_s$ need not be included when $\mathbf{E}_c$ and $\mathbf{H}_c$ from (2.169) and (2.170) are inserted into (2.168). In other words, because $\mathbf{E}_s$ and $\mathbf{H}_s$ are produced by fictitious surface charge and current densities on S_D that arise only if each volume element is separated in free space from all the other volume elements, $\mathbf{E}_s$ and $\mathbf{H}_s$ disappear if the volume elements of integration are chosen without the fictitious surface charge and current densities. Therefore, $\Delta\mathbf{F}_{el}(\mathbf{r},t)$ in (2.168) can be rewritten as

$$\Delta\mathbf{F}_{el}(\mathbf{r},t)=\left[\mathbf{P}\cdot\nabla\mathbf{E}+\frac{\partial\mathbf{P}}{\partial t}\times\mathbf{B}+\mu\mathbf{M}\cdot\nabla\mathbf{H}+\mu\mathbf{M}\times(\nabla\times\mathbf{H})\right.$$
$$\left.-\epsilon\mu\frac{\partial}{\partial t}(\mathbf{M}\times\mathbf{E})\right]\Delta V \tag{2.171}$$

which becomes

$$\Delta\mathbf{F}_{el}(\mathbf{r},t)=\left[\rho\mathbf{E}+\mathbf{P}\cdot\nabla\mathbf{E}+\mu\left(\mathbf{J}+\frac{\partial\mathbf{P}}{\partial t}\right)\times\mathbf{H}\right.$$
$$\left.+\mu\mathbf{M}\cdot\nabla\mathbf{H}-\epsilon\mu\frac{\partial\mathbf{M}}{\partial t}\times\mathbf{E}\right]\Delta V \tag{2.172}$$

when the force on free charge ρ and current $\mathbf{J}$ is added, and the equations $\nabla\times\mathbf{H}=\epsilon\partial\mathbf{E}/\partial t+\partial\mathbf{P}/\partial t+\mathbf{J}$ and $\mathbf{B}=\mu(\mathbf{H}+\mathbf{M})$ are used.

Summing over all the $\Delta\mathbf{F}_{el}(\mathbf{r},t)$ covering a volume V_a, and then letting the ΔV approach zero, converts (2.172) to the volume integral for the total force exerted by the fields on the sources in V_a

$$\mathbf{F}_{el}(t) = \int\limits_{V_a} \left[\rho\mathbf{E} + \mathbf{P}\cdot\nabla\mathbf{E} + \mu\left(\mathbf{J} + \frac{\partial \mathbf{P}}{\partial t}\right)\times\mathbf{H} + \mu\mathbf{M}\cdot\nabla\mathbf{H} - \epsilon\mu\frac{\partial \mathbf{M}}{\partial t}\times\mathbf{E}\right] dV$$
$$= \int\limits_{V_a}\left[\nabla\cdot\bar{\mathbf{T}}_{el} - \epsilon\mu\frac{\partial}{\partial t}(\mathbf{E}\times\mathbf{H})\right]dV = \int\limits_{S_a}\hat{\mathbf{n}}\cdot\bar{\mathbf{T}}_{el}dS - \epsilon\mu\frac{d}{dt}\int\limits_{V_a}\mathbf{E}\times\mathbf{H}dV. \tag{2.173}$$

As explained previously, assume that the surface S_a of V_a is free of polarization densities, so that all equivalent surface charge and current densities are included in the integral over V_a, and the electromagnetic force on the sources in V_a is defined unambiguously.

We have obtained the force equation (2.173) from (2.163), which was derived from first principles for an infinitesimal region of bound electric charge and current (Amperian electric and magnetic dipoles). By similar means, *exactly the same force equation (2.173) is found starting from the force equation (2.166) that applies to a magnetic-charge model of the magnetic dipole.* The power equation corresponding to (2.173) can be similarly derived from either (2.164) or (2.167) in the form

$$P_{el}(t) = \int\limits_{V_a}\left[\left(\mathbf{J} + \frac{\partial \mathbf{P}}{\partial t}\right)\cdot\mathbf{E} + \mu\frac{\partial \mathbf{M}}{\partial t}\cdot\mathbf{H}\right]dV$$
$$= -\int\limits_{V_a}\left[\nabla\cdot(\mathbf{E}\times\mathbf{H}) + \frac{1}{2}\frac{\partial}{\partial t}\left(\epsilon|\mathbf{E}|^2 + \mu|\mathbf{H}|^2\right)\right]dV \tag{2.174}$$
$$= -\int\limits_{S_a}\hat{\mathbf{n}}\cdot(\mathbf{E}\times\mathbf{H})dS - \frac{1}{2}\frac{d}{dt}\int\limits_{V_a}\left(\epsilon|\mathbf{E}|^2 + \mu|\mathbf{H}|^2\right)dV.$$

With S_a not intersecting $\mathbf{P}$ and $\mathbf{M}$, the force and power $\mathbf{F}_{el}$ and P_{el} in (2.173) and (2.174) are the total force and power delivered by the fields to the continuum of source densities ρ, $\mathbf{J}$, $\mathbf{P}$, and $\mathbf{M}$. This total force and power can be obtained from the force and power in (2.138) and (2.139) by adding the volume integrals of $-\epsilon\mu\partial(\mathbf{M}\times\mathbf{E})/\partial t$ and $\partial(\mathbf{M}\cdot\mathbf{B} - \mu|\mathbf{M}|^2/2)/\partial t = -\partial[(\mu|\mathbf{H}|^2 - |\mathbf{B}|^2/\mu)/2]/\partial t$ to (2.138) and (2.139).

Although the same force (2.173) exerted on the source densities in V_a, by the fields produced by all the source densities ρ, $\mathbf{J}$, $\mathbf{P}$, and $\mathbf{M}$, derives from either one of the two different point-dipole forces (2.163) or (2.166), the force expression (2.173) cannot be used, in general, to recover the force in (2.163) on a point Amperian dipole located within the polarization densities of an "external" field. In deriving (2.173) from (2.163) or (2.166), the force on each volume element of electric and magnetic polarization is determined by the cavity fields, which are free of polarization densities, and in deriving (2.173) from (2.163) this meant that the external magnetic induction $\mathbf{B}_{ext}$ in the magnetic dipole terms of (2.163) changed to $\mu\mathbf{H}$ rather than $\mathbf{B}$. This change properly eliminates any spurious self force from $\mathbf{M}dV$ in the derivation of (2.173) from (2.163), but disallows the use of (2.173) for finding the separate forces on two Amperian polarization densities where the two polarization densities overlap. Therefore, (2.163) is not recovered from (2.173) when a point dipole lies within the polarization density of an "external" field (as one can see by letting $\mathbf{M} = \mathbf{m}\delta(\mathbf{r}-\mathbf{r}_0)$ and $\mathbf{H} = \mathbf{H}_{ext}$ in (2.173) and comparing the result with (2.163)). Similarly, the power expression (2.174) cannot be used, in general, to recover the power in (2.164) delivered to a point Amperian magnetic dipole within the magnetic polarization density of an external field.

With S_a not intersecting $\mathbf{P}$ and $\mathbf{M}$, the electromagnetic torque (couple) $\mathbf{C}_{el}(t)$ exerted by the fields on the charge, current, and polarization densities in V_a can be found by applying the same procedures we used to find $\mathbf{F}_{el}$ and P_{el}. The final result can be written as [80, eq. (10.12)]

$$
\begin{aligned}
\mathbf{C}_{el}(t) &= \int\limits_{V_a} \mathbf{r} \times \left[(\rho - \nabla \cdot \mathbf{P})\mathbf{E} + \left(\mathbf{J} + \frac{\partial \mathbf{P}}{\partial t} + \nabla \times \mathbf{M} \right) \times \mathbf{B} \right. \\
&\qquad\qquad \left. - \epsilon\mu \frac{\partial}{\partial t}(\mathbf{M} \times \mathbf{E}) \right] dV \\
&= \int\limits_{V_a} \left\{ \mathbf{P} \times \mathbf{E} + \mu \mathbf{M} \times \mathbf{H} + \mathbf{r} \times \left[\rho \mathbf{E} + \mathbf{P} \cdot \nabla \mathbf{E} + \mu \left(\mathbf{J} + \frac{\partial \mathbf{P}}{\partial t} \right) \times \mathbf{H} \right. \right. \\
&\qquad\qquad \left. \left. + \mu \mathbf{M} \cdot \nabla \mathbf{H} - \epsilon\mu \frac{\partial \mathbf{M}}{\partial t} \times \mathbf{E}) \right] \right\} dV \\
&= \int\limits_{V_a} \mathbf{r} \times \left[\nabla \cdot \bar{\mathbf{T}}_{el} - \epsilon\mu \frac{\partial}{\partial t}(\mathbf{E} \times \mathbf{H}) \right] dV \\
&= \int\limits_{S_a} \mathbf{r} \times \bar{\mathbf{T}}_{el} \cdot \hat{\mathbf{n}} dS - \epsilon\mu \frac{d}{dt} \int\limits_{V_a} \mathbf{r} \times (\mathbf{E} \times \mathbf{H}) dV
\end{aligned}
\tag{2.175}
$$

which can be derived using either the Amperian or magnetic-charge model of the magnetic dipole.

Electromagnetic Momentum and Energy. Having determined the force, power, and torque with polarization densities included, let us return to the force and power equations (2.131) and (2.132) that apply when only free charge and current (ρ and $\mathbf{J}$) are present. It is customary to interpret $\frac{1}{c^2}\int_{V_a} \mathbf{E} \times \mathbf{H} dV$ in (2.131) as the momentum of the electromagnetic fields, or simply the "electromagnetic momentum," contained within V_a, and $\int_{S_a} \hat{\mathbf{n}} \cdot \bar{\mathbf{T}}_{el} dS$ as the electromagnetic momentum per unit time entering the fields of the volume V_a through the surface S_a. Then (2.131) says that the time rate of change of the electromagnetic momentum in the fields of the volume V_a is equal to the electromagnetic momentum per unit time entering the fields of V_a through its surface S_a, minus the force applied by the fields to the charge and current distribution in V_a. (Nineteenth-century physicists viewed the electromagnetic stress tensor $\bar{\mathbf{T}}_{el}(\mathbf{r}, t)$ as determining an actual stress at the time t and position $\mathbf{r}$ within an electromagnetic ether that permeates all space [82, vol. 1, p. 272].)

It is also customary to interpret $\frac{1}{2}\int_{V_a}(\epsilon|\mathbf{E}|^2 + \mu|\mathbf{H}|^2)dV$ in (2.132) as the electromagnetic energy in the fields of V_a (especially because this integral over all space is equal to the energy of formation of a static charge and current distribution [1, secs. 2.8 and 2.15]), and $-\int_{S_a} \hat{\mathbf{n}} \cdot (\mathbf{E} \times \mathbf{H}) dS$ as the electromagnetic power (energy per unit time) entering the fields of V_a through its surface S_a. Then (2.132), which is usually called Poynting's theorem, says that the time rate of change of the electromagnetic energy in the fields of the volume V_a is equal to the electromagnetic power entering the fields of V_a through its surface S_a (via the Poynting vector $\mathbf{E} \times \mathbf{H}$), minus the power delivered by the fields to the charge and current distribution in V_a.

However, caution should be taken when referring to the integrals $\frac{1}{c^2}\int_{V_a} \mathbf{E} \times \mathbf{H} dV$ and $\frac{1}{2}\int_{V_a}(\epsilon|\mathbf{E}|^2 + \mu|\mathbf{H}|^2)dV$ as electromagnetic momentum and energy, respectively, because they do not, in general, transform as a relativistic four-vector [63, sec. 12.10], [103]. Hence, they cannot, in general, represent the total momentum and energy consistently of a charge-current distribution in all inertial frames of reference. Moreover, the electromagnetic force and power on the left-hand sides of (2.131) and (2.132), as well as the time rates of change of the electromagnetic momentum and energy on the right-hand sides of (2.131) and (2.132), do not generally transform as four-vectors [103]. This is true even for causal fields that are zero outside a bounded volume V_a^0, such that the surface integrals in (2.131) and (2.132) are zero, and (2.131) and (2.132) reduce to

$$\mathbf{F}_{el}(t) = \int\limits_{V_a^0} (\rho\mathbf{E} + \mu\mathbf{J} \times \mathbf{H})dV = -\frac{1}{c^2}\frac{d}{dt}\int\limits_{V_a^0} \mathbf{E} \times \mathbf{H}dV \tag{2.176}$$

and

$$P_{el}(t) = \int\limits_{V_a^0} \mathbf{J} \cdot \mathbf{E}dV = -\frac{1}{2}\frac{d}{dt}\int\limits_{V_a^0} (\epsilon|\mathbf{E}|^2 + \mu|\mathbf{H}|^2)dV. \tag{2.177}$$

One way to understand why the force-power in (2.131)–(2.132) or (2.176)–(2.177) does not transform as a four-vector is to realize that the electromagnetic force-power cannot be the only force-power acting on a classical distribution of charge and current. There must be internal binding force-power stabilizing the given charge-current distribution, that is, holding the specified charge-current together or to the bodies on which they reside. For example, this stabilizing force-power could be the short-range molecular force-power from charge-current and fields not included in ρ, $\mathbf{J}$, $\mathbf{E}$, and $\mathbf{H}$, and not necessarily modeled accurately by the classical Maxwell equations. In addition, there can be external force-power applied to the charge-current distribution (by, for example, the Maxwellian fields of charge-current other than the ρ and $\mathbf{J}$ under consideration). Expressing the externally applied force and power as $\mathbf{F}_{ext}$ and P_{ext}, and then writing the binding force $\mathbf{F}_b(t)$ and power $P_b(t)$ acting on the charge-current as the negative time derivative of a binding momentum $\mathbf{G}_b(t)$ and energy $W_b(t)$, that is

$$\mathbf{F}_b(t) = -\frac{d}{dt}\mathbf{G}_b(t) \tag{2.178}$$

$$P_b(t) = -\frac{d}{dt}W_b(t) \tag{2.179}$$

we can add the external and binding force-power to the electromagnetic force-power in (2.176)–(2.177) to get

$$\mathbf{F}_{el}(t) + \mathbf{F}_b(t) + \mathbf{F}_{ext}(t) = \mathbf{F}_{ext}(t) - \frac{d}{dt}\left[\frac{1}{c^2}\int\limits_{V_a^0} \mathbf{E} \times \mathbf{H}dV + \mathbf{G}_b\right] \tag{2.180}$$

$$P_{el}(t) + P_b(t) + P_{ext}(t) = P_{ext}(t) - \frac{d}{dt}\left[\frac{1}{2}\int\limits_{V_a^0} (\epsilon|\mathbf{E}|^2 + \mu|\mathbf{H}|^2)dV + W_b\right]. \tag{2.181}$$

In general, the binding force-power in (2.178)–(2.179) depends in part on the charge-current and electromagnetic fields, and like the electromagnetic force-power in (2.176)–(2.177) does not transform as a four-vector. However, the combined (binding plus electromagnetic) force-power in (2.180)–(2.181) must transform as a relativistic four-vector, because it is the total force-power experienced by the charge-current, except for the externally applied force-power, which we assume transforms as a four-vector. (We are assuming, as a fundamental principle of physics, that the total force-power on a distribution of charge-current transforms as a relativistic four-vector.)

The force and power on the left- or right-hand sides of (2.180) and (2.181) are the total force and power exerted on the charge-current. It is divided into the force-power exerted on the charge-current by externally applied force-power, by the electromagnetic fields of the charge-current, and by the binding force-power that holds the charge-current together or to the bodies on which they reside. Formally, one can set this total force and power exerted on the charge-current equal to the time rate of change of a residual momentum and energy, $d\mathbf{G}_0/dt$ and dW_0/dt, so that (2.180) and (2.181) can be written in the form of a generalized Newtonian second law of motion

$$\mathbf{F}_{ext}(t) = \frac{d}{dt}\left[\frac{1}{c^2}\int\limits_{V_a^0} \mathbf{E} \times \mathbf{H} dV + \mathbf{G}_b + \mathbf{G}_0\right] \tag{2.182}$$

$$P_{ext}(t) = \frac{d}{dt}\left[\frac{1}{2}\int\limits_{V_a^0} (\epsilon|\mathbf{E}|^2 + \mu|\mathbf{H}|^2) dV + W_b + W_0\right]. \tag{2.183}$$

The residual force and power ($d\mathbf{G}_0/dt$ and dW_0/dt) in (2.182)–(2.183) must transform as a four-vector because the left- and right-hand sides of (2.180)–(2.181) transform as a four-vector. Moreover, if $\mathbf{G}_0$ and W_0 are chosen to make the momentum and energy within the brackets of (2.182) and (2.183) equal to the total momentum and energy of the system to which the external force-power is applied, this total system momentum-energy must transform as a four-vector. (Note that arbitrary constants can be added to $\mathbf{G}_b$, $\mathbf{G}_0$, W_b, and W_0 without changing the equations (2.182)–(2.183).) And, of course, the time derivative of this system momentum-energy (right-hand sides of (2.182)–(2.183)) transforms as a four-vector force-power, since the external force-power transforms as a four-vector.

For a system of moving charged particles, each of which can be modeled by a charge distributed uniformly on the surface of a small spherical insulator, the binding force and power ($-d\mathbf{G}_b/dt$ and $-dW_b/dt$) can be found by applying the relativistic version of Newton's second law of motion to both the surface charge and insulator of each charged insulator. The residual force and power ($-d\mathbf{G}_0/dt$ and $-dW_0/dt$) can be determined from the electromagnetic and binding rest masses that are revealed for each charged insulator by evaluating the electromagnetic and binding force-power on the right-hand sides of (2.182)–(2.183). Once the electromagnetic and binding rest masses are found for each charged insulator, the residual force-power is simply chosen to make the sum of the electromagnetic, binding, and residual rest masses equal to the measured (or given) rest mass of each charged insulator. For each of these small, spherical, uniformly charged insulators (classical Lorentz electrons), the self-electromagnetic, binding, and residual forces and powers can be expressed in terms of the total charge, the radius of the sphere, the mass of the uncharged insulator, and the velocity (and its time derivatives) of the center of the sphere [103]. For a more complicated system of charge-current and matter, however, it may not be feasible to explicitly evaluate these forces and powers in terms of the geometry, uncharged mass, and motion of the bodies on which the charge-current resides. (In the context of the Lorentz model of the electron, the residual force-power is traditionally called the material or bare-mass force-power. This traditional terminology is somewhat misleading, however, because, in general, the residual force-power does not equal the force-power of the uncharged system of particles.)

Designating the first terms in the brackets of (2.182) and (2.183) as the electromagnetic momentum and energy ($\mathbf{G}_{el}$ and W_{el}), and the remaining two terms as mechanical momentum and energy ($\mathbf{G}_{mech}$ and W_{mech}), recasts (2.182) and (2.183) in the form

$$\mathbf{F}_{ext}(t) = \frac{d}{dt}\mathbf{G}_{el}(t) + \frac{d}{dt}\mathbf{G}_{mech}(t) \tag{2.184}$$

and

$$P_{ext}(t) = \frac{d}{dt}W_{el}(t) + \frac{d}{dt}W_{mech}(t). \tag{2.185}$$

This division of the total momentum-energy into electromagnetic momentum-energy and mechanical momentum-energy is also somewhat misleading, however, because the so-called mechanical momentum-energy equals the sum of the binding momentum-energy and residual momentum-energy, each of which depends on the charge-current and electromagnetic fields. Moreover, neither the electromagnetic momentum-energy nor the mechanical momentum-energy transforms separately as a four-vector, unless the charge-current is zero (in which case the mechanical momentum-energy and the external force-power are also zero).

Return now to equations (2.176) and (2.177). If the charge and current are zero in V_a^0, (2.176) and (2.177) reduce to

$$0 = \frac{1}{c^2}\frac{d}{dt}\int\limits_{V_a^0} \mathbf{E} \times \mathbf{H} dV \tag{2.186}$$

and

$$0 = \frac{1}{2}\frac{d}{dt}\int\limits_{V_a^0} (\epsilon|\mathbf{E}|^2 + \mu|\mathbf{H}|^2)dV. \tag{2.187}$$

It can then be proved that this conserved electromagnetic momentum and energy ($\frac{1}{c^2}\int_{V_a^0} \mathbf{E} \times \mathbf{H} dV$ and $\frac{1}{2}\int_{V_a^0}(\epsilon|\mathbf{E}|^2 + \mu|\mathbf{H}|^2)dV$) transform as a relativistic four-vector [64, p. 379]. Consequently, when (2.131) and (2.132) are applied to a volume V_a devoid of charge and current, so that (2.131) and (2.132) reduce to

$$\int\limits_{S_a} \hat{\mathbf{n}} \cdot \bar{\mathbf{T}}_{el} dS = \frac{1}{c^2}\frac{d}{dt}\int\limits_{V_a} \mathbf{E} \times \mathbf{H} dV \tag{2.188}$$

and

$$-\int\limits_{S_a} \hat{\mathbf{n}} \cdot (\mathbf{E} \times \mathbf{H}) dS = \frac{1}{2}\frac{d}{dt}\int\limits_{V_a} (\epsilon|\mathbf{E}|^2 + \mu|\mathbf{H}|^2)dV \tag{2.189}$$

it is reasonable to refer to the volume and surface integrals in (2.188) as the electromagnetic momentum within V_a and the electromagnetic momentum per unit time entering V_a through S_a. Similarly, it is reasonable to interpret the volume and surface integrals in (2.189) as the electromagnetic energy within V_a and the electromagnetic power entering V_a through S_a. Still this momentum-energy and force-power in (2.188)–(2.189) will not necessarily transform as relativistic four-vectors unless the volume V_a is large enough to include all the fields; that is, $V_a = V_a^0$ so that the surface integrals are zero and (2.188)–(2.189) reduce to (2.186)–(2.187).

Integrating (2.188) and (2.189) over time from t_0 to t, for fields that are zero in V_a for $t \le t_0$, yields the electromagnetic momentum and energy in V_a at the time t, specifically

$$\int\limits_{t_0}^{t}\int\limits_{S_a} \hat{\mathbf{n}} \cdot \bar{\mathbf{T}}_{el}(\mathbf{r}, t')dSdt' = \frac{1}{c^2}\int\limits_{V_a} \mathbf{E}(\mathbf{r}, t) \times \mathbf{H}(\mathbf{r}, t)dV \tag{2.190}$$

$$-\int\limits_{t_0}^{t}\int\limits_{S_a} \hat{\mathbf{n}} \cdot (\mathbf{E}(\mathbf{r}, t') \times \mathbf{H}(\mathbf{r}, t'))dSdt' = \frac{1}{2}\int\limits_{V_a} (\epsilon|\mathbf{E}(\mathbf{r}, t)|^2 + \mu|\mathbf{H}(\mathbf{r}, t)|^2)dV. \tag{2.191}$$

In Section 5.2.7 we evaluate the left-hand sides of (2.190) and (2.191) with $t = \infty$ for an infinite plane bounding a source-free half space, in order to determine the total electromagnetic momentum and energy radiated into this half space by time-domain sources (located outside this half space) that turn on and off in a finite time interval.

2.2 TIME-DOMAIN ACOUSTIC FIELD EQUATIONS

Consider a compressible nonviscous fluid supporting an acoustic field characterized by the particle velocity $\mathbf{v}(\mathbf{r}, t)$, the total pressure $p_T(\mathbf{r}, t)$, and the total mass density $\rho_{mT}(\mathbf{r}, t)$. The continuity and momentum equations for this continuum of fluid are given, respectively, by [68, ch. 6]

$$\nabla \cdot (\rho_{mT}\mathbf{v}) + \frac{\partial \rho_{mT}}{\partial t} = s(\mathbf{r}, t) \tag{2.192}$$

and

$$\nabla p_T + \rho_{mT}\frac{\partial \mathbf{v}}{\partial t} + \rho_{mT}(\mathbf{v} \cdot \nabla)\mathbf{v} = \mathbf{f}(\mathbf{r}, t) \tag{2.193}$$

where $s(\mathbf{r}, t)$ and $\mathbf{f}(\mathbf{r}, t)$ are the volume sources of mass-flow and external force densities. The equations (2.192)–(2.193) hold for all time t at all observation points $\mathbf{r}$ throughout the fluid. The continuity equation (2.192) holds for all fluids because it is simply a general mathematical statement of mass conservation in a fluid when mass is neither added nor subtracted from the fluid, that is, when $s(\mathbf{r}, t) = 0$. (Both (2.192) and (2.193) assume fluid velocities small enough compared to the speed of light that relativistic effects can be neglected.) The momentum equation (2.193) holds for nonviscous fluids in which the stress tensor has only diagonal elements (normal stress) given by $-p_T$, and thus the divergence of the stress tensor is simply $-\nabla p_T$. Although a perfectly nonviscous fluid is lossless (has no energy dissipation), the momentum equation (2.193) remains valid for fluids that cannot support shear stress and yet exhibit "expansive-friction" losses, provided $-p_T$ refers to the total normal stress; see Section 2.5.2.

Like Maxwell's equations (2.1)–(2.4), the fluid equations (2.192)–(2.193) contain only first-order spatial and time derivatives. Unlike Maxwell's equations (2.1)–(2.4), however, they are nonlinear equations that, in general, are prohibitively difficult to solve. In the next subsection, the nonlinear time-domain acoustic equations are "linearized" for small-amplitude excitations. In the subsequent subsections, the linear time-domain acoustic equations are treated much like Maxwell's equations were treated in the corresponding subsections of Section 2.1. Thus, for the most part, fewer details are given and the derivations are streamlined.

2.2.1 First-Order Linear Differential Equations

Assume the undisturbed fluid (at rest with no sources) has constant uniform mass density and pressure given by ρ_{m0} and p_0, respectively; that is, the undisturbed fluid is quiescent and homogeneous [70, sec. 1-5]. Also, if the changes in velocity, pressure, and density are small, the nonlinear equations (2.192) and (2.193) can be replaced by linear equations that accurately describe the acoustic wave motion. Specifically, $\nabla \cdot (\rho_{mT}\mathbf{v})$ in (2.192) can be approximated by $\rho_{m0}\nabla \cdot \mathbf{v}$, $\rho_{mT}\partial \mathbf{v}/\partial t$ in (2.193) can be approximated by $\rho_{m0}\partial \mathbf{v}/\partial t$, and the $\rho_{mT}(\mathbf{v} \cdot \nabla)\mathbf{v}$ term in (2.193) can be neglected, provided changes in $\mathbf{v}(\mathbf{r}, t)$ and $\rho_{mT}(\mathbf{r}, t)$ from the undisturbed state are small. In addition, for small changes in an ideal, quiescent, homogeneous, isotropic, lossless fluid, the equation of state (constitutive relation) reduces to a proportionality between changes in pressure and mass density, namely

$$\rho_{m_T}(\mathbf{r}, t) - \rho_{m0} = \rho_{m0}\kappa_0(p_T(\mathbf{r}, t) - p_0) \tag{2.194}$$

or

$$\rho_m(\mathbf{r}, t) = \rho_{m0}\kappa_0 p(\mathbf{r}, t). \tag{2.195}$$

The "acoustic mass density" $\rho_m(\mathbf{r}, t) = \rho_{m_T}(\mathbf{r}, t) - \rho_{m0}$ is the difference between the total mass density $\rho_{m_T}(\mathbf{r}, t)$ and the constant uniform mass density ρ_{m0} in the undisturbed fluid. Similarly, the "acoustic pressure" $p(\mathbf{r}, t) = p_T(\mathbf{r}, t) - p_0$ is the difference between the total pressure $p_T(\mathbf{r}, t)$ and the constant uniform pressure p_0 in the undisturbed fluid. Because the fluid is lossless, there is no dispersion, and the compressibility of the fluid can be characterized by a single real, scalar constant κ_0. (In lossy fluids considered in Section 2.5.2, the attenuation must be dispersive, and thus in the frequency domain the compressibility must be a complex function of frequency, and (2.195) is replaced by a convolution integral. Most if not all real fluids are lossy to some degree, but if the loss is negligible over the bandwidth of the acoustic excitation, time-domain acoustic fields in the fluid can often still be characterized to a good approximation by a constant compressibility κ_0.) Differentiating (2.195) with respect to time reveals the simple proportionality between $\partial\rho_m/\partial t$ and $\partial p/\partial t$ given by

$$\frac{\partial \rho_m}{\partial t} = \rho_{m0}\kappa_0\frac{\partial p}{\partial t}. \tag{2.196}$$

In all then, for these small-amplitude acoustic fields, the continuity and momentum equations, (2.192) and (2.193), convert to the first-order linear differential equations [68, ch. 6]

$$\rho_{m0}\nabla\cdot\mathbf{v}(\mathbf{r}, t) + \rho_{m0}\kappa_0\frac{\partial}{\partial t}p(\mathbf{r}, t) = s(\mathbf{r}, t) \tag{2.197}$$

and

$$\nabla p(\mathbf{r}, t) + \rho_{m0}\frac{\partial}{\partial t}\mathbf{v}(\mathbf{r}, t) = \mathbf{f}(\mathbf{r}, t). \tag{2.198}$$

Since the acoustic pressure p and the velocity $\mathbf{v}$ in (2.197) and (2.198) have small amplitudes, the source functions $s(\mathbf{r}, t)$ and $\mathbf{f}(\mathbf{r}, t)$ must also have small amplitudes. (In the context of these linear small-amplitude equations, we shall often refer to the acoustic pressure $p(\mathbf{r}, t)$ as simply the pressure.) Taking the curl of (2.198) gives

$$\rho_{m0}\frac{\partial}{\partial t}(\nabla\times\mathbf{v}) = \nabla\times\mathbf{f} \tag{2.199}$$

and thus the vorticity is given by

$$\nabla\times\mathbf{v}(\mathbf{r}, t) = \frac{1}{\rho_{m0}}\int_{t_a}^{t}\nabla\times\mathbf{f}(\mathbf{r}, t')dt' + \mathbf{w}(\mathbf{r}) \tag{2.200}$$

where $\mathbf{w}(\mathbf{r}) = \nabla\times\mathbf{v}(\mathbf{r}, t_a)$ is the small-amplitude vorticity at arbitrary time $t = t_a$. Note that $\nabla\cdot\mathbf{w}(\mathbf{r}) = 0$. In addition, (2.200) implies that the vorticity is time-varying only in the source region where the curl of the external force density, $\nabla\times\mathbf{f}(\mathbf{r}, t)$, is nonzero. If the vorticity is zero at some time t_a, then $\mathbf{w}(\mathbf{r}) = 0$ and the vorticity remains zero everywhere except in the region where there is a contribution from the curl of the externally applied force density, $\nabla\times\mathbf{f}(\mathbf{r}, t)$. In other words, for small-amplitude disturbances in fluids that are initially at rest, the flows remain irrotational everywhere except possibly where external forces are applied. Within the realm of linear acoustics, it is usually assumed that the time-independent vorticity function $\mathbf{w}(\mathbf{r})$ is zero. For the sake of generality, however, we allow for the possibility of a nonzero time-independent vorticity.

When the sources and fields are zero for $t \leq t_0$, where t_0 is a finite initial time, the equations (2.197) and (2.198) can be integrated over time to give

$$\rho_{m0}\nabla \cdot \int_{t_0}^{t} \mathbf{v}(\mathbf{r}, t')dt' + \rho_{m0}\kappa_0 p(\mathbf{r}, t) = \int_{t_0}^{t} s(\mathbf{r}, t')dt' \tag{2.201}$$

and

$$\nabla \int_{t_0}^{t} p(\mathbf{r}, t')dt' + \rho_{m0}\mathbf{v}(\mathbf{r}, t) = \int_{t_0}^{t} \mathbf{f}(\mathbf{r}, t')dt'. \tag{2.202}$$

(Taking the curl of (2.202) confirms that the time-independent vorticity $\mathbf{w}(\mathbf{r}) = \nabla \times \mathbf{v}(\mathbf{r}, t_0)$ is zero for these fields and sources that are zero for $t \leq t_0$.) For time-domain sources and fields that are zero for $t \leq t_0$, (2.201) and (2.202) are a more general mathematical form of the acoustic equations than (2.197) and (2.198) because they do not require the existence of the time derivatives of p and $\mathbf{v}$. However, (2.201) and (2.202) assume zero time-independent fields, and thus cannot be used to find nonzero time-independent acoustic fields, which, for small amplitudes, are determined by the linear equations obtained from (2.197) and (2.198), specifically

$$\nabla p(\mathbf{r}) = \mathbf{f}(\mathbf{r}) \tag{2.203}$$

and

$$\rho_{m0}\nabla \cdot \mathbf{v}(\mathbf{r}) = s(\mathbf{r}) \tag{2.204}$$

with the time-independent vorticity specified as

$$\nabla \times \mathbf{v}(\mathbf{r}) = \mathbf{w}(\mathbf{r}). \tag{2.205}$$

Taking the curl of (2.203) and the divergence of (2.205) shows that the time-independent force and vorticity source functions must obey $\nabla \times \mathbf{f}(\mathbf{r}) = 0$ and $\nabla \cdot \mathbf{w}(\mathbf{r}) = 0$.

The time-independent linear acoustic equation (2.203) for the pressure field $p(\mathbf{r})$ is uncoupled from the time-independent linear acoustic equations, (2.204) and (2.205), for the velocity field $\mathbf{v}(\mathbf{r})$. This uncoupling parallels that of the electric and magnetic fields in Maxwell's equations, which also uncouple into separate equations, (2.22) and (2.23), for electrostatic and magnetostatic fields when the sources are independent of time. We have obtained (2.203) and (2.204) (and thus (2.205)) for small-amplitude, time-independent fields and sources. However, (2.192) shows that (2.204) also holds for large-amplitude fields and sources, $\mathbf{v}$ and s, in incompressible fluids (ρ_{m_T} independent of $\mathbf{r}$ and t), and (2.193) shows that (2.203) holds also for large-amplitude fields and sources, p and $\mathbf{f}$, if the velocity $\mathbf{v}$ is independent of $\mathbf{r}$ and t.

2.2.2 Second-Order Differential Equations

Taking the time derivative of (2.197) and substituting $\partial\mathbf{v}/\partial t$ from (2.198) shows that $p(\mathbf{r}, t)$ satisfies the inhomogeneous scalar wave equation

$$\nabla^2 p - \frac{1}{c^2}\frac{\partial^2 p}{\partial t^2} = \nabla \cdot \mathbf{f} - \frac{\partial s}{\partial t} \tag{2.206}$$

or simply

$$\nabla^2 p - \frac{1}{c^2}\frac{\partial^2 p}{\partial t^2} = -q \tag{2.207}$$

where the pressure source function $q(\mathbf{r}, t)$ is defined by

$$q(\mathbf{r}, t) = \frac{\partial}{\partial t} s(\mathbf{r}, t) - \nabla \cdot \mathbf{f}(\mathbf{r}, t) \tag{2.208}$$

and c is the constant speed of propagation of small disturbances in the lossless fluid. It is given in terms of the compressibility by

$$c = 1/\sqrt{\rho_{m0}\kappa_0}. \tag{2.209}$$

Taking the time derivative of (2.198) and substituting $\partial p/\partial t$ from (2.197) produces the following vector equation for $\mathbf{v}(\mathbf{r}, t)$

$$\nabla(\nabla \cdot \mathbf{v}) - \frac{1}{c^2}\frac{\partial^2 \mathbf{v}}{\partial t^2} = \frac{1}{\rho_{m0}}\left(\nabla s - \rho_{m0}\kappa_0 \frac{\partial \mathbf{f}}{\partial t}\right). \tag{2.210}$$

With the help of the vector identity

$$\nabla(\nabla \cdot \mathbf{v}) = \nabla \times (\nabla \times \mathbf{v}) + \nabla^2 \mathbf{v} \tag{2.211}$$

and (2.200), which is derivable from (2.198), one finds that

$$\nabla(\nabla \cdot \mathbf{v}) = \nabla^2 \mathbf{v} + \frac{1}{\rho_{m0}} \int_{t_a}^{t} \nabla \times \nabla \times \mathbf{f}(\mathbf{r}, t')dt' + \nabla \times \mathbf{w}(\mathbf{r}). \tag{2.212}$$

Insertion of (2.212) into (2.210) produces the inhomogeneous vector wave equation for $\mathbf{v}$

$$\begin{aligned}\nabla^2 \mathbf{v} - \frac{1}{c^2}\frac{\partial^2 \mathbf{v}}{\partial t^2} = \frac{1}{\rho_{m0}}\Bigg(&\nabla s - \rho_{m0}\kappa_0 \frac{\partial \mathbf{f}}{\partial t} \\ &- \int_{t_a}^{t} \nabla \times \nabla \times \mathbf{f}(\mathbf{r}, t')dt' - \rho_{m0}\nabla \times \mathbf{w}(\mathbf{r})\Bigg)\end{aligned} \tag{2.213}$$

where $\mathbf{w}(\mathbf{r}) = \nabla \times \mathbf{v}(\mathbf{r}, t_a)$ is the given time-independent vorticity source function. Of course, $\mathbf{w}(\mathbf{r}) = 0$ for fields that have zero vorticity everywhere at some point in time t_a.

Equations (2.206) and (2.213) show that p and each rectangular component of $\mathbf{v}$ satisfy the inhomogeneous scalar wave equation. Thus, the analysis in Section 2.1.3 can be used to prove that p and $\mathbf{v}$ satisfy the strong causality condition (the fields are zero beyond a distance $c(t - t_0)$ from a point source that turns on at $t = t_0$) if they obey the weak causality condition (the fields, produced by sources that turn on at $t = t_0$, are zero everywhere for $t \leq t_0$).

2.2.3 Equivalence of First- and Second-Order Differential Equations

The second-order differential equations, (2.206) and (2.213), for $p(\mathbf{r}, t)$ and $\mathbf{v}(\mathbf{r}, t)$, respectively, were derived from the first-order linear differential equations (2.197) and (2.198), assuming the existence of the second-order derivatives and the validity of interchanging space and time derivatives. We shall now show that under the conditions:

1. equation (2.213) can be differentiated and the derivatives interchanged to get (2.214) and (2.215) below,

2. the functions in parentheses in (2.214) and (2.215) have continuous second derivatives with respect to $\mathbf{r}$ and t, and
3. the sources lie in a bounded region of space and are zero for $t \leq t_0$, or, for time-independent fields, approach zero as $r \to \infty$,

the first-order linear differential equations can be derived from the second-order equations (assuming causal fields), and thus the first- and second-order linear acoustic differential equations are equivalent.

Take the time derivative of (2.213) and substitute $\partial s/\partial t$ from (2.206) to get

$$\left(\nabla^2 - \frac{1}{c^2}\frac{\partial^2}{\partial t^2}\right)\left(\nabla p + \rho_{m0}\frac{\partial \mathbf{v}}{\partial t} - \mathbf{f}\right) = 0. \tag{2.214}$$

By use of the argument presented after equation (2.34), this homogeneous scalar wave equation has only the trivial solution, $\nabla p + \rho_{m0}\partial \mathbf{v}/\partial t - \mathbf{f} = 0$, for causal fields excited by time-dependent sources in a bounded region of space that turn on at $t = t_0$. Even for time-independent fields, (2.214) reduces to Laplace's equation, $\nabla^2(\nabla p - \mathbf{f}) = 0$, which has only the trivial solution if $(\nabla p - \mathbf{f})$ is twice continuously differentiable and approaches zero as $r \to \infty$ (see Appendix A). Thus, (2.214) implies (2.198) for time-independent as well as time-dependent fields. And from (2.198), one can derive (2.200).

Having derived (2.198) from (2.206) and (2.213), we take the divergence of (2.213) and substitute $\nabla \cdot \mathbf{f}$ from the divergence of (2.198), and c^2 from (2.209), to get

$$\nabla^2\left(\rho_{m0}\nabla \cdot \mathbf{v} + \rho_{m0}\kappa_0\frac{\partial p}{\partial t} - s\right) = 0. \tag{2.215}$$

Again, the only solution to (2.215) is the trivial solution, $\rho_{m0}\nabla \cdot \mathbf{v} + \rho_{m0}\kappa_0\partial p/\partial t - s = 0$, for fields and sources that are twice continuously differentiable and approach zero as $r \to \infty$. For causal time-dependent fields excited by sources in a bounded region of space that are zero before some initial time, the fields are perfectly zero at infinity. For time-independent fields and sources with $(\rho_{m0}\nabla \cdot \mathbf{v} - s)$ approaching zero as $r \to \infty$, the only solution to (2.215) is again the trivial solution, $\rho_{m0}\nabla \cdot \mathbf{v} - s = 0$. Thus, (2.215) implies (2.197).

In summary, then, under rather general conditions, the first-order linear differential equations, (2.197) and (2.198), and the second-order differential equations, (2.206) and (2.213), are equivalent. Thus, a solution to either set of equations will also be a solution to the other set of equations.

As with the electromagnetic equations, one could also prove the equivalence of the first- and second-order linear acoustic differential equations by proving the uniqueness of solution of the second-order equations, (2.206) and (2.213), under the assumption that a solution to the first-order equations, (2.197) and (2.198), exists (since (2.206) and (2.213) are derivable from (2.197) and (2.198)).

2.2.4 Uniqueness of Solution

This section provides a proof of uniqueness of solution for the time-domain acoustic equations (2.201)–(2.202). The uniqueness proof for these equations, unlike equations (2.197)–(2.198), does not require the existence of the first-time derivatives of pressure and velocity. As with the electromagnetic fields, the proof of uniqueness assumes the fields are continuous functions of $\mathbf{r}$ and t.

Let the difference of two solutions that satisfy (2.201)–(2.202) with the same source distributions $s(\mathbf{r}, t)$ and $\mathbf{f}(\mathbf{r}, t)$ be denoted by $p_d(\mathbf{r}, t)$ and $\mathbf{v}_d(\mathbf{r}, t)$. These difference fields satisfy the homogeneous equations corresponding to (2.201)–(2.202), namely

$$\rho_{m0}\nabla\cdot\int_{t_0}^{t}\mathbf{v}_d(\mathbf{r},t')dt' + \rho_{m0}\kappa_0 p_d(\mathbf{r},t) = 0 \tag{2.216}$$

$$\nabla\int_{t_0}^{t} p_d(\mathbf{r},t')dt' + \rho_{m0}\mathbf{v}_d(\mathbf{r},t) = 0. \tag{2.217}$$

To prove that the only solution to (2.216)–(2.217) is the trivial solution ($p_d = 0, \mathbf{v}_d = 0$), and thus the solution to (2.201)–(2.202) is unique, form the divergence

$$\nabla\cdot\left(\int_{t_0}^{t} p_d dt' \int_{t_0}^{t}\mathbf{v}_d dt'\right) = \int_{t_0}^{t} p_d dt'\;\nabla\cdot\int_{t_0}^{t}\mathbf{v}_d dt' + \nabla\int_{t_0}^{t} p_d dt'\cdot\int_{t_0}^{t}\mathbf{v}_d dt'. \tag{2.218}$$

Then insert the divergence and gradient functions from (2.216)–(2.217) into the right-hand side of (2.218) to obtain

$$\begin{aligned}\nabla\cdot\left(\int_{t_0}^{t} p_d dt' \int_{t_0}^{t}\mathbf{v}_d dt'\right) &= -\left(\kappa_0 p_d\int_{t_0}^{t} p_d dt' + \rho_{m0}\mathbf{v}_d\cdot\int_{t_0}^{t}\mathbf{v}_d dt'\right)\\ &= -\frac{1}{2}\frac{\partial}{\partial t}\left[\kappa_0\left(\int_{t_0}^{t} p_d dt'\right)^2 + \rho_{m0}\left|\int_{t_0}^{t}\mathbf{v}_d dt'\right|^2\right].\end{aligned} \tag{2.219}$$

The last equality in (2.219) holds if $p_d(\mathbf{r},t)$ and $\mathbf{v}_d(\mathbf{r},t)$ are continuous functions of t. The acoustic field equation (2.217) implies that the gradient of $\int_{t_0}^{t} p_d dt'$ exists. Thus, $\int_{t_0}^{t} p_d dt'$ is a continuous function of $\mathbf{r}$. The acoustic field equation (2.216) implies that the divergence of $\int_{t_0}^{t}\mathbf{v}_d dt'$ exists, and if the curl of $\int_{t_0}^{t}\mathbf{v}_d dt'$ also exists, then $\int_{t_0}^{t}\mathbf{v}_d dt'$ is a continuous function of $\mathbf{r}$. If, in addition, p_d, $\mathbf{v}_d$, and $\nabla\times\int_{t_0}^{t}\mathbf{v}_d dt'$ are continuous functions of $\mathbf{r}$ (for each t), (2.216)–(2.217) imply that the first spatial partial derivatives of $\int_{t_0}^{t} p_d dt'$ and $\int_{t_0}^{t}\mathbf{v}_d dt'$ are continuous functions of $\mathbf{r}$, and thus the divergence theorem applies rigorously to transform (2.219) to

$$\begin{aligned}\frac{1}{2}\frac{d}{dt}\int_{V_a}\left[\kappa_0\left(\int_{t_0}^{t} p_d dt'\right)^2 + \rho_{m0}\left|\int_{t_0}^{t}\mathbf{v}_d dt'\right|^2\right]dV\\ = -\int_{S_a}\left(\int_{t_0}^{t} p_d dt' \int_{t_0}^{t}\mathbf{v}_d dt'\right)\cdot\hat{\mathbf{n}}dS\end{aligned} \tag{2.220}$$

where S_a is the closed surface bounding any volume V_a. Leibnitz's rule [90, sec. 4.9] has allowed the time derivative to be brought outside the volume integral in (2.220) because the integrand and its time derivative are continuous functions of $\mathbf{r}$ and t.

For sources s and $\mathbf{f}$ that are located in a bounded volume and are zero for $t \le t_0$, the causality condition demands that the surface integral on the right-hand side of (2.220) approaches zero as the volume V_a becomes larger and larger and the surface S_a recedes to an infinite distance from the sources. Therefore, (2.220) becomes

$$\frac{d}{dt}\int_{V_\infty}\left[\kappa_0\left(\int_{t_0}^{t} p_d dt'\right)^2 + \rho_{m0}\left|\int_{t_0}^{t}\mathbf{v}_d dt'\right|^2\right]dV = 0 \tag{2.221}$$

or

$$\int_{V_\infty}\left[\kappa_0\left(\int_{t_0}^{t} p_d dt'\right)^2 + \rho_{m0}\left|\int_{t_0}^{t} \mathbf{v}_d dt'\right|^2\right] dV = \beta \tag{2.222}$$

where V_∞ denotes all space, and β is a constant independent of time. Since (2.222) holds for all times, it holds, in particular, for $t = t_0$. Thus, β in (2.222) is zero, and (2.222) implies

$$\int_{t_0}^{t} p_d dt' = \int_{t_0}^{t} \mathbf{v}_d dt' = 0 \tag{2.223}$$

for all t. Then (2.216) and (2.217) demand that p_d and $\mathbf{v}_d$ are zero.

Consequently, the uniqueness of solution to (2.201)–(2.202) is assured for causal acoustic fields that are continuous functions of $\mathbf{r}$ and t, provided the curl of $\int_{t_0}^{t} \mathbf{v}_d dt'$ is a continuous function of $\mathbf{r}$ (for each t), the sources are located in a bounded volume, and the sources and fields are zero before a finite initial time t_0.

The homogeneous equations (2.216)–(2.217) for the difference fields hold not only if the fields and sources are zero for $t \le t_0$, but also if the fields and sources are simply specified at $t = t_0$. Therefore, the proof of uniqueness given in this section applies also to the following generalizations of (2.201)–(2.202) for sources and fields specified at $t = t_0$ and zero everywhere for $t < t_{00} \le t_0$

$$\rho_{m0}\nabla\cdot\int_{t_0}^{t} \mathbf{v}(\mathbf{r}, t')dt' + \rho_{m0}\kappa_0(p(\mathbf{r}, t) - p(\mathbf{r}, t_0)) = \int_{t_0}^{t} s(\mathbf{r}, t')dt' \tag{2.224}$$

$$\nabla\int_{t_0}^{t} p(\mathbf{r}, t')dt' + \rho_{m0}(\mathbf{v}(\mathbf{r}, t) - \mathbf{v}(\mathbf{r}, t_0)) = \int_{t_0}^{t} \mathbf{f}(\mathbf{r}, t')dt'. \tag{2.225}$$

For time-independent fields, the pressure is given by the single equation (2.203), which has a unique solution to within a uniform constant pressure. The time-independent velocity field is determined by the equations (2.204) and (2.205), whose uniqueness of solution follows from the uniqueness of solution to Laplace's equation for twice continuously differentiable time-independent potentials that approach zero as $r \to \infty$.

2.2.5 Existence of Solution

The uniqueness theorem of Section 2.2.4 implies that if a causal solution to (2.201)–(2.202) exists, it is the only causal solution. We shall now construct causal "velocity-potential" expressions for $p(\mathbf{r}, t)$ and $\mathbf{v}(\mathbf{r}, t)$, and then determine sufficient conditions on the acoustic source functions, $s(\mathbf{r}, t)$ and $\mathbf{f}(\mathbf{r}, t)$, that ensure the velocity-potential expressions obey the time-domain acoustic equations (2.201)–(2.202) throughout all space.

To construct these velocity-potential expressions, start by assuming what we will eventually prove, that a solution for p and $\mathbf{v}$ to (2.201)–(2.202) exists, and write the pressure as the time derivative of a function $\Phi(\mathbf{r}, t)$, specifically

$$p(\mathbf{r}, t) = \rho_{m0}\frac{\partial}{\partial t}\Phi(\mathbf{r}, t). \tag{2.226}$$

If the pressure $p(\mathbf{r}, t)$ is zero for $t \le t_0$, $\Phi(\mathbf{r}, t)$ can be chosen zero for $t \le t_0$, and (2.226) integrates over time to yield

$$\Phi(\mathbf{r},t) = \frac{1}{\rho_{m0}} \int_{t_0}^{t} p(\mathbf{r},t')dt'. \tag{2.227}$$

Using (2.227) in (2.202) allows $\mathbf{v}$ to be expressed in terms of Φ as

$$\mathbf{v}(\mathbf{r},t) = -\nabla\Phi(\mathbf{r},t) + \frac{1}{\rho_{m0}} \int_{t_0}^{t} \mathbf{f}(\mathbf{r},t')dt' \tag{2.228}$$

that is, as the gradient of a velocity potential function $\Phi(\mathbf{r},t)$ plus a source function.

Substituting $\mathbf{v}(\mathbf{r},t)$ from (2.228) and $p(\mathbf{r},t)$ from (2.226) into (2.197), which is the time derivative of (2.201), shows that the velocity potential $\Phi(\mathbf{r},t)$ satisfies the inhomogeneous scalar wave equation

$$\nabla^2\Phi(\mathbf{r},t) - \frac{1}{c^2}\frac{\partial^2}{\partial t^2}\Phi(\mathbf{r},t) = \frac{1}{\rho_{m0}}\left[\nabla\cdot\int_{t_0}^{t}\mathbf{f}(\mathbf{r},t')dt' - s(\mathbf{r},t)\right] \tag{2.229}$$

or

$$\nabla^2\Phi(\mathbf{r},t) - \frac{1}{c^2}\frac{\partial^2}{\partial t^2}\Phi(\mathbf{r},t) = -j(\mathbf{r},t) \tag{2.230}$$

where the velocity-potential source function $j(\mathbf{r},t)$ is defined in terms of the mass-flow and external force functions (s and $\mathbf{f}$) by

$$j(\mathbf{r},t) = \frac{1}{\rho_{m0}}\left[s(\mathbf{r},t) - \nabla\cdot\int_{t_0}^{t}\mathbf{f}(\mathbf{r},t')dt'\right]. \tag{2.231}$$

Note that the pressure source function $q(\mathbf{r},t)$ defined in (2.208) is simply related to the velocity-potential source function in (2.231) by

$$q(\mathbf{r},t) = \rho_{m0}\frac{\partial}{\partial t}j(\mathbf{r},t). \tag{2.232}$$

The well-known causal solution to (2.230) for sources $j(\mathbf{r},t)$ in a bounded volume V is given by [1, sec. 8.1]; see Section 2.1.3

$$\Phi(\mathbf{r},t) = \frac{1}{4\pi}\lim_{\delta\to 0}\int_{V-V_\delta}\frac{j(\mathbf{r}',t-|\mathbf{r}-\mathbf{r}'|/c)}{|\mathbf{r}-\mathbf{r}'|}dV' \tag{2.233}$$

with $p(\mathbf{r},t)$ and $\mathbf{v}(\mathbf{r},t)$ found from (2.226) and (2.228).

The expression for the pressure given in (2.226) also applies to time-independent acoustic pressure that is zero outside the source region. To see this, insert Φ from (2.233) into (2.226) to get

$$p(\mathbf{r}) = -\frac{1}{4\pi}\lim_{\delta\to 0}\int_{V-V_\delta}\frac{\nabla'\cdot\mathbf{f}(\mathbf{r}')}{|\mathbf{r}-\mathbf{r}'|}dV' = -\frac{1}{4\pi}\nabla\cdot\lim_{\delta\to 0}\int_{V-V_\delta}\frac{\mathbf{f}(\mathbf{r}')}{|\mathbf{r}-\mathbf{r}'|}dV' \tag{2.234}$$

when the sources s and $\mathbf{f}$ are independent of time. Then take the gradient of (2.234), rewrite $\nabla\nabla\cdot$ as $\nabla^2 + \nabla\times\nabla\times$, and note from (2.203) that the source function $\nabla\times\mathbf{f}(\mathbf{r})$ must be zero for time-independent fields, to obtain $\nabla p(\mathbf{r}) = \mathbf{f}(\mathbf{r})$, the time-independent equation (2.203) for the pressure field. The acoustic pressure given by (2.234) is zero outside the source region because it predicts that $p(\mathbf{r}) \to 0$ as $r \to \infty$ and $\nabla p(\mathbf{r}) = 0$ outside the source region. A simpler expression than (2.234) is found in Section 2.2.6 for the time-independent acoustic pressure by integrating (2.203) along any path between two points in space.

The expression for the velocity given in (2.228) also remains valid for time-independent fields but only if the vorticity is zero. To see this, take the divergence of (2.228), substitute $\nabla^2\Phi$ from (2.230), and use (2.231) to obtain the time-independent equation (2.204), that is, $\rho_{m0}\nabla \cdot \mathbf{v} = s(\mathbf{r})$. However, taking the curl of (2.228) and again noting that $\nabla \times \mathbf{f}$ must be zero for time-independent fields, one obtains $\nabla \times \mathbf{v} = 0$, that is, equation (2.205) with the time-independent vorticity $\mathbf{w}(\mathbf{r})$ equal to zero. Expressions for the time-independent velocity are given in Section 2.2.6. However, here we want to determine sufficient conditions on the time-dependent source functions ($j(\mathbf{r}, t)$, $s(\mathbf{r}, t)$, and $\mathbf{f}(\mathbf{r}, t)$) in order for $p(\mathbf{r}, t)$ in (2.226) and $\mathbf{v}(\mathbf{r}, t)$ in (2.228), with $\Phi(\mathbf{r}, t)$ given in (2.233), to rigorously satisfy the acoustic time-domain equations (2.201)–(2.202) throughout all space.

Begin by noting that Leibnitz's rule [90, pp. 266 and 273] and Theorem 2-I of Section 2.1.6 ensures that the first time and spatial derivatives of $\Phi(\mathbf{r}, t)$ in (2.233) exist and are continuous functions of $\mathbf{r}$ and t, if $j(\mathbf{r}, t)$ and $\partial j(\mathbf{r}, t)/\partial t$ are continuous functions of $\mathbf{r}$ and t. Thus, $p(\mathbf{r}, t)$ given in (2.226) and $\mathbf{v}(\mathbf{r}, t)$ given in (2.228) exist and are continuous functions of $(\mathbf{r}, t)$, if $\int_{t_0}^{t} \mathbf{f}(\mathbf{r}', t)dt'$ is a continuous function of $(\mathbf{r}, t)$. Then (2.226) can be integrated over time to obtain (2.227). Substitution of the time integral of pressure from (2.227) and the velocity from (2.228) into (2.202) then shows immediately that (2.202) is satisfied. (The definition of $j(\mathbf{r}, t)$ in (2.231) implies that $\int_{t_0}^{t} \mathbf{f}(\mathbf{r}, t')dt'$ exists if $j(\mathbf{r}, t)$ exists.)

Next use (2.233) and Theorem 2-I of Section 2.1.6 to write $\nabla\Phi$ explicitly as

$$\begin{aligned}\nabla\Phi &= \frac{1}{4\pi}\lim_{\delta\to 0}\int\limits_{V-V_\delta} \nabla\left[\frac{j(\mathbf{r}', t - |\mathbf{r}-\mathbf{r}'|/c)}{|\mathbf{r}-\mathbf{r}'|}\right]dV' \\ &= -\frac{1}{4\pi}\lim_{\delta\to 0}\int\limits_{V-V_\delta}\left[\frac{j(\mathbf{r}', t - |\mathbf{r}-\mathbf{r}'|/c)}{|\mathbf{r}-\mathbf{r}'|^3}\right. \\ &\qquad \left. + \frac{\frac{\partial}{\partial t} j(\mathbf{r}', t - |\mathbf{r}-\mathbf{r}'|/c)}{c|\mathbf{r}-\mathbf{r}'|^2}\right](\mathbf{r}-\mathbf{r}')dV'\end{aligned} \tag{2.235}$$

where again $j(\mathbf{r}, t)$ and $\partial j(\mathbf{r}, t)/\partial t$ are assumed to be continuous functions of $(\mathbf{r}, t)$, so that $\nabla\Phi$ is a continuous function of $(\mathbf{r}, t)$. Integrating (2.235) over time and using Theorem 2-I of Section 2.1.6, we get

$$\begin{aligned}\int\limits_{t_0}^{t} \nabla\Phi(\mathbf{r}, t')dt' &= -\frac{1}{4\pi}\lim_{\delta\to 0}\int\limits_{V-V_\delta}\left[\frac{c}{|\mathbf{r}-\mathbf{r}'|}\int\limits_{t_0}^{t} j(\mathbf{r}', t' - |\mathbf{r}-\mathbf{r}'|/c)dt'\right. \\ &\qquad \left. + j(\mathbf{r}', t - |\mathbf{r}-\mathbf{r}'|/c)\right]\frac{(\mathbf{r}-\mathbf{r}')}{c|\mathbf{r}-\mathbf{r}'|^2}dV' \\ &= \frac{1}{4\pi}\nabla\lim_{\delta\to 0}\int\limits_{V-V_\delta}\left[\frac{1}{|\mathbf{r}-\mathbf{r}'|}\int\limits_{t_0}^{t} j(\mathbf{r}', t' - |\mathbf{r}-\mathbf{r}'|/c)dt'\right]dV'.\end{aligned} \tag{2.236}$$

The interchange of the time and volume integrations needed to get the first equality in (2.236) is justified by writing the volume integration in spherical coordinates centered on $\mathbf{r}$ (to get a continuous integrand), and applying a theorem of integral calculus [90, p. 238] that lets the order of integrations be interchanged when the integrand is continuous in all the integration variables.

We can now integrate (2.228) over time, insert $\int_{t_0}^{t} \nabla\Phi(\mathbf{r}, t')dt'$ from (2.236), take the divergence of the resulting equation, and apply Theorem 2-II of Section 2.1.6 to find

$$\begin{aligned}\nabla \cdot \int_{t_0}^{t} \mathbf{v}(\mathbf{r}, t')dt' = &-\frac{1}{4\pi}\lim_{\delta\to 0}\int_{V-V_\delta^s} \nabla^2\left[\frac{1}{|\mathbf{r}-\mathbf{r}'|}\int_{t_0}^{t} j(\mathbf{r}', t' - |\mathbf{r}-\mathbf{r}'|/c)dt'\right]dV' \\ &+ \int_{t_0}^{t} j(\mathbf{r}, t')dt' + \frac{1}{\rho_{m0}}\nabla\cdot\int_{t_0}^{t}\int_{t_0}^{t'} \mathbf{f}(\mathbf{r}, t'')dt''dt'.\end{aligned} \tag{2.237}$$

To apply Theorem 2-II, we have had to assume that $j(\mathbf{r}, t)$ and $\partial j(\mathbf{r}, t)/\partial t$ are continuous functions of $\mathbf{r}$ and t, and that $\int_{t_0}^{t} j(\mathbf{r}, t')dt'$ is Hölder continuous in $\mathbf{r}$ for all t. (If $j(\mathbf{r}, t)$ is continuous for all $(\mathbf{r}, t)$, and Hölder continuous in $\mathbf{r}$ for all t, so is $\int_{t_0}^{t} j(\mathbf{r}, t')dt'$.) The last term in (2.237) exists and is a continuous function of $(\mathbf{r}, t)$ equal to $\int_{t_0}^{t}[\nabla\cdot\int_{t_0}^{t'} \mathbf{f}(\mathbf{r}, t'')dt'']dt'$ if $\int_{t_0}^{t} \mathbf{f}(\mathbf{r}, t')dt'$ and $\nabla\cdot\int_{t_0}^{t} \mathbf{f}(\mathbf{r}, t')dt'$ are continuous functions of $\mathbf{r}$ and t, or, alternatively, if $\mathbf{f}(\mathbf{r}, t)$ and $\nabla\cdot\mathbf{f}(\mathbf{r}, t)$ are continuous functions of $\mathbf{r}$ and t [90, pp. 266 and 273]. Equation (2.231) and the formula (2.67) with j inserted for $\mathbf{J}$ reduce (2.237) to

$$\begin{aligned}\nabla \cdot \int_{t_0}^{t} \mathbf{v}(\mathbf{r}, t')dt' = &-\frac{1}{4\pi}\lim_{\delta\to 0}\int_{V-V_\delta} \frac{1}{c^2|\mathbf{r}-\mathbf{r}'|}\frac{\partial}{\partial t} j(\mathbf{r}', t - |\mathbf{r}-\mathbf{r}'|/c)dV' \\ &+ \frac{1}{\rho_{m0}}\int_{t_0}^{t} s(\mathbf{r}, t')dt'\end{aligned} \tag{2.238}$$

or

$$\nabla \cdot \int_{t_0}^{t} \mathbf{v}(\mathbf{r}, t')dt' = -\frac{1}{c^2}\frac{\partial}{\partial t}\Phi(\mathbf{r}, t) + \frac{1}{\rho_{m0}}\int_{t_0}^{t} s(\mathbf{r}, t')dt'. \tag{2.239}$$

The relationship (2.231) between j, s, and $\mathbf{f}$ shows that $s(\mathbf{r}, t)$ is a continuous function of $\mathbf{r}$ and t if $j(\mathbf{r}, t)$ and $\nabla\cdot\int_{t_0}^{t} \mathbf{f}(\mathbf{r}, t')dt'$ are continuous functions of $\mathbf{r}$ and t. Substitution of $\nabla\cdot\int_{t_0}^{t} \mathbf{v}(\mathbf{r}, t')dt'$ from (2.239) and $p(\mathbf{r}, t)$ from (2.226) into (2.201) proves that (2.201) is also satisfied.

In summary, we have just proven that $p(\mathbf{r}, t)$ and $\mathbf{v}(\mathbf{r}, t)$ given in (2.226) and (2.228) with Φ defined in (2.233) are the unique, causal, continuous acoustic fields that rigorously satisfy the time-domain linear acoustic equations (2.201)–(2.202) for source functions $j(\mathbf{r}, t)$ and $\mathbf{f}(\mathbf{r}, t)$ that satisfy the conditions:

(i) $j(\mathbf{r}, t)$ and $\mathbf{f}(\mathbf{r}, t)$ are located within a bounded volume V,

(ii) $j(\mathbf{r}, t)$, $\partial j(\mathbf{r}, t)/\partial t$, $\int_{t_0}^{t} \mathbf{f}(\mathbf{r}, t')dt'$, and $\nabla\cdot\int_{t_0}^{t} \mathbf{f}(\mathbf{r}, t')dt'$ are continuous functions of $\mathbf{r}$ and t, and

(iii) $j(\mathbf{r}, t)$ is Hölder continuous in $\mathbf{r}$ for all t.

The definition (2.231) of j in terms of the volume mass-flow and external force densities, s and $\mathbf{f}$, reveals that these sufficient conditions for the existence of solution to (2.201)–(2.202) can be replaced by separate sufficient conditions on s and $\mathbf{f}$, namely (i′) $s(\mathbf{r},t)$ and $\mathbf{f}(\mathbf{r},t)$ are located within a bounded volume V,

(ii′) $s(\mathbf{r},t)$ and $\nabla\cdot\int_{t_0}^{t}\mathbf{f}(\mathbf{r},t')dt'$, their time derivatives, and $\int_{t_0}^{t}\mathbf{f}(\mathbf{r},t')dt'$ are continuous functions of $\mathbf{r}$ and t, and

(iii′) $s(\mathbf{r},t)$ and $\nabla\cdot\int_{t_0}^{t}\mathbf{f}(\mathbf{r},t')dt'$ are Hölder continuous in $\mathbf{r}$ for all t.

The conditions (ii′) and (iii′) can in turn be replaced by

(ii″) $s(\mathbf{r},t)$, $\partial s(\mathbf{r},t)/\partial t$, $\mathbf{f}(\mathbf{r},t)$ and $\nabla\cdot\mathbf{f}(\mathbf{r},t)$ are continuous functions of $\mathbf{r}$ and t, and

(iii″) $s(\mathbf{r},t)$ and $\nabla\cdot\mathbf{f}(\mathbf{r},t)$ are Hölder continuous in $\mathbf{r}$ for all t.

With all sources zero for $t < t_0$, continuity of $j(\mathbf{r},t)$, $q(\mathbf{r},t)$, and $\nabla\cdot\int_{t_0}^{t}\mathbf{f}(\mathbf{r},t')dt'$ in (ii), (ii′), and (ii″) is implied by the continuity of their time derivatives.

The conditions (i′), (ii″), and (iii″) are also sufficient to prove from (2.233) and (2.226), using Theorem 2-II, that Φ and p satisfy the equations[10]

$$\nabla^2\int_{t_0}^{t}\Phi(\mathbf{r},t')dt' - \frac{1}{c^2}\frac{\partial}{\partial t}\Phi(\mathbf{r},t) = -\int_{t_0}^{t} j(\mathbf{r},t')dt' \tag{2.242}$$

and

$$\begin{aligned}\nabla^2\int_{t_0}^{t}\int_{t_0}^{t'} p(\mathbf{r},t'')dt''dt' - \frac{1}{c^2}p(\mathbf{r},t) &= -\rho_{m0}\int_{t_0}^{t} j(\mathbf{r},t')dt' \\ &= -\int_{t_0}^{t}\int_{t_0}^{t'} q(\mathbf{r},t'')dt''dt'.\end{aligned} \tag{2.243}$$

At the end of the next section, where general expressions are determined for the time-independent acoustic pressure and velocity, sufficient conditions on $s(\mathbf{r})$, $\mathbf{f}(\mathbf{r})$, and $\mathbf{w}(\mathbf{r})$ are given for those expressions to obey the time-independent linear acoustic equations (2.203)–(2.205).

2.2.6 Volume Integral Expressions for the Pressure and Velocity Fields

The acoustic pressure and particle velocity can be expressed explicitly in terms of volume sources by substituting the velocity potential from (2.233) into (2.226) and (2.228). For the

[10] Taking the time derivative once of (2.242) and twice of (2.243) produces the familiar wave equations for the velocity potential and acoustic pressure, respectively

$$\nabla^2\Phi - \frac{1}{c^2}\frac{\partial^2\Phi}{\partial t^2} = -j \tag{2.240}$$

and

$$\nabla^2 p - \frac{1}{c^2}\frac{\partial^2 p}{\partial t^2} = -q \tag{2.241}$$

provided the second derivatives of Φ and p exist, and the time derivatives can be interchanged with the ∇^2 operators. The conditions (i′), (ii″), and (iii″) are not sufficient to allow these differentiations and interchanges. Theorem 2-II tells us that sufficient conditions for Φ and p in (2.233) and (2.226) to obey the familiar scalar wave equations (2.240) and (2.241) everywhere are that $j(\mathbf{r},t)$ and $q(\mathbf{r},t)$ are located within a bounded volume V, that $j(\mathbf{r},t)$ and $q(\mathbf{r},t)$ along with their first and second time derivatives are continuous functions of $(\mathbf{r},t)$, and that $j(\mathbf{r},t)$ and $q(\mathbf{r},t)$ are Hölder continuous in $\mathbf{r}$ for all t.

pressure we get

$$p(\mathbf{r},t) = \frac{\rho_{m0}}{4\pi}\lim_{\delta\to 0}\int\limits_{V-V_\delta}\frac{1}{R}\frac{\partial}{\partial t}j(\mathbf{r}',t-R/c)dV' \\ = \frac{1}{4\pi}\lim_{\delta\to 0}\int\limits_{V-V_\delta}\frac{q(\mathbf{r}',t-R/c)}{R}dV' \tag{2.244}$$

or

$$p(\mathbf{r},t) = \frac{1}{4\pi}\lim_{\delta\to 0}\int\limits_{V-V_\delta}\frac{1}{R}\left[\frac{\partial}{\partial t}s(\mathbf{r}',t-R/c) - \{\nabla'\cdot\mathbf{f}\}(\mathbf{r}',t-R/c)\right]dV' \tag{2.245}$$

after substituting j or q from their definition (2.231) or (2.208) in terms of s and $\mathbf{f}$. As usual, the vector $\mathbf{R}$ is defined as $\mathbf{r}-\mathbf{r}'$ with $R = |\mathbf{r}-\mathbf{r}'|$. Since

$$\{\nabla'\cdot\mathbf{f}\}(\mathbf{r}',t-R/c) = \nabla'\cdot\mathbf{f}(\mathbf{r}',t-R/c) - \frac{\mathbf{R}}{cR}\cdot\frac{\partial}{\partial t}\mathbf{f}(\mathbf{r}',t-R/c) \tag{2.246}$$

(2.245) can be rewritten as

$$p(\mathbf{r},t) = \frac{1}{4\pi}\lim_{\delta\to 0}\int\limits_{V-V_\delta}\left[\frac{1}{R}\frac{\partial}{\partial t}s(\mathbf{r}',t-R/c) \right. \\ \left. + \left(\frac{1}{c}\frac{\partial}{\partial t}\mathbf{f}(\mathbf{r}',t-R/c) + \frac{\mathbf{f}(\mathbf{r}',t-R/c)}{R}\right)\cdot\frac{\mathbf{R}}{R^2}\right]dV'. \tag{2.247}$$

The divergence theorem has been applied to obtain the last term in the integrand of (2.247), and the meaning of the curly brackets in (2.246) can be found in (2.62).

For the particle velocity we get

$$\mathbf{v}(\mathbf{r},t) = -\frac{1}{4\pi}\nabla\lim_{\delta\to 0}\int\limits_{V-V_\delta}\frac{j(\mathbf{r}',t-R/c)}{R}dV' + \frac{1}{\rho_{m0}}\int\limits_{t_0}^{t}\mathbf{f}(\mathbf{r},t')dt' \\ = \frac{1}{4\pi}\lim_{\delta\to 0}\int\limits_{V-V_\delta}\left[\frac{1}{c}\frac{\partial}{\partial t}j(\mathbf{r}',t-R/c) + \frac{j(\mathbf{r}',t-R/c)}{R}\right]\frac{\mathbf{R}}{R^2}dV' \\ + \frac{1}{\rho_{m0}}\int\limits_{t_0}^{t}\mathbf{f}(\mathbf{r},t')dt'. \tag{2.248}$$

Inserting j from (2.231) into the first equation of (2.248), we can make use of the relation $\nabla'\cdot(\mathbf{F}(\mathbf{r}',t-R/c)/R) = \{\nabla'\cdot\mathbf{F}\}(\mathbf{r}',t-R/c)/R - \nabla\cdot(\mathbf{F}(\mathbf{r}',t-R/c)/R)$, the divergence theorem, Theorem 2-II of Section 2.1.6, and the vector identity

$$\nabla\nabla\cdot\left[\frac{\mathbf{F}\left(t-\frac{|\mathbf{r}-\mathbf{r}'|}{c}\right)}{|\mathbf{r}-\mathbf{r}'|}\right] = \frac{1}{c^2R}\frac{\partial^2\mathbf{F}}{\partial t^2}\cdot\frac{\mathbf{RR}}{R^2} \\ + \frac{1}{R^2}\left(\frac{1}{c}\frac{\partial\mathbf{F}}{\partial t} + \frac{\mathbf{F}}{R}\right)\cdot\left(\frac{3\mathbf{RR}}{R^2} - \bar{\mathbf{I}}\right),\ \mathbf{r}\neq\mathbf{r}' \tag{2.249}$$

where the $\mathbf{F}$'s on the right-hand side of (2.249) are also evaluated at the retarded time

$t - |\mathbf{r} - \mathbf{r}'|/c$, to convert (2.248) to

$$\mathbf{v}(\mathbf{r}, t) = \frac{1}{4\pi\rho_{m0}} \lim_{\delta \to 0} \int\limits_{V - V_\delta} \Bigg[\left(\frac{1}{c}\frac{\partial}{\partial t} s(\mathbf{r}', t - R/c) + \frac{1}{R} s(\mathbf{r}', t - R/c) \right) \frac{\mathbf{R}}{R^2}$$
$$+ \left(\frac{1}{c^2}\frac{\partial}{\partial t} \mathbf{f}(\mathbf{r}', t - R/c) \right) \cdot \frac{\mathbf{RR}}{R^3}$$
$$+ \frac{1}{R^2} \left(\frac{1}{c} \mathbf{f}(\mathbf{r}', t - R/c) + \frac{1}{R} \int\limits_{t_0}^{t} \mathbf{f}(\mathbf{r}', t - R/c) dt' \right) \tag{2.250}$$
$$\cdot \left(\frac{3\mathbf{RR}}{R^2} - \bar{\mathbf{I}} \right) \Bigg] dV' - \frac{1}{\rho_{m0}} (\bar{\mathbf{L}} - \bar{\mathbf{I}}) \cdot \int\limits_{t_0}^{t} \mathbf{f}(\mathbf{r}, t') dt'.$$

The spherical principal volume V_δ^s in Theorem 2-II has been changed to an arbitrarily shaped principal volume V_δ in (2.250) with the help of the dyadic $\bar{\mathbf{L}}$ defined in (2.84).

The operations required to derive (2.244), (2.245), and (2.248) are rigorously valid under the sufficient conditions (i′), (ii″), and (iii″) on s and $\mathbf{f}$ stated in Section 2.2.5. The derivations of (2.247) and (2.250) are also rigorously allowed, if, in addition to the conditions stated in Section 2.2.5, the first partial derivatives of $\mathbf{f}(\mathbf{r}, t)$ (with respect to x, y, z, and t) are continuous in $(\mathbf{r}, t)$. Consequently, the pressure and velocity fields in (2.247) and (2.250) determine the unique, causal, continuous solution throughout space to the time-domain linear acoustic equations (2.201)–(2.202), under the conditions:

- (i′) $s(\mathbf{r}, t)$ and $\mathbf{f}(\mathbf{r}, t)$ are located within a bounded volume V,
- (ii‴) $s(\mathbf{r}, t)$, $\partial s(\mathbf{r}, t)/\partial t$, $\mathbf{f}(\mathbf{r}, t)$ and its first partial derivatives are continuous functions of $\mathbf{r}$ and t, and
- (iii‴) $s(\mathbf{r}, t)$ and $\nabla \cdot \mathbf{f}(\mathbf{r}, t)$ are Hölder continuous in $\mathbf{r}$ for all t.

Again, with $s(\mathbf{r}, t)$ zero for $t \le t_0$, its continuity in $(\mathbf{r}, t)$ is implied by the continuity of its partial time derivatives in $(\mathbf{r}, t)$. And, of course, the continuity of the first partial derivatives of $\mathbf{f}(\mathbf{r}, t)$ ensures the Hölder continuity of $\mathbf{f}(\mathbf{r}, t)$.

The sufficient conditions on $\mathbf{f}$ in (i′), (ii‴), and (iii‴) for the existence of solution to the acoustic equations (2.201)–(2.202) are identical to those on $\mathbf{J}$ in (i), (ii), and (iii) of Section 2.1.6 for the existence of solution to the electromagnetic equations (2.18)–(2.21). This result could be expected from the similarity between the expressions ((2.247),(2.250)) for the pressure and velocity fields in terms of $\mathbf{f}$ and the expressions ((2.79),(2.83)) for the magnetic and electric fields in terms of $\mathbf{J}$.

Cavity-defined acoustic fields in the source region can be determined from (2.247) and (2.250) similarly to the way they were determined for the electromagnetic fields from (2.79) and (2.83). Specifically, (2.247) and (2.250) show that the acoustic pressure and velocity are unchanged by the removal of source densities $s(\mathbf{r}, t)$ and $\mathbf{f}(\mathbf{r}, t)$ from an infinitesimal volume containing $\mathbf{r}$ during an infinitesimal time interval about t.

We showed in Section 2.2.5 that the expression (2.226) for acoustic pressure remains valid, for time-independent acoustic pressure that is zero outside the source region, and reduces to (2.234). Therefore, for time-independent sources and fields, (2.244), (2.245), and (2.247) still hold and reduce to (2.234) for acoustic pressure that is zero outside the source region. However, a simpler expression for the time-independent pressure can be found by integrating

(2.203) along any regular curve between two points in space to get

$$p(\mathbf{r}) = \int_{\mathbf{r}_0}^{\mathbf{r}} \mathbf{f}(\mathbf{r}') \cdot d\mathbf{r}' \tag{2.251}$$

for $\mathbf{r}_0$ chosen outside the source region where $p(\mathbf{r}_0) = 0$.

It was also shown in Section 2.2.5 that (2.228) remains a valid expression for time-independent velocity fields with $\mathbf{w}(\mathbf{r}) = 0$. Thus, (2.248) and (2.250) also hold for time-independent velocity fields with $\mathbf{w}(\mathbf{r}) = 0$, and reduce to

$$\begin{aligned} \mathbf{v}(\mathbf{r}) &= -\nabla \lim_{\delta\to 0} \int_{V-V_\delta} \frac{s(\mathbf{r}')}{4\pi\rho_{m0}R} dV' \\ &= \frac{1}{4\pi\rho_{m0}} \lim_{\delta\to 0} \int_{V-V_\delta} s(\mathbf{r}')\frac{\mathbf{R}}{R^3} dV', \quad \mathbf{w}(\mathbf{r}) = 0. \end{aligned} \tag{2.252}$$

(This equation is found straightforwardly from the first equation in (2.248) by substituting j from (2.231), expressing $\nabla\nabla\cdot$ as $\nabla^2 + \nabla\times\nabla\times$, and using Stoke's theorem with $\nabla\times\mathbf{f}(\mathbf{r}) = 0$.) A general expression for time-independent $\mathbf{v}(\mathbf{r})$ is given by a mathematical form of Helmholtz's theorem [84, secs. 83–86]

$$\begin{aligned} \mathbf{v}(\mathbf{r}) &= -\nabla \lim_{\delta\to 0} \int_{V-V_\delta} \frac{s(\mathbf{r}')}{4\pi\rho_{m0}R} dV' + \nabla\times \lim_{\delta\to 0} \int_{V-V_\delta} \frac{\mathbf{w}(\mathbf{r}')}{4\pi R} dV' \\ &= \frac{1}{4\pi} \lim_{\delta\to 0} \int_{V-V_\delta} \left[\frac{s(\mathbf{r}')}{\rho_{m0}}\mathbf{R} + \mathbf{w}(\mathbf{r}')\times\mathbf{R}\right] \frac{1}{R^3} dV'. \end{aligned} \tag{2.253}$$

This equation can also be derived from the time-independent solution to (2.213) with $\nabla\times\mathbf{f}(\mathbf{r}) = 0$.

Sufficient conditions on the external force $\mathbf{f}(\mathbf{r})$ for $p(\mathbf{r})$ in (2.251) to obey (2.203) are that $\mathbf{f}(\mathbf{r})$ be continuous and that $\nabla\times\mathbf{f}(\mathbf{r})$ be zero [76, ch. III, secs. 1 and 9]. Sufficient conditions on the mass-flow and vorticity source functions, $s(\mathbf{r})$ and $\mathbf{w}(\mathbf{r})$, for the velocity in (2.253) to obey the time-independent equations (2.204)–(2.205), are that $s(\mathbf{r})$ and $\mathbf{w}(\mathbf{r})$ be located in a bounded volume, that $s(\mathbf{r})$ and $\mathbf{w}(\mathbf{r})$ be Hölder continuous, and that $\nabla\cdot\mathbf{w}(\mathbf{r})$ be zero. Obviously, since $\nabla p(\mathbf{r})$ and both $\nabla\cdot\mathbf{v}$ and $\nabla\times\mathbf{v}$ exist under these conditions, $p(\mathbf{r})$ and $\mathbf{v}(\mathbf{r})$ are continuous functions of $\mathbf{r}$.

2.2.7 Surface Integral Expressions for the Pressure and Velocity Fields (Huygens' Sources)

The acoustic pressure and velocity fields outside the source region can be written in terms of the pressure and normal velocity (Huygens' sources) on a surface S that encloses the sources located in a bounded volume V; see Figure 2.1. These surface integral expressions are readily derived by applying Green's second identity in the infinite source-free region V_0 outside the surface S [1, sec. 8.1]. In that source-free region, the pressure satisfies the homogeneous scalar wave equation

$$\nabla^2 p - \frac{1}{c^2}\frac{\partial^2 p}{\partial t^2} = 0, \quad \mathbf{r}\in V_0 \tag{2.254}$$

and thus p can be inserted into [1, p. 427, eq. 22] to obtain

$$p(\mathbf{r},t) = -\frac{1}{4\pi}\int_S \left[\frac{\hat{\mathbf{n}}'\cdot\{\nabla' p\}(\mathbf{r}',t-|\mathbf{r}-\mathbf{r}'|/c)}{|\mathbf{r}-\mathbf{r}'|} - \frac{\hat{\mathbf{n}}'\cdot(\mathbf{r}-\mathbf{r}')}{|\mathbf{r}-\mathbf{r}'|^2}\left(\frac{1}{c}\frac{\partial}{\partial t}p(\mathbf{r}',t-|\mathbf{r}-\mathbf{r}'|/c) + \frac{p(\mathbf{r}',t-|\mathbf{r}-\mathbf{r}'|/c)}{|\mathbf{r}-\mathbf{r}'|}\right)\right] dS' \tag{2.255}$$

where $\hat{\mathbf{n}}$ is the unit normal to S pointing away from S into V_0, and $\{\nabla' p\}$ indicates that ∇' operates only on the first $\mathbf{r}'$ variable in $p(\mathbf{r}',t-|\mathbf{r}-\mathbf{r}'|/c)$. Causal fields have been assumed that are zero before an initial time t_0, so that the surface integral at infinity in Green's second identity is zero. From (2.198), we have in the source-free region outside S

$$\{\nabla' p\}(\mathbf{r}',t-|\mathbf{r}-\mathbf{r}'|/c) = -\rho_{m0}\frac{\partial}{\partial t}\mathbf{v}(\mathbf{r}',t-|\mathbf{r}-\mathbf{r}'|/c) \tag{2.256}$$

and (2.255) becomes

$$p(\mathbf{r},t) = \frac{1}{4\pi}\int_S \hat{\mathbf{n}}'\cdot\left[\frac{\mathbf{R}}{R^2}\left(\frac{1}{c}\frac{\partial}{\partial t}p(\mathbf{r}',t-R/c) + \frac{1}{R}p(\mathbf{r}',t-R/c)\right) + \frac{\rho_{m0}}{R}\frac{\partial}{\partial t}\mathbf{v}(\mathbf{r}',t-R/c)\right] dS' \tag{2.257}$$

where, as usual, $\mathbf{R} = \mathbf{r}-\mathbf{r}'$ and $R = |\mathbf{r}-\mathbf{r}'|$. With the help of the vector identity

$$\nabla\left[\frac{f(t-R/c)}{R}\right] = -\frac{\mathbf{R}}{R^2}\left[\frac{1}{c}\frac{\partial}{\partial t}f(t-R/c) + \frac{1}{R}f(t-R/c)\right] \tag{2.258}$$

(2.257) can be rewritten as

$$p(\mathbf{r},t) = -\nabla\cdot\int_S \frac{\hat{\mathbf{n}}' p(\mathbf{r}',t-R/c)}{4\pi R} dS' + \frac{\partial}{\partial t}\int_S \frac{\rho_{m0}\hat{\mathbf{n}}'\cdot\mathbf{v}(\mathbf{r}',t-R/c)}{4\pi R} dS'. \tag{2.259}$$

The time-domain particle velocity $\mathbf{v}(\mathbf{r},t)$ in the source-free region outside S can be found from the homogeneous version of (2.202), namely

$$\mathbf{v}(\mathbf{r},t) = -\frac{1}{\rho_{m0}}\nabla\int_{t_0}^{t} p(\mathbf{r},t')dt'. \tag{2.260}$$

Substitution of $p(\mathbf{r},t)$ from (2.259) into (2.260) yields

$$\mathbf{v}(\mathbf{r},t) = \nabla\nabla\cdot\int_S \frac{\hat{\mathbf{n}}'\int_{t_0}^{t} p(\mathbf{r}',t'-R/c)dt'}{4\pi\rho_{m0}R} dS' - \nabla\int_S \frac{\hat{\mathbf{n}}'\cdot\mathbf{v}(\mathbf{r}',t-R/c)}{4\pi R} dS'. \tag{2.261}$$

Using the identities (2.249) and (2.258), we can bring the ∇ operators inside the integrals to convert (2.261) to

$$
\begin{aligned}
\mathbf{v}(\mathbf{r},t) = \frac{1}{4\pi}\int_S \hat{\mathbf{n}}' \cdot \Bigg[& \frac{1}{\rho_{m0}c^2 R}\frac{\partial}{\partial t} p(\mathbf{r}',t-R/c)\frac{\mathbf{R}\mathbf{R}}{R^2} \\
& + \frac{1}{\rho_{m0}R^2}\left(\frac{1}{c}p(\mathbf{r}',t-R/c) + \frac{1}{R}\int_{t_0}^{t} p(\mathbf{r}',t'-R/c)dt'\right) \\
& \cdot \left(\frac{3\mathbf{R}\mathbf{R}}{R^2} - \bar{\mathbf{I}}\right) + \left(\frac{1}{c}\frac{\partial}{\partial t}\mathbf{v}(\mathbf{r}',t-R/c)\right. \\
& \left. + \frac{1}{R}\mathbf{v}(\mathbf{r}',t-R/c)\right)\frac{\mathbf{R}}{R^2}\Bigg] dS'.
\end{aligned}
\tag{2.262}
$$

With the help of (2.67), the identity $\nabla\nabla = \nabla^2 + \nabla \times \nabla\times$, and Stokes theorem on a closed surface [83, app. 2, eq. 52], we can also express the velocity in (2.261) as

$$
\begin{aligned}
\mathbf{v}(\mathbf{r},t) = & \frac{\partial}{\partial t}\int_S \frac{\hat{\mathbf{n}}' p(\mathbf{r}',t-R/c)}{4\pi\rho_{m0}c^2 R} dS' \\
& + \nabla \times \int_S \frac{\hat{\mathbf{n}}' \times \mathbf{v}(\mathbf{r}',t-R/c)}{4\pi R} dS' - \nabla \int_S \frac{\hat{\mathbf{n}}' \cdot \mathbf{v}(\mathbf{r}',t-R/c)}{4\pi R} dS'
\end{aligned}
\tag{2.263}
$$

where use has been made of the surface relationship

$$
\rho_{m0}\hat{\mathbf{n}} \times \mathbf{v}(\mathbf{r},t) = -\hat{\mathbf{n}} \times \nabla_s \int_{t_0}^{t} p(\mathbf{r},t')dt' = \nabla_s \times \int_{t_0}^{t} \hat{\mathbf{n}} p(\mathbf{r},t')dt', \quad \mathbf{r} \in S \tag{2.264}
$$

which can be derived from the momentum equation (2.202) in a source-free region using [83, app. 2, eq. 31]. In other words, to derive (2.263) from (2.261), one shows that the first term on the right-hand side of (2.261) equals the sum of the first and second terms on the right-hand side of (2.263). Since $\nabla \times \mathbf{v}(\mathbf{r},t) = 0$ outside the source region, the curl of the sum of the first and second terms on the right-hand side of (2.263) is zero. Equation (2.263) for the velocity field closely resembles the time-domain analogs of the Stratton-Chu formulas (2.110)–(2.111) for the electric and magnetic fields. The acoustic surface relation (2.264) replaces the electromagnetic relations (2.112)–(2.113).

Comparing (2.257) and (2.262) with (2.247) and (2.250) confirms that $\hat{\mathbf{n}}p$ and $\rho_{m0}\hat{\mathbf{n}} \cdot \mathbf{v}$ in (2.257) and (2.262) can be interpreted as applied surface-force and mass-flow densities, respectively. They are the time-domain Huygens' sources of the pressure-velocity fields in the source-free region outside S.

The time-dependent pressure and velocity fields given in (2.259) or (2.257), and (2.261) or (2.262), are continuous functions of $\mathbf{r}$ and t that satisfy the acoustic equations (2.201)–(2.202) in the source-free region outside S, under the sufficient conditions that $\partial[\hat{\mathbf{n}}p(\mathbf{r},t)]/\partial t$ and $\partial[\hat{\mathbf{n}} \cdot \mathbf{v}(\mathbf{r},t)]/\partial t$ are continuous functions of $(\mathbf{r},t)$ for $\mathbf{r}$ on S. If, in addition to these conditions, the surface S has continuous curvature, and $p(\mathbf{r},t)$ is Hölder continuous on S at each time t, then the right-hand side of (2.257) or (2.259) equals the pressure on S as $\mathbf{r}$ approaches the surface S from outside S.

If $\partial(\hat{\mathbf{n}}p)/\partial t$, $\partial(\hat{\mathbf{n}} \cdot \mathbf{v}/)\partial t$ and $\nabla_s \times (\hat{\mathbf{n}}p)$ are continuous functions of $(\mathbf{r},t)$ for $\mathbf{r}$ on S (which implies through (2.264) that $\partial(\hat{\mathbf{n}} \times \mathbf{v})/\partial t$ is a continuous function of $(\mathbf{r},t)$ on S), then p and $\mathbf{v}$ given in (2.257) or (2.259), and (2.263)–(2.264) are continuous functions of $(\mathbf{r},t)$ and satisfy the acoustic equations (2.201)–(2.202) for $\mathbf{r}$ outside S. If, in addition, $\mathbf{v}(\mathbf{r}, t)$ is Hölder continuous on a continuously curved surface S at each t, the velocity on S is recovered from (2.263) as $\mathbf{r}$ approaches S from outside S and is also recovered from (2.261) or (2.262) (because of the surface relationship (2.264)). These conclusions are proven for these time-dependent acoustic surface integral expressions in the same way the similar conclusions were proven for the time-dependent electromagnetic surface integral expressions in Section 2.1.9.

For time-independent fields, the expressions (2.257) and (2.259) for the pressure reduce to

$$p(\mathbf{r}) = -\nabla \cdot \int_S \frac{\hat{\mathbf{n}}' p(\mathbf{r}')}{4\pi |\mathbf{r} - \mathbf{r}'|} dS'. \tag{2.265}$$

The pressure in (2.265) approaches zero as $r \to \infty$. Moreover, outside the source region the time-independent acoustic pressure satisfies $\nabla p(\mathbf{r}) = 0$, so that $p(\mathbf{r})$ is zero everywhere outside the source region, and, in particular, on S, if it is zero as $\mathbf{r} \to \infty$. Thus, (2.265) merely gives the trivial result $p(\mathbf{r}) = 0$ outside the source region.

Because the time-dependent expressions (2.261) and (2.262) were derived directly from (2.260), which does not apply to time-independent fields, (2.261) and (2.262) are not valid expressions for time-independent velocity. However, it can be shown that (2.263) remains valid for time-independent velocity fields and reduces to

$$\mathbf{v}(\mathbf{r}) = \nabla \times \int_S \frac{\hat{\mathbf{n}}' \times \mathbf{v}(\mathbf{r}')}{4\pi |\mathbf{r} - \mathbf{r}'|} dS' - \nabla \int_S \frac{\hat{\mathbf{n}}' \cdot \mathbf{v}(\mathbf{r}')}{4\pi |\mathbf{r} - \mathbf{r}'|} dS'. \tag{2.266}$$

This equation is a mathematical statement of Helmholtz's theorem [84, secs. 83–86] for fields that are both irrotational and solenoidal outside S.

Sufficient conditions for the continuity of $\mathbf{v}(\mathbf{r})$ in (2.266) and for the divergence and curl of $\mathbf{v}(\mathbf{r})$ to equal zero outside S are that $\hat{\mathbf{n}} \times \mathbf{v}(\mathbf{r})$ and $\hat{\mathbf{n}} \cdot \mathbf{v}(\mathbf{r})$ are continuous on S. (To prove that the curl of $\mathbf{v}(\mathbf{r})$ is zero outside S, one requires the relationship $\nabla_s \cdot [\hat{\mathbf{n}} \times \mathbf{v}(\mathbf{r})] = 0$ on S.) If $\mathbf{v}(\mathbf{r})$ is Hölder continuous on a continuously curved surface S, the velocity on the surface S is recovered from (2.266) by letting $\mathbf{r}$ approach S from outside S.

2.2.8 Wave-Force, Wave-Momentum, Power, and Energy Relations

The momentum density $\rho_{m_T}\mathbf{v}$ in an acoustic field can be expressed as

$$\rho_{m_T}(\mathbf{r}, t)\mathbf{v}(\mathbf{r}, t) = \rho_{m0}\mathbf{v}(\mathbf{r}, t) + \rho_m(\mathbf{r}, t)\mathbf{v}(\mathbf{r}, t) \tag{2.267}$$

or, upon substitution of the acoustic mass density ρ_m from (2.195)

$$\rho_{m_T}(\mathbf{r}, t)\mathbf{v}(\mathbf{r}, t) = \rho_{m0}\mathbf{v}(\mathbf{r}, t) + \rho_{m0}\kappa_0 p(\mathbf{r}, t)\mathbf{v}(\mathbf{r}, t). \tag{2.268}$$

The first term ($\rho_{m0}\mathbf{v}$) on the right-hand side of (2.268) is the "fluid-momentum" density. It is a first-order momentum density that is present in both incompressible and compressible fluids. The second term ($\rho_{m0}\kappa_0 p\mathbf{v}$) in (2.268) is the "wave-momentum" density. It is a second-order momentum density that vanishes for incompressible fluids (since $\rho_m = 0$ in incompressible fluids) and thus is intimately connected with acoustic wave propagation in compressible fluids.

A vector continuity equation for the wave-momentum density $\rho_{m0}\kappa_0 p\mathbf{v}$ is derived straightforwardly by multiplying (2.197) and (2.198) by $\mathbf{v}$ and $\kappa_0 p$, respectively, summing the resulting equations, using the vector identity

$$\mathbf{v}(\nabla\cdot\mathbf{v}) = \nabla\cdot(\mathbf{v}\mathbf{v}) - \frac{1}{2}\nabla|\mathbf{v}|^2 + \mathbf{v}\times(\nabla\times\mathbf{v}) \tag{2.269}$$

and substituting $\nabla\times\mathbf{v}$ from (2.200) to get

$$\kappa_0\mathbf{f}p + s\mathbf{v} + \left(\int_{t_0}^{t}\nabla\times\mathbf{f}dt' + \rho_{m0}\mathbf{w}\right)\times\mathbf{v} = -\nabla\cdot\bar{\mathbf{T}}_{ac} + \frac{1}{c^2}\frac{\partial}{\partial t}(p\mathbf{v}) \tag{2.270}$$

where the acoustic wave-stress dyadic $\bar{\mathbf{T}}_{ac}$ is defined as

$$\bar{\mathbf{T}}_{ac} = -\rho_{m0}\mathbf{v}\mathbf{v} + \frac{1}{2}\left(\rho_{m0}|\mathbf{v}|^2 - \kappa_0 p^2\right)\bar{\mathbf{I}}. \tag{2.271}$$

(Although second-order quantities have been neglected in (2.197) and (2.198), when (2.197) and (2.198) are multiplied by $\mathbf{v}$ and $\kappa_0 p$, the resulting equations neglect only third-order quantities. Therefore, (2.270) is a valid second-order equation.)

Integrating (2.270) over an arbitrary volume V_a with surface S_a and applying the divergence theorem, we obtain

$$\begin{aligned}\mathbf{F}_{ac}(t) &= -\int_{V_a}\left[\kappa_0\mathbf{f}p + s\mathbf{v} + \left(\int_{t_0}^{t}\nabla\times\mathbf{f}dt' + \rho_{m0}\mathbf{w}\right)\times\mathbf{v}\right]dV \\ &= \int_{V_a}\left[\nabla\cdot\bar{\mathbf{T}}_{ac} - \frac{1}{c^2}\frac{\partial}{\partial t}(p\mathbf{v})\right]dV = \int_{S_a}\hat{\mathbf{n}}\cdot\bar{\mathbf{T}}_{ac}dS - \frac{1}{c^2}\frac{d}{dt}\int_{V_a}p\mathbf{v}dV.\end{aligned} \tag{2.272}$$

The "wave force" $\mathbf{F}_{ac}(t)$ is the negative of the integral over V_a of the applied time rate of change of wave-momentum density on the left-hand side of (2.270). That is, $\mathbf{F}_{ac}(t)$ is the wave force applied by the acoustic fields to the source distributions s, $\mathbf{f}$, and $\mathbf{w}$. Thus, (2.272) states that the time rate of change of the wave momentum ($\frac{1}{c^2}\frac{d}{dt}\int_{V_a}p\mathbf{v}dV$) in the volume V_a is equal to the wave momentum per unit time ($\int_{S_a}\hat{\mathbf{n}}\cdot\bar{\mathbf{T}}_{ac}dS$) entering the fields of V_a through its surface S_a, minus the wave force ($\mathbf{F}_{ac}$) applied by the fields to the fluid sources in V_a. (The wave force $\mathbf{F}_{ac}$ is a second-order force that should not be confused with the first-order body force $\int_{V_a}\mathbf{f}dV$ applied to the fluid in V_a.)

A scalar continuity equation for the energy density in the acoustic field can also be derived straightforwardly by multiplying (2.197) by p/ρ_{m0}, taking the scalar product of (2.198) with $\mathbf{v}$, and adding the resulting second-order equations to get

$$\mathbf{f}\cdot\mathbf{v} + \frac{sp}{\rho_{m0}} = \nabla\cdot(p\mathbf{v}) + \frac{1}{2}\frac{\partial}{\partial t}\left(\rho_{m0}|\mathbf{v}|^2 + \kappa_0 p^2\right) \tag{2.273}$$

wherein the acoustic energy density is the sum of the kinetic energy density $\rho_{m0}|\mathbf{v}|^2/2$ and the potential energy density $\kappa_0 p^2/2$.

Integrating (2.273) over an arbitrary volume V_a with surface S_a and applying the divergence theorem gives

$$\begin{aligned}P_{ac}(t) &= -\int_{V_a}\left[\mathbf{f}\cdot\mathbf{v} + \frac{sp}{\rho_{m0}}\right]dV = -\int_{V_a}\left[\nabla\cdot(p\mathbf{v}) + \frac{1}{2}\frac{\partial}{\partial t}\left(\rho_{m0}|\mathbf{v}|^2 + \kappa_0 p^2\right)\right] \\ &= -\int_{S_a}\hat{\mathbf{n}}\cdot p\mathbf{v}dS - \frac{1}{2}\frac{d}{dt}\int_{V_a}\left(\rho_{m0}|\mathbf{v}|^2 + \kappa_0 p^2\right)dV.\end{aligned} \tag{2.274}$$

The applied power $P_{ac}(t)$ is defined as the negative of the integral over V_a of the applied power density on the left-hand side of (2.273). That is, $P_{ac}(t)$ is the acoustic power supplied by the acoustic fields to the source distributions s and $\mathbf{f}$. Thus, (2.274) states that the time rate of change of the acoustic energy $(\frac{1}{2}\frac{d}{dt}\int_{V_a}(\rho_{mo}|\mathbf{v}|^2+\kappa_0 p^2)dV)$ in the acoustic fields of the volume V_a is equal to the acoustic power $(-\int_{S_a}\hat{\mathbf{n}}\cdot p\mathbf{v}dS)$ entering the fields of V_a through its surface S_a, minus the power (P_{ac}) supplied by the fields to the fluid sources in V_a. (The second-order power P_{ac} is not equal to $\mathbf{F}_{ac}\cdot\mathbf{v}$ because $\mathbf{F}_{ac}$ is a second-order "wave force.")

When the source densities are zero in V_a, the functions $\mathbf{F}_{ac}(t)$ and $P_{ac}(t)$ are zero, and (2.272) and (2.274) reduce to

$$\int_{S_a}\hat{\mathbf{n}}\cdot\bar{\mathbf{T}}_{ac}dS=\frac{1}{c^2}\frac{d}{dt}\int_{V_a}p\mathbf{v}dV \tag{2.275}$$

and

$$-\int_{S_a}\hat{\mathbf{n}}\cdot p\mathbf{v}dS=\frac{1}{2}\frac{d}{dt}\int_{V_a}\left(\rho_{mo}|\mathbf{v}|^2+\kappa_0 p^2\right)dV. \tag{2.276}$$

Integrating (2.275) and (2.276) over time from t_0 to t, for fields that are zero in V_a for $t\le t_0$, yields the acoustic wave momentum and energy in V_a at the time t, specifically

$$\int_{t_0}^{t}\int_{S_a}\hat{\mathbf{n}}\cdot\bar{\mathbf{T}}_{ac}(\mathbf{r},t')dSdt'=\frac{1}{c^2}\int_{V_a}p(\mathbf{r},t)\mathbf{v}(\mathbf{r},t)dV \tag{2.277}$$

$$-\int_{t_0}^{t}\int_{S_a}\hat{\mathbf{n}}\cdot p(\mathbf{r},t')\mathbf{v}(\mathbf{r},t')dSdt'=\frac{1}{2}\int_{V_a}\left(\rho_{mo}|\mathbf{v}(\mathbf{r},t)|^2+\kappa_0(p(\mathbf{r},t))^2\right)dV. \tag{2.278}$$

In Section 5.2.7 the left-hand sides of (2.277) and (2.278) are evaluated with $t=\infty$ for an infinite plane bounding a source-free half space, in order to determine the total acoustic wave momentum and energy radiated into this half space by time-domain sources (located outside this half space) that turn on and off in a finite time interval.

2.3 FREQUENCY-DOMAIN ELECTROMAGNETIC FIELD EQUATIONS

Many of the frequency-domain equations will be derived by taking the Fourier transform of the corresponding time-domain equations in Sections 2.1 and 2.2. Thus, we begin here with a brief account of the properties of Fourier transforms that justify the subsequent derivations of the frequency-domain equations.

2.3.1 Fourier Transforms

The Fourier transform f_ω of a function of time $f(t)$ is defined as

$$f_\omega=\frac{1}{2\pi}\int_{-\infty}^{+\infty}f(t)e^{i\omega t}dt. \tag{2.279}$$

The Riemann integral in (2.279) is defined for $-\infty<\omega<\infty$ if $f(t)$ is piecewise continuous for $-\infty<t<\infty$ and $f(t)$ is absolutely integrable, that is, $\int_{-\infty}^{+\infty}|f(t)|dt<\infty$ [104, sec. 5–4]. If, in addition, $f(t)$ is piecewise smooth for $-\infty<t<\infty$, that is, the function $f(t)$ and its first derivative are continuous in any finite interval, except possibly for a finite number of jump discontinuities, then f_ω is a continuous function of ω, and the inverse Fourier transform of f_ω gives $f(t)$ [104, sec. 5–4], [105, vol. 2, pp. 318–323]

$$f(t) = \int_{-\infty}^{+\infty} f_\omega e^{-i\omega t} d\omega. \tag{2.280}$$

The infinite integral with respect to w in (2.280) is taken in the Cauchy principal value sense ($\lim_{a\to\infty}\int_{-a}^{a}$), and the value of the function $f(t)$ at jump discontinuities is assumed equal to the average of the left and right limits, that is, $f(t) = [f(t^+) + f(t^-)]/2$.

The above stated conditions for the validity of (2.279) and (2.280) are sufficient conditions which often are not necessary conditions (even for Riemann integrable functions) for the existence and usefulness of the Fourier transform pair (2.279) and (2.280). Moreover, Fourier transforms can be extended to functions that are square integrable in the Lebesgue sense and to generalized functions [104, p. 251 and sec. 5–16]. Here, however, we shall continue to consider only Riemann integrable functions.

In converting the time-domain equations of Sections 2.1 and 2.2, we need to take the Fourier transform of the time derivatives of functions. A theorem that allows us to do this rigorously states that [104, sec. 5–10]

$$\frac{1}{2\pi}\int_{-\infty}^{+\infty} \frac{df(t)}{dt} e^{i\omega t} dt = -\frac{i\omega}{2\pi}\int_{-\infty}^{+\infty} f(t) e^{i\omega t} dt = -i\omega f_\omega \tag{2.281}$$

if $f(t)$ is continuous and piecewise smooth, and both $f(t)$ and $df(t)/dt$ are absolutely integrable.

When the function f represents a scalar field, or a component of a vector field, throughout space, it is a function of the spatial coordinates $\mathbf{r}$ as well as the time t. Then the Fourier transform pair (2.279) and (2.280) can be written for each $\mathbf{r}$ as

$$f_\omega(\mathbf{r}) = \frac{1}{2\pi}\int_{-\infty}^{+\infty} f(\mathbf{r}, t) e^{i\omega t} dt \tag{2.282}$$

$$f(\mathbf{r}, t) = \int_{-\infty}^{+\infty} f_\omega(\mathbf{r}) e^{-i\omega t} d\omega \tag{2.283}$$

and (2.281) becomes

$$\frac{1}{2\pi}\int_{-\infty}^{+\infty} \frac{\partial}{\partial t} f(\mathbf{r}, t) e^{i\omega t} dt = -\frac{i\omega}{2\pi}\int_{-\infty}^{+\infty} f(\mathbf{r}, t) e^{i\omega t} dt = -i\omega f_\omega(\mathbf{r}). \tag{2.284}$$

In general, the frequency-domain field $f_\omega(\mathbf{r})$ will be complex, even for a real time-domain field. For a real time-domain field, taking the complex conjugate (denoted by *) of (2.282) reveals the reality condition

$$f_{-\omega}(\mathbf{r}) = f_\omega^*(\mathbf{r}) \tag{2.285}$$

which states that the frequency domain field at $-\omega$ equals the complex conjugate of the frequency-domain field at $+\omega$. Thus, it becomes redundant to determine the frequency-domain fields for both negative and positive frequencies when the corresponding time-domain fields are real. Implicit in the reality condition (2.285) is the fact that every nonzero frequency-domain field at frequency ω must be accompanied by its complex conjugate field at $-\omega$.

Note from (2.282) or (2.283) that the units of the time-domain field $f(\mathbf{r}, t)$ and the frequency-domain field $f_\omega(\mathbf{r})$ are not the same. The frequency-domain field $f_\omega(\mathbf{r})$ has the unit of time (seconds) multiplied by the units of the time-domain field. This difference in units must be taken into account throughout the book when checking and comparing the units of frequency-domain and time-domain expressions.

In Fourier transforming the spatial derivatives in the time-domain equations, we will want to interchange differentiation and integration such that

$$\frac{1}{2\pi}\int_{-\infty}^{+\infty} \frac{\partial}{\partial x_i} f(\mathbf{r}, t) e^{i\omega t} dt = \frac{1}{2\pi}\frac{\partial}{\partial x_i}\int_{-\infty}^{+\infty} f(\mathbf{r}, t) e^{i\omega t} dt = \frac{\partial}{\partial x_i} f_\omega(\mathbf{r}) \tag{2.286}$$

where x_i is equal to x, y or z for $i = 1, 2$ or 3, respectively. Sufficient conditions for the validity of (2.286) are that $f(\mathbf{r}, t)$ and $\partial f(\mathbf{r}, t)/\partial x_i$ are continuous functions of $(\mathbf{r}, t)$, the integral on the right converges (that is, the Fourier transform of $f(\mathbf{r}, t)$ exists), and the integral on the left converges uniformly in $\mathbf{r}$ (for example, if $|\partial f(\mathbf{r}, t)/\partial x_i| \leq M(t)$ where $\int_{-\infty}^{+\infty} M(t')dt' < \infty$) [90, sec. 6.23]. Under these conditions, $\partial f_\omega(\mathbf{r})/\partial x_i$ is a continuous function of $\mathbf{r}$ and ω.

2.3.2 First-Order Differential Equations

Taking the Fourier transform of the time-domain Maxwell differential equations (2.1)–(2.4) and making use of (2.284) and (2.286) produces Maxwell's differential equations in the frequency domain

$$\nabla \times \mathbf{E}_\omega(\mathbf{r}) - i\omega\mu \mathbf{H}_\omega(\mathbf{r}) = 0 \tag{2.287}$$

$$\nabla \times \mathbf{H}_\omega(\mathbf{r}) + i\omega\epsilon \mathbf{E}_\omega(\mathbf{r}) = \mathbf{J}_\omega(\mathbf{r}) \tag{2.288}$$

$$\nabla \cdot \mathbf{H}_\omega(\mathbf{r}) = 0 \tag{2.289}$$

$$\nabla \cdot \mathbf{E}_\omega(\mathbf{r}) = \rho_\omega(\mathbf{r})/\epsilon \tag{2.290}$$

where μ and ϵ are the real, scalar, constant permeability and permittivity of homogeneous, isotropic, lossless free space. Electromagnetic frequency-domain equations that apply to homogeneous, isotropic, lossy media, and that include electric and magnetic polarization densities ($\mathbf{P}_\omega$ and $\mathbf{M}_\omega$) in addition to the current and charge densities ($\mathbf{J}_\omega$ and ρ_ω), are considered in Section 2.5.1. Also, in Section 2.3.9, $\mathbf{P}_\omega$ and $\mathbf{M}_\omega$ are included in the expressions for the time-average force and power.

The frequency-domain electric and magnetic fields and the frequency-domain charge and current densities are related to the corresponding time-domain fields and sources in (2.1)–(2.4) through the Fourier transformations in (2.282)–(2.283). For example, the frequency-domain and time-domain electric fields are related by the transform pair

$$\mathbf{E}_\omega(\mathbf{r}) = \frac{1}{2\pi}\int_{-\infty}^{+\infty} \mathbf{E}(\mathbf{r}, t) e^{i\omega t} dt \tag{2.291}$$

$$\mathbf{E}(\mathbf{r}, t) = \int_{-\infty}^{+\infty} \mathbf{E}_\omega(\mathbf{r}) e^{-i\omega t} d\omega. \tag{2.292}$$

As noted in Section 2.3.1, the frequency-domain fields have the units of the time-domain fields multiplied by the unit of time (seconds). Thus, the frequency-domain electric field $\mathbf{E}_\omega(\mathbf{r})$ has units of volt-meter^{-1}-second rather than the units of volt-meter^{-1} of the time-domain electric field $\mathbf{E}(\mathbf{r}, t)$. (If desirable, one can always scale the frequency-domain or time-domain equations by a constant frequency or time interval to make their units identical.)

Since all the time-domain fields and source functions in Maxwell's equations (2.1)–(2.4) are real quantities,[11] the frequency-domain fields and source functions obey the reality condition (2.285). For the frequency-domain electric field

$$\mathbf{E}_{-\omega}(\mathbf{r}) = \mathbf{E}^*_\omega(\mathbf{r}) \tag{2.293}$$

and so on for the other frequency-domain fields and source functions. Because the reality condition makes it unnecessary to determine the fields and sources at both positive and negative frequencies, only positive frequencies ($\omega > 0$) are considered in most of the frequency-domain equations throughout the subsequent chapters of the book. However, for the sake of completeness in this chapter, which introduces the basic equations, all expressions will be written to remain valid for both negative and positive frequencies ($|\omega| > 0$).

If ω is exactly equal to zero, and the frequency-domain fields $\mathbf{E}_\omega(\mathbf{r})$ and $\mathbf{H}_\omega(\mathbf{r})$ are bounded functions of ω, the frequency-domain Maxwell equations (2.287)–(2.290) reduce to the static equations (2.22)–(2.23) for the electric and magnetic fields. Since the electrostatic and magnetostatic equations were discussed in Section 2.1 along with the time-domain electromagnetic field equations, the static equations will not be considered here. In other words, the value of ω in (2.287)–(2.290) is allowed to take any value of ω except $\omega = 0$. Of course, if $f_\omega(\mathbf{r})$ is Riemann integrable, there can be no contribution from the ω integration in (2.283) at a single point such as $\omega = 0$. Thus, the point $\omega = 0$ can be excluded from (2.283), that is, from the inverse Fourier transform. On the other hand, if generalized functions of ω are allowed in $f_\omega(\mathbf{r})$ that contribute to the integration in (2.283) at $\omega = 0$ (for example, a delta function at $\omega = 0$ that produces a static field), these contributions to $f(\mathbf{r}, t)$ can be considered separately.

Taking the divergence of (2.288) and substituting $\nabla \cdot \mathbf{E}_\omega$ from (2.290) produce the continuity equation

$$\nabla \cdot \mathbf{J}_\omega(\mathbf{r}) - i\omega\rho_\omega(\mathbf{r}) = 0. \tag{2.294}$$

Equation (2.289) results from taking the divergence of (2.287), and (2.290) results from taking the divergence of (2.288) and substituting $\nabla \cdot \mathbf{J}_\omega$ from the continuity equation (2.294). However, we retain the divergence equations (2.289) and (2.290) explicitly to emphasize that the traditional Maxwell differential equations in the frequency domain require the existence of the divergence of $\mathbf{E}_\omega(\mathbf{r})$ and $\mathbf{H}_\omega(\mathbf{r})$. The existence of the divergence of $\mathbf{E}_\omega(\mathbf{r})$ and $\mathbf{H}_\omega(\mathbf{r})$ is not implied by (2.287) and (2.288). The existence of the curls and divergences in (2.287)–(2.290) imply the existence of all the partial derivatives of the frequency-domain electric and magnetic fields, $\mathbf{E}_\omega(\mathbf{r})$ and $\mathbf{H}_\omega(\mathbf{r})$, and thus the continuity of $\mathbf{E}_\omega(\mathbf{r})$ and $\mathbf{H}_\omega(\mathbf{r})$ with respect to $\mathbf{r}$.

With the continuity equation (2.294), the charge density $\rho_\omega(\mathbf{r})$ can be eliminated from Maxwell's frequency-domain equation (2.290). Also, if the current lies within a bounded

[11] One can think of the frequency-domain Maxwell equations (2.287)–(2.290) as simply the time-dependent Maxwell equations (2.1)–(2.4) with complex time-harmonic fields and source functions having $e^{-i\omega t}$ time dependence which cancels to produce (2.287)–(2.290). However, the physically realizable fields corresponding to these single-frequency complex time-harmonic fields can be found by taking the real part of the complex time-harmonic fields or by adding the complex time-harmonic fields to their complex conjugate fields. Either of these procedures to extract the real time-harmonic fields from the complex time-harmonic fields can be viewed as a consequence of the reality condition (2.285), which demands that every frequency-domain field at frequency ω must be accompanied by its complex conjugate field at frequency $-\omega$.

volume V, so that the surface S_V of the volume lies in free space outside the source region, the divergence theorem can be applied to the continuity equation to prove

$$\int_V \rho_\omega(\mathbf{r})dV = \frac{1}{i\omega}\int_{S_V} \mathbf{J}_\omega(\mathbf{r})\cdot\hat{\mathbf{n}}dS = 0. \tag{2.295}$$

That is, the total frequency-domain charge is zero ($\omega \neq 0$).

2.3.3 Second-Order Differential Equations

The inhomogeneous vector wave equations for $\mathbf{E}_\omega(\mathbf{r})$ and $\mathbf{H}_\omega(\mathbf{r})$ can be found by taking the curl of (2.287) and (2.288), then substituting from (2.288) and (2.287), respectively, to get

$$\nabla\times\nabla\times\mathbf{E}_\omega(\mathbf{r}) - k^2\mathbf{E}_\omega(\mathbf{r}) = i\omega\mu\mathbf{J}_\omega(\mathbf{r}) \tag{2.296}$$

$$\nabla\times\nabla\times\mathbf{H}_\omega(\mathbf{r}) - k^2\mathbf{H}_\omega(\mathbf{r}) = \nabla\times\mathbf{J}_\omega(\mathbf{r}) \tag{2.297}$$

where $k = \omega\sqrt{\mu\epsilon} = \omega/c = 2\pi/\lambda$ is the propagation constant in free space and λ is the free-space wavelength. Use of the vector operator identity $\nabla\times\nabla\times = -\nabla^2 + \nabla\nabla\cdot$ and the divergence equations (2.289)–(2.290) converts (2.296) and (2.297) to the inhomogeneous vector Helmholtz equations

$$\nabla^2\mathbf{E}_\omega(\mathbf{r}) + k^2\mathbf{E}_\omega(\mathbf{r}) = -i\omega\mu\mathbf{J}_\omega(\mathbf{r}) + \nabla\rho_\omega(\mathbf{r})/\epsilon \tag{2.298}$$

$$\nabla^2\mathbf{H}_\omega(\mathbf{r}) + k^2\mathbf{H}_\omega(\mathbf{r}) = -\nabla\times\mathbf{J}_\omega(\mathbf{r}). \tag{2.299}$$

2.3.4 Equivalence of First- and Second-Order Differential Equations

The second-order differential equations (2.296)–(2.297) and (2.298)–(2.299) were derived from the first-order Maxwell equations (2.287)–(2.290), assuming the existence of the second-order derivatives of $\mathbf{E}_\omega(\mathbf{r})$ and $\mathbf{H}_\omega(\mathbf{r})$. To prove the equivalence of these first- and second-order frequency-domain differential equations, we shall derive the first-order Maxwell equations (2.287)–(2.290) from both (2.296)–(2.297) and (2.298)–(2.299) (and the continuity equation (2.294)). Incidentally, the existence of $\nabla\cdot\mathbf{J}_\omega$ in the continuity equation, and $\nabla\times\mathbf{J}_\omega$ in (2.297) and (2.299), implies that $\mathbf{J}_\omega(\mathbf{r})$ is a continuous function of $\mathbf{r}$. Also, the existence of $\nabla\rho_\omega$ in (2.298) implies that $\rho_\omega(\mathbf{r})$ is a continuous function of $\mathbf{r}$.

The divergence of (2.296) and (2.297), with substitution of $\nabla\cdot\mathbf{J}_\omega$ from (2.294), immediately gives the divergence equations (2.290) and (2.289).

The divergence of (2.298) and (2.299), with substitution of $\nabla\cdot\mathbf{J}_\omega$ from (2.294), produces the homogeneous scalar Helmholtz equations

$$(\nabla^2 + k^2)(\nabla\cdot\mathbf{E}_\omega - \rho_\omega/\epsilon) = 0 \tag{2.300}$$

and

$$(\nabla^2 + k^2)(\nabla\cdot\mathbf{H}_\omega) = 0. \tag{2.301}$$

In Appendix C it is proven that the only solution $\psi(\mathbf{r})$ to the homogeneous scalar Helmholtz equation $\nabla^2\psi + k^2\psi = 0$ is the trivial solution ($\psi(\mathbf{r}) = 0$), provided $\psi(\mathbf{r})$ has continuous second derivatives and satisfies the outgoing wave condition given in (C.2). Therefore, under these conditions, the only solutions to (2.300) and (2.301) are the divergence equations (2.290) and (2.289).

Having derived Maxwell's divergence equations from (2.296)–(2.297) and (2.298)–(2.299), it suffices to derive Maxwell's curl equations (2.287)–(2.288) from (2.298)–(2.299), since (2.298)–(2.299) were derived in Section 2.3.3 from (2.296)–(2.297) and the divergence equations. To do this, take the curl of (2.298) and substitute $\nabla \times \mathbf{J}_\omega$ from (2.299) to get

$$(\nabla^2 + k^2)(\nabla \times \mathbf{E}_\omega - i\omega\mu\mathbf{H}_\omega) = 0 \tag{2.302}$$

which again, under the conditions stated after (2.301), has only the trivial solution (2.287).

To derive (2.288), take the curl of (2.299), substitute $\mathbf{H}_\omega$ from (2.287), and use (2.290) and (2.294) to obtain

$$\nabla^2(\nabla \times \mathbf{H}_\omega + i\omega\epsilon\mathbf{E}_\omega - \mathbf{J}_\omega) = 0 \tag{2.303}$$

which has only the trivial solution (2.288), if $(\nabla \times \mathbf{H}_\omega + i\omega\epsilon\mathbf{E}_\omega - \mathbf{J}_\omega)$ has continuous second derivatives and approaches zero as $r \to \infty$ (see Appendix A).

In summary, the first-order differential Maxwell equations (2.287)–(2.290) were used to derive the continuity equation (2.294) and the second-order differential equations (2.296)–(2.299). Moreover, either (2.296)–(2.297) or (2.298)–(2.299), along with the continuity equation (2.294), were used to derive the first-order Maxwell equations (2.287)–(2.290). Thus, under quite general conditions (stated above as part of the derivations), the first- and second-order electromagnetic differential equations in the frequency domain are equivalent.

One can also prove the equivalence of the first- and second-order differential equations by proving the uniqueness of solution to each of the four equations (2.296)–(2.299) for twice continuously differentiable fields, all components of which satisfy the outgoing wave condition (C.2) in Appendix C. If, then, a solution to (2.287)–(2.290) exists, a unique solution exists to (2.296)–(2.297), (2.298)–(2.299), and (2.287)–(2.290). Thus, the equations (2.296)–(2.297) or (2.298)–(2.299) are equivalent to (2.287)–(2.290).

The equivalence of the first- and second-order frequency-domain electromagnetic field equations has been proven assuming all sources are specified by $\rho_\omega(\mathbf{r})$ and $\mathbf{J}_\omega(\mathbf{r})$. If perfectly conducting electric and magnetic bodies are allowed, the second-order differential equations (2.298)–(2.299) would need the first-order divergence equations to derive the first-order Maxwell equations. Thus, the second-order frequency-domain equations (2.298)–(2.299) and the first-order Maxwell equations would no longer be equivalent [72]. Inside the cavities formed by perfectly conducting surfaces, both the first- and second-order frequency-domain equations allow source-free fields at discrete resonant frequencies.

2.3.5 Uniqueness of Solution

In Section 2.1.5 the uniqueness of solution to the time-domain Maxwell equations was proven with enough generality to include the fields of electromagnetic missiles, that is, pulses with energy that decays slower than $1/r^2$ as $r \to \infty$. One may ask if the frequency-domain Maxwell equations (2.287)–(2.290) allow solutions that decay slower than $1/r$ (for some direction) and yet behave as an outward traveling wave as $r \to \infty$. The usual outward-radiating vector potential solution (given in Section 2.3.6) satisfies Maxwell's frequency-domain equations (2.287)–(2.290), and gives fields that decay as $1/r$ (or faster) as $r \to \infty$ for sources located in a bounded volume. Thus, if one can prove uniqueness of solution to Maxwell's equations (2.287)–(2.290), for outward traveling waves, one also proves that outward traveling fields decay as $1/r$ of faster. However, the usual proof of uniqueness of solution to the frequency-domain Maxwell equations (2.287)–(2.290) assumes as an initial postulate that the fields decay as $1/r$ or faster [66, p. 57] as $r \to \infty$. Therefore, in this section, we shall prove the uniqueness of solution to Maxwell's frequency-domain equations (2.287)–(2.290) with initial postulates that allow the fields to decay slower than $1/r$ as $r \to \infty$. The proof proceeds along the lines of the corresponding proof of uniqueness given in Appendix C for the scalar Helmholtz equation, except that the electromagnetic outgoing wave condition (2.308) is applied just to the fields transverse to $\mathbf{r}$. Once uniqueness is proven under these weaker conditions, and the usual vector potential solution is shown to satisfy these weaker conditions, one immediately recovers the usual result from the vector potential solution that outward traveling frequency-domain fields produced by sources in a bounded volume decay as $1/r$ or faster as $r \to \infty$. Consequently, once uniqueness of solution is proven under a weak outward wave condition, Maxwell's equations do not allow time-harmonic electromagnetic missiles for any finite frequency.

Begin the uniqueness proof by assuming there are two solutions to (2.287)–(2.290) with the same source distributions $\mathbf{J}_\omega(\mathbf{r})$ and $\rho_\omega(\mathbf{r})$, which, of course, is given in terms of $\mathbf{J}_\omega$ by the continuity equation (2.294) for $\omega \neq 0$. The difference of the fields of the two solutions, $\mathbf{E}_{\omega d}(\mathbf{r})$ and $\mathbf{H}_{\omega d}(\mathbf{r})$, satisfies the following homogeneous equations corresponding to (2.287)–(2.290)

$$\nabla \times \mathbf{E}_{\omega d}(\mathbf{r}) - i\omega\mu\mathbf{H}_{\omega d}(\mathbf{r}) = 0 \tag{2.304}$$

$$\nabla \times \mathbf{H}_{\omega d}(\mathbf{r}) + i\omega\epsilon\mathbf{E}_{\omega d}(\mathbf{r}) = 0 \tag{2.305}$$

$$\nabla \cdot \mathbf{H}_{\omega d}(\mathbf{r}) = 0 \tag{2.306}$$

$$\nabla \cdot \mathbf{E}_{\omega d}(\mathbf{r}) = 0. \tag{2.307}$$

We want to prove that the only solution to (2.304)–(2.307) is the trivial solution, and thus establish uniqueness of solution to (2.287)–(2.290). Note that the equations (2.304)–(2.307) require the existence of the curl and divergence of $\mathbf{E}_{\omega d}$ and $\mathbf{H}_{\omega d}$, and thus $\mathbf{E}_{\omega d}$ and $\mathbf{H}_{\omega d}$ are continuous functions of $\mathbf{r}$.

We shall assume, in addition to (2.304)–(2.307), that the fields are twice continuously differentiable and that the fields transverse to $\mathbf{r}$ satisfy the following outgoing wave condition as $r \to \infty$

$$\hat{\mathbf{r}} \times \mathbf{H}_{\omega d}(\mathbf{r}) - \sqrt{\epsilon/\mu}\,\hat{\mathbf{r}} \times (\hat{\mathbf{r}} \times \mathbf{E}_{\omega d}(\mathbf{r})) = \mathbf{o}(1/r) \tag{2.308}$$

for all directions of $\mathbf{r}$ (and for any chosen origin of the coordinate system). The $\mathbf{o}(1/r)$ on the right-hand side of (2.308) denotes a vector function of $\mathbf{r}$ that approaches zero faster than $1/r$ as $r \to \infty$, that is, $\lim_{r\to\infty}[r\mathbf{o}(1/r)] = 0$. The condition (2.308) does not require that either $\mathbf{E}_{\omega d}$ or $\mathbf{H}_{\omega d}$ decays as fast as $1/r$ as $r \to \infty$. However, since $\mathbf{E}_{\omega d}$ and $\mathbf{H}_{\omega d}$ are continuous functions of $\mathbf{r}$, (2.308) implies that the $\mathbf{o}(1/r)$ on the right-hand side of (2.308) is a continuous function of $\mathbf{r}$. In spherical coordinates (r, θ, ϕ), (2.308) can be expressed as

$$\hat{\mathbf{r}} \times \mathbf{H}_{\omega d}(r, \theta, \phi) - \sqrt{\epsilon/\mu}\,\hat{\mathbf{r}} \times (\hat{\mathbf{r}} \times \mathbf{E}_{\omega d}(r, \theta, \phi)) = \mathbf{o}(1/r) \tag{2.309}$$

or, crossing $\hat{\mathbf{r}}$ into (2.309)

$$\hat{\mathbf{r}} \times \mathbf{E}_{\omega d}(r, \theta, \phi) + \sqrt{\mu/\epsilon}\,\hat{\mathbf{r}} \times (\hat{\mathbf{r}} \times \mathbf{H}_{\omega d}(r, \theta, \phi)) = \mathbf{o}(1/r) \tag{2.310}$$

for all (θ, ϕ) (and for any chosen origin of the coordinate system). Saunders [106] proved the uniqueness of solution to Maxwell's homogeneous equations outside scatterers under a similar outgoing wave condition.

Begin the uniqueness proof by applying the divergence theorem to

$$\nabla \cdot (\mathbf{E}_{\omega d} \times \mathbf{H}^*_{\omega d}) = \mathbf{H}^*_{\omega d} \cdot \nabla \times \mathbf{E}_{\omega d} - \mathbf{E}_{\omega d} \cdot \nabla \times \mathbf{H}^*_{\omega d} = i\omega(\mu|\mathbf{H}_{\omega d}|^2 - \epsilon|\mathbf{E}_{\omega d}|^2) \tag{2.311}$$

throughout a spherical volume V_R with surface S_R of radius R, to obtain Poynting's theorem for homogeneous fields in the form

$$\int_{S_R} (\mathbf{E}_{\omega d} \times \mathbf{H}^*_{\omega d}) \cdot \hat{\mathbf{r}}\, dS = i\omega \int_{V_R} (\mu|\mathbf{H}_{\omega d}|^2 - \epsilon|\mathbf{E}_{\omega d}|^2)\, dV. \tag{2.312}$$

Noting that $(\mathbf{E}_{\omega d} \times \mathbf{H}^*_{\omega d}) \cdot \hat{\mathbf{r}} = -[\hat{\mathbf{r}} \times (\hat{\mathbf{r}} \times \mathbf{H}^*_{\omega d})] \cdot (\hat{\mathbf{r}} \times \mathbf{E}_{\omega d})$, substitute $\hat{\mathbf{r}} \times \mathbf{H}^*_{\omega d}$ from (2.309) into the surface integral of (2.312); then take the real part of (2.312) and square the resulting equation to get

$$\left(\int_{S_R} |\hat{\mathbf{r}} \times \mathbf{E}_{\omega d}(R,\theta,\phi)|^2 dS\right)^2 = \left(\int_{S_R} \mathrm{Re}[(\hat{\mathbf{r}} \times \mathbf{E}_{\omega d}(R,\theta,\phi)) \cdot \mathbf{o}(1/R)]dS\right)^2$$
$$\leq \left(\int_{S_R} |\hat{\mathbf{r}} \times \mathbf{E}_{\omega d}(R,\theta,\phi)||\mathbf{o}(1/R)|dS\right)^2 \tag{2.313}$$
$$\leq \int_{S_R} |\hat{\mathbf{r}} \times \mathbf{E}_{\omega d}(R,\theta,\phi)|^2 dS \int_{S_R} |\mathbf{o}(1/R)|^2 dS$$

where (R, θ, ϕ) in (2.313) are the spherical coordinates of $\mathbf{r}$. The last inequality in (2.313) is obtained from the Schwarz inequality for integrals. From (2.313) we can write

$$\left(\int_{S_R} |\hat{\mathbf{r}} \times \mathbf{E}_{\omega d}|^2 dS - \int_{S_R} |\mathbf{o}(1/R)|^2 dS\right) \int_{S_R} |\hat{\mathbf{r}} \times \mathbf{E}_{\omega d}|^2 dS \leq 0 \tag{2.314}$$

which implies that

$$\int_{S_R} |\hat{\mathbf{r}} \times \mathbf{E}_{\omega d}(R,\theta,\phi)|^2 dS \leq \int_{S_R} |\mathbf{o}(1/R)|^2 dS. \tag{2.315}$$

Taking the limit of (2.315) as $R \to \infty$ shows that

$$\lim_{R\to\infty} \int_{4\pi} R^2 |\hat{\mathbf{r}} \times \mathbf{E}_{\omega d}(R,\theta,\phi)|^2 d\Omega = 0 \tag{2.316}$$

where dS has been written in spherical coordinates as $dS = R^2 \sin\theta d\theta d\phi = R^2 d\Omega$.

Substituting $\hat{\mathbf{r}} \times \mathbf{E}_{\omega d}$ from (2.310) (instead of $\hat{\mathbf{r}} \times \mathbf{H}^*_{\omega d}$ from (2.309)) into the surface integral of (2.312), and repeating the above steps, produces

$$\lim_{R\to\infty} \int_{4\pi} R^2 |\hat{\mathbf{r}} \times \mathbf{H}_{\omega d}(R,\theta,\phi)|^2 d\Omega = 0. \tag{2.317}$$

Having derived the strong asymptotic far-field results (2.316) and (2.317) from the homogeneous Maxwell frequency-domain equations and the asymptotic outgoing wave condition in (2.308), we want to express the fields everywhere in terms of integrations over a surface that can be withdrawn to an infinite radius. Toward this end, apply the vector analog of Green's theorem to $\mathbf{E}_{\omega d}(\mathbf{r}')$ and the function $\hat{\mathbf{a}} \exp(ik|\mathbf{r}-\mathbf{r}'|)/|\mathbf{r}-\mathbf{r}'|$ ($\hat{\mathbf{a}}$ is a unit vector in an arbitrary direction) to obtain [1, sec. 8.14] for $\mathbf{r}$ inside S_a

$$\mathbf{E}_{\omega d}(\mathbf{r}) = -\int_{S_a} [i\omega\mu(\hat{\mathbf{n}}' \times \mathbf{H}_{\omega d})G_\omega + (\hat{\mathbf{n}}' \times \mathbf{E}_{\omega d}) \times \nabla' G_\omega + (\hat{\mathbf{n}}' \cdot \mathbf{E}_{\omega d})\nabla' G_\omega]dS' \tag{2.318}$$

where $G_\omega(\mathbf{r}, \mathbf{r}')$ is the free-space scalar Green's function given by

$$G_\omega(\mathbf{r}, \mathbf{r}') = \frac{e^{ik|\mathbf{r}-\mathbf{r}'|}}{4\pi|\mathbf{r}-\mathbf{r}'|} \tag{2.319}$$

and S_a is an arbitrary closed surface. Equation (2.318) is an identity for all twice continuously differentiable fields satisfying the homogeneous Maxwell equations (2.304)–(2.307). Take the curl of (2.318), bring the curl operator under the integral sign (allowable by Leibnitz's rule because $G_\omega(\mathbf{r}, \mathbf{r}')$ and its derivatives, in addition to the fields, are continuous functions of $\mathbf{r}$ and $\mathbf{r}'$ for $\mathbf{r}'$ on S_a and $\mathbf{r}$ inside S_a), and use (2.304) to convert (2.318) to

$$\mathbf{H}_{\omega d}(\mathbf{r}) = -\int\limits_{S_a} [(\hat{\mathbf{n}}' \times \mathbf{H}_{\omega d}) \times \nabla' G_\omega - i\omega\epsilon(\hat{\mathbf{n}}' \times \mathbf{E}_{\omega d}) \cdot \bar{\mathbf{G}}_\omega] dS' \tag{2.320}$$

where $\bar{\mathbf{G}}_\omega(\mathbf{r}, \mathbf{r}')$ is the free-space electric dyadic Green's function given by

$$\begin{aligned}\bar{\mathbf{G}}_\omega(\mathbf{r}, \mathbf{r}') &= \left(\bar{\mathbf{I}} + \frac{1}{k^2}\nabla\nabla\right) G_\omega(\mathbf{r}, \mathbf{r}') \\ &= \frac{e^{ikR}}{4\pi ikR^2}\left[\left(\bar{\mathbf{I}} - \hat{\mathbf{R}}\hat{\mathbf{R}}\right) ikR - \left(\bar{\mathbf{I}} - 3\hat{\mathbf{R}}\hat{\mathbf{R}}\right)\left(1 - \frac{1}{ikR}\right)\right]\end{aligned} \tag{2.321}$$

with $R = |\mathbf{r} - \mathbf{r}'|$ and $\hat{\mathbf{R}} = (\mathbf{r} - \mathbf{r}')/R$.

By letting S_a be the surface of the sphere S_R centered at the origin ($\mathbf{r} = 0$), and taking the absolute value, (2.320) becomes for $\mathbf{r} = 0$

$$\begin{aligned}|\mathbf{H}_{\omega d}(\mathbf{r} = 0)| &= \left| \int\limits_{4\pi} [(\hat{\mathbf{R}} \times \mathbf{H}_{\omega d}) \times \nabla' G_\omega - i\omega\epsilon(\hat{\mathbf{R}} \times \mathbf{E}_{\omega d}) \cdot \bar{\mathbf{G}}_\omega] R^2 d\Omega' \right| \\ &\le \frac{(1/R^2 + k^2)^{\frac{1}{2}}}{4\pi} \int\limits_{4\pi} R|\hat{\mathbf{R}} \times \mathbf{H}_{\omega d}| d\Omega' \\ &\quad + \frac{|kg(R)|}{4\pi}\sqrt{\frac{\epsilon}{\mu}} \int\limits_{4\pi} R|\hat{\mathbf{R}} \times \mathbf{E}_{\omega d}| d\Omega'\end{aligned} \tag{2.322}$$

where $g(R)$ is a function of R that approaches a constant as $R \to \infty$. The Schwarz integral inequality applied to $|\hat{\mathbf{R}} \times \mathbf{E}_{\omega d}|$ and R, or $|\hat{\mathbf{R}} \times \mathbf{H}_{\omega d}|$ and R reveals that

$$\int\limits_{4\pi} R|\hat{\mathbf{R}} \times \mathbf{E}_{\omega d}| d\Omega' \le \left(4\pi R^2 \int\limits_{4\pi} |\hat{\mathbf{R}} \times \mathbf{E}_{\omega d}|^2 d\Omega'\right)^{\frac{1}{2}} \tag{2.323}$$

and

$$\int\limits_{4\pi} R|\hat{\mathbf{R}} \times \mathbf{H}_{\omega d}| d\Omega' \le \left(4\pi R^2 \int\limits_{4\pi} |\hat{\mathbf{R}} \times \mathbf{H}_{\omega d}|^2 d\Omega'\right)^{\frac{1}{2}}. \tag{2.324}$$

Utilizing (2.323) and (2.324) in (2.322) produces the inequality

$$\begin{aligned}|\mathbf{H}_{\omega d}(\mathbf{r} = 0)| &\le \left(\frac{(1/R^2 + k^2)}{4\pi} \int\limits_{4\pi} R^2|\hat{\mathbf{R}} \times \mathbf{H}_{\omega d}|^2 d\Omega'\right)^{\frac{1}{2}} \\ &\quad + |kg(R)|\sqrt{\frac{\epsilon}{\mu}} \left(\frac{1}{4\pi} \int\limits_{4\pi} R^2|\hat{\mathbf{R}} \times \mathbf{E}_{\omega d}|^2 d\Omega'\right)^{\frac{1}{2}}.\end{aligned} \tag{2.325}$$

Taking the limit of (2.325) as $R \to \infty$, and using (2.316) and (2.317), shows that $\mathbf{H}_{\omega d}(\mathbf{r} = 0)$ equals zero. Since the origin ($\mathbf{r} = 0$) of the coordinate system can be chosen at any point in space, $\mathbf{H}_{\omega d}(\mathbf{r}) = 0$ for all $\mathbf{r}$. Then (2.305) demands that $\mathbf{E}_{\omega d}(\mathbf{r}) = 0$ also for all $\mathbf{r}$, and the uniqueness of solution to the frequency-domain Maxwell equations (2.287)–(2.290) is proven.

It is not surprising that this uniqueness theorem should hold because the separable vector spherical-wave solutions that satisfy the frequency-domain Maxwell equations outside the source region and the outgoing wave condition (2.308) have a radial dependence given by spherical Hankel functions of the first kind. However, the spherical Hankel functions of the first kind, or any linear combination of them, do not satisfy Maxwell's homogeneous equations (2.304)–(2.307) at $r = 0$, unless their coefficients and thus the fields are zero.

2.3.6 Existence of Solution

We want to prove that a solution exists to the frequency-domain Maxwell equations (2.287)–(2.290) with far fields that satisfy the outgoing wave condition (2.308). To do this we can first construct vector-scalar potential expressions for $\mathbf{E}_\omega(\mathbf{r})$ and $\mathbf{H}_\omega(\mathbf{r})$, then determine sufficient conditions on the current $\mathbf{J}_\omega(\mathbf{r})$ for these expressions to obey the frequency-domain Maxwell equations (2.287)–(2.290). However, this frequency-domain proof of existence by construction is similar to and simpler than the corresponding proof of existence by construction given in detail in Section 2.1.6 for the time-domain Maxwell equations. Therefore, we shall give only the final results here for the frequency-domain equations.

The frequency-domain magnetic and electric fields $\mathbf{H}_\omega(\mathbf{r})$ and $\mathbf{E}_\omega(\mathbf{r})$ given by

$$\mathbf{H}_\omega(\mathbf{r}) = \frac{1}{\mu}\nabla \times \mathbf{A}_\omega(\mathbf{r}) \tag{2.326}$$

$$\mathbf{E}_\omega(\mathbf{r}) = -\frac{c^2}{i\omega}\nabla \times \nabla \times \mathbf{A}_\omega(\mathbf{r}) + \frac{1}{i\omega\epsilon}\mathbf{J}_\omega(\mathbf{r}) = i\omega\mathbf{A}_\omega(\mathbf{r}) - \nabla\Psi_\omega(\mathbf{r}) \tag{2.327}$$

with the vector and scalar potentials defined as

$$\mathbf{A}_\omega(\mathbf{r}) = \frac{\mu}{4\pi}\lim_{\delta\to 0}\int\limits_{V-V_\delta} \mathbf{J}_\omega(\mathbf{r}')\frac{e^{ik|\mathbf{r}-\mathbf{r}'|}}{|\mathbf{r}-\mathbf{r}'|}dV' \tag{2.328}$$

and

$$\Psi_\omega(\mathbf{r}) = \frac{1}{4\pi\epsilon}\lim_{\delta\to 0}\int\limits_{V-V_\delta} \rho_\omega(\mathbf{r}')\frac{e^{ik|\mathbf{r}-\mathbf{r}'|}}{|\mathbf{r}-\mathbf{r}'|}dV' \tag{2.329}$$

are continuous functions of $\mathbf{r}$ that satisfy Maxwell's frequency-domain equations (2.287)–(2.290) under the sufficient conditions:

(i) $\mathbf{J}_\omega(\mathbf{r})$ is located within a bounded volume V,

(ii) $\mathbf{J}_\omega(\mathbf{r})$ and its first partial derivatives (with respect to x, y, and z) are continuous functions of $\mathbf{r}$, and

(iii) $\nabla \cdot \mathbf{J}_\omega(\mathbf{r})$ is a Hölder continuous function of $\mathbf{r}$.

The continuity of the first partial derivatives of $\mathbf{J}_\omega$ in (ii) is needed, rather than merely the Hölder continuity of $\mathbf{J}_\omega$, because in proving that (2.326)–(2.329) satisfies Maxwell's equations (2.287)–(2.290), one applies the divergence theorem to $\mathbf{J}_\omega \exp(ik|\mathbf{r}-\mathbf{r}'|)/|\mathbf{r}-\mathbf{r}'|$. Continuity of the first partial derivatives of $\mathbf{J}_\omega$ is a sufficient condition for application of the divergence theorem,[12] whereas Hölder continuity of $\mathbf{J}_\omega$ and $\nabla \cdot \mathbf{J}_\omega$ is not sufficient when using the usual derivative definition of the divergence operator. If one uses Müller's integral definitions of the ∇ operators, instead of the usual more restrictive derivative definitions assumed herein, the condition (iii) can be omitted and (ii) can be replaced by the weaker condition that $\mathbf{J}_\omega(\mathbf{r})$ and $\nabla \cdot \mathbf{J}_\omega(\mathbf{r})$ be continuous functions of $\mathbf{r}$ [91, secs. 1 and 11]. (Although these integral definitions of the ∇ operators are applicable to a larger class of functions than the derivative definitions, they are difficult to use in practice.)

[12] According to Günter [107, p. 16], the divergence theorem can be proved under the weaker sufficient conditions that the partial derivatives of $\mathbf{J}_\omega$ be bounded and integrable.

The fields in (2.326)–(2.327) decay as $1/r$ or faster as $r \to \infty$, and satisfy the outgoing wave condition (2.308). Therefore, the uniqueness theorem of Section 2.3.5 implies that (2.326)–(2.327) is the unique outgoing wave solution to the frequency-domain Maxwell equations (2.287)–(2.290). Indeed, the electric and magnetic fields in (2.326)–(2.327) satisfy the strong radiation conditions [66, p. 57]

$$|\mathbf{E}_\omega(\mathbf{r})| = O(1/r), \quad |\mathbf{H}_\omega(\mathbf{r})| = O(1/r) \tag{2.330}$$

$$\hat{\mathbf{r}} \times \mathbf{H}_\omega(\mathbf{r}) + \sqrt{\epsilon/\mu}\,\mathbf{E}_\omega(\mathbf{r}) = \mathbf{o}(1/r) \tag{2.331}$$

$$\hat{\mathbf{r}} \times \mathbf{E}_\omega(\mathbf{r}) - \sqrt{\mu/\epsilon}\,\mathbf{H}_\omega(\mathbf{r}) = \mathbf{o}(1/r). \tag{2.332}$$

Theorem 2-II of Section 2.1.6 can be used to prove that (i) along with the Hölder continuity of $\mathbf{J}_\omega$ and ρ_ω are sufficient conditions for the vector and scalar potentials in (2.328) and (2.329) to satisfy the inhomogeneous Helmholtz equations

$$\nabla^2 \mathbf{A}_\omega + k^2 \mathbf{A}_\omega = -\mu \mathbf{J}_\omega \tag{2.333}$$

and

$$\nabla^2 \Psi_\omega + k^2 \Psi_\omega = -\rho_\omega/\epsilon \tag{2.334}$$

with ρ_ω given in terms of $\nabla \cdot \mathbf{J}_\omega$ through (2.294). Again, if one uses Müller's integral definitions of the ∇ operators, ordinary continuity of $\mathbf{J}_\omega$ and ρ_ω can replace Hölder continuity in the sufficient conditions for (2.333) and (2.334) [91, th. 7].

2.3.7 Volume Integral Expressions for the Electric and Magnetic Fields

The sufficient conditions (i), (ii), and (iii) in Section 2.3.6, for the vector-scalar potential expressions (2.326)–(2.329) to satisfy the frequency-domain Maxwell equations, also allow the differential operators in (2.326) and (2.327) to be brought inside the integrals of the vector and scalar potentials in (2.328) and (2.329). Again, we give only the final results for these volume integral expressions of $\mathbf{H}_\omega(\mathbf{r})$ and $\mathbf{E}_\omega(\mathbf{r})$, which can also be found by taking the Fourier transform of the corresponding time-domain volume integral expressions (2.79), (2.83), and (2.81):

$$\mathbf{H}_\omega(\mathbf{r}) = -\frac{1}{4\pi} \lim_{\delta \to 0} \int_{V-V_\delta} (ikR - 1)\mathbf{J}_\omega(\mathbf{r}') \times \hat{\mathbf{R}} \frac{e^{ikR}}{R^2} dV' \tag{2.335}$$

$$\begin{aligned} \mathbf{E}_\omega(\mathbf{r}) = &-\frac{1}{4\pi}\sqrt{\frac{\mu}{\epsilon}} \lim_{\delta \to 0} \int_{V-V_\delta} \mathbf{J}_\omega(\mathbf{r}') \\ &\cdot \left[(\hat{\mathbf{R}}\hat{\mathbf{R}} - \bar{\mathbf{I}})ikR - (3\hat{\mathbf{R}}\hat{\mathbf{R}} - \bar{\mathbf{I}})\left(1 - \frac{1}{ikR}\right) \right] \frac{e^{ikR}}{R^2} dV' + \frac{\bar{\mathbf{L}} \cdot \mathbf{J}_\omega(\mathbf{r})}{i\omega\epsilon} \\ = &\frac{1}{4\pi} \lim_{\delta \to 0} \int_{V-V_\delta} \left[i\omega\mu \mathbf{J}_\omega(\mathbf{r}') - \frac{\rho_\omega(\mathbf{r}')}{\epsilon}\left(ik - \frac{1}{R}\right)\hat{\mathbf{R}} \right] \frac{e^{ikR}}{R} dV' \end{aligned} \tag{2.336}$$

where $\mathbf{R} = \mathbf{r} - \mathbf{r}'$ and the source (or depolarization) dyadic $\bar{\mathbf{L}}$ is defined in (2.84). Equation (2.335) and the first equation in (2.336) are commonly written in shorter form as [72]

$$\mathbf{H}_\omega(\mathbf{r}) = \lim_{\delta \to 0} \int_{V-V_\delta} \mathbf{J}_\omega(\mathbf{r}') \times \nabla' G_\omega(\mathbf{r}, \mathbf{r}') dV' \tag{2.337}$$

$$\mathbf{E}_\omega(\mathbf{r}) = i\omega\mu \lim_{\delta\to 0} \int\limits_{V-V_\delta} \mathbf{J}_\omega(\mathbf{r}')\cdot\bar{\mathbf{G}}_\omega(\mathbf{r},\mathbf{r}')dV' + \frac{\bar{\mathbf{L}}\cdot\mathbf{J}_\omega(\mathbf{r})}{i\omega\epsilon} \tag{2.338}$$

with $G_\omega(\mathbf{r},\mathbf{r}')$ and $\bar{\mathbf{G}}_\omega(\mathbf{r},\mathbf{r}')$ denoting the free-space scalar and electric dyadic Green's functions in (2.319) and (2.321), respectively.

As with the corresponding time-domain expression (2.83), the value of both the volume integral and the $\bar{\mathbf{L}}$ term on the right-hand side of (2.338) depend on the shape and position with respect to the singularity of the principal volume V_δ, but add to produce the unique electric field on the left-hand side of (2.338). For observation points $\mathbf{r}$ outside the source current $\mathbf{J}_\omega$, the integrand of the volume integral is no longer singular, the limiting principal volume can be omitted, and the $\mathbf{L}$ dyadic term vanishes.

Equations (2.335)–(2.338) determine the frequency-domain "Maxwellian fields," that is, the magnetic and electric fields defined mathematically by Maxwell's equations, in terms of the frequency-domain volume current density. They are also in a form that shows that the frequency-domain "cavity-defined" magnetic field within a current distribution, like the time-domain cavity-defined magnetic field within a current distribution, is identical to the Maxwellian magnetic field. In contrast, the frequency-domain cavity-defined electric field within a current distribution differs from the Maxwellian electric field by $\bar{\mathbf{L}}\cdot\mathbf{J}_\omega(\mathbf{r})/(i\omega\epsilon)$. These results follow directly from (2.337) and (2.338) once cavity-defined fields within a frequency-domain current distribution are defined as the fields measured at a point $\mathbf{r}$ within that current distribution when an infinitesimally small volume of frequency-domain current that contains the point $\mathbf{r}$ is removed [73]. By removing $\mathbf{J}_\omega(\mathbf{r})$ from a small volume of current containing the point $\mathbf{r}$, the magnetic field in (2.337) does not change, whereas the electric field in (2.338) changes by $\bar{\mathbf{L}}\cdot\mathbf{J}_\omega(\mathbf{r})/(i\omega\epsilon)$ because $\mathbf{J}_\omega(\mathbf{r})$ in the $\bar{\mathbf{L}}$ term of (2.338) is no longer present in the cavity containing the point $\mathbf{r}$.

In the time domain, cavity-defined fields are defined by removing the current in the infinitesimal volume at $\mathbf{r}$ for an infinitesimal amount of time about each t. Thus, we showed in Section 2.1.7 that both the electric and magnetic cavity-defined and Maxwellian fields in the time domain were identical. In the frequency domain, however, the cavity-defined fields are found by removing $\mathbf{J}_\omega$ from a small volume around $\mathbf{r}$, and this means that the time-harmonic current is removed from the small volume for all time. The removal of the time-harmonic current at $\mathbf{r}$ for all time, rather than for an infinitesimal time interval about each t, creates a time-harmonic surface charge on the surface of the cavity. This surface charge does not affect the frequency-domain magnetic field, as (2.335) or (2.337) shows, but changes the frequency-domain electric field at $\mathbf{r}$, as (2.336) or (2.338) shows, by the amount $\bar{\mathbf{L}}\cdot\mathbf{J}_\omega(\mathbf{r})/(i\omega\epsilon)$.

If the real time-harmonic current (equal to the real part of the frequency-domain current multiplied by $e^{-i\omega t}$) is removed from the infinitesimal volume at $\mathbf{r}$ for an infinitesimal amount of time Δt about each t, then, like all time-dependent fields within volume current density, both the cavity-defined real time-harmonic electric and magnetic fields are identical to the real time-harmonic Maxwellian fields. (See Footnote 6 in Section 2.1.7.)

2.3.8 Surface Integral Expressions for the Electric and Magnetic Fields (Huygens' Sources)

The frequency-domain electric and magnetic fields, like the time-domain electric and magnetic fields, can be written outside a surface S in terms of integrals of the tangential electric and magnetic fields (Huygens' sources) on the surface S in free space that surrounds the sources located within a bounded volume V, as shown in Figure 2.1. These frequency-domain formulas are commonly referred to as the Stratton-Chu formulas or sometimes as the Larmor-Tedone formulas [1, sec. 8.14], and a second version as the Kottler-Franz formulas [86], [94], [95].

The Stratton-Chu formulas can be found by Fourier transforming (2.110) and (2.111) and bringing the ∇ operators inside the integrals to obtain (for $\mathbf{r}$ outside S)

$$\mathbf{E}_\omega(\mathbf{r}) = \int_S [i\omega\mu\hat{\mathbf{n}}' \times \mathbf{H}_\omega(\mathbf{r}')G_\omega(\mathbf{r},\mathbf{r}') + (\hat{\mathbf{n}}' \times \mathbf{E}_\omega(\mathbf{r}')) \times \nabla' G_\omega(\mathbf{r},\mathbf{r}') + \hat{\mathbf{n}}' \cdot \mathbf{E}_\omega(\mathbf{r}')\nabla' G_\omega(\mathbf{r},\mathbf{r}')]dS' \tag{2.339}$$

$$\mathbf{H}_\omega(\mathbf{r}) = \int_S [-i\omega\epsilon\hat{\mathbf{n}}' \times \mathbf{E}_\omega(\mathbf{r}')G_\omega(\mathbf{r},\mathbf{r}') + (\hat{\mathbf{n}}' \times \mathbf{H}_\omega(\mathbf{r}')) \times \nabla' G_\omega(\mathbf{r},\mathbf{r}') + \hat{\mathbf{n}}' \cdot \mathbf{H}_\omega(\mathbf{r}')\nabla' G_\omega(\mathbf{r},\mathbf{r}')]dS' \tag{2.340}$$

with the free-space Green's function $G_\omega(\mathbf{r},\mathbf{r}')$ given in (2.319). The normal components of $\mathbf{E}_\omega$ and $\mathbf{H}_\omega$ on S can be expressed from (2.288) and (2.287) in terms of the tangential components on S, namely

$$\hat{\mathbf{n}} \cdot \mathbf{E}_\omega(\mathbf{r}) = \frac{\nabla_s \cdot (\hat{\mathbf{n}} \times \mathbf{H}_\omega(\mathbf{r}))}{i\omega\epsilon} \tag{2.341}$$

$$\hat{\mathbf{n}} \cdot \mathbf{H}_\omega(\mathbf{r}) = -\frac{\nabla_s \cdot (\hat{\mathbf{n}} \times \mathbf{E}_\omega(\mathbf{r}))}{i\omega\mu}. \tag{2.342}$$

Thus, the surface integrals in (2.339) and (2.340) can be expressed solely in terms of the tangential components of $\mathbf{E}_\omega$ and $\mathbf{H}_\omega$. If the normal components, $\epsilon\hat{\mathbf{n}} \cdot \mathbf{E}_\omega$ and $\mu\hat{\mathbf{n}} \cdot \mathbf{H}_\omega$, are interpreted as equivalent electric and magnetic surface charge densities, and the tangential components, $\hat{\mathbf{n}} \times \mathbf{H}_\omega$ and $-\hat{\mathbf{n}} \times \mathbf{E}_\omega$, are interpreted as equivalent electric and magnetic surface current densities, (2.341) and (2.342) are merely restatements of the continuity equations, which require that the electric and magnetic surface charge densities equal, respectively, the divergence of the electric and magnetic surface current densities divided by $i\omega$.

Taking the curl of equations (2.339) and (2.340), then substituting from Maxwell's equations (2.287) and (2.288) produces the Kottler-Franz formulas

$$\mathbf{H}_\omega(\mathbf{r}) = \nabla \times \int_S \hat{\mathbf{n}}' \times \mathbf{H}_\omega(\mathbf{r}')G_\omega(\mathbf{r},\mathbf{r}')dS' + \frac{1}{i\omega\mu}\nabla \times \nabla \times \int_S \hat{\mathbf{n}}' \times \mathbf{E}_\omega(\mathbf{r}')G_\omega(\mathbf{r},\mathbf{r}')dS' \tag{2.343}$$

$$\mathbf{E}_\omega(\mathbf{r}) = \nabla \times \int_S \hat{\mathbf{n}}' \times \mathbf{E}_\omega(\mathbf{r}')G_\omega(\mathbf{r},\mathbf{r}')dS' - \frac{1}{i\omega\epsilon}\nabla \times \nabla \times \int_S \hat{\mathbf{n}}' \times \mathbf{H}_\omega(\mathbf{r}')G_\omega(\mathbf{r},\mathbf{r}')dS' \tag{2.344}$$

which explicitly involve only the tangential electric and magnetic fields on the surface S outside the source region. The equivalent electric and magnetic surface currents, $\hat{\mathbf{n}} \times \mathbf{H}_\omega$ and $-\hat{\mathbf{n}} \times \mathbf{E}_\omega$, can be referred to as frequency-domain Huygens' sources of the electromagnetic fields outside S, and thus (2.343) and (2.344) represent a Hertz potential [1, sec. 1.11] formulation of Maxwell's equations in terms of equivalent surface currents.

The curl operators can be brought inside the surface integrals to convert (2.343) and (2.344) to

$$\mathbf{H}_\omega(\mathbf{r}) = -\frac{1}{4\pi}\int_S \Bigg\{(ikR-1)(\hat{\mathbf{n}}'\times\mathbf{H}_\omega)\times\hat{\mathbf{R}} \\ -\sqrt{\frac{\epsilon}{\mu}}(\hat{\mathbf{n}}'\times\mathbf{E}_\omega)\cdot\left[(\hat{\mathbf{R}}\hat{\mathbf{R}}-\bar{\mathbf{I}})ikR-(3\hat{\mathbf{R}}\hat{\mathbf{R}}-\bar{\mathbf{I}})\left(1-\frac{1}{ikR}\right)\right]\Bigg\}\frac{e^{ikR}}{R^2}dS' \tag{2.345}$$

$$\mathbf{E}_\omega(\mathbf{r}) = -\frac{1}{4\pi}\int_S \Bigg\{(ikR-1)(\hat{\mathbf{n}}'\times\mathbf{E}_\omega)\times\hat{\mathbf{R}} \\ +\sqrt{\frac{\mu}{\epsilon}}(\hat{\mathbf{n}}'\times\mathbf{H}_\omega)\cdot\left[(\hat{\mathbf{R}}\hat{\mathbf{R}}-\bar{\mathbf{I}})ikR-(3\hat{\mathbf{R}}\hat{\mathbf{R}}-\bar{\mathbf{I}})\left(1-\frac{1}{ikR}\right)\right]\Bigg\}\frac{e^{ikR}}{R^2}dS' \tag{2.346}$$

where, as usual, $\mathbf{R} = \mathbf{r} - \mathbf{r}'$. If $\hat{\mathbf{n}} \times \mathbf{H}_\omega$ is interpreted as equivalent electric surface current, one sees that the surface integrals over electric current in (2.345) and (2.346) agree with the volume integrals over electric current in (2.335) and (2.336), respectively. Similarly, interpreting $-\hat{\mathbf{n}} \times \mathbf{E}_\omega$ as equivalent magnetic surface current reveals that the surface integrals in (2.345) and (2.346) over magnetic current are duals of the integrals over the electric current.

The frequency-domain electric and magnetic fields given in (2.339)–(2.342), or (2.343)–(2.344), or (2.345)–(2.346) satisfy Maxwell's homogeneous equations (2.287)–(2.290) outside S if $\hat{\mathbf{n}} \times \mathbf{E}_\omega$, $\hat{\mathbf{n}} \times \mathbf{H}_\omega$, $\nabla_s \cdot (\hat{\mathbf{n}} \times \mathbf{E}_\omega) = -i\omega\mu\hat{\mathbf{n}} \cdot \mathbf{H}_\omega$ and $\nabla_s \cdot (\hat{\mathbf{n}} \times \mathbf{H}_\omega) = i\omega\epsilon\hat{\mathbf{n}} \cdot \mathbf{E}_\omega$ are continuous functions of $\mathbf{r}$ on S. If, in addition, $\mathbf{E}_\omega$ and $\mathbf{H}_\omega$ are Hölder continuous on S,[13] and S is a continuously curved surface, then (2.339)–(2.340), or (2.343)–(2.344), or (2.345)–(2.346) recover the electric and magnetic fields on S as r approaches the surface S from outside S (because of the surface relationships (2.341)–(2.342)). The proof of these results parallels the proof of the corresponding results for the time-domain fields in Section 2.1.9. Moreover, repeated differentiation of (2.339)–(2.340), (2.343)–(2.344) or (2.345)–(2.346) with respect to x, y, and z shows that all derivatives of $\mathbf{E}_\omega(\mathbf{r})$ and $\mathbf{H}_\omega(\mathbf{r})$ exist, and that $\mathbf{E}_\omega(\mathbf{r})$ and $\mathbf{H}_\omega(\mathbf{r})$ are analytic functions of complex x, y, and z in a complex neighborhood of real x, y, and z outside the source region; see Section 3.2.4.

2.3.9 Time-Average Force and Power

The electromagnetic force and power per unit volume delivered to a distribution of charge and current density by the electromagnetic fields, which are produced by the entire distribution of charge and current, are given in (2.128) and (2.129). These equations, and their integral

[13]If the fields obey Maxwell's frequency-domain differential equations (2.287)–(2.290) across the surface of integration S, the existence of the curls and divergences of $\mathbf{E}_\omega(\mathbf{r})$ and $\mathbf{H}_\omega(\mathbf{r})$ in (2.287)–(2.290) implies the Hölder continuity (as well as the ordinary continuity) of $\mathbf{E}_\omega(\mathbf{r})$ and $\mathbf{H}_\omega(\mathbf{r})$. If $\hat{\mathbf{n}} \times \mathbf{E}_\omega$ and $\hat{\mathbf{n}} \times \mathbf{H}_\omega$, that is, the Huygens' sources of equivalent electric and magnetic surface currents, in (2.343)–(2.344) or (2.339)–(2.342) were chosen independently so that they were incompatible with Maxwell's equations in free space, the fields on the left-hand sides of these equations would still obey Maxwell's homogeneous equations for $\mathbf{r}$ outside S, but the chosen fields would not be recovered from (2.343)–(2.344) or (2.339)–(2.342) as $\mathbf{r}$ approaches S from outside S. In other words, the surface integral expressions show that additional electric and magnetic surface currents must reside on S (or just in front of S) if the chosen equivalent electric and magnetic surface currents are incompatible with Maxwell's equations in free space [96, ch. II, sec. 4].

counterparts, (2.131) and (2.132), hold for all real time-dependent charge-current and fields, and, in particular, for the real time-harmonic charge-current and fields obtained by taking the real part of the frequency-domain charge-current and fields multiplied by $e^{-i\omega t}$. We can insert these real time-harmonic charge-current and fields into (2.128)–(2.132), average over time, and use the result [1, p. 136]

$$\frac{\omega}{2\pi}\int_0^{2\pi/\omega} \mathrm{Re}\left(A_\omega e^{-i\omega t}\right)\mathrm{Re}\left(B_\omega e^{-i\omega t}\right)dt = \frac{1}{2}\mathrm{Re}(A_\omega B_\omega^*) \tag{2.347}$$

where A_ω and B_ω are any two complex frequency-domain functions. This procedure leads to a time-average electromagnetic force and power, delivered to the charge and current distribution in the volume V_a with surface S_a. Moreover, electric and magnetic polarization densities can be included by applying this time-averaging procedure to (2.173) and (2.174) to obtain

$$\begin{aligned}\mathbf{F}_{el}^{ave} &= \frac{1}{2}\mathrm{Re}\int_{V_a}[(\nabla\cdot(\mathbf{J}_\omega/(i\omega)-\mathbf{P}_\omega))\mathbf{E}_\omega^* + \mu(\mathbf{J}_\omega - i\omega\mathbf{P}_\omega)\times\mathbf{H}_\omega^* \\ &\qquad -\mu(\nabla\cdot\mathbf{M}_\omega)\mathbf{H}_\omega^* + i\omega\epsilon\mu\mathbf{M}_\omega\times\mathbf{E}_\omega^*]dV \\ &= \frac{1}{2}\mathrm{Re}\int_{V_a}\nabla\cdot\bar{\mathbf{T}}_{el}^c dV = \frac{1}{2}\mathrm{Re}\int_{S_a}\hat{\mathbf{n}}\cdot\bar{\mathbf{T}}_{el}^c dS\end{aligned} \tag{2.348}$$

$$\begin{aligned}P_{el}^{ave} &= \frac{1}{2}\mathrm{Re}\int_{V_a}[(\mathbf{J}_\omega - i\omega\mathbf{P}_\omega)\cdot\mathbf{E}_\omega^* - i\omega\mu\mathbf{M}_\omega\cdot\mathbf{H}_\omega^*]dV \\ &= -\frac{1}{2}\mathrm{Re}\int_{V_a}\nabla\cdot(\mathbf{E}_\omega\times\mathbf{H}_\omega^*)dV = -\frac{1}{2}\mathrm{Re}\int_{S_a}\hat{\mathbf{n}}\cdot(\mathbf{E}_\omega\times\mathbf{H}_\omega^*)dS\end{aligned} \tag{2.349}$$

where $\mathbf{P}_\omega$ and $\mathbf{M}_\omega$ denote the frequency-domain electric and magnetic polarization densities, and the complex electromagnetic stress dyadic $\bar{\mathbf{T}}_{el}^c$ is defined as

$$\bar{\mathbf{T}}_{el}^c = \epsilon\left(\mathbf{E}_\omega\mathbf{E}_\omega^* - \frac{1}{2}\bar{\mathbf{I}}|\mathbf{E}_\omega|^2\right) + \mu\left(\mathbf{H}_\omega\mathbf{H}_\omega^* - \frac{1}{2}\bar{\mathbf{I}}|\mathbf{H}_\omega|^2\right). \tag{2.350}$$

Note that the time average of (2.138)–(2.139) (and thus the time average of (2.134)–(2.135)) also yields (2.348)–(2.349) because the difference between the force-power in (2.173)–(2.174) and (2.138)–(2.139) is a time derivative and therefore its time average is zero.

The power relations in (2.349) can also be obtained by taking the real part of the following complex Poynting's theorem [1, sec. 2.20], which is derived straightforwardly from the Maxwell frequency-domain equations (2.287) and (2.288) generalized to include $\mathbf{P}_\omega$ and $\mathbf{M}_\omega$

$$\begin{aligned}&\frac{1}{2}\int_{V_a}[(\mathbf{J}_\omega^* + i\omega\mathbf{P}_\omega^*)\cdot\mathbf{E}_\omega - i\omega\mu\mathbf{M}_\omega\cdot\mathbf{H}_\omega^*]dV + 2i\omega\int_{V_a}\left(\frac{\epsilon}{4}|\mathbf{E}_\omega|^2 - \frac{\mu}{4}|\mathbf{H}_\omega|^2\right)dV \\ &\qquad = -\frac{1}{2}\int_{V_a}\nabla\cdot(\mathbf{E}_\omega\times\mathbf{H}_\omega^*)dV = -\frac{1}{2}\int_{S_a}\hat{\mathbf{n}}\cdot(\mathbf{E}_\omega\times\mathbf{H}_\omega^*)dS.\end{aligned} \tag{2.351}$$

The vector $\mathbf{E}_\omega\times\mathbf{H}_\omega^*/2$ is generally called the complex Poynting vector, and the left-hand side of (2.351), the complex power delivered to the sources in V_a.

One-half of the real part of the integrand in the first volume integral of (2.348) is in the form of a time-average, generalized Lorentz force per unit volume given by $\rho_\omega^e\mathbf{E}_\omega^* + \mu\mathbf{J}_\omega^e\times\mathbf{H}_\omega^* + \rho_\omega^m\mathbf{H}_\omega^* - \epsilon\mathbf{J}_\omega^m\times\mathbf{E}_\omega^*$, where the equivalent electric and magnetic charges and currents are given by $\rho_\omega^e = \nabla\cdot(\mathbf{J}_\omega/(i\omega)-\mathbf{P}_\omega)$, $\rho_\omega^m = -\mu\nabla\cdot\mathbf{M}_\omega$, $\mathbf{J}_\omega^e = \mathbf{J}_\omega - i\omega\mathbf{P}_\omega$, and $\mathbf{J}_\omega^m = -i\omega\mu\mathbf{M}_\omega$.

Similarly, the first volume integral in (2.349) contains the time-average, generalized Lorentz power per unit volume given by $\mathbf{J}^e_\omega \cdot \mathbf{E}^*_\omega + \mathbf{J}^m_\omega \cdot \mathbf{H}^*_\omega$.

As noted above, the time-average force and power on the charge, current, and polarization densities within V_a can also be determined by taking the time average of (2.134) and (2.135) to get

$$\begin{aligned}\mathbf{F}^{ave}_{el} &= \frac{1}{2}\mathrm{Re}\int_{V_a}[(\nabla\cdot(\mathbf{J}_\omega/(i\omega)-\mathbf{P}_\omega))\mathbf{E}^*_\omega + (\mathbf{J}_\omega - i\omega\mathbf{P}_\omega + \nabla\times\mathbf{M}_\omega)\times\mathbf{B}^*_\omega]dV \\ &= \frac{1}{2}\mathrm{Re}\int_{V_a}\nabla\cdot\bar{\mathbf{T}}^c_{el}dV = \frac{1}{2}\mathrm{Re}\int_{S_a}\hat{\mathbf{n}}\cdot\bar{\mathbf{T}}^c_{el}dS\end{aligned} \tag{2.352}$$

and

$$\begin{aligned}P^{ave}_{el} &= \frac{1}{2}\mathrm{Re}\int_{V_a}(\mathbf{J}_\omega - i\omega\mathbf{P}_\omega + \nabla\times\mathbf{M}_\omega)\cdot\mathbf{E}^*_\omega dV \\ &= -\frac{1}{2}\mathrm{Re}\int_{V_a}\nabla\cdot(\mathbf{E}_\omega\times\mathbf{H}^*_\omega)dV = -\frac{1}{2}\mathrm{Re}\int_{S_a}\hat{\mathbf{n}}\cdot(\mathbf{E}_\omega\times\mathbf{H}^*_\omega)dS\end{aligned} \tag{2.353}$$

with $\mathbf{B}_\omega = \mu(\mathbf{H}_\omega + \mathbf{M}_\omega)$.

In general, as explained in Section 2.1.10, only if the surface S_a of the volume V_a does not intersect the polarization densities $\mathbf{P}_\omega$ and $\mathbf{M}_\omega$, do the integrals in (2.348) or (2.352), and (2.349) or (2.353), unambiguously define the time-average electromagnetic force and power, $\mathbf{F}^{ave}_{el}$ and P^{ave}_{el}, delivered to all the sources in V_a. Moreover, if S_a does intersect $\mathbf{P}_\omega$ or $\mathbf{M}_\omega$, the first volume integrals in (2.348) and (2.349) do not necessarily equal those in (2.352) and (2.353), respectively. (The difference can be traced to (2.348) neglecting equivalent magnetic surface charge if S_a intersects $\mathbf{M}_\omega$, whereas (2.352)–(2.353) neglects equivalent electric surface current if S_a intersects $\mathbf{M}_\omega$.)

To ensure that (2.348) or (2.352), and (2.349) or (2.353), are valid expressions for the time-average electromagnetic force and power delivered to all the sources in V_a, one can remove an infinitesimally thin shell of the polarization densities $\mathbf{P}_\omega$ and $\mathbf{M}_\omega$ containing the closed surface S_a. Note that in the power expression (2.349), $\hat{\mathbf{n}}\cdot(\mathbf{E}_\omega\times\mathbf{H}^*_\omega)$ on S_a, and thus P^{ave}_{el}, remain the same with or without the removal of the infinitesimally thin shell of polarization densities containing S_a, because $\hat{\mathbf{n}}\times\mathbf{E}_\omega$ and $\hat{\mathbf{n}}\times\mathbf{H}_\omega$ are continuous across the thin shell. In contrast, in the force expression (2.348), $\hat{\mathbf{n}}\cdot\bar{\mathbf{T}}^c_{el}$ on S_a, and thus $\mathbf{F}^{ave}_{el}$, are changed, in general, by the removal of the infinitesimally thin shell of polarization densities, because $\hat{\mathbf{n}}\cdot\bar{\mathbf{T}}^c_{el}$ involves normal components of $\mathbf{E}_\omega$ and $\mathbf{H}_\omega$ that are not continuous across the thin shell. However, both the force and power integrals in (2.352) and (2.353) are changed by the removal of the infinitesimally thin shell of polarization densities containing S_a, such that the force integrals in (2.352) equal those in (2.348) (with the thin shell of polarization densities removed) and the power integrals in (2.353) equal those in (2.349), respectively.

With S_a devoid of polarization densities, as explained above, the equations in (2.348) or (2.352) show that the volume integral $\mathrm{Re}\int_{V_a}\nabla\cdot\bar{\mathbf{T}}^c_{el}dV/2$ gives the time-average electromagnetic force delivered by the electromagnetic fields to the charge, current, and polarization densities in V_a, and equals the surface integral $\mathrm{Re}\int_{S_a}\hat{\mathbf{n}}\cdot\bar{\mathbf{T}}^c_{el}dS/2$, which can be interpreted as the time-average momentum flux entering the fields of the volume V_a through its surface S_a. However, because $\hat{\mathbf{n}}\cdot\bar{\mathbf{T}}^c_{el}$ is discontinuous across the polarization-free thin shell containing S_a, $\nabla\cdot\bar{\mathbf{T}}^c_{el}(\mathbf{r})$ is not necessarily equal to the time-average force per unit volume exerted by the electromagnetic fields on the source densities at $\mathbf{r}$ if polarization densities are present at $\mathbf{r}$.

Similarly, the equations in (2.349) or (2.353) show that $-\text{Re}\int_{V_a} \nabla\cdot(\mathbf{E}_\omega \times \mathbf{H}_\omega^*)dV/2$ gives the time-average power delivered by the electromagnetic fields to the current and polarization densities and equals $-\text{Re}\int_{S_a} \hat{\mathbf{n}}\cdot(\mathbf{E}_\omega \times \mathbf{H}_\omega^*)dS/2$, which can be interpreted as the time-average electromagnetic power entering the fields of the volume V_a through its surface S_a. However, in contrast to $\hat{\mathbf{n}}\cdot\bar{\mathbf{T}}_{el}^c$, the normal component of the complex Poynting vector $\hat{\mathbf{n}}\cdot(\mathbf{E}_\omega \times \mathbf{H}_\omega^*)/2$ is continuous across a polarization-free thin shell containing S_a. Therefore, the divergence theorem can be applied to an arbitrarily small region V_D (at the point $\mathbf{r}$), with polarization-free surface S_D and volume ΔV, to show that $\int_{S_D} \hat{\mathbf{n}}\cdot(\mathbf{E}_\omega \times \mathbf{H}_\omega^*)dS = \nabla\cdot(\mathbf{E}_\omega(\mathbf{r}) \times \mathbf{H}_\omega^*(\mathbf{r}))\Delta V + o(\Delta V)$. Thus, $-\text{Re}[\nabla\cdot(\mathbf{E}_\omega(\mathbf{r}) \times \mathbf{H}_\omega^*(\mathbf{r}))]/2$ equals the time-average power per unit volume supplied to the current and polarization densities at $\mathbf{r}$. (It is evaluated in (2.416) for the lossy media of Section 2.5.1.) The real part of the complex Poynting vector $\text{Re}(\mathbf{E}_\omega \times \mathbf{H}_\omega^*)/2$ is generally assumed to give the magnitude and direction of the time-average electromagnetic power flow per unit area, although this pointwise interpretation of the real part of the complex Poynting vector does not follow rigorously from $\text{Re}\int_{S_a} \hat{\mathbf{n}}\cdot(\mathbf{E}_\omega \times \mathbf{H}_\omega^*)dS/2$ equaling the time-average power passing through the closed-surface S_a.

The imaginary part of (2.351) produces expressions for the difference between the time-average electric and magnetic energies in V_a multiplied by 2ω; specifically

$$
\begin{aligned}
2\omega \int_{V_a} \frac{1}{4}\{(\epsilon|\mathbf{E}_\omega|^2 - \mu|\mathbf{H}_\omega|^2) - \text{Im}[(\mathbf{J}_\omega/\omega - i\mathbf{P}_\omega)\cdot\mathbf{E}_\omega^* + i\mu\mathbf{M}_\omega\cdot\mathbf{H}_\omega^*]\}dV \\
= -\frac{1}{2}\text{Im}\int_{V_a} \nabla\cdot(\mathbf{E}_\omega \times \mathbf{H}_\omega^*)dV = -\frac{1}{2}\text{Im}\int_{S_a} \hat{\mathbf{n}}\cdot(\mathbf{E}_\omega \times \mathbf{H}_\omega^*)dS.
\end{aligned}
\tag{2.354}
$$

If (2.354) is applied to a surface S in free space enclosing the sources, and if V_0 refers to the infinite free-space volume outside S, we obtain

$$
\frac{1}{2}\text{Im}\int_S \hat{\mathbf{n}}\cdot(\mathbf{E}_\omega \times \mathbf{H}_\omega^*)dS = 2\omega\int_{V_0} \frac{1}{4}(\epsilon|\mathbf{E}_\omega|^2 - \mu|\mathbf{H}_\omega|^2)dV \tag{2.355}
$$

where in (2.355) $\hat{\mathbf{n}}$ is the unit normal to S pointing into V_0. It can be shown [63, sec. 6.9] (see also Section 3.2.9) that the right-hand side of (2.355) can be considered as the contribution of the fields in V_0 to the reactance (imaginary part of the input impedance) of an antenna enclosed by S. In other words, the left-hand side of (2.355) is a measure of the reactance contributed by the fields in the infinite free-space region outside S.

This reactance is not necessarily a measure of the reactive or stored energy in the fields outside S, that is, in V_0, because the right-hand side of (2.355) involves the difference of the electric and magnetic energy densities. Moreover, the sum of the electric and magnetic energy densities integrated over the infinite volume V_0 diverges and does not determine the reactive energy in V_0, because the total fields $\mathbf{E}_\omega$ and $\mathbf{H}_\omega$ include the radiation fields. By subtracting the energy in the radiation fields, various definitions of reactive energy in an exterior volume V_0 can be given in terms of a volume integration over V_0, in which orthogonal eigenfunctions can be utilized to facilitate its evaluation [108]–[112]. There appears to be no one definition of reactive energy of a radiator that is satisfactory for all applications, and the values of reactive energy obtained from the various definitions generally differ. Usually, however, reactive energy is important and useful only when its value is large (for example, when it is used to define a quality factor Q that can be related to the bandwidth of an antenna). For large values of reactive energy, the various definitions give comparable results.

2.4 FREQUENCY-DOMAIN ACOUSTIC FIELD EQUATIONS

Most of the frequency-domain acoustic equations, like the frequency-domain electromagnetic equations, are found by taking the Fourier transform of the corresponding time-domain equations. The reader is referred to Section 2.3.1 for the theorems that rigorously allow the Fourier transformations of the time-domain equations.

2.4.1 First-Order Linear Differential Equations

Taking the Fourier transform of the time-domain, small-amplitude acoustic differential equations (2.197)–(2.198), and making use of (2.284) and (2.286), we obtain the linear acoustic field equations in the frequency domain

$$\rho_{m0}\nabla\cdot\mathbf{v}_\omega(\mathbf{r}) - i\omega\rho_{m0}\kappa_0 p_\omega(\mathbf{r}) = s_\omega(\mathbf{r}) \tag{2.356}$$

$$\nabla p_\omega(\mathbf{r}) - i\omega\rho_{m0}\mathbf{v}_\omega(\mathbf{r}) = \mathbf{f}_\omega(\mathbf{r}) \tag{2.357}$$

for the acoustic pressure and velocity fields, $p_\omega(\mathbf{r})$ and $\mathbf{v}_\omega(\mathbf{r})$, with volume sources of mass-flow and external force densities, $s_\omega(\mathbf{r})$ and $\mathbf{f}_\omega(\mathbf{r})$, in an ideal, quiescent, homogeneous, isotropic, lossless fluid. In this ideal lossless fluid, the constitutive relation (2.195) between the acoustic mass density and pressure transforms to the frequency domain as

$$\rho_{m\omega}(\mathbf{r}) = \rho_{m0}\kappa_0 p_\omega(\mathbf{r}) \tag{2.358}$$

where κ_0 is the real, scalar constant compressibility of the fluid. In the lossy fluids of Section 2.5.2, the compressibility of the fluid becomes a complex function of frequency.

Taking the curl of (2.357) shows that the frequency-domain vorticity is given by

$$\nabla\times\mathbf{v}_\omega(\mathbf{r}) = -\frac{\nabla\times\mathbf{f}_\omega(\mathbf{r})}{i\omega\rho_{m0}} \tag{2.359}$$

which is zero everywhere except where the curl of the externally applied force is nonzero. Equations (2.356), (2.357), and (2.359) imply the existence of the first derivatives of $p_\omega(\mathbf{r})$ and $\mathbf{v}_\omega(\mathbf{r})$, and thus the continuity of these acoustic fields with respect to $\mathbf{r}$.

2.4.2 Second-Order Differential Equations

The inhomogeneous Helmholtz equations for p_ω and $\mathbf{v}_\omega$ can be found by Fourier transforming the second-order differential time-domain acoustic equations (2.207), (2.208), and (2.213) (with $\mathbf{w}(\mathbf{r}) = 0$) to get

$$\nabla^2 p_\omega(\mathbf{r}) + k^2 p_\omega(\mathbf{r}) = -q_\omega(\mathbf{r}) \tag{2.360}$$

and

$$\nabla^2\mathbf{v}_\omega(\mathbf{r}) + k^2\mathbf{v}_\omega(\mathbf{r}) = \frac{1}{\rho_{m0}}\left(\nabla s_\omega(\mathbf{r}) + i\omega\rho_{m0}\kappa_0\mathbf{f}_\omega(\mathbf{r}) + \frac{1}{i\omega}\nabla\times\nabla\times\mathbf{f}_\omega(\mathbf{r})\right) \tag{2.361}$$

where the pressure source function $q_\omega(\mathbf{r})$ is given in terms of the mass-flow and external force densities by

$$q_\omega(\mathbf{r}) = -i\omega s_\omega(\mathbf{r}) - \nabla\cdot\mathbf{f}_\omega(\mathbf{r}). \tag{2.362}$$

The propagation constant is denoted by

$$k = \omega\sqrt{\rho_{m0}\kappa_0} = \omega/c = 2\pi/\lambda \tag{2.363}$$

where c is the constant speed of propagation and λ is the wavelength in the ideal lossless fluid.

2.4.3 Equivalence of First- and Second-Order Differential Equations

The second-order acoustic differential equations (2.360)–(2.361) can be derived from the first-order linear acoustic equations (2.356)–(2.357), assuming the existence of the second derivatives of $p_\omega(\mathbf{r})$ and $\mathbf{v}_\omega(\mathbf{r})$. To prove the equivalence of the first- and second-order acoustic differential equations, it remains to derive the first-order equations (2.356)–(2.357) from the second-order equations (2.360)–(2.361) (with $q_\omega(\mathbf{r})$ defined in (2.362)).

Substitute q_ω from (2.362) into (2.360), then insert s_ω from the resulting equation into (2.361) to get

$$(\nabla^2 + k^2)(\nabla p_\omega - i\omega\rho_{m0}\mathbf{v}_\omega - \mathbf{f}_\omega) = 0. \tag{2.364}$$

Appendix C shows that the only twice continuously differentiable solution to (2.364) satisfying the outgoing wave condition (C.2) is the trivial solution

$$\nabla p_\omega - i\omega\rho_{m0}\mathbf{v}_\omega - \mathbf{f}_\omega = 0. \tag{2.365}$$

Having derived (2.357) from (2.360)–(2.361), take the divergence of (2.361) and substitute $\nabla \cdot \mathbf{f}_\omega$ from the divergence (2.365) into (2.361) to obtain

$$\nabla^2(\rho_{m0}\nabla \cdot \mathbf{v}_\omega - i\omega\rho_{m0}\kappa_0 p_\omega - s_\omega) = 0. \tag{2.366}$$

Appendix A proves that the only twice continuously differentiable solution to Laplace's equation (2.366) that approaches zero as $r \to \infty$ is the trivial solution

$$\rho_{m0}\nabla \cdot \mathbf{v}_\omega - i\omega\rho_{m0}\kappa_0 p_\omega - s_\omega = 0. \tag{2.367}$$

Thus, the equivalence of the first- and second-order set of linear acoustic equations has been proven under the assumptions that the derivatives in (2.364) and (2.366) exist, the field in (2.366) approaches zero as $r \to \infty$, and the field in (2.364) obeys the outgoing wave condition (C.2) as $r \to \infty$. In other words, a solution to either set of equations will also be a solution to the other set of equations.

As with the electromagnetic equations in the frequency domain, one can also prove the equivalence of the first- and second-order acoustic differential equations by assuming a solution to the first-order equations exist, and then proving uniqueness of solution to the second-order equations for twice continuously differentiable fields that satisfy the outgoing wave condition (C.2) of Appendix C.

2.4.4 Uniqueness of Solution

The uniqueness of solution for the frequency-domain linear acoustic equations (2.356)–(2.357) follows from the uniqueness of solution proven in Appendix C for the scalar Helmholtz equation. To see this, form the difference of two solutions to (2.356)–(2.357) with the same sources. These difference acoustic fields, $p_{\omega d}$ and $\mathbf{v}_{\omega d}$, then satisfy the homogeneous equations corresponding to (2.356)–(2.357), namely

$$\rho_{m0}\nabla \cdot \mathbf{v}_{\omega d}(\mathbf{r}) - i\omega\rho_{m0}\kappa_0 p_{\omega d}(\mathbf{r}) = 0 \tag{2.368}$$

$$\nabla p_{\omega d}(\mathbf{r}) - i\omega\rho_{m0}\mathbf{v}_{\omega d}(\mathbf{r}) = 0. \tag{2.369}$$

We want to prove that the only solution to (2.368)–(2.369) is the trivial solution, and thus establish uniqueness of solution to (2.356)–(2.357). Assume, in addition to (2.368)–(2.369), that the fields are twice continuously differentiable and satisfy the following outgoing wave condition a $r \to \infty$

$$p_{\omega d}(\mathbf{r}) - \rho_{m0}c\hat{\mathbf{r}} \cdot \mathbf{v}_{\omega d}(\mathbf{r}) = o(1/r) \tag{2.370}$$

for all directions of $\mathbf{r}$ (and for any chosen origin of the coordinate system). The condition (2.370) does not demand that either $p_{\omega d}$ or $\hat{\mathbf{r}} \cdot \mathbf{v}_{\omega d}$ decays as fast as $1/r$ as $r \to \infty$. However, since $p_{\omega d}$ and $\mathbf{v}_{\omega d}$ are continuous functions of $\mathbf{r}$, (2.370) implies that the function $o(1/r)$ on the right-hand side of (2.370) is a continuous function of $\mathbf{r}$.

Taking the divergence of (2.369) and substituting $\nabla \cdot \mathbf{v}_{\omega d}$ from (2.368) into the resulting equation reveals that $p_{\omega d}$ obeys the homogeneous Helmholtz equation

$$\nabla^2 p_{\omega d}(\mathbf{r}) + k^2 p_{\omega d}(\mathbf{r}) = 0. \tag{2.371}$$

Moreover, substitution of $\hat{\mathbf{r}} \cdot \mathbf{v}_{\omega d}$ from (2.369) into (2.370) shows that $p_{\omega d}$ satisfies the outgoing wave condition

$$\frac{\partial p_{\omega d}}{\partial r} - ikp_{\omega d} = o(1/r) \tag{2.372}$$

for all directions of $\mathbf{r}$ (and for any chosen origin of the coordinate system).

The equations (2.371) and (2.372) are identical in form to (C.1) and (C.2) of Appendix C. Therefore, the results of Appendix C implies that the only solution to (2.371) and (2.372) is the trivial solution $p_{\omega d} = 0$. It then follows from (2.369) that $\mathbf{v}_{\omega d} = 0$ for all $\mathbf{r}$, and the uniqueness of solution to the frequency-domain acoustic field equations (2.356)–(2.357) is proven.

It is not surprising that this uniqueness theorem should hold because the separable spherical-wave solutions that satisfy the equations (2.356)–(2.357) outside the source region and the outgoing wave condition (2.370) have a radial dependence given by spherical Hankel functions of the first kind. However, the spherical Hankel functions of the first kind, or any linear combination of them, do not satisfy the homogeneous equations (2.368)–(2.369) at $r = 0$, unless their coefficients and thus the fields are zero.

2.4.5 Existence of Solution

To prove that a solution exists to the frequency-domain linear acoustic equations (2.356)–(2.357) with far fields that satisfy the outgoing wave condition (2.370), we first construct velocity-potential expressions for p_ω and $\mathbf{v}_\omega$. Next we determine sufficient conditions on the acoustic source functions, s_ω and $\mathbf{f}_\omega$, for these expressions to obey the frequency-domain acoustic equations (2.356)–(2.357) everywhere including the source region. Only the final results will be given here for the frequency-domain acoustic equations.

The frequency-domain acoustic pressure and velocity fields p_ω and $\mathbf{v}_\omega$ given by

$$p_\omega(\mathbf{r}) = -i\omega\rho_{m0}\Phi_\omega(\mathbf{r}) \tag{2.373}$$

$$\mathbf{v}_\omega(\mathbf{r}) = -\nabla\Phi_\omega(\mathbf{r}) - \frac{\mathbf{f}_\omega(\mathbf{r})}{i\omega\rho_{m0}} \tag{2.374}$$

with the velocity potential $\Phi_\omega(\mathbf{r})$ defined as

$$\Phi_\omega(\mathbf{r}) = -\frac{1}{4\pi i\omega\rho_{m0}} \lim_{\delta\to 0} \int\limits_{V-V_\delta} q_\omega(\mathbf{r}')\frac{e^{ik|\mathbf{r}-\mathbf{r}'|}}{|\mathbf{r}-\mathbf{r}'|}dV' \tag{2.375}$$

and q_ω given in (2.362), are continuous functions of $\mathbf{r}$ that satisfy the acoustic frequency-domain equations (2.356)–(2.357) under the sufficient conditions:

(i) $s_\omega(\mathbf{r})$ and $\mathbf{f}_\omega(\mathbf{r})$ are located within a bounded volume V,
(ii) $\mathbf{f}_\omega(\mathbf{r})$ is a continuous function of $\mathbf{r}$, and
(iii) $s_\omega(\mathbf{r})$ and $\nabla \cdot \mathbf{f}_\omega(\mathbf{r})$ are Hölder continuous functions of $\mathbf{r}$.

The acoustic fields in (2.373)–(2.375) decay as $1/r$ or faster as $r \to \infty$ and satisfy the outgoing wave condition (2.370). Therefore, the uniqueness theorem of Section 2.4.4 implies that (2.373)–(2.375) is the unique outgoing wave solution to the frequency-domain acoustic equations (2.356)–(2.357). Indeed, the acoustic pressure in (2.373) satisfies the Sommerfeld radiation conditions [66, p. 56]

$$|p_\omega(\mathbf{r})| = O(1/r) \tag{2.376}$$

$$\frac{\partial}{\partial r} p_\omega(\mathbf{r}) - ikp_\omega(\mathbf{r}) = o(1/r). \tag{2.377}$$

The condition (iii) is also sufficient for Φ_ω to satisfy the inhomogeneous Helmholtz equation

$$\nabla^2\Phi_\omega + k^2\Phi_\omega = \frac{q_\omega(\mathbf{r})}{i\omega\rho_{m0}} - \frac{1}{\rho_{m0}}\left(g_\omega(\mathbf{r}) + \frac{\nabla\cdot\mathbf{f}_\omega(\mathbf{r})}{i\omega}\right). \tag{2.378}$$

2.4.6 Volume Integral Expressions for the Pressure and Velocity Fields

The frequency-domain acoustic volume integral expressions for $p_\omega(\mathbf{r})$ and $\mathbf{v}_\omega(\mathbf{r})$ can be found by taking the Fourier transform of the corresponding time-domain volume integral expressions (2.247) and (2.250) to get

$$\begin{aligned} p_\omega(\mathbf{r}) &= \frac{1}{4\pi}\lim_{\delta\to 0}\int\limits_{V-V_\delta} q_\omega(\mathbf{r}')\frac{e^{ik|\mathbf{r}-\mathbf{r}'|}}{|\mathbf{r}-\mathbf{r}'|}dV' \\ &= -\frac{1}{4\pi}\lim_{\delta\to 0}\int\limits_{V-V_\delta}\left[i\omega s_\omega(\mathbf{r}') + (ikR-1)\mathbf{f}_\omega(\mathbf{r}')\cdot\frac{\hat{\mathbf{R}}}{R}\right]\frac{e^{ikR}}{R}dV' \end{aligned} \tag{2.379}$$

$$\begin{aligned} \mathbf{v}_\omega(\mathbf{r}) &= \frac{1}{4\pi i\omega\rho_{m0}}\lim_{\delta\to 0}\int\limits_{V-V_\delta} q_\omega(\mathbf{r}')(ikR-1)\hat{\mathbf{R}}\frac{e^{ikR}}{R^2}dV' - \frac{\mathbf{f}_\omega(\mathbf{r})}{i\omega\rho_{m0}} \\ &= -\frac{1}{4\pi\rho_{m0}}\lim_{\delta\to 0}\int\limits_{V-V_\delta}\left\{(ikR-1)s_\omega(\mathbf{r}')\hat{\mathbf{R}} + \frac{\mathbf{f}_\omega(\mathbf{r}')}{c}\left[\hat{\mathbf{R}}\hat{\mathbf{R}}ikR\right.\right. \\ &\qquad \left.\left. -(3\hat{\mathbf{R}}\hat{\mathbf{R}} - \bar{\mathbf{I}})\left(1 - \frac{1}{ikR}\right)\right]\right\}\frac{e^{ikR}}{R^2}dV' \\ &\quad + \frac{1}{i\omega\rho_{m0}}(\bar{\mathbf{L}} - \bar{\mathbf{I}})\cdot\mathbf{f}_\omega(\mathbf{r}) \end{aligned} \tag{2.380}$$

with the dyadic $\bar{\mathbf{L}}$ defined in (2.84) and $\mathbf{R} = \mathbf{r} - \mathbf{r}'$. The frequency-domain expressions (2.379) and (2.380) for p_ω and $\mathbf{v}_\omega$ form a rigorously valid unique solution to the acoustic equations (2.356)–(2.357), if, in addition to the conditions (i), (ii), and (iii), of Section 2.4.5, the first partial derivatives of $\mathbf{f}_\omega(\mathbf{r})$ are continuous functions of $\mathbf{r}$.

Cavity-defined frequency-domain acoustic fields can be determined from (2.379) and (2.380) in the same way that the cavity-defined frequency-domain electromagnetic fields were determined from (2.335) and (2.336). In particular, (2.380) shows that the frequency-domain velocity field changes by $(\bar{\mathbf{L}} - \bar{\mathbf{I}})\cdot\mathbf{f}_\omega(\mathbf{r})/(i\omega\rho_{m0})$ if the external force density $\mathbf{f}_\omega$ is removed (and thus the real time-harmonic external force density is removed for all time) from an infinitesimal volume containing $\mathbf{r}$. However, if the real time-harmonic external force density is removed from the infinitesimal volume for an infinitesimal amount of time Δt about each t, rather than for all time, then the velocity field remains unchanged, as explained in Section 2.2.6.

2.4.7 Surface Integral Expressions for the Pressure and Velocity Fields (Huygens' Sources)

The frequency-domain acoustic pressure and velocity fields, like the time-domain acoustic fields, can be written outside a surface S in terms of integrals of the pressure and normal velocity (Huygens' sources) on the surface S that surrounds the sources located within a bounded volume V, as shown in Figure 2.1. These frequency-domain surface integral expressions can be found simply by Fourier transforming the corresponding time-domain expressions (2.259) and (2.261) to get

$$p_\omega(\mathbf{r}) = -\nabla \cdot \int_S \hat{\mathbf{n}}' p_\omega(\mathbf{r}') G_\omega(\mathbf{r}, \mathbf{r}') dS' - i\omega\rho_{m0} \int_S \hat{\mathbf{n}}' \cdot \mathbf{v}_\omega(\mathbf{r}') G_\omega(\mathbf{r}, \mathbf{r}') dS' \tag{2.381}$$

or simply

$$p_\omega(\mathbf{r}) = \int_S \left[p_\omega(\mathbf{r}') \frac{\partial}{\partial n'} G_\omega(\mathbf{r}, \mathbf{r}') - G_\omega(\mathbf{r}, \mathbf{r}') \frac{\partial}{\partial n'} p_\omega(\mathbf{r}') \right] dS' \tag{2.382}$$

and

$$\mathbf{v}_\omega(\mathbf{r}) = -\frac{1}{i\omega\rho_{m0}} \nabla\nabla \cdot \int_S \hat{\mathbf{n}}' p_\omega(\mathbf{r}') G_\omega(\mathbf{r}, \mathbf{r}') dS' - \nabla \int_S \hat{\mathbf{n}}' \cdot \mathbf{v}_\omega(\mathbf{r}') G_\omega(\mathbf{r}, \mathbf{r}') dS' \tag{2.383}$$

with the free-space Green's function $G_\omega(\mathbf{r}, \mathbf{r}')$ given in (2.319).

The del operators can be brought inside the surface integrals to convert (2.381) and (2.383) to the expressions corresponding to (2.257) and (2.262), namely

$$p_\omega(\mathbf{r}) = -\frac{1}{4\pi} \int_S \hat{\mathbf{n}}' \cdot \left[\frac{\hat{\mathbf{R}}}{R} (ikR - 1) p_\omega(\mathbf{r}') + i\omega\rho_{m0} \mathbf{v}_\omega(\mathbf{r}') \right] \frac{e^{ikR}}{R} dS' \tag{2.384}$$

and

$$\begin{aligned} \mathbf{v}_\omega(\mathbf{r}) = -\frac{1}{4\pi\rho_{m0}} \int_S \hat{\mathbf{n}}' \cdot \Bigg\{ & \left[\hat{\mathbf{R}}\hat{\mathbf{R}} ikR - (3\hat{\mathbf{R}}\hat{\mathbf{R}} - \mathbf{I}) \left(1 - \frac{1}{ikR} \right) \right] \frac{p_\omega(\mathbf{r}')}{c} \\ & + \rho_{m0}(ikR - 1) \mathbf{v}_\omega(\mathbf{r}') \hat{\mathbf{R}} \Bigg\} \frac{e^{ikr}}{R^2} dS' \end{aligned} \tag{2.385}$$

with $\mathbf{R} = \mathbf{r} - \mathbf{r}'$.

The frequency-domain acoustic pressure and velocity fields given in (2.381) or (2.384) and (2.383) or (2.385) are continuous functions of $\mathbf{r}$ that satisfy the homogeneous acoustic equations (2.356)–(2.357) outside S if $\hat{\mathbf{n}} p_\omega(\mathbf{r})$ and $\hat{\mathbf{n}} \cdot \mathbf{v}_\omega(\mathbf{r})$ are continuous functions of $\mathbf{r}$ on S. If, in addition, the surface S has continuous curvature, and $p_\omega(\mathbf{r})$ is Hölder continuous on S, then the right-hand side of (2.381) or (2.384) equals the pressure on S as $\mathbf{r}$ approaches the surface S from outside S. Repeated differentiation of (2.381)–(2.383) or (2.384)–(2.385) with respect to x, y, and z shows that all derivatives of $p_\omega(\mathbf{r})$ and $\mathbf{v}_\omega(\mathbf{r})$ exist and that $p_\omega(\mathbf{r})$ and $\mathbf{v}_\omega(\mathbf{r})$ are analytic functions of complex x, y, and z in a complex neighborhood of real x, y, and z outside the source region; see Section 3.2.4.

Finally, one can also Fourier transform the time-domain acoustic formula (2.263) to obtain the formula for the frequency-domain acoustic velocity that is analogous to the Stratton-

Chu formulas for frequency-domain electric and magnetic fields

$$\mathbf{v}_\omega(\mathbf{r}) = -\frac{ik}{\rho_{m0}c}\int\limits_S \hat{\mathbf{n}}' p_\omega(\mathbf{r}')G_\omega(\mathbf{r},\mathbf{r}')dS' + \nabla\times\int\limits_S \hat{\mathbf{n}}'\times\mathbf{v}_\omega(\mathbf{r}')G_\omega(\mathbf{r},\mathbf{r}')dS' \\ -\nabla\int\limits_S \hat{\mathbf{n}}'\cdot\mathbf{v}_\omega(\mathbf{r}')G_\omega(\mathbf{r},\mathbf{r}')dS'. \tag{2.386}$$

The tangential acoustic velocity on the surface S can be written in terms of the acoustic pressure through the surface relationship

$$\rho_{m0}\hat{\mathbf{n}}\times\mathbf{v}_\omega(\mathbf{r}) = \frac{\hat{\mathbf{n}}\times\nabla_s p_\omega(\mathbf{r})}{i\omega} = -\frac{\nabla_s\times\hat{\mathbf{n}}p_\omega(\mathbf{r})}{i\omega}, \quad \mathbf{r}\in S. \tag{2.387}$$

If $\hat{\mathbf{n}}p_\omega$, $\hat{\mathbf{n}}\cdot\mathbf{v}_\omega$, and $\nabla_s\times(\hat{\mathbf{n}}p_\omega)$ are continuous functions of $\mathbf{r}$ for $\mathbf{r}$ on S (which implies through (2.387) that $\hat{\mathbf{n}}\times\mathbf{v}_\omega$ is a continuous function of $\mathbf{r}$ on S), then p_ω and $\mathbf{v}_\omega$ given in (2.381) or (2.384), and (2.386)–(2.387) are continuous functions of $\mathbf{r}$ that satisfy the acoustic equations (2.356)–(2.357) for $\mathbf{r}$ outside S. If, in addition, $\mathbf{v}_\omega(\mathbf{r})$ is Hölder continuous on a continuously curved surface S, the velocity on S is recovered from (2.386) as $\mathbf{r}$ approaches S from outside S, and also recovered from (2.383) or (2.385) (because of the surface relationship (2.387)).

2.4.8 Time-Average Wave Force and Power

The acoustic wave force and acoustic power per unit volume are given in (2.270) and (2.273). These equations, and their integral counterparts, (2.272) and (2.274), hold for all real time-dependent acoustic sources and fields, and, in particular, for the real time-harmonic sources and fields obtained by taking the real part of the frequency-domain sources and fields multiplied by $e^{-i\omega t}$. We can insert these real time-harmonic sources and fields into (2.270), (2.272), (2.273), and (2.274), average over time, and use the result (2.347) to obtain a time-average acoustic wave force and power, delivered by the acoustic fields to the fluid sources in a volume V_a with surface S_a, namely

$$\mathbf{F}_{ac}^{ave} = -\frac{1}{2}\mathrm{Re}\int\limits_{V_a}\left(\kappa_0\mathbf{f}_\omega p_\omega^* + s_\omega\mathbf{v}_\omega^* - \frac{(\nabla\times\mathbf{f}_\omega)\times\mathbf{v}_\omega^*}{i\omega}\right)dV \\ = \frac{1}{2}\mathrm{Re}\int\limits_{V_a}\nabla\cdot\bar{\mathbf{T}}_{ac}^c dV = \frac{1}{2}\mathrm{Re}\int\limits_{S_a}\hat{\mathbf{n}}\cdot\bar{\mathbf{T}}_{ac}^c dS \tag{2.388}$$

$$P_{ac}^{ave} = -\frac{1}{2}\mathrm{Re}\int\limits_{V_a}\left(\mathbf{f}_\omega\cdot\mathbf{v}_\omega^* + \frac{s_\omega p_\omega^*}{\rho_{m0}}\right)dV \\ = -\frac{1}{2}\mathrm{Re}\int\limits_{V_a}\nabla\cdot(p_\omega\mathbf{v}_\omega^*)dV = -\frac{1}{2}\mathrm{Re}\int\limits_{S_a}\hat{\mathbf{n}}\cdot(p_\omega\mathbf{v}_\omega^*)dS \tag{2.389}$$

where the complex acoustic wave-stress dyadic $\bar{\mathbf{T}}_{ac}^c$ is defined as

$$\bar{\mathbf{T}}_{ac}^c = -\rho_{m0}\mathbf{v}_\omega\mathbf{v}_\omega^* + \frac{1}{2}\left(\rho_{m0}|\mathbf{v}_\omega|^2 - \kappa_0|p_\omega|^2\right)\bar{\mathbf{I}}. \tag{2.390}$$

The force equations in (2.388) show that the vector $\mathrm{Re}(\nabla\cdot\bar{\mathbf{T}}_{ac}^c/2)$ is the time-average wave force per unit volume delivered by the acoustic fields to the fluid sources and that the surface integral $\mathrm{Re}\int_{S_a}\hat{\mathbf{n}}\cdot\bar{\mathbf{T}}_{ac}^c dS/2$ is the time-average wave-momentum flux entering the fields of the volume V_a through its surface S_a.

The power relations in (2.389) can also be obtained by taking the real part of a complex acoustic power theorem, which can be derived from the frequency-domain acoustic equations

(2.356) and (2.357). Forming $\nabla\cdot(p_\omega \mathbf{v}_\omega^*)$, substituting from (2.356) and (2.357), and integrating over V_a, we find

$$-\frac{1}{2}\int\limits_{V_a}\left(\mathbf{f}_\omega\cdot\mathbf{v}_\omega^* + \frac{s_\omega^* p_\omega}{\rho_{m0}}\right)dV + 2i\omega\int\limits_{V_a}\frac{1}{4}\left(\kappa_0|p_\omega|^2 - \rho_{m0}|\mathbf{v}_\omega|^2\right)dV$$
$$= -\frac{1}{2}\int\limits_{V_a}\nabla\cdot(p_\omega\mathbf{v}_\omega^*)dV = -\frac{1}{2}\int\limits_{S_a}\hat{\mathbf{n}}\cdot(p_\omega\mathbf{v}_\omega^*)dS. \tag{2.391}$$

The real parts of $-\mathbf{f}_\omega\cdot\mathbf{v}_\omega^*$ and $-s_\omega^* p_\omega/\rho_{m0}$ are the time-average acoustic power per unit volume delivered by the fields to the fluid source densities $\mathbf{f}_\omega$ and s_ω. Thus, the integral in (2.389) over the volume V_a of the sum of these power densities equals the time-average acoustic power P_{ac}^{ave} delivered by the fields to all the sources in V_a.

The second equation in (2.389) shows that $\text{Re}[-\nabla\cdot(p_\omega\mathbf{v}_\omega^*)/2]$ equals the time-average acoustic power per unit volume delivered by the acoustic fields to the fluid sources. (It is evaluated in (2.449) for the lossy fluids of Section 2.5.2.) The last equation in (2.389) shows that $-\text{Re}\int_{S_a}\hat{\mathbf{n}}\cdot(p_\omega\mathbf{v}_\omega^*)dS/2$ equals the time-average acoustic power entering the fields of the volume V_a through its surface S_a.

The imaginary part of (2.391) produces expressions for the difference between the time-average acoustic velocity and pressure energies in V_a multiplied by 2ω, specifically

$$2\omega\int\limits_{V_a}\frac{1}{4}\left\{\left(\kappa_0|p_\omega|^2 - \rho_{m0}|\mathbf{v}_\omega|^2\right) + \text{Im}\frac{1}{\omega}\left[\mathbf{f}_\omega^*\cdot\mathbf{v}_\omega + \frac{s_\omega p_\omega^*}{\rho_{m0}}\right]\right\}dV$$
$$= -\frac{1}{2}\text{Im}\int\limits_{V_a}\nabla\cdot(p_\omega\mathbf{v}_\omega^*)dV = -\frac{1}{2}\text{Im}\int\limits_{S_a}\hat{\mathbf{n}}\cdot(p_\omega\mathbf{v}_\omega^*)dS. \tag{2.392}$$

If (2.392) is applied to a surface S in the fluid enclosing the sources, and V_0 refers to the infinite fluid volume outside S, one obtains

$$\frac{1}{2}\text{Im}\int\limits_{S}\hat{\mathbf{n}}\cdot(p_\omega\mathbf{v}_\omega^*)dS = 2\omega\int\limits_{V_0}\frac{1}{4}\left(\kappa_0|p_\omega|^2 - \rho_{m0}|\mathbf{v}_\omega|^2\right)dV \tag{2.393}$$

where in (2.393) $\hat{\mathbf{n}}$ is the unit normal to S pointing into V_0. The right-hand side of (2.393) can be considered as the contribution of the fields in V_0 to the acoustic reactance (imaginary part of the acoustic input impedance) of an acoustic radiator enclosed by S. In other words, the left-hand side of (2.393) is a measure of the acoustic reactance contributed by the fields in the infinite fluid outside S.

2.5 EQUATIONS FOR LOSSY MEDIA

Throughout most of the book, it is assumed that the electromagnetic sources are located in free space and that the acoustic sources are located in a lossless fluid; thus, the fields propagate without loss. However, many of the results can be readily extended to lossy media. For example, in Section 3.3 it is explained how the frequency-domain formulas can be modified to hold in lossy media. Hence, we provide here a brief introduction to the electromagnetic and acoustic parameters and equations useful for describing simple lossy media in both the frequency and time domains. For lossy media, it is advantageous to begin in the frequency domain and then to find the corresponding time-domain parameters and equations by taking the inverse Fourier transform of the frequency-domain parameters and equations. Electromagnetic fields in lossy materials are considered first, and we end the section with a briefer account of the acoustic fields in lossy fluids.

2.5.1 Electromagnetic Fields

Frequency Domain. In the frequency domain, losses manifest themselves in Maxwell's equations through the real part of the complex conductivity and through the imaginary parts of the complex electric and magnetic susceptibilities. In particular, for linear, isotropic, homogeneous, dispersive materials, the conduction current density $\mathbf{J}_\omega^{in}(\mathbf{r})$, electric polarization density $\mathbf{P}_\omega^{in}(\mathbf{r})$, and magnetic polarization density $\mathbf{M}_\omega^{in}(\mathbf{r})$ induced in the material are related to the electric and magnetic fields by the simple proportionalities

$$\mathbf{J}_\omega^{in}(\mathbf{r}) = \sigma_\omega \mathbf{E}_\omega(\mathbf{r}) \tag{2.394}$$

$$\mathbf{P}_\omega^{in}(\mathbf{r}) = \epsilon \chi_\omega^e \mathbf{E}_\omega(\mathbf{r}) \tag{2.395}$$

$$\mathbf{M}_\omega^{in}(\mathbf{r}) = \chi_\omega^m \mathbf{H}_\omega(\mathbf{r}) \tag{2.396}$$

where σ_ω, χ_ω^e, and χ_ω^m are the conductivity, and electric and magnetic susceptibilities. They are independent of position (because the medium is assumed homogeneous), and, in general, they are complex functions of frequency. (For our purposes, it is unnecessary to introduce the electric displacement vector $\mathbf{D}_\omega = \epsilon\mathbf{E}_\omega + \mathbf{P}_\omega$, and the magnetic induction vector $\mathbf{B}_\omega = \mu(\mathbf{H}_\omega + \mathbf{M}_\omega)$, because we want to express the electromagnetic field equations for lossy media in terms of the $\mathbf{E}_\omega$ and $\mathbf{H}_\omega$ fields alone, as we did for free space.)

The constitutive relations (2.394)–(2.396) allow Maxwell's differential equations in the frequency domain to be written as

$$\nabla \times \mathbf{E}_\omega(\mathbf{r}) - i\omega\mu_\omega \mathbf{H}_\omega(\mathbf{r}) = i\omega\mu \mathbf{M}_\omega^{ap}(\mathbf{r}) \tag{2.397}$$

$$\nabla \times \mathbf{H}_\omega(\mathbf{r}) + i\omega\epsilon_\omega \mathbf{E}_\omega(\mathbf{r}) = \mathbf{J}_\omega^{ap}(\mathbf{r}) - i\omega \mathbf{P}_\omega^{ap}(\mathbf{r}) \tag{2.398}$$

$$\nabla \cdot \mathbf{H}_\omega(\mathbf{r}) = -\mu \nabla \cdot \mathbf{M}_\omega^{ap}(\mathbf{r})/\mu_\omega \tag{2.399}$$

$$\nabla \cdot \mathbf{E}_\omega(\mathbf{r}) = (\rho_\omega^{ap}(\mathbf{r}) - \nabla \cdot \mathbf{P}_\omega^{ap}(\mathbf{r}))/\epsilon_\omega \tag{2.400}$$

where the complex permittivity ϵ_ω and complex permeability μ_ω are defined as

$$\epsilon_\omega = \epsilon[1 + \chi_\omega^e - \sigma_\omega/(i\omega\epsilon)] \tag{2.401}$$

$$\mu_\omega = \mu(1 + \chi_\omega^m) \tag{2.402}$$

and, as usual, ϵ and μ are the free-space permittivity and permeability. In equations (2.397)–(2.400), $\mathbf{J}_\omega^{ap}$, $\mathbf{P}_\omega^{ap}$, $\mathbf{M}_\omega^{ap}$, and ρ_ω^{ap} are the applied current, electric polarization, magnetic polarization, and charge densities in excess of the induced conduction current, polarization, and charge densities. The applied current and charge densities are related by the continuity equation

$$\nabla \cdot \mathbf{J}_\omega^{ap}(\mathbf{r}) - i\omega\rho_\omega^{ap}(\mathbf{r}) = 0 \tag{2.403}$$

which can be derived from (2.400) and the divergence of (2.398). The total current, polarization, and charge densities are given by

$$\mathbf{J}_\omega = \mathbf{J}_\omega^{ap} + \mathbf{J}_\omega^{in}, \quad \mathbf{P}_\omega = \mathbf{P}_\omega^{ap} + \mathbf{P}_\omega^{in}, \quad \mathbf{M}_\omega = \mathbf{M}_\omega^{ap} + \mathbf{M}_\omega^{in}, \quad \rho_\omega = \rho_\omega^{ap} + \rho_\omega^{in}. \tag{2.404}$$

Since the total and applied charge and current densities satisfy the continuity equation, the induced current and charge densities also satisfy the continuity equation

$$\rho_\omega^{in} = \nabla \cdot \mathbf{J}_\omega^{in}/(i\omega) = \sigma_\omega \nabla \cdot \mathbf{E}_\omega/(i\omega) = \sigma_\omega(\rho_\omega^{ap} - \nabla \cdot \mathbf{P}_\omega^{ap})/(i\omega\epsilon_\omega) \tag{2.405}$$

where the last two equations in (2.405) follow from (2.394) and (2.400). Note from (2.405) that the conduction charge $\rho_\omega^{in}(\mathbf{r})$ as well as $\nabla \cdot \mathbf{E}_\omega(\mathbf{r})$ is zero outside the region of the applied sources $(\rho_\omega^{ap} - \nabla \cdot \mathbf{P}_\omega^{ap})$ (assuming $\epsilon_\omega \neq 0$). Also, note that from a macroscopic view of the frequency-domain Maxwell equations (2.397)–(2.400), it is impossible to distinguish between the complex electric susceptibility χ_ω^e and the complex conductivity function $i\sigma_\omega/(\omega\epsilon)$. When

σ_ω, χ^e_ω, and χ^m_ω are zero, the induced sources are zero, and (2.397)–(2.400) reduce to the free-space frequency-domain equations (2.287)–(2.290), generalized to include applied polarization densities.

Substitution of $\mathbf{H}_\omega$ from (2.397) into (2.398), or $\mathbf{E}_\omega$ from (2.398) into (2.397), produces the second-order differential equations

$$\nabla\times\nabla\times\mathbf{E}_\omega - k_\ell^2\mathbf{E}_\omega = i\omega\mu_\omega(\mathbf{J}^{ap}_\omega - i\omega\mathbf{P}^{ap}_\omega) + i\omega\mu\nabla\times\mathbf{M}^{ap}_\omega \tag{2.406}$$

$$\nabla\times\nabla\times\mathbf{H}_\omega - k_\ell^2\mathbf{H}_\omega = \omega^2\mu\epsilon_\omega\mathbf{M}^{ap}_\omega + \nabla\times(\mathbf{J}^{ap}_\omega - i\omega\mathbf{P}^{ap}_\omega). \tag{2.407}$$

The square of the complex propagation constant is given in terms of μ_ω and ϵ_ω as

$$k_\ell^2 = \omega^2\mu_\omega\epsilon_\omega. \tag{2.408}$$

With the help of the identity $\nabla\times\nabla\times = -\nabla^2 + \nabla\nabla\cdot$, and the divergence equations (2.399)–(2.400), the inhomogeneous "curlcurl" equations (2.406)–(2.407) convert to the inhomogeneous Helmholtz equations

$$\nabla^2\mathbf{E}_\omega + k_\ell^2\mathbf{E}_\omega = -i\omega\mu_\omega(\mathbf{J}^{ap}_\omega - i\omega\mathbf{P}^{ap}_\omega) - i\omega\mu\nabla\times\mathbf{M}^{ap}_\omega + \nabla(\rho^{ap}_\omega - \nabla\cdot\mathbf{P}^{ap}_\omega)/\epsilon_\omega \tag{2.409}$$

$$\nabla^2\mathbf{H}_\omega + k_\ell^2\mathbf{H}_\omega = -\omega^2\mu\epsilon_\omega\mathbf{M}^{ap}_\omega - \nabla\times(\mathbf{J}^{ap}_\omega - i\omega\mathbf{P}^{ap}_\omega) - \mu\nabla(\nabla\cdot\mathbf{M}^{ap}_\omega)/\mu_\omega. \tag{2.410}$$

The complex conductivity and susceptibilities must obey the reality condition

$$\sigma_{-\omega} = \sigma^*_\omega, \quad \chi^e_{-\omega} = \chi^{e*}_\omega, \quad \chi^m_{-\omega} = \chi^{m*}_\omega \tag{2.411}$$

because $\mathbf{J}^{in}_\omega$, $\mathbf{P}^{in}_\omega$, $\mathbf{M}^{in}_\omega$, $\mathbf{E}_\omega$, and $\mathbf{H}_\omega$ obey the reality condition. Dividing σ_ω, χ^e_ω, and χ^m_ω into real and imaginary parts

$$\sigma_\omega = \sigma'_\omega + i\sigma''_\omega, \quad \chi^e_\omega = \chi^{e\prime}_\omega + i\chi^{e\prime\prime}_\omega, \quad \chi^m_\omega = \chi^{m\prime}_\omega + i\chi^{m\prime\prime}_\omega \tag{2.412}$$

and applying (2.411) shows that the real and imaginary parts in (2.412) satisfy the even and odd relations

$$\sigma'_{-\omega} = \sigma'_\omega, \quad \chi^{e\prime}_{-\omega} = \chi^{e\prime}_\omega, \quad \chi^{m\prime}_{-\omega} = \chi^{m\prime}_\omega \tag{2.413}$$

$$\sigma''_{-\omega} = -\sigma''_\omega, \quad \chi^{e\prime\prime}_{-\omega} = -\chi^{e\prime\prime}_\omega, \quad \chi^{m\prime\prime}_{-\omega} = -\chi^{m\prime\prime}_\omega. \tag{2.414}$$

In lossy media the complex Poynting's theorem

$$\begin{aligned}\frac{1}{2}\int_{V_a} [(\mathbf{J}^{ap*}_\omega + i\omega\mathbf{P}^{ap*}_\omega)\cdot\mathbf{E}_\omega - i\omega\mu\mathbf{M}^{ap}_\omega\cdot\mathbf{H}^*_\omega]dV \\ + 2i\omega\int_{V_a}\left(\frac{\epsilon^*_\omega}{4}|\mathbf{E}_\omega|^2 - \frac{\mu_\omega}{4}|\mathbf{H}_\omega|^2\right)dV \\ = -\frac{1}{2}\int_{V_a}\nabla\cdot(\mathbf{E}_\omega\times\mathbf{H}^*_\omega)dV = -\frac{1}{2}\int_{S_a}\hat{\mathbf{n}}\cdot(\mathbf{E}_\omega\times\mathbf{H}^*_\omega)dS\end{aligned} \tag{2.415}$$

can be derived from the complex Poynting's theorem (2.351) in lossless free space by substituting into (2.351) the total source densities from (2.404) with $\mathbf{J}^{in}_\omega$, $\mathbf{P}^{in}_\omega$, and $\mathbf{M}^{in}_\omega$ given in (2.394)–(2.396). The time-average power delivered by the electromagnetic fields to the induced sources in the arbitrary volume V_a with surface S_a is given by the real part of the second volume integral in (2.415).

The time-average force equations (2.348) and (2.352) also remain valid in lossy media, if the total sources in (2.404) are inserted into (2.348) and (2.352). Then with S_a chosen within an infinitesimally thin shell devoid of polarization densities, as explained in Section 2.3.9, $\mathbf{F}^{ave}_{el}$ in both (2.348) and (2.352) give the force delivered by the electromagnetic fields to the applied and induced sources in V_a.

At every point in the medium, the time-average power per unit volume (denoted by $\mathcal{Q}_{el}$), delivered by the electromagnetic fields to the total (applied plus induced) current and polarization densities, is determined by the negative real part of the divergence of the complex Poynting vector ($\text{Re}[-\nabla \cdot (\mathbf{E}_\omega \times \mathbf{H}^*_\omega)]/2$; see Section 2.3.9). In particular, at every point in the medium where there are no applied source densities (only induced current and polarization densities given by (2.394)–(2.396)), the second equation in (2.349) (or the real part of the first equation in (2.415)) shows that

$$\mathcal{Q}_{el} = -\frac{1}{2}\text{Re}[\nabla \cdot (\mathbf{E}_\omega \times \mathbf{H}^*_\omega)] = \frac{1}{2}[(\omega\epsilon\chi^{e\prime\prime}_\omega + \sigma'_\omega)|\mathbf{E}_\omega|^2 + \omega\mu\chi^{m\prime\prime}_\omega|\mathbf{H}_\omega|^2]. \tag{2.416}$$

For passive media in which this electromagnetic power is converted irreversibly into heat (by means of frictional forces on the conduction current and induced polarization densities), $\mathcal{Q}_{el} \geq 0$. Thus, from (2.416), which holds for all $\mathbf{E}_\omega$ and $\mathbf{H}_\omega$, $\omega\epsilon\chi^{e\prime\prime}_\omega + \sigma'_\omega \geq 0$ and $\omega\chi^{m\prime\prime}_\omega \geq 0$. If the mechanisms for heat loss from the conduction current and the electric polarization are independent, $\omega\chi^{e\prime\prime}_\omega \geq 0$ and $\sigma'_\omega \geq 0$, and we have [65, p. 274], [10, p. 37]

$$\sigma'_\omega \geq 0, \quad \omega\chi^{e\prime\prime}_\omega \geq 0, \quad \omega\chi^{m\prime\prime}_\omega \geq 0. \tag{2.417}$$

Also, because $\chi^{e\prime\prime}_\omega$ and $\chi^{m\prime\prime}_\omega$ are odd functions of ω, no material can have electric or magnetic polarization loss that is independent of frequency for $-\infty < \omega < \infty$. In the words of E.J. Post [113, p. 162] "The imaginary elements, therefore, if different from zero, should be functions of frequency under all circumstances. One can express this in another way by saying that the imaginary elements are associated with effects which are essentially dispersive." Of course, lossy materials are possible with real conductivity $\sigma'_\omega = \sigma_0$ that is independent of frequency.

Time Domain. The general constitutive relations for time-domain fields in linear, isotropic, homogeneous, dispersive materials can be found by taking the inverse Fourier transform of (2.394)–(2.396) and then applying the convolution theorem to get

$$\mathbf{J}^{in}(\mathbf{r}, t) = \frac{1}{2\pi}\int_{-\infty}^{t} \sigma(t - t')\mathbf{E}(\mathbf{r}, t')dt' \tag{2.418}$$

$$\mathbf{P}^{in}(\mathbf{r}, t) = \frac{\epsilon}{2\pi}\int_{-\infty}^{t} \chi^e(t - t')\mathbf{E}(\mathbf{r}, t')dt' \tag{2.419}$$

$$\mathbf{M}^{in}(\mathbf{r}, t) = \frac{1}{2\pi}\int_{-\infty}^{t} \chi^m(t - t')\mathbf{H}(\mathbf{r}, t')dt' \tag{2.420}$$

where the time-domain constitutive parameters $\sigma(t)$, $\chi^e(t)$, and $\chi^m(t)$ are the inverse Fourier transforms of the frequency-domain constitutive parameters σ_ω, χ^e_ω, and χ^m_ω, respectively. The time-domain constitutive parameters are real functions of t because the frequency-domain constitutive parameters obey the reality condition (2.411). Causality implies that the fields in the future cannot affect the present values of the induced conduction current or electric and magnetic polarization. Therefore, the time-domain constitutive parameters, $\sigma(t)$, $\chi^e(t)$, and $\chi^m(t)$, are zero for $t < 0$, and the upper limits of integration in (2.418)–(2.420) are shown as t rather than $+\infty$. If the electric and magnetic fields are zero for $t \leq t_0$, the lower limits of integration in (2.418)–(2.420) can be changed from $-\infty$ to t_0.

At each point in time t, the conduction-current and polarization densities on the left-hand sides of (2.418)–(2.420) are determined by the fields on the right-hand sides of (2.418)–(2.420) at all previous times. However, at each point in space $\mathbf{r}$, the conduction-current and polarization densities are determined by the fields at the same point $\mathbf{r}$. Thus, the materials are said to exhibit time dispersion or frequency dispersion but not spatial dispersion [65, p. 358].

Taking the inverse Fourier transforms of (2.397)–(2.400) with ϵ_ω and μ_ω inserted from (2.401) and (2.402) gives Maxwell's time-domain differential equations in a medium of conductive, polarizable material

$$\nabla \times \mathbf{E}(\mathbf{r}, t) + \mu \frac{\partial}{\partial t} \left[\mathbf{H}(\mathbf{r}, t) + \frac{1}{2\pi} \int_{-\infty}^{t} \chi^m(t - t')\mathbf{H}(\mathbf{r}, t')dt' \right] = -\mu \frac{\partial}{\partial t} \mathbf{M}^{ap}(\mathbf{r}, t) \tag{2.421}$$

$$\nabla \times \mathbf{H}(\mathbf{r}, t) - \epsilon \frac{\partial}{\partial t} \left[\mathbf{E}(\mathbf{r}, t) + \frac{1}{2\pi} \int_{-\infty}^{t} \chi^e(t - t')\mathbf{E}(\mathbf{r}, t')dt' \right] - \frac{1}{2\pi} \int_{-\infty}^{t} \sigma(t - t')\mathbf{E}(\mathbf{r}, t')dt' = \mathbf{J}^{ap}(\mathbf{r}, t) + \frac{\partial}{\partial t} \mathbf{P}^{ap}(\mathbf{r}, t) \tag{2.422}$$

$$\nabla \cdot \left[\mathbf{H}(\mathbf{r}, t) + \frac{1}{2\pi} \int_{-\infty}^{t} \chi^m(t - t')\mathbf{H}(\mathbf{r}, t')dt' \right] = -\nabla \cdot \mathbf{M}^{ap}(\mathbf{r}, t) \tag{2.423}$$

$$\nabla \cdot \left[\mathbf{E}(\mathbf{r}, t) + \frac{1}{2\pi} \int_{-\infty}^{t} \chi^e(t - t')\mathbf{E}(\mathbf{r}, t')dt' + \frac{1}{2\pi\epsilon} \int_{-\infty}^{t} \int_{-\infty}^{t'} \sigma(t' - t'')\mathbf{E}(\mathbf{r}, t'')dt''dt' \right] = [\rho^{ap}(\mathbf{r}, t) - \nabla \cdot \mathbf{P}^{ap}(\mathbf{r}, t)]/\epsilon. \tag{2.424}$$

Taking the inverse Fourier transform of (2.405) outside the source region shows that $\rho^{in}(\mathbf{r}, t)$ as well as $\nabla \cdot \mathbf{E}(\mathbf{r}, t)$ are zero outside the source region, that is, outside the region in which the applied sources $\rho^{ap}(\mathbf{r}, t) - \nabla \cdot \mathbf{P}^{ap}(\mathbf{r}, t)$ are located. Also, the inverse Fourier transform of (2.399) outside the source region shows that $\nabla \cdot \mathbf{H}(\mathbf{r}, t)$ is zero outside the source region in which the applied magnetic polarization $\mathbf{M}(\mathbf{r}, t)$ is located.

Dispersion Relations. We mentioned above that causality requires that the time-domain constitutive parameters are zero for $t < 0$. We can use this fact to relate the real and imaginary parts of each frequency-domain constitutive parameter by means of the Hilbert transform. Consider, for example, the frequency-domain complex electric susceptibility χ^e_ω, which can be written in terms of the Fourier transform of the time-domain electric susceptibility $\chi^e(t)$ as

$$\chi^e_\omega = \frac{1}{2\pi} \int_{0}^{+\infty} \chi^e(t)e^{i\omega t}dt = \frac{1}{2\pi} \int_{-\infty}^{+\infty} \frac{t}{|t|} \chi^e(t)e^{i\omega t}dt = \frac{1}{2\pi} \lim_{\delta \to 0} \int_{-\infty}^{+\infty} \frac{t}{|t|} e^{-\delta|t|} \chi^e(t)e^{i\omega t}dt. \tag{2.425}$$

The last equality in (2.425) can be proven valid for $\delta > 0$ using the theorem in [71, vol. 2, sec. 225] that allows the limit to be brought inside the integral if $\int_{-\infty}^{+\infty} |\chi^e(t)|dt < \infty$. Assume that χ^e_ω is a Hölder continuous function of ω, and for sufficiently large $|\omega|$, $|\chi^e_\omega| \le C/|\omega|^\alpha$, where C and α are positive constants [114, sec. 43]. Then one can apply the convolution theorem to the last Fourier transform in (2.425), use the result

$$\frac{1}{2\pi}\int_{-\infty}^{+\infty}\frac{t}{|t|}e^{-\delta|t|}e^{i\omega t}dt = \frac{i\omega}{\pi(\omega^2+\delta^2)} \tag{2.426}$$

rewrite the convolution integral as a Cauchy principal value integral, and take the limit as $\delta \to 0$ (using [71, vol. 2, sec. 225] and the fact that χ_ω^e is Hölder continuous), to obtain

$$\chi_\omega^e = \frac{i}{\pi}\,⨍_{-\infty}^{+\infty}\frac{\chi_\nu^e}{\omega-\nu}d\nu. \tag{2.427}$$

The bar through the integral sign in (2.427) denotes the Cauchy principal value integration [115, p. 368]. Equating the real and imaginary parts of (2.427), we obtain the "dispersion relations"

$$\chi_\omega^{e\prime} = \frac{1}{\pi}\,⨍_{-\infty}^{+\infty}\frac{\chi_\nu^{e\prime\prime}}{\nu-\omega}d\nu = \frac{2}{\pi}\,⨍_{0}^{+\infty}\frac{\nu\chi_\nu^{e\prime\prime}}{\nu^2-\omega^2}d\nu \tag{2.428}$$

$$\chi_\omega^{e\prime\prime} = -\frac{1}{\pi}\,⨍_{-\infty}^{+\infty}\frac{\chi_\nu^{e\prime}}{\nu-\omega}d\nu = -\frac{2\omega}{\pi}\,⨍_{0}^{+\infty}\frac{\chi_\nu^{e\prime}}{\nu^2-\omega^2}d\nu. \tag{2.429}$$

Equations (2.428)–(2.429) show that the real and imaginary parts of χ_ω^e are Hilbert transforms of each other [115, p. 372]. These dispersion relations could be obtained directly from the theory of Hilbert transforms, given that the first equation in (2.425) implies that χ_ω^e is an analytic function in the upper half of the complex ω-plane. Here, however, we have derived the dispersion relations (2.428)–(2.429) straightforwardly without relying on the theory of Hilbert transforms or integrating in the complex ω-plane.

The corresponding dispersion relations hold for the magnetic susceptibility χ_ω^m and conductivity σ_ω, provided they too are Hölder continuous functions of ω that approach zero as $C/|\omega|^\alpha$ or faster ($\alpha > 0$) as $|\omega| \to \infty$. For frequency-domain functions like the complex permittivity $\epsilon_\omega = \epsilon_\omega' + i\epsilon_\omega'' = \epsilon(1+\chi_\omega^e) + i\sigma_\omega/\omega$, which approaches the free-space value ϵ as $|\omega| \to \infty$, the dispersion relations (2.428)–(2.429) can be applied to $(\epsilon_\omega - \epsilon - i\sigma_0/\omega)$ to yield the Kramers-Kronig dispersion relations [63, p. 334]

$$\epsilon_\omega' - \epsilon = \frac{1}{\pi}\,⨍_{-\infty}^{+\infty}\frac{\epsilon_\nu''}{\nu-\omega}d\nu = \frac{2}{\pi}\,⨍_{0}^{+\infty}\frac{\nu\epsilon_\nu''}{\nu^2-\omega^2}d\nu \tag{2.430}$$

$$\epsilon_\omega'' = -\frac{1}{\pi}\,⨍_{-\infty}^{+\infty}\frac{\epsilon_\nu'-\epsilon}{\nu-\omega}d\nu + \frac{\sigma_0}{\omega} = -\frac{2\omega}{\pi}\,⨍_{0}^{+\infty}\frac{\epsilon_\nu'-\epsilon}{\nu^2-\omega^2}d\nu + \frac{\sigma_0}{\omega} \tag{2.431}$$

generalized to include a possible real constant conductivity $\sigma_0 = \sigma_{\omega\to 0}$ [63, p. 348, problem 7.23]. When σ_0/ω occurs in ϵ_ω'', the integration in the first integral of (2.430) about $\nu = 0$ is evaluated as a Cauchy principal value.

Formula (2.428) (or (2.430)) is especially useful because approximate values of the odd function $\chi_\omega^{e\prime\prime}$, such that $\chi_\omega^{e\prime\prime} \geq 0$ for $\omega > 0$, can be inserted into (2.428) to obtain presumably a permissible even function $\chi_\omega^{e\prime}$. However, approximate values of the even function $\chi_\omega^{e\prime}$ inserted into formula (2.429) may yield an odd function $\chi_\omega^{e\prime\prime}$ that does not satisfy the condition $\chi_\omega^{e\prime\prime} \geq 0$ for $\omega > 0$ [65, p. 281].

Each of the dispersion relations (2.428) and (2.429) (or (2.430) and (2.431)) holds exactly only if the imaginary and real parts, respectively, of the constitutive parameter are known and integrated over all frequencies. Nonetheless, approximate dispersion relations, corresponding to (2.428) and (2.429) (or (2.430) and (2.431)), can sometimes be found that produce sufficiently accurate values of the real and imaginary parts of the constitutive parameter from the local variation of its imaginary and real parts, respectively [116].

2.5.2 Acoustic Fields

Frequency Domain. Losses can occur in compressible fluids, even those that have zero shear viscosity (cannot support shear stress), through "frictional resistance" that occurs during expansion and compression of the fluid [115, pp. 159–160], [68, p. 273], [69, sec. 81]. For small-amplitude, time-harmonic fields, this "expansive friction" alters the normal stress in the fluid by an amount proportional to the rate of expansion of the fluid, or, equivalently, proportional to $i\omega\rho_{m\omega}$. Therefore, in the frequency domain, the frictional normal stress can be taken into account by simply generalizing the linear constitutive relation (2.358) to

$$p_\omega(\mathbf{r}) = \frac{1}{\rho_{m0}\kappa_0}(1 - i\omega\kappa_0\alpha_\omega)\rho_{m\omega}(\mathbf{r}) \tag{2.432}$$

provided $p_\omega(\mathbf{r})$ in the mass conservation equation (2.356) and the momentum equation (2.357) denotes the negative of the normal acoustic stress. The expansive-friction coefficient α_ω is a complex function of frequency. The expansive-friction part $i\omega\alpha_\omega\rho_{m\omega}/\rho_{m0}$ of the normal acoustic stress in (2.432) also equals $\alpha_\omega\nabla\cdot\mathbf{v}_\omega$ wherever $s_\omega^{ap} = 0$. Alternatively, the frequency-domain constitutive relation in linear, quiescent, isotropic, homogeneous, lossy fluids (with zero shear viscosity) can be written as

$$\rho_{m\omega}(\mathbf{r}) = \rho_{m0}\kappa_\omega p_\omega(\mathbf{r}) \tag{2.433}$$

where

$$\kappa_\omega = \frac{\kappa_0}{1 - i\omega\kappa_0\alpha_\omega} = \kappa_\omega' + i\kappa_\omega'' \tag{2.434}$$

is the complex compressibility, and $p_\omega(\mathbf{r})$ is the negative of the normal acoustic stress.[14]

With the constitutive relation (2.433), and $p_\omega(\mathbf{r})$ referring to the negative of the normal acoustic stress, the linear, frequency-domain continuity and momentum equations in lossy fluids (with zero shear viscosity) can be written as

$$\rho_{m0}\nabla\cdot\mathbf{v}_\omega(\mathbf{r}) - i\omega\rho_{m0}\kappa_\omega p_\omega(\mathbf{r}) = s_\omega^{ap}(\mathbf{r}) \tag{2.435}$$

$$\nabla p_\omega(\mathbf{r}) - i\omega\rho_{m0}\mathbf{v}_\omega(\mathbf{r}) = \mathbf{f}_\omega^{ap}(\mathbf{r}). \tag{2.436}$$

Taking the curl of (2.436) shows that

$$\nabla\times\mathbf{v}_\omega(\mathbf{r}) = -\frac{\nabla\times\mathbf{f}_\omega^{ap}(\mathbf{r})}{i\omega\rho_{m0}}. \tag{2.437}$$

Taking the divergence of (2.436) and substituting $\nabla\cdot\mathbf{v}_\omega$ from the continuity equation (2.435) produces the inhomogeneous Helmholtz equation for the acoustic pressure p_ω in lossy fluids

$$\nabla^2 p_\omega(\mathbf{r}) + k_\ell^2 p_\omega(\mathbf{r}) = -q_\omega^{ap}(\mathbf{r}) \tag{2.438}$$

---

[14] Here, in lossy fluids with zero shear viscosity, we let $p_\omega(\mathbf{r})$ (and $p(\mathbf{r},t)$ in the time-domain equations of this section) refer to the negative of the normal acoustic stress to keep the lossy frequency-domain equations (2.435)–(2.436) in the same form as the lossless frequency domain equations (2.356)–(2.357), and to maintain $\mathrm{Re}[-\nabla\cdot(p_\omega\mathbf{v}_\omega^*)]$ as a measure of time-average acoustic power per unit volume delivered by the acoustic fields. At low frequencies where $\kappa_\omega \approx \kappa_0$ and the expansive-friction losses are negligible ($\alpha_\omega \approx 0$), the negative of the normal acoustic stress becomes approximately equal to the ordinary acoustic pressure.

To make the lossy acoustic equation (2.435) entirely analogous to the lossy electromagnetic equations (2.397)–(2.398), one could let $\kappa_\omega = \kappa_0(1+\chi_\omega)$ and then refer to $s_\omega^{in} = i\omega\rho_{m0}\kappa_0\chi_\omega p_\omega$ as the induced mass-flow density, so that the total mass-flow density is given by $s_\omega = s_\omega^{ap} + s_\omega^{in}$. In (2.436) there is no explicit body force induced by the dispersion in the fluid ($\mathbf{f}_\omega^{in} = 0$), because p_ω refers to the negative of the normal acoustic stress, and thus $\mathbf{f}_\omega = \mathbf{f}_\omega^{ap}$.

where the square of the complex propagation constant k_ℓ is given in terms of κ_ω by

$$k_\ell^2 = \omega^2 \rho_{m0} \kappa_\omega \tag{2.439}$$

and the applied pressure source function q_ω^{ap} is given by

$$q_\omega^{ap}(\mathbf{r}) = -i\omega s_\omega^{ap}(\mathbf{r}) - \nabla \cdot \mathbf{f}_\omega^{ap}(\mathbf{r}). \tag{2.440}$$

Similarly, taking the gradient of the continuity equation (2.435) and substituting $\mathbf{v}_\omega$ from (2.436) gives

$$\nabla(\nabla \cdot \mathbf{v}_\omega(\mathbf{r})) + k_\ell^2 \mathbf{v}_\omega(\mathbf{r}) = \frac{1}{\rho_{m0}}(\nabla s_\omega^{ap}(\mathbf{r}) + i\omega\rho_{m0}\kappa_\omega \mathbf{f}_\omega^{ap}(\mathbf{r})). \tag{2.441}$$

Inserting $\nabla \times \mathbf{v}_\omega(\mathbf{r})$ from (2.437) into the identity $\nabla(\nabla \cdot \mathbf{v}_\omega) = \nabla \times (\nabla \times \mathbf{v}_\omega) + \nabla^2 \mathbf{v}_\omega$, we find

$$\nabla(\nabla \cdot \mathbf{v}_\omega) = \nabla^2 \mathbf{v}_\omega - \frac{1}{i\omega\rho_{m0}} \nabla \times \nabla \times \mathbf{f}_\omega^{ap} \tag{2.442}$$

which, when inserted into (2.441), produces the inhomogeneous vector Helmholtz equation for $\mathbf{v}_\omega$ in lossy fluids

$$\nabla^2 \mathbf{v}_\omega + k_\ell^2 \mathbf{v}_\omega = \frac{1}{\rho_{m0}} \left(\nabla s_\omega^{ap} + i\omega\rho_{m0}\kappa_\omega \mathbf{f}_\omega^{ap} + \frac{1}{i\omega} \nabla \times \nabla \times \mathbf{f}_\omega^{ap} \right). \tag{2.443}$$

As we might expect, the form of the wave equations (2.438) and (2.443) for the frequency-domain pressure and velocity in lossy fluids (with zero shear viscosity) remains the same as in lossless fluids. The complex propagation constant k_ℓ for lossy fluids merely replaces the real propagation constant k for lossless fluids, and, of course, $p_\omega(\mathbf{r})$ refers to the negative of the normal stress.

The complex compressibility κ_ω defined in (2.433)–(2.434) must obey the reality condition

$$\kappa_{-\omega} = \kappa_\omega^* \tag{2.444}$$

because both $\rho_{m\omega}(\mathbf{r})$ and $p_\omega(\mathbf{r})$ obey the reality condition. Thus, the real and imaginary parts of κ_ω satisfy the even and odd relations

$$\kappa'_{-\omega} = \kappa'_\omega, \quad \kappa''_{-\omega} = -\kappa''_\omega. \tag{2.445}$$

In lossy fluids the complex acoustic power theorem

$$\begin{aligned} -\frac{1}{2}\int_{V_a} \left(\mathbf{f}_\omega^{ap} \cdot \mathbf{v}_\omega^* + \frac{s_\omega^{ap*} p_\omega}{\rho_{m0}} \right) dV + 2i\omega \int_{V_a} \left(\frac{\kappa_\omega^*}{4}|p_\omega|^2 - \frac{\rho_{m0}}{4}|\mathbf{v}_\omega|^2 \right) dV \\ = -\frac{1}{2}\int_{V_a} \nabla \cdot (p_\omega \mathbf{v}_\omega^*) dV = -\frac{1}{2}\int_{S_a} \hat{\mathbf{n}} \cdot (p_\omega \mathbf{v}_\omega^*) dS \end{aligned} \tag{2.446}$$

can be obtained from the complex power theorem (2.391) in lossless fluids by inserting into (2.391) the total source densities given in Footnote 14 as

$$\mathbf{f}_\omega = \mathbf{f}_\omega^{ap} + \mathbf{f}_\omega^{in} = \mathbf{f}_\omega^{ap} \tag{2.447}$$

$$s_\omega = s_\omega^{ap} + s_\omega^{in} = s_\omega^{ap} + i\omega\rho_{m0}(\kappa_\omega - \kappa_0) p_\omega. \tag{2.448}$$

The time-average acoustic power supplied by the acoustic fields to the induced sources in V_a is given by the real part of the second volume integral in (2.446).

The time-average acoustic wave-force equation (2.388) also remains valid in lossy fluids if the sources $\mathbf{f}_\omega$ and s_ω in (2.447)–(2.448) are inserted into (2.388). Then $\mathbf{F}_{ac}^{ave}$ in (2.388) gives the time-average wave force delivered by the acoustic fields to the applied and induced sources in V_a.

At every point in the fluid, the time-average acoustic power per unit volume (denoted by $\mathcal{Q}_{ac}$) delivered by the acoustic fields to the fluid sources (applied plus induced) is determined by the real part of the divergence of $-p_\omega \mathbf{v}_\omega^*$; see Section 2.4.8. In particular, at every point in the fluid where there are no applied sources (only the induced mass-flow density defined in Footnote 14 and (2.448)), the second equation in (2.389) (or the real part of the first equation in (2.446)) shows that

$$\mathcal{Q}_{ac} = -\frac{1}{2}\text{Re}[\nabla \cdot (p_\omega \mathbf{v}_\omega^*)] = \frac{1}{2}\omega\kappa_\omega''|p_\omega|^2. \tag{2.449}$$

For passive fluids in which this acoustic power is converted irreversibly into heat by the frictional losses during expansion and compression, $\mathcal{Q}_{ac} \geq 0$. Thus, from (2.449)

$$\omega\kappa_\omega'' \geq 0. \tag{2.450}$$

Also, since κ_ω'' is an odd function of ω, no fluid can have expansive-friction loss that is independent of frequency for $-\infty < \omega < \infty$.

Time Domain. The constitutive relation corresponding to (2.433) for time-domain acoustic fields in linear, quiescent, isotropic, homogeneous, lossy fluids (with zero shear viscosity) can be found by applying the inverse Fourier transform and convolution theorem to (2.433) to get

$$\rho_m(\mathbf{r}, t) = \frac{\rho_{m0}}{2\pi}\int_{-\infty}^{t} \kappa(t - t')p(\mathbf{r}, t')dt' \tag{2.451}$$

with the time-domain compressibility function $\kappa(t)$ defined as the inverse Fourier transform of the frequency-domain compressibility, that is

$$\kappa(t) = \int_{-\infty}^{+\infty} \kappa_\omega e^{-i\omega t} d\omega. \tag{2.452}$$

Causality implies that the real time-domain compressibility $\kappa(t)$ is zero for $t < 0$, and thus the upper limit of integration in (2.451) is given as t instead of $+\infty$. For fields that are zero for $t \leq t_0$, the lower limit of integration in (2.451) can be changed from $-\infty$ to t_0. Also, since $\kappa(t) = 0$ for $t < 0$, the real and imaginary parts of κ_ω are related by the Kronig-Kramers dispersion relations (2.430)–(2.431) with $\sigma_0 = 0$ and κ_ω', κ_ω'', and κ_∞ replacing ϵ_ω', ϵ_ω'', and ϵ, respectively [116].

The inverse Fourier transform of (2.435) and (2.436) gives the linear time-domain continuity and momentum equations for lossy fluids that do not support shear stress, namely

$$\rho_{m0}\nabla \cdot \mathbf{v}(\mathbf{r}, t) + \frac{\rho_{m0}}{2\pi}\frac{\partial}{\partial t}\int_{-\infty}^{t} \kappa(t - t')p(\mathbf{r}, t')dt' = s^{ap}(\mathbf{r}, t) \tag{2.453}$$

$$\nabla p(\mathbf{r}, t) + \rho_{m0}\frac{\partial}{\partial t}\mathbf{v}(\mathbf{r}, t) = \mathbf{f}^{ap}(\mathbf{r}, t). \tag{2.454}$$

The form of the linear momentum equation (2.454) in lossy fluids free of shear stress remains the same as the linear momentum equation (2.198) in the corresponding lossless fluids. Recall, however, that for lossy fluids, $p(\mathbf{r}, t)$ in (2.451), (2.453), and (2.454) refers to the negative of the normal acoustic stress, as explained in Footnote 14.

3

Frequency-Domain Representations

Numerous expansions exist for acoustic and electromagnetic fields radiated by time-harmonic sources that are located in a bounded volume. For example, the fields can be expanded in terms of free-space Green's functions, plane waves, cylindrical waves, or spherical waves [1], [2], [67]. Other expansions that employ directionally localized basis fields have also been proposed [19], [117]–[120]. Each expansion has its advantages and disadvantages, and the choice of expansion depends on the particular problem under consideration.

In this book we are concerned with propagation of fields in homogeneous regions, and with planar near-field measurement techniques [8]. The expansions best suited for our applications employ free-space Green's functions or plane waves. In the present chapter, we derive the frequency-domain versions of these Green's function and plane-wave expansions in a form that is suitable for the subsequent translation into the time domain in Chapter 5. In particular, we derive basic acoustic and electromagnetic formulas that express the fields in a source free half space in terms of the fields in a scan plane, and we introduce the plane-wave notation that will be used throughout the book.

The problem under consideration is shown in Figure 3.1, where V is the finite volume that contains the sources of the acoustic or electromagnetic fields, S is a finite, closed surface that contains V and lies entirely in the source-free region outside V, and $z = z_0$ is the infinite scan plane. In addition to S being a "regular" surface, as explained in the introduction to Chapter 2, we shall assume here in Chapter 3 that S is a continuously curved (smooth) surface, in order to ensure that the normal $\hat{\mathbf{n}}$ to the surface is a continuous function on S. It is assumed that the acoustic frequency-domain fields satisfy the small-amplitude (linearized) acoustic equations of Section 2.4.1 (or Section 2.5.2 for lossy fluids) and that the frequency-domain electromagnetic fields satisfy the Maxwell equations of Section 2.3.2 (or Section 2.5.1 for lossy media). For the acoustic fields of Sections 3.1 and 3.2, the source-free region outside V is a linear, quiescent, homogeneous, isotropic, lossless fluid characterized by the constant mass density ρ_{m0} and constant compressibility κ_0, resulting in a constant wave speed $c = 1/\sqrt{\rho_{m0}\kappa_0}$. For the electromagnetic fields of Sections 3.1 and 3.2, the source-free region is lossless free space with constant permeability μ and constant permittivity ϵ, resulting in a constant wave speed of light $c = 1/\sqrt{\mu\epsilon}$. Section 3.3 deals with lossy acoustic fluids and lossy electromagnetic media.

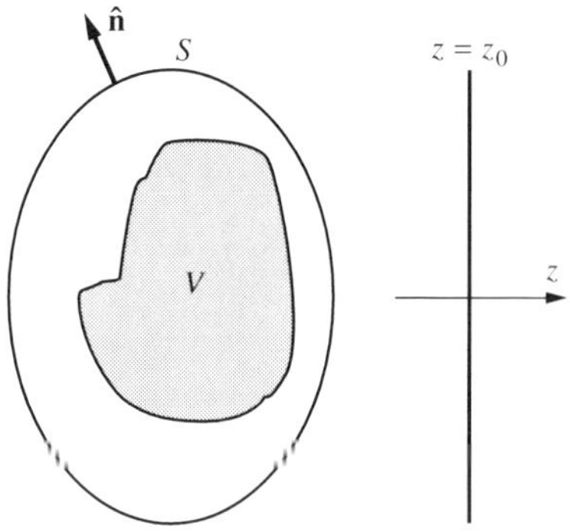

Figure 3.1 The source region V, the surrounding surface S, and the plane $z = z_0$.

The chapter is organized as follows. In Section 3.1 we derive Green's function representations for the fields in the half space $z > z_0$ in terms of their values in the scan plane $z = z_0$. We express far fields in terms of the volume sources, the Huygens' sources on the closed surface S, and the fields in the scan plane $z = z_0$. The volume-source expressions for the far fields are used to relate the low-frequency behavior of the far fields to the low-frequency behavior of the sources.

In Section 3.2 we derive plane-wave spectrum formulas that express the fields in a half space in terms of a spectrum of propagating and evanescent plane waves. These formulas are first derived heuristically and then verified rigorously in Appendix D. For the acoustic point-source field, explicit expressions are derived for the separate contributions from the propagating and evanescent plane waves. For an arbitrary acoustic or electromagnetic source region, the spectrum is expressed in terms of the volume sources, the Huygens' sources on the closed surface S, and the fields in the plane $z = z_0$. The singularities and the asymptotic behavior of the spectrum are determined, and it is shown that the spectrum, the far-field pattern, and the fields outside the source region are analytic functions of their arguments in certain complex regions. Asymptotic analysis is used to derive far-field expressions from the plane-wave spectrum representations, and the separate far-field contributions from the propagating and evanescent plane waves are determined.

The fields everywhere in the half space $z \geq z_0$ are then written in terms of the far-field pattern evaluated at real and complex angles of observation. Homogeneous fields are introduced that satisfy the homogeneous Helmholtz equation throughout space, and it is shown that the homogeneous fields can be expressed throughout space in terms of the far-field pattern evaluated at real angles of observation only. Equating alternative expressions for the homogeneous fields reveals the relationships that form the basis of many inverse scattering techniques. It is also shown how a minimum source region (the minimum region in which the sources can reside) can be determined from the asymptotic behavior of the analytic continuation of the far-field pattern. Finally, the acoustic and electromagnetic power and radiation force transmitted through the plane $z = z_1 > z_0$ are expressed in terms of the spectra and the far-field patterns. In Section 3.3 we explain how the formulas derived in this chapter change when the medium outside the finite source region V is lossy.

Throughout this chapter, $e^{-i\omega t}$ time dependence is suppressed in all the time-harmonic equations, and *the frequency ω is assumed positive*. These frequency-domain field representations will be derived in detail for the case where the source region V is located to the left of the scan plane $z = z_0$, as in Figure 3.1. The necessary modifications are then briefly stated for the sources located to the right of the scan plane $z = z_0$.

3.1 GREEN'S FUNCTION REPRESENTATIONS

We shall obtain Green's function representations that express the acoustic and electromagnetic fields in the source-free half space $z > z_0$ in terms of their values in the plane $z = z_0$. First consider the acoustic field determined by its pressure $p_\omega(\mathbf{r})$ and its velocity $\mathbf{v}_\omega(\mathbf{r})$. The finite source region V that generates the acoustic field is located in the half space $z < z_0$, as shown in Figure 3.1. From the second-order differential equation (2.360) for the acoustic pressure, it follows that this pressure satisfies the homogeneous Helmholtz equation in the source-free half space, that is

$$\nabla^2 p_\omega(\mathbf{r}) + k^2 p_\omega(\mathbf{r}) = 0, \quad z \geq z_0 \tag{3.1}$$

where $k = \omega/c$ and $\mathbf{r} = x\hat{\mathbf{x}} + y\hat{\mathbf{y}} + z\hat{\mathbf{z}}$. The Dirichlet Green's function $G_\omega^D(\mathbf{r}, \mathbf{r}')$ for the half space $z \geq z_0$ satisfies the equation

$$\nabla^2 G_\omega^D(\mathbf{r}, \mathbf{r}') + k^2 G_\omega^D(\mathbf{r}, \mathbf{r}') = -\delta(\mathbf{r} - \mathbf{r}'), \quad z > z_0 \tag{3.2}$$

and the boundary condition $G_\omega^D(\mathbf{r}, \mathbf{r}') = 0$ for $z = z_0$. This Dirichlet Green's function can be expressed in terms of the free-space Green's function

$$G_\omega(\mathbf{r}, \mathbf{r}') = \frac{e^{ik|\mathbf{r}-\mathbf{r}'|}}{4\pi|\mathbf{r} - \mathbf{r}'|} \tag{3.3}$$

as

$$G_\omega^D(\mathbf{r}, \mathbf{r}') = G_\omega(\mathbf{r}, \mathbf{r}') - G_\omega(\mathbf{r}, \mathbf{r}_i'), \quad z \geq z_0 \tag{3.4}$$

where $\mathbf{r}_i' = x'\hat{\mathbf{x}} + y'\hat{\mathbf{y}} + (2z_0 - z')\hat{\mathbf{z}}$ is the image of $\mathbf{r}'$ with respect to the plane $z = z_0$. As shown in Figure 3.2, we let the surface $C_0(R_0)$ be the circular disk defined by $x^2 + y^2 \leq R_0^2$ at $z = z_0$, and $S_0(R_0)$ be the spherical surface defined by $x^2 + y^2 + (z - z_0)^2 = R_0^2$ for $z \geq z_0$. Then we apply Green's second identity to the hemisphere $V_0(R_0)$ bounded by the closed surface $S_0(R_0) \cup C_0(R_0)$, and find that

$$-p_\omega(\mathbf{r}') = \int_{S_0(R_0)\cup C_0(R_0)} \left[p_\omega(\mathbf{r}) \frac{\partial}{\partial n} G_\omega^D(\mathbf{r}, \mathbf{r}') - G_\omega^D(\mathbf{r}, \mathbf{r}') \frac{\partial}{\partial n} p_\omega(\mathbf{r}) \right] dS \tag{3.5}$$

where $\mathbf{r}'$ is inside the surface $S_0(R_0) \cup C_0(R_0)$, and $\hat{\mathbf{n}}$ is the outward normal of this surface. On the disk $C_0(R_0)$, we have $\partial G_\omega^D/\partial n = -2\partial G_\omega/\partial z$ and $G_\omega^D = 0$, and the Sommerfeld radiation conditions for the acoustic pressure (2.376)–(2.377) show that the expression in the square brackets in (3.5) goes to zero faster than R_0^{-2} on the spherical surface $S_0(R_0)$. Consequently, the contribution to the integral in (3.5) from $S_0(R_0)$ vanishes as $R_0 \to +\infty$, and (3.5) reveals

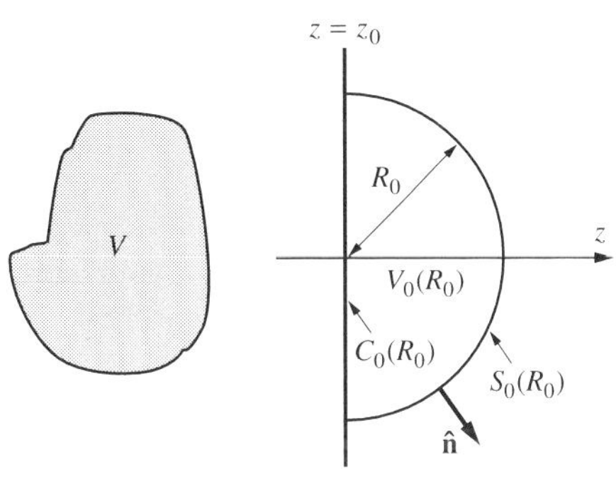

Figure 3.2 The hemisphere $V_0(R_0)$ bounded by the spherical surface $S_0(R_0)$ and the circular disk $C_0(R_0)$.

that

$$
\begin{aligned}
p_\omega(\mathbf{r}) &= 2 \int_{-\infty}^{+\infty}\int_{-\infty}^{+\infty} p_\omega(\mathbf{r}_0) \frac{\partial}{\partial z_0} G_\omega(\mathbf{r}, \mathbf{r}_0) dx_0 dy_0, \quad z > z_0 \\
&= -2 \frac{\partial}{\partial z} \int_{-\infty}^{+\infty}\int_{-\infty}^{+\infty} p_\omega(\mathbf{r}_0) G_\omega(\mathbf{r}, \mathbf{r}_0) dx_0 dy_0, \quad z > z_0
\end{aligned}
\tag{3.6}
$$

where we have replaced $\mathbf{r}'$ with $\mathbf{r}$ and introduced $\mathbf{r}_0 = x_0\hat{\mathbf{x}} + y_0\hat{\mathbf{y}} + z_0\hat{\mathbf{z}}$. Taking the z_0 derivative of the Green's function, we may write (3.6) as

$$
p_\omega(\mathbf{r}) = \frac{z_0 - z}{2\pi} \int_{-\infty}^{+\infty}\int_{-\infty}^{+\infty} p_\omega(\mathbf{r}_0) \left[\frac{ik}{R^2} - \frac{1}{R^3} \right] e^{ikR} \, dx_0 dy_0, \quad z > z_0 \tag{3.7}
$$

with $R = |\mathbf{r} - \mathbf{r}_0|$. The formula (3.7) is the Green's function representation for the pressure in the half space $z > z_0$ in terms of the pressure in the plane $z = z_0$. It was first derived by Rayleigh [121] and will prove valuable in Chapter 5 for obtaining the corresponding time-domain representation. To get an expression for the acoustic velocity in terms of the pressure on $z = z_0$, we substitute (3.7) into the expression $\mathbf{v}_\omega = \nabla p_\omega / (i\omega\rho_{m0})$ obtained from (2.357) and find that

$$
\begin{aligned}
\mathbf{v}_\omega(\mathbf{r}) = {} & \frac{z_0 - z}{2\pi i \omega \rho_{m0}} \int_{-\infty}^{+\infty}\int_{-\infty}^{+\infty} p_\omega(\mathbf{r}_0) \left[\frac{3}{R^5} - \frac{3ik}{R^4} - \frac{k^2}{R^3} \right] e^{ikR} \, \mathbf{R} \, dx_0 dy_0 \\
& - \frac{\hat{\mathbf{z}}}{2\pi i \omega \rho_{m0}} \int_{-\infty}^{+\infty}\int_{-\infty}^{+\infty} p_\omega(\mathbf{r}_0) \left[\frac{ik}{R^2} - \frac{1}{R^3} \right] e^{ikR} \, dx_0 dy_0, \quad z > z_0
\end{aligned}
\tag{3.8}
$$

where $\mathbf{R} = \mathbf{r} - \mathbf{r}_0$ and ρ_{m0} is the constant mass density of the undisturbed fluid.

Now consider the electromagnetic field. Because each of the rectangular components of the electric field $\mathbf{E}_\omega$ satisfies the scalar Helmholtz equation (3.1) and the scalar radiation conditions (2.376)–(2.377), the scalar formula (3.7) immediately gives

$$
\begin{aligned}
\mathbf{E}_\omega(\mathbf{r}) &= -2 \frac{\partial}{\partial z} \int_{-\infty}^{+\infty}\int_{-\infty}^{+\infty} \mathbf{E}_\omega(\mathbf{r}_0) G_\omega(\mathbf{r}, \mathbf{r}_0) dx_0 dy_0, \quad z > z_0 \\
&= \frac{z_0 - z}{2\pi} \int_{-\infty}^{+\infty}\int_{-\infty}^{+\infty} \mathbf{E}_\omega(\mathbf{r}_0) \left[\frac{ik}{R^2} - \frac{1}{R^3} \right] e^{ikR} \, dx_0 dy_0, \quad z > z_0
\end{aligned}
\tag{3.9}
$$

which expresses the electric field in the half space $z > z_0$ in terms of all three rectangular components of the electric field in the plane $z = z_0$.

Next we will derive a formula for the electric field in the half space $z > z_0$ in terms of only the tangential components of the electric field in the plane $z = z_0$. To do this, we apply the dyadic version of Green's second identity [83, p. 509, eq. 56] to the hemisphere $V_0(R_0)$, which is bounded by the closed surface $S_0(R_0) \cup C_0(R_0)$ shown in Figure 3.2, and find that

$$
\begin{aligned}
& \int_{V_0(R_0)} \left[(\nabla \times \nabla \times \mathbf{E}_\omega) \cdot \bar{\mathbf{G}}^D_\omega - \mathbf{E}_\omega \cdot \left(\nabla \times \nabla \times \bar{\mathbf{G}}^D_\omega \right) \right] dV \\
& = - \int_{S_0(R_0) \cup C_0(R_0)} \left[\mathbf{E}_\omega \cdot \left(\hat{\mathbf{n}} \times \nabla \times \bar{\mathbf{G}}^D_\omega \right) + (\nabla \times \mathbf{E}_\omega) \cdot \left(\hat{\mathbf{n}} \times \bar{\mathbf{G}}^D_\omega \right) \right] dS
\end{aligned}
\tag{3.10}
$$

where the principal volume is omitted since $V_0(R_0)$ is source-free, and

$$\bar{\mathbf{G}}_\omega^D(\mathbf{r},\mathbf{r}') = \left[\bar{\mathbf{I}} - \frac{1}{k^2}\nabla\nabla'\right]\left(G_\omega(\mathbf{r},\mathbf{r}') - G_\omega(\mathbf{r},\mathbf{r}_i')\right) + 2\hat{\mathbf{z}}\hat{\mathbf{z}}G_\omega(\mathbf{r},\mathbf{r}_i') \tag{3.11}$$

is the Dirichlet dyadic Green's function for the half space $z \geq z_0$ satisfying

$$\nabla\times\nabla\times\bar{\mathbf{G}}_\omega^D(\mathbf{r},\mathbf{r}') - k^2\bar{\mathbf{G}}_\omega^D(\mathbf{r},\mathbf{r}') = \bar{\mathbf{I}}\delta(\mathbf{r}-\mathbf{r}'), \quad z > z_0 \tag{3.12}$$

and the boundary condition $\hat{\mathbf{z}}\times\bar{\mathbf{G}}_\omega^D = 0$ in the plane $z = z_0$ [122, p. 94]. Here $\bar{\mathbf{I}} = \hat{\mathbf{x}}\hat{\mathbf{x}}+\hat{\mathbf{y}}\hat{\mathbf{y}}+\hat{\mathbf{z}}\hat{\mathbf{z}}$ is the identity dyadic. When $\mathbf{r}'$ is in the region $V_0(R_0)$, the homogeneous vector-wave equation (2.296) satisfied by the electric field in the source-free half space

$$\nabla\times\nabla\times\mathbf{E}_\omega(\mathbf{r}) - k^2\mathbf{E}_\omega(\mathbf{r}) = 0, \quad z \geq z_0 \tag{3.13}$$

together with (3.12) show that the left-hand side of (3.10) is equal to $-\mathbf{E}_\omega(\mathbf{r}')$. Inserting the Dirichlet boundary condition $\hat{\mathbf{z}}\times\bar{\mathbf{G}}_\omega^D = 0$ for $z = z_0$ and the identity

$$\mathbf{E}_\omega(\mathbf{r})\cdot\left(\hat{\mathbf{z}}\times\nabla\times\bar{\mathbf{G}}_\omega^D(\mathbf{r},\mathbf{r}')\right)\Big|_{z=z_0} = -2\nabla' G_\omega(\mathbf{r},\mathbf{r}')\times\left(\hat{\mathbf{z}}\times\mathbf{E}_\omega(\mathbf{r})\right)\Big|_{z=z_0} \tag{3.14}$$

into (3.10), we find

$$\begin{aligned}-\mathbf{E}_\omega(\mathbf{r}') = &-2\int\limits_{C_0(R_0)} \nabla' G_\omega(\mathbf{r},\mathbf{r}')\times(\hat{\mathbf{z}}\times\mathbf{E}_\omega(\mathbf{r}))dS \\ &-\int\limits_{S_0(R_0)}\left[\mathbf{E}_\omega(\mathbf{r})\cdot\left(\hat{\mathbf{n}}\times\nabla\times\bar{\mathbf{G}}_\omega^D(\mathbf{r},\mathbf{r}')\right)\right. \\ &\left.+(\nabla\times\mathbf{E}_\omega(\mathbf{r}))\cdot\left(\hat{\mathbf{n}}\times\bar{\mathbf{G}}_\omega^D(\mathbf{r},\mathbf{r}')\right)\right]dS.\end{aligned} \tag{3.15}$$

Using the result in [83, p. 507, eq. 2], we may write the integrand in the last integral of (3.15) as $\mathbf{E}_\omega\cdot\left(\hat{\mathbf{n}}\times\nabla\times\bar{\mathbf{G}}_\omega^D\right) - \hat{\mathbf{n}}\times(\nabla\times\mathbf{E}_\omega)\cdot\bar{\mathbf{G}}_\omega^D$, which, according to the radiation conditions for the dyadic Green's function (see [66, p. 63]) and the radiation conditions for the electric field (see (2.330)–(2.331)), goes to zero faster than R_0^{-2} as $R_0 \to \infty$. Thus, the contribution from the last integral in (3.15) vanishes as $R_0 \to \infty$, proving that

$$\begin{aligned}\mathbf{E}_\omega(\mathbf{r}) &= 2\int\limits_{-\infty}^{+\infty}\int\limits_{-\infty}^{+\infty}\nabla G_\omega(\mathbf{r},\mathbf{r}_0)\times\left(\hat{\mathbf{z}}\times\mathbf{E}_\omega(\mathbf{r}_0)\right)dx_0dy_0, \quad z > z_0 \\ &= 2\nabla\times\int\limits_{-\infty}^{+\infty}\int\limits_{-\infty}^{+\infty}\hat{\mathbf{z}}\times\mathbf{E}_\omega(\mathbf{r}_0)G_\omega(\mathbf{r},\mathbf{r}_0)dx_0dy_0, \quad z > z_0\end{aligned} \tag{3.16}$$

where we have replaced $\mathbf{r}'$ with $\mathbf{r}$ and introduced $\mathbf{r}_0 = x_0\hat{\mathbf{x}} + y_0\hat{\mathbf{y}} + z_0\hat{\mathbf{z}}$. Upon calculating ∇G_ω, this equation becomes

$$\mathbf{E}_\omega(\mathbf{r}) = \frac{1}{2\pi}\int\limits_{-\infty}^{+\infty}\int\limits_{-\infty}^{+\infty}e^{ikR}\left[\frac{ik}{R^2} - \frac{1}{R^3}\right]\mathbf{R}\times\left(\hat{\mathbf{z}}\times\mathbf{E}_\omega(\mathbf{r}_0)\right)dx_0dy_0, \quad z > z_0 \tag{3.17}$$

where $R = |\mathbf{r}-\mathbf{r}_0|$ and $\mathbf{R} = \mathbf{r}-\mathbf{r}_0$. Equation (3.17) is the Green's function representation for the electric field in the half space $z > z_0$ in terms of the tangential electric field in the plane $z = z_0$. It was first derived by Smythe [123] using a double current sheet in the plane $z = z_0$. The expression (3.17) also holds for the magnetic field, provided $\mathbf{E}_\omega$ is replaced with $\mathbf{H}_\omega$.

Let us finally express the magnetic field $\mathbf{H}_\omega$ in the half space $z > z_0$ in terms of the x and y components of the electric field in the plane $z = z_0$. Since the induction law (2.287) shows that $\mathbf{H}_\omega = -(i/(\omega\mu))\nabla\times\mathbf{E}_\omega$, the magnetic field can be found by taking the curl of

(3.16) to get

$$\mathbf{H}_\omega(\mathbf{r}) = -\frac{2i}{\omega\mu} \int_{-\infty}^{+\infty}\int_{-\infty}^{+\infty} \left(\hat{\mathbf{z}} \times \mathbf{E}_\omega(\mathbf{r}_0)\right) \cdot \left[\left(\bar{\mathbf{I}}k^2 + \nabla\nabla\right) G_\omega(\mathbf{r}, \mathbf{r}_0)\right] dx_0 dy_0, \quad z > z_0 \tag{3.18}$$

or

$$\mathbf{H}_\omega(\mathbf{r}) = -2i\omega\epsilon \int_{-\infty}^{+\infty}\int_{-\infty}^{+\infty} \left(\hat{\mathbf{z}} \times \mathbf{E}_\omega(\mathbf{r}_0)\right) \cdot \bar{\mathbf{G}}_\omega(\mathbf{r}, \mathbf{r}_0) dx_0 dy_0, \quad z > z_0 \tag{3.19}$$

where $\bar{\mathbf{G}}_\omega(\mathbf{r}, \mathbf{r}_0)$ is the free-space electric dyadic Green's function in (2.321). With the help of (2.321), the expression (3.19) can be written in expanded form as

$$\mathbf{H}_\omega(\mathbf{r}) = \frac{1}{2\pi}\sqrt{\frac{\epsilon}{\mu}} \int_{-\infty}^{+\infty}\int_{-\infty}^{+\infty} \left(\hat{\mathbf{z}} \times \mathbf{E}_\omega(\mathbf{r}_0)\right) \cdot \left[\left(\hat{\mathbf{R}}\hat{\mathbf{R}} - \bar{\mathbf{I}}\right) ikR - \left(3\hat{\mathbf{R}}\hat{\mathbf{R}} - \bar{\mathbf{I}}\right)\left(1 - \frac{1}{ikR}\right)\right] \frac{e^{ikR}}{R^2}\, dx_0 dy_0, \quad z > z_0 \tag{3.20}$$

which has the same $\hat{\mathbf{n}}' \times \mathbf{E}_\omega$ dependence as (2.345). Duality shows that an expression for the electric field in $z > z_0$ in terms of the tangential magnetic field on $z = z_0$ can be obtained from (3.20) by replacing $\mathbf{E}_\omega$ with $\mathbf{H}_\omega$ and $\mathbf{H}_\omega$ with $-(\epsilon/\mu)\mathbf{E}_\omega$. The resulting expression for $\mathbf{H}_\omega(\mathbf{r})$ has the same $\hat{\mathbf{n}}' \times \mathbf{H}_\omega$ dependence as (2.346).

When the sources are located in the half space $z > z_0$, the acoustic pressure and velocity in $z < z_0$ are given by the negative of (3.7) and (3.8), and the electromagnetic fields in $z < z_0$ are given by the negative of (3.17) and (3.20).

3.1.1 Far-Field Expressions

We shall derive alternative expressions for the acoustic and electromagnetic far fields in terms of the volume sources, the Huygens' sources on S, and the fields in the plane $z = z_0$. The far-field expressions in terms of the acoustic and electromagnetic volume sources, $q_\omega(\mathbf{r})$ and $\mathbf{J}_\omega(\mathbf{r})$, are easily obtained from the formulas (2.379)–(2.380) and (2.335)–(2.336) of Chapter 2, which show that as $r \to \infty$, the acoustic and electromagnetic far fields are

$$p_\omega(\mathbf{r}) = \frac{e^{ikr}}{r}\frac{1}{4\pi}\int_V q_\omega(\mathbf{r}')e^{-ik\hat{\mathbf{r}}\cdot\mathbf{r}'}dV' + O\left(\frac{1}{r^2}\right) \tag{3.21}$$

$$\begin{aligned} \mathbf{v}_\omega(\mathbf{r}) &= \frac{\hat{\mathbf{r}}\, p_\omega(\mathbf{r})}{c\rho_{m0}} + O\left(\frac{1}{r^2}\right) \\ &= \frac{e^{ikr}}{r}\frac{\hat{\mathbf{r}}}{4\pi c\rho_{m0}}\int_V q_\omega(\mathbf{r}')e^{-ik\hat{\mathbf{r}}\cdot\mathbf{r}'}dV' + O\left(\frac{1}{r^2}\right) \end{aligned} \tag{3.22}$$

$$\mathbf{E}_\omega(\mathbf{r}) = -\frac{e^{ikr}}{r}\sqrt{\frac{\mu}{\epsilon}}\frac{ik}{4\pi}\hat{\mathbf{r}} \times \left(\hat{\mathbf{r}} \times \int_V \mathbf{J}_\omega(\mathbf{r}')e^{-ik\hat{\mathbf{r}}\cdot\mathbf{r}'}dV'\right) + O\left(\frac{1}{r^2}\right) \tag{3.23}$$

and

$$\mathbf{H}_\omega(\mathbf{r}) = \frac{e^{ikr}}{r}\frac{ik}{4\pi}\hat{\mathbf{r}} \times \int_V \mathbf{J}_\omega(\mathbf{r}')e^{-ik\hat{\mathbf{r}}\cdot\mathbf{r}'}dV' + O\left(\frac{1}{r^2}\right). \tag{3.24}$$

The electric and magnetic far fields satisfy the relations

$$\mathbf{H}_\omega(\mathbf{r}) \sim \sqrt{\frac{\epsilon}{\mu}}\,\hat{\mathbf{r}} \times \mathbf{E}_\omega(\mathbf{r}), \qquad \mathbf{E}_\omega(\mathbf{r}) \sim -\sqrt{\frac{\mu}{\epsilon}}\,\hat{\mathbf{r}} \times \mathbf{H}_\omega(\mathbf{r}), \quad r \to \infty. \tag{3.25}$$

The interchange of the volume integral with the limit as $r \to \infty$ required to derive (3.21)–(3.22), and with the curl to obtain (3.23)–(3.24) is allowed as shown in Hobson [71, vol. 2, pp. 324 and 355], under the assumption that the source functions q_ω and $\mathbf{J}_\omega$ are absolutely integrable over the finite source volume V.

We may also express the acoustic far fields in terms of the Huygens' sources $p_\omega(\mathbf{r})$ and $\hat{\mathbf{n}} \cdot \mathbf{v}_\omega(\mathbf{r})$ on the closed surface S in Figure 3.1. From (2.384)–(2.385) of Chapter 2, we find immediately that

$$p_\omega(\mathbf{r}) = -\frac{e^{ikr}}{r}\frac{1}{4\pi}\int_S \left[ik\hat{\mathbf{r}}\cdot\hat{\mathbf{n}}' p_\omega(\mathbf{r}') + i\omega\rho_{m0}\hat{\mathbf{n}}'\cdot\mathbf{v}_\omega(\mathbf{r}')\right] e^{-ik\hat{\mathbf{r}}\cdot\mathbf{r}'} dS' + O\left(\frac{1}{r^2}\right) \tag{3.26}$$

and

$$\begin{aligned}\mathbf{v}_\omega(\mathbf{r}) = -\frac{e^{ikr}}{r}\frac{\hat{\mathbf{r}}}{4\pi c\rho_{m0}}\int_S \Big[&ik\hat{\mathbf{r}}\cdot\hat{\mathbf{n}}' p_\omega(\mathbf{r}') \\ &+ i\omega\rho_{m0}\,\hat{\mathbf{n}}'\cdot\mathbf{v}_\omega(\mathbf{r}')\Big] e^{-ik\hat{\mathbf{r}}\cdot\mathbf{r}'} dS' + O\left(\frac{1}{r^2}\right).\end{aligned} \tag{3.27}$$

The normal component of the velocity is related to the normal derivative of the pressure through $\partial p_\omega/\partial n = i\omega\rho_{m0}\,\hat{\mathbf{n}}\cdot\mathbf{v}_\omega$. Equations (2.345)–(2.346) show that the electromagnetic far fields can be expressed in terms of the Huygens' sources $\hat{\mathbf{n}} \times \mathbf{E}_\omega(\mathbf{r})$ and $\hat{\mathbf{n}} \times \mathbf{H}_\omega(\mathbf{r})$ on the closed surface S as

$$\begin{aligned}\mathbf{E}_\omega(\mathbf{r}) = -\frac{e^{ikr}}{r}\frac{ik}{4\pi}\,\hat{\mathbf{r}} \times \int_S \Big[&-\hat{\mathbf{n}}' \times \mathbf{E}_\omega(\mathbf{r}') \\ &+ \sqrt{\frac{\mu}{\epsilon}}\,\hat{\mathbf{r}} \times \left(\hat{\mathbf{n}}' \times \mathbf{H}_\omega(\mathbf{r}')\right)\Big] e^{-ik\hat{\mathbf{r}}\cdot\mathbf{r}'} dS' + O\left(\frac{1}{r^2}\right)\end{aligned} \tag{3.28}$$

and

$$\begin{aligned}\mathbf{H}_\omega(\mathbf{r}) = \frac{e^{ikr}}{r}\frac{ik}{4\pi}\hat{\mathbf{r}} \times \int_S \Big[&\sqrt{\frac{\epsilon}{\mu}}\,\hat{\mathbf{r}} \times \left(\hat{\mathbf{n}}' \times \mathbf{E}_\omega(\mathbf{r}')\right) \\ &+ \hat{\mathbf{n}}' \times \mathbf{H}_\omega(\mathbf{r}')\Big] e^{-ik\hat{\mathbf{r}}\cdot\mathbf{r}'} dS' + O\left(\frac{1}{r^2}\right)\end{aligned} \tag{3.29}$$

where $\hat{\mathbf{n}}$ is the outward normal to S and the observation point $\mathbf{r}$ is outside S. The interchange of the surface integral with the limit as $r \to \infty$ and with the curl required to derive (3.26)–(3.29) is allowed as shown in Hobson [71, vol. 2, pp. 324 and 355] under the assumption that the fields on the surface S are absolutely integrable. This assumption is always satisfied because the fields satisfy the homogeneous Helmholtz equations in the region outside the source region V. (We are assuming throughout that S lies a finite [nonzero] distance outside the sources.)

We have confirmed that the acoustic and electromagnetic frequency-domain far fields always decay as $1/r$ or faster as $r \to \infty$. Moreover, the formulas (3.21)–(3.29) show that one can introduce far-field patterns $\mathcal{F}_\omega(\theta, \phi)$, $\boldsymbol{\mathcal{F}}_\omega(\theta, \phi)$, and $\boldsymbol{\mathcal{F}}_\omega^H(\theta, \phi)$ defined by

$$\mathcal{F}_\omega(\theta, \phi) = \lim_{r\to\infty} re^{-ikr}\, p_\omega(\mathbf{r}) \tag{3.30}$$

and

$$\boldsymbol{\mathcal{F}}_\omega(\theta,\phi) = \lim_{r\to\infty} re^{-ikr}\,\mathbf{E}_\omega(\mathbf{r}), \quad \boldsymbol{\mathcal{F}}^H_\omega(\theta,\phi) = \lim_{r\to\infty} re^{-ikr}\,\mathbf{H}_\omega(\mathbf{r}) \tag{3.31}$$

so that the far fields can be written as

$$p_\omega(\mathbf{r}) \sim \mathcal{F}_\omega(\theta,\phi)\frac{e^{ikr}}{r}, \quad \mathbf{v}_\omega(\mathbf{r}) \sim \hat{\mathbf{r}}\,\frac{\mathcal{F}_\omega(\theta,\phi)\,e^{ikr}}{c\rho_{m0}\,r} \tag{3.32}$$

and

$$\mathbf{E}_\omega(\mathbf{r}) \sim \boldsymbol{\mathcal{F}}_\omega(\theta,\phi)\frac{e^{ikr}}{r}, \quad \mathbf{H}_\omega(\mathbf{r}) \sim \boldsymbol{\mathcal{F}}^H_\omega(\theta,\phi)\frac{e^{ikr}}{r} \tag{3.33}$$

with

$$\boldsymbol{\mathcal{F}}^H_\omega(\theta,\phi) = \sqrt{\frac{\epsilon}{\mu}}\,\hat{\mathbf{r}}\times\boldsymbol{\mathcal{F}}_\omega(\theta,\phi), \quad \boldsymbol{\mathcal{F}}_\omega(\theta,\phi) = -\sqrt{\frac{\mu}{\epsilon}}\,\hat{\mathbf{r}}\times\boldsymbol{\mathcal{F}}^H_\omega(\theta,\phi) \tag{3.34}$$

and $\hat{\mathbf{r}} = \hat{\mathbf{x}}\cos\phi\sin\theta + \hat{\mathbf{y}}\sin\phi\sin\theta + \hat{\mathbf{z}}\cos\theta$. Here (r,θ,ϕ) are the usual spherical coordinates defined by $x = r\cos\phi\sin\theta$, $y = r\sin\phi\sin\theta$, and $z = r\cos\theta$. Equations (3.21)–(3.24) show that the far-field patterns can be expressed in terms of the volume sources as

$$\mathcal{F}_\omega(\theta,\phi) = \frac{1}{4\pi}\int_V q_\omega(\mathbf{r}')e^{-ik(x'\cos\phi\sin\theta + y'\sin\phi\sin\theta + z'\cos\theta)}dV' \tag{3.35}$$

$$\boldsymbol{\mathcal{F}}_\omega(\theta,\phi) = -\sqrt{\frac{\mu}{\epsilon}}\,\frac{ik}{4\pi}\,\hat{\mathbf{r}}\times\left(\hat{\mathbf{r}}\times\int_V \mathbf{J}_\omega(\mathbf{r}')e^{-ik(x'\cos\phi\sin\theta + y'\sin\phi\sin\theta + z'\cos\theta)}dV'\right) \tag{3.36}$$

and

$$\boldsymbol{\mathcal{F}}^H_\omega(\theta,\phi) = \frac{ik}{4\pi}\,\hat{\mathbf{r}}\times\int_V \mathbf{J}_\omega(\mathbf{r}')e^{-ik(x'\cos\phi\sin\theta + y'\sin\phi\sin\theta + z'\cos\theta)}dV'. \tag{3.37}$$

Using the expressions (3.26)–(3.29), we can also write the far-field patterns in terms of the Huygens' sources on S, as follows

$$\mathcal{F}_\omega(\theta,\phi) = -\frac{1}{4\pi}\int_S \left[ik\hat{\mathbf{r}}\cdot\hat{\mathbf{n}}'p_\omega(\mathbf{r}') + i\omega\rho_{m0}\,\hat{\mathbf{n}}'\cdot\mathbf{v}_\omega(\mathbf{r}')\right]e^{-ik\hat{\mathbf{r}}\cdot\mathbf{r}'}dS' \tag{3.38}$$

$$\boldsymbol{\mathcal{F}}_\omega(\theta,\phi) = -\frac{ik}{4\pi}\,\hat{\mathbf{r}}\times\int_S\left[-\hat{\mathbf{n}}'\times\mathbf{E}_\omega(\mathbf{r}') + \sqrt{\frac{\mu}{\epsilon}}\,\hat{\mathbf{r}}\times\left(\hat{\mathbf{n}}'\times\mathbf{H}_\omega(\mathbf{r}')\right)\right]e^{-ik\hat{\mathbf{r}}\cdot\mathbf{r}'}dS' \tag{3.39}$$

and

$$\boldsymbol{\mathcal{F}}^H_\omega(\theta,\phi) = \frac{ik}{4\pi}\,\hat{\mathbf{r}}\times\int_S\left[\sqrt{\frac{\epsilon}{\mu}}\,\hat{\mathbf{r}}\times\left(\hat{\mathbf{n}}'\times\mathbf{E}_\omega(\mathbf{r}')\right) + \hat{\mathbf{n}}'\times\mathbf{H}_\omega(\mathbf{r}')\right]e^{-ik\hat{\mathbf{r}}\cdot\mathbf{r}'}dS'. \tag{3.40}$$

Next, we shall derive expressions for the far fields in the half space $z > z_0$ in terms of the fields in the plane $z = z_0$. Start with the acoustic pressure p_ω given by (3.7). Inserting the far-field approximations $R = r - \hat{\mathbf{r}}\cdot\mathbf{r}_0 + O(r^{-1})$ and $(z_0 - z)/R = -\cos\theta + O(r^{-1})$ into the Green's function representation (3.7), we find that the far-field pressure is given by

$$p_\omega(\mathbf{r}) = -\frac{e^{ikr}}{r}\,\frac{ik\cos\theta}{2\pi}\int_{-\infty}^{+\infty}\int_{-\infty}^{+\infty} p_\omega(\mathbf{r}_0)e^{-ik\hat{\mathbf{r}}\cdot\mathbf{r}_0}dx_0dy_0 + O\left(\frac{1}{r^2}\right), \quad z > z_0 \tag{3.41}$$

which proves that the acoustic far-field pattern is

$$\mathcal{F}_\omega(\theta,\phi) = -\frac{ik\cos\theta}{2\pi}\int_{-\infty}^{+\infty}\int_{-\infty}^{+\infty} p_\omega(\mathbf{r}_0)e^{-ik\hat{\mathbf{r}}\cdot\mathbf{r}_0}dx_0dy_0, \quad z > z_0. \tag{3.42}$$

Inserting $\mathbf{R} = \mathbf{r} + O(1)$ along with the asymptotic relations above (3.41) into the Green's function representations (3.9) and (3.17) for the electric field, we find the following far-field formulas for the electric field

$$\begin{aligned}\mathbf{E}_\omega(\mathbf{r}) &= \frac{e^{ikr}}{r}\frac{ik\cos\theta}{2\pi}\int_{-\infty}^{+\infty}\int_{-\infty}^{+\infty}\mathbf{E}_\omega(\mathbf{r}_0)e^{-ik\hat{\mathbf{r}}\cdot\mathbf{r}_0}dx_0dy_0 + O\left(\frac{1}{r^2}\right), \quad z > z_0 \\ &= \frac{e^{ikr}}{r}\frac{ik}{2\pi}\hat{\mathbf{r}}\times\int_{-\infty}^{+\infty}\int_{-\infty}^{+\infty}\hat{\mathbf{z}}\times\mathbf{E}_\omega(\mathbf{r}_0)e^{-ik\hat{\mathbf{r}}\cdot\mathbf{r}_0}dx_0dy_0 + O\left(\frac{1}{r^2}\right), \quad z > z_0.\end{aligned} \tag{3.43}$$

The electric far-field pattern can then be expressed in terms of the tangential electric field in the plane $z = z_0$ as

$$\begin{aligned}\mathcal{F}_\omega(\theta,\phi) &= \frac{ik\cos\theta}{2\pi}\int_{-\infty}^{+\infty}\int_{-\infty}^{+\infty}\mathbf{E}_\omega(\mathbf{r}_0)e^{-ik\hat{\mathbf{r}}\cdot\mathbf{r}_0}dx_0dy_0, \quad z > z_0 \\ &= \frac{ik}{2\pi}\hat{\mathbf{r}}\times\int_{-\infty}^{+\infty}\int_{-\infty}^{+\infty}\hat{\mathbf{z}}\times\mathbf{E}_\omega(\mathbf{r}_0)e^{-ik\hat{\mathbf{r}}\cdot\mathbf{r}_0}dx_0dy_0, \quad z > z_0.\end{aligned} \tag{3.44}$$

Dotting $\hat{\mathbf{r}}$ into (3.44) reveals that

$$\hat{\mathbf{r}}\cdot\int_{-\infty}^{+\infty}\int_{-\infty}^{+\infty}\mathbf{E}_\omega(\mathbf{r}_0)e^{-ik\hat{\mathbf{r}}\cdot\mathbf{r}_0}dx_0dy_0 = 0. \tag{3.45}$$

Similar relations hold for the magnetic far-field pattern in terms of the tangential magnetic field in the plane $z = z_0$. The formula (3.34) shows that the magnetic far field and far-field pattern can also be expressed in terms of the tangential electric field as

$$\begin{aligned}\mathbf{H}_\omega(\mathbf{r}) = &\frac{e^{ikr}}{r}\sqrt{\frac{\epsilon}{\mu}}\frac{ik}{2\pi}\hat{\mathbf{r}}\times\left(\hat{\mathbf{r}}\times\int_{-\infty}^{+\infty}\int_{-\infty}^{+\infty}\hat{\mathbf{z}}\times\mathbf{E}_\omega(\mathbf{r}_0)e^{-ik\hat{\mathbf{r}}\cdot\mathbf{r}_0}dx_0dy_0\right) \\ &+ O\left(\frac{1}{r^2}\right), \quad z > z_0\end{aligned} \tag{3.46}$$

and

$$\mathcal{F}^H_\omega(\theta,\phi) = \sqrt{\frac{\epsilon}{\mu}}\frac{ik}{2\pi}\hat{\mathbf{r}}\times\left(\hat{\mathbf{r}}\times\int_{-\infty}^{+\infty}\int_{-\infty}^{+\infty}\hat{\mathbf{z}}\times\mathbf{E}_\omega(\mathbf{r}_0)e^{-ik\hat{\mathbf{r}}\cdot\mathbf{r}_0}dx_0dy_0\right), \quad z > z_0. \tag{3.47}$$

It should be noted that the derivation of the far-field formulas (3.41)–(3.47) required that the limit as $r \to \infty$ be interchanged with the infinite (x_0, y_0) integration. We cannot use standard theorems of calculus to justify this interchange because p_ω and $\mathbf{E}_\omega$ are not absolutely integrable in the infinite (x_0, y_0) plane. However, these far-field formulas can be proven valid by first expressing the fields in the plane $z = z_0$ in terms of their volume sources, or equivalent surface sources on a closed surface S, and then proceeding as in Appendix D. The results of Appendix D show that when $\theta = 0$ the integrations in the far-field formulas (3.41)–(3.47) may diverge by oscillation. For $0 < \theta < \pi/2$ they converge provided the integrations are performed as described in Appendix D (see also the discussion following equation (3.71)).

When the sources are located in the half space $z > z_0$, the far fields in the half space $z < z_0$ are given by the negative of (3.41)–(3.47). The formulas (3.21)–(3.40) remain unchanged.

3.1.2 Low-Frequency Behavior of the Far-Field Patterns

The integral over all time of a time-dependent function equals the function's frequency spectrum evaluated at ω equal to zero. Thus, it is of interest for both frequency-domain and time-domain representations to determine the behavior of the far fields as ω approaches zero.

From (3.36) we see that the frequency dependence of the electric far-field pattern as $\omega \to 0$ is ω times the frequency dependence of the integrated current over the bounded volume. Equation (3.37) shows that the magnetic far-field pattern has the same frequency behavior. This means that the electric and magnetic far-field patterns in the time domain will depend on the time derivative of the spatially integrated current. A further implication, proven in Section 5.1.5, is that the time-domain electric and magnetic far-field patterns integrated over all time will be zero for currents in a finite region that turn on and off in a finite time period. Only if the current stays on forever, so that the fields, for example, eventually become static, can the electric and magnetic far-field patterns integrated over all time have nonzero values.

The expression (2.379) in Chapter 2 shows that the acoustic far-field pattern can be written as

$$\mathcal{F}_\omega(\theta,\phi) = -\frac{1}{4\pi}\int\limits_V \left[i\omega s_\omega(\mathbf{r}') + ik\hat{\mathbf{r}}\cdot\mathbf{f}_\omega(\mathbf{r}')\right]e^{-ik\hat{\mathbf{r}}\cdot\mathbf{r}'}dV' \tag{3.48}$$

where $s_\omega(\mathbf{r})$ and $\mathbf{f}_\omega(\mathbf{r})$ are the volume sources of mass-flow density and force density, respectively. For low frequencies, $e^{-ik\hat{\mathbf{r}}\cdot\mathbf{r}'} = 1 - ik\hat{\mathbf{r}}\cdot\mathbf{r}' + O(\omega^2)$ and (3.48) shows that the first term of the low-frequency expansion for $\mathcal{F}_\omega(\theta,\phi)$ is

$$\mathcal{F}_\omega(\theta,\phi) \sim -\frac{i\omega}{4\pi}\int\limits_V \left[s_\omega(\mathbf{r}') + \frac{\hat{\mathbf{r}}\cdot\mathbf{f}_\omega(\mathbf{r}')}{c}\right]dV'. \tag{3.49}$$

Thus, for low frequencies the frequency dependence of the acoustic far-field pattern is ω times the frequency dependence of the integrated volume-source functions $s_\omega(\mathbf{r})$ and $\mathbf{f}_\omega(\mathbf{r})$.

3.2 PLANE-WAVE SPECTRUM REPRESENTATIONS

In Section 3.1, Green's function formulas for the frequency-domain fields in the half space $z > z_0$ were derived in terms of the fields on the infinite scan plane $z = z_0$. In principle, one could measure the electric field, for example, tangent to the scan plane $z = z_0$, and then compute the electric and magnetic fields for $z > z_0$ from the formulas (3.17) and (3.20). In practice, however, the integration of (3.17) and (3.20) for many observation points in the half space $z > z_0$ can be prohibitively time consuming. Fortunately, a reformulation in terms of plane-wave spectrum representations enables fast Fourier transform algorithms to greatly reduce the time needed to compute the fields. Moreover, plane-wave spectrum representations facilitate probe correction for both frequency-domain and time-domain measurements, as shown in Chapters 6 and 7.

The plane-wave spectrum representations for the acoustic and electromagnetic fields in the half space $z \geq z_0$ are derived heuristically in this section and are then validated rigorously in Appendix D. We begin by expressing the acoustic pressure in terms of the spatial Fourier integral

$$p_\omega(\mathbf{r}) = \frac{1}{2\pi} \int_{-\infty}^{+\infty} \int_{-\infty}^{+\infty} b_\omega(k_x, k_y, z) e^{i(k_x x + k_y y)} dk_x dk_y, \quad z \geq z_0 \tag{3.50}$$

where b_ω is the Fourier transform of $p_\omega(\mathbf{r})$ with respect to x and y given by

$$b_\omega(k_x, k_y, z) = \frac{1}{2\pi} \int_{-\infty}^{+\infty} \int_{-\infty}^{+\infty} p_\omega(\mathbf{r}) e^{-i(k_x x + k_y y)} dx dy, \quad z \geq z_0. \tag{3.51}$$

Inserting the Fourier integral (3.50) into the Helmholtz equation (3.1), and taking the inverse Fourier transform, we find that the function b_ω satisfies

$$\left[\frac{\partial^2}{\partial z^2} + \gamma^2\right] b_\omega(k_x, k_y, z) = 0, \quad \gamma^2 = k^2 - k_x^2 - k_y^2 \tag{3.52}$$

so that

$$b_\omega(k_x, k_y, z) = T_\omega^A(k_x, k_y) e^{i\gamma z} + T_\omega^B(k_x, k_y) e^{-i\gamma z} \tag{3.53}$$

where T_ω^A and T_ω^B are independent of z, and γ is defined by

$$\gamma = \begin{cases} |\sqrt{k^2 - k_x^2 - k_y^2}|, & k_x^2 + k_y^2 \leq k^2 \\ i|\sqrt{k_x^2 + k_y^2 - k^2}|, & k_x^2 + k_y^2 > k^2 \end{cases} . \tag{3.54}$$

The pressure decays as $z \to \infty$. Thus, from the Fourier representation (3.50) we see that T_ω^B must be zero for imaginary γ. For real γ, an asymptotic evaluation (see Section 3.2.5) of (3.50) with b_ω inserted from (3.53) shows that T_ω^B must be zero if no energy is coming in from $z = +\infty$. Therefore, (3.53) reduces to

$$b_\omega(k_x, k_y, z) = T_\omega(k_x, k_y) e^{i\gamma z}, \quad z \geq z_0 \tag{3.55}$$

where T_ω is independent of z and is called the plane-wave spectrum of p_ω.

Inserting b_ω from (3.55) into the Fourier representation (3.50), we find that

$$p_\omega(\mathbf{r}) = \frac{1}{2\pi} \int_{-\infty}^{+\infty} \int_{-\infty}^{+\infty} T_\omega(k_x, k_y) e^{i(k_x x + k_y y + \gamma z)} dk_x dk_y, \quad z \geq z_0 \tag{3.56}$$

with the spectrum given by

$$T_\omega(k_x, k_y) = \frac{e^{-i\gamma z}}{2\pi} \int_{-\infty}^{+\infty} \int_{-\infty}^{+\infty} p_\omega(\mathbf{r}) e^{-i(k_x x + k_y y)} dx dy, \quad z \geq z_0. \tag{3.57}$$

Note that the left-hand side of (3.57) is independent of z, and thus the right-hand side of (3.57) must also be independent of z. If z is set equal to z_0 in (3.57), the formula (3.56) gives the pressure for $z \geq z_0$ in terms of the pressure in the plane $z = z_0$. The formulas (3.56) and (3.57) comprise the plane-wave spectrum representation of the acoustic pressure in the half space $z \geq z_0$. The definition (3.54) of γ shows that for propagation in the $+z$ direction, the plane wave $e^{i(k_x x+k_y y+\gamma z)}$ is nondecaying when $k_x^2 + k_y^2 < k^2$ and decaying in the z direction when $k_x^2 + k_y^2 > k^2$. Therefore, $e^{i(k_x x+k_y y+\gamma z)}$ is called a propagating plane wave when $k_x^2 + k_y^2 < k^2$ and an evanescent plane wave when $k_x^2 + k_y^2 > k^2$. The formula (3.56) for the pressure also remains valid for observation points in z planes that lie between the scan plane and the acoustic sources.

Using the relation $\mathbf{v}_\omega = \nabla p_\omega/(i\omega\rho_{m0})$ in conjunction with the plane-wave spectrum representation (3.56) shows that the velocity can be expressed in terms of plane waves as

$$\mathbf{v}_\omega(\mathbf{r}) = \frac{1}{2\pi\omega\rho_{m0}} \int\limits_{-\infty}^{+\infty}\int\limits_{-\infty}^{+\infty} \mathbf{k}\, T_\omega(k_x, k_y) e^{i(k_x x+k_y y+\gamma z)} dk_x dk_y, \quad z \geq z_0 \tag{3.58}$$

where

$$\mathbf{k} = k_x\hat{\mathbf{x}} + k_y\hat{\mathbf{y}} + \gamma\hat{\mathbf{z}} \tag{3.59}$$

is the propagation vector.

The Green's function representation (3.7) can be derived from the plane-wave representations (3.56) and (3.57) by using the identity

$$2\pi \frac{\partial}{\partial z} \frac{e^{ik|\mathbf{r}-\mathbf{r}_0|}}{|\mathbf{r}-\mathbf{r}_0|} = -\int\limits_{-\infty}^{+\infty}\int\limits_{-\infty}^{+\infty} e^{i[k_x(x-x_0)+k_y(y-y_0)+\gamma(z-z_0)]} dk_x dk_y, \quad z > z_0 \tag{3.60}$$

which may be proven directly by expanding the free-space Green's function in a plane-wave spectrum or by taking the z derivative of the Weyl identity [124]

$$\frac{e^{ik|\mathbf{r}-\mathbf{r}_0|}}{4\pi|\mathbf{r}-\mathbf{r}_0|} = \frac{i}{8\pi^2} \int\limits_{-\infty}^{+\infty}\int\limits_{-\infty}^{+\infty} \frac{e^{i(k_x(x-x_0)+k_y(y-y_0)+\gamma|z-z_0|)}}{\gamma} dk_x dk_y \tag{3.61}$$

which holds for $z > z_0$ as well as for $z < z_0$. In fact, a simple evaluation shows that it holds for all $\mathbf{r}$ except $\mathbf{r} = \mathbf{r}_0$. (For $z = z_0$ the $k_x k_y$ integration in (3.61) must be performed like the xy integration in Appendix D.)

Now consider the electric field $\mathbf{E}_\omega$. Since each of its rectangular components satisfies the scalar Helmholtz equation (3.1) and the scalar radiation conditions (2.376)–(2.377), the plane-wave spectrum representation of the electric field is immediately found from (3.56) and (3.57) to be

$$\mathbf{E}_\omega(\mathbf{r}) = \frac{1}{2\pi} \int\limits_{-\infty}^{+\infty}\int\limits_{-\infty}^{+\infty} \mathbf{T}_\omega(k_x, k_y) e^{i(k_x x+k_y y+\gamma z)} dk_x dk_y, \quad z \geq z_0 \tag{3.62}$$

and the spectrum $\mathbf{T}_\omega$ for the electric field is given by

$$\mathbf{T}_\omega(k_x, k_y) = \frac{e^{-i\gamma z}}{2\pi} \int\limits_{-\infty}^{+\infty}\int\limits_{-\infty}^{+\infty} \mathbf{E}_\omega(\mathbf{r}) e^{-i(k_x x+k_y y)} dx dy, \quad z \geq z_0. \tag{3.63}$$

If z is set equal to z_0 in (3.63), the formula (3.62) gives the electric field for $z \geq z_0$ in terms of the electric field in the plane $z = z_0$. The formula (3.62) also remains valid for observation points in z planes that lie between the scan plane and the sources. Since the electric field has zero divergence, the spectrum for the electric field satisfies

$$\mathbf{k} \cdot \mathbf{T}_\omega(k_x, k_y) = 0 \tag{3.64}$$

where $\mathbf{k}$ is defined in (3.59). That is, $\mathbf{T}_\omega$ is perpendicular to the propagation direction of the plane wave $e^{i(k_x x+k_y y+\gamma z)} = e^{i\mathbf{k}\cdot\mathbf{r}}$. Consequently, for $\gamma \neq 0$ the z component of $\mathbf{T}_\omega$ can be found from the x and y components of $\mathbf{T}_\omega$. Thus, the electric field in the free-space region $z \geq z_0$ is determined uniquely by specifying the x and y components of the electric field in the measurement plane $z = z_0$, as confirmed by the Green's function representation (3.17). Note that inserting $\mathbf{T}_\omega$ from (3.63) into (3.64), replacing $\mathbf{r}$ by $\mathbf{r}_0$ and $\mathbf{k}$ by $k\hat{\mathbf{r}}$, produces equation (3.45).

The z component of the electric field can also be found directly from the x and y components by integrating $\nabla \cdot \mathbf{E}_\omega = 0$ with respect to z, to obtain the formula

$$\hat{\mathbf{z}} \cdot \mathbf{E}_\omega(\mathbf{r}) = -\int_{z_0}^{z} \left[\frac{\partial}{\partial x} \hat{\mathbf{x}} \cdot \mathbf{E}_\omega(x, y, z') + \frac{\partial}{\partial y} \hat{\mathbf{y}} \cdot \mathbf{E}_\omega(x, y, z') \right] dz' + C_\omega(x, y) \tag{3.65}$$

where $z \geq z_0$ and $C_\omega(x, y)$ is a function of x and y only. Since the sources of the electromagnetic field are contained in a finite region of space, the amplitude of the electric field is of order r^{-1} at infinity, so $\hat{\mathbf{z}} \cdot \mathbf{E}_\omega \to 0$ as $r \to \infty$. Therefore

$$\hat{\mathbf{z}} \cdot \mathbf{E}_\omega(\mathbf{r}) = \int_{z}^{+\infty} \left[\frac{\partial}{\partial x} \hat{\mathbf{x}} \cdot \mathbf{E}_\omega(x, y, z') + \frac{\partial}{\partial y} \hat{\mathbf{y}} \cdot \mathbf{E}_\omega(x, y, z') \right] dz', \quad z \geq z_0. \tag{3.66}$$

Thus, the z component of the electric field is determined uniquely from the x and y components of the electric field in the source-free half space $z \geq z_0$.

Using the induction law (2.287), $\nabla \times \mathbf{E}_\omega = i\omega\mu\mathbf{H}_\omega$, which gives the magnetic field in terms of the electric field, we find that the magnetic field is given by (3.62) with $\mathbf{T}_\omega$ replaced by

$$\mathbf{T}_\omega^H(k_x, k_y) = \sqrt{\frac{\epsilon}{\mu}}\, \hat{\mathbf{k}} \times \mathbf{T}_\omega(k_x, k_y) = \frac{1}{\omega\mu}\, \mathbf{k} \times \mathbf{T}_\omega(k_x, k_y) \tag{3.67}$$

that is,

$$\mathbf{H}_\omega(\mathbf{r}) = \frac{1}{2\pi} \int_{-\infty}^{+\infty} \int_{-\infty}^{+\infty} \mathbf{T}_\omega^H(k_x, k_y) e^{i(k_x x + k_y y + \gamma z)} dk_x dk_y, \quad z \geq z_0 \tag{3.68}$$

where $\mathbf{T}_\omega^H$ is the spectrum for the magnetic field also given by

$$\mathbf{T}_\omega^H(k_x, k_y) = \frac{e^{-i\gamma z}}{2\pi} \int_{-\infty}^{+\infty} \int_{-\infty}^{+\infty} \mathbf{H}_\omega(\mathbf{r}) e^{-i(k_x x + k_y y)} dx dy, \quad z \geq z_0 \tag{3.69}$$

and $\hat{\mathbf{k}} = \mathbf{k}/k$ with $\mathbf{k}$ given in (3.59). Note that $\mathbf{k}$ is a complex vector for γ imaginary (evanescent spectrum) and thus $\hat{\mathbf{k}}$ is a complex unit vector for γ imaginary because $k = \omega/c = \sqrt{\mathbf{k} \cdot \mathbf{k}}$ is a real number.

As with the electric field, the tangential components of the magnetic field in the plane $z = z_0$ uniquely determine all the field components in the half space $z \geq z_0$. Also, the z component of the magnetic field can be determined from the x and y components of the magnetic field in the half space $z \geq z_0$ by the formula (3.66) with $\mathbf{E}_\omega$ replaced by $\mathbf{H}_\omega$.

In general, any two of the six rectangular components of $\mathbf{E}_\omega$ and $\mathbf{H}_\omega$ in the plane $z = z_0$ determine all fields in the half space $z \geq z_0$. This more general result is proven in the following way. First note that the spectra $\mathbf{T}_\omega$ and $\mathbf{T}_\omega^H$, through (3.62) and (3.68), determine all fields in the half space $z \geq z_0$, so all we have to prove is that any two of the six rectangular components of $\mathbf{E}_\omega$ and $\mathbf{H}_\omega$ in the plane $z = z_0$ determine $\mathbf{T}_\omega$ and $\mathbf{T}_\omega^H$.

If any two rectangular components of $\mathbf{E}_\omega$ are known for $z = z_0$, the formula (3.63) determines the corresponding two rectangular components of $\mathbf{T}_\omega$. The third component of $\mathbf{T}_\omega$ can then be determined from (3.64), and $\mathbf{T}_\omega^H$ can be determined from (3.67). Similarly, if any two rectangular components of $\mathbf{H}_\omega$ are known for $z = z_0$, the formula (3.68) determines the corresponding two rectangular components of $\mathbf{T}_\omega^H$. The third component of $\mathbf{T}_\omega^H$ can then be determined from

$$\mathbf{k} \cdot \mathbf{T}_\omega^H(k_x, k_y) = 0 \tag{3.70}$$

and $\mathbf{T}_\omega$ can be determined from

$$\mathbf{T}_\omega(k_x, k_y) = -\sqrt{\frac{\mu}{\epsilon}}\, \hat{\mathbf{k}} \times \mathbf{T}_\omega^H(k_x, k_y) = -\frac{1}{\omega\epsilon}\, \mathbf{k} \times \mathbf{T}_\omega^H(k_x, k_y). \tag{3.71}$$

Assume now that one rectangular component of $\mathbf{E}_\omega$ and another rectangular component $\mathbf{H}_\omega$ are known for $z = z_0$ (for example, the x component of $\mathbf{E}_\omega$ and the z component of $\mathbf{H}_\omega$). Then the corresponding rectangular components of $\mathbf{T}_\omega$ and $\mathbf{T}_\omega^H$ are determined from (3.63) and (3.69), and the remaining four rectangular components of $\mathbf{T}_\omega$ and $\mathbf{T}_\omega^H$ can be determined directly from (3.67) and (3.71). Finally, assume that the same rectangular components of $\mathbf{E}_\omega$ and $\mathbf{H}_\omega$ are known for $z = z_0$ (for example, the x components of $\mathbf{E}_\omega$ and $\mathbf{H}_\omega$). Again, the corresponding rectangular components of $\mathbf{T}_\omega$ and $\mathbf{T}_\omega^H$ are determined from (3.63) and (3.69), and the remaining four components are solutions to a set of linear equations obtained from (3.67) and (3.71). These equations are easily solved by invoking (3.64) and (3.70). Thus, we have proven that all fields in $z \geq z_0$ are determined by any two of the six rectangular components of $\mathbf{E}_\omega$ and $\mathbf{H}_\omega$ on $z = z_0$. Using the identity (3.60), we can derive both the Green's function expressions (3.9) and (3.17) for the electric field from the plane-wave representations (3.62)–(3.64). Similarly, the corresponding expressions for the magnetic field can be derived from (3.68)–(3.70).

In Appendix D we prove that the plane-wave spectrum representations (3.56)–(3.57), (3.62)–(3.63), and (3.68)–(3.69) produce the correct fields in the region $z \geq z_0$. The integrands in the plane-wave spectrum expressions (3.57), (3.63), and (3.69) are not absolutely integrable, and it is found in Appendix D that the integrations over the plane $z = z_0$ should be performed as follows. First, introduce a scan area D_R, which is part of the plane $z = z_0$ and lies within the circle of radius R. In addition, require that the scan area D_R has a fixed shape and, as $R \to \infty$, it covers the entire plane $z = z_0$. The spectra (3.57), (3.63), and (3.69) are then determined by first integrating over D_R and then letting $R \to \infty$. For example, this integration can be performed in polar coordinates by first integrating with respect to the angular variable.

In Appendix D it is also shown that the plane-wave spectra obtained from (3.57), (3.63), and (3.69), when calculated in the manner described above, equal the plane-wave spectra obtained from the Huygens' source expressions (3.105), (3.110), and (3.111) in Section 3.2.2, except possibly at $(k_x, k_y) = (0, 0)$. At $(k_x, k_y) = (0, 0)$, the expressions (3.57), (3.63), and (3.69) may diverge by oscillation, thereby producing a removable singularity. However, this removable singularity does not affect the validity of the plane-wave expansions (3.56), (3.62), and (3.68) for the fields. Moreover, the removable singularity occurs when the scan area D_R is circular but does not occur when D_R is rectangular.

Since the far fields are proportional to the spectra (see Section 3.2.5), the on-axis far field (corresponding to $(k_x, k_y) = (0, 0)$) evaluated from (3.57), (3.63), or (3.69) diverges in general by oscillation no matter how large a radius is chosen for a circular scan area. For the main beam of directive radiators, the oscillation is usually negligible. However, for broadbeam radiators the oscillation can severely limit the accuracy of the far-field pattern near the z axis computed from plane-polar, measured near-field data.

For a rectangular scan area, the integrals (3.57), (3.63), and (3.69) converge for all (k_x, k_y) including the on-axis direction $(k_x, k_y) = (0, 0)$, as the size of the rectangular scan area approaches infinity. However, for broadbeam radiators and practical finite-size rectangular scan areas, the computed far fields near the z axis may be in error by several decibels [125]. This limitation in the measurement of broadbeam radiators by planar near-field scanning in the frequency domain was discussed in [42] and [46]. In Section 8.3 we show that these finite-scan errors can be avoided in the time domain.

If the sources are located in the half space $z > z_0$, the fields in the region $z \le z_0$ are determined by the formulas (3.56), (3.57), (3.58), (3.62), (3.63), (3.64), (3.67), (3.68), (3.69), (3.70), and (3.71) with $e^{i(k_x x+k_y y+\gamma z)}$ replaced by $e^{i(k_x x+k_y y-\gamma z)}$, $e^{-i(k_x x+k_y y+\gamma z)}$ replaced by $e^{-i(k_x x+k_y y-\gamma z)}$, $\mathbf{k}$ replaced by $\mathbf{k}^- = k_x\hat{\mathbf{x}}+k_y\hat{\mathbf{y}}-\gamma\hat{\mathbf{z}}$, and $\hat{\mathbf{k}}$ replaced by $\hat{\mathbf{k}}^- = \mathbf{k}^-/k$. Moreover, the z component of the electric field in the half space $z \le z_0$ is given by (3.66) with the upper limit of integration changed to $-\infty$.

3.2.1 Propagating and Evanescent Fields of a Point Source

According to the Weyl formula (3.61), the scalar field $G_\omega(\mathbf{r},\mathbf{r}')$ of a point source located at $\mathbf{r}' = x'\hat{\mathbf{x}} + y'\hat{\mathbf{y}} + z'\hat{\mathbf{z}}$ can be expressed as

$$G_\omega(\mathbf{r},\mathbf{r}') = \frac{e^{ikR}}{4\pi R} = G_\omega^p(\mathbf{r},\mathbf{r}') + G_\omega^e(\mathbf{r},\mathbf{r}'), \quad R = |\mathbf{r}-\mathbf{r}'| \tag{3.72}$$

where

$$G_\omega^p(\mathbf{r},\mathbf{r}') = \frac{i}{8\pi^2}\iint\limits_{k_x^2+k_y^2<k^2} \frac{e^{i\mathbf{k}\cdot\mathbf{R}}}{\gamma}\, dk_x dk_y, \quad z > z' \tag{3.73}$$

is referred to as the propagating field, and

$$G_\omega^e(\mathbf{r},\mathbf{r}') = \frac{i}{8\pi^2}\iint\limits_{k_x^2+k_y^2>k^2} \frac{e^{i\mathbf{k}\cdot\mathbf{R}}}{\gamma}\, dk_x dk_y, \quad z > z' \tag{3.74}$$

is referred to as the evanescent field. Here $\mathbf{R} = \mathbf{r} - \mathbf{r}'$ is the vector from the source point to the observation point, and $\mathbf{k} = k_x\hat{\mathbf{x}} + k_y\hat{\mathbf{y}} + \gamma\hat{\mathbf{z}}$ is the propagation vector of the plane wave $e^{i\mathbf{k}\cdot\mathbf{R}}$. The region of convergence of (3.73) is actually $(z \ge z')$, and the region of convergence of (3.74) is $(z \ge z', R \ne 0)$. In this section we shall evaluate the propagating and evanescent fields and investigate their asymptotic behavior for large R. For a fixed observation point $\mathbf{r}$ with $z > z'$, we define the spherical source angles (θ_s, ϕ_s) with respect to the source point $\mathbf{r}'$ such that $\hat{\mathbf{R}}(\theta_s,\phi_s) = \mathbf{R}/R = \sin\theta_s\cos\phi_s\hat{\mathbf{x}} + \sin\theta_s\sin\phi_s\hat{\mathbf{y}} + \cos\theta_s\hat{\mathbf{z}}$, with $0 \le \theta_s < \pi/2$. Thus, an observation point $\mathbf{r}$ can be described both in terms of the source coordinates (R,θ_s,ϕ_s) and in terms of the standard spherical coordinates (r,θ,ϕ). We first consider observation points for which $\theta = 0$.

Observation Points with $\theta = 0$. When $\theta = 0$ the observation point is $\mathbf{r} = z\hat{\mathbf{z}}$ with $z > z'$. Assume first that the point source is located on the z axis so that $x' = y' = 0$, $\theta_s = 0$, $R = |z - z'|$, and $\mathbf{R} = (z - z')\hat{\mathbf{z}}$. Then the integral (3.73) for the propagating field reduces to

$$\begin{aligned} G_\omega^p(\mathbf{r},\mathbf{r}') &= \frac{i}{8\pi^2}\iint\limits_{k_x^2+k_y^2<k^2} \frac{e^{i\gamma(z-z')}}{\gamma}\, dk_x dk_y \\ &= \frac{e^{ik(z-z')}}{4\pi(z-z')} - \frac{1}{4\pi(z-z')}, \quad \theta_s = 0 \end{aligned} \tag{3.75}$$

where we have performed the (k_x, k_y) integration in polar coordinates. Combining (3.75) with (3.72) shows that the evanescent field is

$$G_\omega^e(\mathbf{r},\mathbf{r}') = \frac{1}{4\pi(z-z')}, \quad \theta_s = 0. \tag{3.76}$$

Assume now that $(x', y') \ne (0, 0)$. Introducing the integration variables (γ, k_ϕ), defined such that $k_x = (k^2 - \gamma^2)^{1/2}\cos k_\phi$ and $k_y = (k^2 - \gamma^2)^{1/2}\sin k_\phi$, into the integral (3.73) then yields

$$G^p_\omega(\mathbf{r},\mathbf{r}') = \frac{i}{8\pi^2}\int_0^{2\pi}\int_0^{k} e^{-i(k^2-\gamma^2)^{1/2}(x'\cos k_\phi + y'\sin k_\phi)}\, e^{-i\gamma z'}\, e^{i\gamma z} d\gamma\, dk_\phi, \quad \theta = 0 \tag{3.77}$$

which can be integrated by parts with respect to γ to show that

$$\begin{aligned} G^p_\omega(\mathbf{r},\mathbf{r}') = {} & \frac{e^{ik(z-z')}}{4\pi(z-z')} - \frac{1}{8\pi^2(z-z')}\int_0^{2\pi} e^{-ik(x'\cos k_\phi + y'\sin k_\phi)} dk_\phi \\ & + O\left(\frac{1}{(z-z')^2}\right), \quad \theta = 0. \end{aligned} \tag{3.78}$$

By making the substitution $x' = \rho'\cos\phi'$ and $y' = \rho'\sin\phi'$ with $\rho' = \sqrt{x'^2 + y'^2}$, and using the representation

$$J_0(k\rho') = \frac{1}{2\pi}\int_0^{2\pi} e^{-ik\rho'\cos(k_\phi - \phi')} dk_\phi \tag{3.79}$$

for the Bessel function of order zero, we find that the propagating far field can also be expressed as

$$G^p_\omega(\mathbf{r},\mathbf{r}') = \frac{e^{ik(z-z')}}{4\pi(z-z')} - \frac{J_0\left(k\sqrt{x'^2+y'^2}\right)}{4\pi(z-z')} + O\left(\frac{1}{(z-z')^2}\right), \quad \theta = 0. \tag{3.80}$$

The corresponding evanescent far field is

$$\begin{aligned} G^e_\omega(\mathbf{r},\mathbf{r}') &= \frac{1}{8\pi^2(z-z')}\int_0^{2\pi} e^{-ik(x'\cos k_\phi + y'\sin k_\phi)} dk_\phi + O\left(\frac{1}{(z-z')^2}\right) \\ &= \frac{J_0\left(k\sqrt{x'^2+y'^2}\right)}{4\pi(z-z')} + O\left(\frac{1}{(z-z')^2}\right), \quad \theta = 0. \end{aligned} \tag{3.81}$$

The formulas (3.80) and (3.81) show that, unless $k\sqrt{x'^2+y'^2}$ equals one of the zeros of the Bessel function of order zero, the evanescent field is of the same order ($1/z,\ \ z \to \infty$) as the propagating field.

Observation Points with $\boldsymbol{0 < \theta < \pi/2}$. When $0 < \theta < \pi/2$, the source angle θ_s lies in the interval $0 < \theta_s < \pi/2$. In order to perform the integration in (3.73), we introduce new integration variables (u, v) as follows. The unit propagation vector $\hat{\mathbf{k}} = \mathbf{k}/k$ describes a point on the unit hemisphere given by $0 \le \theta < \pi/2$, and we introduce the new integration variable u by the equation $u = \hat{\mathbf{k}}\cdot\hat{\mathbf{R}}(\theta_s,\phi_s)$. Then $-\sin\theta_s < u \le 1$, as seen from Figure 3.3. For a fixed (θ_s,ϕ_s) the spherical unit vectors $\hat{\mathbf{R}}(\theta_s,\phi_s)$, $\hat{\boldsymbol{\theta}}(\theta_s,\phi_s)$, and $\hat{\boldsymbol{\phi}}(\theta_s,\phi_s)$ are orthogonal and

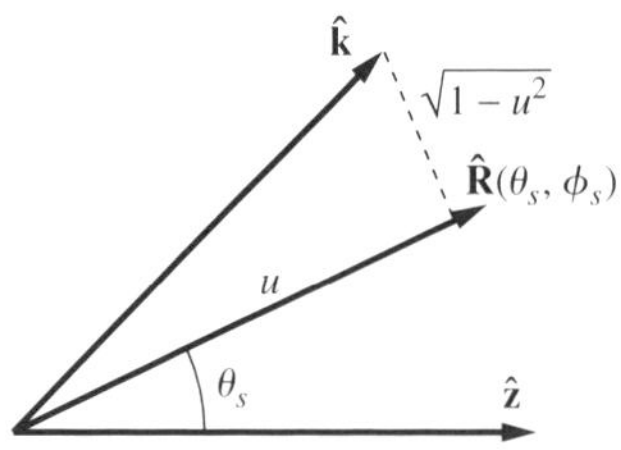

Figure 3.3 The vectors $\hat{\mathbf{k}}$ and $\hat{\mathbf{R}}$ defining the variable $u = \hat{\mathbf{k}}\cdot\hat{\mathbf{R}}$.

can be used to expand the unit vector $\hat{\mathbf{k}}$. We define the second new integration variable v such that this expansion becomes

$$\hat{\mathbf{k}}(u, v) = u\hat{\mathbf{R}}(\theta_s, \phi_s) + \sqrt{1-u^2}\left(\cos v\hat{\boldsymbol{\phi}}(\theta_s, \phi_s) + \sin v\hat{\boldsymbol{\theta}}(\theta_s, \phi_s)\right) \tag{3.82}$$

where the factor $\sqrt{1-u^2}$ ensures that the right-hand side of (3.82) is a unit vector. The well-known expressions for the spherical unit vectors in terms of the rectangular unit vectors show that

$$\begin{aligned} k_x &= ku \sin\theta_s \cos\phi_s + k\sqrt{1-u^2}(\sin v \cos\theta_s \cos\phi_s - \cos v \sin\phi_s) \\ k_y &= ku \sin\theta_s \sin\phi_s + k\sqrt{1-u^2}(\sin v \cos\theta_s \sin\phi_s + \cos v \cos\phi_s) \\ \gamma &= ku \cos\theta_s - k\sqrt{1-u^2} \sin v \sin\theta_s. \end{aligned} \tag{3.83}$$

These equations express the old integration variables (k_x, k_y) in terms of the new variables (u, v) and show that the Jacobian for the coordinate transformation is given by

$$\left|\frac{\partial(k_x, k_y)}{\partial(u, v)}\right| = k^2 \left|u\cos\theta_s - \sqrt{1-u^2}\sin\theta_s \sin v\right| = k\gamma \tag{3.84}$$

so that $dk_x\, dk_y = k\,\gamma\, du\, dv$.

We have already shown that $-\sin\theta_s < u \le 1$ and will now determine the limits of integration for v. Since $\gamma \ge 0$, the expression (3.83) shows that the range of the v integration is that part of the interval from 0 to 2π for which $\sin v < (u\cos\theta_s)/(\sqrt{1-u^2}\sin\theta_s)$. Thus, for $\sin\theta_s < u \le 1$, the v integration is over the entire interval $0 \le v < 2\pi$. This can also be seen from Figure 3.3 by noting that when $\sin\theta_s < u \le 1$, the angle between $\hat{\mathbf{k}}$ and $\hat{\mathbf{R}}$ is less than $\pi/2 - \theta_s$. For $-\sin\theta_s < u < \sin\theta_s$ we find that the length of the v integration interval is $\pi + 2\arcsin\left[(u\cos\theta_s)/(\sqrt{1-u^2}\sin\theta_s)\right]$, where $-\pi/2 \le \arcsin(x) \le \pi/2$ when $-1 \le x \le 1$. Using (u, v) as integration variables in the integral (3.73) yields the following formula for the propagating field of the point source

$$\begin{aligned} G_\omega^p(\mathbf{r}, \mathbf{r}') &= \frac{ik}{8\pi^2}\int_{\sin\theta_s}^{1} 2\pi e^{ikRu} du \\ &\quad + \frac{ik}{8\pi^2}\int_{-\sin\theta_s}^{\sin\theta_s}\left(\pi + 2\arcsin\left[\frac{u\cos\theta_s}{\sqrt{1-u^2}\sin\theta_s}\right]\right) e^{ikRu} du \\ &= \frac{e^{ikR}}{4\pi R} - \frac{\cos\theta_s}{4\pi^2 R}\int_{-\sin\theta_s}^{\sin\theta_s}\frac{e^{ikRu}}{(1-u^2)\sqrt{\sin^2\theta_s - u^2}} du, \quad 0 < \theta_s < \pi/2 \end{aligned} \tag{3.85}$$

where we have integrated by parts once. The formula (3.72) for the total field of the point source shows that the evanescent field is given by

$$G_\omega^e(\mathbf{r}, \mathbf{r}') = \frac{\cos\theta_s}{4\pi^2 R}\int_{-\sin\theta_s}^{\sin\theta_s}\frac{e^{ikRu}}{(1-u^2)\sqrt{\sin^2\theta_s - u^2}} du, \quad 0 < \theta_s < \pi/2. \tag{3.86}$$

Far-field expressions that give the propagating and evanescent fields for large R will now be derived. From the asymptotic formulas of Bleistein and Handelsman [126, p. 91] we find that when $0 < u_0 < 1$

$$\int_{-u_0}^{u_0}\frac{e^{iXu}}{(1-u^2)\sqrt{u_0^2 - u^2}} du = \sqrt{\frac{2\pi}{Xu_0}}\,\frac{\cos(Xu_0 - \pi/4)}{1-u_0^2} + O\left(\frac{1}{X^{3/2}}\right), \quad X \to \infty \tag{3.87}$$

which, together with (3.85) and (3.86), shows that the propagating field for large R behaves as

$$G_\omega^p(\mathbf{r}, \mathbf{r}') = \frac{e^{ikR}}{4\pi R} - \frac{\cos(kR\sin\theta_s - \pi/4)}{2\pi\cos\theta_s\sqrt{2\pi k\sin\theta_s}}\frac{1}{R^{3/2}} + O\left(\frac{1}{R^{5/2}}\right), \quad 0 < \theta_s < \pi/2 \tag{3.88}$$

and that the evanescent field for large R behaves as

$$G_\omega^e(\mathbf{r}, \mathbf{r}') = \frac{\cos(kR\sin\theta_s - \pi/4)}{2\pi\cos\theta_s\sqrt{2\pi k\sin\theta_s}}\frac{1}{R^{3/2}} + O\left(\frac{1}{R^{5/2}}\right), \quad 0 < \theta_s < \pi/2. \tag{3.89}$$

Hence, for $0 < \theta_s < \pi/2$ the evanescent field is of the order $R^{-3/2}$ and is therefore negligible compared to the propagating far field, which is of order R^{-1}.

For $r \gg r'$ we have $R = r - \mathbf{r}'\cdot\hat{\mathbf{r}} + O(r^{-1})$, $R^{-1} = r^{-1} + O(r^{-2})$, $R\sin\theta_s = \sqrt{(x-x')^2 + (y-y')^2} = r\sin\theta - x'\cos\phi - y'\sin\phi + O(r^{-1})$, $\sin\theta_s = \sin\theta + O(r^{-1})$, and $\cos\theta_s = \cos\theta + O(r^{-1})$, where (r, θ, ϕ) are the standard spherical coordinates centered at the origin. Inserting these asymptotic relations into (3.88) and (3.89), we obtain

$$\begin{aligned} G_\omega^p(\mathbf{r}, \mathbf{r}') = {} & \frac{e^{ikr}\, e^{-ik\mathbf{r}'\cdot\hat{\mathbf{r}}}}{4\pi r} - \frac{\cos(kr\sin\theta - kx'\cos\phi - ky'\sin\phi - \pi/4)}{2\pi\cos\theta\sqrt{2\pi k\sin\theta}\; r^{3/2}} \\ & + O\left(\frac{1}{r^2}\right), \quad 0 < \theta < \pi/2 \end{aligned} \tag{3.90}$$

and

$$\begin{aligned} G_\omega^e(\mathbf{r}, \mathbf{r}') = \frac{\cos(kr\sin\theta - kx'\cos\phi - ky'\sin\phi - \pi/4)}{2\pi\cos\theta\sqrt{2\pi k\sin\theta}\; r^{3/2}} + O\left(\frac{1}{r^{5/2}}\right), \\ 0 < \theta < \pi/2 \end{aligned} \tag{3.91}$$

which express the propagating and evanescent far fields from the point source located at $\mathbf{r}'$ in terms of the spherical coordinates (r, θ, ϕ).

Observation Points with $\theta = \pi/2$. Note that $\theta_s < \pi/2$ for all observation points in the plane $z = z_0 > z'$, so the formulas (3.85) and (3.86) hold. In the asymptotic formulas (3.88) and (3.89), however, it is assumed that θ_s is a constant less than $\pi/2$, so these asymptotic formulas do not hold for far-field observation points in the plane $z = z_0$. To derive far-field formulas that hold for $z = z_0$, we introduce the polar integration variables (k_ρ, k_ϕ), defined such that $k_x = k_\rho\cos k_\phi$ and $k_y = k_\rho\sin k_\phi$, into (3.73) and use the representation (3.79) for the Bessel function of order zero to get[1]

$$G_\omega^p(\mathbf{r}, \mathbf{r}') = \frac{i}{4\pi}\int_0^k \frac{e^{i\gamma(z-z')}}{\gamma} J_0\left(k_\rho\sqrt{(x-x')^2 + (y-y')^2}\right) k_\rho\, dk_\rho \tag{3.95}$$

---

[1] We can also write from (3.61) that

$$\frac{e^{ikR}}{4\pi R} = \frac{i}{4\pi}\int_0^\infty \frac{e^{i\gamma(z-z')}}{\gamma} J_0\left(k_\rho\sqrt{(x-x')^2 + (y-y')^2}\right) k_\rho\, dk_\rho. \tag{3.92}$$

Thus,

$$G_\omega^p(\mathbf{r}, \mathbf{r}') = \frac{e^{ikR}}{4\pi R} - \frac{i}{4\pi}\int_k^\infty \frac{e^{i\gamma(z-z')}}{\gamma} J_0\left(k_\rho\sqrt{(x-x')^2 + (y-y')^2}\right) k_\rho\, dk_\rho \tag{3.93}$$

and

$$G_\omega^e(\mathbf{r}, \mathbf{r}') = \frac{i}{4\pi}\int_k^\infty \frac{e^{i\gamma(z-z')}}{\gamma} J_0\left(k_\rho\sqrt{(x-x')^2 + (y-y')^2}\right) k_\rho\, dk_\rho. \tag{3.94}$$

These general formulas (3.93) and (3.94), which hold for all θ in the domain $0 \le \theta \le \pi/2$ and give $G_\omega^p(\mathbf{r}, \mathbf{r}')$ and $G_\omega^e(\mathbf{r}, \mathbf{r}')$ in terms of a radial integration, could also be evaluated asymptotically to find far-field expressions for $G_\omega^p(\mathbf{r}, \mathbf{r}')$ and $G_\omega^e(\mathbf{r}, \mathbf{r}')$ at $\theta = 0$ and $0 < \theta < \pi/2$ as well as at $\theta = \pi/2$.

which holds for all (x, y, z), that is, for all θ in the domain $0 \le \theta \le \pi/2$. Inserting the asymptotic approximation

$$J_0(x) \sim \frac{1}{\sqrt{2\pi x}}\left(e^{i(x-\pi/4)} + e^{-i(x-\pi/4)}\right) \tag{3.96}$$

into (3.95), and again using the asymptotic formulas of Bleistein and Handelsman [126, p. 91], we find that the propagating far field in the plane $z = z_0$ is given by

$$\begin{aligned} G^p_\omega(\mathbf{r}, \mathbf{r}') &= \frac{e^{ik\sqrt{(x-x')^2+(y-y')^2}}}{8\pi\sqrt{(x-x')^2+(y-y')^2}} - \frac{e^{-ik\sqrt{(x-x')^2+(y-y')^2}}}{8\pi\sqrt{(x-x')^2+(y-y')^2}} \\ &\quad + O\left(\frac{1}{(x-x')^2+(y-y')^2}\right) \\ &= \frac{i}{4\pi R_\rho}\sin kR_\rho + O\left(\frac{1}{R_\rho^2}\right) \end{aligned} \tag{3.97}$$

where $R_\rho = \sqrt{(x-x')^2+(y-y')^2}$. Combining this result with (3.72) shows that the corresponding evanescent far field in the plane $z = z_0$ is

$$\begin{aligned} G^e_\omega(\mathbf{r}, \mathbf{r}') &= \frac{e^{ik\sqrt{(x-x')^2+(y-y')^2}}}{8\pi\sqrt{(x-x')^2+(y-y')^2}} + \frac{e^{-ik\sqrt{(x-x')^2+(y-y')^2}}}{8\pi\sqrt{(x-x')^2+(y-y')^2}} \\ &\quad + O\left(\frac{1}{(x-x')^2+(y-y')^2}\right) \\ &= \frac{1}{4\pi R_\rho}\cos kR_\rho + O\left(\frac{1}{R_\rho^2}\right). \end{aligned} \tag{3.98}$$

These expressions show that the propagating as well as the evanescent far field contains one incoming and one outgoing wave both of order r^{-1}. Consequently, when the observation point is in a plane perpendicular to the z axis, the propagating and evanescent waves contribute equally to the total far field.

Assuming that $r \gg r'$, we find that the far fields in (3.97) and (3.98) become

$$\begin{aligned} G^p_\omega(\mathbf{r}, \mathbf{r}') &= \frac{e^{ik\sqrt{x^2+y^2}}\, e^{-ik(x'\cos\phi+y'\sin\phi)}}{8\pi\sqrt{x^2+y^2}} \\ &\quad - \frac{e^{-ik\sqrt{x^2+y^2}}\, e^{ik(x'\cos\phi+y'\sin\phi)}}{8\pi\sqrt{x^2+y^2}} + O\left(\frac{1}{x^2+y^2}\right), \quad \theta = \pi/2 \end{aligned} \tag{3.99}$$

and

$$\begin{aligned} G^e_\omega(\mathbf{r}, \mathbf{r}') &= \frac{e^{ik\sqrt{x^2+y^2}}\, e^{-ik(x'\cos\phi+y'\sin\phi)}}{8\pi\sqrt{x^2+y^2}} \\ &\quad + \frac{e^{-ik\sqrt{x^2+y^2}}\, e^{ik(x'\cos\phi+y'\sin\phi)}}{8\pi\sqrt{x^2+y^2}} + O\left(\frac{1}{x^2+y^2}\right), \quad \theta = \pi/2. \end{aligned} \tag{3.100}$$

Also, we can prove from (3.95) and [127, eq. 6.554(2)] that when $z = z' = 0$, so that $\theta = \pi/2$ for all $R_\rho = \sqrt{(x-x')^2+(y-y')^2} > 0$, the first-order equations in (3.97) and (3.98) hold exactly; that is

$$G^p_\omega(\mathbf{r}, \mathbf{r}') = \frac{i}{4\pi R_\rho}\sin kR_\rho, \quad z = z' = 0, \quad \theta = \pi/2 \tag{3.101}$$

and

$$G_\omega^e(\mathbf{r}, \mathbf{r}') = \frac{1}{4\pi R_\rho} \cos kR_\rho, \quad z = z' = 0, \quad \theta = \pi/2. \tag{3.102}$$

Equations (3.101) and (3.102) can be derived simply by noting that G_ω^p in (3.95) is imaginary and G_ω^e in (3.94) is real for $(z - z') = 0$. Therefore, G_ω^p and G_ω^e equal the imaginary and real parts of $e^{ikR_\rho}/(4\pi R_\rho)$, respectively.

The analysis of this section has shown that, except for on-axis observation points ($\theta = 0$) and for observation points in a plane perpendicular to the z axis ($\theta = \pi/2$), the evanescent far field of a point source is negligible compared to the propagating far field of that point source. In Section 3.2.5 we shall determine the far-field contributions from the propagating and evanescent fields of a general source distribution of finite extent. Moreover, time-domain analogs of the point-source formulas of this section will be derived in Section 5.2.1.

3.2.2 Spectrum Given in Terms of the Sources

We shall express the acoustic spectrum $T_\omega(k_x, k_y)$ in terms of the acoustic source function $q_\omega(\mathbf{r})$ and in terms of Huygens' sources on the smooth surface S shown in Figure 3.1. Similarly, the spectra $\mathbf{T}_\omega(k_x, k_y)$ and $\mathbf{T}_\omega^H(k_x, k_y)$ for the electromagnetic field will be expressed in terms of the current source $\mathbf{J}_\omega(\mathbf{r})$ and in terms of Huygens' sources on S.

Inserting the Weyl formula (3.61) for the free-space Green's function into the formula (2.379) for the acoustic pressure gives

$$p_\omega(\mathbf{r}) = \frac{i}{8\pi^2} \int_V q_\omega(\mathbf{r}') \int_{-\infty}^{+\infty}\int_{-\infty}^{+\infty} \frac{e^{i[k_x(x-x')+k_y(y-y')+\gamma(z-z')]}}{\gamma} dk_x dk_y dV', \quad z \geq z_0. \tag{3.103}$$

The finite source volume V is located in the region $z < z_0$ so that $z' < z_0 \leq z$ throughout the integration in (3.103). The integrand in (3.103) therefore decays exponentially at infinity in the (k_x, k_y) plane, and the integration over V can be brought inside the (k_x, k_y) integration. Comparing the resulting expression for p_ω with the plane-wave spectrum formula (3.56) shows that the spectrum can be expressed in terms of the volume sources as [29]

$$T_\omega(k_x, k_y) = \frac{i}{4\pi\gamma} \int_V q_\omega(\mathbf{r}) e^{-i(k_x x + k_y y + \gamma z)} dV. \tag{3.104}$$

To derive an expression for the spectrum in terms of the Huygens' sources on the surface S, note that $(\nabla^2 + k^2)p_\omega = -q_\omega$ and $(\nabla^2 + k^2)e^{-i(k_x x + k_y y + \gamma z)} = 0$. Then the formula (3.104) along with Green's second identity applied to the region bounded by S immediately show that the spectrum can be found from the Huygens' sources on S as

$$T_\omega(k_x, k_y) = \frac{i}{4\pi\gamma} \int_S \Big[-i\mathbf{k} \cdot \hat{\mathbf{n}} p_\omega(\mathbf{r}) - i\omega\rho_{m0}\, \hat{\mathbf{n}} \cdot \mathbf{v}_\omega(\mathbf{r}) \Big] e^{-i(k_x x + k_y y + \gamma z)} dS \tag{3.105}$$

where we have used $\partial p_\omega / \partial n = i\omega\rho_{m0}\, \hat{\mathbf{n}} \cdot \mathbf{v}_\omega$ and $\hat{\mathbf{n}}$ is the outward normal to S. Alternatively, (3.105) could be derived by first inserting the Weyl identity (3.61) into the Huygens' source expression (2.381), then interchanging the orders of integration, and finally comparing the resulting equation to the plane-wave spectrum formula (3.56).

Now consider the magnetic field, which can be expressed in terms of the current as

$$\mathbf{H}_\omega(\mathbf{r}) = \nabla \times \int_V \mathbf{J}_\omega(\mathbf{r}') \frac{e^{ik|\mathbf{r}-\mathbf{r}'|}}{4\pi|\mathbf{r}-\mathbf{r}'|} dV'. \tag{3.106}$$

Substituting the Weyl expression (3.61) into (3.106) and interchanging the orders of integration (which, as in the acoustic case, is valid because of the exponential decay of the integrand at infinity in the (k_x, k_y) plane) gives for $z \geq z_0$ the formula

$$\mathbf{H}_\omega(\mathbf{r}) = \frac{i}{8\pi^2}\nabla \times \int_{-\infty}^{+\infty}\int_{-\infty}^{+\infty} \frac{e^{i(k_x x+k_y y+\gamma z)}}{\gamma} \int_V \mathbf{J}_\omega(\mathbf{r}')e^{-i(k_x x'+k_y y'+\gamma z')}dV'dk_x dk_y. \tag{3.107}$$

Because of the exponential decay of the integrand, we can move the curl operator inside the (k_x, k_y) integration in (3.107). Then, by using $\nabla \times (e^{i\mathbf{k}\cdot\mathbf{r}}\mathbf{J}_\omega(\mathbf{r}')) = i\mathbf{k}e^{i\mathbf{k}\cdot\mathbf{r}} \times \mathbf{J}_\omega(\mathbf{r}')$, and comparing to the plane-wave spectrum formula (3.68), we obtain the desired relation between the spectrum and the current

$$\mathbf{T}^H_\omega(k_x, k_y) = -\frac{1}{4\pi\gamma}\mathbf{k} \times \int_V \mathbf{J}_\omega(\mathbf{r})e^{-i(k_x x+k_y y+\gamma z)}dV. \tag{3.108}$$

The spectrum for the electric field is found from (3.71) to be

$$\mathbf{T}_\omega(k_x, k_y) = \frac{1}{4\pi k\gamma}\sqrt{\frac{\mu}{\epsilon}}\,\mathbf{k} \times \left(\mathbf{k} \times \int_V \mathbf{J}_\omega(\mathbf{r})e^{-i(k_x x+k_y y+\gamma z)}dV\right) \tag{3.109}$$

which is in agreement with Kerns' result [8, ch. 3, eqs. (2.2–7a) and (2.2–7b)].

The expressions for the electromagnetic spectra in terms of the Huygens' sources are obtained in the following way. First substitute the Weyl identity (3.61) into the formulas (2.343)–(2.344) giving the electromagnetic fields in terms of the Huygens' sources. Then, interchange the orders of integration and bring the curl operators inside the (k_x, k_y) integration. Finally, compare the resulting expressions for the electromagnetic fields with the plane-wave spectrum formulas (3.62) and (3.68) to get

$$\begin{aligned}\mathbf{T}^H_\omega(k_x, k_y) = &-\frac{1}{4\pi\gamma}\mathbf{k} \times \int_S \hat{\mathbf{n}} \times \mathbf{H}_\omega(\mathbf{r})e^{-i(k_x x+k_y y+\gamma z)}\,dS \\ &-\frac{1}{4\pi k\gamma}\sqrt{\frac{\epsilon}{\mu}}\,\mathbf{k} \times \left(\mathbf{k} \times \int_S \hat{\mathbf{n}} \times \mathbf{E}_\omega(\mathbf{r})e^{-i(k_x x+k_y y+\gamma z)}\,dS\right)\end{aligned} \tag{3.110}$$

and

$$\begin{aligned}\mathbf{T}_\omega(k_x, k_y) = &\frac{1}{4\pi k\gamma}\sqrt{\frac{\mu}{\epsilon}}\,\mathbf{k} \times \left(\mathbf{k} \times \int_S \hat{\mathbf{n}} \times \mathbf{H}_\omega(\mathbf{r})e^{-i(k_x x+k_y y+\gamma z)}\,dS\right) \\ &-\frac{1}{4\pi\gamma}\mathbf{k} \times \int_S \hat{\mathbf{n}} \times \mathbf{E}_\omega(\mathbf{r})e^{-i(k_x x+k_y y+\gamma z)}dS.\end{aligned} \tag{3.111}$$

Because the expressions for the spectra in this section involve integrations only over *finite* regions, they will prove valuable in the following sections, where the analytic properties of the plane-wave spectra are determined.

If the sources are located in the half space $z > z_0$, the fields in the half space $z \leq z_0$ can be determined from the spectra given in (3.104), (3.105), (3.108), (3.109), (3.110), and (3.111) with $e^{-i(k_x x+k_y y+\gamma z)}$ replaced by $e^{-i(k_x x+k_y y-\gamma z)}$, and $\mathbf{k}$ replaced by $\mathbf{k}^- = k_x\hat{\mathbf{x}} + k_y\hat{\mathbf{y}} - \gamma\hat{\mathbf{z}}$.

3.2.3 Singularities and Asymptotic Behavior of the Spectrum

From the expressions in Section 3.2.2, we can determine the singularities and asymptotic behavior of the spectra. The formula (3.104) shows that the acoustic spectrum can be written as

$$T_\omega(k_x, k_y) = \frac{i}{\gamma} F_\omega(k_x, k_y) \tag{3.112}$$

for all (k_x, k_y), where

$$F_\omega(k_x, k_y) = \frac{1}{4\pi} \int\limits_V q_\omega(x, y, z) e^{-i(k_x x + k_y y + \gamma z)} dV \tag{3.113}$$

is a continuous bounded function for all (k_x, k_y), when the source function $q_\omega(x, y, z)$ is integrable. For reasons that become apparent in Section 3.2.5, $F_\omega(k_x, k_y)$ is named the far-field function.

Equation (3.105) proves that the far-field function can also be written in terms of Huygens' sources as

$$F_\omega(k_x, k_y) = \frac{1}{4\pi} \int\limits_S \left[-i\mathbf{k} \cdot \hat{\mathbf{n}} p_\omega(\mathbf{r}) - i\omega\rho_{m0}\, \hat{\mathbf{n}} \cdot \mathbf{v}_\omega(\mathbf{r})\right] e^{-i(k_x x + k_y y + \gamma z)} dS. \tag{3.114}$$

Since the pressure satisfies the homogeneous Helmholtz equation outside the source volume V, the Huygens' sources p_ω and $\partial p_\omega / \partial n = i\omega\rho_{m0}\, \hat{\mathbf{n}} \cdot \mathbf{v}_\omega$ are continuous on the smooth surface S. Hence, (3.114) shows that $F_\omega(k_x, k_y)$ is always a continuous function of (k_x, k_y). Without any explicit assumptions on the source function q_ω, we have now shown that *the acoustic spectrum $T_\omega(k_x, k_y)$ is continuous for all real (k_x, k_y), except possibly at $k_x^2 + k_y^2 = k^2$ (corresponding to $\gamma = 0$), where it may have a $1/\gamma$ singularity.*

Similarly, in the electromagnetic case we have

$$\mathbf{T}_\omega(k_x, k_y) = \frac{i}{\gamma} \mathbf{F}_\omega(k_x, k_y), \quad \mathbf{T}_\omega^H(k_x, k_y) = \frac{i}{\gamma} \mathbf{F}_\omega^H(k_x, k_y) \tag{3.115}$$

where $\mathbf{F}_\omega(k_x, k_y)$ and $\mathbf{F}_\omega^H(k_x, k_y)$ are the electric and magnetic far-field functions satisfying

$$\mathbf{F}_\omega(k_x, k_y) = -\sqrt{\frac{\mu}{\epsilon}}\, \hat{\mathbf{k}} \times \mathbf{F}_\omega^H(k_x, k_y). \tag{3.116}$$

From (3.108) and (3.110) it is seen that the far-field function for the magnetic field can be expressed in terms of volume and Huygens' sources on S as

$$\mathbf{F}_\omega^H(k_x, k_y) = \frac{i}{4\pi} \mathbf{k} \times \int\limits_V \mathbf{J}_\omega(\mathbf{r}) e^{-i(k_x x + k_y y + \gamma z)} dV \tag{3.117}$$

and

$$\begin{aligned} \mathbf{F}_\omega^H(k_x, k_y) = {} & \frac{i}{4\pi} \mathbf{k} \times \int\limits_S \hat{\mathbf{n}} \times \mathbf{H}_\omega(\mathbf{r}) e^{-i(k_x x + k_y y + \gamma z)} dS \\ & + \frac{i}{4\pi k} \sqrt{\frac{\epsilon}{\mu}}\, \mathbf{k} \times \left(\mathbf{k} \times \int\limits_S \hat{\mathbf{n}} \times \mathbf{E}_\omega(\mathbf{r}) e^{-i(k_x x + k_y y + \gamma z)} dS \right). \end{aligned} \tag{3.118}$$

The volume and Huygens' source expressions for the electric far-field function can be determined from (3.116)–(3.118). Since the electric and magnetic fields satisfy the homogeneous vector-wave equation outside the source region, the Huygens' sources $\hat{\mathbf{n}} \times \mathbf{E}_\omega(\mathbf{r})$ and $\hat{\mathbf{n}} \times \mathbf{H}_\omega(\mathbf{r})$ occurring in (3.118) are continuous functions on the smooth surface S. Thus, the electromagnetic far-field functions are continuous for all (k_x, k_y), and we have shown that *the electromagnetic spectra* $\mathbf{T}_\omega(k_x, k_y)$ *and* $\mathbf{T}^H_\omega(k_x, k_y)$ *are continuous for all real* (k_x, k_y), *except possibly at* $k_x^2 + k_y^2 = k^2$ *(corresponding to* $\gamma = 0$*), where they may have* $1/\gamma$ *singularities.* The $1/\gamma$ singularity in the spectrum $\mathbf{T}_\omega$ is also exhibited by the formulas of Kerns [8, ch. 3, eqs. (2.2–7a) and (2.2–7b)] relating spectra to the prescribed currents.

Next, we shall determine the asymptotic behavior of the acoustic and electromagnetic spectra in the limit $k_x^2 + k_y^2 \to \infty$, which is equivalent to $|\gamma| \to \infty$. Let z_s be the maximum (more precisely, the supremum) value of z for which $q_\omega(x, y, z)$ is nonzero. Then for $k_x^2 + k_y^2 > k^2$, take the absolute value of both sides of (3.113) and multiply by $e^{-|\gamma|z}$ to get

$$|F_\omega(k_x, k_y)|e^{-|\gamma|z} = \frac{e^{-|\gamma|(z-z_s)}}{4\pi}\left|\int_V q_\omega(x', y', z')e^{-i(k_x x'+k_y y')}e^{-|\gamma|(z_s-z')}dV'\right|. \tag{3.119}$$

The value of the integral in (3.119) is finite, even as $|\gamma| \to \infty$, because $z' \le z_s$ throughout the integration over V. Hence for $z > z_s$, the formula (3.119) proves the asymptotic relation

$$\lim_{|\gamma|\to\infty} |F_\omega(k_x, k_y)|e^{-|\gamma|z} = 0, \quad z > z_s. \tag{3.120}$$

The result in (3.120) holds no matter how k_x and k_y approach infinity in the (k_x, k_y) plane. Moreover, (3.119) shows that the far-field function can be expressed as

$$F_\omega(k_x, k_y) = C_\omega(k_x, k_y)e^{|\gamma|z_s}, \quad k_x^2 + k_y^2 > k^2 \tag{3.121}$$

where $C_\omega(k_x, k_y)$ is bounded as $|\gamma| \to \infty$ and given by

$$C_\omega(k_x, k_y) = \frac{1}{4\pi}\int_V q_\omega(x', y', z')e^{-i(k_x x'+k_y y')}e^{-|\gamma|(z_s-z')}dV'. \tag{3.122}$$

Note that equations (3.119)–(3.122) are valid even if the source function $q_\omega(x, y, z)$ contains one-, two-, or three-dimensional delta functions (that is, surface, line, or point sources) and their derivatives. Also, z_s can be a negative, zero, or positive real number, depending on the position of the origin of the xyz coordinate system with respect to the source region. From (3.112) we see that the plane-wave spectrum $T_\omega(k_x, k_y)$ is simply $iF_\omega(k_x, k_y)/\gamma$. Thus, (3.119)–(3.122) determine the asymptotic behavior of the plane-wave spectrum. Specifically,

$$\lim_{|\gamma|\to\infty} |T_\omega(k_x, k_y)|e^{-|\gamma|z} = 0, \quad z \ge z_s \tag{3.123}$$

and

$$T_\omega(k_x, k_y) = C^T_\omega(k_x, k_y)e^{|\gamma|z_s}, \quad k_x^2 + k_y^2 > k^2 \tag{3.124}$$

where

$$\lim_{|\gamma|\to\infty} C^T_\omega(k_x, k_y) = 0. \tag{3.125}$$

The electromagnetic equations analogous to (3.120)–(3.122) can be written as

$$\lim_{|\gamma|\to\infty} |\mathbf{F}_\omega(k_x, k_y)|e^{-|\gamma|z} = 0, \quad z > z_s \tag{3.126}$$

and

$$\mathbf{F}_\omega(k_x, k_y) = \mathbf{C}_\omega(k_x, k_y)e^{|\gamma|z_s}, \quad k_x^2 + k_y^2 > k^2 \tag{3.127}$$

where $\mathbf{C}_\omega(k_x, k_y)$ is bounded as $|\gamma| \to \infty$ and given by

$$\mathbf{C}_\omega(k_x, k_y) = -\frac{i}{4\pi\epsilon\omega}\mathbf{k} \times \left(\mathbf{k} \times \int_V \mathbf{J}_\omega(x', y', z')e^{-i(k_x x'+k_y y')}e^{-|\gamma|(z_s-z')}dV'\right). \tag{3.128}$$

From (3.123)–(3.125) it is seen that the spectrum for the electric field satisfies the relations

$$\lim_{|\gamma|\to\infty} |\mathbf{T}_\omega(k_x, k_y)|e^{-|\gamma|z} = 0, \quad z \geq z_s \tag{3.129}$$

and

$$\mathbf{T}_\omega(k_x, k_y) = \mathbf{C}^T_\omega(k_x, k_y)e^{|\gamma|z_s}, \quad k_x^2 + k_y^2 > k^2 \tag{3.130}$$

where

$$\lim_{|\gamma|\to\infty} \mathbf{C}^T_\omega(k_x, k_y) = 0. \tag{3.131}$$

Here z_s is the maximum (more precisely, the supremum) value of z for which the source current $\mathbf{J}_\omega(x, y, z)$ is nonzero. The asymptotic expressions for the far-field functions will be used in Section 3.2.8 to determine the minimum region in which the sources can be located.

If the sources are located in the half space $z > z_0$, and the source-free half space of interest is $z \leq z_0$, the far-field functions are still defined by (3.112) and (3.115), and are given by (3.114), (3.116), (3.117), and (3.118) with $e^{-i(k_x x + k_y y + \gamma z)}$ replaced by $e^{-i(k_x x + k_y y - \gamma z)}$, $\mathbf{k}$ replaced by $\mathbf{k}^- = k_x\hat{\mathbf{x}} + k_y\hat{\mathbf{y}} - \gamma\hat{\mathbf{z}}$, and $\hat{\mathbf{k}}$ replaced by $\hat{\mathbf{k}}^- = \mathbf{k}^-/k$. Also, with z_s being the minimum (more precisely, the infimum) value of z for which the source function $q_\omega(x, y, z)$ is nonzero, the far-field functions can be written as $F_\omega(k_x, k_y) = C_\omega(k_x, k_y)e^{-|\gamma|z_s}$ and $\mathbf{F}_\omega(k_x, k_y) = \mathbf{C}_\omega(k_x, k_y)e^{-|\gamma|z_s}$ for $k_x^2 + k_y^2 > k^2$, where $C_\omega(k_x, k_y)$ and $\mathbf{C}_\omega(k_x, k_y)$ are bounded as $k_x^2 + k_y^2 \to \infty$. Similarly the spectra can be written as $T_\omega(k_x, k_y) = C^T_\omega(k_x, k_y)e^{-|\gamma|z_s}$ and $\mathbf{T}_\omega(k_x, k_y) = \mathbf{C}^T_\omega(k_x, k_y)e^{-|\gamma|z_s}$ for $k_x^2 + k_y^2 > k^2$, where $C^T_\omega(k_x, k_y)$ and $\mathbf{C}^T_\omega(k_x, k_y)$ approach zero as $k_x^2 + k_y^2 \to \infty$.

3.2.4 Analyticity of Spectrum, Far-Field Pattern, and Fields

To prove that the spectra, the far-field patterns, and the fields are analytic functions of their arguments in certain complex regions, we use the Huygens' source expressions and the following theorem.

Theorem 3-I Let S be a closed smooth surface with a finite area. (For example, S can be a sphere of finite radius.) Assume that for all $w \in A$, the complex function $f(\mathbf{r}', w)$ is analytic with respect to the complex variable w for all $\mathbf{r}' \in S$. Moreover, assume that $f(\mathbf{r}', w)$ and $\partial f(\mathbf{r}', w)/\partial w$ are continuous functions of $(\mathbf{r}', w)$ for all $w \in A$ and $\mathbf{r}' \in S$. Then $\int_S f(\mathbf{r}', w)dS'$ is analytic for $w \in A$.

To prove this theorem, let $w = u + iv$, with u and v real, and use the result of Hobson [71, vol. 2, sec. 246, p. 355] to show that $(\partial/\partial u)\int_S f(\mathbf{r}', u + iv)dS' = \int_S(\partial/\partial u)f(\mathbf{r}', u + iv)dS'$ and $(\partial/\partial v)\int_S f(\mathbf{r}', u+iv)dS' = \int_S(\partial/\partial v)f(\mathbf{r}', u+iv)dS'$. These equations show that $\int_S f(\mathbf{r}', w)dS'$ satisfies the Cauchy-Riemann conditions since these conditions are satisfied by the analytic function $f(\mathbf{r}', w)$. Consequently, $\int_S f(\mathbf{r}', w)dS'$ is an analytic function for $w \in A$, and the theorem is proven. The theorem is analogous to the result of Whittaker and Watson [88, sec. 5.31, p. 92], which holds for line integrals.

Analyticity of the Spectrum and the Far-Field Function. Let us first consider the acoustic spectrum $T_\omega(k_x, k)$ given by the Huygens' source expression (3.105). Since the surface S is smooth and located outside the source region V (see Figure 3.1), the pressure and the normal component of the velocity are continuous on S. Applying Theorem 3-I to the expression (3.105), and using the fact that $\hat{\mathbf{n}}$ is continuous on S, proves that the spectrum T_ω is an analytic function of complex k_x and k_y in the regions of the complex k_x and k_y planes where $\gamma^{-1}e^{-i(k_x x + k_y y + \gamma z)}$ is analytic. To further determine the region of analyticity, we specify that the branch cut of the square root defining γ extend from 0 to $i\infty$. Then we have the following result. *The spectrum* $T_\omega(k_x, k_y)$ *is an analytic function of complex* k_x *and* k_y *around the real* k_x *and* k_y *axes, except possibly at* $k_x^2 + k_y^2 = k^2$ *(corresponding to* $\gamma = 0$*). Moreover, it follows that the far-field function* $F_\omega(k_x, k_y) = -i\gamma T_\omega(k_x, k_y)$ *is also an analytic function of complex* k_x *and* k_y *around the real* k_x *and* k_y *axes, except possibly at* $k_x^2 + k_y^2 = k^2$.

Similarly, from the Huygens' source expressions (3.110)–(3.111) it is found that since the electric and magnetic fields are continuous on S, *the rectangular components of the spectra* $\mathbf{T}_\omega(k_x, k_y)$ *and* $\mathbf{T}_\omega^H(k_x, k_y)$*, as well as the rectangular components of the far-field functions* $\mathbf{F}_\omega(k_x, k_y)$ *and* $\mathbf{F}_\omega^H(k_x, k_y)$*, are analytic functions of* k_x *and* k_y *around the real* k_x *and* k_y *axes, except possibly at* $k_x^2 + k_y^2 = k^2$.

These analytical properties of the spectra and far-field functions show that $T_\omega(k_x, k_y)$ *and* $F_\omega(k_x, k_y)$*, as well as the rectangular components of* $\mathbf{T}_\omega(k_x, k_y)$, $\mathbf{F}_\omega(k_x, k_y)$, $\mathbf{T}_\omega^H(k_x, k_y)$, *and* $\mathbf{F}_\omega^H(k_x, k_y)$*, are infinitely differentiable with respect to real* k_x *and* k_y*, except possibly at* $k_x^2 + k_y^2 = k^2$.

We will now determine some of the properties of the spectrum and far-field function when they are expressed in terms of (γ, k_ϕ), where k_ϕ is defined such that $k_x = k_\rho \cos k_\phi$ and $k_y = k_\rho \sin k_\phi$, with $k_\rho = \sqrt{k^2 - \gamma^2}$. In terms of these variables, $\mathbf{k} = \hat{\mathbf{x}}k_\rho \cos k_\phi + \hat{\mathbf{y}}k_\rho \sin k_\phi + \hat{\mathbf{z}}\gamma$, which is analytic for all complex k_ϕ and for γ in a complex region surrounding the line segments $0 \le \gamma \le k$ and $0 \le -i\gamma \le \infty$, except possibly at $\gamma = k$ (since $dk_\rho/d\gamma = -\gamma/k_\rho = -\gamma/\sqrt{k^2 - \gamma^2}$). Thus, the expression (3.114) along with Theorem 3-I show that the far-field function $F_\omega(\sqrt{k^2 - \gamma^2}\cos k_\phi, \sqrt{k^2 - \gamma^2}\sin k_\phi)$ is analytic for all complex k_ϕ and for γ in a complex region surrounding these line segments. As a result, the function $F_\omega(\sqrt{k^2 - \gamma^2}\cos k_\phi, \sqrt{k^2 - \gamma^2}\sin k_\phi)$ is infinitely differentiable for all real k_ϕ and for $0 \le \gamma \le k$ and $0 \le -i\gamma \le \infty$, except possibly at $\gamma = k$. Because of the singularity at $\gamma = 0$, the spectrum $T_\omega(\sqrt{k^2 - \gamma^2}\cos k_\phi, \sqrt{k^2 - \gamma^2}\sin k_\phi)$ is infinitely differentiable for all real k_ϕ and for $0 \le \gamma \le k$ and $0 \le -i\gamma \le \infty$, except possibly at $\gamma = k$ and $\gamma = 0$. Similar results hold for the rectangular components of the electromagnetic spectra and far-field functions.

Analyticity of the Far-Field Pattern. Consider now the acoustic far-field pattern given by (3.38). Note that $\hat{\mathbf{r}} = \hat{\mathbf{x}}\cos\phi\sin\theta + \hat{\mathbf{y}}\sin\phi\sin\theta + \hat{\mathbf{z}}\cos\theta$ is analytic for all complex θ and ϕ, (that is, $\hat{\mathbf{r}}$ is an entire function of complex θ and ϕ). Since the pressure and the normal component of the velocity are continuous on S, it follows immediately from (3.38) and Theorem 3-I that *the far-field pattern is an entire function of complex* θ *and* ϕ. Similarly, the Huygens' source expressions (3.39) and (3.40) can be used to show that *the rectangular components of the electric and magnetic far-field patterns are entire functions of complex* θ *and* ϕ. As a consequence, if a far-field pattern is zero for $(\theta, \phi) \in B$, where B has a nonzero area, then the pattern is zero everywhere. For example, if the far-field pattern is zero for $\theta_1 < \theta < \theta_2$ and $\phi_1 < \phi < \phi_2$, then the far-field pattern is zero everywhere.

Also, if the far-field pattern is known for $(\theta, \phi) \in B$, where B has a nonzero area, then, by analytic continuation, the far-field pattern is known for all (θ, ϕ). Therefore, because the far-field pattern determines the field everywhere outside the source region (as will be demonstrated in Section 3.2.6), the far field over a solid angular region with a nonzero area determines the field everywhere outside the source region. These results on analyticity of the far-field patterns hold only for sources of finite extent; they have also been proven in [91, th. 26] and [9, sec. 4.3].

Comparing the acoustic expressions (3.38) and (3.105) shows that the acoustic far-field pattern and the spectrum are related through the following well-known expression, which will also be proven by asymptotic analysis in Section 3.2.5

$$\mathcal{F}_\omega(\theta, \phi) = -ik\cos\theta\, T_\omega(k\cos\phi\sin\theta, k\sin\phi\sin\theta), \quad 0 \le \theta < \pi/2 \tag{3.132}$$

and similarly for the electromagnetic field

$$\boldsymbol{\mathcal{F}}_\omega(\theta, \phi) = -ik\cos\theta\, \mathbf{T}_\omega(k\cos\phi\sin\theta, k\sin\phi\sin\theta), \quad 0 \le \theta < \pi/2. \tag{3.133}$$

Since $\sin(\pi/2 - i\alpha) = \cosh\alpha$ and $\cos(\pi/2 - i\alpha) = i\sinh\alpha$, it is seen that the spectrum in the entire evanescent region $k_x^2 + k_y^2 > k^2$ can be determined from the far-field pattern in the complex angular region $\theta = \pi/2 - i\alpha$, with $0 \le \alpha \le +\infty$, and $0 \le \phi \le 2\pi$. Thus, if the spectrum is zero throughout the evanescent region $k_x^2 + k_y^2 > k^2$, the far-field pattern is zero for $\theta = \pi/2 - i\alpha$, with $0 \le \alpha \le +\infty$ and $0 \le \phi \le 2\pi$. Because it is an entire function, the far-field pattern is then zero for all θ and ϕ, and the spectrum is seen from (3.132) to be zero for all k_x and k_y. We have now proven that for sources in a finite region of space, the spectrum is zero for all k_x and k_y if it is zero throughout the evanescent region $k_x^2 + k_y^2 > k^2$. Thus, a finite acoustic or electromagnetic source region cannot radiate a field that contains only propagating plane waves. (This result was first proven for scalar fields by Wolf and Nieto-Vesperinas [128] under the assumption that the volume-source function is continuous. Our proof relies on the continuity of the Huygens' sources on S and does not require continuity of the volume-source function.)

Analyticity of the Fields. Having dealt with the spectra and the far-field patterns, consider the fields themselves. Begin with the Huygens' source expression (2.384) for the acoustic pressure and require that the branch cut for the square root defining $R = [(x - x')^2 + (y - y')^2 + (z - z')^2]^{1/2}$ extend from 0 to $i\infty$. Let $\mathbf{r} = x\hat{\mathbf{x}} + y\hat{\mathbf{y}} + z\hat{\mathbf{z}}$ be a fixed real observation point outside the surface S. Then R is an analytic function with respect to x in a complex region around the real value of x attained at the fixed observation point. Similar results hold for the variables y and z. Consequently, R is analytic with respect to x, y, and z in complex regions that surround real observation points outside the surface S.

The continuity on S of the pressure and the normal component of the velocity can now be used in conjunction with Theorem 3-I to show that *the acoustic field is analytic with respect to x, y, and z in complex regions that surround real observation points outside the surface S.* Similarly, one finds that *the acoustic field is analytic with respect to cylindrical or spherical coordinates in complex regions surrounding real observation points outside S.*

Also, the Huygens' source expressions (2.345) and (2.346) for the magnetic and electric fields can be used to show that *the electromagnetic fields are analytic with respect to rectangular, cylindrical, or spherical coordinates in complex regions surrounding real observation points outside S.* Thus, if the acoustic or electromagnetic fields are identically zero in a region with nonzero volume located outside S, then they must be zero everywhere outside S. Also, by analytic continuation, the near fields in a free-space region with nonzero volume determine the fields everywhere outside the source region.

3.2.5 Far-Field Expressions

We shall express the acoustic and electromagnetic far fields in terms of the spectra and far-field functions. We evaluate the separate far-field contributions from the propagating and evanescent plane waves for directions of observation in the entire range $0 \le \theta \le \pi/2$, which includes observation points in the scan plane $z = z_0$. Some of the acoustic far-field results have been derived previously by Sherman, Stamnes, and Lalor [129] and Nieto-Vesperinas [9, sec. 2.12]. Traditionally, far-field formulas are obtained by applying the method of stationary phase to the Fourier integrals (3.56), (3.62), and (3.68). In this section, however, we obtain the far-field formulas with the help of the asymptotic results for the point source in Section 3.2.1 and the expressions for the fields in terms of their volume sources.

For the acoustic pressure with volume source $q_\omega(\mathbf{r})$, (2.379) and (3.56) show that

$$p_\omega(\mathbf{r}) = \frac{1}{2\pi} \int_{-\infty}^{+\infty}\int_{-\infty}^{+\infty} T_\omega(k_x, k_y) e^{i(k_x x + k_y y + \gamma z)} dk_x dk_y = \int_V G_\omega(\mathbf{r}, \mathbf{r}') q_\omega(\mathbf{r}') dV' \tag{3.134}$$

where $G_\omega(\mathbf{r}, \mathbf{r}') = e^{ik|\mathbf{r}-\mathbf{r}'|}/(4\pi|\mathbf{r} - \mathbf{r}'|)$ is the free-space Green's function.

The propagating field $p_\omega^p(\mathbf{r})$ corresponding to the acoustic pressure $p_\omega(\mathbf{r})$ is simply that contribution to $p_\omega(\mathbf{r})$ which comes from the propagating plane waves, that is

$$p_\omega^p(\mathbf{r}) = \frac{1}{2\pi} \iint\limits_{k_x^2+k_y^2<k^2} T_\omega(k_x, k_y) e^{i(k_x x + k_y y + \gamma z)} dk_x dk_y = \int\limits_V G_\omega^p(\mathbf{r}, \mathbf{r}') q_\omega(\mathbf{r}') dV' \tag{3.135}$$

where $G_\omega^p(\mathbf{r}, \mathbf{r}')$ is the propagating point-source field given by (3.73). Similarly, the contribution from the evanescent plane waves is referred to as the evanescent field and is given by

$$p_\omega^e(\mathbf{r}) = \frac{1}{2\pi} \iint\limits_{k_x^2+k_y^2>k^2} T_\omega(k_x, k_y) e^{i(k_x x + k_y y + \gamma z)} dk_x dk_y = \int\limits_V G_\omega^e(\mathbf{r}, \mathbf{r}') q_\omega(\mathbf{r}') dV' \tag{3.136}$$

where $G_\omega^e(\mathbf{r}, \mathbf{r}')$ is the evanescent point-source field in (3.74). The total pressure is then given by $p_\omega(\mathbf{r}) = p_\omega^p(\mathbf{r}) + p_\omega^e(\mathbf{r})$.

For the electric field we can introduce the vector source function $\mathbf{Q}_\omega(\mathbf{r})$, which is given in terms of ρ_ω and $\mathbf{J}_\omega$ by (3.205) below, so that the electric field is

$$\mathbf{E}_\omega(\mathbf{r}) = \frac{1}{2\pi} \int_{-\infty}^{+\infty}\int_{-\infty}^{+\infty} \mathbf{T}_\omega(k_x, k_y) e^{i(k_x x + k_y y + \gamma z)} dk_x dk_y = \int\limits_V G_\omega(\mathbf{r}, \mathbf{r}') \mathbf{Q}_\omega(\mathbf{r}') dV'. \tag{3.137}$$

Analogous to the definitions for the acoustic pressure, we define the propagating and evanescent electric fields by

$$\mathbf{E}_\omega^p(\mathbf{r}) = \frac{1}{2\pi} \iint\limits_{k_x^2+k_y^2<k^2} \mathbf{T}_\omega(k_x, k_y) e^{i(k_x x + k_y y + \gamma z)} dk_x dk_y = \int\limits_V G_\omega^p(\mathbf{r}, \mathbf{r}') \mathbf{Q}_\omega(\mathbf{r}') dV' \tag{3.138}$$

and

$$\mathbf{E}_\omega^e(\mathbf{r}) = \frac{1}{2\pi} \iint\limits_{k_x^2+k_y^2>k^2} \mathbf{T}_\omega(k_x, k_y) e^{i(k_x x + k_y y + \gamma z)} dk_x dk_y = \int\limits_V G_\omega^e(\mathbf{r}, \mathbf{r}') \mathbf{Q}_\omega(\mathbf{r}') dV'. \tag{3.139}$$

The total electric field is then $\mathbf{E}_\omega(\mathbf{r}) = \mathbf{E}_\omega^p(\mathbf{r}) + \mathbf{E}_\omega^e(\mathbf{r})$. The corresponding magnetic fields are given by similar expressions. Far-field formulas for the propagating and evanescent fields can now be derived by substituting asymptotic formulas for $G_\omega^p(\mathbf{r}, \mathbf{r}')$ and $G_\omega^e(\mathbf{r}, \mathbf{r}')$ into (3.135), (3.136), (3.138), and (3.139).

Far Fields for $\theta = 0$. We first consider the special case where the observation point is along the z axis, that is, $\theta = 0$ and $\mathbf{r} = z\hat{\mathbf{z}}$. Substituting the expression (3.78) into (3.135) shows that the propagating acoustic far field is

$$\begin{aligned} p_\omega^p(\mathbf{r}) = {} & \frac{e^{ikz}}{4\pi z} \int\limits_V q_\omega(\mathbf{r}') e^{-ikz'} dV' \\ & - \frac{1}{8\pi^2 z} \int_0^{2\pi} \int\limits_V q_\omega(\mathbf{r}') e^{-ik(x' \cos k_\phi + y' \sin k_\phi)} dV' \, dk_\phi + O\left(\frac{1}{z^2}\right). \end{aligned} \tag{3.140}$$

From the expressions (3.104) and (3.113) for $T_\omega(k_x, k_y)$ and $F_\omega(k_x, k_y)$ it is seen that (3.140) implies

$$p_\omega^p(\mathbf{r}) = -ikT_\omega(0, 0) \frac{e^{ikz}}{z} - \frac{1}{2\pi z} \int_0^{2\pi} F_\omega(k \cos k_\phi, k \sin k_\phi) dk_\phi + O\left(\frac{1}{z^2}\right) \tag{3.141}$$

which expresses the propagating far field in terms of the spectrum and the far-field function. Similarly, by substituting (3.81) into (3.136) we find that the evanescent far field is

$$p_\omega^e(\mathbf{r}) = \frac{1}{2\pi z}\int_0^{2\pi} F_\omega(k\cos k_\phi, k\sin k_\phi)dk_\phi + O\left(\frac{1}{z^2}\right) \tag{3.142}$$

which shows that, unless the integral in (3.142) is zero, the evanescent plane-wave contribution to the on-axis far field is of order z^{-1}. The integrand in (3.142) can also be expressed in terms of the far-field pattern evaluated at $\theta = \pi/2$ as $F_\omega(k\cos k_\phi, k\sin k_\phi) = \mathcal{F}_\omega(\pi/2, k_\phi)$. Adding (3.141) and (3.142) shows that the total on-axis far field is simply

$$p_\omega(\mathbf{r}) = -ikT_\omega(0,0)\frac{e^{ikz}}{z} + O\left(\frac{1}{z^2}\right) \tag{3.143}$$

which is in agreement with the Green's function far-field expressions (3.21), (3.26), and (3.41).

For the electric field we find similarly that the propagating far field is

$$\mathbf{E}_\omega^p(\mathbf{r}) = -ik\mathbf{T}_\omega(0,0)\frac{e^{ikz}}{z} - \frac{1}{2\pi z}\int_0^{2\pi}\mathbf{F}_\omega(k\cos k_\phi, k\sin k_\phi)dk_\phi + O\left(\frac{1}{z^2}\right) \tag{3.144}$$

and that the evanescent far field is

$$\mathbf{E}_\omega^e(\mathbf{r}) = \frac{1}{2\pi z}\int_0^{2\pi}\mathbf{F}_\omega(k\cos k_\phi, k\sin k_\phi)dk_\phi + O\left(\frac{1}{z^2}\right) \tag{3.145}$$

which shows that the propagating and evanescent on-axis far fields are both of order z^{-1}. The integrand in (3.145) can alternatively be expressed in terms of the vector far-field pattern for the electric field evaluated at $\theta = \pi/2$ as $\mathbf{F}_\omega(k\cos k_\phi, k\sin k_\phi) = \boldsymbol{\mathcal{F}}_\omega(\pi/2, k_\phi)$. The total on-axis electric far field is

$$\mathbf{E}_\omega(\mathbf{r}) = -ik\mathbf{T}_\omega(0,0)\frac{e^{ikz}}{z} + O\left(\frac{1}{z^2}\right) \tag{3.146}$$

which is in agreement with the Green's function formulas (3.23), (3.28), and (3.43). The formulas for the magnetic field are given by (3.144), (3.145), and (3.146) with $\mathbf{T}_\omega(k_x, k_y)$ and $\mathbf{F}_\omega(k_x, k_y)$ replaced by $(\epsilon/\mu)^{1/2}\hat{\mathbf{k}} \times \mathbf{T}_\omega(k_x, k_y)$ and $(\epsilon/\mu)^{1/2}\hat{\mathbf{k}} \times \mathbf{F}_\omega(k_x, k_y)$, respectively.

The results of this section show that the far-field patterns for the acoustic and electromagnetic fields for $\theta = 0$ can be expressed in terms of the far-field functions and spectra as

$$\mathcal{F}_\omega(0,0) = F_\omega(0,0) = -ikT_\omega(0,0) \tag{3.147}$$

$$\boldsymbol{\mathcal{F}}_\omega(0,0) = \mathbf{F}_\omega(0,0) = -ik\mathbf{T}_\omega(0,0) \tag{3.148}$$

and

$$\boldsymbol{\mathcal{F}}_\omega^H(0,0) = \mathbf{F}_\omega^H(0,0) = -ik\mathbf{T}_\omega^H(0,0). \tag{3.149}$$

Far Fields for 0 < θ < π/2. To obtain an expression for the propagating far-field pressure, insert the asymptotic formula (3.90) for $G_\omega^p(\mathbf{r}, \mathbf{r}')$ into (3.135) to get

$$p_\omega^p(\mathbf{r}) = \frac{e^{ikr}}{4\pi r}\int_V e^{-ik\mathbf{r}'\cdot\hat{\mathbf{r}}}q_\omega(\mathbf{r}')dV' - \frac{1}{2\pi\cos\theta\sqrt{2\pi k\sin\theta}\; r^{3/2}} \cdot \int_V \cos(kr\sin\theta - kx'\cos\phi - ky'\sin\phi - \pi/4)q_\omega(\mathbf{r}')dV' + O\left(\frac{1}{r^2}\right). \tag{3.150}$$

Writing $\cos x = (e^{ix} + e^{-ix})/2$ and introducing the expressions (3.104) and (3.113) for $T_\omega(k_x, k_y)$ and $F_\omega(k_x, k_y)$, we see that (3.150) becomes

$$
\begin{aligned}
p_\omega^p(\mathbf{r}) = & -ik\cos\theta T_\omega(k\cos\phi\sin\theta, k\sin\phi\sin\theta)\frac{e^{ikr}}{r} \\
& -\frac{e^{i(kr\sin\theta-\pi/4)}F_\omega(k\cos\phi, k\sin\phi)}{\cos\theta\sqrt{2\pi k\sin\theta}\; r^{3/2}} \\
& -\frac{e^{-i(kr\sin\theta-\pi/4)}F_\omega(-k\cos\phi, -k\sin\phi)}{\cos\theta\sqrt{2\pi k\sin\theta}\; r^{3/2}} + O\left(\frac{1}{r^2}\right).
\end{aligned}
\tag{3.151}
$$

Similarly, by substituting the formula (3.91) for $G_\omega^e(\mathbf{r}, \mathbf{r}')$ into (3.136), we find that the evanescent far field is

$$
\begin{aligned}
p_\omega^e(\mathbf{r}) = & \frac{e^{i(kr\sin\theta-\pi/4)}F_\omega(k\cos\phi, k\sin\phi)}{\cos\theta\sqrt{2\pi k\sin\theta}\; r^{3/2}} \\
& +\frac{e^{-i(kr\sin\theta-\pi/4)}F_\omega(-k\cos\phi, -k\sin\phi)}{\cos\theta\sqrt{2\pi k\sin\theta}\; r^{3/2}} + O\left(\frac{1}{r^{5/2}}\right).
\end{aligned}
\tag{3.152}
$$

Hence, for $0 < \theta < \pi/2$, the evanescent far field is of order $r^{-3/2}$ and is negligible compared to the propagating far field, which is of order r^{-1}. By adding (3.151) and (3.152), the total far field is found to be

$$
p_\omega(\mathbf{r}) = -ik\cos\theta T_\omega(k\cos\phi\sin\theta, k\sin\phi\sin\theta)\frac{e^{ikr}}{r} + O\left(\frac{1}{r^2}\right) \tag{3.153}
$$

which is identical to the far-field formulas (3.21), (3.26), and (3.41) obtained from the Green's function representation.

The analogous formulas for the propagating, evanescent, and total electric far fields are

$$
\begin{aligned}
\mathbf{E}_\omega^p(\mathbf{r}) = & -ik\cos\theta \mathbf{T}_\omega(k\cos\phi\sin\theta, k\sin\phi\sin\theta)\frac{e^{ikr}}{r} \\
& -\frac{e^{i(kr\sin\theta-\pi/4)}\mathbf{F}_\omega(k\cos\phi, k\sin\phi)}{\cos\theta\sqrt{2\pi k\sin\theta}\; r^{3/2}} \\
& -\frac{e^{-i(kr\sin\theta-\pi/4)}\mathbf{F}_\omega(-k\cos\phi, -k\sin\phi)}{\cos\theta\sqrt{2\pi k\sin\theta}\; r^{3/2}} + O\left(\frac{1}{r^2}\right)
\end{aligned}
\tag{3.154}
$$

$$
\begin{aligned}
\mathbf{E}_\omega^e(\mathbf{r}) = & \frac{e^{i(kr\sin\theta-\pi/4)}\mathbf{F}_\omega(k\cos\phi, k\sin\phi)}{\cos\theta\sqrt{2\pi k\sin\theta}\; r^{3/2}} \\
& +\frac{e^{-i(kr\sin\theta-\pi/4)}\mathbf{F}_\omega(-k\cos\phi, -k\sin\phi)}{\cos\theta\sqrt{2\pi k\sin\theta}\; r^{3/2}} + O\left(\frac{1}{r^{5/2}}\right)
\end{aligned}
\tag{3.155}
$$

and

$$
\mathbf{E}_\omega(\mathbf{r}) = -ik\cos\theta \mathbf{T}_\omega(k\cos\phi\sin\theta, k\sin\phi\sin\theta)\frac{e^{ikr}}{r} + O\left(\frac{1}{r^2}\right). \tag{3.156}
$$

The formulas for the magnetic field are given by (3.154), (3.155), and (3.156) with $\mathbf{T}_\omega(k_x, k_y)$ and $\mathbf{F}_\omega(k_x, k_y)$ replaced by $(\epsilon/\mu)^{1/2}\hat{\mathbf{k}} \times \mathbf{T}_\omega(k_x, k_y)$ and $(\epsilon/\mu)^{1/2}\hat{\mathbf{k}} \times \mathbf{F}_\omega(k_x, k_y)$, respectively.

According to the results of this section, the far-field patterns defined in (3.30) and (3.31) for the acoustic and electromagnetic fields can be expressed in terms of the far-field functions (see Sections 3.2.3 and 3.2.4) and the spectra for $0 < \theta < \pi/2$ as

$$\begin{aligned}\mathcal{F}_\omega(\theta, \phi) &= F_\omega(k \cos\phi \sin\theta, k \sin\phi \sin\theta) \\ &= -ik \cos\theta\, T_\omega(k \cos\phi \sin\theta, k \sin\phi \sin\theta)\end{aligned} \tag{3.157}$$

$$\begin{aligned}\boldsymbol{\mathcal{F}}_\omega(\theta, \phi) &= \mathbf{F}_\omega(k \cos\phi \sin\theta, k \sin\phi \sin\theta) \\ &= -ik \cos\theta\, \mathbf{T}_\omega(k \cos\phi \sin\theta, k \sin\phi \sin\theta)\end{aligned} \tag{3.158}$$

and

$$\begin{aligned}\boldsymbol{\mathcal{F}}^H_\omega(\theta, \phi) &= \mathbf{F}^H_\omega(k \cos\phi \sin\theta, k \sin\phi \sin\theta) \\ &= -ik \cos\theta\, \mathbf{T}^H_\omega(k \cos\phi \sin\theta, k \sin\phi \sin\theta)\end{aligned} \tag{3.159}$$

where $\boldsymbol{\mathcal{F}}_\omega(\theta, \phi)$ and $\boldsymbol{\mathcal{F}}^H_\omega(\theta, \phi)$ are related through (3.34).

Far Fields for $\theta = \pi/2$. Finally, let us consider the far fields in the scan plane $z = z_0$ where $\mathbf{r} = x\hat{\mathbf{x}} + y\hat{\mathbf{y}} + z_0\hat{\mathbf{z}}$. Inserting the asymptotic expansion (3.99) for $G^p_\omega(\mathbf{r}, \mathbf{r}')$ into (3.135) and employing the expression (3.113) for $F_\omega(k_x, k_y)$ shows that the propagating acoustic far field is

$$\begin{aligned}p^p_\omega(\mathbf{r}) = {} & F_\omega(k\cos\phi, k\sin\phi)\frac{e^{ik\sqrt{x^2+y^2}}}{2\sqrt{x^2+y^2}} \\ & - F_\omega(-k\cos\phi, -k\sin\phi)\frac{e^{-ik\sqrt{x^2+y^2}}}{2\sqrt{x^2+y^2}} + O\left(\frac{1}{x^2+y^2}\right).\end{aligned} \tag{3.160}$$

Similarly, the evanescent far field is found to be

$$\begin{aligned}p^e_\omega(\mathbf{r}) = {} & F_\omega(k\cos\phi, k\sin\phi)\frac{e^{ik\sqrt{x^2+y^2}}}{2\sqrt{x^2+y^2}} \\ & + F_\omega(-k\cos\phi, -k\sin\phi)\frac{e^{-ik\sqrt{x^2+y^2}}}{2\sqrt{x^2+y^2}} + O\left(\frac{1}{x^2+y^2}\right).\end{aligned} \tag{3.161}$$

Hence, for $\theta = \pi/2$ both propagating and evanescent plane waves contribute to the far field with a term of order r^{-1}, and both contribute with incoming as well as outgoing waves. Adding the contributions from the propagating and evanescent plane waves, we find that the total far field is an outgoing wave given by

$$p_\omega(\mathbf{r}) = F_\omega(k\cos\phi, k\sin\phi)\frac{e^{ik\sqrt{x^2+y^2}}}{\sqrt{x^2+y^2}} + O\left(\frac{1}{x^2+y^2}\right) \tag{3.162}$$

which agrees with the result obtained by letting $\theta \to \pi/2$ in the formula (3.153). This expression shows that the incoming waves in the expressions (3.160) and (3.161) cancel. Thus, the incoming waves result from dividing the spectrum into propagating and evanescent parts. Consequently, when the sources are in a finite region of space and radiate in a direction for which $\theta = \pi/2$, the evanescent part of the spectrum cannot be zero. (In Section 3.2.4 it was shown that a finite source region cannot radiate a field for which the evanescent part of the spectrum is zero, unless the spectrum is zero everywhere.) Furthermore, by neglecting the contribution from the evanescent part of the spectrum, we obtain the erroneous result that waves are coming in from infinity.

For the electric field, one finds similarly that the contribution to the far field in the scan plane from the propagating plane waves is

$$\begin{aligned}\mathbf{E}^p_\omega(\mathbf{r}) &= \mathbf{F}_\omega(k\cos\phi, k\sin\phi)\frac{e^{ik\sqrt{x^2+y^2}}}{2\sqrt{x^2+y^2}}\\ &\quad -\mathbf{F}_\omega(-k\cos\phi, -k\sin\phi)\frac{e^{-ik\sqrt{x^2+y^2}}}{2\sqrt{x^2+y^2}} + O\left(\frac{1}{x^2+y^2}\right)\end{aligned} \tag{3.163}$$

and that the contribution from the evanescent plane waves is

$$\begin{aligned}\mathbf{E}^e_\omega(\mathbf{r}) &= \mathbf{F}_\omega(k\cos\phi, k\sin\phi)\frac{e^{ik\sqrt{x^2+y^2}}}{2\sqrt{x^2+y^2}}\\ &\quad +\mathbf{F}_\omega(-k\cos\phi, -k\sin\phi)\frac{e^{-ik\sqrt{x^2+y^2}}}{2\sqrt{x^2+y^2}} + O\left(\frac{1}{x^2+y^2}\right).\end{aligned} \tag{3.164}$$

The total electric far field is then

$$\mathbf{E}_\omega(\mathbf{r}) = \mathbf{F}_\omega(k\cos\phi, k\sin\phi)\frac{e^{ik\sqrt{x^2+y^2}}}{\sqrt{x^2+y^2}} + O\left(\frac{1}{x^2+y^2}\right) \tag{3.165}$$

which agrees with the result (3.156) with $\theta \to \pi/2$. The formulas for the magnetic field are given by (3.163), (3.164), and (3.165), with $\mathbf{F}_\omega(k_x, k_y)$ replaced by $(\epsilon/\mu)^{1/2}\hat{\mathbf{k}} \times \mathbf{F}_\omega(k_x, k_y)$.

For $\theta = \pi/2$ the far-field patterns for the acoustic and electromagnetic fields can conveniently be expressed in terms of the far-field functions as

$$\mathcal{F}_\omega(\pi/2, \phi) = F_\omega(k\cos\phi, k\sin\phi) \tag{3.166}$$

$$\boldsymbol{\mathcal{F}}_\omega(\pi/2, \phi) = \mathbf{F}_\omega(k\cos\phi, k\sin\phi) \tag{3.167}$$

and

$$\boldsymbol{\mathcal{F}}^H_\omega(\pi/2, \phi) = \mathbf{F}^H_\omega(k\cos\phi, k\sin\phi). \tag{3.168}$$

Since the spectra have singularities at $k_x^2 + k_y^2 = k^2$, corresponding to $\theta = \pi/2$, it is not possible to express the far fields at $\theta = \pi/2$ in terms of the spectra $T_\omega(k_x, k_y)$, $\mathbf{T}_\omega(k_x, k_y)$, and $\mathbf{T}^H_\omega(k_x, k_y)$.

Power Radiated into the Hemisphere. We have just determined the separate asymptotic far-field contributions of the propagating and evanescent fields. We shall now determine the power radiated by the propagating and evanescent fields into the hemisphere $0 \le \theta \le \pi/2$.

For the acoustic field this radiated power is given by (see Section 2.4.8)

$$\begin{aligned}\mathcal{P}^h_{ac} &= \lim_{r\to\infty}\frac{r^2}{2}\int_0^{2\pi}\int_0^{\pi/2}\mathrm{Re}\left[p_\omega(\mathbf{r})\,\hat{\mathbf{r}}\cdot\mathbf{v}^*_\omega(\mathbf{r})\right]\sin\theta d\theta d\phi\\ &= -\lim_{r\to\infty}\frac{r^2}{2\omega\rho_{m0}}\int_0^{2\pi}\int_0^{\pi/2}\mathrm{Im}\left[p_\omega(\mathbf{r})\frac{\partial}{\partial r}p^*_\omega(\mathbf{r})\right]\sin\theta d\theta d\phi\end{aligned} \tag{3.169}$$

where the relation $\mathbf{v}^*_\omega = -\nabla p^*_\omega/(i\omega\rho_{m0})$ has been used. From $p_\omega(\mathbf{r}) = p^p_\omega(\mathbf{r}) + p^e_\omega(\mathbf{r})$ we find

that

$$p_\omega(\mathbf{r})\frac{\partial}{\partial r}p^*_\omega(\mathbf{r}) = p^p_\omega(\mathbf{r})\frac{\partial}{\partial r}p^{p*}_\omega(\mathbf{r}) + p^p_\omega(\mathbf{r})\frac{\partial}{\partial r}p^{e*}_\omega(\mathbf{r}) + p^e_\omega(\mathbf{r})\frac{\partial}{\partial r}p^{p*}_\omega(\mathbf{r}) + p^e_\omega(\mathbf{r})\frac{\partial}{\partial r}p^{e*}_\omega(\mathbf{r}). \tag{3.170}$$

For $0 < \theta < \pi/2$, the formula (3.152) for p^e_ω shows that all the terms in (3.170) that contain p^e_ω decay as $r^{-5/2}$ or faster as $r \to \infty$, so that

$$p_\omega(\mathbf{r})\frac{\partial}{\partial r}p^*_\omega(\mathbf{r}) = p^p_\omega(\mathbf{r})\frac{\partial}{\partial r}p^{p*}_\omega(\mathbf{r}) + O\left(\frac{1}{r^{5/2}}\right), \quad 0 < \theta < \pi/2. \tag{3.171}$$

For $\theta = 0$ and $\theta = \pi/2$, it is found from (3.142) and (3.161) that the terms in (3.170) involving p^e_ω are, in general, of order r^{-2} as $r \to \infty$. However, the values of the integrand in (3.169) at the isolated points $\theta = 0$ and $\theta = \pi/2$ do not affect the value of the integral over θ, so the total power can be expressed solely in terms p^p_ω as

$$\mathcal{P}^h_{ac} = -\lim_{r\to\infty}\frac{r^2}{2\omega\rho_{m0}}\int_0^{2\pi}\int_0^{\pi/2}\operatorname{Im}\left[p^p_\omega(\mathbf{r})\frac{\partial}{\partial r}p^{p*}_\omega(\mathbf{r})\right]\sin\theta d\theta d\phi. \tag{3.172}$$

Consequently, the evanescent plane waves do *not* contribute to the power radiated into the far-field hemisphere $0 \le \theta \le \pi/2$, as will be confirmed in Section 3.2.9 where the radiated power is expressed in terms of $T_\omega(k_x, k_y)$.

For the electromagnetic field, the power radiated through the hemisphere $0 \le \theta \le \pi/2$ is found by integrating Poynting's vector to get (see Section 2.3.9)

$$\mathcal{P}^h_{el} = \lim_{r\to\infty}\frac{r^2}{2}\int_0^{2\pi}\int_0^{\pi/2}\operatorname{Re}\left(\hat{\mathbf{r}}\cdot\left[\mathbf{E}_\omega(\mathbf{r})\times\mathbf{H}^*_\omega(\mathbf{r})\right]\right)\sin\theta d\theta d\phi. \tag{3.173}$$

Writing $\mathbf{E}_\omega(\mathbf{r}) = \mathbf{E}^p_\omega(\mathbf{r}) + \mathbf{E}^e_\omega(\mathbf{r})$, we have

$$\hat{\mathbf{r}}\cdot\left[\mathbf{E}_\omega(\mathbf{r})\times\mathbf{H}^*_\omega(\mathbf{r})\right] = \hat{\mathbf{r}}\cdot\Big[\mathbf{E}^p_\omega(\mathbf{r})\times\mathbf{H}^{p*}_\omega(\mathbf{r}) + \mathbf{E}^p_\omega(\mathbf{r})\times\mathbf{H}^{e*}_\omega(\mathbf{r}) + \mathbf{E}^e_\omega(\mathbf{r})\times\mathbf{H}^{p*}_\omega(\mathbf{r}) + \mathbf{E}^e_\omega(\mathbf{r})\times\mathbf{H}^{e*}_\omega(\mathbf{r})\Big]. \tag{3.174}$$

For $0 < \theta < \pi/2$, the formula (3.155) for $\mathbf{E}^e_\omega$, and the corresponding formula for $\mathbf{H}^e_\omega$, show that all the terms in (3.174) that contain evanescent fields decay as $r^{-5/2}$ or faster as $r \to \infty$. Hence

$$\hat{\mathbf{r}}\cdot\left[\mathbf{E}_\omega(\mathbf{r})\times\mathbf{H}^*_\omega(\mathbf{r})\right] = \hat{\mathbf{r}}\cdot\left[\mathbf{E}^p_\omega(\mathbf{r})\times\mathbf{H}^{p*}_\omega(\mathbf{r})\right] + O\left(\frac{1}{r^{5/2}}\right), \quad 0 < \theta < \pi/2. \tag{3.175}$$

For $\theta = 0$ and $\theta = \pi/2$ it is found from (3.145) and (3.164) that, in general, all terms in (3.174) are of order r^{-2} as $r \to \infty$. Since the values of the integrand in (3.173) at the isolated points $\theta = 0$ and $\theta = \pi/2$ do not affect the value of the integral over θ, the total power can be expressed solely in terms of propagating fields as

$$\mathcal{P}^h_{el} = \lim_{r\to\infty}\frac{r^2}{2}\int_0^{2\pi}\int_0^{\pi/2}\operatorname{Re}\left(\hat{\mathbf{r}}\cdot\left[\mathbf{E}^p_\omega(\mathbf{r})\times\mathbf{H}^{p*}_\omega(\mathbf{r})\right]\right)\sin\theta d\theta d\phi \tag{3.176}$$

and thus the evanescent plane waves do *not* contribute to the power radiated into the far-field hemisphere $0 \le \theta \le \pi/2$. In Section 3.2.9 this fact is confirmed by using Parseval's theorem to express the radiated power in terms of the spectrum $\mathbf{T}_\omega(k_x, k_y)$.

We have now shown that the evanescent fields can contribute with r^{-1} terms to the far fields at $\theta = 0$ and $\theta = \pi/2$, but do not contribute to the power that is radiated into the hemisphere $0 \le \theta \le \pi/2$.

The result that the evanescent spectrum contributes to the far field at $\theta = 0$ and at $\theta = \pi/2$ with terms of order r^{-1} has important practical consequences. When computing the fields of radiators (or the coupling between radiators) from the plane-wave spectrum formulas, usually the evanescent spectrum is unknown. Thus, only the propagating part of the spectrum is used in the numerical integration of the formulas [45]. We have just shown that, in general, this neglect of the evanescent spectrum will produce errors in the computed fields (or coupling) near the $\theta = 0$ and $\theta = \pi/2$ directions, regardless of how large the distance from the radiator (or distance between radiators). This error will usually be negligible for radiators with relatively low far fields at $\theta = \pi/2$. However, for broadbeam radiators, the errors caused by integrating only the propagating part of the spectrum, that is, by neglecting the evanescent part of the spectrum, can be large when computing fields (or coupling) near the $\theta = 0$ direction, as well as the $\theta = \pi/2$ directions.

The contribution from the evanescent spectrum results from dividing the fields into propagating and evanescent parts. The above asymptotic evaluations of the entire plane-wave spectrum representations showed that the far fields for $0 \le \theta < \pi/2$ are ultimately determined by the propagating spectra alone [130].

Finally, we mention that if the sources are located in the half space $z > z_0$, the total far fields for $\pi/2 < \theta \le \pi$ are given by the negative of (3.153) and (3.156).

3.2.6 Near Fields in Terms of the Far Fields in Spherical Coordinates

By expressing the spectra in terms of the far-field patterns, we derive formulas for the fields in a half space in terms of the far-field patterns and their analytic continuations. We begin by using (3.157) (valid for $0 < \theta < \pi/2$) and (3.147) (valid for $\theta = 0$) to show that the acoustic far-field pattern is related to the spectrum through the well-known expression

$$\mathcal{F}_\omega(\theta, \phi) = -ik\cos\theta\, T_\omega(k\cos\phi\sin\theta, k\sin\phi\sin\theta), \qquad 0 \le \theta < \pi/2,\ 0 \le \phi \le 2\pi \tag{3.177}$$

which was derived previously as (3.132) from the Huygens' source expressions (3.38) and (3.105). To find the spectrum in terms of the far-field pattern, we need to express θ and ϕ in terms of $k_x = k\cos\phi\sin\theta$ and $k_y = k\sin\phi\sin\theta$. For $k_x^2 + k_y^2 < k^2$ and $0 \le \theta < \pi/2$ it is found that $\theta = \arccos(\gamma/k)$ and $\phi = \arctan_2(k_x, k_y)$, where the function $\arctan_2$ is defined such that $(k_x^2+k_y^2)^{1/2}\cos(\arctan_2(k_x, k_y)) = k_x$, $(k_x^2+k_y^2)^{1/2}\sin(\arctan_2(k_x, k_y)) = k_y$, and $0 \le \arctan_2(k_x, k_y) < 2\pi$. (For example, if $k_x > 0$ and $k_y > 0$, $\arctan_2(k_x, k_y) = \arctan(k_y/k_x)$.)

With this definition of $\arctan_2$, the spectrum is given by

$$\begin{aligned} T_\omega(k_x, k_y) &= \frac{i}{\gamma} F_\omega(k_x, k_y) \\ &= \frac{i}{\gamma}\mathcal{F}_\omega\left(\arccos\frac{\gamma}{k}, \arctan_2(k_x, k_y)\right), \quad k_x^2 + k_y^2 < k^2 \end{aligned} \tag{3.178}$$

where the first equation in (3.178) is taken from (3.112). For $k_x^2 + k_y^2 > k^2$, the far-field pattern $\mathcal{F}_\omega$ does not immediately give us the spectrum since its values for complex observation angles are required. (The function $\arccos(\gamma/k)$ is complex when γ is purely imaginary.) However, as shown in Section 3.2.4 for sources in a finite region, the far-field pattern $\mathcal{F}_\omega(\theta, \phi)$ is an entire function of (θ, ϕ). In principle, it is therefore possible to analytically continue the far-field pattern $\mathcal{F}_\omega(\theta, \phi)$ from real angles of observation to complex values of θ, so that the expression (3.178) can be used to calculate the spectrum for all real (k_x, k_y).

Similarly, the spectrum (3.63) and far-field pattern (3.31) for the electric field are related by the expressions

$$\boldsymbol{\mathcal{F}}_\omega(\theta, \phi) = -ik\cos\theta\, \mathbf{T}_\omega(k\cos\phi\sin\theta, k\sin\phi\sin\theta), \qquad 0 \le \theta < \pi/2,\ 0 \le \phi \le 2\pi \tag{3.179}$$

$$\begin{aligned}\mathbf{T}_\omega(k_x, k_y) &= \frac{i}{\gamma}\mathbf{F}_\omega(k_x, k_y) \\ &= \frac{i}{\gamma}\boldsymbol{\mathcal{F}}_\omega\left(\arccos\frac{\gamma}{k}, \arctan_2(k_x, k_y)\right), \quad k_x^2 + k_y^2 < k^2.\end{aligned} \tag{3.180}$$

The expression (3.179), which was derived previously as (3.133) from the Huygens' source expressions (3.39) and (3.111), follows from (3.148) and (3.158). Also, Section 3.2.4 shows that the electric far-field pattern, for sources in a finite region of space, is an entire function of complex (θ, ϕ). In principle, we can therefore analytically continue the electric far-field pattern $\boldsymbol{\mathcal{F}}_\omega$ to complex values of θ, and (3.180) can be used to calculate the spectrum for all real (k_x, k_y).

In practice, numerical errors usually prevent the analytic continuation of a measured far-field pattern. When a closed-form expression for the acoustic far-field pattern $\mathcal{F}_\omega(\theta, \phi)$ or the electromagnetic far-field pattern $\boldsymbol{\mathcal{F}}_\omega(\theta, \phi)$ is known for real (θ, ϕ), one way to obtain the analytic continuation to complex values of θ is to expand the far field in a series of spherical harmonics [131, eqs. (3.9) and (4.10)], [37, eq. (6)]. With such an analytic continuation, the near fields can be computed from the far fields through the acoustic formula (3.56) or the electromagnetic formulas (3.62) and (3.68).

To derive a formula for the acoustic pressure in the half space $z \ge z_0$ in terms of the acoustic far-field pattern, insert $T_\omega(k_x, k_y)$ from (3.178) into (3.56), and separate the propagating and evanescent parts of the spectrum, to get

$$\begin{aligned}p_\omega(\mathbf{r}) = &\frac{1}{2\pi}\iint\limits_{k_x^2+k_y^2<k^2} \frac{i}{\gamma}\mathcal{F}_\omega\left(\arccos\frac{\gamma}{k}, \arctan_2(k_x, k_y)\right) e^{i(k_x x + k_y y + \gamma z)} dk_x dk_y \\ &+ \frac{1}{2\pi}\iint\limits_{k_x^2+k_y^2>k^2} \frac{i}{\gamma}\mathcal{F}_\omega\left(\arccos\frac{\gamma}{k}, \arctan_2(k_x, k_y)\right) e^{i(k_x x + k_y y + \gamma z)} dk_x dk_y.\end{aligned} \tag{3.181}$$

Then introduce the new integration variables $(\sigma_\theta, \sigma_\phi)$ given by

$$k_x = k\cos\sigma_\phi\sin\sigma_\theta, \quad k_y = k\sin\sigma_\phi\sin\sigma_\theta, \quad \gamma = k\cos\sigma_\theta \tag{3.182}$$

with $0 \le \sigma_\phi \le 2\pi$ and $\sigma_\theta \in \Gamma$, where Γ is the contour in the complex σ_θ plane consisting of the line segment from 0 to $\pi/2$ and the half line from $\pi/2$ to $\pi/2 - i\infty$, as shown in Figure 3.4. Since $\sin(\pi/2 - i\alpha) = \cosh\alpha$ and $\cos(\pi/2 - i\alpha) = i\sinh\alpha$, it is seen that the entire real

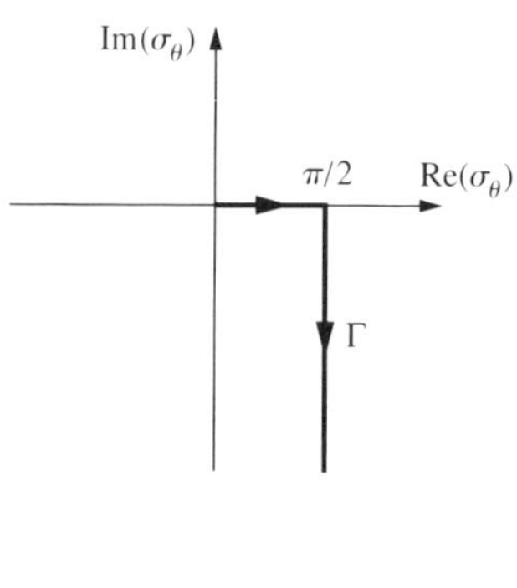

Figure 3.4 The contour Γ in the complex σ_θ plane.

(k_x, k_y) plane is covered through the transformation (3.182). Furthermore, with these new variables we have $dk_x dk_y = k\gamma \sin\sigma_\theta d\sigma_\theta d\sigma_\phi$. Inserting this transformation into the spectral expression (3.181), we obtain

$$
\begin{aligned}
p_\omega(\mathbf{r}) = {} & \frac{1}{2\pi}\int_0^{2\pi}\int_0^{\pi/2} ik\mathcal{F}_\omega(\sigma_\theta,\sigma_\phi)e^{ik\mathbf{r}\cdot\hat{\boldsymbol{\sigma}}_{\theta\phi}}\sin\sigma_\theta d\sigma_\theta d\sigma_\phi \\
& + \frac{1}{2\pi}\int_0^{2\pi}\int_0^{+\infty} k\mathcal{F}_\omega(\pi/2 - i\alpha,\sigma_\phi)e^{ik\cosh\alpha\,\mathbf{r}\cdot\hat{\boldsymbol{\sigma}}_\phi}e^{-kz\sinh\alpha}\cosh\alpha\,d\alpha\,d\sigma_\phi
\end{aligned}
\tag{3.183}
$$

where

$$\hat{\boldsymbol{\sigma}}_{\theta\phi} = \hat{\mathbf{x}}\cos\sigma_\phi\sin\sigma_\theta + \hat{\mathbf{y}}\sin\sigma_\phi\sin\sigma_\theta + \hat{\mathbf{z}}\cos\sigma_\theta \tag{3.184}$$

$$\hat{\boldsymbol{\sigma}}_\phi = \hat{\mathbf{x}}\cos\sigma_\phi + \hat{\mathbf{y}}\sin\sigma_\phi \tag{3.185}$$

and we have used the fact that $d\sigma_\theta = -id\alpha$ for $k_x^2 + k_y^2 > k^2$. Equation (3.183) is the desired expression (valid for $\omega > 0$) for the acoustic pressure in the half space $z \geq z_0$ in terms of the acoustic far-field pattern and its analytic continuation to complex values of σ_θ.

Inserting the pressure in (3.183) into $\mathbf{v}_\omega = \nabla p_\omega/(i\omega\rho_{m0})$, we can express the acoustic velocity in the half space $z \geq z_0$ in terms of the far-field pattern as

$$
\begin{aligned}
\mathbf{v}_\omega(\mathbf{r}) = {} & \frac{1}{2\pi c\rho_{m0}}\int_0^{2\pi}\int_0^{\pi/2} \hat{\boldsymbol{\sigma}}_{\theta\phi}\, ik\,\mathcal{F}_\omega(\sigma_\theta,\sigma_\phi)e^{ik\mathbf{r}\cdot\hat{\boldsymbol{\sigma}}_{\theta\phi}}\sin\sigma_\theta d\sigma_\theta d\sigma_\phi \\
& + \frac{1}{2\pi c\rho_{m0}}\int_0^{2\pi}\int_0^{+\infty} \left(\hat{\boldsymbol{\sigma}}_\phi\cosh\alpha + \hat{\mathbf{z}}i\sinh\alpha\right) k\,\mathcal{F}_\omega(\pi/2 - i\alpha,\sigma_\phi) \\
& \cdot e^{ik\cosh\alpha\,\mathbf{r}\cdot\hat{\boldsymbol{\sigma}}_\phi}e^{-kz\sinh\alpha}\cosh\alpha\,d\alpha\,d\sigma_\phi .
\end{aligned}
\tag{3.186}
$$

For the electric and magnetic fields, $\mathbf{T}_\omega(k_x, k_y)$ can be substituted from (3.180) into (3.62) and (3.68), and the transformation (3.182) can again be applied to give the electromagnetic expressions (valid for $\omega > 0$) analogous to (3.183)

$$
\begin{aligned}
\mathbf{E}_\omega(\mathbf{r}) = {} & \frac{1}{2\pi}\int_0^{2\pi}\int_0^{\pi/2} ik\boldsymbol{\mathcal{F}}_\omega(\sigma_\theta,\sigma_\phi)e^{ik\mathbf{r}\cdot\hat{\boldsymbol{\sigma}}_{\theta\phi}}\sin\sigma_\theta d\sigma_\theta d\sigma_\phi \\
& + \frac{1}{2\pi}\int_0^{2\pi}\int_0^{+\infty} k\boldsymbol{\mathcal{F}}_\omega(\pi/2 - i\alpha,\sigma_\phi)e^{ik\cosh\alpha\,\mathbf{r}\cdot\hat{\boldsymbol{\sigma}}_\phi}e^{-kz\sinh\alpha}\cosh\alpha\,d\alpha\,d\sigma_\phi
\end{aligned}
\tag{3.187}
$$

and

$$
\begin{aligned}
\mathbf{H}_\omega(\mathbf{r}) = {} & \frac{1}{2\pi}\sqrt{\frac{\epsilon}{\mu}}\int_0^{2\pi}\int_0^{\pi/2} ik\hat{\boldsymbol{\sigma}}_{\theta\phi}\times\boldsymbol{\mathcal{F}}_\omega(\sigma_\theta,\sigma_\phi)e^{ik\mathbf{r}\cdot\hat{\boldsymbol{\sigma}}_{\theta\phi}}\sin\sigma_\theta d\sigma_\theta d\sigma_\phi \\
& + \frac{1}{2\pi}\sqrt{\frac{\epsilon}{\mu}}\int_0^{2\pi}\int_0^{+\infty} k(\hat{\boldsymbol{\sigma}}_\phi\cosh\alpha + \hat{\mathbf{z}}i\sinh\alpha) \\
& \times\boldsymbol{\mathcal{F}}_\omega(\pi/2 - i\alpha,\sigma_\phi)e^{ik\cosh\alpha\,\mathbf{r}\cdot\hat{\boldsymbol{\sigma}}_\phi}e^{-kz\sinh\alpha}\cosh\alpha\,d\alpha\,d\sigma_\phi
\end{aligned}
\tag{3.188}
$$

where we have used (3.34) and

$$\mathcal{F}^H_\omega(\pi/2 - i\alpha, \sigma_\phi) = \sqrt{\frac{\epsilon}{\mu}}\,(\hat{\boldsymbol{\sigma}}_\phi \cosh\alpha + \hat{\mathbf{z}} i \sinh\alpha) \times \mathcal{F}_\omega(\pi/2 - i\alpha, \sigma_\phi) \qquad (3.189)$$

which relates the electric and magnetic far-field patterns. Alternatively, (3.188) follows from (3.187) and the Maxwell equation $\mathbf{H}_\omega = \nabla \times \mathbf{E}_\omega/(i\omega\mu)$.

When the sources are located in the half space $z > z_0$, the formulas of this section are valid for the fields in $z \le z_0$ provided the following changes are made. In (3.178), (3.180), and (3.181), $\arccos(\gamma/k)$ must be replaced by $\arccos(-\gamma/k)$, and $e^{i(k_x x + k_y y + \gamma z)}$ must be replaced by $e^{i(k_x x + k_y y - \gamma z)}$. In (3.183), (3.186), (3.187), and (3.188), $\int_0^{\pi/2}$ must be replaced by $\int_{\pi/2}^{\pi}$, and α must be replaced by $-\alpha$, while keeping the signs of $d\sigma_\theta$ and $d\alpha$ the same.

3.2.7 Homogeneous Fields

The expressions (3.183), (3.186), (3.187), and (3.188) for the acoustic and electromagnetic fields are valid only outside the source region in the half space $z \ge z_0$. They involve both the propagating spectra (far-field patterns for real angles on the hemisphere defined by $0 \le \sigma_\theta \le \pi/2$) and the evanescent spectra (far-field patterns for complex angles, $\sigma_\theta = \pi/2 - i\alpha$ with $0 \le \alpha < \infty$). In this section we use the advanced Green's function to define "homogeneous fields," which are closely related to the original fields but which satisfy the homogeneous wave equation throughout all space. It is shown that these homogeneous fields can be expressed in terms of the original far-field patterns for real angles on the far-field sphere defined by $0 \le \sigma_\theta \le \pi$. Thus, the homogeneous fields do not require evanescent spectra.

These expressions for the homogeneous fields are used in Chapter 5 to determine the analogous expressions in the time domain. The analogous expressions in the time domain are especially useful because in the time domain the homogeneous fields are equal to the original fields throughout all space after all sources are turned off.

We begin by deriving an alternative representation for the acoustic pressure $p_\omega(\mathbf{r})$. Let S_0 be a closed surface that contains the observation point $\mathbf{r}$ and the source volume V. Then insert the advanced Green's function

$$G^a_\omega(\mathbf{r}, \mathbf{r}') = \frac{e^{-ik|\mathbf{r}-\mathbf{r}'|}}{4\pi|\mathbf{r}-\mathbf{r}'|} \qquad (3.190)$$

and the acoustic pressure $p_\omega(\mathbf{r}')$ into Green's second identity with volume $V - V_\delta$ bounded by $S - S_\delta$, where V_δ is an infinitesimal volume with surface S_δ surrounding the singularity at $\mathbf{r}' = \mathbf{r}$ [1, sec. 3.4]. Then use the facts that

$$\nabla^2 G^a_\omega(\mathbf{r}, \mathbf{r}') + k^2 G^a_\omega(\mathbf{r}, \mathbf{r}') = 0, \quad \mathbf{r}' \ne \mathbf{r} \qquad (3.191)$$

and

$$\nabla^2 p_\omega(\mathbf{r}) + k^2 p_\omega(\mathbf{r}) = -q_\omega(\mathbf{r}) \qquad (3.192)$$

to get

$$\begin{aligned} p_\omega(\mathbf{r}) = {} & \frac{1}{4\pi} \lim_{\delta\to 0} \int_{V-V_\delta} \frac{q_\omega(\mathbf{r}') e^{-ik|\mathbf{r}-\mathbf{r}'|}}{|\mathbf{r}-\mathbf{r}'|} dV' \\ & + \frac{1}{4\pi} \int_{S_0} \left[\frac{e^{-ik|\mathbf{r}-\mathbf{r}'|}}{|\mathbf{r}-\mathbf{r}'|} \frac{\partial}{\partial n'} p_\omega(\mathbf{r}') - p_\omega(\mathbf{r}') \frac{\partial}{\partial n'} \frac{e^{-ik|\mathbf{r}-\mathbf{r}'|}}{|\mathbf{r}-\mathbf{r}'|} \right] dS' \end{aligned} \qquad (3.193)$$

which is valid for all observation points $\mathbf{r}$ inside S_0. (Throughout the remainder of this section, the limits on the volume integrals will be omitted because the values of the integrals are all independent of the shape of V_δ.) Now let the integration surface S_0 in (3.193) be the sphere of radius r' and express the surface integration in terms of the spherical coordinates $(r', \sigma_\theta, \sigma_\phi)$, defined such that $\hat{\mathbf{r}}' = \hat{\boldsymbol{\sigma}}_{\theta\phi}$ with $\hat{\boldsymbol{\sigma}}_{\theta\phi}$ given by (3.184), and $dS' = r'^2 \sin\sigma_\theta d\sigma_\theta d\sigma_\phi$. Letting $r' \to \infty$ and inserting the asymptotic relations

$$\frac{e^{-ik|\mathbf{r}-\mathbf{r}'|}}{|\mathbf{r}-\mathbf{r}'|} \sim e^{ik\hat{\mathbf{r}}'\cdot\mathbf{r}}\,\frac{e^{-ikr'}}{r'}, \quad \frac{\partial}{\partial n'}\frac{e^{-ik|\mathbf{r}-\mathbf{r}'|}}{|\mathbf{r}-\mathbf{r}'|} \sim -ik\, e^{ik\hat{\mathbf{r}}'\cdot\mathbf{r}}\,\frac{e^{-ikr'}}{r'}, \quad r' \to \infty \tag{3.194}$$

and the far-field formula $p_\omega(\mathbf{r}') \sim \mathcal{F}_\omega(\theta', \phi')e^{ikr'}/r'$ into (3.193) proves that the pressure throughout space (including the source region) can be expressed in terms of its source function and far-field pattern as

$$\begin{aligned} p_\omega(\mathbf{r}) = {} & \frac{1}{4\pi}\int_V \frac{q_\omega(\mathbf{r}')e^{-ik|\mathbf{r}-\mathbf{r}'|}}{|\mathbf{r}-\mathbf{r}'|}dV' \\ & + \frac{1}{2\pi}\int_0^{2\pi}\int_0^{\pi} ik\,\mathcal{F}_\omega(\sigma_\theta, \sigma_\phi)e^{ik\mathbf{r}\cdot\hat{\boldsymbol{\sigma}}_{\theta\phi}} \sin\sigma_\theta d\sigma_\theta d\sigma_\phi. \end{aligned} \tag{3.195}$$

The analogous formula for the acoustic velocity is obtained by inserting (3.195) into $\mathbf{v}_\omega = \nabla p_\omega/(i\omega\rho_{m0})$ to get

$$\begin{aligned} \mathbf{v}_\omega(\mathbf{r}) = {} & \frac{1}{4\pi i\omega\rho_{m0}}\int_V q_\omega(\mathbf{r}')\left[ik + \frac{1}{|\mathbf{r}-\mathbf{r}'|}\right]\frac{e^{-ik|\mathbf{r}-\mathbf{r}'|}}{|\mathbf{r}-\mathbf{r}'|^2}\left(\mathbf{r}'-\mathbf{r}\right)dV' \\ & + \frac{ik}{2\pi c\rho_{m0}}\int_0^{2\pi}\int_0^{\pi} \hat{\boldsymbol{\sigma}}_{\theta\phi}\,\mathcal{F}_\omega(\sigma_\theta, \sigma_\phi)e^{ik\mathbf{r}\cdot\hat{\boldsymbol{\sigma}}_{\theta\phi}} \sin\sigma_\theta d\sigma_\theta d\sigma_\phi. \end{aligned} \tag{3.196}$$

We now define the advanced and homogeneous pressure fields. The acoustic pressure can be expressed in terms of the retarded Green's function (3.3) and the source function q_ω as

$$p_\omega(\mathbf{r}) = \frac{1}{4\pi}\int_V \frac{q_\omega(\mathbf{r}')e^{ik|\mathbf{r}-\mathbf{r}'|}}{|\mathbf{r}-\mathbf{r}'|}dV' \tag{3.197}$$

which holds throughout all space, including the source region. The advanced pressure field $p_\omega^a(\mathbf{r})$ is obtained by replacing the causal retarded Green's function in (3.197) by the advanced Green's function (3.190), so that

$$p_\omega^a(\mathbf{r}) = \frac{1}{4\pi}\int_V \frac{q_\omega(\mathbf{r}')e^{-ik|\mathbf{r}-\mathbf{r}'|}}{|\mathbf{r}-\mathbf{r}'|}dV' \tag{3.198}$$

where $p_\omega^a(\mathbf{r})$ satisfies the same Helmholtz equation (3.192) as $p_\omega(\mathbf{r})$. (Of course, $p_\omega^a(\mathbf{r})$, and $p_\omega^h(\mathbf{r})$ below, do not satisfy the outgoing radiation condition.) Defining the homogeneous pressure field p_ω^h such that $p_\omega = p_\omega^a + p_\omega^h$, we find from (3.198) and (3.197) that

$$p_\omega^h(\mathbf{r}) = \frac{1}{4\pi}\int_V \frac{q_\omega(\mathbf{r}')}{|\mathbf{r}-\mathbf{r}'|}\left(e^{ik|\mathbf{r}-\mathbf{r}'|} - e^{-ik|\mathbf{r}-\mathbf{r}'|}\right)dV'. \tag{3.199}$$

The homogeneous pressure satisfies the homogeneous Helmholtz equation throughout space

$$\nabla^2 p_\omega^h(\mathbf{r}) + k^2 p_\omega^h(\mathbf{r}) = 0 \tag{3.200}$$

and its far field is given by

$$p_\omega^h(\mathbf{r}) \sim \frac{e^{ikr}}{r}\mathcal{F}_\omega(\hat{\mathbf{r}}) - \frac{e^{-ikr}}{r}\mathcal{F}_\omega(-\hat{\mathbf{r}}), \quad r \to \infty \tag{3.201}$$

where $\mathcal{F}_\omega$ is the far-field pattern for p_ω. Since $p_\omega = p^a_\omega + p^h_\omega$, we see from (3.193) and (3.195) that the homogeneous pressure field can be expressed as

$$p^h_\omega(\mathbf{r}) = \frac{1}{4\pi}\int_{S_0}\left[\frac{e^{-ik|\mathbf{r}-\mathbf{r}'|}}{|\mathbf{r}-\mathbf{r}'|}\frac{\partial}{\partial n'}p_\omega(\mathbf{r}') - p_\omega(\mathbf{r}')\frac{\partial}{\partial n'}\frac{e^{-ik|\mathbf{r}-\mathbf{r}'|}}{|\mathbf{r}-\mathbf{r}'|}\right]dS' \tag{3.202}$$

and

$$p^h_\omega(\mathbf{r}) = \frac{1}{2\pi}\int_0^{2\pi}\int_0^{\pi} ik\,\mathcal{F}_\omega(\sigma_\theta, \sigma_\phi)e^{ik\mathbf{r}\cdot\hat{\sigma}_{\theta\phi}}\sin\sigma_\theta d\sigma_\theta d\sigma_\phi. \tag{3.203}$$

In contrast to the formula (3.183) for p_ω, the formula (3.203) for p^h_ω holds throughout all space, and it involves an integration of the far-field pattern over real angles only. It does not require the evanescent spectrum, that is, the analytic continuation of the far-field pattern to complex values of σ_θ. However, the integration in (3.203) covers the entire far-field sphere, whereas the real-angle integration in (3.183) covers only the far-field pattern on the hemisphere in which (3.183) is valid, namely, $z \geq z_0$. The representation (3.203) for homogeneous or "source-free" fields was first obtained by Whittaker and Watson [88, sec. 18.6], and it can also be found in [131] and [9, sec. 2.7].

The expressions for the electric field are easily obtained from the acoustic formulas by noting that the electric field satisfies the inhomogeneous vector Helmholtz equation (2.298), which states that

$$\nabla^2\mathbf{E}_\omega(\mathbf{r}) + k^2\mathbf{E}_\omega(\mathbf{r}) = -\mathbf{Q}_\omega(\mathbf{r}) \tag{3.204}$$

where the electromagnetic source function is

$$\mathbf{Q}_\omega(\mathbf{r}) = i\omega\mu\mathbf{J}_\omega(\mathbf{r}) - \frac{1}{\epsilon}\nabla\rho_\omega(\mathbf{r}). \tag{3.205}$$

We find that the electromagnetic analogs of the acoustic formulas (3.193) and (3.195) are

$$\begin{aligned}\mathbf{E}_\omega(\mathbf{r}) &= \frac{1}{4\pi}\int_V \frac{\mathbf{Q}_\omega(\mathbf{r}')e^{-ik|\mathbf{r}-\mathbf{r}'|}}{|\mathbf{r}-\mathbf{r}'|}dV' \\ &+ \frac{1}{4\pi}\int_{S_0}\left[\frac{e^{-ik|\mathbf{r}-\mathbf{r}'|}}{|\mathbf{r}-\mathbf{r}'|}\frac{\partial}{\partial n'}\mathbf{E}_\omega(\mathbf{r}') - \mathbf{E}_\omega(\mathbf{r}')\frac{\partial}{\partial n'}\frac{e^{-ik|\mathbf{r}-\mathbf{r}'|}}{|\mathbf{r}-\mathbf{r}'|}\right]dS'\end{aligned} \tag{3.206}$$

and [24]

$$\begin{aligned}\mathbf{E}_\omega(\mathbf{r}) &= \frac{1}{4\pi}\int_V \frac{\mathbf{Q}_\omega(\mathbf{r}')e^{-ik|\mathbf{r}-\mathbf{r}'|}}{|\mathbf{r}-\mathbf{r}'|}dV' \\ &+ \frac{1}{2\pi}\int_0^{2\pi}\int_0^{\pi} ik\,\boldsymbol{\mathcal{F}}(\sigma_\theta, \sigma_\phi)e^{ik\mathbf{r}\cdot\hat{\sigma}_{\theta\phi}}\sin\sigma_\theta d\sigma_\theta d\sigma_\phi.\end{aligned} \tag{3.207}$$

The formula (3.206) is valid for all observation points inside S_0, and (3.207) is valid throughout space (including the source region).

The advanced electric field given by

$$\mathbf{E}^a_\omega(\mathbf{r}) = \frac{1}{4\pi}\int_V \frac{\mathbf{Q}_\omega(\mathbf{r}')e^{-ik|\mathbf{r}-\mathbf{r}'|}}{|\mathbf{r}-\mathbf{r}'|}dV' \tag{3.208}$$

satisfies the vector Helmholtz equation (3.204), and the homogeneous electric field $\mathbf{E}^h_\omega$, defined such that $\mathbf{E}_\omega = \mathbf{E}^a_\omega + \mathbf{E}^h_\omega$, can be expressed in terms of the source function $\mathbf{Q}_\omega$ as

$$\mathbf{E}^h_\omega(\mathbf{r}) = \frac{1}{4\pi}\int_V \frac{\mathbf{Q}_\omega(\mathbf{r}')}{|\mathbf{r}-\mathbf{r}'|}\left(e^{ik|\mathbf{r}-\mathbf{r}'|} - e^{-ik|\mathbf{r}-\mathbf{r}'|}\right)dV'. \tag{3.209}$$

The homogeneous electric field satisfies the homogeneous vector Helmholtz equation throughout space

$$\nabla^2\mathbf{E}^h_\omega(\mathbf{r}) + k^2\mathbf{E}^h_\omega(\mathbf{r}) = 0 \tag{3.210}$$

and has a far field given by

$$\mathbf{E}^h_\omega(\mathbf{r}) \sim \frac{e^{ikr}}{r}\mathcal{F}_\omega(\hat{\mathbf{r}}) - \frac{e^{-ikr}}{r}\mathcal{F}_\omega(-\hat{\mathbf{r}}), \quad r\to\infty. \tag{3.211}$$

Since $\mathbf{E}_\omega = \mathbf{E}^a_\omega + \mathbf{E}^h_\omega$, (3.206) and (3.207) show that the homogeneous electric field can be expressed as

$$\mathbf{E}^h_\omega(\mathbf{r}) = \frac{1}{4\pi}\int_{S_0}\left[\frac{e^{-ik|\mathbf{r}-\mathbf{r}'|}}{|\mathbf{r}-\mathbf{r}'|}\frac{\partial}{\partial n'}\mathbf{E}_\omega(\mathbf{r}') - \mathbf{E}_\omega(\mathbf{r}')\frac{\partial}{\partial n'}\frac{e^{-ik|\mathbf{r}-\mathbf{r}'|}}{|\mathbf{r}-\mathbf{r}'|}\right]dS' \tag{3.212}$$

and

$$\mathbf{E}^h_\omega(\mathbf{r}) = \frac{1}{2\pi}\int_0^{2\pi}\int_0^{\pi} ik\,\mathcal{F}(\sigma_\theta,\sigma_\phi)e^{ik\mathbf{r}\cdot\hat{\boldsymbol{\sigma}}_{\theta\phi}}\sin\sigma_\theta d\sigma_\theta d\sigma_\phi \tag{3.213}$$

where $\hat{\boldsymbol{\sigma}}_{\theta\phi}$ is given by (3.184).

The homogeneous magnetic field, $\mathbf{H}^h_\omega = \nabla\times\mathbf{E}^h_\omega/(i\omega\mu)$, found by taking the curl of (3.213), is given by

$$\mathbf{H}^h_\omega(\mathbf{r}) = \frac{1}{2\pi}\sqrt{\frac{\epsilon}{\mu}}\int_0^{2\pi}\int_0^{\pi} ik\hat{\boldsymbol{\sigma}}_{\theta\phi}\times\mathcal{F}(\sigma_\theta,\sigma_\phi)e^{ik\mathbf{r}\cdot\hat{\boldsymbol{\sigma}}_{\theta\phi}}\sin\sigma_\theta d\sigma_\theta d\sigma_\phi. \tag{3.214}$$

In the frequency domain, the advanced fields are in general nonzero, so the homogeneous fields do not equal the actual fields. In the time domain on the other hand, the advanced fields are zero throughout space at times after which all sources are turned off, showing that the homogeneous fields equal the actual fields after all sources are turned off. This result will be proven in Section 5.2.6 and justifies the introduction of homogeneous fields.

Inverse Source Formulas. Equating the right-hand sides of (3.199), (3.202), and (3.203) produces the following general expressions for the Fourier integral of the far-field pattern [132], [133], [134, sec. 9.6]

$$\begin{aligned}\frac{1}{2ik}\int_{S_0}&\left[\frac{e^{-ik|\mathbf{r}-\mathbf{r}'|}}{|\mathbf{r}-\mathbf{r}'|}\frac{\partial}{\partial n'}p_\omega(\mathbf{r}') - p_\omega(\mathbf{r}')\frac{\partial}{\partial n'}\frac{e^{-ik|\mathbf{r}-\mathbf{r}'|}}{|\mathbf{r}-\mathbf{r}'|}\right]dS' \\ &= \int_0^{2\pi}\int_0^{\pi}\mathcal{F}_\omega(\sigma_\theta,\sigma_\phi)e^{ik\mathbf{r}\cdot\hat{\boldsymbol{\sigma}}_{\theta\phi}}\sin\sigma_\theta d\sigma_\theta d\sigma_\phi \\ &= \int_V q_\omega(\mathbf{r}')\frac{\sin(k|\mathbf{r}-\mathbf{r}'|)}{k|\mathbf{r}-\mathbf{r}'|}dV'.\end{aligned} \tag{3.215}$$

The first equality in (3.215) is valid only for $\mathbf{r}$ inside S_0. The second equality in (3.215) remains valid for all $\mathbf{r}$ (inside and outside the source region) and can be derived directly from (3.35) recast in the form

$$\mathcal{F}_\omega(\sigma_\theta, \sigma_\phi) = \frac{1}{4\pi} \int_V q_\omega(\mathbf{r}') e^{-ik\mathbf{r}'\cdot\hat{\boldsymbol{\sigma}}_{\theta\phi}} dV'. \tag{3.216}$$

Simply multiply (3.216) by $e^{ik\mathbf{r}\cdot\hat{\boldsymbol{\sigma}}_{\theta\phi}}$ and integrate to get

$$\begin{aligned}\int_0^{2\pi}\int_0^{\pi} &\mathcal{F}_\omega(\sigma_\theta, \sigma_\phi) e^{ik\mathbf{r}\cdot\hat{\boldsymbol{\sigma}}_{\theta\phi}} \sin\sigma_\theta d\sigma_\theta d\sigma_\phi \\ &= \frac{1}{4\pi} \int_V q_\omega(\mathbf{r}') \left[\int_0^{2\pi}\int_0^{\pi} e^{ik(\mathbf{r}-\mathbf{r}')\cdot\hat{\boldsymbol{\sigma}}_{\theta\phi}} \sin\sigma_\theta d\sigma_\theta d\sigma_\phi \right] dV'\end{aligned} \tag{3.217}$$

which reduces to (3.215) with the aid of the equality [135]

$$\int_0^{2\pi}\int_0^{\pi} e^{ik(\mathbf{r}-\mathbf{r}')\cdot\hat{\boldsymbol{\sigma}}_{\theta\phi}} \sin\sigma_\theta d\sigma_\theta d\sigma_\phi = 4\pi \frac{\sin(k|\mathbf{r}-\mathbf{r}'|)}{k|\mathbf{r}-\mathbf{r}'|} \tag{3.218}$$

which can be proven straightforwardly by choosing $\mathbf{r} - \mathbf{r}'$ parallel to $\hat{\mathbf{z}}$.

Similarly, equating the right-hand sides of (3.209), (3.212), and (3.213), we obtain the following expressions for the electric field analogous to (3.215)

$$\begin{aligned}\frac{1}{2ik} \int_{S_0} &\left[\frac{e^{-ik|\mathbf{r}-\mathbf{r}'|}}{|\mathbf{r}-\mathbf{r}'|} \frac{\partial}{\partial n'} \mathbf{E}_\omega(\mathbf{r}') - \mathbf{E}_\omega(\mathbf{r}') \frac{\partial}{\partial n'} \frac{e^{-ik|\mathbf{r}-\mathbf{r}'|}}{|\mathbf{r}-\mathbf{r}'|} \right] dS' \\ &= \int_0^{2\pi}\int_0^{\pi} \boldsymbol{\mathcal{F}}_\omega(\sigma_\theta, \sigma_\phi) e^{ik\mathbf{r}\cdot\hat{\boldsymbol{\sigma}}_{\theta\phi}} \sin\sigma_\theta d\sigma_\theta d\sigma_\phi \\ &= \int_V \mathbf{Q}_\omega(\mathbf{r}') \frac{\sin(k|\mathbf{r}-\mathbf{r}'|)}{k|\mathbf{r}-\mathbf{r}'|} dV'.\end{aligned} \tag{3.219}$$

The far-field integrals in (3.183) and (3.187) show that all these expressions in (3.215) and (3.219) are proportional to the sum of the Fourier transforms of the far-field patterns to the left and right of the sources, but with the evanescent spectra excluded from the far-field patterns. If the evanescent spectra were included, the integrals over the left and right far-field patterns would equal the near field to the left and right of the source region, respectively. Thus, the equalities in (3.215) or (3.219) give three different representations outside the source region for the near field devoid of its evanescent spectra and analytically continued through the source region.

The volume integral representations on the right-hand side of (3.215) or (3.219) reveal that the expressions in (3.215) or (3.219) decrease asymptotically as $1/r$ outside the source region. Thus, when the Fourier integral of the far-field pattern in (3.215) or (3.219) is evaluated for values of $\mathbf{r}$ covering the source region, and these values are appropriately plotted (and possibly filtered), a "fuzzy" image of the source region is produced [136], [137]. If there are "hot spots" in the near field outside the source region caused by focusing or constructive interference of the backward propagated radiated fields (as would occur in scattering from curved surfaces), these hot spots appear as false images.

It may be tempting to interpret the $\sin(k|\mathbf{r}-\mathbf{r}'|)/|\mathbf{r}-\mathbf{r}'|$ convolution function in the integrand on the right-hand side of (3.215) or (3.219) as a point-spread function that behaves approximately as a delta function (times π) and thus allows the volume convolution integral to be evaluated approximately as $\pi q_\omega(\mathbf{r})/k$ in (3.215) or $\pi \mathbf{Q}_\omega(\mathbf{r})/k$ in (3.219). In general, this would be an erroneous interpretation and conclusion for two reasons. First, in three dimensions (volume integration), unlike in one dimension, $\sin(k|\mathbf{r} - \mathbf{r}'|)/(\pi|\mathbf{r} - \mathbf{r}'|)$ does not approach the delta function as $k \to \infty$. This is vividly illustrated by moving the origin of the volume integration in (3.215) or (3.219) to $\mathbf{r} = 0$ and writing dV' in spherical coordinates as $r'^2 \sin\theta' dr' d\theta' d\phi'$. Second, the source functions, $q_\omega(\mathbf{r}')$ and $\mathbf{Q}_\omega(\mathbf{r}')$, for most radiators and scatterers vary as rapidly as or more rapidly than $\sin(k|\mathbf{r} - \mathbf{r}'|)$. Of course, for a source distribution composed of a finite sum of point sources (for example, $q_\omega(\mathbf{r}) = \sum_j q_{\omega j}\delta(\mathbf{r} - \mathbf{r}_j)$ in (3.215)) separated by half a wavelength or more, the resulting fields $\sum_j q_{\omega j} \sin(k|\mathbf{r} - \mathbf{r}_j|)/(k|\mathbf{r} - \mathbf{r}_j|)$ will produce a local maximum at the position of each point source. This result forms the basis of many radar imaging and inverse scattering techniques [136], [138].

3.2.8 Minimum Source Region for a Given Far-Field Pattern

In Section 3.2.4, as well as in [91, th. 26] and [9, sec. 4.3], it is proven that the far-field pattern produced by sources in a bounded volume is an entire function of the complex spherical angles (θ, ϕ). In addition, Müller [91, ths. 26,27,29] has proven that the sources of a given entire far-field pattern must be contained in a sphere with a radius at least as large as R. The value of R is determined from the spherical expansion coefficients of the far-field pattern. For many far-field patterns the value of R is greater than zero, regardless of the position of the origin chosen for the spherical expansion. Outside the sphere of radius R, the field of the entire far-field pattern is also an analytic function of the spherical coordinates. This function can be uniquely continued analytically to a minimum source region V_{min} located inside the sphere of radius R. (This can be done, in principle, by expanding the field in a series of Bessel spherical wave functions that converges uniformly up to the surface of the minimum source region [91, ch. 3].) Outside V_{min}, which can be singly or multiply connected, the field obeys the homogeneous wave equation. In addition, V_{min} is the smallest possible source region in that the sources of the given far-field pattern must extend at least as far as the surface of V_{min}, or else the field could be further analytically continued inside the surface.

In this section, we derive a simple method, based on the plane-wave spectrum formulation, for determining the minimum convex closed surface S_{min} in which the sources of a given acoustic or electromagnetic far-field pattern can be located, that is, the minimum convex surface enclosing the minimum source region V_{min}. Unlike the value of R in the method based on a spherical wave expansion, the shape and size of S_{min} found here from the far-field pattern is independent of the position of the origin of the coordinate system in which the far-field pattern is defined. The results of this section were first presented in [25].

We begin by recalling the formula (3.120) in Section 3.2.3, which shows that the far-field functions for acoustic sources in a bounded volume V satisfy the asymptotic relation

$$\lim_{|\gamma|\to\infty} |F_\omega(k_x, k_y)|e^{-|\gamma|z} = 0, \quad z > z_s, \quad \text{for all } k_y/k_x \tag{3.220}$$

where z_s equals the maximum (more precisely, the supremum) value of z for which the source function $q_\omega(x, y, z)$ is nonzero. Here, z_s can be a negative, zero, or positive real number, depending on the position of the origin of the xyz coordinate system with respect to the source region.

An infinite number of different source distributions can radiate the same far field and thus produce the same field everywhere outside the source region. For example, any number of spherically symmetric source distributions produce the same constant far-field pattern. Thus, depending on the particular source distribution $q_\omega(x, y, z)$, (3.220) may hold for $z > z_1$ with $z_1 < z_s$. Moreover, unless the source distribution does not radiate at all, there must be a minimum z_1 for which the limit in (3.220) is zero. Letting z_{min} be the maximum value that z

attains in V_{min}, so that $z \leq z_{min}$ when $\mathbf{r} \in V_{min}$, the minimum z_1 is equal to z_{min}. To prove this, assume that z_1 were less than z_{min}, so that the plane $z = z_1$ cuts through the minimum source region V_{min}. Then the spectrum $T_\omega(k_x, k_y) = (i/\gamma)F_\omega(k_x, k_y)$ would decay exponentially according to (3.220) for $z > z_1$, and the field given by the plane-wave expansion

$$p_\omega(\mathbf{r}) = \frac{1}{2\pi}\int_{-\infty}^{+\infty}\int_{-\infty}^{+\infty} T_\omega(k_x, k_y)e^{i(k_x x+k_y y+\gamma z)}dk_x dk_y \tag{3.221}$$

would satisfy the homogeneous wave equation for $z > z_1$ and be an analytic function of the spatial coordinates for $z > z_1$.[2] Consequently, $z = z_1$ cannot cut through the minimum source region, and the minimum value of z_1 is therefore equal to z_{min}.

Hence, for $z < z_{min}$, the limit in (3.220) is indeed not equal to zero at least for one value of k_y/k_x as $|\gamma| = |k_x^2 + k_y^2 - k^2|^{1/2} \to \infty$. Thus, for any far-field function generated by sources in a bounded volume, we have

$$\limsup_{|\gamma|\to\infty} |F_\omega(k_x, k_y)|e^{-|\gamma|z} = \begin{cases} 0, & z > z_{min}, \quad \text{for all } k_y/k_x \\ \infty, & z < z_{min}, \quad \text{for at least one } k_y/k_x. \end{cases} \tag{3.222}$$

The infinite value occurs on the lower right-hand side of (3.222) because if the value of the limit is nonzero for all $z < z_{min}$, the exponential factor on the left-hand side of (3.222) demands that its value be infinite for all $z < z_{min}$. The "limit superior" [93, p. 62] is necessary in (3.222) to allow for oscillating values of $F_\omega(k_x, k_y)$ as $|\gamma| \to \infty$.

The condition (3.222) can be rewritten as follows in terms of a far-field pattern $\mathcal{F}_\omega(\theta, \phi)$ defined by (3.30) and (3.32)

$$\limsup_{\theta\to\frac{\pi}{2}-i\infty} |\mathcal{F}_\omega(\theta, \phi)|e^{-kz|\cos\theta|} = \begin{cases} 0, & z > z_{min}, \quad \text{for all } \phi \\ \infty, & z < z_{min}, \quad \text{for at least one } \phi \end{cases} \tag{3.223}$$

where we have used the fact that

$$\mathcal{F}_\omega(\theta, \phi) = F_\omega(k\cos\phi\sin\theta, k\sin\phi\sin\theta) \tag{3.224}$$

for $0 \leq \phi \leq 2\pi$; $0 \leq \theta \leq \pi/2$ and $\theta = \pi/2 - i\alpha$ with $0 \leq \alpha < \infty$.

Now, suppose we are given an analytic far-field pattern and have determined the value of z_{min} such that (3.223) is satisfied. The minimum source region V_{min} for this far-field pattern must extend in the z direction as far as $z = z_{min}$. If (3.223) holds for all orientations of the z axis (with z_{min} and $\mathcal{F}_\omega(\theta, \phi)$ depending, in general, on the orientation), then the corresponding spectrum $T_\omega(k_x, k_y)$ for all orientations of the z axis can be inserted into (3.221) to generate a function $p_\omega(\mathbf{r})$ that satisfies the homogeneous wave equation outside the minimum convex closed surface S_{min} formed by all the z_{min}, as shown in Figure 3.5. In other words, S_{min} will be the minimum convex surface enclosing the minimum source region. The plane-wave representation (3.221) also gives us the value of $p_\omega(\mathbf{r})$ and all its derivatives on S_{min}. In particular, it gives the Huygens' sources ($p_\omega(\mathbf{r})$ and $(\partial/\partial n)p_\omega(\mathbf{r})$) that generate the fields outside S_{min}.

Consequently, we have proven that a far field satisfying the conditions (3.222) or (3.223) for every orientation of the z axis can be generated by sources that lie within or on the minimum convex closed-surface S_{min} determined by the locus of values of z_{min} for all orientations of the z axis. Moreover, this far field cannot be generated by sources that extend to a radius that is less than the radius R_{min} of the sphere that circumscribes S_{min} (see Figure 3.5). If the origin of this

[2] To prove that the field in (3.221) satisfies the homogeneous wave equation for $z > z_1$, apply the ∇^2 operator to the right-hand side of (3.221) and interchange the integrations and differentiations for $z > z_1$ as allowed by the exponential decay of the integrand [90, sec. 6.23]. Similarly, it follows that the field in (3.221) is an analytic function of x, y, and z in domains that include the real x, y, and z axes.

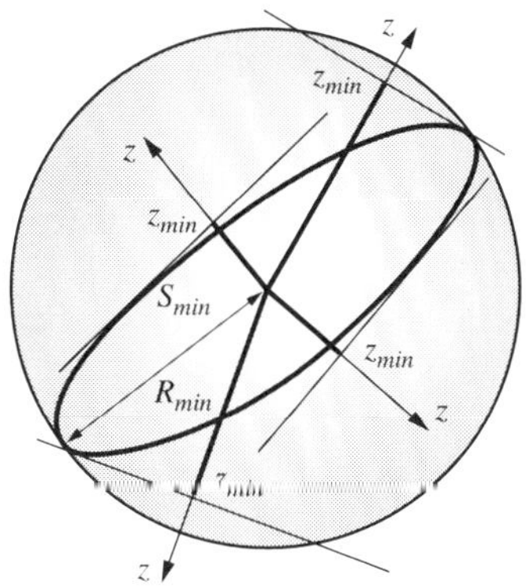

Figure 3.5 Minimum convex closed surface S_{min} formed by the locus of points z_{min} for all orientations of the z axis.

circumscribing sphere is chosen as the origin of the coordinate system that Müller [91] uses to expand the fields in spherical waves, then this R_{min} equals the minimum R in [91, ths. 26 and 27]. For other origins, Müller's minimum R is greater than R_{min}. The sources that generate the far-field pattern $\mathcal{F}_\omega(\theta, \phi)$ are not unique and can be chosen to extend outside S_{min}.

For many far-field patterns, S_{min} will simply be a single point in space so that $R_{min} = 0$. This means that the pressure $p_\omega(\mathbf{r})$ associated with $\mathcal{F}_\omega(\theta, \phi)$ can be produced by a convergent sum of spherical multipoles located at a single point in space (S_{min}). If $R_{min} > 0$, a multipole expansion (located at the center of the sphere with radius R_{min}) of the pressure $p_\omega(\mathbf{r})$ having the far-field pattern $\mathcal{F}_\omega(\theta, \phi)$ will converge for radii $r > R_{min}$. Moreover, at least for some spherical angles (θ, ϕ), it will diverge when $r < R_{min}$. If the infinite sum of spherical multipoles is truncated to a finite mode number N that is just large enough to produce a reasonably accurate approximation to the exact far-field pattern, the resulting approximate field will converge for all $r > 0$ but will have extremely high reactive fields for $r < N/k$ [44], [108].

For electromagnetic fields, the scalar acoustic far field is merely replaced by the vector electromagnetic far field, and (3.222) and (3.223) are replaced by

$$\limsup_{|\gamma|\to\infty} |\mathbf{F}_\omega(k_x, k_y)| e^{-|\gamma| z} = \begin{cases} 0, & z > z_{min}, \quad \text{for all } k_y/k_x \\ \infty, & z < z_{min}, \quad \text{for at least one } k_y/k_x \end{cases} \tag{3.225}$$

$$\limsup_{\theta\to\frac{\pi}{2}-i\infty} |\mathcal{F}_\omega(\theta, \phi)| e^{-kz|\cos\theta|} = \begin{cases} 0, & z > z_{min}, \quad \text{for all } \phi \\ \infty, & z < z_{min}, \quad \text{for at least one } \phi. \end{cases} \tag{3.226}$$

Circular-Disk Acoustic Radiator. Let us use the acoustic formula (3.223) to determine the minimum convex surface S_{min} for the far-field pattern of a circular-disk acoustic radiator of radius a. This far-field pattern as a function of spherical angles (θ, ϕ) measured with respect to a rectangular xyz coordinate system, with the z axis normal to the circular disk and passing through the center of the disk at $z = 0$, is given by [68, eq. 7.4.36]

$$\mathcal{F}_\omega(\theta, \phi) = A\frac{J_1(ka\sin\theta)}{\sin\theta}, \quad 0 \le \theta \le \pi \tag{3.227}$$

where $J_1(x)$ is the first-order Bessel function and A is a constant independent of θ and ϕ. For this particular orientation of the z axis, the far-field pattern is independent of the azimuthal angle ϕ.

Substituting the far-field pattern (3.227) into (3.223) shows that $z_{min} = 0$ for this z axis normal to the center of the disk. For the z axis pointed in the opposite direction, z_{min} also equals zero, because $\sin(\pi - \theta) = \sin\theta$; that is, the far-field pattern in (3.227) is identical to the right and left of the circular-disk radiator. Thus, we have confirmed that the sources of the far-field pattern (3.227) can all lie in the plane of the circular disk that radiates this far-field pattern.

To find the minimum distance that the sources must extend from the center of the disk in a direction in the plane of the disk, rotate the z axis 90° about the y axis. Then rewrite the far-field pattern in terms of the new spherical angles (θ', ϕ') measured with respect to this rotated z axis, which now lies in the plane of the disk. Specifically, we have $\sin\theta = (1 - \sin^2\theta' \cos^2\phi')^{1/2}$, and (3.227) shows that

$$\mathcal{F}_\omega(\pi/2 - i\alpha, \phi') = A\frac{J_1\left(ka\sqrt{1-\cosh^2\alpha\cos^2\phi'}\right)}{\sqrt{1-\cosh^2\alpha\cos^2\phi'}}, \quad 0 \le \alpha < \infty \tag{3.228}$$

where we have also used the identity $\sin(\pi/2 - i\alpha) = \cosh\alpha$. Since

$$(1 - \cosh^2\alpha\cos^2\phi')^{1/2} \sim \pm i|\cos\phi'|\cosh\alpha, \quad \alpha \to \infty, \quad \cos\phi' \neq 0 \tag{3.229}$$

the asymptotic formula

$$\frac{J_1(\pm ix)}{\pm ix} \sim \frac{e^x}{x\sqrt{2\pi x}}, \quad x \to \infty \tag{3.230}$$

can be invoked to get

$$\limsup_{\alpha\to\infty} |\mathcal{F}_\omega(\pi/2 - i\alpha, \phi')|e^{-kz\sinh\alpha} = \limsup_{\alpha\to\infty} \frac{Ae^{ka\cosh\alpha|\cos\phi'|}\,e^{-kz\sinh\alpha}}{\sqrt{2\pi ka}\,(\cosh\alpha|\cos\phi'|)^{3/2}} \tag{3.231}$$

when $\cos\phi' \neq 0$. For $\cos\phi' = 0$, we have $\mathcal{F}_\omega(\pi/2 - i\alpha, \phi') = A\,J_1(ka)$. Combining these results shows that

$$\limsup_{\theta'\to\frac{\pi}{2}-i\infty} |\mathcal{F}_\omega(\theta', \phi')|e^{-kz|\cos\theta'|} = \begin{cases} 0, & z > a, \quad \text{for all } \phi' \\ \infty, & z < a, \quad \text{for at least one } \phi'. \end{cases} \tag{3.232}$$

Comparing (3.232) with (3.223), we see that $z_{min} = a$ for any orientation of the z axis in the plane of the disk. Therefore, the minimum convex surface S_{min} containing sources that can produce the far-field pattern of the circular-disk radiator simply coincides with the actual infinitesimally thin physical disk of radius a. The same far-field pattern could be excited by sources that extend outside this disk. However, the above exercise has proven that the physical disk comprising the circular acoustic radiator is the smallest source region that can exactly produce its far-field pattern. A similar exercise using (3.226) and the far-field pattern (5.341) found in Chapter 5 for a circular-disk antenna with uniform surface current produces the same results.

In summary, when an exact closed-form expression for the acoustic or electromagnetic far field is known, (3.222)–(3.223) or (3.225)–(3.226) determine the minimum convex closed surface in which the sources of the far field can reside.

3.2.9 Power and Momentum-Flux (Radiation-Force) Relations

The time-average electromagnetic and acoustic power and momentum-flux (radiation force) radiated through the plane $z = z_1 \ge z_0$ shall now be expressed in terms of the electromagnetic spectrum $\mathbf{T}_\omega$ in (3.63) and the acoustic spectrum T_ω in (3.57), respectively. Begin with the electromagnetic field for which Poynting's and Parseval's theorems show that the complex power $\mathcal{P}^c_{el}(z_1)$ transmitted across the plane $z = z_1$ is given from (2.351) of Section

2.3.9 as

$$\begin{aligned}\mathcal{P}^c_{el}(z_1) &= \frac{1}{2}\hat{\mathbf{z}}\cdot\int_{-\infty}^{+\infty}\int_{-\infty}^{+\infty}\mathbf{E}_\omega(\mathbf{r}_1)\times\mathbf{H}^*_\omega(\mathbf{r}_1)dx_1dy_1 \\ &= \frac{1}{2}\hat{\mathbf{z}}\cdot\int_{-\infty}^{+\infty}\int_{-\infty}^{+\infty}\mathbf{T}_\omega(k_x,k_y)\times\mathbf{T}^{H*}_\omega(k_x,k_y)e^{iz_1(\gamma-\gamma^*)}dk_xdk_y\end{aligned} \tag{3.233}$$

where $\mathbf{r}_1 = x_1\hat{\mathbf{x}} + y_1\hat{\mathbf{y}} + z_1\hat{\mathbf{z}}$ and $\mathbf{T}^H_\omega$ is the spectrum for the magnetic field given in (3.67) and (3.69). The superscript $*$ denotes complex conjugation. Now, from (3.67) we have

$$\begin{aligned}\mathbf{T}_\omega\times\mathbf{T}^{H*}_\omega &= \frac{1}{\omega\mu}\mathbf{T}_\omega\times\left(\mathbf{k}^*\times\mathbf{T}^*_\omega\right) \\ &= \frac{1}{\omega\mu}|\mathbf{T}_\omega|^2\mathbf{k}^* - \frac{1}{\omega\mu}\mathbf{T}^*_\omega\mathbf{T}_\omega\cdot\mathbf{k}^*\end{aligned} \tag{3.234}$$

and from (3.64) $\mathbf{T}_\omega\cdot\mathbf{k}^* = (\gamma^*-\gamma)\hat{\mathbf{z}}\cdot\mathbf{T}_\omega$, so that

$$\hat{\mathbf{z}}\cdot\mathbf{T}_\omega\times\mathbf{T}^{H*}_\omega = \frac{1}{\omega\mu}|\mathbf{T}_\omega|^2\gamma^* - \frac{1}{\omega\mu}(\gamma^*-\gamma)|\hat{\mathbf{z}}\cdot\mathbf{T}_\omega|^2. \tag{3.235}$$

Then (3.233) becomes

$$\begin{aligned}\mathcal{P}^c_{el}(z_1) = &\frac{1}{2\omega\mu}\iint\limits_{k_x^2+k_y^2<k^2}|\mathbf{T}_\omega|^2\gamma\,dk_xdk_y \\ &+\frac{i}{2\omega\mu}\iint\limits_{k_x^2+k_y^2>k^2}|\gamma|\left[2|\hat{\mathbf{z}}\cdot\mathbf{T}_\omega|^2 - |\mathbf{T}_\omega|^2\right]e^{-2z_1|\gamma|}dk_xdk_y\end{aligned} \tag{3.236}$$

since γ is real for $k_x^2+k_y^2 < k^2$, and imaginary for $k_x^2+k_y^2 > k^2$. The first and second integrals in (3.236) are the time-average real and imaginary power flow, respectively, transmitted across the plane $z = z_1$. The real power (the first term of (3.236)) is carried solely by the propagating part of the spectrum and is independent of z_1. The imaginary power (the second term of (3.236)) involves only the evanescent part of the spectrum and decays exponentially as $z_1 \to +\infty$. No real power is radiated by the evanescent part of the spectrum, as we showed previously in Section 3.2.5.

It can be shown that the real and imaginary parts of the complex power contribute to the real and imaginary parts, respectively, of the input impedance of a transmitting antenna fed by a single propagating waveguide mode [63, sec. 6.9]; see Section 2.3.9. This is proven by applying Poynting's theorem to the volume between two closed surfaces. The inner surface covers the reference plane of the waveguide feed and the surface of the power supply of the antenna (see Figure 6.1). The outer surface surrounds the antenna and is formed by the infinite plane $z = z_1$, and its capping spherical surface of infinite radius in the region $z \le z_1$.

As mentioned previously in Section 2.3.9, the imaginary part of the complex power divided by ω, that is, $\mathrm{Im}(\mathcal{P}^c_{el}(z_1))/\omega$, is not in general equal to the stored antenna energy that oscillates across the plane $z = z_1$. A measure of this so-called reactive energy in the half space $z > z_1$ can be found, however, in terms of a (k_x, k_y) integration over the evanescent spectrum [111], [112].

Introducing the angular integration variables (θ, ϕ), determined by the identities $k_x = k\sin\theta\cos\phi$ and $k_y = k\sin\theta\sin\phi$, into (3.236) shows that the time-average real power radiated through the plane $z = z_1$ is given by

$$\mathcal{P}_{el}^{h} = \text{Re}\left(\mathcal{P}_{el}^{c}(z_1)\right) = \frac{1}{2\omega\mu} \iint\limits_{k_x^2+k_y^2<k^2} |\mathbf{T}_\omega|^2 \gamma \, dk_x dk_y$$
$$= \frac{1}{2}\sqrt{\frac{\epsilon}{\mu}} \int_0^{2\pi}\int_0^{\pi/2} |k\cos\theta \mathbf{T}_\omega(k\sin\theta\cos\phi, k\sin\theta\sin\phi)|^2 \sin\theta d\theta d\phi \tag{3.237}$$
$$= \frac{1}{2}\sqrt{\frac{\epsilon}{\mu}} \int_0^{2\pi}\int_0^{\pi/2} |\boldsymbol{\mathcal{F}}_\omega(\theta,\phi)|^2 \sin\theta d\theta d\phi$$

where we have used $dk_x dk_y = k\gamma \sin\theta d\theta d\phi$ and the relation (3.179) between $\boldsymbol{\mathcal{F}}_\omega$ and $\mathbf{T}_\omega$. The formula (3.237) verifies that the real power radiated through the plane $z = z_1 > z_0$ equals the power radiated through the far-field hemisphere $0 \le \theta \le \pi/2$, and thus the superscript h on $\mathcal{P}_{el}^{h}$ in (3.237).

The time-average electromagnetic momentum flux (radiation force) in the z direction, $\mathcal{N}_{el}^{z}$, transmitted through the plane $z = z_1 \ge z_0$ is given from (2.348) of Section 2.3.9 as

$$\mathcal{N}_{el}^{z} = -\frac{1}{2}\text{Re}\int_{-\infty}^{+\infty}\int_{-\infty}^{+\infty} \hat{\mathbf{z}}\cdot\bar{\mathbf{T}}_{el}^{c}\cdot\hat{\mathbf{z}}dx_1dy_1 = \frac{1}{2}\int_{-\infty}^{+\infty}\int_{-\infty}^{+\infty}\left[\epsilon\left(\frac{|\mathbf{E}_\omega|^2}{2} - |\hat{\mathbf{z}}\cdot\mathbf{E}_\omega|^2\right) + \mu\left(\frac{|\mathbf{H}_\omega|^2}{2} - |\hat{\mathbf{z}}\cdot\mathbf{H}_\omega|^2\right)\right]dx_1dy_1 \tag{3.238}$$

which, from Parseval's theorem, converts to

$$\mathcal{N}_{el}^{z} = \frac{1}{2}\int_{-\infty}^{+\infty}\int_{-\infty}^{+\infty}\left[\epsilon\left(\frac{|\mathbf{T}_\omega|^2}{2} - |\hat{\mathbf{z}}\cdot\mathbf{T}_\omega|^2\right) + \frac{1}{\omega^2\mu}\left(\frac{|\mathbf{k}\times\mathbf{T}_\omega|^2}{2} - |\hat{\mathbf{z}}\cdot(\mathbf{k}\times\mathbf{T}_\omega)|^2\right)\right]dk_xdk_y. \tag{3.239}$$

A fairly lengthy manipulation of the $\mathbf{k}\times\mathbf{T}_\omega$ terms in the integrand of (3.239) shows that they cancel the $\mathbf{T}_\omega$ terms for the evanescent spectrum and that the remaining integral over the propagating spectrum reduces to simply

$$\mathcal{N}_{el}^{z} = \frac{\epsilon}{2k^2}\iint\limits_{k_x^2+k_y^2<k^2} |\mathbf{T}_\omega|^2\gamma^2 dk_xdk_y$$
$$= \frac{\epsilon}{2}\int_0^{2\pi}\int_0^{\pi/2} |\boldsymbol{\mathcal{F}}_\omega(\theta,\phi)|^2 \cos\theta\sin\theta d\theta d\phi. \tag{3.240}$$

This momentum flux $\mathcal{N}_{el}^{z}$ is a positive real number independent of the distance z_1 to the plane of integration, and is equal to the radiation force that would be exerted on an infinite, flat absorbing screen located in the xy plane at any $z = z_1 \ge z_0$.

The time-average total electromagnetic momentum flux $\mathcal{N}_{el}^h$ radiated through the far-field hemisphere $0 \leq \theta \leq \pi/2$ is given by

$$\begin{aligned}\mathcal{N}_{el}^h &= -\lim_{r\to\infty} \frac{1}{2} \int_0^{2\pi}\int_0^{\pi/2} \hat{\mathbf{r}} \cdot \bar{\mathbf{T}}_{el}^c(r,\theta,\phi) \cdot \hat{\mathbf{r}} r^2 \sin\theta d\theta d\phi \\ &= \frac{\epsilon}{2} \int_0^{2\pi}\int_0^{\pi/2} |\boldsymbol{\mathcal{F}}_\omega(\theta,\phi)|^2 \sin\theta d\theta d\phi.\end{aligned} \tag{3.241}$$

Comparing the total power in (3.237) with the total momentum flux in (3.241) confirms that the time-average momentum flux radiated through the far-field hemisphere equals the time-average power radiated through the hemisphere divided by the speed of propagation c, that is

$$\mathcal{N}_{el}^h = \mathcal{P}_{el}^h / c. \tag{3.242}$$

Comparing the integrands of (3.237) and (3.241) also confirms that in any one direction (θ, ϕ), the ratio of the far-field momentum flux per unit area (radiation pressure) in the $\hat{\mathbf{r}}$ direction to the far-field power flow per unit area (radiation intensity) is just $1/c$.

For the acoustic fields we find that the complex time-average power transmitted across the plane $z = z_1 \geq z_0$ is given from (2.391) of Section 2.4.8 as

$$\begin{aligned}\mathcal{P}_{ac}^c(z_1) &= \frac{1}{2} \int_{-\infty}^{+\infty}\int_{-\infty}^{+\infty} p_\omega(\mathbf{r}_1)\, \hat{\mathbf{z}} \cdot \mathbf{v}_\omega^*(\mathbf{r}_1) dx_1 dy_1 \\ &= \frac{1}{2\omega\rho_{m0}} \iint_{k_x^2+k_y^2<k^2} |T_\omega|^2 \gamma dk_x dk_y \\ &\quad - \frac{i}{2\omega\rho_{m0}} \iint_{k_x^2+k_y^2>k^2} |T_\omega|^2 |\gamma| e^{-2z_1|\gamma|} dk_x dk_y\end{aligned} \tag{3.243}$$

where ρ_{m0} is the constant mass density of the undisturbed fluid, and we have used Parseval's theorem and the formulas (3.56) and (3.58) for the pressure and velocity. The first and second integrals in the last equation of (3.243) are the real and imaginary acoustic power, respectively, transmitted through the plane $z = z_1$. As for the electromagnetic field, the real power is carried solely by the propagating part of the spectrum and is independent of z_1. The imaginary power involves only the evanescent part of the spectrum and decays as $z_1 \to +\infty$. We can also show that the real power transmitted through the plane $z = z_1 > z_0$, obtained from (3.243), equals the power radiated through the far-field hemisphere $0 \leq \theta \leq \pi/2$ given by

$$\begin{aligned}\mathcal{P}_{ac}^h = \text{Re}\left(\mathcal{P}_{ac}^c(z_1)\right) &= \frac{1}{2\omega\rho_{m0}} \iint_{k_x^2+k_y^2<k^2} |T_\omega|^2 \gamma dk_x dk_y \\ &= \frac{1}{2c\rho_{m0}} \int_0^{2\pi}\int_0^{\pi/2} |k\cos\theta T_\omega(k\sin\theta\cos\phi, k\sin\theta\sin\phi)|^2 \sin\theta d\theta d\phi \\ &= \frac{1}{2c\rho_{m0}} \int_0^{2\pi}\int_0^{\pi/2} |\mathcal{F}_\omega(\theta,\phi)|^2 \sin\theta d\theta d\phi.\end{aligned} \tag{3.244}$$

For the time-average acoustic wave momentum flux (radiation force) in the z direction transmitted through the plane $z = z_1$, we find from (2.388) of Section 2.4.8 that

$$\begin{aligned}
\mathcal{N}^z_{ac} &= -\frac{1}{2}\mathrm{Re}\int_{-\infty}^{+\infty}\int_{-\infty}^{+\infty} \hat{\mathbf{z}}\cdot\bar{\mathbf{T}}^c_{ac}\cdot\hat{\mathbf{z}}\,dx_1dy_1 \\
&= -\frac{1}{2}\int_{-\infty}^{+\infty}\int_{-\infty}^{+\infty}\left[\rho_{m0}\left(\frac{|\mathbf{v}_\omega|^2}{2} - |\hat{\mathbf{z}}\cdot\mathbf{v}_\omega|^2\right) - \kappa_0\frac{|p_\omega|^2}{2}\right]dx_1dy_1 \\
&= \frac{1}{2\rho_{m0}\omega^2}\iint\limits_{k_x^2+k_y^2<k^2} |T_\omega|^2\gamma^2 dk_x dk_y \\
&= \frac{1}{2\rho_{m0}c^2}\int_0^{2\pi}\int_0^{\pi/2} |\mathcal{F}_\omega(\theta,\phi)|^2\cos\theta\sin\theta d\theta d\phi.
\end{aligned} \tag{3.245}$$

The total acoustic wave momentum flux radiated through a far-field hemisphere is given by

$$\begin{aligned}
\mathcal{N}^h_{ac} &= -\lim_{r\to\infty}\frac{1}{2}\int_0^{2\pi}\int_0^{\pi/2}\hat{\mathbf{r}}\cdot\bar{\mathbf{T}}^c_{ac}(r,\theta,\phi)\cdot\hat{\mathbf{r}}r^2\sin\theta d\theta d\phi \\
&= \frac{1}{2\rho_{m0}c^2}\int_0^{2\pi}\int_0^{\pi/2}|\mathcal{F}_\omega(\theta,\phi)|^2\sin\theta d\theta d\phi
\end{aligned} \tag{3.246}$$

so that

$$\mathcal{N}^h_{ac} = \mathcal{P}^h_{ac}/c \tag{3.247}$$

and in any one direction (θ, ϕ), the ratio of the far-field momentum flux per unit area (radiation pressure) in the $\hat{\mathbf{r}}$ direction to the far-field power flow per unit area (radiation intensity) is just $1/c$.

When the sources are located in the half space $z > z_0$, the complex power transmitted through the plane $z = z_1 \le z_0$ is given by (3.236) and (3.243) with z_1 in the exponentials replaced by $-z_1$. Moreover, in all the expressions containing θ integrations, the integral sign $\int_0^{\pi/2}$ is replaced by $\int_{\pi/2}^{\pi}$.

3.3 LOSSY MEDIA

In Section 2.5, the electromagnetic and acoustic equations are given for applied sources located in linear, isotropic, homogeneous, lossy media. Comparing the frequency-domain electromagnetic equations (2.397)–(2.400) in lossy media with those in lossless media (2.287)–(2.290), we see that the lossy equations emerge from the lossless equations if the real permeability and permittivity μ and ϵ in lossless free space are replaced by the complex permeability and permittivity μ_ω and ϵ_ω given by

$$\mu_\omega = \mu(1 + \chi^m_\omega) \tag{3.248}$$

and

$$\epsilon_\omega = \epsilon[1 + \chi^e_\omega - \sigma_\omega/(i\omega\epsilon)]. \tag{3.249}$$

The applied polarization densities $\mathbf{P}^{ap}_\omega$ and $\mathbf{M}^{ap}_\omega$, which are included in the lossy equations (2.397)–(2.400), will be assumed zero in this section, so as to limit the applied sources to current and charge densities in both the lossy and lossless equations of Chapter 3. The real propagation constant k in free space is replaced by the complex propagation constant k_ℓ for lossy media defined as

$$k_\ell = \omega(\mu_\omega \epsilon_\omega)^{\frac{1}{2}} \tag{3.250}$$

with positive real and imaginary parts for $\omega > 0$. If we also write k_ℓ as ω/c_ω, the constant c_ω is a complex "speed of propagation" that replaces the real speed of propagation c in lossless free space. From (3.250) the complex speed can also be written as

$$c_\omega = \frac{\omega}{k_\ell} = \frac{1}{(\mu_\omega \epsilon_\omega)^{\frac{1}{2}}}. \tag{3.251}$$

Since the real and imaginary parts of k_ℓ are both positive in lossy media for $\omega > 0$, the real part of c_ω is positive and the imaginary part of c_ω is negative in lossy media for $\omega > 0$.

Comparing the frequency domain linear acoustic equations (2.435)–(2.436) and (2.356)–(2.357) in lossy and lossless fluids reveals that the lossy equations can be obtained from the lossless equations by replacing the real compressibility κ_0 with the complex compressibility κ_ω. The complex propagation constant k_ℓ in lossy fluids replaces the real propagation constant k in lossless fluids and is defined as

$$k_\ell = \omega(\rho_{m0} \kappa_\omega)^{\frac{1}{2}} \tag{3.252}$$

with positive real and imaginary parts for $\omega > 0$. The complex speed of propagation c_ω in lossy fluids that replaces the real speed of propagation c in lossless fluids is given by

$$c_\omega = \frac{\omega}{k_\ell} = \frac{1}{(\rho_{m0} \kappa_\omega)^{\frac{1}{2}}} \tag{3.253}$$

which has positive and negative real and imaginary parts, respectively, for $\omega > 0$.

The Helmholtz equations for the electric field $\mathbf{E}_\omega(\mathbf{r})$ in a lossy material, and for the acoustic pressure $p_\omega(\mathbf{r})$ in a lossy fluid, can be written from (2.409) and (2.438) as

$$\nabla^2 \mathbf{E}_\omega(\mathbf{r}) + k_\ell^2 \mathbf{E}_\omega(\mathbf{r}) = -\mathbf{Q}_\omega^{ap}(\mathbf{r}) \tag{3.254}$$

and

$$\nabla^2 p_\omega(\mathbf{r}) + k_\ell^2 p_\omega(\mathbf{r}) = -q_\omega^{ap}(\mathbf{r}) \tag{3.255}$$

with the applied sources given by

$$\mathbf{Q}_\omega^{ap}(\mathbf{r}) = i\omega\mu_\omega \mathbf{J}_\omega^{ap}(\mathbf{r}) - \nabla \rho_\omega^{ap}(\mathbf{r})/\epsilon_\omega \tag{3.256}$$

and

$$q_\omega^{ap}(\mathbf{r}) = -i\omega s_\omega^{ap}(\mathbf{r}) - \nabla \cdot \mathbf{f}_\omega^{ap}(\mathbf{r}). \tag{3.257}$$

With these replacements in the frequency-domain electromagnetic equations for lossless free space, and in the frequency-domain acoustic equations for lossless fluids, namely

$$\mu \to \mu_\omega, \quad \epsilon \to \epsilon_\omega, \quad \kappa_0 \to \kappa_\omega, \quad k \to k_\ell, \quad c \to c_\omega \tag{3.258}$$

and

$$\mathbf{J}_\omega \to \mathbf{J}_\omega^{ap}, \quad \rho_\omega \to \rho_\omega^{ap}, \quad s_\omega \to s_\omega^{ap}, \quad \mathbf{f}_\omega \to \mathbf{f}_\omega^{ap}, \quad \mathbf{Q}_\omega \to \mathbf{Q}_\omega^{ap}, \quad q_\omega \to q_\omega^{ap} \tag{3.259}$$

we can see which of the foregoing lossless equations in Chapter 3 remain valid for lossy media. In particular, all the formulas in Section 3.1 on Green's function representations, in Section 3.1.1 on far-field expressions, and in Section 3.1.2 on the low-frequency behavior of the far-field patterns remain valid for lossy media under the replacements (3.258)–(3.259). The correct values of $\sqrt{\epsilon_\omega/\mu_\omega}$ and $\sqrt{\mu_\omega/\epsilon_\omega}$ can be determined by writing these quantities as $\epsilon_\omega c_\omega$ and $\mu_\omega c_\omega$, respectively. The far-field expressions in Section 3.1.1 with k_ℓ replacing k show that the fields are exponentially attenuated in lossy media as $r \to \infty$.

All the plane-wave spectrum representations in Section 3.2 remain valid for lossy media with the replacements (3.258)–(3.259). The z component (γ_ℓ) of the complex propagation vector

$$\mathbf{k}_\ell = k_\ell \hat{\mathbf{k}}_\ell = k_x \hat{\mathbf{x}} + k_y \hat{\mathbf{y}} + \gamma_\ell \hat{\mathbf{z}} \tag{3.260}$$

is given by

$$\gamma_\ell = (k_\ell^2 - k_x^2 - k_y^2)^{\frac{1}{2}} = \gamma_r + i\gamma_i \tag{3.261}$$

with k_x and k_y real and the square root chosen to make the real and imaginary parts of γ_ℓ positive, so that every plane wave in the entire plane-wave spectrum is attenuated. This means that the plane-wave spectra in lossy media do not divide into propagating and evanescent parts. Thus, Sections 3.2.1 and 3.2.5, as well as the equations in Sections 3.2.6 and 3.2.7 that divide the spectrum into propagating and evanescent parts, are not applicable to lossy media. The other equations in Sections 3.2.6 and 3.2.7 remain valid in lossy media under the replacements (3.258)–(3.259). For example, (3.177)–(3.180), which relate the spectra to the far fields, remain valid in lossy media. In lossy media, however, the constants k_ℓ and γ_ℓ, which are complex for all real (k_x, k_y), replace k and γ, respectively. Thus, to get any part of the spectra, one has to know the far-field patterns $\mathcal{F}_\omega(\theta_\ell, \phi)$ and $\boldsymbol{\mathcal{F}}_\omega(\theta_\ell, \phi)$ for complex θ_ℓ angles of observation defined by

$$\cos\theta_\ell = \frac{\gamma_\ell}{k_\ell}, \quad \sin\theta_\ell = \frac{\sqrt{k_x^2 + k_y^2}}{k_\ell}. \tag{3.262}$$

Also, since $\gamma_\ell \neq 0$, the plane-wave spectra in lossy media have no singularities corresponding to the singularities in lossless media at $\gamma = 0$. Accordingly, the discussions in Sections 3.2.3 on singularities and in Section 3.2.4 on analyticity have to be modified for lossy media. In general, the spectra have no singularities when losses are present because the fields in the scan plane $z = z_0$ decay exponentially at infinity and their integrals over the infinite scan plane converge for all real k_x and k_y; see (3.57) and (3.63).

The formulas in Section 3.2.8 for determining the minimum source region for a given far-field pattern apply to lossy media under the replacements (3.258)–(3.259). However, for lossy media, it is more expedient to use the formulas (3.222) and (3.225) to find the acoustic and electromagnetic minimum source regions instead of (3.223) and (3.226) in which θ would have to be replaced by θ_ℓ given in (3.262). Using (3.222) instead of (3.223) in the example of the circular-disk acoustic radiator in Section 3.2.8 produces the same circular-disk minimum source region in both lossy and lossless fluids.

The formulas for time-average power in Section 3.2.9 that are not divided into propagating and evanescent parts also generalize to lossy media with the replacements (3.258)–(3.259). Specifically, the complex electromagnetic power transmitted across the plane $z = z_1 \geq z_0$ is still determined by integrating the complex Poynting vector in (3.233) (see (2.415) in Section 2.5.1), which yields for a lossy medium

$$\begin{aligned}\mathcal{P}_{el}^c(z_1) = \frac{1}{2\omega\mu_\omega^*} \Bigg[&\int_{-\infty}^{+\infty}\int_{-\infty}^{+\infty} |\mathbf{T}_\omega|^2 \gamma_r e^{-2z_1\gamma_i} dk_x dk_y \\ &+ i \int_{-\infty}^{+\infty}\int_{-\infty}^{+\infty} \gamma_i \left[2|\hat{\mathbf{z}}\cdot\mathbf{T}_\omega|^2 - |\mathbf{T}_\omega|^2\right] e^{-2z_1\gamma_i} dk_x dk_y \Bigg].\end{aligned} \tag{3.263}$$

Similarly, the complex acoustic power in (3.243) (see (2.446) in Section 2.5.2) transmitted across the plane $z = z_1 \geq z_0$ yields in a lossy fluid

$$\mathcal{P}^c_{ac}(z_1) = \frac{1}{2\omega\rho_{m0}} \left[\int_{-\infty}^{+\infty}\int_{-\infty}^{+\infty} |T_\omega|^2 \gamma_r e^{-2z_1\gamma_i} dk_x dk_y \right. \\ \left. -i \int_{-\infty}^{+\infty}\int_{-\infty}^{+\infty} |T_\omega|^2 \gamma_i e^{-2z_1\gamma_i} dk_x dk_y \right]. \tag{3.264}$$

The time-average electromagnetic momentum flux (radiation force) in the z direction transmitted across the plane $z = z_1$ in a lossy material can be found by integrating $\bar{\mathbf{T}}^c_{el}$ in (2.348) or (2.352) over this z_1 plane, if an infinitesimally thin sheet of the polarization densities containing the z_1 plane is removed (as explained in Section 2.3.9; see also Section 2.5.1). With the electric and magnetic fields on the polarization-free z_1 plane denoted by $\mathbf{E}^0_\omega$ and $\mathbf{H}^0_\omega$, (2.348) or (2.352) gives

$$\mathcal{N}^z_{el} = -\frac{1}{2}\text{Re} \int_{-\infty}^{+\infty}\int_{-\infty}^{+\infty} \hat{\mathbf{z}} \cdot \bar{\mathbf{T}}^c_{el} \cdot \hat{\mathbf{z}} dx_1 dy_1 \\ = \frac{1}{2} \int_{-\infty}^{+\infty}\int_{-\infty}^{+\infty} \left[\epsilon \left(\frac{|\mathbf{E}^0_\omega|^2}{2} - |\hat{\mathbf{z}} \cdot \mathbf{E}^0_\omega|^2 \right) + \mu \left(\frac{|\mathbf{H}^0_\omega|^2}{2} - |\hat{\mathbf{z}} \cdot \mathbf{H}^0_\omega|^2 \right) \right] dx_1 dy_1. \tag{3.265}$$

Continuity of the tangential components of the electric and magnetic fields across the infinitesimally thin sheet, and continuity of the normal components of the magnetic induction and electric displacement across this sheet, allow one to express $\mathbf{E}^0_\omega$ and $\mathbf{H}^0_\omega$ in terms of $\mathbf{E}_\omega$ and $\mathbf{H}_\omega$ as follows

$$\hat{\mathbf{z}} \times \mathbf{E}^0_\omega = \hat{\mathbf{z}} \times \mathbf{E}_\omega, \quad \hat{\mathbf{z}} \times \mathbf{H}^0_\omega = \hat{\mathbf{z}} \times \mathbf{H}_\omega \tag{3.266}$$

$$\hat{\mathbf{z}} \cdot \mathbf{E}^0_\omega = (1 + \chi^e_\omega)\hat{\mathbf{z}} \cdot \mathbf{E}_\omega, \quad \hat{\mathbf{z}} \cdot \mathbf{H}^0_\omega = (1 + \chi^m_\omega)\hat{\mathbf{z}} \cdot \mathbf{H}_\omega. \tag{3.267}$$

(Note that the conduction current is not removed from the infinitesimally thin sheet containing the z_1 plane.) By inserting (3.266)–(3.267) into (3.265), and substituting the plane-wave representations for $\mathbf{E}_\omega$ and $\mathbf{H}_\omega$ in lossy media, (3.265) can be rewritten as

$$\mathcal{N}^z_{el} = \frac{1}{4} \int_{-\infty}^{+\infty}\int_{-\infty}^{+\infty} \left[\epsilon \left(|\mathbf{E}_\omega|^2 - (1 + |1 + \chi^e_\omega|^2)|\hat{\mathbf{z}} \cdot \mathbf{E}_\omega|^2 \right) \right. \\ \left. + \mu \left(|\mathbf{H}_\omega|^2 - (1 + |1 + \chi^m_\omega|^2)|\hat{\mathbf{z}} \cdot \mathbf{H}_\omega|^2 \right) \right] dx_1 dy_1 \\ = \frac{1}{4} \int_{-\infty}^{+\infty}\int_{-\infty}^{+\infty} \left[\epsilon \left(|\mathbf{T}_\omega|^2 - (1 + |1 + \chi^e_\omega|^2)|\hat{\mathbf{z}} \cdot \mathbf{T}_\omega|^2 \right) \right. \\ + \frac{\mu}{\omega^2 |\mu_\omega|^2} \left(|\mathbf{k}_\ell \times \mathbf{T}_\omega|^2 - (1 + |1 + \chi^m_\omega|^2) \right. \\ \left. \left. \cdot |\hat{\mathbf{z}} \cdot (\mathbf{k}_\ell \times \mathbf{T}_\omega)|^2 \right) \right] e^{-2z_1\gamma_i} dk_x dk_y. \tag{3.268}$$

Unlike the plane-wave integrals in (3.263) and (3.264) for the time-average power, or the plane-wave integral for the acoustic momentum flux in (3.269) below, the plane-wave integral in (3.268) for the electromagnetic momentum flux in lossy media cannot, in general, be written simply in terms of the spectrum squared ($|\mathbf{T}_\omega|^2$).

The time-average acoustic wave-momentum flux (radiation force) in the z direction transmitted across the plane $z = z_1$ in a lossy fluid can be found from (2.388) (see (3.245) and Sections 2.4.8 and 2.5.2). Specifically, entering the plane-wave representations for p_ω and $\mathbf{v}_\omega$ in a lossy fluid into the integrals of (3.245) containing p_ω and $\mathbf{v}_\omega$, one finds that the acoustic wave-momentum can be expressed in the form

$$\begin{aligned}
\mathcal{N}_{ac}^z &= -\frac{1}{2}\mathrm{Re}\int_{-\infty}^{+\infty}\int_{-\infty}^{+\infty} \hat{\mathbf{z}}\cdot\bar{\mathbf{T}}_{ac}^c\cdot\hat{\mathbf{z}}\,dx_1dy_1 \\
&= -\frac{1}{2}\int_{-\infty}^{+\infty}\int_{-\infty}^{+\infty}\left[\rho_{m0}\left(\frac{|\mathbf{v}_\omega|^2}{2} - |\hat{\mathbf{z}}\cdot\mathbf{v}_\omega|^2\right) - \kappa_0\frac{|p_\omega|^2}{2}\right]dx_1dy_1 \\
&= \frac{1}{2\rho_{m0}\omega^2}\int_{-\infty}^{+\infty}\int_{-\infty}^{+\infty}|T_\omega|^2\left[\gamma_r^2 - \rho_{m0}\omega^2(\kappa_\omega' - \kappa_0)/2\right]e^{-2z_1\gamma_i}\,dk_xdk_y
\end{aligned} \tag{3.269}$$

where κ_ω' is the real part of κ_ω, as defined previously in (2.434) of Section 2.5.2.

When the sources are located in the lossy half space $z > z_0$, the power and radiation force transmitted to the left through the plane $z = z_1 \le z_0$ are given by (3.263), (3.264), (3.268), and (3.269) with z_1 in the exponentials replaced by $-z_1$.

4

Static Electric and Magnetic Fields

We assumed throughout the derivations of the frequency-domain formulas in Chapter 3 that the frequency ω was positive. Some of these formulas derived for $\omega > 0$ remain valid for $\omega = 0$ and therefore apply to static electric and magnetic fields. However, fundamental differences between static fields ($\omega = 0$) and frequency-domain fields ($\omega > 0$) hamper the use of many frequency-domain formulas in the calculation of static fields. Static electric and magnetic fields are uncoupled, whereas the frequency-domain electric and magnetic fields are coupled. For example, formulas derived for $\omega > 0$ that express the spectra in terms of electric and magnetic fields on an arbitrary surface cannot immediately be used for static-field calculations. In addition, static far fields behave differently than frequency-domain far fields, and, in particular, static electric far fields decay as r^{-2} (or faster), whereas frequency-domain far fields with $\omega > 0$ decay as r^{-1}. Therefore, the frequency-domain far-field formulas do not provide any explicit information about the static far fields.

In Chapter 5 we will derive formulas for time-domain electromagnetic fields that are zero for $t < t_0$ and can have an arbitrary time dependence for $t > t_0$. From these formulas one can compute fields that begin at $t = t_0$ and approach static values as $t \to \infty$. For example, if a capacitor is charged starting at $t = t_0$, the time-domain magnetic field approaches zero as $t \to \infty$, and the time-domain electric field approaches a nonzero static electric field as $t \to \infty$. In principle, this static electric field can be calculated from any of the time-domain formulas of Chapter 5 if the entire time histories of the time-domain electric and magnetic fields are known. However, if these time histories are not known, some of the time-domain formulas involving both electric and magnetic fields cannot be used to compute this static electric field. Moreover, the time-domain far-field formulas for electromagnetic fields that begin at $t = t_0$ provide no explicit information about the static far fields.

Consequently, neither the frequency-domain formulas in Chapter 3 nor the time-domain formulas that will be derived in Chapter 5 are particularly useful for static-field calculations. Therefore, in this chapter we derive static formulas directly from the static Maxwell equations and obtain expressions that involve only static electric or static magnetic fields. These expressions do not require any knowledge of the time histories of the time-domain electromagnetic fields from which the static fields originated. Also, far-field formulas will be given that explicitly determine the radial and angular dependencies of the static fields in terms of the total electric charge, the electric dipole moment, and the magnetic dipole moment.

The static electric and static magnetic fields considered in this chapter are generated by impressed sources that lie within the finite volume V, which is located in the half space $z < z_0$, as shown in Figure 3.1. The impressed sources are described by the static current $\mathbf{J}$ and the static charge ρ. As mentioned in Section 2.1.1, one can easily modify the analysis involving $\mathbf{J}$ and ρ to take into account the polarization $\mathbf{P}$ and the magnetization $\mathbf{M}$. However, the important results of this chapter are formulas that give the static fields in source-free regions in terms of their values on a surface. Such formulas are independent of the exact nature of the sources that produce the fields.

The source-free region outside V is assumed homogeneous, isotropic, and lossless, and is characterized by the constant permittivity ϵ and constant permeability μ. The electrostatic and magnetostatic Maxwell equations (2.22)–(2.23) reduce simply to

$$\nabla \times \mathbf{E} = 0, \quad \nabla \cdot \mathbf{E} = 0 \tag{4.1}$$

and

$$\nabla \times \mathbf{H} = 0, \quad \nabla \cdot \mathbf{H} = 0 \tag{4.2}$$

outside the source region V. It should also be mentioned that the Newtonian gravitational field can be treated like the electrostatic fields. Since mass is always positive, however, there are no elementary gravitational dipoles.

In Sections 4.1 and 4.2, both the scalar potentials and the static fields are represented in terms of free-space, Dirichlet, and Neumann Green's functions. These various types of Green's functions produce different field representations, and they lead to alternative expressions for the total electric charge, the electric dipole moment, and the magnetic dipole moment. With S being the general closed surface in Figure 3.1 that contains the source volume V, expressions are derived for the scalar potentials and the static fields outside S in terms of the potentials or fields on S. These expressions valid for a general closed surface S are then used to obtain formulas for the potentials and fields in the half space $z > z_0$ in terms of the potentials or fields in the scan plane $z = z_0$.

In Section 4.3 plane-wave spectrum representations are derived for the static electric and magnetic fields and their scalar potentials. Only evanescent plane waves occur, and unlike the frequency-domain spectra that have singularities for $\omega > 0$, the static spectra are bounded functions. It is shown that the total electric charge, the electric dipole moment, and the magnetic dipole moment can be determined from the spectra and their derivatives evaluated at zero spatial frequencies.

Formulas will be derived in detail for fields in the half space $z > z_0$ that are generated by sources in the half space $z < z_0$. At the end of each section, it is explained how these formulas valid for $z > z_0$ can be modified to determine the fields in the half space $z < z_0$, when the sources are located in the half space $z > z_0$.

4.1 GREEN'S FUNCTION REPRESENTATIONS FOR THE STATIC ELECTRIC FIELD

We begin by deriving representations in terms of Green's functions for the static electric field $\mathbf{E}$. Since $\nabla \cdot \mathbf{E} = \rho/\epsilon$ and $\nabla \times \mathbf{E} = 0$ throughout space, we can represent $\mathbf{E}$ as the gradient of a single-valued scalar potential Ψ, that is

$$\mathbf{E}(\mathbf{r}) = -\nabla \Psi(\mathbf{r}) \tag{4.3}$$

where the scalar potential satisfies the Poisson equation

$$\nabla^2 \Psi(\mathbf{r}) = -\frac{\rho(\mathbf{r})}{\epsilon} \tag{4.4}$$

and ϵ is the constant permittivity. The scalar potential and the static electric field can be expressed from (2.86) as the integrals over the charge distribution

$$\Psi(\mathbf{r}) = \frac{1}{4\pi\epsilon}\lim_{\delta\to 0}\int\limits_{V-V_\delta}\frac{\rho(\mathbf{r}')}{|\mathbf{r}-\mathbf{r}'|}dV', \quad \mathbf{E}(\mathbf{r}) = \frac{1}{4\pi\epsilon}\lim_{\delta\to 0}\int\limits_{V-V_\delta}\rho(\mathbf{r}')\frac{\mathbf{r}-\mathbf{r}'}{|\mathbf{r}-\mathbf{r}'|^3}dV'. \tag{4.5}$$

These integrals can be expanded to prove that the static far fields are given by [63, sec. 4.1]

$$\Psi(\mathbf{r}) = \frac{q^{tot}}{4\pi\epsilon r} + \frac{\mathbf{p}\cdot\hat{\mathbf{r}}}{4\pi\epsilon r^2} + O\left(\frac{1}{r^3}\right), \quad \mathbf{E}(\mathbf{r}) = \frac{q^{tot}\,\hat{\mathbf{r}}}{4\pi\epsilon r^2} + \frac{3(\mathbf{p}\cdot\hat{\mathbf{r}})\hat{\mathbf{r}}-\mathbf{p}}{4\pi\epsilon r^3} + O\left(\frac{1}{r^4}\right) \tag{4.6}$$

where $r = |\mathbf{r}|$, $\hat{\mathbf{r}} = \mathbf{r}/r$, and

$$q^{tot} = \int\limits_V \rho(\mathbf{r}')dV', \quad \mathbf{p} = \int\limits_V \mathbf{r}'\rho(\mathbf{r}')dV' \tag{4.7}$$

are the total charge and the total dipole moment of the charge distribution. Note that the dipole moment $\mathbf{p}$ depends on the choice of origin when the total charge q^{tot} is nonzero. Moreover, when q^{tot} is nonzero, the far-field formulas (4.6) show that $\Psi = O(r^{-1})$ and $|\mathbf{E}| = O(r^{-2})$ as $r\to\infty$. When q^{tot} is zero but the dipole moment $\mathbf{p}$ is nonzero, the fields satisfy the relations $\Psi = O(r^{-2})$ and $|\mathbf{E}| = O(r^{-3})$ as $r\to\infty$. If both q^{tot} and $\mathbf{p}$ are zero, the far fields decay faster, and a general multipole expansion [63, sec. 4.1] can be used to determine their angular and radial dependencies.

4.1.1 Expressions Involving the Static Electric Field on a Closed Surface

Let S be the closed, smooth, finite surface shown in Figure 3.1, which contains the source region V and has $\hat{\mathbf{n}}$ as its outward normal unit vector. Then introduce a static Green's function $G(\mathbf{r},\mathbf{r}')$ that satisfies the differential equation

$$\nabla'^2 G(\mathbf{r},\mathbf{r}') = -\delta(\mathbf{r}-\mathbf{r}') \tag{4.8}$$

where ∇'^2 is the Laplacian operator with respect to the primed coordinates, and both $\mathbf{r}$ and $\mathbf{r}'$ are outside S. Moreover, assume that $G(\mathbf{r},\mathbf{r}') = O[(r')^{-1}]$ as $r'\to\infty$ for all fixed $\mathbf{r}$. So far, we have not specified any boundary conditions for $G(\mathbf{r},\mathbf{r}')$, and in the following derivation $G(\mathbf{r},\mathbf{r}')$ can be any Green's function that satisfies the conditions above. For example, $G(\mathbf{r},\mathbf{r}')$ can be the free-space static Green's function $1/(4\pi|\mathbf{r}-\mathbf{r}'|)$.

Now introduce a sphere S_0 with radius R_0 that encloses the surface S and the point $\mathbf{r}$, which is outside S. Then apply Green's second identity with the functions $\Psi(\mathbf{r}')$ and $G(\mathbf{r},\mathbf{r}')$ to the volume bounded by S and S_0. The asymptotic relations $\Psi(\mathbf{r}') = O[(r')^{-1}]$ and $G(\mathbf{r},\mathbf{r}') = O[(r')^{-1}]$ as $r'\to\infty$ show that the surface integral over S_0 vanishes as $R_0\to\infty$, so that the scalar potential at the point $\mathbf{r}$ outside S is given by

$$\Psi(\mathbf{r}) = \int\limits_S \left[\Psi(\mathbf{r}')\frac{\partial}{\partial n'}G(\mathbf{r},\mathbf{r}') - G(\mathbf{r},\mathbf{r}')\frac{\partial}{\partial n'}\Psi(\mathbf{r}')\right]dS'. \tag{4.9}$$

This expression gives the potential outside S in terms of the potential and its normal derivative on S.

Inserting the free-space static Green's function $1/(4\pi|\mathbf{r}-\mathbf{r}'|)$ into (4.9) gives the well-known formula [63, sec. 1.10]

$$\Psi(\mathbf{r}) = \frac{1}{4\pi}\int\limits_S \left[\Psi(\mathbf{r}')\frac{\partial}{\partial n'}\frac{1}{|\mathbf{r}-\mathbf{r}'|} - \frac{1}{|\mathbf{r}-\mathbf{r}'|}\frac{\partial}{\partial n'}\Psi(\mathbf{r}')\right]dS'. \tag{4.10}$$

Let $G^D(\mathbf{r}, \mathbf{r}')$ and $G^N(\mathbf{r}, \mathbf{r}')$ be the Dirichlet and Neumann Green's functions for the surface S satisfying (4.8) when $\mathbf{r}$ and $\mathbf{r}'$ are outside S and the boundary conditions

$$G^D(\mathbf{r}, \mathbf{r}') = 0, \quad \frac{\partial}{\partial n'} G^N(\mathbf{r}, \mathbf{r}') = 0, \quad \mathbf{r}' \in S \tag{4.11}$$

when $\mathbf{r}$ is outside S. Assuming that these Green's functions can be found and decay as $(r')^{-1}$ at infinity, the formula (4.9), with each of these functions inserted, shows that the potential outside S can be expressed in terms of the potential on S as

$$\Psi(\mathbf{r}) = \int_S \Psi(\mathbf{r}') \frac{\partial}{\partial n'} G^D(\mathbf{r}, \mathbf{r}') dS' \tag{4.12}$$

or in terms of the normal derivative of the potential on S as

$$\Psi(\mathbf{r}) = -\int_S G^N(\mathbf{r}, \mathbf{r}') \frac{\partial}{\partial n'} \Psi(\mathbf{r}') dS'. \tag{4.13}$$

Invoking the relationship (4.3) for the electric field as the gradient of the potential, each of the formulas (4.9), (4.10), (4.12), and (4.13) can be used to obtain the electric field outside S.

The potential on S can be expressed in terms of the tangential electric field on S through the integration of (4.3)

$$\Psi(\mathbf{r}') = -\int_{\mathbf{r}_1}^{\mathbf{r}'} \mathbf{E}(\mathbf{r}'') \cdot d\mathbf{r}'', \quad \mathbf{r}' \in S \tag{4.14}$$

where $\mathbf{r}_1$ is a fixed point on S. The integration in (4.14) is performed along a smooth curve that lies on S, has a differential tangent vector $d\mathbf{r}''$, begins at $\mathbf{r}_1$, and ends at $\mathbf{r}'$. The potential given by the integration (4.14) equals zero at the fixed point $\mathbf{r}_1$ on the surface S. From (4.3) it follows immediately that the normal derivative of the potential on S is related to the normal component of the electric field on S through the formula

$$\frac{\partial}{\partial n'} \Psi(\mathbf{r}') = -\hat{\mathbf{n}}' \cdot \mathbf{E}(\mathbf{r}'), \quad \mathbf{r}' \in S. \tag{4.15}$$

Substituting the formulas (4.14) and (4.15) into (4.9) or (4.10) gives us an expression for the potential outside S in terms of all three components of the electric field on S. Moreover, inserting (4.14) into (4.12) and (4.15) into (4.13) produces expressions for the potential outside S in terms of the tangential and normal components, respectively, of the electric field on S. Again, all these expressions for the potential can be used in conjunction with (4.3) to obtain the corresponding formulas for the electric field outside S.

Alternatively, we can use the free-space Green's function with the vector Green's identity (2.102) to obtain the expression (2.124) for the electric field outside S in terms of all three components of the electric field on S. The del operator in (2.124) can be brought under the integral sign to yield

$$\mathbf{E}(\mathbf{r}) = \frac{1}{4\pi} \int_S \left[\big(\hat{\mathbf{n}}' \times \mathbf{E}(\mathbf{r}')\big) \times \nabla' \frac{1}{|\mathbf{r} - \mathbf{r}'|} + \hat{\mathbf{n}}' \cdot \mathbf{E}(\mathbf{r}') \nabla' \frac{1}{|\mathbf{r} - \mathbf{r}'|} \right] dS' \tag{4.16}$$

in which we can insert

$$\nabla' \frac{1}{|\mathbf{r} - \mathbf{r}'|} = \frac{\mathbf{r} - \mathbf{r}'}{|\mathbf{r} - \mathbf{r}'|^3}. \tag{4.17}$$

Taking the gradient of (4.10) produces the following expression for the electric field outside S

$$\mathbf{E}(\mathbf{r}) = \frac{1}{4\pi} \int_S \left[\Psi(\mathbf{r}') \frac{\partial}{\partial n'} \nabla' \frac{1}{|\mathbf{r} - \mathbf{r}'|} + \hat{\mathbf{n}}' \cdot \mathbf{E}(\mathbf{r}') \nabla' \frac{1}{|\mathbf{r} - \mathbf{r}'|} \right] dS'. \tag{4.18}$$

The second term of (4.18) is seen to agree with the second term of (4.16). Using the fact that $\nabla \times \mathbf{E} = 0$ throughout space, we can also show that the first terms in (4.16) and (4.18) agree. Moreover, from (4.14) the scalar potential occurring in (4.10) can be expressed in terms of the tangential electric field on S, and the two formulas (4.16) and (4.18) both express the electric field outside S in terms of the electric field on S. However, it is worth noticing an important difference between (4.16) and (4.18): The electric field given by (4.18) automatically satisfies $\nabla \times \mathbf{E} = 0$ no matter what values are inserted for the field on S. This is not true for the electric field given by (4.16). Of course, if the exact electric field is known on S, the curl of (4.16) is zero. If approximate values for the electric field on S are inserted into (4.16), the electric field outside S calculated from (4.16) may not satisfy $\nabla \times \mathbf{E} = 0$.

Far from the source region V where $r \gg r'$, we can employ the expansions

$$\frac{1}{|\mathbf{r}-\mathbf{r}'|} = \frac{1}{r} + \frac{\hat{\mathbf{r}}\cdot\mathbf{r}'}{r^2} + O\left(\frac{1}{r^3}\right) \tag{4.19}$$

and

$$\nabla'\frac{1}{|\mathbf{r}-\mathbf{r}'|} = \frac{\mathbf{r}-\mathbf{r}'}{|\mathbf{r}-\mathbf{r}'|^3} = \frac{\hat{\mathbf{r}}}{r^2} + \frac{3(\mathbf{r}'\cdot\hat{\mathbf{r}})\hat{\mathbf{r}} - \mathbf{r}'}{r^3} + O\left(\frac{1}{r^4}\right) \tag{4.20}$$

in (4.10) to show that the total charge q^{tot} and the dipole moment $\mathbf{p}$ occurring in the far-field formulas (4.6) can be expressed in terms of the potential and its normal derivative on S as

$$q^{tot} = -\epsilon\int_S \frac{\partial}{\partial n'}\Psi(\mathbf{r}')dS', \quad \mathbf{p} = \epsilon\int_S \left[\hat{\mathbf{n}}'\Psi(\mathbf{r}') - \mathbf{r}'\frac{\partial}{\partial n'}\Psi(\mathbf{r}')\right]dS'. \tag{4.21}$$

The divergence theorem and the Poisson equation (4.4) for the potential can be used to show that the expressions in (4.21) for q^{tot} and $\mathbf{p}$ are identical to the ones given in (4.7). In particular, the expression for $\mathbf{p}$ in (4.7) is obtained from (4.21) by applying the divergence theorem to the dyadic $(\nabla\Psi)\mathbf{r}$.

Expressions for q^{tot} and $\mathbf{p}$ in terms of the electric field on S can also be derived, beginning with the proof of the relationship

$$\int_S \hat{\mathbf{n}}\Psi(\mathbf{r})dS = \frac{1}{2}\int_S (\hat{\mathbf{n}} \times \nabla\Psi(\mathbf{r})) \times \mathbf{r}dS. \tag{4.22}$$

From the dyadic divergence theorem [83, p. 509, eq. 53] and the gradient form of Gauss's theorem [83, p. 488, eq. 29] it follows that

$$\begin{aligned}\frac{1}{2}\int_S (\hat{\mathbf{n}} \times \nabla\Psi(\mathbf{r})) \times \mathbf{r}dS &= \frac{1}{2}\int_S \left[\hat{\mathbf{n}}\cdot\mathbf{r}\,\nabla\Psi(\mathbf{r}) - \mathbf{r}\cdot\nabla\Psi(\mathbf{r})\,\hat{\mathbf{n}}\right]dS \\ &= \frac{1}{2}\int_{V_S} \left[\nabla\cdot(\mathbf{r}\nabla\Psi(\mathbf{r})) - \nabla(\mathbf{r}\cdot\nabla\Psi(\mathbf{r}))\right]dV \\ &= \int_{V_S} \nabla\Psi(\mathbf{r})dV = \int_S \hat{\mathbf{n}}\Psi(\mathbf{r})dS\end{aligned} \tag{4.23}$$

where V_S is the volume enclosed by S. (The source region V is contained in V_S.) Inserting (4.22) with $\nabla\Psi = -\mathbf{E}$ and the expression (4.15) for the normal derivative of the potential into (4.21), we obtain the following expressions for q^{tot} and $\mathbf{p}$ in terms of the electric field

$$q^{tot} = \epsilon\int_S \hat{\mathbf{n}}'\cdot\mathbf{E}(\mathbf{r}')dS', \quad \mathbf{p} = \epsilon\int_S \left[\hat{\mathbf{n}}'\cdot\mathbf{E}(\mathbf{r}')\mathbf{r}' - \frac{1}{2}\Big(\hat{\mathbf{n}}' \times \mathbf{E}(\mathbf{r}')\Big) \times \mathbf{r}'\right]dS'. \tag{4.24}$$

4.1.2 Expressions Involving the Static Electric Field in a Plane

Having expressed the field outside the closed surface S in terms of the field on S, we shall now derive formulas for the field in the source-free half space $z > z_0$ in terms of the field in the scan plane $z = z_0$. Begin with the formula (4.10) for the potential, and let S consist of the circular disk ($z' = z_0, x'^2 + y'^2 \leq R_0^2$) and the half sphere ($z' < z_0, x'^2 + y'^2 + (z' - z_0)^2 \leq R_0^2$), as shown in Figure 4.1. With the observation point located in the half space $z > z_0$, we let $R_0 \to \infty$ and find that the integral over the half sphere vanishes so that

$$\Psi(\mathbf{r}) = \frac{1}{4\pi} \int_{-\infty}^{+\infty}\int_{-\infty}^{+\infty} \left[\Psi(\mathbf{r}_0) \frac{z - z_0}{|\mathbf{r} - \mathbf{r}_0|^3} - \frac{1}{|\mathbf{r} - \mathbf{r}_0|} \frac{\partial}{\partial z_0} \Psi(\mathbf{r}_0) \right] dx_0 dy_0, \quad z > z_0 \tag{4.25}$$

where $\mathbf{r}_0 = x_0\hat{\mathbf{x}} + y_0\hat{\mathbf{y}} + z_0\hat{\mathbf{z}}$.

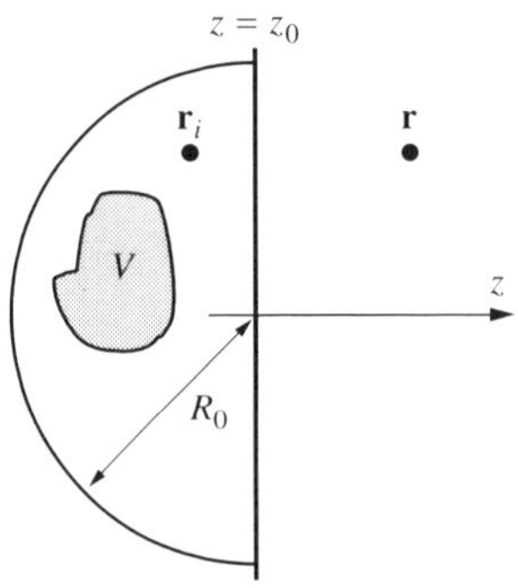

Figure 4.1 The closed surface consisting of a half sphere and a circular disk that both have radius R_0.

The Dirichlet and Neumann Green's functions that satisfy the boundary conditions (4.11) in the plane $z' = z_0$ are given by the well-known expressions

$$G^D(\mathbf{r}, \mathbf{r}') = \frac{1}{4\pi|\mathbf{r} - \mathbf{r}'|} - \frac{1}{4\pi|\mathbf{r}_i - \mathbf{r}'|}, \quad G^N(\mathbf{r}, \mathbf{r}') = \frac{1}{4\pi|\mathbf{r} - \mathbf{r}'|} + \frac{1}{4\pi|\mathbf{r}_i - \mathbf{r}'|} \tag{4.26}$$

where $\mathbf{r}_i = x\hat{\mathbf{x}} + y\hat{\mathbf{y}} + (2z_0 - z)\hat{\mathbf{z}}$ is the image of $\mathbf{r}$ with respect to the scan plane $z = z_0$. When $\mathbf{r}$ is in the half space $z > z_0$, these Green's functions satisfy the differential equation (4.8) for all $\mathbf{r}'$ except at the image point $\mathbf{r}' = \mathbf{r}_i$ located in the half space $z < z_0$. Therefore, for a fixed observation point $\mathbf{r}$ in the half space $z > z_0$, we let S in the formula (4.9) consist of the circular disk ($z' = z_0, x'^2 + y'^2 \leq R_0^2$) and the half sphere ($z' < z_0, x'^2 + y'^2 + (z' - z_0)^2 \leq R_0^2$), where R_0 is large enough so that the image point $\mathbf{r}_i$ is enclosed by S, as shown in Figure 4.1. Substituting each of the Green's functions in (4.26) into (4.9), and proving that the contributions from the half sphere vanish as $R_0 \to \infty$, we obtain the two representations

$$\Psi(\mathbf{r}) = \frac{1}{2\pi} \int_{-\infty}^{+\infty}\int_{-\infty}^{+\infty} \Psi(\mathbf{r}_0) \frac{z - z_0}{|\mathbf{r} - \mathbf{r}_0|^3} dx_0 dy_0, \quad z > z_0 \tag{4.27}$$

and

$$\Psi(\mathbf{r}) = -\frac{1}{2\pi} \int_{-\infty}^{+\infty}\int_{-\infty}^{+\infty} \frac{1}{|\mathbf{r} - \mathbf{r}_0|} \frac{\partial}{\partial z_0} \Psi(\mathbf{r}_0) dx_0 dy_0, \quad z > z_0 \tag{4.28}$$

where

$$\frac{\partial}{\partial z_0} G^D(\mathbf{r}, \mathbf{r}_0) = \frac{z - z_0}{2\pi|\mathbf{r} - \mathbf{r}_0|^3}, \quad G^N(\mathbf{r}, \mathbf{r}_0) = \frac{1}{2\pi|\mathbf{r} - \mathbf{r}_0|} \tag{4.29}$$

have been used. The formulas (4.25), (4.27), and (4.28) express the potential in the half space $z > z_0$ in terms of the potential and its normal derivative, the potential itself, and the normal derivative of the potential, respectively, in the scan plane $z = z_0$. With

$$\rho_0 = \sqrt{x_0^2 + y_0^2} \tag{4.30}$$

being the radial distance in the scan plane, we have $\Psi(\mathbf{r}_0) = O(\rho_0^{-1})$ and $\partial\Psi(\mathbf{r}_0)/\partial z_0 = O(\rho_0^{-3})$ as $\rho_0 \to \infty$, which shows that the integrands in (4.25), (4.27), and (4.28) decay as ρ_0^{-4}. Therefore, the integrals in (4.25), (4.27), and (4.28) converge absolutely.

Equation (4.14) shows that the potential in the scan plane $z = z_0$ can be expressed in terms of the x or y component of the electric field in the scan plane as

$$\Psi(\mathbf{r}_0) = -\int_{-\infty}^{x_0} \hat{\mathbf{x}} \cdot \mathbf{E}(x', y_0, z_0)dx', \quad \Psi(\mathbf{r}_0) = -\int_{-\infty}^{y_0} \hat{\mathbf{y}} \cdot \mathbf{E}(x_0, y', z_0)dy' \tag{4.31}$$

where we have used the asymptotic relations $|\mathbf{E}(x', y_0, z_0)| = O[(x')^{-2}]$ as $|x'| \to \infty$ and $|\mathbf{E}(x_0, y', z_0)| = O[(y')^{-2}]$ as $|y'| \to \infty$. Inserting the first or second expressions for $\Psi(\mathbf{r}_0)$ from (4.31) into (4.27) produces a formula for the potential in the half space $z > z_0$ in terms of the x or y component of the electric field in the scan plane $z = z_0$. Equation (4.15) shows that the normal derivative of the potential in the scan plane can be determined from the normal component of the electric field as

$$\frac{\partial}{\partial z_0}\Psi(\mathbf{r}_0) = -\hat{\mathbf{z}} \cdot \mathbf{E}(\mathbf{r}_0). \tag{4.32}$$

By inserting this result into (4.28), we get a formula for the potential in the half space $z > z_0$ in terms of the z component of the electric field in the scan plane $z = z_0$. To obtain explicit formulas for the electric field, we take the gradient of (4.27) and (4.28) in order to get

$$\mathbf{E}(\mathbf{r}) = \frac{1}{2\pi}\int_{-\infty}^{+\infty}\int_{-\infty}^{+\infty} \Psi(\mathbf{r}_0)\left[3(z - z_0)\frac{\mathbf{r} - \mathbf{r}_0}{|\mathbf{r} - \mathbf{r}_0|^5} - \frac{\hat{\mathbf{z}}}{|\mathbf{r} - \mathbf{r}_0|^3}\right]dx_0dy_0, \quad z > z_0 \tag{4.33}$$

and

$$\mathbf{E}(\mathbf{r}) = \frac{1}{2\pi}\int_{-\infty}^{+\infty}\int_{-\infty}^{+\infty} \hat{\mathbf{z}} \cdot \mathbf{E}(\mathbf{r}_0)\frac{\mathbf{r} - \mathbf{r}_0}{|\mathbf{r} - \mathbf{r}_0|^3}dx_0dy_0, \quad z > z_0. \tag{4.34}$$

By performing a derivation with (4.16), similar to the one that led to (4.25), one can prove that the electric field in the half space $z > z_0$ can be expressed in terms of all three rectangular components of the electric field in the scan plane $z = z_0$ as

$$\mathbf{E}(\mathbf{r}) = \frac{1}{4\pi}\int_{-\infty}^{+\infty}\int_{-\infty}^{+\infty}\left[\left(\hat{\mathbf{z}} \times \mathbf{E}(\mathbf{r}_0)\right) \times \frac{\mathbf{r} - \mathbf{r}_0}{|\mathbf{r} - \mathbf{r}_0|^3} + \hat{\mathbf{z}} \cdot \mathbf{E}(\mathbf{r}_0)\frac{\mathbf{r} - \mathbf{r}_0}{|\mathbf{r} - \mathbf{r}_0|^3}\right]dx_0dy_0, \quad z > z_0 \tag{4.35}$$

which, by use of (4.34), implies that

$$\mathbf{E}(\mathbf{r}) = \frac{1}{2\pi}\int_{-\infty}^{+\infty}\int_{-\infty}^{+\infty}\left(\hat{\mathbf{z}} \times \mathbf{E}(\mathbf{r}_0)\right) \times \frac{\mathbf{r} - \mathbf{r}_0}{|\mathbf{r} - \mathbf{r}_0|^3}dx_0dy_0, \quad z > z_0. \tag{4.36}$$

From the far-field expressions in (4.6), it is seen that in the scan plane $z = z_0$, the integrands in (4.33), (4.35), and (4.36) decay as ρ_0^{-4} and the integrand in (4.34) decays as ρ_0^{-5}. Hence, all the integrals in (4.33)–(4.36) converge absolutely.

We have just presented formulas that determine the static electric field and its potential in the half space $z > z_0$ in terms of an integral of the potential or any one of the rectangular components of the electric field in the scan plane $z = z_0$.

The total charge q^{tot} and the dipole moment $\mathbf{p}$ occurring in the far-field formulas (4.6) will now be expressed in terms of the field in the scan plane $z = z_0$. Begin with the formula in (4.21) giving q^{tot} in terms of the normal derivative of the potential on the surface S. In this formula, let S consist of the disk ($z' = z_0$, $x'^2 + y'^2 \le R_0^2$) and the half sphere ($z' < z_0$, $x'^2 + y'^2 + (z' - z_0)^2 \le R_0^2$). As $R_0 \to \infty$, we can employ the expression (4.15) along with (4.6) for the electric far field to show that the contribution from the half sphere is simply $q^{tot}/2$. Thus, we have the formulas

$$q^{tot} = -2\epsilon \int_{-\infty}^{+\infty}\int_{-\infty}^{+\infty} \frac{\partial}{\partial z_0}\Psi(\mathbf{r}_0)dx_0dy_0 = 2\epsilon \int_{-\infty}^{+\infty}\int_{-\infty}^{+\infty} \hat{\mathbf{z}} \cdot \mathbf{E}(\mathbf{r}_0)dx_0dy_0 \tag{4.37}$$

which express the total charge in terms of the normal derivative of the potential or the z component of the electric field in the scan plane $z = z_0$. As mentioned before, $\partial\Psi(\mathbf{r}_0)/\partial z_0 = O(\rho_0^{-3})$ in the scan plane $z = z_0$, so the integrals in (4.37) converge absolutely.

Next consider the dipole moment $\mathbf{p}$ given by the second equation in (4.21), and let S consist of the circular disk and half sphere of radii R_0 used in the derivation of (4.37). Under the assumption that $q^{tot} = 0$, one can use the far-field expressions (4.6) to show that the contribution from the half sphere is simply $\mathbf{p}/2$, as $R_0 \to \infty$. This result is obtained by use of the formula

$$\int_0^{2\pi}\int_{\pi/2}^{\pi} \hat{\mathbf{r}}\hat{\mathbf{r}} \sin\theta d\theta d\phi = \frac{2\pi}{3}\bar{\mathbf{I}} \tag{4.38}$$

where $\bar{\mathbf{I}}$ is the identity dyadic. Consequently, when $q^{tot} = 0$, the dipole moment $\mathbf{p}$ can be expressed in terms of the potential and its normal derivative in the scan plane $z = z_0$ as

$$\mathbf{p} = 2\epsilon \int_0^{+\infty}\int_0^{2\pi} \left[\hat{\mathbf{z}}\Psi(\mathbf{r}_0) - \mathbf{r}_0\frac{\partial}{\partial z_0}\Psi(\mathbf{r}_0)\right] d\phi_0\rho_0 d\rho_0 \tag{4.39}$$

where the polar integration variables (ρ_0, ϕ_0) determine the point $\mathbf{r}_0$ in the scan plane through the equation $\mathbf{r}_0 = \hat{\mathbf{x}}\rho_0\cos\phi_0 + \hat{\mathbf{y}}\rho_0\sin\phi_0 + \hat{\mathbf{z}}z_0$. When $q^{tot} = 0$, we have $\Psi(\mathbf{r}_0) = O(\rho_0^{-2})$ and $\partial\Psi(\mathbf{r}_0)/\partial z_0 = O(\rho_0^{-3})$ in the scan plane $z = z_0$. Thus, the expression in the square brackets of (4.39) decays as ρ_0^{-2}, and the integral over the scan plane in (4.39) diverges unless it is calculated with the ϕ_0 integration performed before the ρ_0 integration. (This prescription for integrating over the scan plane is similar to the prescriptions used in Appendix D for time-harmonic fields with $\omega > 0$.) However, the derivation shows that it does indeed converge when calculated as in (4.39), but it is not possible, as in the previously derived formulas, to express $\mathbf{p}$ in terms of a general integral over the scan plane. Note that both terms in (4.39) contribute in general to the dipole moment and that the integral in (4.39) diverges when $q^{tot} \neq 0$. Also, when $q^{tot} = 0$, (4.37) shows that the z component of the second term in (4.39) is zero, and thus the first term in (4.39) gives the z component of $\mathbf{p}$, and the second term in (4.39) gives the x and y components of $\mathbf{p}$. In other words, (4.39) can be rewritten as

$$\mathbf{p} = 2\epsilon \int_0^{+\infty}\int_0^{2\pi} \left[\hat{\mathbf{z}}\Psi(\mathbf{r}_0) - \boldsymbol{\rho}_0\frac{\partial}{\partial z_0}\Psi(\mathbf{r}_0)\right] d\phi_0\rho_0 d\rho_0 \tag{4.40}$$

where

$$\boldsymbol{\rho}_0 = \hat{\mathbf{x}}\rho_0\cos\phi_0 + \hat{\mathbf{y}}\rho_0\sin\phi_0. \tag{4.41}$$

To derive an expression for $\mathbf{p}$ in terms of only $\partial\Psi/\partial z_0$, that is, in terms of only the component of the electric field normal to the scan plane, begin by inserting the Neumann Green's function from (4.26) into the formula (4.9) for the potential. Let the closed surface S in (4.9) consist of the circular disk ($z' = z_0$, $x'^2 + y'^2 \le R_0^2$) and the half sphere ($z' < z_0$, $x'^2 + y'^2 + (z' - z_0)^2 \le R_0^2$), so that

$$\begin{aligned}\Psi(\mathbf{r}) = &-\frac{1}{2\pi}\int_0^{R_0}\int_0^{2\pi}\frac{1}{|\mathbf{r}-\mathbf{r}_0|}\frac{\partial}{\partial z_0}\Psi(\mathbf{r}_0)d\phi_0\rho_0 d\rho_0 \\ &-\frac{R_0^2}{4\pi}\int_0^{2\pi}\int_{\pi/2}^{\pi}\left(\left[\frac{1}{|\mathbf{r}-R_0\hat{\mathbf{r}}'|}+\frac{1}{|\mathbf{r}_i-R_0\hat{\mathbf{r}}'|}\right]\frac{\partial}{\partial R_0}\Psi(R_0\hat{\mathbf{r}}')\right. \\ &\qquad\left. -\Psi(R_0\hat{\mathbf{r}}')\frac{\partial}{\partial R_0}\left[\frac{1}{|\mathbf{r}-R_0\hat{\mathbf{r}}'|}+\frac{1}{|\mathbf{r}_i-R_0\hat{\mathbf{r}}'|}\right]\right)\sin\theta' d\theta' d\phi'\end{aligned} \tag{4.42}$$

where (4.29) has been employed, and $\mathbf{r}_i = x\hat{\mathbf{x}} + y\hat{\mathbf{y}} + (2z_0 - z)\hat{\mathbf{z}}$ is the image of $\mathbf{r}$ with respect to the scan plane $z = z_0$. For (4.9) to hold, the Green's function must satisfy (4.8) when $\mathbf{r}$ and $\mathbf{r}'$ are outside S. Thus, (4.42) is valid provided that the image point $\mathbf{r}_i$ is inside S, as shown in Figure 4.1. Therefore, we let the observation point $\mathbf{r} = x\hat{\mathbf{z}} + y\hat{\mathbf{y}} + z\hat{\mathbf{z}}$ in (4.42) satisfy $|\mathbf{r} - z_0\hat{\mathbf{z}}| < R_0$ and $z > z_0$. We further specify that $\hat{\mathbf{r}}$ be fixed and that $\mathbf{r} = \mathcal{R}(R_0)\hat{\mathbf{r}}$ where the function $\mathcal{R}(R_0)$ satisfies the conditions $\mathcal{R}(R_0) \to \infty$ and $\mathcal{R}(R_0)/R_0 \to 0$ as $R_0 \to \infty$. Thus, the observation point moves to infinity in the half space $z > z_0$ as $R_0 \to \infty$, and $\mathbf{r}_i$ is always inside S so that (4.42) holds.

When $q^{tot} = 0$, (4.6) shows that the $\hat{\mathbf{r}}$ component of the dipole moment is given by

$$\mathbf{p}\cdot\hat{\mathbf{r}} = 4\pi\epsilon \lim_{R_0\to\infty}[\mathcal{R}(R_0)]^2\Psi(\hat{\mathbf{r}}\mathcal{R}(R_0)). \tag{4.43}$$

Inserting (4.42) into (4.43), we find that there is no contribution from the half sphere. Therefore, the $\hat{\mathbf{r}}$ component of the dipole moment is given in terms of the normal derivative of the potential in the scan plane $z = z_0$ as

$$\mathbf{p}\cdot\hat{\mathbf{r}} = -2\epsilon \lim_{R_0\to\infty}\int_0^{R_0}\int_0^{2\pi}\frac{[\mathcal{R}(R_0)]^2}{|\hat{\mathbf{r}}\mathcal{R}(R_0)-\mathbf{r}_0|}\frac{\partial}{\partial z_0}\Psi(\mathbf{r}_0)d\phi_0\rho_0 d\rho_0 \tag{4.44}$$

when $q^{tot} = 0$. Letting $\hat{\mathbf{r}} = \hat{\mathbf{z}}$ in (4.44) to get p_z and using (4.40) to get p_x and p_y, we obtain the following expression for $\mathbf{p}$ in terms of $\partial\Psi/\partial z_0$ in the scan plane

$$\mathbf{p} = -2\epsilon \lim_{R_0\to\infty}\int_0^{R_0}\int_0^{2\pi}\left[\hat{\mathbf{z}}\frac{[\mathcal{R}(R_0)]^2}{|\hat{\mathbf{z}}\mathcal{R}(R_0)-\mathbf{r}_0|}+\boldsymbol{\rho}_0\right]\frac{\partial}{\partial z_0}\Psi(\mathbf{r}_0)d\phi_0\rho_0 d\rho_0. \tag{4.45}$$

Inserting $\partial\Psi/\partial z_0$ from (4.32) into (4.45), we obtain an expression for $\mathbf{p}$ in terms of the z component of the electric field in the scan plane. In numerical calculations, we can choose $\mathcal{R}(R_0) = \sqrt{aR_0}$ where a is a positive length.

Similarly, inserting the Dirichlet Green's function from (4.26) into (4.9), we find that $\mathbf{p}\cdot\hat{\mathbf{r}}$ can be determined from the potential in the scan plane as

$$\mathbf{p}\cdot\hat{\mathbf{r}} = 2\epsilon \lim_{R_0\to\infty}\int_0^{R_0}\int_0^{2\pi}\frac{(\hat{\mathbf{z}}\cdot\hat{\mathbf{r}}\mathcal{R}(R_0)-z_0)[\mathcal{R}(R_0)]^2}{|\hat{\mathbf{r}}\mathcal{R}(R_0)-\mathbf{r}_0|^3}\Psi(\mathbf{r}_0)d\phi_0\rho_0 d\rho_0 \tag{4.46}$$

when $q^{tot} = 0$. Substituting from (4.31) into (4.46) produces a formula for $\mathbf{p}\cdot\hat{\mathbf{r}}$ in terms of the x or y component of the electric field in the scan plane $z = z_0$. In (4.46) we can let $\hat{\mathbf{r}} = \hat{\mathbf{z}}$,

$\hat{\mathbf{r}} = \hat{\mathbf{x}}/\sqrt{2} + \hat{\mathbf{z}}/\sqrt{2}$, and $\hat{\mathbf{r}} = \hat{\mathbf{y}}/\sqrt{2} + \hat{\mathbf{z}}/\sqrt{2}$ to get expressions for the rectangular components of $\mathbf{p}$ in terms of the potential in the scan plane.

We can also express $\mathbf{p}$ in terms of all three rectangular components of the electric field in the scan plane. To do this, let S in the formula (4.24) for $\mathbf{p}$ consist of the circular disk and half sphere of radii R_0 used in the derivation of (4.37). By use of (4.38), we can show that as $R_0 \to \infty$, the contribution from the half sphere is simply $\mathbf{p}/2$ when $q^{tot} = 0$. Thus, when $q^{tot} = 0$ the dipole moment is simply

$$\mathbf{p} = 2\epsilon \int_0^{+\infty}\int_0^{2\pi} \left[\hat{\mathbf{z}} \cdot \mathbf{E}(\mathbf{r}_0)\mathbf{r}_0 - \frac{1}{2}\Big(\hat{\mathbf{z}} \times \mathbf{E}(\mathbf{r}_0) \Big) \times \mathbf{r}_0 \right] d\phi_0 \rho_0 d\rho_0. \tag{4.47}$$

Using (4.37) with $q^{tot} = 0$, expressing $\big(\hat{\mathbf{z}} \times \mathbf{E}(\mathbf{r}_0)\big) \times \hat{\mathbf{z}}$ as $-\nabla_s \Psi$ with $\nabla_s = \nabla - \hat{\mathbf{z}}\frac{\partial}{\partial z}$, and applying the two-dimensional divergence theorem to show that this $\nabla_s \Psi$ term in (4.47) integrates to zero, we can write (4.47) as

$$\mathbf{p} = 2\epsilon \int_0^{+\infty}\int_0^{2\pi} \left[\hat{\mathbf{z}} \cdot \mathbf{E}(\mathbf{r}_0)\boldsymbol{\rho}_0 - \frac{1}{2}\Big(\hat{\mathbf{z}} \times \mathbf{E}(\mathbf{r}_0) \Big) \times \boldsymbol{\rho}_0 \right] d\phi_0 \rho_0 d\rho_0. \tag{4.48}$$

As in the expression (4.39) or (4.40), the integral over the scan plane in (4.47) or (4.48) diverges unless it is calculated in the prescribed manner. Equation (4.45), which determines $\mathbf{p}$ from only the z component of the electric field in the scan plane, may be more useful for numerical calculations than (4.47) or (4.48), which uses all three rectangular components of the electric field in the scan plane.

Rigorous derivations have now been presented for the expressions (4.37), (4.39)–(4.40), and (4.44)–(4.48) that give q^{tot} and $\mathbf{p}$ in terms of the fields in the scan plane $z = z_0$. Once q^{tot} and $\mathbf{p}$ are known, (4.6) determines the far fields. One could be tempted to derive a far-field expression for the potential by inserting the far-field formula (4.19) for $1/|\mathbf{r} - \mathbf{r}_0|$ into (4.28), which gives the potential in terms of its normal derivative in the scan plane $z = z_0$. Although this procedure would produce the correct expression (4.37) for q^{tot}, it produces the incorrect result that $\mathbf{p}$ is given by the second term of (4.39).

The reason this nonrigorous procedure succeeds in producing the correct expression for q^{tot} but fails to produce the correct expression for $\mathbf{p}$ can be explained as follows. In the scan plane, the first term in the expansion (4.19) for $1/|\mathbf{r} - \mathbf{r}_0|$ is of order 1, and since $\partial\Psi(\mathbf{r}_0)/\partial z_0 = O(\rho_0^{-3})$, the resulting integrand decays as ρ_0^{-3} at infinity. The resulting integral over the scan plane therefore converges absolutely and uniformly with respect to r, and we can interchange the limit $r \to \infty$ with the infinite $\mathbf{r}_0$ integration in (4.28) to obtain the correct expression for q^{tot}. The second term of the expansion (4.19) of $1/|\mathbf{r} - \mathbf{r}_0|$ is of order ρ_0, and the resulting integrand decays as ρ_0^{-2} at infinity. This ρ_0^{-2} decay results in an integral that is not absolutely convergent, and we cannot interchange the limit $r \to \infty$ with the infinite $\mathbf{r}_0$ integration in (4.28) to obtain the expression for $\mathbf{p}$. Similar erroneous far-field results are obtained if we substitute the far-field expressions (4.19) and (4.20) into the formulas (4.27) and (4.33)–(4.36), and mistakenly interchange the limits as $r \to \infty$ with the infinite (x_0, y_0) integrations.

The formulas of this section show that the static electric field is uniquely determined in the source-free half space $z > z_0$ from the values of one of its rectangular components in the scan plane $z = z_0$. We can also prove this result directly from the static Maxwell equations without deriving any representations for the fields.

Begin by assuming that the x component of the electric field is known in the scan plane $z = z_0$. To show that E_x for $z = z_0$ uniquely determines the electric field for $z > z_0$, we simply prove that if $E_x = 0$ for $z = z_0$, the electric field is identically zero throughout the source-free half space $z > z_0$. If $E_x = 0$ for $z = z_0$, the scalar potential satisfies $\partial\Psi/\partial x = 0$ for $z = z_0$. Since Ψ decays as r^{-1} (or faster) at infinity, this implies that the scalar potential is identically zero over the entire plane $z = z_0$. Moreover, Ψ satisfies Laplace's equation, and the divergence theorem gives

$$\begin{aligned}\int\limits_{z\geq z_0} |\nabla\Psi(\mathbf{r})|^2 dV &= \int\limits_{z\geq z_0} \nabla\cdot(\Psi(\mathbf{r})\nabla\Psi(\mathbf{r}))dV \\ &= -\int\limits_{-\infty}^{+\infty}\int\limits_{-\infty}^{+\infty} \Psi(\mathbf{r}_0)\frac{\partial}{\partial z_0}\Psi(\mathbf{r}_0)dx_0dy_0 = 0\end{aligned} \tag{4.49}$$

since there is no contribution from the far-field hemisphere in the region $z > z_0$. Thus, $-\nabla\Psi = \mathbf{E} = 0$ for $z > z_0$ and the proof is complete. A similar proof can be performed to show that the y component of the electric field in the scan plane uniquely determines the electric field in the half space $z > z_0$.

To show that E_z in the plane $z = z_0$ uniquely determines the electric field in the half space $z > z_0$, we simply prove that if $E_z = 0$ for $z = z_0$, the electric field is zero for $z > z_0$. When $E_z = 0$ for $z = z_0$, Ψ satisfies $\partial\Psi/\partial z = 0$ in the plane $z = z_0$, and (4.49) again shows that $-\nabla\Psi = \mathbf{E}$ is identically zero in the half space $z > z_0$. Thus, specifying any rectangular component of the electric field in the plane $z = z_0$ determines all electric field components in the half space $z > z_0$.

4.2 GREEN'S FUNCTION REPRESENTATIONS FOR THE STATIC MAGNETIC FIELD

Now consider the static magnetic field $\mathbf{H}$ generated by the source current $\mathbf{J}$ located in the bounded volume V. The static magnetic field can be expressed from (2.85) in terms of the source current as

$$\mathbf{H}(\mathbf{r}) = \frac{1}{4\pi}\lim_{\delta\to 0}\int\limits_{V-V_\delta} \mathbf{J}(\mathbf{r}')\times\frac{\mathbf{r}-\mathbf{r}'}{|\mathbf{r}-\mathbf{r}'|^3}dV' \tag{4.50}$$

which can be expanded asymptotically for large r as [63, sec. 5.6]

$$\mathbf{H}(\mathbf{r}) = \frac{3(\mathbf{m}\cdot\hat{\mathbf{r}})\,\hat{\mathbf{r}} - \mathbf{m}}{4\pi r^3} + O\left(\frac{1}{r^4}\right) \tag{4.51}$$

where $\mathbf{m}$ is the magnetic dipole moment given by

$$\mathbf{m} = \frac{1}{2}\int\limits_V \mathbf{r}'\times\mathbf{J}(\mathbf{r}')dV'. \tag{4.52}$$

Since there is no magnetic charge, the static magnetic far field in (4.51) decays as r^{-3} at infinity when $\mathbf{m} \neq 0$. If $\mathbf{m} = 0$, the magnetic far field decays as r^{-4} (or faster), and a multipole expansion [1, sec. 4.5] can be used to determine its radial and angular dependencies.

The static magnetic field satisfies the equation $\nabla\times\mathbf{H} = \mathbf{J}$. Therefore, it cannot be expressed as the gradient of a single-valued scalar potential throughout all space [1, sec. 4.2]. However, here we are interested only in the magnetic field outside a closed surface S that contains all the sources, and we can choose S such that the region bounded by S is simply connected (contains no holes) [90, p. 305]. With this choice of S, we have $\mathbf{J} = 0$ outside S, and for all closed curves C outside S we have $\int_C \mathbf{H}(\mathbf{r})\cdot d\mathbf{r} = \int_{S_C}\mathbf{J}\cdot\hat{\mathbf{n}}dS = 0$, where $d\mathbf{r}$ is the differential of the position vector tangent to C, S_C is any surface located outside S that has C as its boundary, and $\hat{\mathbf{n}}$ is the unit normal vector to S_C. This result, which follows from Stokes' theorem, shows that we can express the static magnetic field *outside* S (which is chosen so that the region enclosed by S is simply connected) in terms of a *single-valued* magnetic scalar potential Ψ^m such that $\mathbf{H} = -\nabla\Psi^m$. Because no magnetic charge exists, the magnetic scalar potential decays as r^{-2} (or faster) at infinity [1, sec. 4.16], and its far field can be expressed in terms of the magnetic dipole moment as

$$\Psi^m(\mathbf{r}) = \frac{\mathbf{m}\cdot\hat{\mathbf{r}}}{4\pi r^2} + O\left(\frac{1}{r^3}\right). \tag{4.53}$$

4.2.1 Expressions Involving the Static Magnetic Field on a Closed Surface

Since $\nabla\cdot\mathbf{H} = 0$, the magnetic scalar potential obeys $\nabla^2\Psi^m = 0$ outside S. Thus, formulas that determine the static magnetic field and its scalar potential in terms of fields on a surface can be obtained by replacing $\mathbf{E}$ and Ψ in the static electric-field formulas above with $\mathbf{H}$ and Ψ^m, respectively. Moreover, the magnetic dipole moment can be found by inspection of (4.21) to be given in terms of the magnetic scalar potential and its normal derivative on S as

$$\mathbf{m} = \int_S \left[\hat{\mathbf{n}}'\Psi^m(\mathbf{r}') - \mathbf{r}'\frac{\partial}{\partial n'}\Psi^m(\mathbf{r}')\right] dS'. \tag{4.54}$$

Since no magnetic charge exists, there is no magnetic equivalent of q^{tot}.

The expression for the electric dipole moment in (4.24) was obtained by use of the identity (4.22), whose derivation involved an integration of the electric scalar potential Ψ over the source region V. Since the magnetic scalar potential Ψ^m is not a single-valued function in V, it is not guaranteed that (4.22) holds for Ψ^m. However, Ψ^m decays as r^{-2} (or faster) at infinity, and by applying the divergence theorem [83, p. 509, eq. 53] to the infinite region *outside* S, and taking into account the $\mathbf{m}/3$ contribution from the far-field sphere, we can show that (4.22) indeed holds for Ψ^m. Thereby we find from (4.54) that the magnetic dipole moment can be expressed in terms of the magnetic field on S as

$$\mathbf{m} = \int_S \left[\hat{\mathbf{n}}'\cdot\mathbf{H}(\mathbf{r}')\mathbf{r}' - \frac{1}{2}\left(\hat{\mathbf{n}}'\times\mathbf{H}(\mathbf{r}')\right)\times\mathbf{r}'\right] dS'. \tag{4.55}$$

4.2.2 Expressions Involving the Static Magnetic Field in a Plane

Performing derivations similar to those that led to (4.39), (4.40), and (4.44)–(4.48) shows that the magnetic dipole moment can be expressed in terms of the potential and magnetic field in the scan plane $z = z_0$ as

$$\begin{aligned}\mathbf{m} &= 2\int_0^{+\infty}\int_0^{2\pi}\left[\hat{\mathbf{z}}\Psi^m(\mathbf{r}_0) - \mathbf{r}_0\frac{\partial}{\partial z_0}\Psi^m(\mathbf{r}_0)\right] d\phi_0\rho_0 d\rho_0 \\ &= 2\int_0^{+\infty}\int_0^{2\pi}\left[\hat{\mathbf{z}}\Psi^m(\mathbf{r}_0) - \boldsymbol{\rho}_0\frac{\partial}{\partial z_0}\Psi^m(\mathbf{r}_0)\right] d\phi_0\rho_0 d\rho_0\end{aligned} \tag{4.56}$$

$$\mathbf{m}\cdot\hat{\mathbf{r}} = -2\lim_{R_0\to\infty}\int_0^{R_0}\int_0^{2\pi}\frac{[\mathcal{R}(R_0)]^2}{|\hat{\mathbf{r}}\mathcal{R}(R_0) - \mathbf{r}_0|}\frac{\partial}{\partial z_0}\Psi^m(\mathbf{r}_0)d\phi_0\rho_0 d\rho_0 \tag{4.57}$$

$$\mathbf{m} = -2 \lim_{R_0 \to \infty} \int_0^{R_0} \int_0^{2\pi} \left[\hat{\mathbf{z}} \frac{[\mathcal{R}(R_0)]^2}{|\hat{\mathbf{z}} \mathcal{R}(R_0) - \mathbf{r}_0|} + \boldsymbol{\rho}_0 \right] \frac{\partial}{\partial z_0} \Psi^m(\mathbf{r}_0) d\phi_0 \rho_0 d\rho_0 \tag{4.58}$$

$$\mathbf{m} \cdot \hat{\mathbf{r}} = 2 \lim_{R_0 \to \infty} \int_0^{R_0} \int_0^{2\pi} \frac{\left(\hat{\mathbf{z}} \cdot \hat{\mathbf{r}} \mathcal{R}(R_0) - z_0\right) [\mathcal{R}(R_0)]^2}{|\hat{\mathbf{r}} \mathcal{R}(R_0) - \mathbf{r}_0|^3} \Psi^m(\mathbf{r}_0) d\phi_0 \rho_0 d\rho_0 \tag{4.59}$$

and

$$\begin{aligned} \mathbf{m} &= 2 \int_0^{+\infty} \int_0^{2\pi} \left[\hat{\mathbf{z}} \cdot \mathbf{H}(\mathbf{r}_0) \mathbf{r}_0 - \frac{1}{2} \Big(\hat{\mathbf{z}} \times \mathbf{H}(\mathbf{r}_0) \Big) \times \mathbf{r}_0 \right] d\phi_0 \rho_0 d\rho_0 \\ &= 2 \int_0^{+\infty} \int_0^{2\pi} \left[\hat{\mathbf{z}} \cdot \mathbf{H}(\mathbf{r}_0) \boldsymbol{\rho}_0 - \frac{1}{2} \Big(\hat{\mathbf{z}} \times \mathbf{H}(\mathbf{r}_0) \Big) \times \boldsymbol{\rho}_0 \right] d\phi_0 \rho_0 d\rho_0 \end{aligned} \tag{4.60}$$

where $\mathbf{r}_0 = \boldsymbol{\rho}_0 + \hat{\mathbf{z}} z_0$, $\boldsymbol{\rho}_0 = \hat{\mathbf{x}} \rho_0 \cos\phi_0 + \hat{\mathbf{y}} \rho_0 \sin\phi_0$, and $-\hat{\mathbf{z}} \cdot \mathbf{H}$ can be inserted for $\partial \Psi^m / \partial z$. As defined previously in Section 4.1.2, the function $\mathcal{R}(R_0)$ satisfies the conditions $\mathcal{R}(R_0) \to \infty$ and $\mathcal{R}(R_0)/R_0 \to 0$ as $R_0 \to \infty$. The convergence properties of the integrals in (4.56)–(4.60) are discussed below equation (4.39).

When the sources are located in the half space $z > z_0$, the formulas (4.25), (4.27)–(4.28), (4.33)–(4.37), (4.39)–(4.40), (4.44)–(4.48), and (4.56)–(4.60) must be multiplied by -1 to hold for the fields in the half space $z < z_0$.

4.3 PLANE-WAVE SPECTRUM REPRESENTATIONS

In Sections 4.1.2 and 4.2.2, Green's function expressions were found for the electrostatic and magnetostatic fields in a source-free half space $z > z_0$ in terms of any one rectangular component of the electrostatic and magnetostatic fields, respectively, in the infinite scan plane $z = z_0$. In this section, we show that the static fields for $z > z_0$ can also be determined from their component values at $z = z_0$ by means of evanescent plane-wave spectrum representations [139]. Although we first have to evaluate the plane-wave spectra from the static fields measured on the $z = z_0$ scan plane, and then integrate the spectra to get the static fields in the half space $z > z_0$, fast Fourier transform algorithms can make this indirect method involving integration of the plane-wave spectra much more efficient than the direct integration of the Green's function representations to evaluate the static fields.

4.3.1 Static Electric Field

Outside the source region, the static electric field is a continuous function of $\mathbf{r}$ that decays monotonically as r^{-2} or faster as r approaches infinity. It is therefore both Riemann and Lebesgue square integrable in any plane $z = z_1 \geq z_0$. Hence, the Fourier integral theorem for Lebesgue square-integrable functions [140], [141], [115, p. 458] shows that the electric field in the half space $z \geq z_0$ can be expressed in terms of the Fourier integral

$$\mathbf{E}(\mathbf{r}) = \frac{1}{2\pi} \int_{-\infty}^{+\infty} \int_{-\infty}^{+\infty} \mathbf{b}(k_x, k_y, z) e^{i(k_x x + k_y y)} dk_x dk_y, \quad z \geq z_0 \tag{4.61}$$

where $\mathbf{r} = x\hat{\mathbf{x}} + y\hat{\mathbf{y}} + z\hat{\mathbf{z}}$ and $\mathbf{b}(k_x, k_y, z)$ is the Fourier transform of $\mathbf{E}(\mathbf{r})$. Because $\nabla^2 \mathbf{E} = 0$, the function **b** satisfies

$$\left[\frac{\partial^2}{\partial z^2} - k_x^2 - k_y^2\right]\mathbf{b}(k_x, k_y, z) = 0. \tag{4.62}$$

Since the electric field in (4.61) decays as $z \to \infty$, we require that $\mathbf{b} \to 0$ as $z \to \infty$. The only solution to (4.62) that satisfies this requirement is

$$\mathbf{b}(k_x, k_y, z) = \mathbf{T}(k_x, k_y)e^{-z\sqrt{k_x^2+k_y^2}}, \quad z \geq z_0 \tag{4.63}$$

where $\mathbf{T}(k_x, k_y)$ is the plane-wave spectrum for the static electric field. By inserting (4.63) into (4.61), we find that the static electric field can be expressed in terms of the plane wave spectrum as

$$\mathbf{E}(\mathbf{r}) = \frac{1}{2\pi}\int_{-\infty}^{+\infty}\int_{-\infty}^{+\infty} \mathbf{T}(k_x, k_y)e^{i(k_x x+k_y y+i\sqrt{k_x^2+k_y^2}z)}dk_x dk_y, \quad z \geq z_0. \tag{4.64}$$

Taking the inverse Fourier transform of (4.64) yields the following formula for the plane-wave spectrum in terms of the static electric field in a plane

$$\mathbf{T}(k_x, k_y) = \frac{e^{z\sqrt{k_x^2+k_y^2}}}{2\pi}\int_{-\infty}^{+\infty}\int_{-\infty}^{+\infty} \mathbf{E}(\mathbf{r})e^{-i(k_x x+k_y y)}dxdy, \quad z \geq z_0. \tag{4.65}$$

Letting $z = z_0$ in (4.65), we find that the formula (4.64) becomes a representation for the electric field in the half space $z \geq z_0$ in terms of the electric field in the scan plane $z = z_0$. Note that the "propagation" vector for the plane waves used in the expansion (4.64) is given by

$$\mathbf{k} = k_x\hat{\mathbf{x}} + k_y\hat{\mathbf{y}} + i\sqrt{k_x^2 + k_y^2}\,\hat{\mathbf{z}} \tag{4.66}$$

and that for $(k_x, k_y) \neq (0, 0)$, the plane waves $e^{i\mathbf{k}\cdot\mathbf{r}}$ are evanescent in the positive z direction.

We have initially relied upon the Fourier integral theorem, which requires Lebesgue integration and convergence in the mean, to assure the validity of the plane-wave spectrum representation (4.64) and (4.65). However, the spectrum $\mathbf{T}(k_x, k_y)$ can be evaluated at each $(k_x, k_y) \neq (0, 0)$ using ordinary Riemann integration in (4.65) because $\mathbf{E}(\mathbf{r})$ is a continuous function of $\mathbf{r}$ that decays monotonically as r^{-2} or faster as $r \to \infty$. We prove below that $\mathbf{T}(k_x, k_y)$ is a continuous function for all real (k_x, k_y) except possibly at $(k_x, k_y) = (0, 0)$ where it is bounded. Moreover, (4.75) can be used to show that for $z > z_0$ the function $\mathbf{T}(k_x, k_y)e^{-z\sqrt{k_x^2+k_y^2}}$ in the integrand of (4.64) decays exponentially as $k_x^2+k_y^2 \to \infty$. Therefore, $\mathbf{E}(\mathbf{r})$ in (4.64) can also be evaluated at each point $\mathbf{r}$ in the half space $z > z_0$ by ordinary Riemann integration.

Taking the divergence and curl of (4.64), and bringing the derivatives inside the integral in (4.64), we find that $\nabla \cdot \mathbf{E} = 0$ implies

$$\mathbf{k} \cdot \mathbf{T}(k_x, k_y) = 0 \tag{4.67}$$

and that $\nabla \times \mathbf{E} = 0$ implies

$$\mathbf{k} \times \mathbf{T}(k_x, k_y) = 0. \tag{4.68}$$

Crossing $\mathbf{k}$ into (4.68) shows that (4.67) is implied by (4.68) because $\mathbf{k} \cdot \mathbf{k} = 0$. Expressing equation (4.68) in terms of the rectangular components of $\mathbf{T}$ and $\mathbf{k}$ proves that the spectrum can be written as

$$\mathbf{T}(k_x, k_y) = -iL(k_x, k_y)\mathbf{k} \tag{4.69}$$

where L is a scalar function determined by any one component of $\mathbf{T}(k_x, k_y)$, specifically

$$L(k_x, k_y) = \frac{iT_x(k_x, k_y)}{k_x} = \frac{iT_y(k_x, k_y)}{k_y} = \frac{T_z(k_x, k_y)}{\sqrt{k_x^2 + k_y^2}}. \tag{4.70}$$

Although (4.69) implies that $\mathbf{T} \cdot \mathbf{T} = 0$ because $\mathbf{k} \cdot \mathbf{k} = 0$, this does not imply that $\mathbf{T} = 0$, because $\mathbf{T}$ is a complex vector. From (4.70), (4.65), and (4.64) we see that the electric field in the half space $z \geq z_0$ is completely determined from one component of the electric field (for example E_z) in the plane $z = z_0$, as was also proven in Section 4.1.2. Equation (4.69) shows that the scalar function $L(k_x, k_y)$ is the spectrum of the scalar potential $\Psi(\mathbf{r})$; that is,

$$\Psi(\mathbf{r}) = \frac{1}{2\pi} \int_{-\infty}^{+\infty} \int_{-\infty}^{+\infty} L(k_x, k_y) e^{i(k_x x + k_y y + i\sqrt{k_x^2 + k_y^2} z)} dk_x dk_y, \quad z \geq z_0. \tag{4.71}$$

The integral in (4.71) can be evaluated by Riemann integration because (4.76) below shows that $L(k_x, k_y)$ is a continuous function of (k_x, k_y), except at $(k_x, k_y) = (0, 0)$ where it has an integrable $1/\sqrt{k_x^2 + k_y^2}$ singularity, and $L(k_x, k_y)e^{-z\sqrt{k_x^2+k_y^2}}$ decays exponentially as $k_x^2 + k_y^2 \to \infty$ for $z > z_0$. Also, $L(k_x, k_y)$ can be evaluated for all $(k_x, k_y) \neq (0, 0)$ by Riemann integration of

$$L(k_x, k_y) = \frac{1}{2\pi} \int_{-\infty}^{+\infty} \int_{-\infty}^{+\infty} \Psi(\mathbf{r}) e^{-i(k_x x + k_y y + i\sqrt{k_x^2 + k_y^2} z)} dx dy, \quad z \geq z_0 \tag{4.72}$$

provided the integration in (4.72) is done in the manner prescribed in Appendix D and Section 3.2.

The Weyl identity (3.61) with $\omega = 0$ shows that the free-space static Green's function can be expressed in terms of plane waves as

$$\frac{1}{4\pi |\mathbf{r} - \mathbf{r}'|} = \frac{1}{8\pi^2} \int_{-\infty}^{+\infty} \int_{-\infty}^{+\infty} \frac{e^{i\mathbf{k}\cdot(\mathbf{r}-\mathbf{r}')}}{\sqrt{k_x^2 + k_y^2}} dk_x dk_y, \quad z > z' \tag{4.73}$$

with $\mathbf{k}$ given in (4.66). Moreover, a plane-wave expansion of the gradient of the free-space Green's function can be found by taking the gradient of (4.73) to get

$$\nabla' \frac{1}{4\pi |\mathbf{r} - \mathbf{r}'|} = \frac{\mathbf{r} - \mathbf{r}'}{4\pi |\mathbf{r} - \mathbf{r}'|^3} = -\frac{i}{8\pi^2} \int_{\infty}^{+\infty} \int_{\infty}^{+\infty} \mathbf{k} \frac{e^{i\mathbf{k}\cdot(\mathbf{r}-\mathbf{r}')}}{\sqrt{k_x^2 + k_y^2}} dk_x dk_y, \quad z > z'. \tag{4.74}$$

To obtain an expression for the spectrum $\mathbf{T}$ in terms of the charge distribution $\rho(\mathbf{r})$, insert (4.74) into the Green's function representation in (4.5) for the electric field. Then interchange the orders of integration, permitted by the exponential decay of the integrand at infinity in the (k_x, k_y) plane, and compare the resulting expression with (4.64) to get

$$\mathbf{T}(k_x, k_y) = -\frac{i\mathbf{k}}{4\pi\epsilon\sqrt{k_x^2 + k_y^2}} \int_V \rho(\mathbf{r}') \, e^{-i\mathbf{k}\cdot\mathbf{r}'} dV'. \tag{4.75}$$

We can also obtain (4.75) by inserting (4.73) into the Green's function representation (4.5) for the scalar potential, and comparing the resulting expression with (4.71) to get

$$L(k_x, k_y) = \frac{1}{4\pi\epsilon\sqrt{k_x^2 + k_y^2}} \int_V \rho(\mathbf{r}') \, e^{-i\mathbf{k}\cdot\mathbf{r}'} dV'. \tag{4.76}$$

Then $\mathbf{T}(k_x, k_y)$ in (4.75) is found from (4.76) through (4.69). Similarly, (4.74) can be inserted into (4.16) to get

$$\mathbf{T}(k_x, k_y) = -\frac{i}{4\pi\sqrt{k_x^2+k_y^2}} \int_S \left[\left(\hat{\mathbf{n}}' \times \mathbf{E}(\mathbf{r}')\right) \times \mathbf{k} + \hat{\mathbf{n}}' \cdot \mathbf{E}(\mathbf{r}')\, \mathbf{k} \right] e^{-i\mathbf{k}\cdot\mathbf{r}'} dS' \tag{4.77}$$

which gives the spectrum $\mathbf{T}$ in terms of the electric field on the closed surface S. The vector $\hat{\mathbf{n}}$ is the unit normal to S pointing out of the volume enclosed by S.

Since the static electric field satisfies the differential equations $\nabla \times \mathbf{E} = 0$ and $\nabla \cdot \mathbf{E} = 0$ outside the source region, it is continuous on the surface S. Hence (4.77) shows that *the spectrum* $\mathbf{T}(k_x, k_y)$ *for the static electric field is continuous for all real* (k_x, k_y), *except possibly at* $(k_x, k_y) = (0, 0)$. Moreover, from (4.66) it is seen that $\mathbf{k}(k_x^2 + k_y^2)^{-1/2}$ is bounded as $(k_x, k_y) \to (0, 0)$, so $\mathbf{T}(k_x, k_y)$ *is bounded as* $(k_x, k_y) \to (0, 0)$. In particular, if the point $(k_x, k_y) = (0, 0)$ is approached along a direction that forms the angle k_ϕ with the k_x axis, we have

$$\frac{\mathbf{k}}{\sqrt{k_x^2+k_y^2}} \to \hat{\mathbf{x}} \cos k_\phi + \hat{\mathbf{y}} \sin k_\phi + \hat{\mathbf{z}} i. \tag{4.78}$$

The result (4.78) can be used along with (4.75) to show that $\mathbf{T}(0, 0) = 0$ if and only if the total charge q^{tot} is zero. By expanding the exponential in (4.76) in a Taylor series around $(k_x, k_y) = (0, 0)$, (4.76) can be written as

$$L(k_x, k_y) = \frac{1}{4\pi\epsilon\sqrt{k_x^2+k_y^2}} \left(q^{tot} - i\mathbf{k}\cdot\mathbf{p}\right) + O\left(\sqrt{k_x^2+k_y^2}\right) \tag{4.79}$$

which shows that the total charge in (4.7) is given in terms of $L(k_x, k_y)$ by the limit

$$q^{tot} = \lim_{(k_x,k_y)\to(0,0)} 4\pi\epsilon\sqrt{k_x^2+k_y^2}\, L(k_x, k_y) \tag{4.80}$$

or in terms of T_z simply by

$$q^{tot} = 4\pi\epsilon\, T_z(0, 0). \tag{4.81}$$

If one of the rectangular components of the electric field is known in the scan plane $z = z_0$, we can determine the function $L(k_x, k_y)$ or $T_z(k_x, k_y)$ by use of (4.65) and (4.70), and the total charge can then be determined from (4.80) or (4.81). We thus have a procedure for computing the total charge, which determines the leading far-field terms in (4.6), from any of the rectangular components of the electric field in the scan plane.

We can also determine the electric dipole moment given in (4.7) from the potential spectrum $L(k_x, k_y)$ or any one component of the electric-field spectrum $\mathbf{T}(k_x, k_y)$, for example $T_z(k_x, k_y)$. By inspection, we find from (4.79) that with

$$U(k_x, k_y) = 4\pi i\epsilon \frac{\partial}{\partial k_x}\left[\sqrt{k_x^2+k_y^2}\, L(k_x, k_y)\right] = 4\pi i\epsilon \frac{\partial}{\partial k_x} T_z(k_x, k_y) \tag{4.82}$$

and

$$V(k_x, k_y) = 4\pi i\epsilon \frac{\partial}{\partial k_y}\left[\sqrt{k_x^2+k_y^2}\, L(k_x, k_y)\right] = 4\pi i\epsilon \frac{\partial}{\partial k_y} T_z(k_x, k_y) \tag{4.83}$$

the rectangular components of the dipole moment $\mathbf{p}$ can be determined from the limits

$$\begin{aligned} p_x &= \lim_{k_y\to 0} U(0, k_y), \quad p_y = \lim_{k_x\to 0} V(k_x, 0) \\ p_z &= i p_x - i \lim_{k_x\to 0^+} U(k_x, 0) = i p_y - i \lim_{k_y\to 0^+} V(0, k_y). \end{aligned} \tag{4.84}$$

Note that k_x and k_y must approach zero from the positive side in the expressions for p_z, whereas they can approach zero arbitrarily in the expressions for p_x and p_y. The formulas in (4.84) allow one to compute the electric dipole moment, which determines the second term in the far-field formulas (4.6), from one rectangular component of the static electric field in the scan plane.

4.3.2 Static Magnetic Field

For the static magnetic field $\mathbf{H}$, we introduce the plane-wave spectrum $\mathbf{T}^H$ and find that the equations (4.64), (4.65), and (4.77) also hold for the magnetic field when $\mathbf{T}$ and $\mathbf{E}$ are replaced by $\mathbf{T}^H$ and $\mathbf{H}$, respectively. Moreover, $\mathbf{T}^H$ satisfies the conditions (4.67) and (4.68), and it can be written in terms of the magnetic-potential spectrum $L^H(k_x, k_y)$ as

$$\mathbf{T}^H(k_x, k_y) = -iL^H(k_x, k_y)\mathbf{k} \tag{4.85}$$

where $\mathbf{k}$ is given in (4.66). The magnetic potential Ψ^m and its plane-wave spectrum $L^H(k_x, k_y)$ are related by the Fourier transform pair (4.71)–(4.72) with Ψ and L replaced by Ψ^m and L^H. The scalar spectrum $L^H(k_x, k_y)$ can also be expressed in terms of any one component of the spectrum for the magnetic field as

$$L^H(k_x, k_y) = \frac{iT_x^H(k_x, k_y)}{k_x} = \frac{iT_y^H(k_x, k_y)}{k_y} = \frac{T_z^H(k_x, k_y)}{\sqrt{k_x^2 + k_y^2}}. \tag{4.86}$$

Inserting (4.74) into (4.50) proves that the spectrum can be expressed in terms of the source current as

$$\mathbf{T}^H(k_x, k_y) = \frac{i\mathbf{k}}{4\pi\sqrt{k_x^2 + k_y^2}} \times \int_V \mathbf{J}(\mathbf{r}')\, e^{-i\mathbf{k}\cdot\mathbf{r}'} dV'. \tag{4.87}$$

Hence, $\mathbf{T}^H(k_x, k_y)$ *is continuous for all real* (k_x, k_y), *and satisfies* $\mathbf{T}^H(0, 0) = 0$ because $\mathbf{k}(k_x^2 + k_y^2)^{-1/2}$ is bounded and $\int_V \mathbf{J}(\mathbf{r}')dV' = 0$ [63, p. 185]. The continuity of $\mathbf{T}^H(k_x, k_y)$ also follows from the r^{-3} decay of the magnetic field, which proves that the double Fourier integral over the scan plane of the magnetic field converges absolutely for all real (k_x, k_y).

Dotting $\hat{\mathbf{z}}$ into (4.87) and using (4.86) shows that the function $L^H(k_x, k_y)$ can be written in terms of the current as

$$L^H(k_x, k_y) = \frac{k_y\hat{\mathbf{x}} - k_x\hat{\mathbf{y}}}{4\pi i(k_x^2 + k_y^2)} \cdot \int_V \mathbf{J}(\mathbf{r}')\, e^{-i\mathbf{k}\cdot\mathbf{r}'} dV'. \tag{4.88}$$

Unlike the function $L(k_x, k_y)$, which is singular at $(k_x, k_y) = (0, 0)$ when $q^{tot} \neq 0$, the function $L^H(k_x, k_y)$ is bounded for all real values of (k_x, k_y). This can be seen from (4.88) and the relation

$$\int_V \mathbf{J}(\mathbf{r}')\, e^{-i\mathbf{k}\cdot\mathbf{r}'} dV' = -i\mathbf{k} \cdot \int_V \mathbf{r}'\mathbf{J}(\mathbf{r}')dV' + O\left(k_x^2 + k_y^2\right), \quad (k_x, k_y) \to (0, 0). \tag{4.89}$$

With the identity [63, p. 185]

$$\mathbf{k} \cdot \int_V \mathbf{r}'\mathbf{J}(\mathbf{r}')dV' = -\frac{1}{2}\mathbf{k} \times \int_V \mathbf{r}' \times \mathbf{J}(\mathbf{r}')dV' \tag{4.90}$$

(4.89) becomes

$$\begin{aligned}\int_V \mathbf{J}(\mathbf{r}')\, e^{-i\mathbf{k}\cdot\mathbf{r}'} dV' &= i\mathbf{k} \times \frac{1}{2}\int_V \mathbf{r}' \times \mathbf{J}(\mathbf{r}')dV' + O\left(k_x^2 + k_y^2\right) \\ &= i\mathbf{k} \times \mathbf{m} + O\left(k_x^2 + k_y^2\right).\end{aligned} \tag{4.91}$$

By substituting from (4.91) into (4.87), using the relation $\mathbf{k} \times (\mathbf{k} \times \mathbf{m}) = (\mathbf{k} \cdot \mathbf{m})\,\mathbf{k}$, and comparing the resulting expression with (4.85), we obtain the following alternative expression for $L^H(k_x, k_y)$

$$L^H(k_x, k_y) = -\frac{i\mathbf{k} \cdot \mathbf{m}}{4\pi\sqrt{k_x^2 + k_y^2}} + O\left(\sqrt{k_x^2 + k_y^2}\right). \tag{4.92}$$

The magnetic dipole term in the formula (4.92) has the same form as the electric dipole term in the formula (4.79). Therefore, the formulas (4.82)–(4.84), which determine the electric dipole moment from the function $L(k_x, k_y)$, imply that the rectangular components of the magnetic dipole moment $\mathbf{m}$ can be determined from the limits

$$\begin{aligned} m_x &= \lim_{k_y \to 0} U^H(0, k_y), \quad m_y = \lim_{k_x \to 0} V^H(k_x, 0) \\ m_z &= i m_x - i \lim_{k_x \to 0^+} U^H(k_x, 0) = i m_y - i \lim_{k_y \to 0^+} V^H(0, k_y) \end{aligned} \tag{4.93}$$

where the functions $U^H(k_x, k_y)$ and $V^H(k_x, k_y)$ are determined from $L^H(k_x, k_y)$ and $T_z^H(k_x, k_y)$ by the expressions

$$U^H(k_x, k_y) = 4\pi i\,\frac{\partial}{\partial k_x}\left[\sqrt{k_x^2 + k_y^2}\,L^H(k_x, k_y)\right] = 4\pi i\,\frac{\partial}{\partial k_x} T_z^H(k_x, k_y) \tag{4.94}$$

and

$$V^H(k_x, k_y) = 4\pi i\,\frac{\partial}{\partial k_y}\left[\sqrt{k_x^2 + k_y^2}\,L^H(k_x, k_y)\right] = 4\pi i\,\frac{\partial}{\partial k_y} T_z^H(k_x, k_y). \tag{4.95}$$

When the sources are located in the half space $z > z_0$, some of the results of this section must be modified to hold for the fields in the half space $z < z_0$. In (4.64)–(4.65), (4.70)–(4.72), and (4.86), $(k_x^2 + k_y^2)^{1/2}$ is replaced by $-(k_x^2 + k_y^2)^{1/2}$. The propagation vector $\mathbf{k}$ in (4.66) is replaced by

$$\mathbf{k}^- = k_x\hat{\mathbf{x}} + k_y\hat{\mathbf{y}} - i\sqrt{k_x^2 + k_y^2}\,\hat{\mathbf{z}}. \tag{4.96}$$

The formulas (4.67)–(4.69), (4.73)–(4.77), (4.79), (4.85), (4.87) and (4.92) hold with $\mathbf{k}$ replaced by $\mathbf{k}^-$. The formula (4.88) is multiplied by -1 and has $\mathbf{k}$ replaced by $\mathbf{k}^-$. The terms with T_z in (4.81)–(4.83), and T_z^H in (4.94)–(4.95), change sign. In the formulas (4.84) and (4.93), the imaginary unit i is replaced by $-i$.

5

Time-Domain Representations

In Chapter 3 we derived frequency-domain representations for acoustic and electromagnetic fields radiated by sources located in a region of finite extent (bounded volume). In the present chapter, these frequency-domain representations will be translated into the time domain. The field expansions in this chapter employ either time-domain Green's functions or time-domain plane waves. These expansions are well suited for studying propagation in homogeneous media and for the probe-corrected formulation in Chapter 7 of planar near-field scanning in the time domain. In addition to presenting field expansions, we also derive general theorems on the behavior of time-domain far fields, and we consider special time-domain far-field phenomena such as electromagnetic "bullets" and "missiles."

As in Chapter 3, the problem under consideration is shown in Figure 3.1, where V is the finite volume that contains the sources, S is a smooth, finite, closed surface that contains V, and $z = z_0$ is the scan plane. The chapter is organized as follows. In Section 5.1 we derive time-domain Green's function representations for acoustic and electromagnetic fields in the half space $z > z_0$ in terms of their values in the scan plane $z = z_0$. Far-field expressions are also derived and conditions on the volume sources, the Huygens' sources on the closed surface S, and the fields in the scan plane $z = z_0$ are presented that ensure that the far fields decay as $1/r$ or faster.

Time-domain plane-wave representations are obtained in Section 5.2 by applying the standard Fourier transform to the frequency-domain plane-wave formulas in Section 3.2. Thereby we obtain formulas that involve time-domain spectra, which can be determined as space-time Radon transforms of the fields in the scan plane $z = 0$. (To simplify the field expressions, the scan plane is placed at $z_0 = 0$ in Sections 5.2 and 5.3.) These expressions form the basis of the probe-corrected formulation in Chapter 7, and many of them were first published in [21], [22]. Simple closed-form expressions are obtained for the propagating and evanescent time-domain plane-wave contributions to the field of an acoustic point source. It is shown that the evanescent contribution is exactly zero throughout a large space-time region, and therefore in this region the point-source field is determined solely by the propagating plane-wave contribution.

The time-domain spectra are then expressed in terms of the volume sources and the Huygens' sources on the surface S, and the singularities and analytical properties of the spectra are determined. Time-domain far-field expressions are also obtained in terms of the time-domain spectra, and alternative expressions for the time-domain fields at all times everywhere in the half space $z > 0$ are derived in terms of the time-domain far-field pattern and a time-domain far-field function. Using the advanced free-space Green's function, we derive exact representations that express the fields for all times throughout all space in terms of the far-field patterns and the volume-source functions. These representations prove that after all sources have been turned off, the time-domain fields everywhere can be expressed in terms of the time-domain far-field patterns alone over the entire far-field sphere. Finally, in Section 5.2 we present expressions in terms of the time-domain spectra and far-field patterns for the total acoustic and electromagnetic energy and momentum radiated through a plane $z = z_1 > 0$.

The analytic Fourier transform is applied in Section 5.3 to the frequency-domain plane-wave formulas of Section 3.2, and plane-wave representations are obtained for the analytic-signal fields in terms of the analytic-signal spectra. The resulting formulas involve complex times and are simpler in form than the corresponding real time-domain formulas of Section 5.2. Various analytic properties and expressions for the analytic-signal spectra and far-field patterns are presented, and differences between the analytic-signal representation and the standard time-domain representation are pointed out.

In Section 5.4 we consider real time-domain electromagnetic pulses with fields that decay slower than $1/r$ and energy that decays slower than $1/r^2$ as $r \to \infty$ (electromagnetic missiles). Conditions necessary to launch an electromagnetic missile are given for accelerating point charges, for a continuum of volume current, and for Huygens' sources. To illustrate electromagnetic missiles, we consider radiation from a circular disk of uniform current. We derive exact closed-form expressions for the on-axis fields of a current disk with arbitrary time dependence, and for the fields everywhere outside a current disk with delta-function time dependence. We also consider the special cases of step-function and rectangular-pulse time dependencies.

All time-domain fields are zero for $t \leq t_0$ and can have arbitrary time dependence for $t > t_0$. We assume throughout Chapter 5 that the source-free region outside V is lossless.[1] The derivations will be performed for the case where the source region V is located to the left of the scan plane $z = z_0$, as shown in Figure 3.1. At the end of each section, it is explained how the formulas change when the sources are located to the right of the scan plane $z = z_0$.

5.1 GREEN'S FUNCTION REPRESENTATIONS

Time-domain formulas that are analogs to the frequency-domain Green's function formulas of Section 3.1 will be derived in this section. The time-domain formulas are first obtained in Section 5.1.1 by Fourier transforming the frequency-domain Green's function formulas, and are rederived in Section 5.1.2 by working directly in the time domain with time-domain Green's functions. In Section 5.1.3, we present formulas that give the time-domain velocity and magnetic field in the half space $z > z_0$ in terms of the time-domain pressure and electric field, respectively, in the plane $z = z_0$, and the class of time functions for which the Fourier transform can be used to calculate the time-domain fields is discussed. Time-domain far-field expressions in terms of the volume sources, the Huygens' sources on S, and the fields in the plane $z = z_0$ are given in Section 5.1.4. In Section 5.1.5, the spatial dependence of the time integral of the far fields is determined.

5.1.1 Derivation from the Fourier Transform

The time-domain acoustic pressure $p(\mathbf{r}, t)$ satisfying the wave equation

$$\nabla^2 p(\mathbf{r}, t) - \frac{1}{c^2}\frac{\partial^2}{\partial t^2} p(\mathbf{r}, t) = 0, \quad z \geq z_0 \tag{5.1}$$

[1] Oughstun and Sherman [10] derive frequency-domain plane-wave representations of the electromagnetic field and find asymptotic expressions for time-domain fields in certain types of lossy dispersive media.

with c being the speed of sound, can be determined from the frequency-domain acoustic pressure $p_\omega(\mathbf{r})$ by the Fourier inversion formula

$$p(\mathbf{r}, t) = \int_{-\infty}^{+\infty} p_\omega(\mathbf{r}) e^{-i\omega t} d\omega. \tag{5.2}$$

Writing the propagation constant as $k = \omega/c$, we observe that the frequency-domain formulas (3.6) and (3.7) become

$$\begin{aligned} p_\omega(\mathbf{r}) &= -\frac{1}{2\pi}\frac{\partial}{\partial z}\int_{-\infty}^{+\infty}\int_{-\infty}^{+\infty} \frac{p_\omega(\mathbf{r}_0)}{R} e^{i\omega R/c} dx_0 dy_0, \quad z > z_0 \\ &= \frac{z_0 - z}{2\pi}\int_{-\infty}^{+\infty}\int_{-\infty}^{+\infty} p_\omega(\mathbf{r}_0)\left[\frac{i\omega}{cR^2} - \frac{1}{R^3}\right] e^{i\omega R/c}\, dx_0 dy_0, \quad z > z_0 \end{aligned} \tag{5.3}$$

with $\mathbf{r}_0 = x_0\hat{\mathbf{x}} + y_0\hat{\mathbf{y}} + z_0\hat{\mathbf{z}}$ and $R = |\mathbf{r} - \mathbf{r}_0|$. Inserting $p_\omega(\mathbf{r})$ from (5.3) into (5.2) and interchanging the orders of integration yields

$$\begin{aligned} p(\mathbf{r}, t) &= -\frac{1}{2\pi}\frac{\partial}{\partial z}\int_{-\infty}^{+\infty}\int_{-\infty}^{+\infty} \frac{p(\mathbf{r}_0, t - R/c)}{R} dx_0 dy_0, \quad z > z_0 \\ &= \frac{z - z_0}{2\pi}\int_{-\infty}^{+\infty}\int_{-\infty}^{+\infty}\left[\frac{1}{cR^2}\frac{\partial}{\partial t} p(\mathbf{r}_0, t - R/c)\right. \\ &\qquad \left. + \frac{p(\mathbf{r}_0, t - R/c)}{R^3}\right] dx_0 dy_0, \quad z > z_0 \end{aligned} \tag{5.4}$$

where the derivative rule

$$\frac{\partial}{\partial t} p(\mathbf{r}, t) = \int_{-\infty}^{+\infty} -i\omega p_\omega(\mathbf{r}) e^{-i\omega t} d\omega \tag{5.5}$$

has been used. Equation (5.4) expresses the time-domain pressure in the region $z > z_0$ in terms of the time-domain pressure and its time derivative on the plane $z = z_0$. In Section 5.1.3, we express the time-domain acoustic velocity $\mathbf{v}(\mathbf{r}, t)$ in $z > z_0$ in terms of the time-domain pressure in the plane $z = z_0$.

Similarly, by Fourier transforming the electromagnetic frequency-domain formulas (3.9), (3.16) and (3.17), we find that the time-domain electric field is given by [142], [143]

$$\begin{aligned} \mathbf{E}(\mathbf{r}, t) &= \frac{z - z_0}{2\pi}\int_{-\infty}^{+\infty}\int_{-\infty}^{+\infty}\left[\frac{1}{cR^2}\frac{\partial}{\partial t}\mathbf{E}(\mathbf{r}_0, t - R/c)\right. \\ &\qquad \left. + \frac{\mathbf{E}(\mathbf{r}_0, t - R/c)}{R^3}\right] dx_0 dy_0, \quad z > z_0 \\ &= \frac{1}{2\pi}\nabla \times \int_{-\infty}^{+\infty}\int_{-\infty}^{+\infty} \frac{\hat{\mathbf{z}} \times \mathbf{E}(\mathbf{r}_0, t - R/c)}{R} dx_0 dy_0, \quad z > z_0 \\ &= -\frac{1}{2\pi}\int_{-\infty}^{+\infty}\int_{-\infty}^{+\infty} \mathbf{R} \times \left[\frac{1}{cR^2}\hat{\mathbf{z}} \times \frac{\partial}{\partial t}\mathbf{E}(\mathbf{r}_0, t - R/c)\right. \\ &\qquad \left. + \frac{\hat{\mathbf{z}} \times \mathbf{E}(\mathbf{r}_0, t - R/c)}{R^3}\right] dx_0 dy_0, \quad z > z_0 \end{aligned} \tag{5.6}$$

with $\mathbf{R} = \mathbf{r} - \mathbf{r}_0$. Although it has been assumed that all fields are zero for $t \le t_0$, the formula (5.6) is also valid for static electric fields, as can be seen from the static formula (4.36). The expression (5.6) also holds for the magnetic field, provided $\mathbf{E}$ is replaced by $\mathbf{H}$. An expression for the magnetic field in the half space $z > z_0$ in terms of the electric field in the plane $z = z_0$ will be derived in Section 5.1.3.

So far, it has been assumed that the time dependence of the field is such that its time Fourier transform exists and the Fourier inversion formula holds. For time signals with infinite energy, such as the unit-step function, the standard Fourier transform cannot be used because the Fourier integral does not exist, and the analysis in this section does not apply. However, we can repair this deficiency in the derivation of (5.4) and (5.6) in a number of ways. One way is to use distribution theory. Another is to give the time signal with infinite energy an exponential decay and then let the decay go to zero after having derived the desired formulas. Still another way is to use the complex Fourier transform, described in Dettman [144, pp. 365–369], which is obtained by letting the real contour from the standard Fourier integration become complex. Use of the complex Fourier transform will be illustrated in Section 5.1.3, where the magnetic field is calculated.

When the sources are located in the half space $z > z_0$, the fields in the half space $z < z_0$ are given by the negative of (5.4) and (5.6).

5.1.2 Derivation from Time-Domain Green's Functions

In this section, we use time-domain Green's functions and Green's second identity to rederive the time-domain formulas (5.4) and (5.6) for the acoustic pressure and the electric field in the half space $z > z_0$. Our method of derivation is similar to the one used by Stratton [1, sec. 8.1] to obtain a Huygens' source expression for scalar time-domain fields.

First consider the acoustic pressure $p(\mathbf{r}, t)$ satisfying the wave equation (5.1) in the source-free region. The time-domain Dirichlet Green's function $G^D(\mathbf{r}, \mathbf{r}', t, t')$, satisfying the wave equation

$$\nabla^2 G^D(\mathbf{r}, \mathbf{r}', t, t') - \frac{1}{c^2}\frac{\partial^2}{\partial t^2} G^D(\mathbf{r}, \mathbf{r}', t, t') = -\delta(\mathbf{r} - \mathbf{r}')\delta(t - t'), \quad z > z_0 \tag{5.7}$$

and the Dirichlet boundary condition $G^D = 0$ for $z = z_0$ or $z' = z_0$ is given by

$$G^D(\mathbf{r}, \mathbf{r}', t, t') = G(\mathbf{r}, \mathbf{r}', t, t') - G(\mathbf{r}, \mathbf{r}'_i, t, t'), \quad z \ge z_0 \tag{5.8}$$

where $\mathbf{r}'_i = x'\hat{\mathbf{x}} + y'\hat{\mathbf{y}} + (2z_0 - z')\hat{\mathbf{z}}$ is the image of $\mathbf{r}'$ with respect to the plane $z = z_0$. Moreover, $G(\mathbf{r}, \mathbf{r}', t, t')$ is the time-domain free-space Green's function, which can be determined from the corresponding frequency-domain Green's function $G_\omega(\mathbf{r}, \mathbf{r}') = e^{ik|\mathbf{r}-\mathbf{r}'|}/(4\pi|\mathbf{r} - \mathbf{r}'|)$ through

$$G(\mathbf{r}, \mathbf{r}', t, t') = \int_{-\infty}^{+\infty} \frac{e^{i\omega t'}}{2\pi} G_\omega(\mathbf{r}, \mathbf{r}')\, e^{-i\omega t} d\omega = \frac{\delta(t' - t + |\mathbf{r} - \mathbf{r}'|/c)}{4\pi|\mathbf{r} - \mathbf{r}'|}. \tag{5.9}$$

Note that $G^D(\mathbf{r}, \mathbf{r}', t, t')$ and $G(\mathbf{r}, \mathbf{r}', t, t')$ are retarded with respect to $(\mathbf{r}, t)$ when $(\mathbf{r}', t')$ is fixed. In the following derivation, we employ the Dirichlet Green's function obtained by interchanging t and t' in $G^D(\mathbf{r}, \mathbf{r}', t, t')$. We find that $G^D(\mathbf{r}, \mathbf{r}', t', t)$ satisfies (5.7) so that

$$\begin{aligned} p(\mathbf{r}, t)\nabla^2 G^D(\mathbf{r}, \mathbf{r}', t', t) &- G^D(\mathbf{r}, \mathbf{r}', t', t)\nabla^2 p(\mathbf{r}, t) \\ = -p(\mathbf{r}, t)\delta(\mathbf{r} - \mathbf{r}')\delta(t - t') &+ \frac{1}{c^2}\frac{\partial}{\partial t}\Bigg[p(\mathbf{r}, t)\frac{\partial}{\partial t} G^D(\mathbf{r}, \mathbf{r}', t', t) \\ &- G^D(\mathbf{r}, \mathbf{r}', t', t)\frac{\partial}{\partial t} p(\mathbf{r}, t)\Bigg]. \end{aligned} \tag{5.10}$$

Applying Green's second identity to the region $V_0(R_0)$ bounded by the circular disk $C_0(R_0)$ and the hemispherical surface $S_0(R_0)$ in Figure 3.2, we find that when $\mathbf{r}' \in V_0(R_0)$

$$\int\limits_{S_0(R_0)\cup C_0(R_0)} \left[p(\mathbf{r},t)\frac{\partial}{\partial n}G^D(\mathbf{r},\mathbf{r}',t',t) - G^D(\mathbf{r},\mathbf{r}',t',t)\frac{\partial}{\partial n}p(\mathbf{r},t)\right]dS$$
$$= -p(\mathbf{r}',t)\delta(t-t') + \frac{1}{c^2}\frac{\partial}{\partial t}\int\limits_{V_0(R_0)}\left[p(\mathbf{r},t)\frac{\partial}{\partial t}G^D(\mathbf{r},\mathbf{r}',t',t)\right. \qquad (5.11)$$
$$\left.-G^D(\mathbf{r},\mathbf{r}',t',t)\frac{\partial}{\partial t}p(\mathbf{r},t)\right]dV.$$

Since all fields are zero for $t \le t_0$ and the source begins to radiate at $t = t_0$, we can choose R_0 large enough so that the acoustic pressure on $S_0(R_0)$ is zero for $t \le t'$. Then the integral over $S_0(R_0)$ in (5.11) equals zero for all t because $G^D(\mathbf{r},\mathbf{r}',t',t) = 0$ for $t > t'$. Integrating (5.11) with respect to t from $-\infty$ to $+\infty$ yields

$$p(\mathbf{r}',t') = \int\limits_{C_0(R_0)}\int\limits_{-\infty}^{+\infty} p(\mathbf{r},t)\frac{\partial}{\partial z}G^D(\mathbf{r},\mathbf{r}',t',t)\,dt\,dS, \quad \mathbf{r}' \in V_0(R_0) \qquad (5.12)$$

where we have used the Dirichlet boundary condition on $G^D(\mathbf{r},\mathbf{r}',t',t)$, and the facts that $p(\mathbf{r},-\infty) = 0$ and $G^D(\mathbf{r},\mathbf{r}',t',+\infty) = 0$.

Letting $R_0 \to +\infty$ and using the relations

$$\left.\frac{\partial}{\partial z}G^D(\mathbf{r},\mathbf{r}',t',t)\right|_{z=z_0} = \frac{z_0 - z'}{2\pi|\mathbf{r}-\mathbf{r}'|^2}\left[\frac{1}{c}\delta'(t-t'+|\mathbf{r}-\mathbf{r}'|/c) - \frac{\delta(t-t'+|\mathbf{r}-\mathbf{r}'|/c)}{|\mathbf{r}-\mathbf{r}'|}\right] \qquad (5.13)$$

and

$$\int\limits_{-\infty}^{+\infty}\delta'(t'-t)p(\mathbf{r},t')dt' = -\frac{\partial}{\partial t}p(\mathbf{r},t) \qquad (5.14)$$

(5.12) becomes

$$p(\mathbf{r},t) = \frac{z-z_0}{2\pi}\int\limits_{-\infty}^{+\infty}\int\limits_{-\infty}^{+\infty}\left[\frac{1}{cR^2}\frac{\partial}{\partial t}p(\mathbf{r}_0,t-R/c) + \frac{p(\mathbf{r}_0,t-R/c)}{R^3}\right]dx_0dy_0, \qquad (5.15)$$
$$z > z_0$$

where we have replaced $(\mathbf{r}',t')$ with $(\mathbf{r},t)$ and introduced the integration vector $\mathbf{r}_0 = x_0\hat{\mathbf{x}} + y_0\hat{\mathbf{y}} + z_0\hat{\mathbf{z}}$ and $R = |\mathbf{r}-\mathbf{r}_0|$. The expression (5.15) for the time-domain pressure in the half space $z > z_0$, obtained directly in the time domain, is seen to agree with the expression (5.4), obtained by Fourier transforming the frequency-domain formula (3.7).

To derive the corresponding formula for the electric field, it is convenient to use the time-domain Dirichlet dyadic Green's function $\bar{\mathbf{G}}^D(\mathbf{r},\mathbf{r}',t,t')$ for the half space $z \ge z_0$. The expression for $\bar{\mathbf{G}}^D(\mathbf{r},\mathbf{r}',t,t')$ is found by Fourier transforming the corresponding frequency-domain dyadic Green's function (3.11) to get

$$\begin{aligned}\bar{\mathbf{G}}^D(\mathbf{r},\mathbf{r}',t,t') &= \int_{-\infty}^{+\infty} \frac{e^{i\omega t'}}{2\pi} \bar{\mathbf{G}}^D_\omega(\mathbf{r},\mathbf{r}')\, e^{-i\omega t} d\omega \\ &= \bar{\mathbf{I}} G^D(\mathbf{r},\mathbf{r}',t,t') + c^2 \int_{-\infty}^{t-t'} \int_{-\infty}^{u} \nabla\nabla' G^D(\mathbf{r},\mathbf{r}',u',0) du' du \\ &\quad + 2\hat{\mathbf{z}}\hat{\mathbf{z}} G(\mathbf{r},\mathbf{r}'_i,t,t')\end{aligned} \tag{5.16}$$

where $G^D(\mathbf{r},\mathbf{r}',t,t')$ is the Dirichlet Green's function (5.8) and $G(\mathbf{r},\mathbf{r}',t,t')$ is the free-space Green's function (5.9). The dyadic Green's function (5.16) is retarded with respect to $(\mathbf{r},t)$ (for fixed $(\mathbf{r}',t')$) and satisfies the dyadic wave equation

$$\nabla\times\nabla\times\bar{\mathbf{G}}^D(\mathbf{r},\mathbf{r}',t,t') + \frac{1}{c^2}\frac{\partial^2}{\partial t^2}\bar{\mathbf{G}}^D(\mathbf{r},\mathbf{r}',t,t') = \bar{\mathbf{I}}\delta(\mathbf{r}-\mathbf{r}')\delta(t-t'), \quad z>z_0 \tag{5.17}$$

as well as the Dirichlet boundary condition $\hat{\mathbf{z}}\times\bar{\mathbf{G}}^D = 0$ on the plane $z = z_0$. In the following derivation of the expression for the electric field, we shall use the Dirichlet dyadic Green's function obtained by interchanging t and t' in $\bar{\mathbf{G}}^D(\mathbf{r},\mathbf{r}',t,t')$. We find that $\bar{\mathbf{G}}^D(\mathbf{r},\mathbf{r}',t',t)$ satisfies the wave equation (5.17) so that

$$\begin{aligned}&(\nabla\times\nabla\times\mathbf{E}(\mathbf{r},t))\cdot\bar{\mathbf{G}}^D(\mathbf{r},\mathbf{r}',t',t) - \mathbf{E}(\mathbf{r},t)\cdot\left(\nabla\times\nabla\times\bar{\mathbf{G}}^D(\mathbf{r},\mathbf{r}',t',t)\right) \\ &= -\mathbf{E}(\mathbf{r},t)\delta(\mathbf{r}-\mathbf{r}')\delta(t-t') + \frac{1}{c^2}\frac{\partial}{\partial t}\left[\mathbf{E}(\mathbf{r},t)\cdot\frac{\partial}{\partial t}\bar{\mathbf{G}}^D(\mathbf{r},\mathbf{r}',t',t)\right. \\ &\qquad \left. - \left(\frac{\partial}{\partial t}\mathbf{E}(\mathbf{r},t)\right)\cdot\bar{\mathbf{G}}^D(\mathbf{r},\mathbf{r}',t',t)\right]\end{aligned} \tag{5.18}$$

and thus the dyadic version of Green's second identity (3.10) applied to the region $V_0(R_0)$ bounded by the circular disk $C_0(R_0)$ and the hemispherical surface $S_0(R_0)$ in Figure 3.2 yields

$$\begin{aligned}&\mathbf{E}(\mathbf{r}',t')\delta(t-t') + \frac{1}{c^2}\frac{\partial}{\partial t}\int_{V_0(R_0)}\left[\left(\frac{\partial}{\partial t}\mathbf{E}(\mathbf{r},t)\right)\cdot\bar{\mathbf{G}}^D(\mathbf{r},\mathbf{r}',t',t)\right. \\ &\qquad\left. -\mathbf{E}(\mathbf{r},t)\cdot\frac{\partial}{\partial t}\bar{\mathbf{G}}^D(\mathbf{r},\mathbf{r}',t',t)\right]dV \\ &= \int_{C_0(R_0)\cup S_0(R_0)}\left[\mathbf{E}(\mathbf{r},t)\cdot\left(\hat{\mathbf{n}}\times\nabla\times\bar{\mathbf{G}}^D(\mathbf{r},\mathbf{r}',t',t)\right)\right. \\ &\qquad\left. +(\nabla\times\mathbf{E}(\mathbf{r},t))\cdot\left(\hat{\mathbf{n}}\times\bar{\mathbf{G}}^D(\mathbf{r},\mathbf{r}',t',t)\right)\right]dS, \quad \mathbf{r}'\in V_0(R_0).\end{aligned} \tag{5.19}$$

The source begins to radiate at $t = t_0$ and all fields are zero for $t \le t_0$. Hence, if R_0 is chosen large enough, the electric field on $S_0(R_0)$ is zero for $t \le t'$. Since $\bar{\mathbf{G}}^D(\mathbf{r},\mathbf{r}',t',t) = 0$ for $t > t'$, the integral over $S_0(R_0)$ in (5.19) equals zero for all t. Integrating (5.19) with respect to t from $-\infty$ to $+\infty$ and using $\mathbf{E}(\mathbf{r},-\infty) = 0$ and $\bar{\mathbf{G}}^D(\mathbf{r},\mathbf{r}',t',+\infty) = 0$, along with the Dirichlet boundary condition on $\bar{\mathbf{G}}^D$, yields

$$\mathbf{E}(\mathbf{r}',t') = 2\int_{-\infty}^{+\infty}\int_{-\infty}^{+\infty}\int_{-\infty}^{+\infty} \nabla' G(\mathbf{r},\mathbf{r}',t,t')\times\left(\hat{\mathbf{z}}\times\mathbf{E}(\mathbf{r}',t)\right) dt\,dx\,dy, \quad z>z_0 \tag{5.20}$$

where the time-domain analog of (3.14) has been used. Now from (5.9)

$$\nabla' G(\mathbf{r}, \mathbf{r}', t, t') = \frac{(\mathbf{r}' - \mathbf{r})}{4\pi |\mathbf{r}' - \mathbf{r}|^2} \left[\frac{1}{c} \delta'(t - t' + |\mathbf{r}' - \mathbf{r}|/c) - \frac{\delta(t - t' + |\mathbf{r}' - \mathbf{r}|/c)}{|\mathbf{r}' - \mathbf{r}|} \right] \tag{5.21}$$

and (5.20) becomes

$$\mathbf{E}(\mathbf{r}, t) = -\frac{1}{2\pi} \int_{-\infty}^{+\infty}\int_{-\infty}^{+\infty} \mathbf{R} \times \left[\frac{1}{cR^2} \hat{\mathbf{z}} \times \frac{\partial}{\partial t} \mathbf{E}(\mathbf{r}_0, t - R/c) + \frac{\hat{\mathbf{z}} \times \mathbf{E}(\mathbf{r}_0, t - R/c)}{R^3} \right] dx_0 dy_0, \quad z > z_0 \tag{5.22}$$

in which we have replaced $(\mathbf{r}', t')$ with $(\mathbf{r}, t)$ and introduced the integration vector $\mathbf{r}_0 = x_0\hat{\mathbf{x}} + y_0\hat{\mathbf{y}} + z_0\hat{\mathbf{z}}$ and the vector $\mathbf{R} = \mathbf{r} - \mathbf{r}_0$ with $R = |\mathbf{r} - \mathbf{r}_0|$. The formula (5.22) for the electric field in the half space $z > z_0$ is seen to agree with the formula (5.6) obtained by Fourier transforming the frequency-domain expression (3.17).

5.1.3 Velocity and Magnetic Field in Terms of Pressure and Electric Field

The frequency-domain formula (3.8) for the velocity in $z > z_0$ in terms of the pressure in the plane $z = z_0$ has a term of order ω^{-1}. The corresponding time-domain relation is therefore obtained by first multiplying (3.8) by $-i\omega$, and then taking the inverse Fourier transform to get an expression for the time derivative of the velocity. The resulting expression is finally integrated with respect to time to get

$$\begin{aligned} \mathbf{v}(\mathbf{r}, t) = &-\frac{\hat{\mathbf{z}}}{2\pi\rho_{m0}} \int_{-\infty}^{+\infty}\int_{-\infty}^{+\infty} \left[\frac{1}{R^3} \int_{t_0}^{t-R/c} p(\mathbf{r}_0, t')dt' + \frac{1}{cR^2} p(\mathbf{r}_0, t - R/c) \right] dx_0 dy_0 \\ &+ \frac{z - z_0}{2\pi\rho_{m0}} \int_{-\infty}^{+\infty}\int_{-\infty}^{+\infty} \left[\frac{3}{R^5} \int_{t_0}^{t-R/c} p(\mathbf{r}_0, t')dt' + \frac{3}{cR^4} p(\mathbf{r}_0, t - R/c) \right. \\ &\left. + \frac{1}{c^2R^3} \frac{\partial}{\partial t} p(\mathbf{r}_0, t - R/c) \right] \mathbf{R}\, dx_0 dy_0, \quad z > z_0 \end{aligned} \tag{5.23}$$

where we have used the fact that the pressure is zero for $t \le t_0$. Below we will discuss ways of generalizing the Fourier transform to deal with the ω^{-1} singularity that may occur when the time-domain velocity is calculated directly from the frequency-domain pressure.

Next, the time-domain magnetic field in the half space $z > z_0$ will be expressed in terms of the time-domain electric field in the plane $z = z_0$. This time-domain result is first derived by taking the inverse Fourier transform of the corresponding frequency-domain result. Then we explain how it can be found directly in the time domain with a time-domain Green's function.

Equation (3.20) shows that there are terms of order ω^{-1} in the frequency-domain formula for the magnetic field. Therefore, as in the acoustic derivation above, we first multiply (3.20) by $-i\omega$ and then take the inverse Fourier transform to get an expression for the time derivative of the magnetic field. (When we multiply (3.20) by $-i\omega$, we recover the time-harmonic form of Maxwell's equation, $i\omega\mu\mathbf{H}_\omega = \nabla \times \mathbf{E}_\omega$, which holds for all ω, including $\omega = 0$.) Finally, we integrate this expression for the time derivative of the magnetic field, and we recall that all fields are zero for $t \leq t_0$, to obtain the following time-domain formula for the magnetic field

$$
\begin{aligned}
\mathbf{H}(\mathbf{r}, t) = \frac{1}{2\pi}\sqrt{\frac{\epsilon}{\mu}} \int_{-\infty}^{+\infty}\int_{-\infty}^{+\infty} \Bigg(& -\frac{\mathbf{R}}{cR^3} \cdot \left[\hat{\mathbf{z}} \times \frac{\partial}{\partial t}\mathbf{E}(\mathbf{r}_0, t - R/c)\right] \mathbf{R} \\
& + \frac{1}{cR}\hat{\mathbf{z}} \times \frac{\partial}{\partial t}\mathbf{E}(\mathbf{r}_0, t - R/c) - \frac{3\mathbf{R}}{R^4} \cdot \left[\hat{\mathbf{z}} \times \mathbf{E}(\mathbf{r}_0, t - R/c)\right] \mathbf{R} \\
& + \frac{1}{R^2}\hat{\mathbf{z}} \times \mathbf{E}(\mathbf{r}_0, t - R/c) - \frac{3c\mathbf{R}}{R^5} \cdot \left[\int_{t_0}^{t-R/c} \hat{\mathbf{z}} \times \mathbf{E}(\mathbf{r}_0, t')dt'\right] \mathbf{R} \\
& + \frac{c}{R^3} \int_{t_0}^{t-R/c} \hat{\mathbf{z}} \times \mathbf{E}(\mathbf{r}_0, t')dt' \Bigg) dx_0 dy_0, \quad z > z_0.
\end{aligned}
\tag{5.24}
$$

This is the time-domain formula that gives the magnetic field in the region $z > z_0$ in terms of the x and y components of the electric field in the plane $z = z_0$. Both this formula for the magnetic field and the formula (5.6) for the electric field were derived previously by Baum [142], [143] by taking the inverse Fourier transform of the Smythe formulas [123]. An expression for the electric field in $z > z_0$ in terms of the tangential magnetic field in the plane $z = z_0$ is obtained by replacing $\mathbf{E}$ with $\mathbf{H}$ and $\mathbf{H}$ with $-(\epsilon/\mu)\mathbf{E}$ in (5.24).

Equation (5.24) could have been derived directly in the time domain in the following way. One can use Maxwell's equations and the initial conditions $\mathbf{E}(\mathbf{r}, t) = 0$, $\mathbf{H}(\mathbf{r}, t) = 0$ for $t \leq t_0$ to prove that $\mathbf{H}(\mathbf{r}, t) = -\mu^{-1} \int_{t_0}^{t} \nabla \times \mathbf{E}(\mathbf{r}, t')dt'$. The time-domain formula (5.22) for the electric field is inserted into this expression for the magnetic field, and (5.24) is recovered. Alternatively, equation (5.24) can be derived in a manner similar to the derivation of the electric-field formula (5.22), using the Neumann dyadic Green's function for the half space $z > z_0$. This shows that (5.24) is valid for all electromagnetic fields for which both $\mathbf{E}$ and $\mathbf{H}$ are identically zero for $t \leq t_0$, even for fields with a time dependence like the unit-step function that cannot be represented by a standard Fourier integral. However, (5.24) cannot represent a purely magnetostatic field, that is, a nonzero magnetic field that remains constant for all time ($-\infty < t < \infty$), unlike (5.22), which remains valid for electrostatic fields. If such a magnetostatic field is present, one has to add its value to the formula (5.24) to get the correct total magnetic field.

The discussion of static fields brings up a subtlety in (5.22) and (5.24) that may easily be overlooked. Suppose all sources and fields are zero for $t \leq t_0$ so that (5.22) and (5.24) apply. At $t = t_0$, close a switch that charges a capacitor and causes current to flow through a resistor, such as when a battery is connected to a capacitor and resistor in parallel. After a time t_1, assume all charge and current have practically reached a constant value, so that for $r \ll c(t - t_1)$ there remain only static electric and magnetic fields. These remnant static electric and magnetic fields, which build up from zero fields at $t = t_0$, are entirely determined from the formulas (5.22) and (5.24), respectively. This is unremarkable for (5.22) which, as mentioned before, remains valid for electrostatic fields ($-\infty < t < \infty$). However, it seems quite remarkable that (5.24) can represent a remnant static magnetic field for $t > t_0$ even though it cannot represent a purely magnetostatic field for all time ($-\infty < t < \infty$). The explanation lies in the fact that the remnant static magnetic field is generated by time-varying sources with electric and magnetic fields intimately related through Maxwell's equations, whereas a purely magnetostatic field throughout all space for all time ($-\infty < t < \infty$) exists independently of an electric field turned on at $t = t_0$.

So far we have not been concerned with the restrictions on the time functions we impose by assuming they can be represented by standard Fourier-transform integrals. The standard theorems for the Fourier transform hold for time functions that are either absolutely integrable or square integrable [140], [141] (see Section 2.3.1) and are therefore not valid for time functions like the unit step. Furthermore, even if a time function $f(t)$ can be represented by a standard Fourier integral, it is not guaranteed that we can calculate the time integral of $f(t)$ directly from its Fourier transform, that is, calculate the integral $\int_{-\infty}^{+\infty}(-i\omega)^{-1}f_\omega e^{-i\omega t}d\omega$ in which the integrand possibly has a singularity at $\omega = 0$. Such an integral with an ω^{-1} singularity has to be evaluated to determine the time-domain magnetic field directly from the frequency domain electric field in the scan plane (see (3.20)). The same problem occurs when the time-domain acoustic velocity is calculated directly from the frequency-domain pressure (see (3.8)). Hence, the standard Fourier transform is insufficient for analyzing certain types of problems that can occur in planar near-field measurements.

One way of dealing with this problem is to use the complex Fourier transform explained in Dettman [144, pp. 365–369] instead of the standard Fourier transform. Let us illustrate the use of the complex Fourier transform by calculating the time integral of the unit-step function directly from its spectrum. The unit-step function $h(t)$ is given by $h(t) = 0, t \le 0$ and $h(t) = 1, t > 0$. According to Dettman [144, p. 365], the complex Fourier transform of the unit-step function is

$$h_\omega = \frac{1}{2\pi}\int_0^{+\infty} h(t)e^{i\omega t}dt = -\frac{1}{2\pi i\omega}, \quad \mathrm{Im}(\omega) > 0 \tag{5.25}$$

and the inversion formula is

$$\int_C h_\omega e^{-i\omega t}d\omega = \frac{h(t^+) + h(t^-)}{2} = \begin{cases} h(t), & t \neq 0 \\ \frac{1}{2}, & t = 0 \end{cases} \tag{5.26}$$

where C is the complex contour above the real axis shown in Figure 5.1.

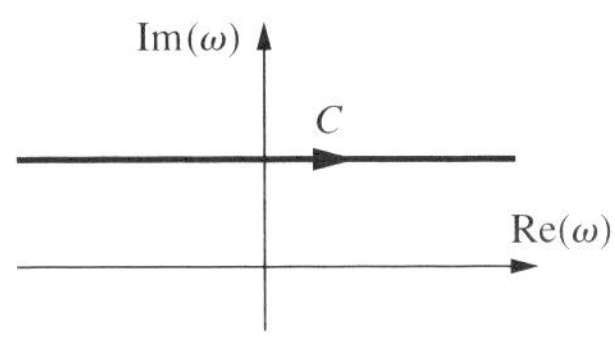

Figure 5.1 The inversion contour C for the complex Fourier transform.

The inverse Fourier transform of $(-i\omega)^{-1}h_\omega = -(2\pi\omega^2)^{-1}$ will now be calculated. When $t < 0$, the function $-(2\pi\omega^2)^{-1}e^{-i\omega t}$ is exponentially decaying on the semicircle $\omega = Re^{i\theta}$ with $0 < \theta < \pi$ as $R \to \infty$. Because this function has no singularities above the contour C, it follows that the inversion integral is zero.

When $t > 0$, the function $-(2\pi\omega^2)^{-1}e^{-i\omega t}$ is exponentially decaying on the semicircle $\omega = Re^{i\theta}$ with $\pi < \theta < 2\pi$ as $R \to \infty$, and the inversion integral equals $-2\pi i$ times the sum of all residues in the half plane below the contour C. The only singularity is the pole of order 2 at $\omega = 0$, and its residue is $i/(2\pi)$, so the inversion integral equals t. Also, for $t = 0$ the value of the inversion integral is zero. Hence

$$-\frac{1}{2\pi}\int_C \frac{e^{-i\omega t}}{\omega^2}d\omega = \begin{cases} 0, & t \le 0 \\ t, & t > 0 \end{cases} \tag{5.27}$$

which is seen to equal $\int_{-\infty}^{t} h(t')dt'$. Thus, one can use the rigorous complex Fourier transform to directly evaluate the time-domain magnetic field from the frequency-domain electric field, even when the time dependence of the electric field is a unit-step function.

If a piecewise smooth time function $f(t)$ is absolutely integrable, that is, if $\int_{-\infty}^{+\infty} |f(t)|dt < \infty$, and satisfies $f(t) = 0$ for $t \le t_0$ and $f(t) = O(t^{-1-\alpha}), 0 < \alpha < 1$ as $t \to \infty$, we may derive a formula that gives the time integral of $f(t)$ in terms of f_ω evaluated at only *real* ω. To obtain this formula, we displace the contour of the complex Fourier transform onto the real axis except near the singularity at $\omega = 0$ and find that

$$\int_{t_0}^{t} f(t')dt' = \int_C \frac{f_\omega e^{-i\omega t}}{-i\omega} d\omega = \fint_{-\infty}^{+\infty} \frac{f_\omega e^{-i\omega t}}{-i\omega} d\omega + \lim_{a\to 0} \int_{C_a} \frac{f_\omega e^{-i\omega t}}{-i\omega} d\omega \tag{5.28}$$

where the bar on the integral sign in (5.28) denotes a Cauchy principal value integration at $\omega = 0$ and C_a is the semicircle of radius a on which $\omega = ae^{i\theta}, 0 < \theta < \pi$. The asymptotic formulas of Bleistein and Handelsman [126, p. 134] can be used to show that $f_\omega = f_0 + O(\omega^\alpha)$ for $|\omega| \to 0$. And since $e^{-i\omega t} = 1 + O(a)$ on the contour C_a, we find that the last integral in (5.28) simply equals πf_0, so that

$$\int_{-\infty}^{t} f(t')dt' = \fint_{-\infty}^{+\infty} \frac{f_\omega e^{-i\omega t}}{-i\omega} d\omega + \pi f_0 = 2\int_0^{+\infty} \mathrm{Re}\left(\frac{f_\omega e^{-i\omega t}}{-i\omega}\right) d\omega + \pi f_0 \tag{5.29}$$

where $f_\omega = f_{-\omega}^*$ has been used. Note that the last integral in (5.29) would not converge if Re were moved outside the integration. Equation (5.29) is a practical formula for directly evaluating the inverse Fourier transform of the problematic integrals in (3.8) and (3.20) that involve $1/\omega$.

5.1.4 Far-Field Expressions

In this section we consider the electromagnetic fields before the acoustic fields. From (2.79) and (2.80) the magnetic and electric fields outside the source region can be expressed in terms of the current as

$$\mathbf{H}(\mathbf{r}, t) = -\frac{1}{4\pi} \int_V \left[\frac{\mathbf{R}}{cR^2} \times \frac{\partial}{\partial t} \mathbf{J}(\mathbf{r}', t - R/c) + \frac{\mathbf{R}}{R^3} \times \mathbf{J}(\mathbf{r}', t - R/c) \right] dV' \tag{5.30}$$

and

$$\begin{aligned} \mathbf{E}(\mathbf{r}, t) = \frac{1}{4\pi\epsilon} \int_V \Bigg[& \frac{1}{c^2 R} \frac{\partial \mathbf{J}(\mathbf{r}', t - R/c)}{\partial t} \cdot \left(\frac{\mathbf{RR}}{R^2} - \bar{\mathbf{I}} \right) \\ & + \frac{1}{R^2} \left(\frac{1}{c} \mathbf{J}(\mathbf{r}', t - R/c) + \frac{1}{R} \int_{t_0}^{t-R/c} \mathbf{J}(\mathbf{r}', t')dt' \right) \cdot \left(\frac{3\mathbf{RR}}{R^2} - \bar{\mathbf{I}} \right) \Bigg] dV' \end{aligned} \tag{5.31}$$

where $\mathbf{R} = \mathbf{r} - \mathbf{r}'$ and $R = |\mathbf{R}|$. We will now show that the far fields decay as $1/r$ or faster as $r \to \infty$, provided the time-domain current $\mathbf{J}(\mathbf{r}', t)$ satisfies the following conditions. (i) $\mathbf{J}(\mathbf{r}', t - R/c)$ as well as its time derivative and time integral are integrable over V for all $(\mathbf{r}, t)$, and (ii) there exists an integrable time-independent function $\alpha(\mathbf{r}')$ such that $|\partial \mathbf{J}(\mathbf{r}', t)/\partial t| \le \alpha(\mathbf{r}')$ (then $|\mathbf{J}(\mathbf{r}', t)| \le |t - t_0|\alpha(\mathbf{r}')$ and $\left|\int_{t_0}^{t} \mathbf{J}(\mathbf{r}', t)dt'\right| \le \frac{1}{2}|t - t_0|^2 \alpha(\mathbf{r}')$, where t_0 is the time at which the sources are turned on). These conditions do not exclude currents with spatially integrable singularities, such as those that appear at the edges of perfect conductors. Moreover, the conditions are trivially satisfied if the time derivative of the current exists and is finite. For currents that satisfy (i) and (ii), equation (5.30) reveals that

$$
\begin{aligned}
|r\mathbf{H}(\mathbf{r}, t + r/c)| &\le \frac{1}{4\pi c}\int_V \frac{r}{R^2}\left|\mathbf{R}\times\frac{\partial}{\partial t}\mathbf{J}(\mathbf{r}', t - R/c + r/c)\right| dV' \\
&\quad + \frac{1}{4\pi}\int_V \frac{r}{R^3}\left|\mathbf{R}\times\mathbf{J}(\mathbf{r}', t - R/c + r/c)\right| dV' \\
&\le \frac{1}{4\pi c}\int_V \alpha(\mathbf{r}')dV' + O\left(\frac{1}{r}\right).
\end{aligned}
\tag{5.32}
$$

To obtain the analogous results for the electric field, we use the expression (5.31) and find that

$$
|r\mathbf{E}(\mathbf{r}, t + r/c)| \le \frac{1}{4\pi c}\sqrt{\frac{\mu}{\epsilon}}\int_V \alpha(\mathbf{r}')dV' + O\left(\frac{1}{r}\right). \tag{5.33}
$$

The inequalities (5.32) and (5.33) prove that the magnetic and electric far fields decay as $1/r$ (or faster) as $r \to \infty$, provided $\mathbf{J}(\mathbf{r}', t - R/c)$ as well as its time derivative and time integral are integrable over V for all $(\mathbf{r}, t)$, and the time derivative of $\mathbf{J}(\mathbf{r}', t)$ is bounded by an integrable time-independent function.

If, in addition to (i) and (ii), the current satisfies the condition (iii) that for all t and almost all $\mathbf{r}'$, $\lim_{r\to\infty} \partial\mathbf{J}(\mathbf{r}', t - R/c + r/c)/\partial t$ exists and equals $\partial\mathbf{J}(\mathbf{r}', t + \mathbf{r}'\cdot\hat{\mathbf{r}}/c)/\partial t$, the theorem on limits of integrals in [71, vol. 2, p. 324] along with (5.30) shows that the magnetic far field is given by

$$
\mathbf{H}(\mathbf{r}, t) = -\frac{1}{4\pi cr}\hat{\mathbf{r}}\times\int_V \frac{\partial}{\partial t}\mathbf{J}(\mathbf{r}', t - r/c + \hat{\mathbf{r}}\cdot\mathbf{r}'/c)dV' + O\left(\frac{1}{r^2}\right). \tag{5.34}
$$

The condition (iii) is trivially satisfied by currents with first time derivatives that are continuous functions of time. Similarly, for currents that satisfy (i), (ii), and (iii), the electric far field is found from (5.31) to be expressible as

$$
\mathbf{E}(\mathbf{r}, t) = \sqrt{\frac{\mu}{\epsilon}}\frac{1}{4\pi cr}\hat{\mathbf{r}}\times\left(\hat{\mathbf{r}}\times\int_V \frac{\partial}{\partial t}\mathbf{J}(\mathbf{r}', t - r/c + \hat{\mathbf{r}}\cdot\mathbf{r}'/c)dV'\right) + O\left(\frac{1}{r^2}\right) \tag{5.35}
$$

and thus the far fields satisfy the relationships

$$
\mathbf{H}(\mathbf{r}, t) \sim \sqrt{\frac{\epsilon}{\mu}}\hat{\mathbf{r}}\times\mathbf{E}(\mathbf{r}, t), \quad \mathbf{E}(\mathbf{r}, t) \sim -\sqrt{\frac{\mu}{\epsilon}}\hat{\mathbf{r}}\times\mathbf{H}(\mathbf{r}, t), \quad r \to \infty. \tag{5.36}
$$

The far-field expressions in the time domain (5.34) and (5.35) check with the far-field expressions in the frequency domain (3.24) and (3.23) in that they are Fourier transforms of each other. However, in this section we have given conditions on the current that assure the validity of the time-domain far-field expressions (5.32)–(5.35).

The electromagnetic fields outside the source region are written in (2.122) and (2.123) in terms of the Huygens' sources $\hat{\mathbf{n}}\times\mathbf{E}$ and $\hat{\mathbf{n}}\times\mathbf{H}$ on the closed surface S with normal $\hat{\mathbf{n}}$ (see Figure 3.1). Using a procedure similar to the one that led to (5.32) proves that *the electromagnetic fields decay as $1/r$ (or faster) as $r \to \infty$, provided $\hat{\mathbf{n}}\times\mathbf{E}(\mathbf{r}', t - R/c)$ and $\hat{\mathbf{n}}\times\mathbf{H}(\mathbf{r}', t - R/c)$ as well as their time derivatives and time integrals are integrable over S for all $(\mathbf{r}, t)$, and the time derivatives of $\hat{\mathbf{n}}\times\mathbf{E}(\mathbf{r}', t)$ and $\hat{\mathbf{n}}\times\mathbf{H}(\mathbf{r}', t)$ are bounded by a time-independent function that is integrable on S.*

If, in addition, the Huygens' sources satisfy the condition that for all t and almost all $\mathbf{r}'$, $\lim_{r\to\infty}\hat{\mathbf{n}}\times\partial\mathbf{E}(\mathbf{r}', t - R/c + r/c)/\partial t$ exists and equals $\hat{\mathbf{n}}\times\partial\mathbf{E}(\mathbf{r}', t + \mathbf{r}'\cdot\hat{\mathbf{r}}/c)/\partial t$ (and similarly for $\hat{\mathbf{n}} \times \mathbf{H}(\mathbf{r}', t)$), the theorem on limits of integrals in [71, vol. 2, p. 324] along with (2.123) shows that the electric far field is given by [59]

$$\mathbf{E}(\mathbf{r}, t) = -\frac{1}{4\pi r}\hat{\mathbf{r}} \times \int\limits_S \left[\frac{1}{c}\hat{\mathbf{n}}' \times \frac{\partial}{\partial t}\mathbf{E}(\mathbf{r}', t - r/c + \hat{\mathbf{r}}\cdot\mathbf{r}'/c) - \mu\hat{\mathbf{r}} \times \left(\hat{\mathbf{n}}' \times \frac{\partial}{\partial t}\mathbf{H}(\mathbf{r}', t - r/c + \hat{\mathbf{r}}\cdot\mathbf{r}'/c)\right)\right] dS' + O\left(\frac{1}{r^2}\right). \tag{5.37}$$

The additional condition is trivially satisfied if the first time derivatives of the Huygens' sources are continuous functions of time. Similarly, the magnetic far field is found to be

$$\mathbf{H}(\mathbf{r}, t) = -\frac{1}{4\pi r}\hat{\mathbf{r}} \times \int\limits_S \left[\frac{1}{c}\hat{\mathbf{n}}' \times \frac{\partial}{\partial t}\mathbf{H}(\mathbf{r}', t - r/c + \hat{\mathbf{r}}\cdot\mathbf{r}'/c) + \epsilon\hat{\mathbf{r}} \times \left(\hat{\mathbf{n}}' \times \frac{\partial}{\partial t}\mathbf{E}(\mathbf{r}', t - r/c + \hat{\mathbf{r}}\cdot\mathbf{r}'/c)\right)\right] dS' + O\left(\frac{1}{r^2}\right). \tag{5.38}$$

The far fields (5.37) and (5.38) satisfy the far-field relationships in (5.36). Expressions for the θ and ϕ components of the electric far field in (5.37) may be found in [145].

Similar results are obtained for the acoustic pressure and velocity from the formulas (2.244) and (2.248) with (2.232) inserted. Specifically, for the acoustic pressure $p(\mathbf{r}, t)$ and the acoustic velocity $\mathbf{v}(\mathbf{r}, t)$ in the lossless fluid of static mass density ρ_{m0}, *the far fields decay as $1/r$ (or faster) as $r \to \infty$, provided the source function $q(\mathbf{r}, t - R/c)$ and its time integral are integrable over V for all $(\mathbf{r}, t)$, and $q(\mathbf{r}', t)$ is bounded by an integrable time-independent function in the finite source region V.*

If, in addition, the source function satisfies the condition that for all t and almost all $\mathbf{r}'$, $\lim_{r\to\infty} q(\mathbf{r}', t - R/c + r/c)$ exists and equals $q(\mathbf{r}', t + \mathbf{r}'\cdot\hat{\mathbf{r}}/c)$ (which is trivially satisfied if the source function is continuous with respect to time), the theorem in [71, vol. 2, p. 324] along with (2.244) show that the far-field pressure is

$$p(\mathbf{r}, t) = \frac{1}{4\pi r}\int\limits_V q(\mathbf{r}', t - r/c + \hat{\mathbf{r}}\cdot\mathbf{r}'/c)dV' + O\left(\frac{1}{r^2}\right) \tag{5.39}$$

and similarly from (2.248) and (2.232) we obtain

$$\begin{aligned}\mathbf{v}(\mathbf{r}, t) &= \frac{\hat{\mathbf{r}}\,p(\mathbf{r}, t)}{c\rho_{m0}} + O\left(\frac{1}{r^2}\right)\\ &= \frac{\hat{\mathbf{r}}}{4\pi c\rho_{m0} r}\int\limits_V q(\mathbf{r}', t - r/c + \hat{\mathbf{r}}\cdot\mathbf{r}'/c)dV' + O\left(\frac{1}{r^2}\right).\end{aligned} \tag{5.40}$$

The far-field expressions (5.39) and (5.40) can be obtained by Fourier transforming (3.21) and (3.22).

Alternatively, we can use the Huygens' source expressions (2.257) and (2.262) to show that *the far fields decay as $1/r$ (or faster) as $r \to \infty$, provided $p(\mathbf{r}', t - R/c)$ as well as its time derivative and time integral are integrable over S for all $(\mathbf{r}, t)$, $\mathbf{v}(\mathbf{r}', t - R/c)$ and its time derivative are integrable over S for all $(\mathbf{r}, t)$, and the time derivatives of $p(\mathbf{r}', t)$ and $\mathbf{v}(\mathbf{r}', t)$ are bounded by an integrable time-independent function on the closed surface S.*

If, in addition, the time derivative of $p(\mathbf{r}', t)$ satisfies the condition that for all t and almost all $\mathbf{r}'$, $\lim_{r\to\infty} \partial p(\mathbf{r}', t - R/c + r/c)/\partial t$ exists and equals $\partial p(\mathbf{r}', t + \mathbf{r}'\cdot\hat{\mathbf{r}}/c)/\partial t$ (and $\hat{\mathbf{n}}\cdot\mathbf{v}(\mathbf{r}', t)$ satisfies a similar condition), the theorem on limits of integrals in [71, vol. 2, p. 324] along with (2.257) and (2.262) show that the far fields can be written as

$$p(\mathbf{r},t)=\frac{1}{4\pi r}\int_S\left[\frac{\hat{\mathbf{r}}\cdot\hat{\mathbf{n}}'}{c}\frac{\partial}{\partial t}p(\mathbf{r}',t-r/c+\hat{\mathbf{r}}\cdot\mathbf{r}'/c)\right.$$
$$\left.+\rho_{m0}\hat{\mathbf{n}}'\cdot\frac{\partial}{\partial t}\mathbf{v}(\mathbf{r}',t-r/c+\hat{\mathbf{r}}\cdot\mathbf{r}'/c)\right]dS'+O\left(\frac{1}{r^2}\right) \tag{5.41}$$

and

$$\mathbf{v}(\mathbf{r},t)=\frac{\hat{\mathbf{r}}}{4\pi c\rho_{m0}r}\int_S\left[\frac{\hat{\mathbf{r}}\cdot\hat{\mathbf{n}}'}{c}\frac{\partial}{\partial t}p(\mathbf{r}',t-r/c+\hat{\mathbf{r}}\cdot\mathbf{r}'/c)\right.$$
$$\left.+\rho_{m0}\hat{\mathbf{n}}'\cdot\frac{\partial}{\partial t}\mathbf{v}(\mathbf{r}',t-r/c+\hat{\mathbf{r}}\cdot\mathbf{r}'/c)\right]dS'+O\left(\frac{1}{r^2}\right) \tag{5.42}$$

which can also be obtained by Fourier transforming the frequency-domain formulas (3.26) and (3.27).

In the remainder of this section, we assume that the volume and Huygens' sources satisfy the conditions stated above so that the far fields decay as $1/r$ (or faster) and the formulas (5.34)–(5.42) hold. (In Section 5.4 we consider sources and fields for which these conditions are not satisfied, and the resulting far fields decay slower than $1/r$.) With these assumptions, the formulas (5.34)–(5.42) show that we can define time-domain far-field patterns as

$$\mathcal{F}(\theta,\phi,t)=\lim_{r\to\infty}rp(\mathbf{r},t+r/c) \tag{5.43}$$

and

$$\boldsymbol{\mathcal{F}}(\theta,\phi,t)=\lim_{r\to\infty}r\mathbf{E}(\mathbf{r},t+r/c),\quad \boldsymbol{\mathcal{F}}^H(\theta,\phi,t)=\lim_{r\to\infty}r\mathbf{H}(\mathbf{r},t+r/c) \tag{5.44}$$

such that the far fields can be written as

$$p(\mathbf{r},t)\sim\frac{\mathcal{F}(\theta,\phi,t-r/c)}{r},\quad \mathbf{v}(\mathbf{r},t)\sim\hat{\mathbf{r}}\frac{\mathcal{F}(\theta,\phi,t-r/c)}{c\rho_{m0}r} \tag{5.45}$$

and

$$\mathbf{E}(\mathbf{r},t)\sim\frac{\boldsymbol{\mathcal{F}}(\theta,\phi,t-r/c)}{r},\quad \mathbf{H}(\mathbf{r},t)\sim\frac{\boldsymbol{\mathcal{F}}^H(\theta,\phi,t-r/c)}{r} \tag{5.46}$$

with

$$\boldsymbol{\mathcal{F}}^H(\theta,\phi,t)=\sqrt{\frac{\epsilon}{\mu}}\hat{\mathbf{r}}\times\boldsymbol{\mathcal{F}}(\theta,\phi,t),\quad \boldsymbol{\mathcal{F}}(\theta,\phi,t)=-\sqrt{\frac{\mu}{\epsilon}}\hat{\mathbf{r}}\times\boldsymbol{\mathcal{F}}^H(\theta,\phi,t) \tag{5.47}$$

where $\hat{\mathbf{r}}=\hat{\mathbf{x}}\cos\phi\sin\theta+\hat{\mathbf{y}}\sin\phi\sin\theta+\hat{\mathbf{z}}\cos\theta$ and (r,θ,ϕ) are the usual spherical coordinates defined by $x=r\cos\phi\sin\theta$, $y=r\sin\phi\sin\theta$, and $z=r\cos\theta$.

We note from the expressions (5.34), (5.35), and (5.39) that the far-field patterns can be expressed in terms of the volume sources as

$$\mathcal{F}(\theta,\phi,t)=\frac{1}{4\pi}\int_V q(\mathbf{r}',t+\hat{\mathbf{r}}\cdot\mathbf{r}'/c)dV' \tag{5.48}$$

$$\boldsymbol{\mathcal{F}}(\theta,\phi,t)=\sqrt{\frac{\mu}{\epsilon}}\frac{1}{4\pi c}\hat{\mathbf{r}}\times\left(\hat{\mathbf{r}}\times\int_V\frac{\partial}{\partial t}\mathbf{J}(\mathbf{r}',t+\hat{\mathbf{r}}\cdot\mathbf{r}'/c)dV'\right) \tag{5.49}$$

and

$$\boldsymbol{\mathcal{F}}^H(\theta,\phi,t)=-\frac{1}{4\pi c}\hat{\mathbf{r}}\times\int_V\frac{\partial}{\partial t}\mathbf{J}(\mathbf{r}',t+\hat{\mathbf{r}}\cdot\mathbf{r}'/c)dV'. \tag{5.50}$$

Similarly, (5.37), (5.38), and (5.41) can be used to express the far-field patterns in terms of the Huygens' sources on S, as follows

$$\mathcal{F}(\theta,\phi,t) = \frac{1}{4\pi}\int_S \left[\frac{\hat{\mathbf{r}}\cdot\hat{\mathbf{n}}'}{c}\frac{\partial}{\partial t}p(\mathbf{r}',t+\hat{\mathbf{r}}\cdot\mathbf{r}'/c) + \rho_{m0}\hat{\mathbf{n}}'\cdot\frac{\partial}{\partial t}\mathbf{v}(\mathbf{r}',t+\hat{\mathbf{r}}\cdot\mathbf{r}'/c)\right]dS' \tag{5.51}$$

$$\boldsymbol{\mathcal{F}}(\theta,\phi,t) = -\frac{1}{4\pi}\hat{\mathbf{r}}\times\int_S \left[\frac{1}{c}\hat{\mathbf{n}}'\times\frac{\partial}{\partial t}\mathbf{E}(\mathbf{r}',t+\hat{\mathbf{r}}\cdot\mathbf{r}'/c) - \mu\hat{\mathbf{r}}\times\left(\hat{\mathbf{n}}'\times\frac{\partial}{\partial t}\mathbf{H}(\mathbf{r}',t+\hat{\mathbf{r}}\cdot\mathbf{r}'/c)\right)\right]dS' \tag{5.52}$$

and

$$\boldsymbol{\mathcal{F}}^H(\theta,\phi,t) = -\frac{1}{4\pi}\hat{\mathbf{r}}\times\int_S \left[\frac{1}{c}\hat{\mathbf{n}}'\times\frac{\partial}{\partial t}\mathbf{H}(\mathbf{r}',t+\hat{\mathbf{r}}\cdot\mathbf{r}'/c) + \epsilon\hat{\mathbf{r}}\times\left(\hat{\mathbf{n}}'\times\frac{\partial}{\partial t}\mathbf{E}(\mathbf{r}',t+\hat{\mathbf{r}}\cdot\mathbf{r}'/c)\right)\right]dS'. \tag{5.53}$$

Having expressed the far fields in terms of the volume and Huygens' sources, we shall now derive expressions for the far fields in terms of the field values in the scan plane $z = z_0$. Because we assume that all fields are zero for $t \le t_0$ and that the sources are in a finite region of space, the $\mathbf{r}_0$ integration in (5.4) is limited to a finite scan plane. Therefore, the expansion $R = r - \hat{\mathbf{r}}\cdot\mathbf{r}_0 + O(r^{-1})$ as $r \to \infty$ can be inserted into (5.4) to obtain the time-domain far-field expression

$$p(\mathbf{r},t) = \frac{\cos\theta}{2\pi cr}\int_{-\infty}^{+\infty}\int_{-\infty}^{+\infty}\frac{\partial}{\partial t}p(\mathbf{r}_0,t-r/c+\mathbf{r}_0\cdot\hat{\mathbf{r}}/c)dx_0dy_0 + O\left(\frac{1}{r^2}\right), \quad z > z_0 \tag{5.54}$$

where $\hat{\mathbf{r}} = \hat{\mathbf{x}}\cos\phi\sin\theta + \hat{\mathbf{y}}\sin\theta\sin\phi + \hat{\mathbf{z}}\cos\theta$. This result could also have been obtained by Fourier transforming the frequency-domain far field (3.41). Note that this derivation shows that the far field decays as $1/r$ (or faster) if the first time derivative of the field in the scan plane $z = z_0$ is a finite function of time. The acoustic far-field pattern is seen from (5.54) to equal

$$\mathcal{F}(\theta,\phi,t) = \frac{\cos\theta}{2\pi c}\int_{-\infty}^{+\infty}\int_{-\infty}^{+\infty}\frac{\partial}{\partial t}p(\mathbf{r}_0,t+\mathbf{r}_0\cdot\hat{\mathbf{r}}/c)dx_0dy_0, \quad 0 \le \theta < \pi/2. \tag{5.55}$$

For the electric field we similarly find from (5.6) that the far fields and the far-field patterns are given by

$$\begin{aligned}\mathbf{E}(\mathbf{r},t) &= \frac{\cos\theta}{2\pi cr}\int_{-\infty}^{+\infty}\int_{-\infty}^{+\infty}\frac{\partial}{\partial t}\mathbf{E}(\mathbf{r}_0,t-r/c+\mathbf{r}_0\cdot\hat{\mathbf{r}}/c)dx_0dy_0 \\ &\quad + O\left(\frac{1}{r^2}\right), \quad z > z_0 \\ &= -\frac{1}{2\pi cr}\hat{\mathbf{r}}\times\int_{-\infty}^{+\infty}\int_{-\infty}^{+\infty}\hat{\mathbf{z}}\times\frac{\partial}{\partial t}\mathbf{E}(\mathbf{r}_0,t-r/c+\mathbf{r}_0\cdot\hat{\mathbf{r}}/c)dx_0dy_0 \\ &\quad + O\left(\frac{1}{r^2}\right), \quad z > z_0\end{aligned} \tag{5.56}$$

and

$$\mathcal{F}(\theta, \phi, t) = \frac{\cos\theta}{2\pi c} \int_{-\infty}^{+\infty}\int_{-\infty}^{+\infty} \frac{\partial}{\partial t}\mathbf{E}(\mathbf{r}_0, t + \mathbf{r}_0 \cdot \hat{\mathbf{r}}/c) dx_0 dy_0, \quad 0 \le \theta < \pi/2$$

$$= -\frac{1}{2\pi c}\hat{\mathbf{r}} \times \int_{-\infty}^{+\infty}\int_{-\infty}^{+\infty} \hat{\mathbf{z}} \times \frac{\partial}{\partial t}\mathbf{E}(\mathbf{r}_0, t + \mathbf{r}_0 \cdot \hat{\mathbf{r}}/c) dx_0 dy_0, \quad 0 \le \theta < \pi/2. \tag{5.57}$$

The far-field result (5.56) was derived previously by Hill [146] by taking the inverse Fourier transform of (3.43). Dotting $\hat{\mathbf{r}}$ into (5.57) and integrating over time from $-\infty$ to t reveals that

$$\hat{\mathbf{r}} \cdot \int_{-\infty}^{+\infty}\int_{-\infty}^{+\infty} \mathbf{E}(\mathbf{r}_0, t + \mathbf{r}_0 \cdot \hat{\mathbf{r}}/c) dx_0 dy_0 = 0. \tag{5.58}$$

The time-domain magnetic far field in the half space $z > z_0$ is found from (5.24) to be

$$\mathbf{H}(\mathbf{r}, t) = -\frac{1}{2\pi c r}\sqrt{\frac{\epsilon}{\mu}}\hat{\mathbf{r}} \times \left[\hat{\mathbf{r}} \times \int_{-\infty}^{+\infty}\int_{-\infty}^{+\infty} \hat{\mathbf{z}} \times \frac{\partial}{\partial t}\mathbf{E}(\mathbf{r}_0, t - r/c + \mathbf{r}_0 \cdot \hat{\mathbf{r}}/c) dx_0 dy_0\right] + O\left(\frac{1}{r^2}\right), \quad z > z_0 \tag{5.59}$$

and the magnetic far-field pattern is

$$\mathcal{F}^H(\theta, \phi, t) = -\frac{1}{2\pi c}\sqrt{\frac{\epsilon}{\mu}}\hat{\mathbf{r}} \times \left[\hat{\mathbf{r}} \times \int_{-\infty}^{+\infty}\int_{-\infty}^{+\infty} \hat{\mathbf{z}} \times \frac{\partial}{\partial t}\mathbf{E}(\mathbf{r}_0, t + \mathbf{r}_0 \cdot \hat{\mathbf{r}}/c) dx_0 dy_0\right], \quad 0 < \theta < \pi/2. \tag{5.60}$$

Thus, the electric and magnetic far fields decay as $1/r$ (or faster) if the first time derivative of the x and y components of the electric (or magnetic) field in the plane $z = z_0$ are finite functions of time.

The time-domain far-field expressions in this section and the corresponding frequency-domain far-field expressions in Section 3.1.1 are related by Fourier transform pairs. In particular, the time-domain and frequency-domain far-field patterns are related by

$$\mathcal{F}(\theta, \phi, t) = \int_{-\infty}^{+\infty} \mathcal{F}_\omega(\theta, \phi)\, e^{-i\omega t} d\omega, \quad \mathcal{F}_\omega(\theta, \phi) = \frac{1}{2\pi}\int_{-\infty}^{+\infty} \mathcal{F}(\theta, \phi, t)\, e^{i\omega t} dt \tag{5.61}$$

and

$$\boldsymbol{\mathcal{F}}(\theta, \phi, t) = \int_{-\infty}^{+\infty} \boldsymbol{\mathcal{F}}_\omega(\theta, \phi)\, e^{-i\omega t} d\omega, \quad \boldsymbol{\mathcal{F}}_\omega(\theta, \phi) = \frac{1}{2\pi}\int_{-\infty}^{+\infty} \boldsymbol{\mathcal{F}}(\theta, \phi, t)\, e^{i\omega t} dt. \tag{5.62}$$

When the sources are located in the half space $z > z_0$, the far fields in the half space $z < z_0$ are given by the negativeof (5.54)–(5.57) and (5.59)–(5.60).

5.1.5 Far Fields Integrated over Time

We shall now show, as mentioned in Section 3.1.2, that there are restrictions on the time dependence that the far fields can have. We will begin by considering the electromagnetic field and assume that the sources $\mathbf{J}$ and ρ are contained in a bounded volume V, and turn on at $t = t_0$. Furthermore, let us assume that the sources approach zero as $t \to \infty$, so that all fields approach zero as $t \to \infty$. These assumptions are satisfied in cases where the sources turn on and off in a finite time interval.

Integrating Maxwell's equations (2.1)–(2.4) with respect to time then shows that (see (2.18)–(2.21))

$$\nabla \times \int_{t_0}^{+\infty} \mathbf{E}(\mathbf{r}, t)dt = 0 \tag{5.63}$$

$$\nabla \times \int_{t_0}^{+\infty} \mathbf{H}(\mathbf{r}, t)dt = \int_{t_0}^{+\infty} \mathbf{J}(\mathbf{r}, t)dt \tag{5.64}$$

$$\nabla \cdot \int_{t_0}^{+\infty} \mathbf{H}(\mathbf{r}, t)dt = 0 \tag{5.65}$$

and

$$\nabla \cdot \int_{t_0}^{+\infty} \mathbf{E}(\mathbf{r}, t)dt = \frac{1}{\epsilon} \int_{t_0}^{+\infty} \rho(\mathbf{r}, t)dt. \tag{5.66}$$

When the source terms $\int_{t_0}^{+\infty} \mathbf{J}(\mathbf{r}, t)dt$, and $\int_{t_0}^{+\infty} \rho(\mathbf{r}, t)dt$ are integrable with respect to $\mathbf{r}$ in the region V where the sources reside, (5.63)–(5.66) are simply the equations that hold for static electric and magnetic fields generated by sources located in the bounded volume V. In addition, the continuity equation (2.5) shows that the time-integrated current has zero divergence, that is, $\nabla \cdot \int_{t_0}^{+\infty} \mathbf{J}(\mathbf{r}, t)dt = 0$. For sources that are integrable in a bounded volume V, and for far fields that approach zero as $r \to \infty$, the electrostatic field decays as $1/r^2$ (or faster) and the magnetostatic field decays as $1/r^3$ (or faster) away from the source regions (see (4.6) and (4.51)) [1, chs. 3 and 4], [63, secs. 4.1 and 5.6]. Thus, the following asymptotic results hold

$$\int_{t_0}^{+\infty} \mathbf{E}(\mathbf{r}, t)dt = O\left(\frac{1}{r^2}\right), \quad \int_{t_0}^{+\infty} \mathbf{H}(\mathbf{r}, t)dt = O\left(\frac{1}{r^3}\right), \quad r \to \infty. \tag{5.67}$$

The integral of the electric field over all time approaches zero as $1/r^2$ (or faster) as the distance r to the source region approaches infinity, and the integral of the magnetic field over all time approaches zero as $1/r^3$ (or faster) as the distance r to the source region approaches infinity. These statements are true when the source terms $\int_{t_0}^{+\infty} \mathbf{J}(\mathbf{r}, t)dt$ and $\int_{t_0}^{+\infty} \rho(\mathbf{r}, t)dt$ are integrable with respect to $\mathbf{r}$ in a bounded volume V, and the time-integrated fields approach zero as $r \to \infty$. They hold even in cases where the far fields themselves decay slower than $1/r$. In particular, equation (5.67) holds for the electromagnetic missiles that will be discussed in Section 5.4.

If the far fields themselves decay as $1/r$ (or faster), so that the far-field patterns (5.44) are finite functions, we can insert the expansions

$$\begin{aligned}\mathbf{E}(\mathbf{r}, t) &= \frac{\mathcal{F}(\theta, \phi, t - r/c)}{r} + O\left(\frac{1}{r^2}\right),\\ \mathbf{H}(\mathbf{r}, t) &= \frac{\mathcal{F}^H(\theta, \phi, t - r/c)}{r} + O\left(\frac{1}{r^2}\right)\end{aligned} \tag{5.68}$$

into (5.67) to prove that the far-field patterns satisfy

$$\int_{-\infty}^{+\infty} \mathcal{F}(\theta, \phi, t)dt = 0, \quad \int_{-\infty}^{+\infty} \mathcal{F}^H(\theta, \phi, t)dt = 0 \tag{5.69}$$

which shows that the far-field patterns integrated over all time equal zero. These results could also have been proven directly by integrating over all time the expressions (5.49) and (5.50) for the far-field patterns in terms of the current.

The asymptotic formulas (5.67) for the time-domain fields were derived under the assumption that $\int_{t_0}^{+\infty} \mathbf{J}(\mathbf{r}, t)dt$ and $\int_{t_0}^{+\infty} \rho(\mathbf{r}, t)dt$ are integrable with respect to $\mathbf{r}$. We shall now consider an example in which these functions are not integrable. Let the source current have a continuous first time derivative, be zero for $t \leq t_0$, and approach a nonzero static value as $t \to \infty$ so that

$$\lim_{t\to\infty} \mathbf{J}(\mathbf{r}, t) = \mathbf{J}_0(\mathbf{r}). \tag{5.70}$$

In this case, $\int_{t_0}^{+\infty} \mathbf{J}(\mathbf{r}, t)dt$ is infinite for all values of $\mathbf{r}$ for which $\mathbf{J}_0(\mathbf{r}) \neq 0$, and the far fields decay as $1/r$ (or faster). The expressions (5.49) and (5.50) for the far-field patterns show that

$$\int_{-\infty}^{+\infty} \mathcal{F}(\theta, \phi, t)dt = \sqrt{\frac{\mu}{\epsilon}} \frac{1}{4\pi c} \hat{\mathbf{r}} \times \left(\hat{\mathbf{r}} \times \int_V \mathbf{J}_0(\mathbf{r}')dV'\right) \tag{5.71}$$

and

$$\int_{\infty}^{+\infty} \mathcal{F}^H(\theta, \phi, t)dt = -\frac{1}{4\pi c} \hat{\mathbf{r}} \times \int_V \mathbf{J}_0(\mathbf{r}')dV' \tag{5.72}$$

which are, in general, nonzero. With $\mathbf{J}_0$ inserted from the vector identity, $\nabla \cdot (\mathbf{J}_0\mathbf{r}) = \mathbf{r}(\nabla \cdot \mathbf{J}_0) + \mathbf{J}_0$, equations (5.71) and (5.72) become

$$\int_{-\infty}^{+\infty} \mathcal{F}(\theta, \phi, t)dt = \sqrt{\frac{\mu}{\epsilon}} \frac{1}{4\pi c} \hat{\mathbf{r}} \times \left(\hat{\mathbf{r}} \times \int_V \mathbf{r}' \frac{\partial \rho_0}{\partial t}(\mathbf{r}', \infty)dV'\right) \tag{5.73}$$

and

$$\int_{-\infty}^{+\infty} \mathcal{F}^H(\theta, \phi, t)dt = -\frac{1}{4\pi c} \hat{\mathbf{r}} \times \int_V \mathbf{r}' \frac{\partial \rho_0}{\partial t}(\mathbf{r}', \infty)dV' \tag{5.74}$$

where use has been made of the divergence theorem and the continuity equation

$$-\frac{\partial \rho_0}{\partial t}(\mathbf{r}, \infty) = -\lim_{t\to\infty} \frac{\partial}{\partial t}\rho_0(\mathbf{r}, t) = \nabla \cdot \mathbf{J}_0(\mathbf{r}). \tag{5.75}$$

Since the divergence of $\mathbf{J}_0(\mathbf{r})$ is not a function of time in (5.75), the charge distribution increases linearly with time as $t \to \infty$; that is

$$\rho_0(\mathbf{r}, t) \sim -\nabla \cdot \mathbf{J}_0(\mathbf{r}) t, \quad t \to \infty. \tag{5.76}$$

Hence, in this case the electromagnetic far-field patterns integrated over all time can be nonzero when the current approaches a static value as $t \to \infty$, but only if the static charge distribution, or more precisely, the electric dipole moment (or higher-order moments) of the charge distribution, grows linearly with time as $t \to \infty$. (The continuity equation, which is valid for all time, implies that the *total* charge for sources in a bounded volume remains zero because it is assumed zero for $t \leq t_0$.)

For acoustics, assume that the fields are zero for $t \leq t_0$, the mass-flow density $s(\mathbf{r}, t)$ and the force density $\mathbf{f}(\mathbf{r}, t)$ turn on at $t = t_0$, they approach zero as $t \to \infty$ (so that the fields also approach zero as $t \to \infty$), and have time integrals $\int_{t_0}^{+\infty} s(\mathbf{r}, t)dt$ and $\int_{t_0}^{+\infty} \mathbf{f}(\mathbf{r}, t)dt$ that are both integrable with respect to $\mathbf{r}$ in the bounded source region V. Then (2.201), (2.202), and (2.252) show that

$$\int_{t_0}^{+\infty} p(\mathbf{r}, t)dt = 0, \quad \mathbf{r} \notin V \tag{5.77}$$

$$\int_{t_0}^{+\infty} \mathbf{v}(\mathbf{r}, t)dt = O\left(\frac{1}{r^2}\right), \quad r \to \infty \tag{5.78}$$

and

$$\int_{-\infty}^{+\infty} \mathcal{F}(\theta, \phi, t)dt = 0. \tag{5.79}$$

Equation (5.77) shows that the integral over all times of the near-field pressure outside the source region is zero. However, in general the time-domain acoustic velocity near field and the electromagnetic near fields at a point in space do not integrate over all time to zero. An integration over all time of the expression (5.199) below for the acoustic velocity near field or (5.200) and (5.203) for the electric and magnetic near fields shows that it is only the evanescent part of the fields that contributes to the integration over all time t, provided the sources (and thus the far fields) are nonzero only during a finite time interval. The integration over all time t of the propagating part of the near fields, that is, the first integral on the right-hand sides of (5.199), (5.200), and (5.203), can be interchanged with the *finite* angular integration to give zero for far fields that are zero outside a finite time interval. The integration over all time t of the evanescent part of the near fields, that is, the second integrals on the right-hand sides of (5.199), (5.200), and (5.203), cannot be rigorously interchanged with the *infinite* (ξ, η) and t' integrations, and is not, in general, equal to zero.

5.2 TIME-DOMAIN ANALOGS OF THE PLANE-WAVE SPECTRUM REPRESENTATIONS

The time-domain analogs of the frequency-domain plane-wave spectrum formulas in Section 3.2 will be derived in this section. Without loss of generality, it is assumed throughout this section that the scan plane is located at $z = z_0 = 0$ because it substantially simplifies the analysis. The time-domain representations express both acoustic and electromagnetic fields in the region $z > 0$ in terms of a space-time Radon transform of the fields in the plane $z = 0$. The derivations will be performed by taking the inverse Fourier transform of the frequency-domain plane-wave spectrum formulas of Section 3.2.

These frequency-domain formulas were derived for $\omega > 0$, and many of them do not hold for $\omega < 0$. To obtain formulas that are valid for all real ω, we use the fact that the frequency-domain field f_ω, corresponding to a *real* time-domain field $f(t)$, satisfies the reality condition (2.285)

$$f_{-\omega} = f_\omega^* \tag{5.80}$$

where $*$ indicates complex conjugation. This equation, which will be used extensively throughout this section, determines the frequency-domain field for negative frequencies, given the frequency-domain field for positive frequencies. (The frequency-domain Green's function formulas used in the derivations of Section 5.1 all satisfied (5.80) and were therefore automatically valid for all real ω.)

We will begin by rewriting the frequency-domain plane-wave formulas for the acoustic pressure p_ω and its spectrum T_ω in (3.56) and (3.57) for $\omega > 0$ as

$$p_\omega(\mathbf{r}) = \frac{1}{2\pi} \int_{-\infty}^{+\infty}\int_{-\infty}^{+\infty} \omega^2 T_\omega(\omega\xi, \omega\eta) e^{i\omega(\xi x + \eta y)} e^{i\gamma(\omega) z} d\xi d\eta, \quad z \geq 0 \tag{5.81}$$

and

$$T_\omega(\omega\xi, \omega\eta) = \frac{1}{2\pi} \int_{-\infty}^{+\infty}\int_{-\infty}^{+\infty} p_\omega(\mathbf{r}_0) e^{-i\omega(\xi x_0 + \eta y_0)} dx_0 dy_0 \tag{5.82}$$

where $\mathbf{r}_0 = x_0\hat{\mathbf{x}} + y_0\hat{\mathbf{y}}$ and we have introduced the spectral variables (ξ, η) such that $(k_x, k_y) = (\omega\xi, \omega\eta)$. Since the reality condition (5.80) is satisfied by the frequency-domain acoustic pressure $p_\omega(\mathbf{r})$, it is also satisfied by the right-hand side of (5.82). Thus, $T_\omega(\omega\xi, \omega\eta)$ defined by (5.82) for all real ω satisfies the reality condition

$$T_{-\omega}(-\omega\xi, -\omega\eta) = T_\omega^*(\omega\xi, \omega\eta) \tag{5.83}$$

where $T_{-\omega}(-\omega\xi, -\omega\eta)$ denotes the right-hand side of (5.82) with ω replaced by $-\omega$.

Taking the complex conjugate of (5.81) shows that p_ω in (5.81) satisfies the reality condition (5.80) if $\gamma(-\omega) = \gamma^*(\omega)$. This result can also be obtained from first principles by repeating the derivation of (3.56) and (3.57) in Section 3.2 for $\omega < 0$. Hence, the following expression for $p_\omega(\mathbf{r})$ is valid for all real ω

$$\begin{aligned} p_\omega(\mathbf{r}) = & \frac{1}{2\pi} \iint_{\xi^2+\eta^2<c^{-2}} \omega^2 T_\omega(\omega\xi, \omega\eta) e^{i\omega(\xi x + \eta y + \zeta z)} d\xi d\eta \\ & + \frac{1}{2\pi} \iint_{\xi^2+\eta^2>c^{-2}} \omega^2 T_\omega(\omega\xi, \omega\eta) e^{i\omega(\xi x + \eta y)} e^{-|\omega||\zeta| z} d\xi d\eta, \quad z \geq 0 \end{aligned} \tag{5.84}$$

where ζ is a new spectral variable given by

$$\zeta = \begin{cases} \left|\sqrt{c^{-2} - \xi^2 - \eta^2}\right|, & \xi^2 + \eta^2 \leq c^{-2} \\ i\left|\sqrt{\xi^2 + \eta^2 - c^{-2}}\right|, & \xi^2 + \eta^2 > c^{-2}. \end{cases} \tag{5.85}$$

Note that $\gamma(\omega) = \omega\zeta$ for $\omega > 0$ and $\gamma(\omega) = \omega\zeta^*$ for $\omega < 0$.

We can now take the inverse Fourier transform of (5.84) to get

$$
\begin{aligned}
p(\mathbf{r},t) = &-\frac{1}{2\pi}\iint\limits_{\xi^2+\eta^2<c^{-2}} \frac{\partial^2}{\partial t^2}T(\xi,\eta,t-\xi x-\eta y-\zeta z)d\xi d\eta \\
&-\frac{1}{2\pi^2}\iint\limits_{\xi^2+\eta^2>c^{-2}}\int_{-\infty}^{+\infty}\frac{\partial^2}{\partial t^2}\frac{|\zeta|z\,T(\xi,\eta,t-t'-\xi x-\eta y)}{|\zeta|^2z^2+t'^2}dt'd\xi d\eta, \quad z>0
\end{aligned}
\tag{5.86}
$$

where $T(\xi,\eta,t)$ is just the inverse Fourier transform of $T_\omega(\omega\xi,\omega\eta)$

$$
\begin{aligned}
T(\xi,\eta,t) &= \int_{-\infty}^{+\infty} T_\omega(\omega\xi,\omega\eta)\,e^{-i\omega t}d\omega \\
&= \frac{1}{2\pi}\int_{-\infty}^{+\infty}\int_{-\infty}^{+\infty}\int_{-\infty}^{+\infty} p_\omega(\mathbf{r}_0)e^{-i\omega(t+\xi x_0+\eta y_0)}d\omega dx_0 dy_0 \\
&= \frac{1}{2\pi}\int_{-\infty}^{+\infty}\int_{-\infty}^{+\infty} p(\mathbf{r}_0,t+\xi x_0+\eta y_0)dx_0 dy_0
\end{aligned}
\tag{5.87}
$$

and use has been made of the identity

$$
\int_{-\infty}^{+\infty} e^{-|\omega||\zeta|z}e^{-i\omega t}d\omega = \frac{2|\zeta|z}{|\zeta|^2z^2+t^2}, \quad z>0
\tag{5.88}
$$

and the convolution theorem

$$
\int_{-\infty}^{+\infty} f_\omega g_\omega e^{-i\omega t}d\omega = \frac{1}{2\pi}\int_{-\infty}^{+\infty} f(t-t')g(t')dt'.
\tag{5.89}
$$

The time-domain function $T(\xi,\eta,t)$ is a space-time Radon transform of the time-domain pressure $p(\mathbf{r},t)$ in the plane $z=0$. We shall refer to $T(\xi,\eta,t)$ as the time-domain plane-wave spectrum (or just time-domain spectrum).

The formula (5.86), which is a time-domain analog of the frequency-domain plane-wave spectrum formula (3.56), gives the pressure in the half space $z>0$ in terms of the time-domain spectrum determined in (5.87) from the pressure in the plane $z=0$. We see from (5.86) that the second-time derivative of the spectrum, and therefore also the second-time derivative of the pressure in the plane $z=0$, are required to calculate the pressure for $z>0$.

For the subsequent analysis it is convenient to introduce a time-domain propagation vector $\widetilde{\boldsymbol{\chi}}$, which is analogous to the frequency-domain propagation vector $\mathbf{k}=k_x\hat{\mathbf{x}}+k_y\hat{\mathbf{y}}+\gamma\hat{\mathbf{z}}$. We define a normalized propagation vector

$$
\boldsymbol{\chi}_\omega(\xi,\eta) = \xi\hat{\mathbf{x}}+\eta\hat{\mathbf{y}}+\zeta\hat{\mathbf{z}}, \quad \omega>0
\tag{5.90}
$$

so that $\omega\boldsymbol{\chi}_\omega=\mathbf{k}$ for positive frequencies. By requiring $\boldsymbol{\chi}_\omega=\boldsymbol{\chi}^*_{-\omega}$ we find the following expression for the normalized propagation vector that is valid for all real ω

$$
\boldsymbol{\chi}_\omega(\xi,\eta) = \begin{cases} \xi\hat{\mathbf{x}}+\eta\hat{\mathbf{y}}+\zeta\hat{\mathbf{z}}, & \xi^2+\eta^2\le c^{-2} \\ \xi\hat{\mathbf{x}}+\eta\hat{\mathbf{y}}+i\dfrac{\omega}{|\omega|}|\zeta|\hat{\mathbf{z}}, & \xi^2+\eta^2>c^{-2} \end{cases}.
\tag{5.91}
$$

Taking the Fourier transform of (5.91), multiplied by a frequency-domain function f_ω, shows that $\widetilde{\boldsymbol{\chi}}(\xi, \eta)$ is a time-domain vector operator given by

$$\widetilde{\boldsymbol{\chi}}(\xi, \eta) = \begin{cases} \xi\hat{\mathbf{x}} + \eta\hat{\mathbf{y}} + \zeta\hat{\mathbf{z}}, & \xi^2 + \eta^2 \le c^{-2} \\ \xi\hat{\mathbf{x}} + \eta\hat{\mathbf{y}} + |\zeta|\widetilde{\mathcal{H}}\hat{\mathbf{z}}, & \xi^2 + \eta^2 > c^{-2} \end{cases} \tag{5.92}$$

where $\widetilde{\mathcal{H}}$ denotes the Hilbert transform

$$\widetilde{\mathcal{H}} f(t) = \fint_{-\infty}^{+\infty} \frac{f(t')}{\pi(t - t')} dt' \tag{5.93}$$

and the bar on the integral sign indicates a Cauchy principal-value integral. Since $\widetilde{\mathcal{H}}^2 = -1$, that is, $\widetilde{\mathcal{H}}[\widetilde{\mathcal{H}} f(t)] = -f(t)$, the vector operator $\widetilde{\boldsymbol{\chi}}$ satisfies $\widetilde{\boldsymbol{\chi}} \cdot \widetilde{\boldsymbol{\chi}} = c^{-2}$, that is, $\widetilde{\boldsymbol{\chi}} \cdot (\widetilde{\boldsymbol{\chi}} f(t)) = c^{-2} f(t)$, for all (ξ, η). To derive (5.92) we have used the identity

$$\int_{-\infty}^{+\infty} i \frac{\omega}{|\omega|} f_\omega e^{-i\omega t} d\omega = -2\mathrm{Im} \int_{0}^{+\infty} f_\omega e^{-i\omega t} d\omega = \widetilde{\mathcal{H}} f(t) \tag{5.94}$$

which can be proven from the following three facts.

1. $f(t) = 2\mathrm{Re} \int_0^{+\infty} f_\omega e^{-i\omega t} d\omega.$
2. $\int_0^{+\infty} f_\omega e^{-i\omega t} d\omega$ is analytic for $\mathrm{Im}(t) < 0$.
3. The real and imaginary parts of an analytic function are related on the real axis through the Hilbert transform [115, pp. 370–372].

(Sufficient conditions for the existence of the Hilbert transform of $f(t)$ is that $f(t)$ be Hölder continuous and that $f(t)$ approaches the definite value $f(\infty)$ as $t \to \pm\infty$ so that for sufficiently large $|t|$, $|f(t) - f(\infty)| \le C/|t|^\alpha$, where C and α are positive constants [114, sec. 43]. This latter condition plays the role of the Hölder condition at $t = \pm\infty$, and of course, $f(\infty) = 0$ for the functions of time we are considering that are zero for $t \le t_0$. As shown in Section 2.5.1, the Hilbert-transform relationships in (5.94) can also be proven by first multiplying the integrand of the first integral in (5.94) by $e^{-\delta|\omega|}$ ($\delta > 0$), second applying the convolution theorem, third expressing the convolution as a principal-value integral, and finally taking the limit of the integral as $\delta \to 0$ by means of the theorem in Hobson [71, vol. 2, sec. 225].)

Introducing the vector $\boldsymbol{\chi}_\omega$ into the frequency-domain formula (3.58) for the velocity and using (5.80), we obtain the following formula valid for all real ω

$$\begin{aligned} \mathbf{v}_\omega(\mathbf{r}) = & \frac{1}{2\pi \rho_{m0}} \iint_{\xi^2+\eta^2<c^{-2}} \omega^2 \boldsymbol{\chi}_\omega(\xi, \eta) T_\omega(\omega\xi, \omega\eta) e^{i\omega(\xi x + \eta y + \zeta z)} d\xi d\eta \\ & + \frac{1}{2\pi \rho_{m0}} \iint_{\xi^2+\eta^2>c^{-2}} \omega^2 \boldsymbol{\chi}_\omega(\xi, \eta) T_\omega(\omega\xi, \omega\eta) e^{i\omega(\xi x + \eta y)} e^{-|\omega||\zeta|z} d\xi d\eta, \quad z \ge 0. \end{aligned} \tag{5.95}$$

The Fourier transform applied to (5.95) proves that the velocity for $z > 0$ can be expressed in terms of the acoustic time-domain spectrum as

$$\mathbf{v}(\mathbf{r},t) = -\frac{1}{2\pi\rho_{m0}} \iint\limits_{\xi^2+\eta^2<c^{-2}} \widetilde{\chi}(\xi,\eta)\frac{\partial^2}{\partial t^2}T(\xi,\eta,t-\xi x-\eta y-\zeta z)d\xi d\eta$$
$$-\frac{1}{2\pi^2\rho_{m0}} \iint\limits_{\xi^2+\eta^2>c^{-2}} \int_{-\infty}^{+\infty} \widetilde{\chi}(\xi,\eta)\frac{\partial^2}{\partial t^2}\frac{|\zeta|z\,T(\xi,\eta,t-t'-\xi x-\eta y)}{|\zeta|^2z^2+t'^2}dt'd\xi d\eta, \quad z>0 \tag{5.96}$$

where the identity (5.88) and the convolution theorem (5.89) have been used.

Before considering the electromagnetic fields, we will discuss the time-domain basis functions that are used implicitly in the expansions (5.86) and (5.96). Begin by noting that the basis functions in the frequency-domain expansions (3.56) and (3.58) are the frequency-domain plane waves $(2\pi)^{-1}e^{i(k_x x+k_y y+\gamma z)}$. Since the time-domain expansions (5.86) and (5.96) are obtained by Fourier transforming (3.56) and (3.58), the basis functions used implicitly in (5.86) and (5.96) are the time-domain analogs of the frequency-domain plane waves. We shall explicitly write down these time-domain basis functions and thereby obtain alternative expressions for the time-domain expansions.

The frequency-domain plane waves $(2\pi)^{-1}e^{i(k_x x+k_y y+\gamma z)}$ can be written in the following form, which satisfies (5.80), and therefore is valid for all real ω

$$\Phi^B_\omega(\mathbf{r},\xi,\eta) = \frac{1}{2\pi}\begin{cases} e^{i\omega(\xi x+\eta y+\zeta z)}, & \omega>0 \\ e^{i\omega(\xi x+\eta y+\zeta^* z)}, & \omega<0 \end{cases} \tag{5.97}$$

where ζ is given by (5.85) and $z>0$. Taking the inverse Fourier transform of equation (5.97) shows that the time-domain basis functions are given by

$$\Phi^B(\mathbf{r},t,\xi,\eta) = \begin{cases} \delta(t-\xi x-\eta y-\zeta z), & \xi^2+\eta^2\le c^{-2} \\ \dfrac{\pi^{-1}z|\zeta|}{z^2|\zeta|^2+(t-\xi x-\eta y)^2}, & \xi^2+\eta^2>c^{-2} \end{cases} \tag{5.98}$$

where $z>0$, and the identity (5.88) has been used.

Since these time-domain basis fields are obtained from frequency-domain plane waves, we will refer to them as time-domain plane waves. The propagating plane wave $\Phi^B(\mathbf{r},t,\xi,\eta)$, $\xi^2+\eta^2\le c^{-2}$ is a delta function, which is nonzero only in the plane $t=\xi x+\eta y+\zeta z$ and is propagating in the direction given by the unit vector $\hat{\mathbf{k}}(\xi,\eta)=c\widetilde{\chi}(\xi,\eta)=c\xi\hat{\mathbf{x}}+c\eta\hat{\mathbf{y}}+c\zeta\hat{\mathbf{z}}$, as shown in Figure 5.2. The evanescent plane wave $\Phi^B(\mathbf{r},t,\xi,\eta)$, $\xi^2+\eta^2>c^{-2}$ is more complicated. In the plane $\xi x+\eta y=t_0$, which is parallel to the z axis, the amplitude of the evanescent plane wave has a maximum at $z=|t-t_0|/|\zeta|$ and decays as z^{-1} at infinity, as shown in Figure 5.3. Furthermore, from the analysis of Morse and Feshbach [115, pp. 813–814] it follows that

$$\lim_{z|\zeta|\to 0^+}\frac{\pi^{-1}z|\zeta|}{z^2|\zeta|^2+(t-\xi x-\eta y)^2} = \delta(t-\xi x-\eta y) \tag{5.99}$$

so that, in agreement with (5.97), all time-domain plane waves are delta functions in the plane $z=0$. The expansion (5.86) may now be written in the alternative form

$$p(\mathbf{r},t) = -\frac{1}{2\pi}\int_{-\infty}^{+\infty}\int_{-\infty}^{+\infty}\int_{-\infty}^{+\infty}\Phi^B(\mathbf{r},t-t',\xi,\eta)\frac{\partial^2}{\partial t'^2}T(\xi,\eta,t')dt'd\xi d\eta, \quad z>0 \tag{5.100}$$

expressing the time-domain pressure in the half space $z>0$ in terms of the time-domain plane waves (5.98). This form of the expansion will prove useful in Chapter 7 for deriving the probe-corrected formulas in the time domain.

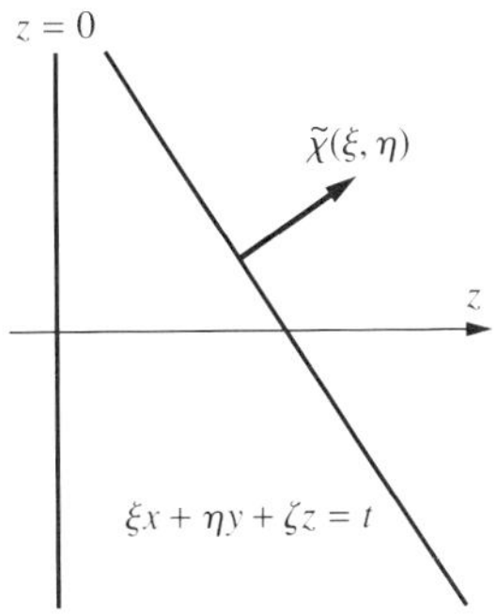

Figure 5.2 Propagating time-domain plane wave.

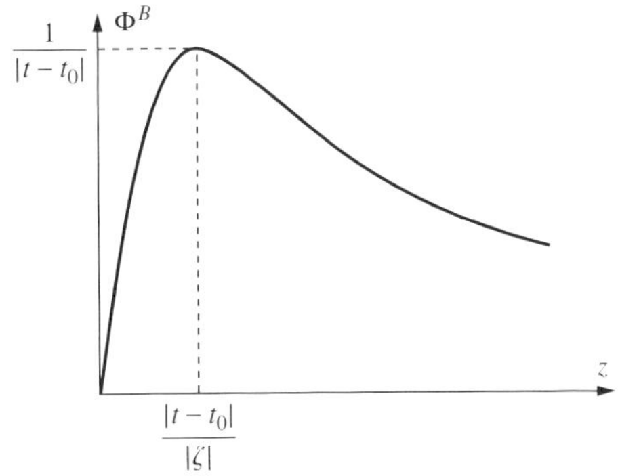

Figure 5.3 Evanescent time-domain plane wave as a function of z for fixed t and $\xi x + \eta y = t_0$.

In the plane $z = 0$, (5.98)–(5.99) reduce (5.100) (or, equivalently, (5.86)) to

$$p(\mathbf{r}_0, t) = -\frac{1}{2\pi} \int_{-\infty}^{+\infty} \int_{-\infty}^{+\infty} \frac{\partial^2}{\partial t^2} T(\xi, \eta, t - \xi x_0 - \eta y_0) d\xi d\eta, \quad \mathbf{r}_0 = x_0 \hat{\mathbf{x}} + y_0 \hat{\mathbf{y}} \tag{5.101}$$

which is the inverse space-time Radon transform of $T(\xi, \eta, t)$ [148] corresponding to the space-time Radon transform of $p(\mathbf{r}_0, t)$ given from (5.87) as

$$T(\xi, \eta, t) = \frac{1}{2\pi} \int_{-\infty}^{+\infty} \int_{-\infty}^{+\infty} p(\mathbf{r}_0, t + \xi x_0 + \eta y_0) dx_0 dy_0, \quad \mathbf{r}_0 = x_0 \hat{\mathbf{x}} + y_0 \hat{\mathbf{y}}. \tag{5.102}$$

Now consider the electromagnetic fields. The formulas derived in this section hold for each rectangular component of the electric and magnetic fields, and we find by taking the Fourier transform of (3.62) and (3.63), rewritten here as

$$\mathbf{E}_\omega(\mathbf{r}) = \frac{1}{2\pi} \int_{-\infty}^{+\infty} \int_{-\infty}^{+\infty} \omega^2 \, \mathbf{T}_\omega(\omega\xi, \omega\eta) \, e^{i\omega(\xi x + \eta y)} \, e^{i\gamma(\omega) z} d\xi d\eta \tag{5.103}$$

and

$$\mathbf{T}_\omega(\omega\xi, \omega\eta) = \frac{1}{2\pi} \int_{-\infty}^{+\infty} \int_{-\infty}^{+\infty} \mathbf{E}_\omega(\mathbf{r}_0) e^{-i\omega(\xi x_0 + \eta y_0)} dx_0 dy_0, \quad \mathbf{r}_0 = x_0 \hat{\mathbf{x}} + y_0 \hat{\mathbf{y}} \tag{5.104}$$

that the time-domain electric field is given by the time-domain plane-wave expansion

$$\mathbf{E}(\mathbf{r}, t) = -\frac{1}{2\pi} \iint_{\xi^2+\eta^2<c^{-2}} \frac{\partial^2}{\partial t^2} \mathbf{T}(\xi, \eta, t - \xi x - \eta y - \zeta z) d\xi d\eta$$
$$-\frac{1}{2\pi^2} \iint_{\xi^2+\eta^2>c^{-2}} \int_{-\infty}^{+\infty} \frac{\partial^2}{\partial t^2} \frac{z|\zeta| \, \mathbf{T}(\xi, \eta, t - t' - \xi x - \eta y)}{|\zeta|^2 z^2 + t'^2} dt' d\xi d\eta, \quad z > 0 \tag{5.105}$$

which can also be expressed explicitly in terms of the time-domain plane waves (5.98) as

$$\mathbf{E}(\mathbf{r}, t) = -\frac{1}{2\pi} \int_{-\infty}^{+\infty} \int_{-\infty}^{+\infty} \int_{-\infty}^{+\infty} \Phi^B(\mathbf{r}, t - t', \xi, \eta) \frac{\partial^2}{\partial t'^2} \mathbf{T}(\xi, \eta, t') dt' d\xi d\eta, \quad z > 0. \tag{5.106}$$

The time-domain plane-wave spectrum for the electric field is given by

$$\begin{aligned}\mathbf{T}(\xi,\eta,t) &= \int_{-\infty}^{+\infty} \mathbf{T}_\omega(\omega\xi,\omega\eta)\, e^{-i\omega t} d\omega \\ &= \frac{1}{2\pi}\int_{-\infty}^{+\infty}\int_{-\infty}^{+\infty} \mathbf{E}(\mathbf{r}_0, t+\xi x_0+\eta y_0) dx_0 dy_0, \quad \mathbf{r}_0 = x_0\hat{\mathbf{x}} + y_0\hat{\mathbf{y}}\end{aligned} \tag{5.107}$$

which is a space-time Radon transform of the electric field in the plane $z = 0$. With the help of (5.98), (5.99), (5.106) reduces in the $z = 0$ plane to the inverse space-time Radon transform [148]

$$\mathbf{E}(\mathbf{r}_0, t) = -\frac{1}{2\pi}\int_{-\infty}^{+\infty}\int_{-\infty}^{+\infty} \frac{\partial^2}{\partial t^2}\mathbf{T}(\xi,\eta,t-\xi x_0-\eta y_0) d\xi d\eta, \quad \mathbf{r}_0 = x_0\hat{\mathbf{x}} + y_0\hat{\mathbf{y}}. \tag{5.108}$$

The formulas (5.105)–(5.107) express the time-domain electric field in the half space $z > 0$ in terms of the electric field in the scan plane $z = 0$. We note that the second time derivative of the electric field in scan plane $z = 0$ is required to compute the electric field for $z > 0$.

The frequency-domain formula (3.64), showing that the spectrum $\mathbf{T}_\omega$ is perpendicular to the complex propagation direction $\mathbf{k}$, can be written as

$$\boldsymbol{\chi}_\omega(\xi,\eta)\cdot\mathbf{T}_\omega(\omega\xi,\omega\eta) = 0 \tag{5.109}$$

where $\boldsymbol{\chi}_\omega$ is given by (5.91). Taking the Fourier transform of (5.109) shows that

$$\widetilde{\boldsymbol{\chi}}(\xi,\eta)\cdot\mathbf{T}(\xi,\eta,t) = 0 \tag{5.110}$$

for all (ξ, η). Hence

$$\mathbf{T}(\xi,\eta,t)\cdot(\xi\hat{\mathbf{x}}+\eta\hat{\mathbf{y}}+\zeta\hat{\mathbf{z}}) = 0, \quad \xi^2+\eta^2 < c^{-2} \tag{5.111}$$

so for $\xi^2+\eta^2 < c^{-2}$ the time-domain plane-wave spectrum of the electric field is perpendicular to the direction $\xi\hat{\mathbf{x}}+\eta\hat{\mathbf{y}}+\zeta\hat{\mathbf{z}}$, and completely determined from two components of the electric field in the plane $z = 0$. For $\xi^2+\eta^2 > c^{-2}$ the formula (5.92) for $\widetilde{\boldsymbol{\chi}}(\xi,\eta)$ gives

$$\mathbf{T}(\xi,\eta,t)\cdot(\xi\hat{\mathbf{x}}+\eta\hat{\mathbf{y}}) + |\zeta|\widetilde{\mathcal{H}}\mathbf{T}(\xi,\eta,t)\cdot\hat{\mathbf{z}} = 0, \quad \xi^2+\eta^2 > c^{-2}. \tag{5.112}$$

Taking the Hilbert transform of (5.112), and using $\widetilde{\mathcal{H}}^2 = -1$, gives us the z component of $\mathbf{T}(\xi,\eta,t)$ in terms of the x and y components of $\mathbf{T}(\xi,\eta,t)$, namely

$$T_z(\xi,\eta,t) = \widetilde{\mathcal{H}}\mathbf{T}(\xi,\eta,t)\cdot(\xi\hat{\mathbf{x}}+\eta\hat{\mathbf{y}})/|\zeta|, \quad \xi^2+\eta^2 > c^{-2}. \tag{5.113}$$

Equations (5.111) and (5.113) enable us to calculate the z component of the time-domain spectrum for the electric field from its x and y components. Note that inserting $\mathbf{T}$ from (5.107) into (5.111) and replacing $\xi\hat{\mathbf{x}}+\eta\hat{\mathbf{y}}+\zeta\hat{\mathbf{z}}$ by $\hat{\mathbf{r}}/c$ yields equation (5.58).

The formulas derived in this section also hold for the magnetic field provided that $\mathbf{E}$ is replaced by $\mathbf{H}$ and $\mathbf{T}$ is replaced by $\mathbf{T}^H$, where $\mathbf{T}^H$ is the time-domain plane-wave spectrum for the magnetic field. The relation (3.67) between the spectrum $\mathbf{T}_\omega$ for the electric field and the spectrum $\mathbf{T}^H_\omega$ for the magnetic field shows that

$$\mathbf{T}^H_\omega(\omega\xi,\omega\eta) = \frac{1}{\mu}\boldsymbol{\chi}_\omega(\xi,\eta)\times\mathbf{T}_\omega(\omega\xi,\omega\eta) \tag{5.114}$$

and therefore the time-domain spectra for the electric and magnetic fields satisfy

$$\mathbf{T}^H(\xi, \eta, t) = \frac{1}{\mu}\widetilde{\boldsymbol{\chi}}(\xi, \eta) \times \mathbf{T}(\xi, \eta, t) \tag{5.115}$$

for all (ξ, η). From the expression (5.92) for $\widetilde{\boldsymbol{\chi}}$ we find that

$$\mathbf{T}^H(\xi, \eta, t) = \frac{1}{\mu}(\xi\hat{\mathbf{x}} + \eta\hat{\mathbf{y}} + \zeta\hat{\mathbf{z}}) \times \mathbf{T}(\xi, \eta, t), \quad \xi^2 + \eta^2 < c^{-2} \tag{5.116}$$

and that

$$\begin{aligned}\mathbf{T}^H(\xi, \eta, t) &= \frac{1}{\mu}(\xi\hat{\mathbf{x}} + \eta\hat{\mathbf{y}}) \times \mathbf{T}(\xi, \eta, t) \\ &+ \frac{|\zeta|}{\mu}\hat{\mathbf{z}} \times \widetilde{\mathcal{H}}\mathbf{T}(\xi, \eta, t), \quad \xi^2 + \eta^2 > c^{-2}.\end{aligned} \tag{5.117}$$

The time-domain plane-wave spectrum for the magnetic field is thereby expressed in terms of the time-domain plane-wave spectrum for the electric field. Equation (5.116) shows that $\mathbf{T}(\xi, \eta, t) \cdot \mathbf{T}^H(\xi, \eta, t)$ equals zero for propagating plane waves, whereas (5.117) reveals that, in general, $\mathbf{T}(\xi, \eta, t) \cdot \mathbf{T}^H(\xi, \eta, t)$ is not equal to zero for evanescent plane waves.

The formulas derived in this section are time-domain analogs to the frequency-domain formulas of Section 3.2, and it is seen that a space-time Radon transform in the time-domain formulas plays the role of the spatial Fourier transform in the frequency-domain formulas. Furthermore, the time-domain spectra for the electric and magnetic fields in the plane $z = 0$ satisfy the relations (5.111)–(5.112) and (5.116)–(5.117), which are similar to but more complicated than the relations (3.64) and (3.67) satisfied by the frequency-domain plane-wave spectra for the electric and magnetic fields. If the evanescent fields are negligible, the second integrations in (5.86) and (5.105) are negligible, and the time-domain plane-wave spectrum relations take the simpler form of the analogous frequency-domain plane-wave spectra relations.

When the sources are located in $z > 0$, the fields in $z < 0$ are determined from (5.86), (5.87), (5.96), (5.98), (5.100), (5.105), (5.106), (5.107), (5.110), and (5.115) with z replaced by $-z$ and $\widetilde{\boldsymbol{\chi}}$ replaced by $\widetilde{\boldsymbol{\chi}}^- = \widetilde{\boldsymbol{\chi}} - 2(\hat{\mathbf{z}} \cdot \widetilde{\boldsymbol{\chi}})\hat{\mathbf{z}}$.

5.2.1 Propagating and Evanescent Fields of a Point Source

We next consider the plane-wave expansion of the time-domain free-space Green's function $G(\mathbf{r}, \mathbf{r}', t, t')$ in (5.9). The contribution from the propagating plane waves (called the propagating field) and the contribution from the evanescent plane waves (called the evanescent field) will be computed explicitly, and we shall determine in which space-time regions the evanescent field is exactly zero. Tygel and Hubral [16] and Heyman [17] have also derived expressions for the evanescent field of this free-space Green's function.

The time-domain free-space Green's function $G(\mathbf{r}, \mathbf{r}', t, t')$ is related to the frequency-domain free-space Green's function $G_\omega(\mathbf{r}, \mathbf{r}')$ through the relation

$$G(\mathbf{r}, \mathbf{r}', t, t') = \int_{-\infty}^{+\infty} \frac{e^{i\omega t'}}{2\pi} G_\omega(\mathbf{r}, \mathbf{r}')\, e^{-i\omega t}\, d\omega. \tag{5.118}$$

In Section 3.2.1 we derived expressions for the propagating and evanescent fields of $G_\omega(\mathbf{r}, \mathbf{r}')$. Hence, the expressions for the propagating and evanescent fields of $G(\mathbf{r}, \mathbf{r}', t, t')$ are readily obtained by applying (5.118) to the formulas of Section 3.2.1. In the present section, we assume that the source point is at the origin ($\mathbf{r}' = 0$) and that the initial time is zero ($t' = 0$), so that the source coordinates (R, θ_s, ϕ_s) defined in Section 3.2.1 equal the standard spherical coordinates (r, θ, ϕ).[2]

Introducing the variables (ξ, η, ζ) into the frequency-domain Weyl formulas (3.72)–(3.74) with $\mathbf{r}' = 0$, applying (5.118), and performing a derivation similar to the one that led to (5.86) yield the time-domain Weyl formulas

$$G(\mathbf{r}, 0, t, 0) = \frac{\delta(t - r/c)}{4\pi r} = G^p(\mathbf{r}, 0, t, 0) + G^e(\mathbf{r}, 0, t, 0) \tag{5.119}$$

where

$$G^p(\mathbf{r}, 0, t, 0) = -\frac{1}{8\pi^2} \iint\limits_{\xi^2+\eta^2<c^{-2}} \frac{\delta'(t - \xi x - \eta y - \zeta z)}{\zeta} d\xi d\eta, \quad z > 0 \tag{5.120}$$

is the propagating time-domain field, and

$$\begin{aligned} G^e(\mathbf{r}, 0, t, 0) &= \frac{1}{8\pi^3} \frac{\partial}{\partial t} \iint\limits_{\xi^2+\eta^2>c^{-2}} \frac{1}{|\zeta|} \frac{t - \xi x - \eta y}{|\zeta|^2 z^2 + (t - \xi x - \eta y)^2} d\xi d\eta, \quad z > 0 \\ &= \frac{1}{8\pi^3} \iint\limits_{\xi^2+\eta^2>c^{-2}} \frac{1}{|\zeta|} \frac{|\zeta|^2 z^2 - (t - \xi x - \eta y)^2}{\left(|\zeta|^2 z^2 + (t - \xi x - \eta y)^2\right)^2} d\xi d\eta, \quad z > 0 \end{aligned} \tag{5.121}$$

is the evanescent time-domain field. We will now derive closed-form expressions for $G^p(\mathbf{r}, 0, t, 0)$ and $G^e(\mathbf{r}, 0, t, 0)$ for observation points in the region $0 \le \theta \le \pi/2$.

***Observation Points with* $\boldsymbol{\theta = 0}$.** For this angle of observation, the observation point is given by $\mathbf{r} = z\hat{\mathbf{z}}$ with $z > 0$. Inserting the expressions (3.75) and (3.76) into (5.118) with $\mathbf{r}' = 0$ and $t' = 0$ immediately gives

$$G^p(\mathbf{r}, 0, t, 0) = \frac{\delta(t - z/c)}{4\pi z} - \frac{\delta(t)}{4\pi z} \tag{5.122}$$

and

$$G^e(\mathbf{r}, 0, t, 0) = \frac{\delta(t)}{4\pi z} \tag{5.123}$$

which could easily have been obtained directly from (5.120) and (5.121). Hence, along the z axis the evanescent field is of the same order (z^{-1}) as the propagating field, and the evanescent field is identically zero for $t > 0$.

***Observation Points with* $\boldsymbol{0 < \theta < \pi/2}$.** Inserting (3.85) into (5.118) with $\mathbf{r}' = 0$ and $t' = 0$ yields the following expression for the propagating time-domain field

$$G^p(\mathbf{r}, 0, t, 0) = \frac{\delta(t - r/c)}{4\pi r} - \frac{c\cos\theta}{4\pi^2 r^2} \frac{I\left(t; -rc^{-1}\sin\theta, rc^{-1}\sin\theta\right)}{(1 - (ct/r)^2)\sqrt{\sin^2\theta - (ct/r)^2}} \tag{5.124}$$

where $I(t; a, b) = 1$ for $a < t < b$ and zero otherwise. Combining the formulas (5.119) and (5.124) shows that the evanescent time-domain field is given by

$$G^e(\mathbf{r}, 0, t, 0) = \frac{c\cos\theta}{4\pi^2 r^2} \frac{I\left(t; -rc^{-1}\sin\theta, rc^{-1}\sin\theta\right)}{(1 - (ct/r)^2)\sqrt{\sin^2\theta - (ct/r)^2}}. \tag{5.125}$$

[2] In Section 3.2.1 we derived field expressions for frequency-domain point sources that were not necessarily located at the origin. These field expressions (valid for $\mathbf{r}' \neq 0$) were subsequently used in Section 3.2.5 to obtain simple closed-form expressions for the propagating and evanescent frequency-domain far fields of general source distributions. However, simple closed-form expressions for the propagating and evanescent far fields of general source distributions do not exist in the time domain, so in the present section we deal only with time-domain point sources that are located at the origin.

The evanescent field for all points in the half space $z > 0$ begins at $t = -(r/c)\sin\theta \le 0$, so the evanescent field is noncausal because the source does not turn on until $t = 0$. This noncausality arises from dividing the field into propagating and evanescent fields. However, the sum of the propagating and evanescent fields in (5.119) is indeed causal. Since $r\sin\theta = (x^2 + y^2)^{1/2}$, the formulas (5.123) and (5.125) show that the evanescent field is nonzero only for $|t| \le (x^2 + y^2)^{1/2}/c$. That is, for a fixed time $t > 0$, the evanescent field is zero everywhere inside a circular cylinder of radius ct whose axis passes through the source point and is parallel to the z axis. Inside this cylinder the propagating field equals the exact field. For fixed r and θ, the evanescent field (5.125) is shown in Figure 5.4, where the singularities of $(\sin^2\theta - (ct/r)^2)^{-1/2}$ at $t = \pm rc^{-1}\sin\theta$ are clearly visible.

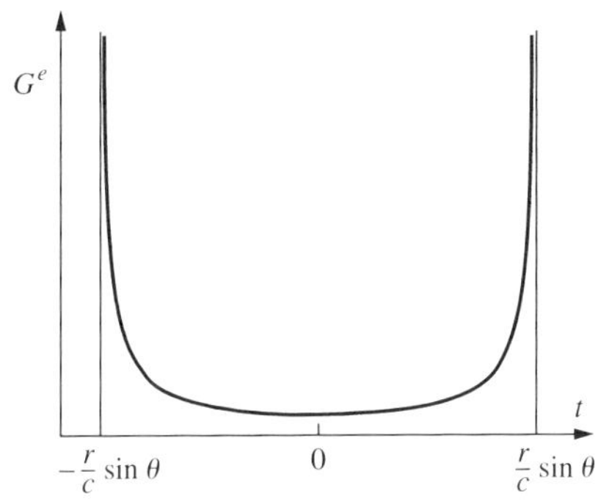

Figure 5.4 The evanescent field for $\delta(t - r/c)/(4\pi r)$.

Observation Points with $\theta = \pi/2$. For observation points in the $z = 0$ plane where the point source resides, we can substitute the frequency-domain formulas (3.101) and (3.102) into (5.118) to get the time-domain propagating and evanescent fields of the point source for $\theta = \pi/2$ $(\rho \ne 0)$

$$G^p(\mathbf{r}, 0, t, 0) = \frac{\delta(t - \rho/c)}{8\pi\rho} - \frac{\delta(t + \rho/c)}{8\pi\rho} \tag{5.126}$$

$$G^e(\mathbf{r}, 0, t, 0) = \frac{\delta(t - \rho/c)}{8\pi\rho} + \frac{\delta(t + \rho/c)}{8\pi\rho} \tag{5.127}$$

in which $\rho = \sqrt{x^2 + y^2}$. In this $\theta = \pi/2$ plane, both the evanescent and propagating fields of the point source decay as $1/\rho$, and both possess incoming and outgoing waves of equal magnitude. The sum of the propagating and evanescent fields equals the outward propagating field, $\delta(t - \rho/c)/(4\pi\rho)$, of the point source.

The Rectangular-Pulse Point Source. To further illustrate the properties of propagating and evanescent fields, we shall consider the acoustic point source whose source function is $q(\mathbf{r}, t) = I(t; 0, T)\delta(\mathbf{r})$ with $T > 0$. This source turns on at $t = 0$ and turns off at $t = T$, and its exact pressure is simply[3]

$$p(\mathbf{r}, t) = \frac{I(t - r/c; 0, T)}{4\pi r}. \tag{5.128}$$

For $\theta = 0$ the corresponding evanescent field is found by convolving the rectangular pulse

[3]Because $\int_{V_\infty}\int_0^\infty q(\mathbf{r}, t)\,dt\,dV = T \ne 0$, where V_∞ denotes all space, it can be shown from (2.208) that $s(\mathbf{r}, t)$ or $\mathbf{f}(\mathbf{r}, t)$ cannot reside in a bounded volume and approach zero as t→ ∞. Thus, the integral of the pressure and velocity over all time need not obey (5.77) and (5.78).

$I(t; 0, T)$ with the expression (5.123) to get

$$p^e(\mathbf{r}, t) = \frac{I(t; 0, T)}{4\pi z}, \quad \theta = 0. \tag{5.129}$$

Similarly, for $\theta = \pi/2$ ($z = z' = 0, \rho \neq 0$), the evanescent pressure field of the rectangular-pulse point source is found from (5.127) as

$$p^e(\mathbf{r}, t) = \frac{I(t - \rho/c; 0, T)}{8\pi\rho} + \frac{I(t + \rho/c; 0, T)}{8\pi\rho}. \tag{5.130}$$

For $0 < \theta < \pi/2$ the evanescent field is found by convolving the rectangular pulse $I(t; 0, T)$ with the expression in (5.125) to get

$$p^e(\mathbf{r}, t) = \frac{c\cos\theta}{4\pi^2 r^2} \int_0^T \frac{I(t - t'; -rc^{-1}\sin\theta, rc^{-1}\sin\theta)}{(1 - (c(t - t')/r)^2)\sqrt{\sin^2\theta - (c(t - t')/r)^2}} dt', \tag{5.131}$$
$$0 < \theta < \pi/2.$$

To derive a closed-form expression for this evanescent field, we define the function

$$\begin{aligned} \Omega(v; \theta) &= \frac{\cos\theta}{\pi} \int_{-1}^{v} \frac{du}{(1 - u^2\sin^2\theta)\sqrt{1 - u^2}} \\ &= \frac{1}{2} + \mathrm{sign}(v)\left(\frac{1}{2} - \frac{1}{\pi}\arctan\left[\frac{\sqrt{1 - v^2}}{|v|\cos\theta}\right]\right), \quad -1 \leq v \leq 1 \end{aligned} \tag{5.132}$$

which is increasing with respect to v and satisfies $\Omega(-1; \theta) = 0$, $\Omega(0; \theta) = \frac{1}{2}$, and $\Omega(1; \theta) = 1$. For $T < 2rc^{-1}\sin\theta$ the evanescent field can now be written as

$$p^e(\mathbf{r}, t) = \begin{cases} 0, & t \leq -\frac{r\sin\theta}{c} \\ \frac{1}{4\pi r}\Omega\left(\frac{ct}{r\sin\theta}; \theta\right), & -\frac{r\sin\theta}{c} < t \leq T - \frac{r\sin\theta}{c} \\ \frac{1}{4\pi r}\Omega\left(\frac{ct}{r\sin\theta}; \theta\right) - \frac{1}{4\pi r}\Omega\left(\frac{c[t-T]}{r\sin\theta}; \theta\right), & T - \frac{r\sin\theta}{c} < t \leq \frac{r\sin\theta}{c} \\ \frac{1}{4\pi r} - \frac{1}{4\pi r}\Omega\left(\frac{c[t-T]}{r\sin\theta}; \theta\right), & \frac{r\sin\theta}{c} < t < T + \frac{r\sin\theta}{c} \\ 0, & t \geq T + \frac{r\sin\theta}{c} \end{cases} \tag{5.133}$$

and for $T > 2rc^{-1}\sin\theta$ the evanescent field is

$$p^e(\mathbf{r}, t) = \begin{cases} 0, & t \leq -\frac{r\sin\theta}{c} \\ \frac{1}{4\pi r}\Omega\left(\frac{ct}{r\sin\theta}; \theta\right), & -\frac{r\sin\theta}{c} < t \leq \frac{r\sin\theta}{c} \\ \frac{1}{4\pi r}, & \frac{r\sin\theta}{c} < t \leq T - \frac{r\sin\theta}{c} \\ \frac{1}{4\pi r} - \frac{1}{4\pi r}\Omega\left(\frac{c[t-T]}{r\sin\theta}; \theta\right), & T - \frac{r\sin\theta}{c} < t < T + \frac{r\sin\theta}{c} \\ 0, & t \geq T + \frac{r\sin\theta}{c}. \end{cases} \tag{5.134}$$

The propagating fields of the rectangular-pulse point source are found by subtracting the evanescent fields in (5.129)–(5.130) and (5.133)–(5.134) from the total acoustic field in (5.128). In Figure 5.5 the propagating field is shown for $r = 5cT$ and four different values of θ. At these observation points, the total point-source pressure equals $1/(20\pi cT)$ for $5T < t < 6T$ and is zero otherwise. When $\theta = 0$, the evanescent field is simply the rectangular pulse (5.129) that starts at $t = 0$ and ends at $t = T$. For $0 < \theta < 5.7^0$ the evanescent field is given by (5.134),

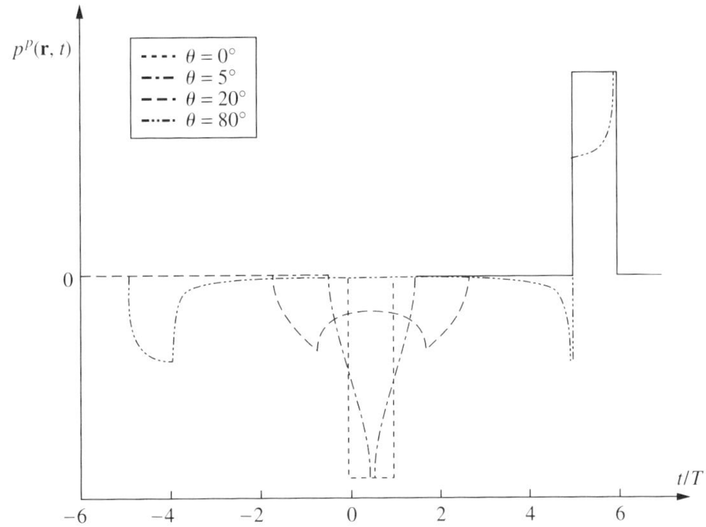

Figure 5.5 The propagating field for the rectangular-pulse point source at observation points where $r = 5cT$.

and for $5.7^0 \leq \theta < 90^0$ the evanescent field is given by (5.133). The formulas (5.133) and (5.134) show that for $0 \leq \theta < 53.1^0$ the evanescent field is zero in the interval $5T \leq t \leq 6T$ where the total field is nonzero. Thus, for these values of θ, by time gating one can recover the total field from the propagating field. When $53.1^0 \leq \theta < 90^0$, the evanescent field is no longer zero throughout the interval $5T \leq t \leq 6T$, and the total field can no longer be recovered from the propagating field. This phenomenon is illustrated by the curve with $\theta = 80^0$ in Figure 5.5.

Asymptotic formulas, valid for large r, for the maximum value of the evanescent field can be derived from (5.133). We find that the maximum value of the evanescent field (5.133) is attained at $t = T - rc^{-1}\sin\theta$ and at $t = rc^{-1}\sin\theta$. Since $\Omega(\delta - 1; \theta) = 1 - \Omega(1 - \delta; \theta) \sim \sqrt{2\delta}/(\pi\cos\theta)$ as $\delta \to 0$, the maximum value of the evanescent field for large r is given by

$$\max_t p^e(\mathbf{r}, t) \sim \frac{1}{2\pi^2\cos\theta}\sqrt{\frac{cT}{2\sin\theta}}\frac{1}{r^{3/2}}, \quad 0 < \theta < \pi/2, \quad r \to \infty. \tag{5.135}$$

For $\theta = 0$ and $\pi/2$ the maximum value of the evanescent field is seen from (5.129) and (5.130) to be simply $1/(4\pi z)$ and $1/(8\pi\rho)$, respectively. This analysis shows that the evanescent field of the rectangular-pulse point-source far field is of order $r^{-3/2}$ for $0 < \theta < \pi/2$, and the evanescent field is therefore negligible compared to the propagating field, which is of order r^{-1}. However, for $\theta = 0$ and $\pi/2$, the evanescent field is not negligible because it is of the same order (r^{-1}) as the propagating field. Moreover, for this rectangular-pulse point-source field, the far-field pattern (defined in (5.43)) for the evanescent field is nonzero only for $\theta = 0$ and $\pi/2$.

5.2.2 Spectrum Given in Terms of the Sources

In this section, the time-domain spectra $T(\xi, \eta, t)$, $\mathbf{T}(\xi, \eta, t)$, and $\mathbf{T}^H(\xi, \eta, t)$ will be expressed in terms of both the volume sources and the Huygens' sources on the surface S shown in Figure 3.1. The results will be obtained by Fourier transforming the frequency-domain formulas of Section 3.2.2. Some of these frequency-domain formulas, which were derived for $\omega > 0$, do not hold for $\omega < 0$, and we obtain formulas that are valid for all real ω by using (5.80), as was done in the derivation of the time-domain plane-wave spectrum formulas. (Analytic-signal analogs of the volume-source expressions (5.138), (5.141), and (5.144) for scalar fields have been obtained previously by Heyman [17]. See also the discussion in Section 5.3.1.)

Let us begin with the acoustic field and recall the relation (3.104) showing that the frequency-domain spectrum $T_\omega(k_x, k_y)$ can be expressed in terms of the frequency-domain volume-source function $q_\omega(\mathbf{r})$ as

$$T_\omega(k_x, k_y) = \frac{i}{4\pi\gamma} \int_V q_\omega(\mathbf{r}) e^{-i(k_x x + k_y y + \gamma z)} dV. \tag{5.136}$$

With ζ given by (3.83), we find for propagating plane waves that

$$-i\omega T_\omega(\omega\xi, \omega\eta) = \frac{1}{4\pi\zeta} \int_V q_\omega(\mathbf{r}) e^{-i\omega(\xi x + \eta y + \zeta z)} dV, \quad \xi^2 + \eta^2 < c^{-2} \tag{5.137}$$

satisfies (5.80) and therefore holds for all ω. Taking the Fourier transform of (5.137) gives the following relation between the time-domain spectrum and the time-domain source function

$$\frac{\partial}{\partial t} T(\xi, \eta, t) = \frac{1}{4\pi\zeta} \int_V q(\mathbf{r}, t + \xi x + \eta y + \zeta z) dV, \quad \xi^2 + \eta^2 < c^{-2}. \tag{5.138}$$

For evanescent plane waves we find that

$$\begin{aligned} -i\omega T_\omega(\omega\xi, \omega\eta) = {} & \frac{1}{4\pi|\zeta|} \int_V q_\omega(\mathbf{r}) e^{-i\omega(\xi x + \eta y)} \\ & \cdot \left(-i\frac{\omega}{|\omega|} e^{|\omega||\zeta|z} \right) dV, \quad \xi^2 + \eta^2 > c^{-2} \end{aligned} \tag{5.139}$$

satisfies (5.80) and therefore holds for all ω. Note that the sources are located in the region $z < 0$ so that the value of z is negative throughout the integration in (5.139). Since

$$\int_{-\infty}^{+\infty} \left(-i\frac{\omega}{|\omega|} e^{|\omega||\zeta|z} \right) e^{-i\omega t} d\omega = -\frac{2t}{|\zeta|^2 z^2 + t^2}, \quad z < 0 \tag{5.140}$$

equation (5.139) and the convolution theorem (5.89) show that the time-domain spectrum for evanescent plane waves is given in terms of the volume-source function by

$$\frac{\partial}{\partial t} T(\xi, \eta, t) = -\frac{1}{4\pi^2|\zeta|} \int_V \int_{-\infty}^{+\infty} \frac{(t - t') q(\mathbf{r}, t' + \xi x + \eta y)}{|\zeta|^2 z^2 + (t - t')^2} dt' dV, \quad \xi^2 + \eta^2 > c^{-2}. \tag{5.141}$$

Alternatively, we can move the factor $-i\omega/|\omega|$ in (5.139) in front of q_ω and then use the identities

$$\int_{-\infty}^{+\infty} e^{|\omega||\zeta|z} e^{-i\omega t} d\omega = -\frac{2|\zeta|z}{|\zeta|^2 z^2 + t^2}, \quad z < 0 \tag{5.142}$$

$$\int_{-\infty}^{+\infty} -i\frac{\omega}{|\omega|} q_\omega(\mathbf{r}) e^{-i\omega t} d\omega = -\mathcal{H} q(\mathbf{r}, t) \tag{5.143}$$

to get the following formula

$$\frac{\partial}{\partial t}T(\xi,\eta,t)=\frac{1}{4\pi^2}\int_V\int_{-\infty}^{+\infty}\frac{z\widetilde{\mathcal{H}}q(\mathbf{r},t')}{|\zeta|^2z^2+(t-t'+\xi x+\eta y)^2}dt'dV,\quad \xi^2+\eta^2>c^{-2} \tag{5.144}$$

which expresses the spectrum in terms of the Hilbert transform of the volume-source function.

We can also write the spectrum in terms of the Huygens' sources on the smooth surface S enclosing the source region. The frequency-domain formula (3.105) shows that for propagating plane waves

$$T_\omega(\xi\omega,\eta\omega)=\frac{1}{4\pi\zeta}\int_S\left[\hat{\mathbf{n}}\cdot\boldsymbol{\chi}_\omega(\xi,\eta)p_\omega(\mathbf{r})+\rho_{m0}\,\hat{\mathbf{n}}\cdot\mathbf{v}_\omega(\mathbf{r})\right]e^{-i\omega(\xi x+\eta y+\zeta z)}dS,\quad \xi^2+\eta^2<c^{-2} \tag{5.145}$$

where $\hat{\mathbf{n}}$ is the outward normal to S, and $\boldsymbol{\chi}_\omega$ is given in (5.91). Since this expression satisfies (5.80), it is valid for all real ω, and we can immediately take the Fourier transform to get

$$\begin{aligned}T(\xi,\eta,t)=\frac{1}{4\pi\zeta}\int_S\Big[&\hat{\mathbf{n}}\cdot\widetilde{\boldsymbol{\chi}}(\xi,\eta)p(\mathbf{r},t+\xi x+\eta y+\zeta z)\\&+\rho_{m0}\hat{\mathbf{n}}\cdot\mathbf{v}(\mathbf{r},t+\xi x+\eta y+\zeta z)\Big]dS,\quad \xi^2+\eta^2<c^{-2}.\end{aligned} \tag{5.146}$$

Similarly, for the evanescent plane waves, with the help of (5.80), we find that for all real ω

$$\begin{aligned}T_\omega(\xi\omega,\eta\omega)=\frac{1}{4\pi|\zeta|}\int_S&\left[\hat{\mathbf{n}}\cdot\boldsymbol{\chi}_\omega(\xi,\eta)p_\omega(\mathbf{r})+\rho_{m0}\,\hat{\mathbf{n}}\cdot\mathbf{v}_\omega(\mathbf{r})\right]\\&\cdot\left[-i\frac{\omega}{|\omega|}e^{-i\omega(\xi x+\eta y)}e^{|\omega||\zeta|z}\right]dS,\quad \xi^2+\eta^2>c^{-2}\end{aligned} \tag{5.147}$$

so that

$$\begin{aligned}T(\xi,\eta,t)=-\frac{1}{4\pi^2|\zeta|}\int_S\int_{-\infty}^{+\infty}&\Big[\hat{\mathbf{n}}\cdot\widetilde{\boldsymbol{\chi}}(\xi,\eta)p(\mathbf{r},t'+\xi x+\eta y)\\&+\rho_{m0}\hat{\mathbf{n}}\cdot\mathbf{v}(\mathbf{r},t'+\xi x+\eta y)\Big]\\&\cdot\frac{t-t'}{|\zeta|^2z^2+(t-t')^2}dt'dS,\quad \xi^2+\eta^2>c^{-2}\end{aligned} \tag{5.148}$$

where the convolution theorem (5.89) and the formula (5.140) have been used. The time-domain spectrum for the acoustic field has thus been expressed in terms of the Huygens' sources for both propagating and evanescent plane waves.

The analogous formulas for the time-domain plane-wave spectra of the electric and magnetic fields in terms of volume current density can be found in a similar way. The formula (3.108) shows that

$$\mathbf{T}^H_\omega(\xi\omega,\eta\omega)=-\frac{1}{4\pi\zeta}\boldsymbol{\chi}_\omega(\xi,\eta)\times\int_V\mathbf{J}_\omega(\mathbf{r})e^{-i\omega(\xi x+\eta y+\zeta z)}dV,\quad \xi^2+\eta^2<c^{-2} \tag{5.149}$$

so that

$$\mathbf{T}^H(\xi,\eta,t) = -\frac{1}{4\pi\zeta}\widetilde{\boldsymbol{\chi}}(\xi,\eta)\times\int_V \mathbf{J}(\mathbf{r},t+\xi x+\eta y+\zeta z)dV,\quad \xi^2+\eta^2<c^{-2}. \tag{5.150}$$

For the evanescent plane waves we find that

$$\mathbf{T}^H_\omega(\xi\omega,\eta\omega) = \frac{1}{4\pi|\zeta|}\boldsymbol{\chi}_\omega(\xi,\eta)\times\int_V \mathbf{J}_\omega(\mathbf{r})e^{-i\omega(\xi x+\eta y)}\left(i\frac{\omega}{|\omega|}e^{|\omega||\zeta|z}\right)dV, \quad \xi^2+\eta^2>c^{-2} \tag{5.151}$$

satisfies (5.80). Thus, the convolution theorem (5.89) and the formula (5.140) transforms (5.151) to

$$\mathbf{T}^H(\xi,\eta,t) = \frac{1}{4\pi^2|\zeta|}\widetilde{\boldsymbol{\chi}}(\xi,\eta)\times\int_V\int_{-\infty}^{+\infty}\frac{(t-t')\mathbf{J}(\mathbf{r},t'+\xi x+\eta y)}{|\zeta|^2z^2+(t-t')^2}dt'dV, \quad \xi^2+\eta^2>c^{-2}. \tag{5.152}$$

The analogous formulas for the spectrum of the electric field are determined from (5.150), (5.152), and the equation

$$\mathbf{T}(\xi,\eta,t) = -\frac{1}{\epsilon}\widetilde{\boldsymbol{\chi}}(\xi,\eta)\times\mathbf{T}^H(\xi,\eta,t) \tag{5.153}$$

which is valid for all (ξ,η) and results from (5.115) or the inverse Fourier transform of the frequency-domain relationship $\mathbf{T}_\omega = -\boldsymbol{\chi}_\omega\times\mathbf{T}^H_\omega/\epsilon$. Similarly, from (3.110) we find that the spectrum for the magnetic field can be expressed in terms of the Huygens' sources as

$$\begin{aligned}\mathbf{T}^H(\xi,\eta,t) = &-\frac{1}{4\pi\zeta}\widetilde{\boldsymbol{\chi}}(\xi,\eta)\times\int_S \hat{\mathbf{n}}\times\mathbf{H}(\mathbf{r},t+\xi x+\eta y+\zeta z)dS\\ &-\frac{1}{4\pi\mu\zeta}\widetilde{\boldsymbol{\chi}}(\xi,\eta)\times\left(\widetilde{\boldsymbol{\chi}}(\xi,\eta)\times\int_S \hat{\mathbf{n}}\times\mathbf{E}(\mathbf{r},t+\xi x+\eta y+\zeta z)dS\right), \quad \xi^2+\eta^2<c^{-2}\end{aligned} \tag{5.154}$$

and

$$\begin{aligned}\mathbf{T}^H(\xi,\eta,t) = &\frac{1}{4\pi^2|\zeta|}\widetilde{\boldsymbol{\chi}}(\xi,\eta)\times\int_S\int_{-\infty}^{+\infty}\hat{\mathbf{n}}\times\frac{(t-t')\mathbf{H}(\mathbf{r},t'+\xi x+\eta y)}{|\zeta|^2z^2+(t-t')^2}dt'dS\\ &+\frac{1}{4\pi^2\mu|\zeta|}\widetilde{\boldsymbol{\chi}}(\xi,\eta)\times\left(\widetilde{\boldsymbol{\chi}}(\xi,\eta)\times\int_S\int_{-\infty}^{+\infty}\hat{\mathbf{n}}\times\frac{(t-t')\mathbf{E}(\mathbf{r},t'+\xi x+\eta y)}{|\zeta|^2z^2+(t-t')^2}dt'dS\right), \quad \xi^2+\eta^2>c^{-2}.\end{aligned} \tag{5.155}$$

The formulas for the electric-field spectrum $\mathbf{T}(\xi,\eta,t)$ are found from (5.153). The results obtained in this section will prove valuable for the determination in the following section of the singularities of the spectra.

When the sources are located in $z > 0$, the fields in $z < 0$ can be determined from the spectra given by (5.138), (5.141), (5.144), (5.146), (5.148), (5.150), (5.152), (5.153), (5.154), and (5.155) with z replaced by $-z$ and $\widetilde{\chi}$ replaced by $\widetilde{\chi}^- = \widetilde{\chi} - 2(\hat{\mathbf{z}} \cdot \widetilde{\chi})\hat{\mathbf{z}}$.

5.2.3 Singularities of the Spectrum

The possible singularities (with respect to ξ and η) of the first time derivatives of the spectra will now be determined. We start by defining the acoustic time-domain far-field function as

$$F(\xi, \eta, t) = \int_{-\infty}^{+\infty} F_\omega(\xi\omega, \eta\omega) e^{-i\omega t} d\omega \tag{5.156}$$

where F_ω is the frequency-domain far-field function, which is given in terms of the frequency-domain plane-wave spectrum by (3.112) for all ω. Since $T_\omega(\xi\omega, \eta\omega)$ and $i\gamma(\omega)$ satisfy the reality condition (5.80), equation (3.112) shows that $F_\omega(\xi\omega, \eta\omega)$ satisfies (5.80). Expressing $\gamma(\omega)$ as $\omega\zeta$ for $\omega > 0$ and $\omega\zeta^*$ for $\omega < 0$ allows (3.112) to be rewritten for all ω as

$$-i\omega T_\omega(\xi\omega, \eta\omega) = \frac{1}{|\zeta|} \begin{cases} F_\omega(\xi\omega, \eta\omega), & \xi^2 + \eta^2 < c^{-2} \\ -i\dfrac{\omega}{|\omega|} F_\omega(\xi\omega, \eta\omega), & \xi^2 + \eta^2 > c^{-2} \end{cases} \tag{5.157}$$

where ζ is defined by (5.85). Taking the Fourier transform of (5.157) and using the Hilbert transform relation (5.94) shows that the time derivative of the spectrum can be expressed in terms of the far-field function as

$$\frac{\partial}{\partial t} T(\xi, \eta, t) = \frac{1}{|\zeta|} \begin{cases} F(\xi, \eta, t), & \xi^2 + \eta^2 < c^{-2} \\ -\widetilde{\mathcal{H}} F(\xi, \eta, t), & \xi^2 + \eta^2 > c^{-2} \end{cases} \tag{5.158}$$

where $\widetilde{\mathcal{H}}$ is the Hilbert transform (5.93). With the help of the formula $\widetilde{\mathcal{H}}^2 = -1$, this relation can be used to express the time-domain far-field function in terms of the time derivative of the spectrum and its Hilbert transform. Next we shall derive source expressions for the time-domain far-field function using the frequency-domain source expressions of Section 3.2.3.

The frequency-domain expression (3.113) for the far-field function in terms of the volume sources is valid for $\omega > 0$. It can be used along with the relation (5.80) to obtain the following frequency-domain formulas for the far-field function that are valid for all real ω

$$F_\omega(\xi\omega, \eta\omega) = \frac{1}{4\pi} \int_V q_\omega(\mathbf{r}) e^{-i\omega(\xi x + \eta y + \zeta z)} dV, \quad \xi^2 + \eta^2 < c^{-2} \tag{5.159}$$

and

$$F_\omega(\xi\omega, \eta\omega) = \frac{1}{4\pi} \int_V q_\omega(\mathbf{r}) e^{-i\omega(\xi x + \eta y + i\frac{\omega}{|\omega|}|\zeta| z)} dV, \quad \xi^2 + \eta^2 > c^{-2}. \tag{5.160}$$

(These equations also follow from (3.113) made valid for all ω by inserting $\gamma(\omega) = \omega\zeta$ for $\omega > 0$ and $\gamma(\omega) = \omega\zeta^*$ for $\omega < 0$.) Taking the Fourier transform of these equations gives us

$$F(\xi, \eta, t) = \frac{1}{4\pi} \int_V q(\mathbf{r}, t + \xi x + \eta y + \zeta z) dV, \quad \xi^2 + \eta^2 < c^{-2} \tag{5.161}$$

and

$$F(\xi,\eta,t) = -\frac{1}{4\pi^2}\int\limits_V\int\limits_{-\infty}^{+\infty}\frac{|\zeta|z\,q(\mathbf{r},t'+\xi x+\eta y)}{|\zeta|^2z^2+(t-t')^2}dt'dV,\quad \xi^2+\eta^2>c^{-2} \tag{5.162}$$

where the convolution theorem (5.89) and the formula (5.142) have been used.

The expressions (5.161) and (5.162) show that when the source function $q(\mathbf{r},t)$ is continuous (see Footnote 4, Chapter 2), the far-field function is continuous in each of the regions $\xi^2+\eta^2<c^{-2}$ and $\xi^2+\eta^2>c^{-2}$. Moreover, the relation (5.99) with z replaced by $-z$ inserted into (5.162) shows that the time-domain far-field function is continuous across the circle $\xi^2+\eta^2=c^{-2}$. Thus, *for continuous source functions, the time-domain far-field function is a continuous function for all real (ξ,η,t) (including $\xi^2+\eta^2=c^{-2}$).* The expression (5.158) also shows that *when the source function $q(\mathbf{r},t)$ is continuous, the time derivative of the spectrum $T(\xi,\eta,t)$ is a continuous function of real (ξ,η,t), for $\xi^2+\eta^2<c^{-2}$, and at $\zeta=0$ it may have a $1/\zeta$ singularity.*

We can use (3.114) (valid for $\omega>0$) along with (5.80) (determining the frequency-domain field for $\omega<0$) to derive formulas that give the frequency-domain far-field pattern in terms of Huygens' sources for all real ω. Taking the inverse Fourier transform of these formulas then gives

$$\begin{aligned}F(\xi,\eta,t) = \frac{1}{4\pi}\int\limits_S\Big[&\hat{\mathbf{n}}\cdot\widetilde{\boldsymbol{\chi}}(\xi,\eta)\frac{\partial}{\partial t}p(\mathbf{r},t+\xi x+\eta y+\zeta z)\\ &+\rho_{m0}\hat{\mathbf{n}}\cdot\frac{\partial}{\partial t}\mathbf{v}(\mathbf{r},t+\xi x+\eta y+\zeta z)\Big]dS,\quad \xi^2+\eta^2<c^{-2}\end{aligned} \tag{5.163}$$

and

$$\begin{aligned}F(\xi,\eta,t) = -\frac{1}{4\pi^2}\int\limits_S\int\limits_{-\infty}^{+\infty}\Big[&\hat{\mathbf{n}}\cdot\widetilde{\boldsymbol{\chi}}(\xi,\eta)\frac{\partial}{\partial t'}p(\mathbf{r},t'+\xi x+\eta y)\\ &+\rho_{m0}\hat{\mathbf{n}}\cdot\frac{\partial}{\partial t'}\mathbf{v}(\mathbf{r},t'+\xi x+\eta y)\Big]\frac{z|\zeta|}{|\zeta|^2z^2+(t-t')^2}dt'dS,\quad \xi^2+\eta^2>c^{-2}.\end{aligned} \tag{5.164}$$

Thus the time-domain far-field function is expressed in terms of the time-domain Huygens' sources $\partial p/\partial t$ and $\rho_{m0}(\partial/\partial t)\hat{\mathbf{n}}\cdot\mathbf{v}=-\partial p/\partial n$ on the smooth surface S, which is located in the region $z<0$. Use of the relation (5.99) in (5.164) proves that *when the time-domain Huygens' sources are continuous on S, the far-field function is continuous for all real (ξ,η,t).* Equation (5.158) then proves that *when the Huygens' sources $\partial p/\partial t$ and $\rho_{m0}(\partial/\partial t)\hat{\mathbf{n}}\cdot\mathbf{v}$ are continuous on S, the time derivative of the spectrum $T(\xi,\eta,t)$ is a continuous function of real (ξ,η,t) for $\xi^2+\eta^2<c^{-2}$ and at $\zeta=0$ it may have a $1/\zeta$ singularity.* This result with Huygens' sources also holds in cases where the source function $q(\mathbf{r},t)$ is discontinuous or has singularities with respect to $\mathbf{r}$ and is therefore more general.

The analogous results for the electric and magnetic fields are derived in a similar straightforward manner. One can use (5.153) or (3.116) to show that the far-field function for the electric field can be determined from the far-field function for the magnetic field by the formula

$$\mathbf{F}(\xi,\eta,t) = -\frac{1}{\epsilon}\widetilde{\boldsymbol{\chi}}(\xi,\eta)\times\mathbf{F}^H(\xi,\eta,t) \tag{5.165}$$

or equivalently

$$\mathbf{F}^H(\xi,\eta,t) = \frac{1}{\mu}\widetilde{\boldsymbol{\chi}}(\xi,\eta) \times \mathbf{F}(\xi,\eta,t). \tag{5.166}$$

In the following we shall present only the expressions for the magnetic field.

The far-field function for the magnetic field is related to the spectrum for the magnetic field through

$$\frac{\partial}{\partial t}\mathbf{T}^H(\xi,\eta,t) = \frac{1}{|\zeta|}\begin{cases} \mathbf{F}^H(\xi,\eta,t), & \xi^2+\eta^2 < c^{-2} \\ -\widetilde{\mathcal{H}}\mathbf{F}^H(\xi,\eta,t), & \xi^2+\eta^2 > c^{-2} \end{cases} \tag{5.167}$$

where the magnetic far-field function is found from (3.117) to be

$$\mathbf{F}^H(\xi,\eta) = -\frac{1}{4\pi}\widetilde{\boldsymbol{\chi}}(\xi,\eta) \times \int_V \frac{\partial}{\partial t}\mathbf{J}(\mathbf{r}, t+\xi x+\eta y+\zeta z)dV, \quad \xi^2+\eta^2 < c^{-2} \tag{5.168}$$

$$\mathbf{F}^H(\xi,\eta) = \frac{1}{4\pi^2}\widetilde{\boldsymbol{\chi}}(\xi,\eta) \times \int_V \int_{-\infty}^{+\infty} \frac{|\zeta| z(\partial/\partial t')\mathbf{J}(\mathbf{r}, t'+\xi x+\eta y)}{|\zeta|^2 z^2 + (t-t')^2} dt' dV, \quad \xi^2+\eta^2 > c^{-2}. \tag{5.169}$$

From the expressions (5.168)–(5.169) we can prove the following results. *When the time derivative of the source current* $\mathbf{J}(\mathbf{r},t)$ *is continuous, the time-domain far-field functions for both the electric and magnetic fields are continuous functions for all real* (ξ,η,t) *(including* $\xi^2+\eta^2 = c^{-2}$*).* Moreover, (5.167) then shows that *when the time derivative of the source current* $\mathbf{J}(\mathbf{r},t)$ *is continuous, the first time derivatives of the electric and magnetic spectra* $\mathbf{T}(\xi,\eta,t)$ *and* $\mathbf{T}^H(\xi,\eta,t)$ *are continuous functions of real* (ξ,η,t) *for* $\xi^2+\eta^2 < c^{-2}$*, and at* $\zeta = 0$ *they may have* $1/\zeta$ *singularities.*

From the formula (3.118) it can be proven that the far-field functions can be expressed in terms of the time-domain Huygens' sources $\hat{\mathbf{n}} \times \mathbf{E}(\mathbf{r},t)$ and $\hat{\mathbf{n}} \times \mathbf{H}(\mathbf{r},t)$ as

$$\begin{aligned}\mathbf{F}^H(\xi,\eta,t) = &-\frac{1}{4\pi}\widetilde{\boldsymbol{\chi}}(\xi,\eta) \times \int_S \hat{\mathbf{n}} \times \frac{\partial}{\partial t}\mathbf{H}(\mathbf{r}, t+\xi x+\eta y+\zeta z)dS \\ &-\frac{1}{4\pi\mu}\widetilde{\boldsymbol{\chi}}(\xi,\eta) \times \left(\widetilde{\boldsymbol{\chi}}(\xi,\eta) \times \int_S \hat{\mathbf{n}} \times \frac{\partial}{\partial t}\mathbf{E}(\mathbf{r}, t+\xi x+\eta y+\zeta z)dS\right), \\ &\xi^2+\eta^2 < c^{-2}\end{aligned} \tag{5.170}$$

and

$$\begin{aligned}\mathbf{F}^H(\xi,\eta,t) = &\frac{1}{4\pi^2}\widetilde{\boldsymbol{\chi}}(\xi,\eta) \times \int_S \int_{-\infty}^{+\infty} \hat{\mathbf{n}} \times \frac{|\zeta| z(\partial/\partial t')\mathbf{H}(\mathbf{r}, t'+\xi x+\eta y)}{|\zeta|^2 z^2 + (t-t')^2} dt' dS \\ &+\frac{1}{4\pi^2\mu}\widetilde{\boldsymbol{\chi}}(\xi,\eta) \times \left(\widetilde{\boldsymbol{\chi}}(\xi,\eta) \times \int_S \int_{-\infty}^{+\infty} \hat{\mathbf{n}} \times \frac{|\zeta| z(\partial/\partial t')\mathbf{E}(\mathbf{r}, t'+\xi x+\eta y)}{|\zeta|^2 z^2 + (t-t')^2} dt' dS\right), \\ &\xi^2+\eta^2 > c^{-2}.\end{aligned} \tag{5.171}$$

Equations (5.170)–(5.171) can be used to show that *when the first time derivatives of the Huygens' sources* $\hat{\mathbf{n}} \times \mathbf{E}$ *and* $\hat{\mathbf{n}} \times \mathbf{H}$ *are continuous on S, the time-domain far-field functions for both the electric and magnetic fields are continuous functions for all real* (ξ, η, t) *(including* $\xi^2 + \eta^2 = c^{-2}$*).* In addition, (5.167) then shows that *when the first time derivatives of the Huygens' sources* $\hat{\mathbf{n}} \times \mathbf{E}$ *and* $\hat{\mathbf{n}} \times \mathbf{H}$ *are continuous on S, the first time derivatives of the electric and magnetic spectra* $\mathbf{T}(\xi, \eta, t)$ *and* $\mathbf{T}^H(\xi, \eta, t)$ *are continuous functions of* (ξ, η, t) *for* $\xi^2 + \eta^2 < c^{-2}$*, and at* $\zeta = 0$ *they may have* $1/\zeta$ *singularities.*

When the sources are located in $z > 0$, the spectra are still related to the far-field functions through (5.158) and (5.167), and the far-field functions are given by (5.161), (5.162), (5.163), (5.164), (5.165), (5.168), (5.169), (5.170), and (5.171), with z replaced by $-z$ and χ replaced by $\widetilde{\chi}^- = \widetilde{\chi} - 2(\hat{\mathbf{z}} \cdot \widetilde{\chi})\hat{\mathbf{z}}$.

5.2.4 Analyticity of the Far-Field Pattern and Its Relation to the Far-Field Function

In Section 3.2.4 it was shown that the frequency-domain far-field pattern $\mathcal{F}_\omega(\theta, \phi)$ is an entire function of complex (θ, ϕ) and that the frequency-domain far-field function $F_\omega(k_x, k_y)$ is an analytic function in a neighborhood of real k_x and k_y, except possibly at $k_x^2 + k_y^2 = k^2$, corresponding to $\gamma = 0$. Moreover, it was shown that $\mathcal{F}_\omega(\theta, \phi) = F_\omega(k \cos\phi \sin\theta, k \sin\phi \sin\theta)$ for $0 \le \phi \le 2\pi$ and $\theta \in \Gamma$, where Γ is the contour in Figure 3.4. These results hold for $\omega > 0$ and were used in Section 3.2.6 to express the near field in terms of the far-field pattern and its analytic continuation. In the present section, we shall investigate the possibility of analytically continuing the time-domain far-field pattern to complex angles of observation, and we shall determine how the analytically continued far-field pattern is related to the time-domain far-field function. It is assumed that the sources satisfy the conditions stated in Section 5.1.4, so that the far fields decay as $1/r$ (or faster) as $r \to \infty$. In Section 5.4 on electromagnetic missiles (and in the last sentence of this section), we consider far fields that decay slower than $1/r$. The analogous results for the analytic-signal far-field pattern are presented in Section 5.3.2 and are significantly different from the results in this section.

Properties of the Far-Field Pattern. Let us begin by examining the far-field pattern. From (3.35) we see that the frequency-domain far-field pattern can be expressed in terms of the source function q_ω as

$$\mathcal{F}_\omega(\theta, \phi) = \frac{1}{4\pi} \int_V q_\omega(\mathbf{r}) e^{-i(\omega/c)(x \cos\phi \sin\theta + y \sin\phi \sin\theta + z \cos\theta)} dV \tag{5.172}$$

which, for real (θ, ϕ), satisfies (5.80) and is valid for all real ω. Moreover, (5.172) is an entire function of (θ, ϕ) for all real ω. We therefore define the frequency-domain far-field pattern by (5.172) for all complex (θ, ϕ) and all real ω, even though this formula does not, in general, satisfy the reality condition (5.80) when (θ, ϕ) is complex. As a consequence, the corresponding time-domain far-field pattern is in general no longer real for complex values of (θ, ϕ).

Although physically observable time-domain fields must be real, the far-field pattern at complex angles of observation is not a physically observable field and can therefore be complex. If a closed-form expression is known for the time-domain far-field pattern of a given source function for real angles of observation, when that expression is analytically continuable to complex angles of observation, it will, in general, be complex. Moreover, it will equal the time-domain far-field pattern in (5.173) below, as illustrated in the example below where the source function is given in (5.180).

A closed-form expression for the time-domain far-field pattern of a given source function can, in principle, be obtained by Fourier transforming (5.172) to get

$$\mathcal{F}(\theta, \phi, t) = \frac{1}{4\pi} \int_{-\infty}^{+\infty} \int_V q_\omega(\mathbf{r}) e^{-i(\omega/c)(x\cos\phi\sin\theta + y\sin\phi\sin\theta + z\cos\theta)} e^{-i\omega t} dV\, d\omega. \tag{5.173}$$

To find out if a closed-form expression for the time-domain far-field pattern can be analytically continued to complex θ, we insert a complex value of θ into (5.173), so that $\sin\theta$ and $\cos\theta$ are complex. The factor $e^{-i(\omega/c)(x\cos\phi\sin\theta+y\sin\phi\sin\theta+z\cos\theta)}$ grows exponentially as $|\omega| \to \infty$ provided $(x, y, z) \neq (0, 0, 0)$, and the Fourier integral in (5.173) diverges unless the source function q_ω decays exponentially as $|\omega| \to \infty$. For example, consider the rectangular-pulse point-source function $q(\mathbf{r}, t) = \delta(\mathbf{r} + a\hat{\mathbf{z}}) I(t; 0, T)$, where $a > 0$, and $I(t; 0, T) = 1$ for $0 \le t \le T$ and zero otherwise. The acoustic far-field pattern is then given by $\mathcal{F}(\theta, \phi, t) = I(t - (a/c)\cos\theta; 0, T)/(4\pi)$, which is clearly a discontinuous function of θ that cannot be analytically continued to complex values of θ. In this case, $q_\omega = O(\omega^{-1})$, and the integrand in (5.173) behaves as $\omega^{-1}\exp[-\omega(a/c)\mathrm{Im}(\cos\theta)]$ for $|\omega| \to \infty$. Thus, the Fourier integral diverges for complex values of $\cos\theta$, and the far-field pattern cannot be analytically continued.

However, if the following two conditions hold for the frequency-domain source function $q_\omega(\mathbf{r})$, the time-domain far-field pattern $\mathcal{F}(\theta, \phi, t)$ is an analytic function of (θ, ϕ) in a complex neighborhood of the real θ and ϕ axes.

1. The function $q_\omega(\mathbf{r})$ is integrable over the bounded source volume V.
2. There exists a $T > 0$, such that $q_\omega(\mathbf{r}) = O(e^{-|\omega|T})$ for all $\mathbf{r}$ as $|\omega| \to \infty$.

This theorem follows directly from the results of Whittaker and Watson [88, sec. 5.32] and shows that for source functions q_ω that decay exponentially or faster with respect to ω, the time-domain far-field pattern must be an analytic function around the real (θ, ϕ) axes.

Similarly, the Huygens' source expression (3.38) for $\mathcal{F}_\omega(\theta, \phi)$ can be used to show that the time-domain far-field pattern is an analytic function of complex (θ, ϕ) in a complex neighborhood of the real θ and ϕ axes, provided the following condition is satisfied. On the surface S, there exists a $T > 0$, such that for all $\mathbf{r} \in S$, $p_\omega(\mathbf{r}) = O(e^{-|\omega|T})$ as $|\omega| \to \infty$. (The frequency-domain Huygens' sources are always continuous on S, because they obey the homogeneous acoustic differential equations across the surface S, and are therefore automatically integrable on S.) Unlike the theorem based on the volume-source expressions, this theorem does not require any explicit condition on the volume-source function q_ω.

Although the time-domain far-field pattern cannot, in general, be analytically continued to complex angles of observation, we can still prove an important theorem for its angular dependence. This theorem holds under the extremely weak condition that the time-domain and frequency-domain far-field patterns, as well as the time-domain and frequency-domain fields outside the source region, are related by Fourier transform pairs. Thus, we have

$$\mathcal{F}(\theta, \phi, t) = \int_{-\infty}^{+\infty} \mathcal{F}_\omega(\theta, \phi) e^{-i\omega t} d\omega \tag{5.174}$$

and

$$\mathcal{F}_\omega(\theta, \phi) = \frac{1}{2\pi} \int_{-\infty}^{+\infty} \mathcal{F}(\theta, \phi, t) e^{i\omega t} dt. \tag{5.175}$$

These relations hold for all physically realizable fields, and even for hypothetical fields such as the rectangular-pulse point source considered above, where the time-domain far-field pattern cannot be analytically continued to complex angles of observation. If the time-domain far-field pattern is zero for all times and for $(\theta, \phi) \in B$, where B has a nonzero area, then (5.175) proves the frequency-domain far-field pattern is zero for all real ω for $(\theta, \phi) \in B$. (For example, the region B can have the form $\theta_1 < \theta < \theta_2$ and $\phi_1 < \phi < \phi_2$.) However, as explained in Section 3.2.4, the frequency-domain far-field pattern is always an entire function of complex (θ, ϕ) for all real ω. Consequently, $\mathcal{F}_\omega(\theta, \phi) \equiv 0$ for all frequencies and all angles of observation. This, in turn, implies from (5.174) that the time-domain far-field pattern is zero for all angles of observation at all times, and we have the following theorem. *If the time-domain acoustic far-field pattern of a finite source region is zero at all times for $(\theta, \phi) \in B$, where B has a nonzero area, then the time-domain far-field pattern is zero for all angles of observation at all times.* Hence, the field generated by time-domain sources in a finite region of space cannot have zero sidelobes, unless the far-field pattern is zero everywhere.

Relations Between the Far-Field Pattern and the Far-Field Function. The time-domain far-field function $F(\xi, \eta, t)$ is given in terms of the volume source $q(\mathbf{r}, t)$ by (5.161) and (5.162), which were obtained by Fourier transforming the frequency-domain formulas

$$F_\omega(\xi\omega, \eta\omega) = \frac{1}{4\pi} \int_V q_\omega(\mathbf{r}) e^{-i\omega(\xi x + \eta y + \zeta z)} dV, \quad \xi^2 + \eta^2 < c^{-2} \tag{5.176}$$

and

$$F_\omega(\xi\omega, \eta\omega) = \frac{1}{4\pi} \int_V q_\omega(\mathbf{r}) e^{-i\omega(\xi x + \eta y + i\frac{\omega}{|\omega|}|\zeta| z)} dV, \quad \xi^2 + \eta^2 > c^{-2}. \tag{5.177}$$

These formulas satisfy the reality condition (5.80) and are valid for all real ω. Since the absolute value of ζ occurs in (5.177), the far-field function is in general not an analytic function of (ξ, η), and we shall consider only real (ξ, η).

Comparing (5.176) with (5.172) shows that the time-domain far-field pattern is related to the time-domain far-field function by the formula

$$\mathcal{F}(\theta, \phi, t) = F(c^{-1} \sin\theta \cos\phi, c^{-1} \sin\theta \sin\phi, t), \qquad 0 \le \theta \le \pi/2, \quad 0 \le \phi \le 2\pi. \tag{5.178}$$

Thus, the time-domain far-field function $F(\xi, \eta, t)$ for $\xi^2 + \eta^2 \le c^{-2}$ is completely determined by the far-field pattern $\mathcal{F}(\theta, \phi, t)$ for $0 \le \theta \le \pi/2$ and $0 \le \phi \le 2\pi$.

Having shown that the time-domain far-field pattern in some cases can be analytically continued to complex angles of observation, we shall now see how this analytically continued far-field pattern is related to the time-domain far-field function. Letting $\theta = \pi/2 - i\alpha$, we have $\cos\theta = i \sinh\alpha$ and $\sin\theta = \cosh\alpha$, and every point (ξ, η) with $\xi^2 + \eta^2 > c^{-2}$ can then be written as $(\xi, \eta) = (c^{-1} \cosh\alpha \cos\phi, c^{-1} \cosh\alpha \sin\phi)$ with $0 \le \alpha < \infty$ and $0 \le \phi \le 2\pi$. In terms of α, the variable ζ is given by $\zeta = ic^{-1} \sinh\alpha$. With these new variables, the formulas (5.172) and (5.177) show that

$$F_\omega(\omega c^{-1} \cosh\alpha \cos\phi, \omega c^{-1} \cosh\alpha \sin\phi) = \mathcal{F}_\omega\left(\frac{\pi}{2} - i\frac{\omega}{|\omega|}\alpha, \phi\right) \tag{5.179}$$

which is valid for all real ω. Because of the factor $\omega/|\omega|$ in (5.179), even in cases where the time-domain far-field pattern can be analytically continued, its values for complex θ are not equal to the values of the time-domain far-field function. Hence, *in the time domain, the far-field function for $\xi^2 + \eta^2 > c^{-2}$ cannot be determined directly from the far-field pattern, even in cases where the far-field pattern can be analytically continued to complex θ.*

EXAMPLE

To illustrate these time-domain results, we consider the example of the point source given by $q_\omega(\mathbf{r}) = \delta(\mathbf{r}+a\hat{\mathbf{z}})e^{-|\omega|T}$, where a and T are positive constants. This source is exponentially decaying for $|\omega| \to \infty$, and the corresponding time-domain far-field pattern can be analytically continued to complex regions around the real θ and ϕ axes. The time-domain source function is in this case (see (5.88))

$$q(\mathbf{r}, t) = \delta(\mathbf{r} + a\hat{\mathbf{z}})\frac{2T}{T^2 + t^2} \tag{5.180}$$

and the far-field pattern is found from (5.48) to be

$$\mathcal{F}(\theta, \phi, t) = \frac{1}{2\pi}\frac{T}{T^2 + (t - (a/c)\cos\theta)^2}. \tag{5.181}$$

Moreover, the formulas (5.161) and (5.162) show that the far-field function is given by

$$F(\xi, \eta, t) = \frac{1}{2\pi}\frac{T}{T^2 + (t - a\zeta)^2}, \quad \xi^2 + \eta^2 < c^{-2} \tag{5.182}$$

and

$$F(\xi, \eta, t) = \frac{1}{2\pi}\frac{T + a|\zeta|}{t^2 + (T + a|\zeta|)^2}, \quad \xi^2 + \eta^2 > c^{-2}. \tag{5.183}$$

These expressions satisfy the formula (5.178), and it is seen that $F(\xi, \eta, t)$ given by (5.183) is not an analytic function because $|\zeta|$ occurs in the expression. Moreover, equation (5.181) confirms that the far-field pattern, for all real t, is analytic with respect to θ when $|\mathrm{Im}(\cos\theta)| < cT/a$, since then the denominator in (5.181) is nonzero for all real t. For $\theta = \pi/2 - i\alpha$, with $\alpha \geq 0$, we find that

$$\mathcal{F}(\pi/2 - i\alpha, \phi, t) = \frac{1}{2\pi}\frac{T}{T^2 + (t - i(a/c)\sinh\alpha)^2} \tag{5.184}$$

and

$$F(c^{-1}\sin(\pi/2 - i\alpha)\cos\phi, c^{-1}\sin(\pi/2 - i\alpha)\sin\phi, t) = \frac{1}{2\pi}\frac{T + (a/c)\sinh\alpha}{t^2 + (T + (a/c)\sinh\alpha)^2} \tag{5.185}$$

which shows that the analytic continuation of the far-field pattern to complex θ does not equal the far-field function at the corresponding point in the (ξ, η) plane. In particular, the analytically continued far-field pattern (5.184) is complex for $\alpha > 0$, whereas the far-field function (5.185) is real for all $\alpha \geq 0$.

Electromagnetic Fields. The foregoing results of this section apply also to the electromagnetic fields. In particular, the formulas (5.178) and (5.179) hold for the electromagnetic fields when the scalar far-field pattern $\mathcal{F}$ and the scalar far-field function F are replaced by the vector far-field pattern $\boldsymbol{\mathcal{F}}$ and vector far-field function $\mathbf{F}$, respectively. Also, the following theorem holds for all electromagnetic fields whose time and frequency-domain far-field patterns can be represented by Fourier integrals. *If the electric or magnetic time-domain far-field pattern of a finite source region is zero at all times for $(\theta, \phi) \in B$, where B has a nonzero area, then both the electric and magnetic time-domain far-field patterns are zero for all angles of observation at all times.*

Thus, "electromagnetic bullets" (fields with zero far fields outside an angular cone) [149], [150] cannot be excited by physically realizable sources in a finite region of space, even though time-domain far fields with very low sidelobes are permissible. This is true even for the fields of "electromagnetic missiles" (which have far fields that decay slower than $1/r$ as $r \to \infty$) considered in Section 5.4, because the frequency-domain far-field patterns of electromagnetic missiles (produced by sources in a bounded volume) are entire functions of complex (θ, ϕ), and their far-field patterns in the time and frequency-domains are Fourier transforms of each other except possibly in the isolated direction of the electromagnetic missile.

5.2.5 Far-Field Expressions

This section expresses the time-domain far fields in terms of the time-domain spectra $T(\xi, \eta, t)$ and $\mathbf{T}(\xi, \eta, t)$. It is assumed that the sources satisfy the conditions stated in Section 5.1.4, so that the far fields decay as $1/r$, or faster, as $r \to \infty$. All formulas will be derived by Fourier transforming the frequency-domain expressions of Section 3.2.5, and we shall consider only the total far fields obtained by adding the contributions from propagating and evanescent plane waves. Thus, unlike Section 3.2.5, the present section does not include separate expressions for the far-field contributions from the propagating and evanescent plane waves. However, in Section 5.2.1 we derived the separate time-domain far-field contributions from propagating and evanescent plane waves for a scalar point source.

Far Fields for $0 \le \theta < \pi/2$. Taking the Fourier transform of the frequency-domain formula (3.153) (which, according to (3.143), is also valid for $\theta = 0$) shows that

$$p(\mathbf{r}, t) = \frac{\cos\theta}{cr}\frac{\partial}{\partial t}T(c^{-1}\cos\phi\sin\theta, c^{-1}\sin\phi\sin\theta, t - r/c) + O\left(\frac{1}{r^2}\right) \tag{5.186}$$

which, upon inserting the definition of T from (5.87), is seen to equal the far field (5.54) found from a Green's function representation. Also, the time-domain far-field pattern defined by (5.43) can be expressed in terms of T as

$$\mathcal{F}(\theta, \phi, t) = \frac{\cos\theta}{c}\frac{\partial}{\partial t}T(c^{-1}\cos\phi\sin\theta, c^{-1}\sin\phi\sin\theta, t). \tag{5.187}$$

Similarly, from (3.156) (which, according to (3.146), is valid also for $\theta = 0$), we find that the electric far field is

$$\mathbf{E}(\mathbf{r}, t) = \frac{\cos\theta}{cr}\frac{\partial}{\partial t}\mathbf{T}(c^{-1}\cos\phi\sin\theta, c^{-1}\sin\phi\sin\theta, t - r/c) + O\left(\frac{1}{r^2}\right) \tag{5.188}$$

and the electric far-field pattern is

$$\boldsymbol{\mathcal{F}}(\theta, \phi, t) = \frac{\cos\theta}{c}\frac{\partial}{\partial t}\mathbf{T}(c^{-1}\cos\phi\sin\theta, c^{-1}\sin\phi\sin\theta, t). \tag{5.189}$$

With the help of (5.107) and (5.110), it can be shown that the formulas (5.188) and (5.189) are identical to the Green's function formulas (5.56) and (5.57). The magnetic far field can be expressed in terms of $\mathbf{T}$ as

$$\mathbf{H}(\mathbf{r}, t) = \sqrt{\frac{\epsilon}{\mu}}\frac{\cos\theta}{cr}\hat{\mathbf{r}} \times \frac{\partial}{\partial t}\mathbf{T}(c^{-1}\cos\phi\sin\theta, c^{-1}\sin\phi\sin\theta, t - r/c) + O\left(\frac{1}{r^2}\right) \tag{5.190}$$

which confirms that the electric and magnetic far-field patterns satisfy (5.47).

Far Fields for $\theta = \pi/2$. Taking the Fourier transform of the far-field formula (3.162) shows that the time-domain far field and the far-field pattern at $\theta = \pi/2$ can be expressed in terms of the time-domain far-field function $F(\xi, \eta, t)$ as

$$p(\mathbf{r}, t) = F(c^{-1}\cos\phi, c^{-1}\sin\phi, t - r/c)\frac{1}{r} + O\left(\frac{1}{r^2}\right) \tag{5.191}$$

and

$$\mathcal{F}(\pi/2, \phi, t) = F(c^{-1}\cos\phi, c^{-1}\sin\phi, t). \tag{5.192}$$

Similarly, for the electric far field, (3.165) shows that

$$\mathbf{E}(\mathbf{r},t) = \mathbf{F}(c^{-1}\cos\phi, c^{-1}\sin\phi, t - r/c)\frac{1}{r} + O\left(\frac{1}{r^2}\right) \tag{5.193}$$

and

$$\mathcal{F}(\pi/2,\phi,t) = \mathbf{F}(c^{-1}\cos\phi, c^{-1}\sin\phi, t). \tag{5.194}$$

Again, the magnetic far-field pattern can be determined from the electric far-field pattern through (5.47).

When the sources are located in the half space $z > 0$, the far fields and the far-field patterns in the half space $z < 0$ are given by the negative of the right-hand sides of (5.186)–(5.190).

5.2.6 Near Fields in Terms of the Far Fields in Spherical Coordinates

This section derives formulas that express the time-domain field in terms of its far-field pattern and its far-field function. We find that the time-domain field for all times everywhere in the region $z > 0$ can be determined from its far-field pattern in the hemisphere $0 \le \theta \le \pi/2$ and its far-field function for $\xi^2 + \eta^2 > c^{-2}$. At times after all sources have been turned off, the time-domain field can be expressed throughout space in terms of its far-field pattern on the entire far-field sphere $0 \le \theta \le \pi$. The formulas of this section are time-domain analogs of the frequency-domain formulas in Sections 3.2.6 and 3.2.7. Throughout this section, we assume that the sources satisfy the conditions stated in Section 5.1.4, ensuring that the time-domain far-fields decay as $1/r$, or faster.

First, consider the acoustic pressure $p(\mathbf{r},t)$. In Section 3.2.6 it was shown that the frequency-domain acoustic pressure for $\omega > 0$ is given by (3.183), which states that for $z > 0$,

$$p_\omega(\mathbf{r}) = \frac{1}{2\pi c}\int_0^{2\pi}\int_0^{\pi/2} i\omega\mathcal{F}_\omega(\sigma_\theta,\sigma_\phi)e^{i(\omega/c)\mathbf{r}\cdot\hat{\boldsymbol{\sigma}}_{\theta\phi}}\sin\sigma_\theta\, d\sigma_\theta\, d\sigma_\phi$$
$$+\frac{1}{2\pi c}\int_0^{2\pi}\int_0^{+\infty}\omega\mathcal{F}_\omega(\pi/2 - i\alpha,\sigma_\phi)e^{i(\omega/c)\cosh\alpha\,\mathbf{r}\cdot\hat{\boldsymbol{\sigma}}_\phi}e^{-(\omega/c)z\sinh\alpha}\cosh\alpha\, d\alpha\, d\sigma_\phi \tag{5.195}$$

where $\hat{\boldsymbol{\sigma}}_{\theta\phi} = \hat{\mathbf{x}}\cos\sigma_\phi\sin\sigma_\theta + \hat{\mathbf{y}}\sin\sigma_\phi\sin\sigma_\theta + \hat{\mathbf{z}}\cos\sigma_\theta$ and $\hat{\boldsymbol{\sigma}}_\phi = \hat{\mathbf{x}}\cos\sigma_\phi + \hat{\mathbf{y}}\sin\sigma_\phi$. In the formula (5.195), the frequency-domain plane-wave spectrum for the evanescent plane waves is expressed in terms of the frequency-domain far-field pattern for complex θ (second term in (5.195)). In Section 5.2.4 we showed that the time-domain plane-wave spectrum for evanescent plane waves cannot be determined directly from the time-domain far-field pattern for complex θ. As a consequence, it is not possible to derive a formula (valid for all times) that expresses the time-domain pressure directly in terms of its time-domain far-field pattern.

Instead we use the relation (3.178) between the far-field function and the far-field pattern in the second term of the formula (3.181) to show that the frequency-domain pressure for $z > 0$

and $\omega > 0$ is given by

$$
\begin{aligned}
p_\omega(\mathbf{r}) = {} & \frac{1}{2\pi c} \int_0^{2\pi} \int_0^{\pi/2} i\omega \mathcal{F}_\omega(\sigma_\theta, \sigma_\phi) e^{i(\omega/c)\mathbf{r}\cdot\hat{\boldsymbol{\sigma}}_{\theta\phi}} \sin\sigma_\theta d\sigma_\theta d\sigma_\phi \\
& + \frac{1}{2\pi} \iint_{\xi^2+\eta^2>c^{-2}} \frac{\omega}{|\zeta|} F_\omega(\xi\omega, \eta\omega) e^{i\omega(\xi x+\eta y)} e^{-\omega|\zeta|z} d\eta d\xi
\end{aligned}
\tag{5.196}
$$

where we have introduced (ξ, η) such that $(k_x, k_y) = (\omega\xi, \omega\eta)$ and ζ is given in (5.95). To determine the frequency-domain pressure for $\omega < 0$, we simply use the reality condition (5.80) and find that the following relation is valid for all real ω

$$
\begin{aligned}
p_\omega(\mathbf{r}) = {} & \frac{1}{2\pi c} \int_0^{2\pi} \int_0^{\pi/2} i\omega \mathcal{F}_\omega(\sigma_\theta, \sigma_\phi) e^{i(\omega/c)\mathbf{r}\cdot\hat{\boldsymbol{\sigma}}_{\theta\phi}} \sin\sigma_\theta d\sigma_\theta d\sigma_\phi \\
& + \frac{1}{2\pi} \iint_{\xi^2+\eta^2>c^{-2}} \left(-\frac{i\omega}{|\zeta|}\right) F_\omega(\xi\omega, \eta\omega) e^{i\omega(\xi x+\eta y)} \left(\frac{i\omega}{|\omega|} e^{-|\omega||\zeta|z}\right) d\eta d\xi .
\end{aligned}
\tag{5.197}
$$

Taking the Fourier transform of (5.197) and using the convolution theorem (5.89) and the formula (5.140) gives us the time-domain acoustic pressure for $z > 0$ as

$$
\begin{aligned}
p(\mathbf{r}, t) = {} & -\frac{1}{2\pi c} \int_0^{2\pi} \int_0^{\pi/2} \frac{\partial}{\partial t} \mathcal{F}(\sigma_\theta, \sigma_\phi, t - \mathbf{r}\cdot\hat{\boldsymbol{\sigma}}_{\theta\phi}/c) \sin\sigma_\theta d\sigma_\theta d\sigma_\phi \\
& + \frac{1}{2\pi^2} \iint_{\xi^2+\eta^2>c^{-2}} \int_{-\infty}^{+\infty} \frac{t' (\partial/\partial t) F(\xi, \eta, t - t' - \xi x - \eta y)}{|\zeta| \left(|\zeta|^2 z^2 + t'^2 \right)} dt' d\eta d\xi
\end{aligned}
\tag{5.198}
$$

which is valid for all times. An alternative form of the second term in (5.198), which involves the Hilbert transform of the far-field function, can be derived by inserting the relation (5.158) into the second integral of (5.86).

Using the relation $\mathbf{v}_\omega = \nabla p_\omega/(i\omega\rho_{m0})$ along with (5.197) gives a formula for the frequency-domain velocity, which can be Fourier transformed to get

$$
\begin{aligned}
\mathbf{v}(\mathbf{r}, t) = {} & -\frac{1}{2\pi c^2 \rho_{m0}} \int_0^{2\pi} \int_0^{\pi/2} \hat{\boldsymbol{\sigma}}_{\theta\phi} \frac{\partial}{\partial t} \mathcal{F}(\sigma_\theta, \sigma_\phi, t - \mathbf{r}\cdot\hat{\boldsymbol{\sigma}}_{\theta\phi}/c) \sin\sigma_\theta d\sigma_\theta d\sigma_\phi \\
& + \frac{1}{2\pi^2 \rho_{m0}} \iint_{\xi^2+\eta^2>c^{-2}} \int_{-\infty}^{+\infty} \widetilde{\boldsymbol{\chi}}(\xi, \eta) \frac{t' (\partial/\partial t) F(\xi, \eta, t - t' - \xi x - \eta y)}{|\zeta| \left(|\zeta|^2 z^2 + t'^2 \right)} dt' d\eta d\xi .
\end{aligned}
\tag{5.199}
$$

Thus, the time-domain velocity in the half space $z > 0$ is expressed in terms of the far-field pattern and far-field function.

For the electric field we find similarly

$$\mathbf{E}(\mathbf{r},t) = -\frac{1}{2\pi c}\int_0^{2\pi}\int_0^{\pi/2} \frac{\partial}{\partial t}\mathcal{F}(\sigma_\theta,\sigma_\phi,t-\mathbf{r}\cdot\hat{\boldsymbol{\sigma}}_{\theta\phi}/c)\sin\sigma_\theta d\sigma_\theta d\sigma_\phi + \frac{1}{2\pi^2}\iint_{\xi^2+\eta^2>c^{-2}}\int_{-\infty}^{+\infty} \frac{t'(\partial/\partial t)\mathbf{F}(\xi,\eta,t-t'-\xi x-\eta y)}{|\zeta|\left(|\zeta|^2z^2+t'^2\right)}dt'd\eta d\xi \tag{5.200}$$

where $\mathcal{F}$ and $\mathbf{F}$ are the time-domain far-field pattern and far-field function, respectively, of the electric field. The formula (5.200) also holds for the magnetic field, provided $\mathcal{F}$ and $\mathbf{F}$ are replaced with $\mathcal{F}^H$ and $\mathbf{F}^H$.

The far-field patterns for the electric and magnetic fields are related through (5.47), which states that

$$\mathcal{F}^H(\sigma_\theta,\sigma_\phi,t) = \sqrt{\frac{\epsilon}{\mu}}\hat{\boldsymbol{\sigma}}_{\theta\phi}\times\mathcal{F}(\sigma_\theta,\sigma_\phi,t) \tag{5.201}$$

and (5.166) shows that the far-field functions of the electric and magnetic fields are related through

$$\mathbf{F}^H(\xi,\eta,t) = \frac{1}{\mu}\widetilde{\boldsymbol{\chi}}(\xi,\eta)\times\mathbf{F}(\xi,\eta,t) \tag{5.202}$$

where $\widetilde{\boldsymbol{\chi}}(\xi,\eta)$ is given by (5.92). Inserting these formulas into (5.200) (with $\mathcal{F}$ replaced by $\mathcal{F}^H$, $\mathbf{F}$ replaced by $\mathbf{F}^H$, and $\mathbf{E}$ replaced by $\mathbf{H}$) gives us the following expression for the magnetic field in terms of the electric far-field pattern and far-field function

$$\mathbf{H}(\mathbf{r},t) = -\frac{\epsilon}{2\pi}\int_0^{2\pi}\int_0^{\pi/2} \hat{\boldsymbol{\sigma}}_{\theta\phi}\times\frac{\partial}{\partial t}\mathcal{F}(\sigma_\theta,\sigma_\phi,t-\mathbf{r}\cdot\hat{\boldsymbol{\sigma}}_{\theta\phi}/c)\sin\sigma_\theta d\sigma_\theta d\sigma_\phi + \frac{1}{2\pi^2\mu}\iint_{\xi^2+\eta^2>c^{-2}}\int_{-\infty}^{+\infty} \widetilde{\boldsymbol{\chi}}(\xi,\eta)\times\frac{t'(\partial/\partial t)\mathbf{F}(\xi,\eta,t-t'-\xi x-\eta y)}{|\zeta|\left(|\zeta|^2z^2+t'^2\right)}dt'd\eta d\xi. \tag{5.203}$$

The formulas (5.198), (5.200), and (5.203) express the time-domain fields everywhere in the half space $z>0$ for all times in terms of the far-field patterns over the hemisphere $0\le\theta\le\pi/2$ and the far-field functions over the evanescent region $\xi^2+\eta^2>c^{-2}$.

We will now derive formulas, which hold throughout space after all sources have been turned off, for the fields in terms of the far-field patterns over the entire far-field sphere $0\le\theta\le\pi$. These formulas were first derived by Moses, Nagem, and Sandri [149]. However, we present a greatly simplified derivation using the formulas involving the homogeneous fields that were derived in Section 3.2.7. Taking the Fourier transform of the formula (3.195), which expresses the frequency-domain acoustic pressure in terms of the advanced Green's function, gives

$$p(\mathbf{r},t) = \frac{1}{4\pi}\int_V \frac{q(\mathbf{r}',t+|\mathbf{r}-\mathbf{r}'|/c)}{|\mathbf{r}-\mathbf{r}'|}dV' - \frac{1}{2\pi c}\int_0^{2\pi}\int_0^{\pi} \frac{\partial}{\partial t}\mathcal{F}(\sigma_\theta,\sigma_\phi,t-\mathbf{r}\cdot\hat{\boldsymbol{\sigma}}_{\theta\phi}/c)\sin\sigma_\theta d\sigma_\theta d\sigma_\phi \tag{5.204}$$

which is valid for all times throughout all space. This formula expresses the time-domain acoustic pressure in terms of its far-field pattern over the entire far-field sphere and its source function evaluated at an advanced time. Alternatively, (5.204) can be derived directly in the time domain using the advanced time-domain free-space Green's function and Green's second identity. The first term on the right-hand side of (5.204) is the advanced pressure, which satisfies the same inhomogeneous wave equation as $p(\mathbf{r}, t)$, namely, $[\nabla^2 - \partial^2/(c^2\partial t^2)]p(\mathbf{r}, t) = -q(\mathbf{r}, t)$, and the second term is the homogeneous pressure, which satisfies the homogeneous wave equation throughout space, as discussed in Section 3.2.7.

After all sources are turned off, the source function $q(\mathbf{r}, t)$ equals zero, and the first integral in (5.204) is zero. That is, the advanced pressure is zero, so the total pressure equals the homogeneous pressure

$$p(\mathbf{r}, t) = -\frac{1}{2\pi c}\int_0^{2\pi}\int_0^{\pi}\frac{\partial}{\partial t}\mathcal{F}(\sigma_\theta, \sigma_\phi, t - \mathbf{r}\cdot\hat{\boldsymbol{\sigma}}_{\theta\phi}/c)\sin\sigma_\theta d\sigma_\theta d\sigma_\phi, \quad t > t_1 \tag{5.205}$$

where t_1 is the time at which the sources turn off so that $q(\mathbf{r}, t) = 0$ for $t > t_1$ and all $\mathbf{r}$. One can also insert (5.48) into (5.205) to get an expression for the pressure in terms of the source function $q(\mathbf{r}, t)$. The formula (5.205) gives the pressure throughout space, at times after which all sources have been turned off, in terms of the far-field pattern over the entire far-field hemisphere. Equating (5.198) and (5.205) shows that after all sources are turned off, and for $z > 0$, the integral over the evanescent region of the time-domain far-field function (second term on the right-hand side of (5.198)) equals the integral of the time-domain far-field pattern over the opposite hemisphere $\pi/2 < \sigma_\theta \le \pi$.

Equation (5.204) shows that (5.205), in fact, is valid throughout the time interval $t > t_{10}(\mathbf{r})$, where

$$t_{10}(\mathbf{r}) = t_1 - \min_{\mathbf{r}'\in V}|\mathbf{r} - \mathbf{r}'|/c. \tag{5.206}$$

The expression for $t_{10}(\mathbf{r})$ shows that when the observation point is outside the source volume V, the formula (5.205) also holds in the time interval $t_{10}(\mathbf{r}) < t < t_1$ in which the sources still radiate.

With the Causality Trick (CT) [16], solutions to the wave equation are obtained by adding retarded and advanced fields. CT solutions involve spectral integrations over *finite* regions, whereas the standard causal solutions usually involve spectral integrations over *infinite* regions. CT solutions are therefore easier to compute, and they equal the standard causal solutions after a certain time period. The expression (5.205) for the acoustic pressure corresponds to a CT solution. The usefulness of the Causality Trick is clearly illustrated by the fact that (5.205) is much easier to use than (5.198) in numerical calculations.

Taking the inverse Fourier transform of the electromagnetic formula (3.207) gives [24]

$$\begin{aligned}\mathbf{E}(\mathbf{r}, t) = {} & \frac{1}{4\pi}\int_V \frac{\mathbf{Q}(\mathbf{r}', t + |\mathbf{r} - \mathbf{r}'|/c)}{|\mathbf{r} - \mathbf{r}'|}dV' \\ & - \frac{1}{2\pi c}\int_0^{2\pi}\int_0^{\pi}\frac{\partial}{\partial t}\boldsymbol{\mathcal{F}}(\sigma_\theta, \sigma_\phi, t - \mathbf{r}\cdot\hat{\boldsymbol{\sigma}}_{\theta\phi}/c)\sin\sigma_\theta d\sigma_\theta d\sigma_\phi\end{aligned} \tag{5.207}$$

throughout all space and time. In (5.207), the first term is the advanced electric field, and the second term is the homogeneous electric field. The time-domain source function $\mathbf{Q}(\mathbf{r}, t)$ can be found in terms of the time-domain current and charge densities by taking the inverse Fourier transform of (3.205) to get

$$\mathbf{Q}(\mathbf{r}, t) = -\mu \frac{\partial}{\partial t}\mathbf{J}(\mathbf{r}, t) - \frac{1}{\epsilon}\nabla\rho(\mathbf{r}, t). \tag{5.208}$$

The first term on the right-hand side of (5.207) is the advanced electric field, which satisfies the same inhomogeneous vector-wave equation as $\mathbf{E}(\mathbf{r}, t)$, namely, $[\nabla^2 - \partial^2/(c^2\partial t^2)]\mathbf{E}(\mathbf{r}, t) = -\mathbf{Q}(\mathbf{r}, t)$, and the second term is the homogeneous field, which satisfies the homogeneous vector-wave equation throughout all space, as discussed in Section 3.2.7.

Let t_1 be the time at which all sources are turned off so that $\mathbf{J}(\mathbf{r}, t)$ and $\rho(\mathbf{r}, t)$ equal zero for $t > t_1$ and all $\mathbf{r}$. Then (5.208) shows that $\mathbf{Q}(\mathbf{r}, t) = 0$ for all $\mathbf{r}$ and $t > t_1$, and the advanced source function $\mathbf{Q}(\mathbf{r}', t + |\mathbf{r} - \mathbf{r}'|/c)$ equals zero throughout the integration in (5.207) for $t > t_1$. Consequently, for $t > t_1$ the volume integral of (5.207) equals zero and [24]

$$\mathbf{E}(\mathbf{r}, t) = -\frac{1}{2\pi c}\int_0^{2\pi}\int_0^{\pi} \frac{\partial}{\partial t}\boldsymbol{\mathcal{F}}(\sigma_\theta, \sigma_\phi, t - \mathbf{r}\cdot\hat{\boldsymbol{\sigma}}_{\theta\phi}/c)\sin\sigma_\theta d\sigma_\theta d\sigma_\phi, \quad t > t_1. \tag{5.209}$$

Similarly, the magnetic field corresponding to (5.209) is found from (5.201) to be

$$\mathbf{H}(\mathbf{r}, t) = -\frac{\epsilon}{2\pi}\int_0^{2\pi}\int_0^{\pi} \hat{\boldsymbol{\sigma}}_{\theta\phi} \times \frac{\partial}{\partial t}\boldsymbol{\mathcal{F}}(\sigma_\theta, \sigma_\phi, t - \mathbf{r}\cdot\hat{\boldsymbol{\sigma}}_{\theta\phi}/c)\sin\sigma_\theta d\sigma_\theta d\sigma_\phi, \quad t > t_1. \tag{5.210}$$

The expressions in (5.209) and (5.210) hold for $t > t_{10}(\mathbf{r})$ with $t_{10}(\mathbf{r})$ given in (5.206). (Equations (5.209) and (5.210) can also be derived directly in the time domain using the advanced free-space time-domain Green's function. This alternate derivation avoids the use of the Fourier transform and does not require the existence of the Fourier transform of the time-domain functions.)

In summary, both acoustic and electromagnetic near fields, for all observation points in the region $z > 0$ and for every instant of time, can be expressed as integrals of the far-field patterns over the hemisphere $0 \le \theta \le \pi/2$ and the far-field functions over the evanescent region $\xi^2 + \eta^2 > c^{-2}$. Moreover, the near fields, at all points of observation and at times after all sources have been turned off, can be expressed in terms of integrals of the far fields over the entire far-field sphere $0 \le \theta \le \pi$. This latter, rather surprising result was first derived by Moses et al. [149] for both acoustic and electromagnetic fields.

The formulas (5.205), (5.209), and (5.210) allow us to calculate the time-domain near fields from a given time-domain far-field pattern. However, as we showed in Sections 5.1.5 and 5.2.4, the time-domain far-field patterns must satisfy two conditions: first, if all sources reside in a bounded volume and are nonzero only during a finite time interval, the time-domain far-field pattern integrated over all time must be zero; second, if all sources reside in a bounded volume, the time-domain far-field pattern cannot be zero for all time in an angular region of the unit sphere with nonzero area. If one inserts into (5.205), (5.209), or (5.210) a far-field pattern that violates this second condition, for example, a far-field pattern that is zero in the region $(\theta_1 < \theta < \theta_2, \phi_1 < \phi < \phi_2)$, the computed near fields can only be generated by sources that are not confined to a bounded region of space. That is, the sources of these fields (called bullets in [149], [150]) extend an infinite distance and cannot be enclosed by a sphere of finite radius.

When the sources are located in $z > 0$, the fields in $z < 0$ are determined from (5.198), (5.199), (5.200), (5.202), and (5.203) with $\int_0^{\pi/2}$ replaced by $\int_{\pi/2}^{\pi}$ and $\widetilde{\chi}$ replaced by $\widetilde{\chi}^- = \widetilde{\chi} - 2(\hat{\mathbf{z}}\cdot\widetilde{\chi})\hat{\mathbf{z}}$. The formulas (5.201), (5.204), (5.205), (5.207), (5.209), and (5.210) remain unchanged.

5.2.7 Energy and Momentum Relations

In this section, the total electromagnetic and acoustic energy and momentum radiated through the plane $z = z_1 \geq z_0$, by sources that are nonzero only during a finite time interval, will be expressed in terms of the time-domain spectra and far-field patterns. The energy expressions will be derived using the frequency-domain power relations of Section 3.2.9 and Parseval's theorem, which states that

$$\int_{-\infty}^{+\infty} f(t)g(t)dt = 2\pi \int_{-\infty}^{+\infty} f_\omega g_\omega^* d\omega = 2\pi \int_{-\infty}^{+\infty} \mathrm{Re}\left(f_\omega g_\omega^*\right) d\omega \tag{5.211}$$

where $f(t)$ and $g(t)$ are real-time functions, and f_ω and g_ω are their Fourier transforms. The last equality in (5.211) is obtained by taking the real part of the first equation.

According to the time-domain Poynting's theorem, the total electromagnetic energy radiated through the plane $z = z_1 \geq z_0$ is given from (2.191) and (5.211) as

$$\begin{aligned}\mathcal{E}_{el} &= \int_{-\infty}^{+\infty}\int_{-\infty}^{+\infty}\int_{-\infty}^{+\infty} \hat{\mathbf{z}} \cdot (\mathbf{E}(\mathbf{r}_1, t) \times \mathbf{H}(\mathbf{r}_1, t))\, dt dx_1 dy_1 \\ &= 2\pi \int_{-\infty}^{+\infty}\int_{-\infty}^{+\infty}\int_{-\infty}^{+\infty} \mathrm{Re}\left[\hat{\mathbf{z}} \cdot \left(\mathbf{E}_\omega(\mathbf{r}_1) \times \mathbf{H}_\omega^*(\mathbf{r}_1)\right)\right] d\omega dx_1 dy_1\end{aligned} \tag{5.212}$$

where $\mathbf{r}_1 = x_1\hat{\mathbf{x}} + y_1\hat{\mathbf{y}} + z_1\hat{\mathbf{z}}$. From the frequency-domain power relations (3.233) and (3.236) we find that

$$\begin{aligned}2\pi &\int_{-\infty}^{+\infty}\int_{-\infty}^{+\infty} \mathrm{Re}\left[\hat{\mathbf{z}} \cdot \left(\mathbf{E}_\omega(\mathbf{r}_1) \times \mathbf{H}_\omega^*(\mathbf{r}_1)\right)\right] dx_1 dy_1 \\ &= \frac{2\pi}{\omega\mu} \iint_{k_x^2+k_y^2<k^2} |\mathbf{T}_\omega(k_x, k_y)|^2 \gamma\, dk_x dk_y \\ &= \frac{2\pi}{\mu} \iint_{\xi^2+\eta^2<c^{-2}} |\omega \mathbf{T}_\omega(\xi\omega, \eta\omega)|^2 \zeta\, d\xi d\eta\end{aligned} \tag{5.213}$$

where (ξ, η) has been introduced such that $(k_x, k_y) = (\xi\omega, \eta\omega)$. The last integral in (5.213) satisfies the reality condition (5.80) and is therefore valid for all real ω. Inserting (5.213) into (5.212) gives

$$\mathcal{E}_{el} = \frac{1}{\mu} \iint_{\xi^2+\eta^2<c^{-2}} \zeta \int_{-\infty}^{+\infty} \left|\frac{\partial}{\partial t}\mathbf{T}(\xi, \eta, t)\right|^2 dt d\xi d\eta \tag{5.214}$$

where Parseval's theorem (5.211) with $f = g$ has been used. The expression (5.214) shows that the total energy $\mathcal{E}_{el}$ is independent of z_1 and radiated solely by the propagating plane waves. The total energy carried by the evanescent fields across a plane $z = z_1 \geq z_0$ is zero, for sources that are turned on only during a finite time interval. During part of the time interval the energy in the evanescent fields may flow in the positive z direction across the $z = z_1 \geq z_0$ plane, and in the negative z direction during the rest of the time interval, as long as the net energy transferred by evanescent fields is zero. In Section 5.1.5 we found that just the opposite is true of the near fields themselves. The integral over all time of the propagating part of the near field is zero, whereas the integral over all time of the evanescent part of the near field is not necessarily zero.

Introducing the angular integration variables (θ, ϕ), determined by the relations $\xi = c^{-1}\sin\theta\cos\phi$ and $\eta = c^{-1}\sin\theta\sin\phi$, into (5.214) reveals that $\mathcal{E}_{el}$ also equals

$$\begin{aligned}\mathcal{E}_{el} &= \frac{1}{c\mu}\int_0^{2\pi}\int_0^{\pi/2}\int_{-\infty}^{+\infty}\left|\frac{\cos\theta}{c}\frac{\partial}{\partial t}\mathbf{T}(c^{-1}\sin\theta\cos\phi,\ c^{-1}\sin\theta\sin\phi, t)\right|^2\sin\theta\,dt\,d\theta\,d\phi \\ &= \sqrt{\frac{\epsilon}{\mu}}\int_0^{2\pi}\int_0^{\pi/2}\int_{-\infty}^{+\infty}|\boldsymbol{\mathcal{F}}(\theta,\phi,t)|^2\sin\theta\,dt\,d\theta\,d\phi\end{aligned} \tag{5.215}$$

where we have used the relation (5.189) between $\boldsymbol{\mathcal{F}}(\theta, \phi, t)$ and $\mathbf{T}(\xi, \eta, t)$. Equation (5.215) confirms that the total energy radiated through the plane $z = z_1 \geq z_0$ is equal to the total energy radiated into the far-field hemisphere $0 \leq \theta \leq \pi/2$.

To determine the electromagnetic momentum in the z direction, $\mathcal{M}_{el}^z$, radiated through the plane $z = z_1 \geq z_0$, we can integrate the electromagnetic stress dyadic in (2.190) to get

$$\begin{aligned}\mathcal{M}_{el}^z &= -\int_{-\infty}^{+\infty}\int_{-\infty}^{+\infty}\int_{-\infty}^{+\infty}\hat{\mathbf{z}}\cdot\bar{\mathbf{T}}_{el}\cdot\hat{\mathbf{z}}\,dt dx_1 dy_1 \\ &= \int_{-\infty}^{+\infty}\int_{-\infty}^{+\infty}\int_{-\infty}^{+\infty}\left[\epsilon\left(\frac{|\mathbf{E}|^2}{2}-(\hat{\mathbf{z}}\cdot\mathbf{E})^2\right)+\mu\left(\frac{|\mathbf{H}|^2}{2}-(\hat{\mathbf{z}}\cdot\mathbf{H})^2\right)\right]dt dx_1 dy_1 \\ &= 2\pi\int_{-\infty}^{+\infty}\int_{-\infty}^{+\infty}\int_{-\infty}^{+\infty}\left[\epsilon\left(\frac{|\mathbf{E}_\omega|^2}{2}-|\hat{\mathbf{z}}\cdot\mathbf{E}_\omega|^2\right)+\mu\left(\frac{|\mathbf{H}_\omega|^2}{2}-|\hat{\mathbf{z}}\cdot\mathbf{H}_\omega|^2\right)\right]d\omega dx_1 dy_1.\end{aligned} \tag{5.216}$$

The frequency-domain formulas (3.238), (3.239), and (3.240) convert (5.216) to

$$\mathcal{M}_{el}^z = \frac{2\pi}{\mu}\iint_{\xi^2+\eta^2<c^{-2}}\zeta^2\int_{-\infty}^{+\infty}\left|\omega^2\mathbf{T}_\omega(\xi\omega, \eta\omega)\right|^2 d\omega\,d\xi\,d\eta \tag{5.217}$$

or

$$\begin{aligned}\mathcal{M}_{el}^z &= \frac{1}{\mu}\iint_{\xi^2+\eta^2<c^{-2}}\zeta^2\int_{-\infty}^{+\infty}\left|\frac{\partial}{\partial t}\mathbf{T}(\xi,\eta,t)\right|^2 dt\,d\xi\,d\eta \\ &= \epsilon\int_0^{2\pi}\int_0^{\pi/2}\int_{-\infty}^{+\infty}|\boldsymbol{\mathcal{F}}(\theta,\phi,t)|^2\cos\theta\sin\theta\,dt\,d\theta\,d\phi.\end{aligned} \tag{5.218}$$

This momentum $\mathcal{M}_{el}^z$ is independent of the location of the plane of integration $z = z_1 \geq z_0$, and is equal to the momentum that would be imparted to an infinite, flat absorbing screen located in the plane of integration.

The total electromagnetic momentum $\mathcal{M}_{el}^h$ radiated through the far-field hemisphere $0 \le \theta \le \pi/2$ is given by

$$\begin{aligned}\mathcal{M}_{el}^h &= -\lim_{r\to\infty}\int_0^{2\pi}\int_0^{\pi/2}\int_{-\infty}^{+\infty} \hat{\mathbf{r}}\cdot\bar{\mathbf{T}}_{el}\cdot\hat{\mathbf{r}}\sin\theta\, dt\, d\theta\, d\phi \\ &= \epsilon\int_0^{2\pi}\int_0^{\pi/2}\int_{-\infty}^{+\infty} |\mathcal{F}(\theta,\phi,t)|^2\sin\theta\, dt\, d\theta\, d\phi.\end{aligned} \tag{5.219}$$

Equations (5.215) and (5.219) confirm that $\mathcal{E}_{el} = \mathcal{M}_{el}^h c$, or that the total energy radiated is equal to the wave propagation speed c times the total momentum radiated through a far-field hemisphere. The integrands of the last integrals of (5.215) and (5.218) confirm that the z-directed momentum per unit area radiated through a point on a far-field hemisphere is equal to the energy radiated through that point multiplied by $\hat{\mathbf{z}}\cdot\hat{\mathbf{r}}/c = \cos\theta/c$.

Similarly, for the acoustic field we find that the total energy and z-directed momentum radiated through the plane $z = z_1 \ge z_0$ are given from (2.278) and (2.277) as

$$\begin{aligned}\mathcal{E}_{ac} &= \int_{-\infty}^{+\infty}\int_{-\infty}^{+\infty}\int_{-\infty}^{+\infty} p(\mathbf{r}_1,t)\hat{\mathbf{z}}\cdot\mathbf{v}(\mathbf{r}_1,t)dt dx_1 dy_1 \\ &= \frac{1}{\rho_{m0}}\iint_{\xi^2+\eta^2<c^{-2}} \zeta\int_{-\infty}^{+\infty}\left|\frac{\partial}{\partial t}T(\xi,\eta,t)\right|^2 dt d\xi d\eta \\ &= \frac{1}{c\rho_{m0}}\int_0^{2\pi}\int_0^{\pi/2}\int_{-\infty}^{+\infty} |\mathcal{F}(\theta,\phi,t)|^2\sin\theta\, dt\, d\theta\, d\phi\end{aligned} \tag{5.220}$$

and

$$\begin{aligned}\mathcal{M}_{ac}^z &= -\int_{-\infty}^{+\infty}\int_{-\infty}^{+\infty}\int_{-\infty}^{+\infty} \hat{\mathbf{z}}\cdot\bar{\mathbf{T}}_{ac}\cdot\hat{\mathbf{z}}\, dt dx_1 dy_1 \\ &= -\frac{1}{2}\int_{-\infty}^{+\infty}\int_{-\infty}^{+\infty}\int_{-\infty}^{+\infty}\left[\rho_{m0}\left(\frac{|\mathbf{v}|^2}{2} - |\hat{\mathbf{z}}\cdot\mathbf{v}|^2\right) - \kappa_0\frac{p^2}{2}\right] dt dx_1 dy_1 \\ &= \frac{1}{\rho_{m0}}\iint_{\xi^2+\eta^2<c^{-2}} \zeta^2\int_{-\infty}^{+\infty}\left|\frac{\partial}{\partial t}T(\xi,\eta,t)\right|^2 dt\, d\xi\, d\eta \\ &= \frac{1}{c^2\rho_{m0}}\int_0^{2\pi}\int_0^{\pi/2}\int_{-\infty}^{+\infty} |\mathcal{F}(\theta,\phi,t)|^2\cos\theta\sin\theta\, dt\, d\theta\, d\phi\end{aligned} \tag{5.221}$$

where we have made use of the frequency-domain formulas (3.243) and (3.245), and Parseval's theorem (5.211). Thus, the total energy and z-directed wave momentum radiated through $z = z_1$ are determined solely by propagating plane waves and are independent of z_1. As in (5.219), the total acoustic wave momentum $\mathcal{M}_{ac}^h$ radiated into the far-field hemisphere

$0 \le \theta \le \pi/2$ is given by

$$\begin{aligned} \mathcal{M}_{ac}^{h} &= -\lim_{r\to\infty} \int_0^{2\pi}\int_0^{\pi/2}\int_{-\infty}^{+\infty} \hat{\mathbf{r}}\cdot\bar{\mathbf{T}}_{ac}\cdot\hat{\mathbf{r}} \sin\theta \, dt\, d\theta\, d\phi \\ &= \frac{1}{c^2\rho_{m0}} \int_0^{2\pi}\int_0^{\pi/2}\int_{-\infty}^{+\infty} |\mathcal{F}(\theta,\phi,t)|^2 \sin\theta\, dt\, d\theta\, d\phi. \end{aligned} \tag{5.222}$$

Time-Domain Antenna Parameters. In the frequency domain, a number of fundamental parameters, such as directivity, gain, effective area, and radiation efficiency, have been defined to characterize antennas at each frequency [151, ch. 2]. These definitions of frequency-domain antenna parameters can be generalized to also characterize time-domain antennas. Although the generalizations are not unique, most of the definitions of the frequency-domain parameters can be extended in a way that seems most straightforward and reasonable. Since these "logical extensions" of the definitions of frequency-domain antenna parameters have been given in [152], we shall limit our discussion to the definition of directivity.

The frequency-domain directivity $D_\omega(\theta,\phi)$ of an antenna in the direction given by (θ,ϕ) is defined as the ratio of the radiation intensity (power radiated per unit solid angle) in the given far-field direction from the antenna to the radiation intensity averaged over all directions [153, p. 260]. The average radiation intensity equals the total power radiated by the antenna divided by 4π. Thus, the frequency-domain directivity can be expressed in terms of the far-field pattern as

$$D_\omega(\theta,\phi) = \frac{4\pi|\boldsymbol{\mathcal{F}}_\omega(\theta,\phi)|^2}{\int_0^{2\pi}\int_0^{\pi} |\boldsymbol{\mathcal{F}}_\omega(\theta,\phi)|^2 \sin\theta d\theta d\phi}. \tag{5.223}$$

An analogous definition of directivity $D(\theta,\phi)$ for antennas operating in the time domain is the ratio of the energy radiated per unit solid angle in the given far-field direction (θ,ϕ) to the energy radiated per unit solid angle averaged over all directions. The average energy radiated per unit solid angle equals the total energy radiated by the antenna divided by 4π. Thus, the time-domain directivity can be written in terms of the time-domain far-field pattern as

$$\begin{aligned} D(\theta,\phi) &= \frac{4\pi \int_{-\infty}^{+\infty} |\boldsymbol{\mathcal{F}}(\theta,\phi,t)|^2 dt}{\int_0^{2\pi}\int_0^{\pi}\int_{-\infty}^{+\infty} |\boldsymbol{\mathcal{F}}(\theta,\phi,t)|^2 \sin\theta dt d\theta d\phi} \\ &= \frac{4\pi \int_{-\infty}^{+\infty} |\boldsymbol{\mathcal{F}}_\omega(\theta,\phi)|^2 d\omega}{\int_0^{2\pi}\int_0^{\pi}\int_{-\infty}^{+\infty} |\boldsymbol{\mathcal{F}}_\omega(\theta,\phi)|^2 \sin\theta d\omega d\theta d\phi} \end{aligned} \tag{5.224}$$

where the second equation in (5.224) is obtained from Parseval's theorem. This time-domain definition (5.224) of directivity exhibits the drawback common to all the "logically extended" time-domain definitions of the fundamental frequency-domain antenna parameters. Namely, they depend on the shape of the time-domain waveform that excites the antenna. This drawback leads one to consider specifying a set of common reference pulses to be used with the definitions of the time-domain parameters. Nonetheless, from their experience with using the time-domain antenna parameters, Allen, Hill, and Ondrejka [152] conclude that "Even without standard pulse shapes, time-domain antenna characterizations offer a viable alternative to their frequency-domain counterparts for describing antennas excited by transient waveforms."

When the sources are located in the half space $z > z_0$, and the plane of integration is given by $z = z_1 \le z_0$, all the numbered equations of this section remain the same except for $\int_{\pi/2}^{\pi}$ replacing $\int_0^{\pi/2}$.

5.3 ANALYTIC-SIGNAL ANALOGS OF THE PLANE-WAVE SPECTRUM REPRESENTATIONS

The analytic Fourier transform [154], [4, sec. 3.1] will now be used in conjunction with the frequency-domain plane-wave spectrum formulas of Section 3.2 to obtain analytic-signal representations of the field in the region $z > 0$. The analytic-signal representations are simpler in form than the time-domain representations of Section 5.2, but they involve complex times and analytic-signal fields that are not directly measurable and are therefore less attractive for measurement purposes. However, all the time-domain formulas of Section 5.2 can be obtained by taking the real part of the analytic-signal representations of the present section. As in Section 5.2, we assume, without loss of generality, that the scan plane is given by $z = z_0 = 0$ because it substantially simplifies the analysis.

Steinberg, Heyman, and Felsen [18] obtained results that are two-dimensional analogs of the acoustic formulas of the present section, and Melamed and Heyman [19] derived the three-dimensional acoustic formula (5.233) below.

The analytic-signal acoustic pressure $\overset{+}{p}(\mathbf{r}, t)$, which corresponds to the real measurable acoustic pressure $p(\mathbf{r}, t)$, is given by the inverse analytic Fourier transform of the frequency-domain pressure $p_\omega(\mathbf{r})$ as

$$\overset{+}{p}(\mathbf{r}, t) = 2 \int_0^{+\infty} p_\omega(\mathbf{r}) e^{-i\omega t} d\omega, \quad \mathrm{Im}(t) \le 0 \tag{5.225}$$

where the usual frequency and time-domain pressure fields are related through the standard Fourier transforms

$$p(\mathbf{r}, t) = \int_{-\infty}^{+\infty} p_\omega(\mathbf{r}) e^{-i\omega t} d\omega, \quad p_\omega(\mathbf{r}) = \frac{1}{2\pi} \int_{-\infty}^{+\infty} p(\mathbf{r}, t) e^{i\omega t} dt. \tag{5.226}$$

Thus, the analytic-signal pressure is simply obtained by integrating the frequency-domain pressure p_ω over positive ω only. From the definition (5.225) it follows that $\overset{+}{p}(\mathbf{r}, t)$ is analytic in the lower half of the complex t plane. The analytic-signal pressure may also be expressed directly in terms of the real time-domain pressure as

$$\overset{+}{p}(\mathbf{r}, t) = \begin{cases} p(\mathbf{r}, t) - i\widetilde{\mathcal{H}} p(\mathbf{r}, t), & \mathrm{Im}(t) = 0 \\ \dfrac{1}{i\pi} \displaystyle\int_{-\infty}^{+\infty} \frac{p(\mathbf{r}, t')}{t - t'} dt', & \mathrm{Im}(t) < 0 \end{cases} \tag{5.227}$$

where $\widetilde{\mathcal{H}}$ is the Hilbert transform defined in (5.93). The minus sign in (5.227) occurs because $\overset{+}{p}$ is analytic in the "lower" half plane rather than in the upper half as in Morse and Feshbach [115]. The first part of (5.227) follows from the fact that on the real axis, the real and imaginary parts of an analytic function are related through the Hilbert transform [115, pp. 370–372]. The second part of (5.227) can be proven by inserting $p_\omega(\mathbf{r})$ from (5.226) into (5.225), interchanging the order of integration, and performing the integration over ω. The first part of (5.227) can also be derived by expressing (5.225) for real t as

$$\begin{aligned}\overset{+}{p}(\mathbf{r},t) &= \int_{-\infty}^{+\infty} p_\omega(\mathbf{r})e^{-i\omega t}d\omega + 2i\,\mathrm{Im}\int_0^{+\infty} p_\omega(\mathbf{r})e^{-i\omega t}d\omega \\ &= p(\mathbf{r},t) + 2i\,\mathrm{Im}\int_0^{+\infty} p_\omega(\mathbf{r})e^{-i\omega t}d\omega, \quad \mathrm{Im}(t)=0\end{aligned} \tag{5.228}$$

and applying the formula (5.94).

The formula (5.227) reveals that the real pressure is recovered from the analytic-signal pressure by the equation

$$p(\mathbf{r},t) = \mathrm{Re}\,[\,\overset{+}{p}(\mathbf{r},t)], \quad \mathrm{Im}(t)=0. \tag{5.229}$$

To obtain a formula for $\overset{+}{p}$ in the region $z > 0$ in terms of $\overset{+}{p}$ in the plane $z = 0$, we define the analytic-signal spectrum $\overset{+}{T}(\xi,\eta,t)$ corresponding to the real time-domain spectrum $T(\xi,\eta,t)$ in (5.87)

$$\begin{aligned}\overset{+}{T}(\xi,\eta,t) &= \frac{1}{\pi}\int_{-\infty}^{+\infty}\int_{-\infty}^{+\infty}\int_0^{+\infty} p_\omega(\mathbf{r}_0)e^{-i\omega(t+\xi x_0+\eta y_0)}d\omega dx_0 dy_0 \\ &= \frac{1}{2\pi}\int_{-\infty}^{+\infty}\int_{-\infty}^{+\infty} \overset{+}{p}(\mathbf{r}_0, t+\xi x_0+\eta y_0)dx_0 dy_0, \quad \mathrm{Im}(t)\le 0.\end{aligned} \tag{5.230}$$

The formula (5.230) shows that the analytic-signal spectrum is determined from a space-time Radon transform of the analytic-signal pressure in the scan plane $z = 0$. Moreover, the relation (5.227) satisfied by all analytic-signal fields shows that

$$\overset{+}{T}(\xi,\eta,t) = \begin{cases} T(\xi,\eta,t) - i\widetilde{\mathcal{H}}T(\xi,\eta,t), & \mathrm{Im}(t)=0 \\ \dfrac{1}{i\pi}\displaystyle\int_{-\infty}^{+\infty}\frac{T(\xi,\eta,t')}{t-t'}dt', & \mathrm{Im}(t)<0\end{cases} \tag{5.231}$$

where $T(\xi,\eta,t)$ is the real time-domain spectrum given by (5.87).

The formula (5.81) for the frequency-domain acoustic pressure in the region $z > 0$ shows that for positive frequencies

$$p_\omega(\mathbf{r}) = \frac{1}{2\pi}\int_{-\infty}^{+\infty}\int_{-\infty}^{+\infty}\omega^2 T_\omega(\omega\xi,\omega\eta)e^{i\omega(\xi x+\eta y+\zeta z)}d\xi d\eta, \quad z\ge 0 \tag{5.232}$$

where ζ is given in (5.85). Taking the inverse analytic Fourier transform (which employs only positive frequencies) of (5.232), we find that

$$\overset{+}{p}(\mathbf{r},t) = -\frac{1}{2\pi}\int_{-\infty}^{+\infty}\int_{-\infty}^{+\infty}\frac{\partial^2}{\partial t^2}\overset{+}{T}(\xi,\eta,t-\xi x-\eta y-\zeta z)d\xi d\eta, \quad z>0,\ \mathrm{Im}(t)\le 0 \tag{5.233}$$

which gives the analytic-signal pressure in the half space $z > 0$ in terms of the analytic-signal spectrum. This formula was also derived by Melamed and Heyman [19]. In (5.233) no time convolution is necessary to determine the contribution from the evanescent region

$(\xi^2 + \eta^2 > c^{-2})$. The formula (5.233) for the analytic-signal pressure is therefore much simpler than the analogous formula (5.86) for the real pressure.

Substitution of $\overset{+}{T}$ from (5.231) into (5.233) and taking the real part to get $p(\mathbf{r}, t)$, we find

$$
\begin{aligned}
p(\mathbf{r}, t) = & -\frac{1}{2\pi} \iint\limits_{\xi^2+\eta^2<c^{-2}} \frac{\partial^2}{\partial t^2} T(\xi, \eta, t - \xi x - \eta y - \zeta z) d\xi d\eta \\
& - \mathrm{Re}\left(\frac{1}{2\pi^2 i} \iint\limits_{\xi^2+\eta^2>c^{-2}} \int\limits_{-\infty}^{+\infty} \frac{\partial^2}{\partial t^2} \frac{T(\xi, \eta, t')}{t - \xi x - \eta y + i|\zeta| z - t'} dt' d\xi d\eta \right).
\end{aligned}
\tag{5.234}
$$

Changing the t' integration variable in (5.234) to $t' - \xi x - \eta y$ recovers the formula (5.86), which was derived from the standard Fourier transform. To calculate the real pressure $p = \mathrm{Re}(\overset{+}{p})$ for real times from (5.233), we have to know the analytic-signal spectrum $\overset{+}{T}$ for complex times because ζ is complex when $\xi^2 + \eta^2 > c^{-2}$. Equation (5.230) shows that to get $\overset{+}{T}$ for complex times, the analytic-signal pressure $\overset{+}{p}$ must be known in the plane $z = 0$ for complex times. Consequently, to use (5.233) to calculate the real pressure p for real times, one has to know the analytic-signal pressure $\overset{+}{p}$ in the plane $z = 0$ for both complex and real times. (Of course, $\overset{+}{p}(\mathbf{r}, t)$ for both complex and real times can be computed from the analytic Fourier transform (5.225) of the frequency-domain pressure.)

Inserting $\mathbf{k} = \omega \boldsymbol{\chi}_\omega$ into (3.58) and taking the inverse analytic Fourier transform shows that the analytic-signal velocity is given in terms of the spectrum as

$$
\overset{+}{\mathbf{v}}(\mathbf{r}, t) = -\frac{1}{2\pi \rho_{m0}} \int\limits_{-\infty}^{+\infty} \int\limits_{-\infty}^{+\infty} \overset{+}{\boldsymbol{\chi}}(\xi, \eta) \frac{\partial^2}{\partial t^2} \overset{+}{T}(\xi, \eta, t - \xi x - \eta y - \zeta z) d\xi d\eta, \qquad z > 0, \quad \mathrm{Im}(t) \le 0
\tag{5.235}
$$

where the propagation vector is

$$
\overset{+}{\boldsymbol{\chi}}(\xi, \eta) = \xi \hat{\mathbf{x}} + \eta \hat{\mathbf{y}} + \zeta \hat{\mathbf{z}}
\tag{5.236}
$$

and thus $\overset{+}{\boldsymbol{\chi}} \cdot \overset{+}{\boldsymbol{\chi}} = c^{-2}$. Note that the vector $\overset{+}{\boldsymbol{\chi}}(\xi, \eta)$ is a simple complex function of (ξ, η), whereas the real vector $\widetilde{\boldsymbol{\chi}}(\xi, \eta)$ in (5.92) is an operator involving the Hilbert transform.

To derive electromagnetic analytic-signal formulas, we can begin with the analytic-signal electric field $\overset{+}{\mathbf{E}}(\mathbf{r}, t)$, which is determined from the frequency-domain electric field by

$$
\overset{+}{\mathbf{E}}(\mathbf{r}, t) = 2 \int\limits_0^{+\infty} \mathbf{E}_\omega(\mathbf{r}) e^{-i\omega t} d\omega, \quad \mathrm{Im}(t) \le 0.
\tag{5.237}
$$

Corresponding to (5.227), we have the analytic-signal electric field determined from the real time-domain electric field by

$$
\overset{+}{\mathbf{E}}(\mathbf{r}, t) = \begin{cases} \mathbf{E}(\mathbf{r}, t) - i\widetilde{\mathcal{H}}\mathbf{E}(\mathbf{r}, t), & \mathrm{Im}(t) = 0 \\ \dfrac{1}{i\pi} \displaystyle\int\limits_{-\infty}^{+\infty} \frac{\mathbf{E}(\mathbf{r}, t')}{t - t'} dt', & \mathrm{Im}(t) < 0. \end{cases}
\tag{5.238}
$$

The analytic-signal spectrum for the electric field is given by the space-time Radon transform of the analytic-signal electric field in the plane $z = 0$

$$
\overset{+}{\mathbf{T}}(\xi, \eta, t) = \frac{1}{2\pi} \int\limits_{-\infty}^{+\infty} \int\limits_{-\infty}^{+\infty} \overset{+}{\mathbf{E}}(\mathbf{r}_0, t + \xi x_0 + \eta y_0) dx_0 dy_0, \quad \mathrm{Im}(t) \le 0.
\tag{5.239}
$$

The analytic-signal spectrum can also be expressed in terms of the real time-domain spectrum as

$$\overset{+}{\mathbf{T}}(\xi,\eta,t) = \begin{cases} \mathbf{T}(\xi,\eta,t) - i\widetilde{\mathcal{H}}\mathbf{T}(\xi,\eta,t), & \operatorname{Im}(t) = 0 \\ \dfrac{1}{i\pi}\displaystyle\int_{-\infty}^{+\infty} \frac{\mathbf{T}(\xi,\eta,t')}{t-t'}dt', & \operatorname{Im}(t) < 0. \end{cases} \tag{5.240}$$

From (5.233) it is then found that the time-domain electric field is given in terms of the analytic-signal spectrum $\overset{+}{\mathbf{T}}(\xi,\eta,t)$ as

$$\overset{+}{\mathbf{E}}(\mathbf{r},t) = -\frac{1}{2\pi}\int_{-\infty}^{+\infty}\int_{-\infty}^{+\infty} \frac{\partial^2}{\partial t^2}\overset{+}{\mathbf{T}}(\xi,\eta,t-\xi x-\eta y-\zeta z)d\xi d\eta, \qquad z>0, \quad \operatorname{Im}(t)\le 0. \tag{5.241}$$

Substituting $\overset{+}{\mathbf{T}}$ from (5.240) into (5.241) and taking the real part to get $\mathbf{E}(\mathbf{r},t)$ yields the real formula (5.105), which was obtained from the standard Fourier transform.

The frequency-domain relation (5.109) shows that $\mathbf{T}_\omega$ is orthogonal to the vector $\boldsymbol{\chi}_\omega$, which is defined such that $\omega\boldsymbol{\chi}_\omega = \mathbf{k}$ for positive frequencies. Taking the inverse analytic Fourier transform of (5.109) proves that for all real (ξ,η)

$$\overset{+}{\boldsymbol{\chi}}(\xi,\eta)\cdot\overset{+}{\mathbf{T}}(\xi,\eta,t) = 0 \tag{5.242}$$

where the vector $\overset{+}{\boldsymbol{\chi}}$ is given in (5.236). Equation (5.242) shows that for all (ξ,η) the analytic-signal spectrum is orthogonal to the complex propagation direction $\xi\hat{\mathbf{x}}+\eta\hat{\mathbf{y}}+\zeta\hat{\mathbf{z}}$, so this spectrum is completely determined from two components of the electric field in the plane $z=0$. Specifically, the z component of $\overset{+}{\mathbf{T}}$ is given in terms of the x and y components by

$$\overset{+}{T}_z(\xi,\eta,t) = -\frac{1}{\zeta}\left(\xi\hat{\mathbf{x}}+\eta\hat{\mathbf{y}}\right)\cdot\overset{+}{\mathbf{T}}(\xi,\eta,t). \tag{5.243}$$

The formulas (5.237)–(5.243) also hold for the analytic-signal magnetic field $\overset{+}{\mathbf{H}}(\mathbf{r},t)$ and its analytic-signal spectrum $\overset{+}{\mathbf{T}}{}^H(\xi,\eta,t)$. The relation (5.114) between the frequency-domain spectra for the electric and magnetic fields along with the analytic Fourier transform show that for all real (ξ,η) the analytic-signal spectra are related through

$$\overset{+}{\mathbf{T}}{}^H(\xi,\eta,t) = \frac{1}{\mu}\overset{+}{\boldsymbol{\chi}}(\xi,\eta)\times\overset{+}{\mathbf{T}}(\xi,\eta,t). \tag{5.244}$$

The formulas (5.233) and (5.241) for the analytic-signal fields in the region $z>0$ are more compact than the corresponding formulas (5.86) and (5.105) for the real fields. However, to compute $\overset{+}{T}$ and $\overset{+}{\mathbf{T}}$ required in (5.233) and (5.241) directly from the time-domain measurements of $p(\mathbf{r},t)$ and $\mathbf{E}(\mathbf{r},t)$, we would use (5.227) or (5.231) and (5.238) or (5.240), respectively. As determined above, this procedure is equivalent to using (5.86) and (5.105) to compute $p(\mathbf{r},t)$ and $\mathbf{E}(\mathbf{r},t)$, respectively. Thus, the Green's function representations of Section 5.1 and the plane-wave time-domain spectrum formulas of Section 5.2 obtained from the standard Fourier transform are ultimately the more useful expressions for numerical calculations since they involve only real fields that are obtainable directly from time-domain measurements. Moreover, these real time-domain plane-wave spectrum formulas reduce in simplicity to that of the corresponding plane-wave spectrum formulas in the frequency domain when the evanescent fields (second integrals in (5.86) and (5.105)) are negligible.

When the sources are located in $z>0$, the fields in $z<0$ are determined by (5.230), (5.233), (5.235), (5.239), (5.241), (5.242), and (5.244) with z replaced by $-z$ and $\overset{+}{\boldsymbol{\chi}}$ replaced by $\overset{+}{\boldsymbol{\chi}}{}^- = \xi\hat{\mathbf{x}}+\eta\hat{\mathbf{y}}-\zeta\hat{\mathbf{z}}$.

5.3.1 Spectrum Given in Terms of the Sources

The analytic-signal spectra $\overset{+}{T}$, $\overset{+}{\mathbf{T}}$, and $\overset{+}{\mathbf{T}}^H$ can be expressed in terms of the analytic-signal volume sources and Huygens' sources on the surface S shown in Figure 3.1 by applying the inverse analytic Fourier transform to the frequency-domain formulas in Section 3.2.2.

The expression (3.104) for the frequency-domain spectrum of the acoustic field shows that for all $\omega > 0$ and all real (ξ, η) we have

$$-i\omega T_\omega(\omega\xi, \omega\eta) = \frac{1}{4\pi\zeta}\int_V q_\omega(\mathbf{r})e^{-i\omega(\xi x+\eta y+\zeta z)}dV \tag{5.245}$$

where $q_\omega(\mathbf{r})$ is the acoustic source function and ζ is given in (5.85). Taking the inverse analytic Fourier transform of (5.245) gives us the following volume-source expression, valid for all real (ξ, η), for the analytic-signal spectrum

$$\frac{\partial}{\partial t}\overset{+}{T}(\xi, \eta, t) = \frac{1}{4\pi\zeta}\int_V \overset{+}{q}(\mathbf{r}, t + \xi x + \eta y + \zeta z)dV, \quad \text{Im}(t) \le 0 \tag{5.246}$$

where

$$\overset{+}{q}(\mathbf{r}, t) = 2\int_0^{+\infty} q_\omega(\mathbf{r})e^{-i\omega t}d\omega, \quad \text{Im}(t) \le 0 \tag{5.247}$$

is the analytic-signal acoustic source function. (Inserting (5.246) into (5.233) produces an expression that is in agreement with Heyman's result [17, eqs. (4.13)–(4.14)]. Note that in [17] the spectral variables (ξ_1, ξ_2, ζ) are dimensionless.) Since the sources are located in the region $z < 0$ and $\text{Im}(\zeta) \ge 0$, we have $\text{Im}(t + \xi x + \eta y + \zeta z) \le 0$ throughout the integration in (5.246). For propagating plane waves $(\xi^2 + \eta^2 < c^{-2})$, the real time-domain spectrum is given by the expression (5.138), which is identical to the expression (5.246) for the analytic-signal spectrum. However, for evanescent plane waves $(\xi^2 + \eta^2 > c^{-2})$, the real time-domain spectrum is given by the expressions (5.141) or (5.144), which involve time convolutions and are much more complicated than the expression (5.246) for the analytic-signal spectrum. This is another example showing that the analytic Fourier transform can lead to simpler expressions than the standard real Fourier transform. Taking the real part of (5.246) and using the relation

$$\overset{+}{q}(\mathbf{r}, t) = \begin{cases} q(\mathbf{r}, t) - i\widetilde{\mathcal{H}}q(\mathbf{r}, t), & \text{Im}(t) = 0 \\ \dfrac{1}{i\pi}\displaystyle\int_{-\infty}^{+\infty} \frac{q(\mathbf{r}, t')}{t - t'}dt', & \text{Im}(t) < 0 \end{cases} \tag{5.248}$$

produces the expressions (5.138), (5.141), and (5.144) for the real time-domain spectrum.

In a similar manner the Huygens' source expression (3.105) shows that for positive frequencies and all real (ξ, η)

$$T_\omega(\xi\omega, \eta\omega) = \frac{1}{4\pi\zeta}\int_S \left[\hat{\mathbf{n}} \cdot \chi_\omega(\xi, \eta)p_\omega(\mathbf{r}) + \rho_{m0}\hat{\mathbf{n}} \cdot \mathbf{v}_\omega(\mathbf{r})\right]e^{-i\omega(\xi x+\eta y+\zeta z)}dS \tag{5.249}$$

where $\hat{\mathbf{n}}$ is the outward normal to the surface S, which encloses the sources and is located in the region $z < 0$. Taking the inverse analytic Fourier transform of (5.249) shows that for all real (ξ, η)

$$\overset{+}{T}(\xi,\eta,t) = \frac{1}{4\pi\zeta}\int_S \left[\hat{\mathbf{n}}\cdot\overset{+}{\chi}(\xi,\eta)\overset{+}{p}(\mathbf{r}, t+\xi x+\eta y+\zeta z) + \rho_{mo}\hat{\mathbf{n}}\cdot\overset{+}{\mathbf{v}}(\mathbf{r}, t+\xi x+\eta y+\zeta z)\right]dS, \quad \mathrm{Im}(t)\le 0. \tag{5.250}$$

For propagating plane waves ($\xi^2+\eta^2 < c^{-2}$), the real time-domain spectrum is given by the Huygens' source expression (5.146), which is identical to the expression (5.250) for the analytic-signal spectrum. For evanescent plane waves ($\xi^2+\eta^2 > c^{-2}$), the real time-domain spectrum is given by the Huygens' source expression (5.148), which involves time convolutions and is more complicated than the expression (5.250) for the analytic-signal spectrum. Taking the real part of (5.250) and using the relation (5.227) for $\overset{+}{p}$ and the corresponding expression for $\overset{+}{\mathbf{v}}$, produces the expressions (5.146) and (5.148) for the real time-domain spectrum in terms of Huygens' sources.

Using the volume and Huygens' source expressions (3.108) and (3.110) for the spectrum of the magnetic field, and performing derivations similar to those that led to (5.246) and (5.250), we find the following two expressions that are valid for all real (ξ, η)

$$\overset{+}{\mathbf{T}}{}^H(\xi,\eta,t) = -\frac{1}{4\pi\zeta}\overset{+}{\chi}(\xi,\eta)\times\int_V \overset{+}{\mathbf{J}}(\mathbf{r}, t+\xi x+\eta y+\zeta z)dV, \quad \mathrm{Im}(t)\le 0 \tag{5.251}$$

and

$$\overset{+}{\mathbf{T}}{}^H(\xi,\eta,t) = -\frac{1}{4\pi\zeta}\overset{+}{\chi}(\xi,\eta)\times\int_S \hat{\mathbf{n}}\times\overset{+}{\mathbf{H}}(\mathbf{r}, t+\xi x+\eta y+\zeta z)dS - \frac{1}{4\pi\mu\zeta}\overset{+}{\chi}(\xi,\eta)\times\left(\overset{+}{\chi}(\xi,\eta)\times\int_S \hat{\mathbf{n}}\times\overset{+}{\mathbf{E}}(\mathbf{r}, t+\xi x+\eta y+\zeta z)dS\right), \quad \mathrm{Im}(t)\le 0 \tag{5.252}$$

where $\overset{+}{\mathbf{J}}$ is the analytic-signal source current. Again, we note that these volume and Huygens' source expressions for the analytic-signal spectrum of the magnetic field are identical to the corresponding real expressions (5.150) and (5.154) for propagating plane waves ($\xi^2+\eta^2 < c^{-2}$). For evanescent plane waves ($\xi^2+\eta^2 > c^{-2}$), the real time-domain spectrum for the magnetic field is given by the volume and Huygens' source expressions (5.152) and (5.155), which involves time convolutions and are more complicated than the expressions (5.251) and (5.252) for the corresponding analytic-signal spectrum. We can use the relation (5.238), and the analogous relations for the magnetic field and the current, in taking the real part of (5.251) and (5.252) to get the corresponding formulas (5.152) and (5.155) for the real time-domain spectrum of the magnetic field.

The formulas for the analytic-signal spectrum $\overset{+}{\mathbf{T}}(\xi,\eta,t)$ of the electric field in terms of volume and Huygens' sources can be obtained from (5.251) and (5.252) by noting that

$$\overset{+}{\mathbf{T}}(\xi,\eta,t) = -\frac{1}{\epsilon}\overset{+}{\chi}(\xi,\eta)\times\overset{+}{\mathbf{T}}{}^H(\xi,\eta,t), \quad \mathrm{Im}(t)\le 0 \tag{5.253}$$

which holds for all real (ξ, η). This relation between $\overset{+}{\mathbf{T}}(\xi,\eta,t)$ and $\overset{+}{\mathbf{T}}{}^H(\xi,\eta,t)$ follows from (5.244) and $\overset{+}{\chi}\cdot\overset{+}{\chi} = c^{-2}$.

When the sources are located in the half space $z>0$, the fields in $z<0$ can be determined from the spectra given in (5.246), (5.250), (5.251), (5.252), and (5.253), with z replaced by $-z$ and $\overset{+}{\chi}$ replaced by $\overset{+}{\chi}{}^- = \xi\hat{\mathbf{x}}+\eta\hat{\mathbf{y}}-\zeta\hat{\mathbf{z}}$.

5.3.2 Singularities of the Spectrum

The possible singularities with respect to (ξ, η) of the first time derivatives of the analytic-signal spectra $\overset{+}{T}$, $\overset{+}{\mathbf{T}}$, and $\overset{+}{\mathbf{T}}^H$ will now be determined using the volume and Huygens' source expressions of Section 5.3.1. We start by defining the analytic-signal far-field function for the acoustic field by

$$\overset{+}{F}(\xi, \eta, t) = 2 \int_0^{+\infty} F_\omega(\xi\omega, \eta\omega) e^{-i\omega t} d\omega, \quad \text{Im}(t) \leq 0 \tag{5.254}$$

where F_ω is the frequency-domain far-field function. Equation (3.112) relates the frequency-domain far-field function to the spectrum and shows that for $\omega > 0$ and all real (ξ, η) we have

$$-i\omega T_\omega(\xi\omega, \eta\omega) = \frac{F_\omega(\xi\omega, \eta\omega)}{\zeta}. \tag{5.255}$$

Taking the inverse analytic Fourier transform of (5.255) gives the following relation (valid for all real (ξ, η)) between the analytic-signal far-field function and the analytic-signal spectrum

$$\frac{\partial}{\partial t} \overset{+}{T}(\xi, \eta, t) = \frac{\overset{+}{F}(\xi, \eta, t)}{\zeta}, \quad \text{Im}(t) \leq 0. \tag{5.256}$$

The expression (5.246) for the analytic-signal spectrum in terms of the analytic-signal volume source $\overset{+}{q}(\mathbf{r}, t)$ along with (5.256) immediately gives us

$$\overset{+}{F}(\xi, \eta, t) = \frac{1}{4\pi} \int_V \overset{+}{q}(\mathbf{r}, t + \xi x + \eta y + \zeta z) dV, \quad \text{Im}(t) \leq 0 \tag{5.257}$$

which is valid for all real (ξ, η) and expresses the analytic-signal far-field function in terms of the analytic-signal volume-source function. This formula shows that when the source function $\overset{+}{q}(\mathbf{r}, t)$ is continuous, the far-field function $\overset{+}{F}(\xi, \eta, t)$ is continuous for all (ξ, η, t). The relation (5.256) then proves that *when the source function* $\overset{+}{q}(\mathbf{r}, t)$ *is continuous, the time derivative of the spectrum* $\overset{+}{T}(\xi, \eta, t)$ *is continuous for all* (ξ, η, t)*, except possibly at* $\xi^2 + \eta^2 = c^{-2}$ *(corresponding to* $\zeta = 0$*), where it may have a* $1/\zeta$ *singularity.*

We can also combine the Huygens' source expression (5.250) for the spectrum with (5.256) to show that for all real (ξ, η) the analytic-signal far-field function is related to the Huygens' sources through

$$\begin{aligned} \overset{+}{F}(\xi, \eta, t) = \frac{1}{4\pi} \int_S \Big[&\hat{\mathbf{n}} \cdot \overset{+}{\boldsymbol{\chi}}(\xi, \eta) \frac{\partial}{\partial t} \overset{+}{p}(\mathbf{r}, t + \xi x + \eta y + \zeta z) \\ &+ \rho_{m0} \hat{\mathbf{n}} \cdot \frac{\partial}{\partial t} \overset{+}{\mathbf{v}}(\mathbf{r}, t + \xi x + \eta y + \zeta z) \Big] dS, \quad \text{Im}(t) \leq 0. \end{aligned} \tag{5.258}$$

Consequently, the far-field function $\overset{+}{F}(\xi, \eta, t)$ is a continuous function when the Huygens' sources $\partial \overset{+}{p}/\partial t$ and $\rho_{m0}(\partial/\partial t)\hat{\mathbf{n}} \cdot \overset{+}{\mathbf{v}} = -\partial \overset{+}{p}/\partial n$ are continuous on S. Moreover, (5.256) shows that *when the Huygens' sources* $\partial \overset{+}{p}/\partial t$ *and* $\rho_{m0}(\partial/\partial t)\hat{\mathbf{n}} \cdot \overset{+}{\mathbf{v}}$ *are continuous on S, the time derivative of the spectrum* $\overset{+}{T}(\xi, \eta, t)$ *is continuous for all* (ξ, η, t)*, except possibly at* $\xi^2 + \eta^2 = c^{-2}$ *(corresponding to* $\zeta = 0$*), where it may have a* $1/\zeta$ *singularity.*

Corresponding results for the electromagnetic fields can be derived in a similar fashion. The frequency-domain far-field functions for the electric and magnetic fields are given in

(3.115) in terms of the spectra for the electric and magnetic fields, and analogous to (5.256) we find that for all real (ξ, η)

$$\frac{\partial}{\partial t}\overset{+}{\mathbf{T}}(\xi, \eta, t) = \frac{\overset{+}{\mathbf{F}}(\xi, \eta, t)}{\zeta}, \quad \frac{\partial}{\partial t}\overset{+}{\mathbf{T}}{}^{H}(\xi, \eta, t) = \frac{\overset{+}{\mathbf{F}}{}^{H}(\xi, \eta, t)}{\zeta}, \quad \mathrm{Im}(t) \leq 0. \tag{5.259}$$

Moreover, the relation (5.253) between $\overset{+}{\mathbf{T}}$ and $\overset{+}{\mathbf{T}}{}^{H}$ yields for all real (ξ, η)

$$\overset{+}{\mathbf{F}}(\xi, \eta, t) = -\frac{1}{\epsilon}\overset{+}{\chi}(\xi, \eta) \times \overset{+}{\mathbf{F}}{}^{H}(\xi, \eta, t), \quad \mathrm{Im}(t) \leq 0. \tag{5.260}$$

The expression (5.251) giving the analytic-signal spectrum in terms of the current $\overset{+}{\mathbf{J}}(\mathbf{r}, t)$ shows that for all real (ξ, η)

$$\overset{+}{\mathbf{F}}{}^{H}(\xi, \eta, t) = -\frac{1}{4\pi}\overset{+}{\chi}(\xi, \eta) \times \int_V \frac{\partial}{\partial t}\overset{+}{\mathbf{J}}(\mathbf{r}, t + \xi x + \eta y + \zeta z) dV, \quad \mathrm{Im}(t) \leq 0. \tag{5.261}$$

From this formula we see that when the time derivative of the current $\overset{+}{\mathbf{J}}(\mathbf{r}, t)$ is continuous, the far-field function $\overset{+}{\mathbf{F}}{}^{H}$ is continuous for all (ξ, η, t). The formulas (5.259) and (5.260) then prove that *when the time derivative of the current $\overset{+}{\mathbf{J}}(\mathbf{r}, t)$ is continuous, the first time derivatives of the spectra $\overset{+}{\mathbf{T}}(\xi, \eta, t)$ and $\overset{+}{\mathbf{T}}{}^{H}(\xi, \eta, t)$ are continuous for all (ξ, η, t), except possibly at $\xi^2 + \eta^2 = c^{-2}$ (corresponding to $\zeta = 0$), where they may have $1/\zeta$ singularities.*

Similarly, from (5.252) we find that for all real (ξ, η) the analytic-signal far-field function for the magnetic field may be expressed in terms of Huygens' sources as

$$\begin{aligned} \overset{+}{\mathbf{F}}{}^{H}(\xi, \eta, t) = &-\frac{1}{4\pi}\overset{+}{\chi}(\xi, \eta) \times \int_S \hat{\mathbf{n}} \times \frac{\partial}{\partial t}\overset{+}{\mathbf{H}}(\mathbf{r}, t + \xi x + \eta y + \zeta z) dS \\ &- \frac{1}{4\pi\mu}\overset{+}{\chi}(\xi, \eta) \times \left(\overset{+}{\chi}(\xi, \eta) \times \int_S \hat{\mathbf{n}} \times \frac{\partial}{\partial t}\overset{+}{\mathbf{E}}(\mathbf{r}, t + \xi x + \eta y + \zeta z) dS\right), \\ &\mathrm{Im}(t) \leq 0. \end{aligned} \tag{5.262}$$

Thus, $\overset{+}{\mathbf{F}}{}^{H}(\xi, \eta, t)$ is continuous when the first time derivatives of the Huygens' sources $\hat{\mathbf{n}} \times \overset{+}{\mathbf{E}}(\mathbf{r}, t)$ and $\hat{\mathbf{n}} \times \overset{+}{\mathbf{H}}(\mathbf{r}, t)$ are continuous on S. Combining this result with (5.259) proves that *when the first time derivatives of the Huygens' sources $\hat{\mathbf{n}} \times \overset{+}{\mathbf{E}}(\mathbf{r}, t)$ and $\hat{\mathbf{n}} \times \overset{+}{\mathbf{H}}(\mathbf{r}, t)$ are continuous on S, the first time derivatives of the spectra $\overset{+}{\mathbf{T}}(\xi, \eta, t)$ and $\overset{+}{\mathbf{T}}{}^{H}(\xi, \eta, t)$ are continuous for all (ξ, η, t), except possibly at $\xi^2 + \eta^2 = c^{-2}$ (corresponding to $\zeta = 0$), where they may have $1/\zeta$ singularities.*

When the sources are located in the half space $z > 0$, the far-field functions for the fields in $z < 0$ are given by (5.257), (5.258), (5.260), (5.261), and (5.262), with z replaced by $-z$ and $\overset{+}{\chi}$ replaced by $\overset{+}{\chi}{}^{-} = \xi\hat{\mathbf{x}} + \eta\hat{\mathbf{y}} - \zeta\hat{\mathbf{z}}$. The formulas (5.256) and (5.259) remain the same.

5.3.3 Analyticity of the Far-Field Pattern and the Far-Field Function

We shall investigate the possible analytic continuation of the analytic-signal far-field pattern and far-field function with respect to (θ, ϕ) and (ξ, η), respectively. Moreover, we shall determine a relation between the analytic-signal far-field pattern and far-field function. The results presented in this section are the analytic-signal analogs of the time-domain results of Section 5.2.4. It is found that the properties of the analytic-signal fields are significantly different from the properties described in Section 5.2.4 for the real time-domain fields.

Throughout this section, we assume that the sources satisfy the conditions stated in Section 5.1.4, so that the analytic-signal far fields decay as $1/r$, or faster, as $r \to \infty$. Far fields that decay slower than $1/r$ are considered in Section 5.4.

Properties of the Far-Field Pattern. The frequency-domain far-field pattern is given in terms of the volume-source function in (3.35), which shows that the analytic-signal far-field pattern is

$$\begin{aligned}\overset{+}{\mathcal{F}}(\theta,\phi,t) &= \frac{1}{2\pi}\int_0^{+\infty}\int_V q_\omega(\mathbf{r})e^{-i(\omega/c)(x\cos\phi\sin\theta+y\sin\phi\sin\theta+z\cos\theta)}e^{-i\omega t}dV\,d\omega \\ &= \frac{1}{4\pi}\int_V \overset{+}{q}(\mathbf{r}, t+(x\cos\phi\sin\theta+y\sin\phi\sin\theta+z\cos\theta)/c)dV, \\ &\qquad \mathrm{Im}(t)\le 0.\end{aligned} \tag{5.263}$$

In the volume integrations of (5.263), x and y can take on any real value, whereas z is always negative because the sources are located in the half space $z < 0$. Moreover, the analytic-signal source function $\overset{+}{q}(\mathbf{r}, t)$ is analytic for all t with $\mathrm{Im}(t) \le 0$, but it is not, in general, defined for t with $\mathrm{Im}(t) > 0$. Consequently, we see from the last part of (5.263) that the analytic-signal far-field pattern cannot, in general, be analytically continued to complex (θ, ϕ) because this would require that $\overset{+}{q}(\mathbf{r}, t)$ be evaluated at times for which $\mathrm{Im}(t) > 0$.

However, the analytic-signal far-field pattern is always defined for $0 \le \phi \le 2\pi$ and $\theta \in \Gamma$, where Γ is the contour in Figure 3.4 that consists of a line segment from 0 to $\pi/2$ and a line segment from $\pi/2$ to $\pi/2 - i\infty$. This is shown by noting that $\sin(\pi/2 - i\alpha) = \cosh\alpha$, $\cos(\pi/2 - i\alpha) = i\sinh\alpha$, and $z < 0$ throughout the integration over V, so that $\mathrm{Im}(t + (x\cos\phi\sin\theta + y\sin\phi\sin\theta + z\cos\theta)/c) \le 0$ when $\mathrm{Im}(t) \le 0$, $0 \le \phi \le 2\pi$ and $\theta \in \Gamma$. This result will prove important when we derive a relation between the far-field pattern and the far-field function below.

As we did for the real time-domain far-field pattern considered in Section 5.2.4, we can show that the analytic-signal far-field pattern can be analytically continued to complex (θ, ϕ) around the real θ and ϕ axes if the source function $q_\omega(\mathbf{r})$ satisfies the following two conditions.

1. The function $q_\omega(\mathbf{r})$ is integrable over the bounded source volume V.
2. There exists a $T > 0$, such that $q_\omega(\mathbf{r}) = O(e^{-\omega T})$ for all $\mathbf{r}$ as $\omega \to \infty$.

To prove this, begin by noting that with $q_\omega(\mathbf{r})$ satisfying 1. and 2., the analytic-signal source function $\overset{+}{q}(\mathbf{r}, t)$ is analytic for all t with $\mathrm{Im}(t) < T$, as can be seen from (5.247). The expression for the far-field pattern in terms of $\overset{+}{q}$ in (5.263) is therefore analytic for (θ, ϕ) in a complex region around the real θ and ϕ axes for $\mathrm{Im}(t) \le 0$. Alternatively, we can use the Huygens' source expression (3.38) to show that the far-field pattern can be analytically continued if the Huygens' sources decay exponentially as $\omega \to \infty$. Thereby conditions 1. and 2. are replaced by the requirement that on the surface S, there exists a $T > 0$, such that for all $\mathbf{r} \in S$, $p_\omega(\mathbf{r}) = O(e^{-\omega T})$ as $\omega \to \infty$. (The frequency-domain Huygens' sources are continuous on S, because they obey the homogeneous acoustic differential equations across the surface S and are therefore automatically integrable on S.)

Although the analytic-signal far-field pattern cannot always be analytically continued to a complex region around the real θ and ϕ axes, we can still show an important theorem for its angular dependence. The theorem holds under the extremely weak condition that the real time-domain far-field pattern and the frequency-domain far-field pattern are related through the Fourier transform pairs (5.174) and (5.175). It states that *if the analytic-signal acoustic far-field pattern $\overset{+}{\mathcal{F}}(\theta, \phi, t)$ of a finite source region is zero for all real times for $(\theta, \phi) \in B$,* *where B has a nonzero area, then the analytic-signal far-field pattern is zero for all angles of observation at all times with* $\operatorname{Im}(t) \le 0$. To prove this theorem, we begin by noting that if $\overset{+}{\mathcal{F}}(\theta, \phi, t) = 0$ for all real t and $(\theta, \phi) \in B$, then the first part of

$$\overset{+}{\mathcal{F}}(\theta, \phi, t) = \begin{cases} \mathcal{F}(\theta, \phi, t) - i\widetilde{\mathcal{H}}\mathcal{F}(\theta, \phi, t), & \operatorname{Im}(t) = 0 \\ \dfrac{1}{i\pi} \displaystyle\int_{-\infty}^{+\infty} \dfrac{\mathcal{F}(\theta, \phi, t')}{t - t'} dt', & \operatorname{Im}(t) < 0 \end{cases} \tag{5.264}$$

implies that $\mathcal{F}(\theta, \phi, t) = 0$ for all real t and $(\theta, \phi) \in B$. According to Section 5.2.4, this, in turn, implies that $\mathcal{F}(\theta, \phi, t) = 0$ for all real t and all complex (θ, ϕ). With this result inserted into (5.264), we have completed the proof of the theorem.

Properties of the Far-Field Function. The frequency-domain expressions (5.176) and (5.177) show that for $\omega > 0$ and all real (ξ, η) the frequency-domain far-field function can be written as

$$F_\omega(\xi\omega, \eta\omega) = \frac{1}{4\pi} \int_V q_\omega(\mathbf{r}) e^{-i\omega(\xi x + \eta y + \zeta z)} dV \tag{5.265}$$

so that, for all real (ξ, η), the analytic-signal far-field function is

$$\begin{aligned} \overset{+}{F}(\xi, \eta, t) &= \frac{1}{2\pi} \int_0^{+\infty} \int_V q_\omega(\mathbf{r}) e^{-i\omega(\xi x + \eta y + \zeta z)} e^{-i\omega t} dV d\omega \\ &= \frac{1}{4\pi} \int_V \overset{+}{q}(\mathbf{r}, t + \xi x + \eta y + \zeta z) dV, \quad \operatorname{Im}(t) \le 0. \end{aligned} \tag{5.266}$$

Like the far-field pattern, we find that the analytic-signal far-field function cannot, in general, be analytically continued to complex (ξ, η) because the analytic-signal source function $\overset{+}{q}(\mathbf{r}, t)$ is defined only for $\operatorname{Im}(t) \le 0$.

However, we can show that if the source function $q_\omega(\mathbf{r})$ satisfies conditions 1. and 2. above, the analytic-signal far-field function can be analytically continued to a complex region around the real ξ and η axes, except possibly at $\xi^2 + \eta^2 = c^{-2}$. The proof goes as follows. By specifying that the branch cut for the square root defining ζ in (5.85) extends from 0 to $i\infty$, the function ζ is analytic for complex values of (ξ, η) in regions around the real ξ and η axes, except at $\xi^2 + \eta^2 = c^{-2}$, where ζ has a branch point. With this specification and the fact that $\overset{+}{q}(\mathbf{r}, t)$ is analytic for all t with $\operatorname{Im}(t) < T$, we see that the far-field function given by (5.266) indeed is analytic with respect to (ξ, η) in a complex region around the real ξ and η axes, except possibly at $\xi^2 + \eta^2 = c^{-2}$, where ζ has a branch point. Thus, unlike the real time-domain far-field function that could not be analytically continued, the analytic-signal far-field function can be analytically continued to complex ξ and η when the source function decays exponentially as $\omega \to \infty$. Like the analytic-signal far-field pattern, the analytic-signal far-field function can also be analytically continued if the frequency-domain Huygens' sources decay exponentially as $\omega \to \infty$. This latter result is proven from the Huygens' source expression (5.258).

Relations Between the Far-Field Pattern and the Far-Field Function. Comparing (5.266) and (5.263) produces the following relation, which is valid for all source functions, between the analytic-signal far-field pattern and the analytic-signal far-field function

$$\begin{aligned} &\overset{+}{\mathcal{F}}(\theta, \phi, t) = \overset{+}{F}(c^{-1} \sin\theta \cos\phi, c^{-1} \sin\theta \sin\phi, t), \\ &\qquad\qquad 0 \le \phi \le 2\pi, \quad \theta \in \Gamma, \quad \operatorname{Im}(t) \le 0 \end{aligned} \tag{5.267}$$

where Γ is the contour shown in Figure 3.4. This relation holds even when the general analytic continuation of the far-field pattern is not possible. With $\xi = c^{-1}\sin\theta\cos\phi$ and $\eta = c^{-1}\sin\theta\sin\phi$, the region given by $0 \le \phi \le 2\pi$ and $\theta \in \Gamma$ corresponds to the entire real (ξ, η) plane. Thus, *the analytic-signal far-field function* $\overset{+}{F}(\xi, \eta, t)$ *for all real* (ξ, η) *is completely determined from the analytic-signal far-field pattern* $\overset{+}{\mathcal{F}}(\theta, \phi, t)$ *for* $0 \le \phi \le 2\pi$ *and* $\theta \in \Gamma$. Once the analytic-signal far-field function is determined, the analytic-signal spectrum is determined from (5.256).

Consequently, like the frequency-domain spectrum (Section 3.2.4), but unlike the real time-domain spectrum (Section 5.2.4), the analytic-signal spectrum is always completely determined from the values of its far-field pattern in the region $0 \le \phi \le 2\pi$ and $\theta \in \Gamma$.

EXAMPLE

To illustrate these properties of the analytic-signal far-field pattern and far-field function, we consider the example where the source function is $q_\omega(\mathbf{r}) = \delta(\mathbf{r} + a\hat{\mathbf{z}})e^{-\omega T}$, with a and T being positive constants. This source was also considered in Section 5.2.4, and it satisfies conditions 1. and 2. above, so that its corresponding analytic-signal far-field pattern and far-field function can be analytically continued to complex regions around the real axes for their arguments. For this source the analytic-signal source function in (5.247) becomes

$$\overset{+}{q}(\mathbf{r}, t) = \delta(\mathbf{r} + a\hat{\mathbf{z}})\frac{2}{T + it}, \quad \mathrm{Im}(t) \le 0 \tag{5.268}$$

so the analytic-signal far-field pattern is found from (5.263) to be

$$\overset{+}{\mathcal{F}}(\theta, \phi, t) = \frac{1}{2\pi}\frac{1}{T + i(t - (a/c)\cos\theta)}, \quad \mathrm{Im}(t) \le 0. \tag{5.269}$$

The analytic-signal far-field function is found from (5.257) to be

$$\overset{+}{F}(\xi, \eta, t) = \frac{1}{2\pi}\frac{1}{T + i(t - a\zeta)}, \quad \mathrm{Im}(t) \le 0 \tag{5.270}$$

for all real (ξ, η). Taking the real part of these analytic-signal expressions produces, in accordance with (5.229), the real time-domain source function (5.180), the real time-domain far-field pattern (5.181), and the real time-domain far-field function (5.182)–(5.183). The far-field pattern in (5.269) and the far-field function in (5.270) are seen to satisfy the relation (5.267). Also, for all t with $\mathrm{Im}(t) \le 0$, the far-field pattern in (5.269) is an analytic function of complex θ throughout the complex region around the real θ axis for which $\mathrm{Im}(\cos\theta) > -cT/a$. In particular, the far-field pattern is analytic with respect to θ around the contour Γ on which $\mathrm{Im}(\cos\theta) \ge 0$. Similarly, for all t with $\mathrm{Im}(t) \le 0$, the far-field function in (5.270) is an analytic function of complex (ξ, η) in complex regions around the real ξ and η axes, for which $\mathrm{Im}(\zeta) > -T/a$, except at $\xi^2 + \eta^2 = c^{-2}$, where ζ has a branch point.

Electromagnetic Fields. The results of this section also hold for the electromagnetic fields, and the corresponding proofs are identical to the proofs for the acoustic field. In particular, we find that *the analytic-signal far-field functions* $\overset{+}{\mathbf{F}}(\xi, \eta, t)$ *and* $\overset{+}{\mathbf{F}}{}^H(\xi, \eta, t)$ *for the electric and magnetic fields for all real* (ξ, η) *are completely determined from the analytic-signal far-field pattern* $\overset{+}{\boldsymbol{\mathcal{F}}}(\theta, \phi, t)$ *or* $\overset{+}{\boldsymbol{\mathcal{F}}}{}^H(\theta, \phi, t)$ *for* $0 \le \phi \le 2\pi$ *and* $\theta \in \Gamma$. Moreover, *if the analytic-signal far-field pattern* $\overset{+}{\boldsymbol{\mathcal{F}}}(\theta, \phi, t)$ *or* $\overset{+}{\boldsymbol{\mathcal{F}}}{}^H(\theta, \phi, t)$ *for the electric or magnetic field of a finite source region is zero for all real times for* $(\theta, \phi) \in B$, *where B has a nonzero area, then the analytic-signal far-field patterns for both electric and magnetic fields are zero for all angles of observation at all times with* $\mathrm{Im}(t) \le 0$. Both these theorems hold for all source functions and do not require a general analytical continuation of the far-field pattern to complex (θ, ϕ).

5.3.4 Far-Field Expressions

To express the analytic-signal far fields in terms of the analytic-signal spectra $\overset{+}{T}(\xi, \eta, t)$ and $\overset{+}{\mathbf{T}}(\xi, \eta, t)$ for the acoustic and electric fields, let us assume that the sources satisfy the conditions stated in Section 5.1.4, so that the far fields decay as $1/r$, or faster, as $r \to \infty$. The formulas will be derived by applying the analytic Fourier transform to the frequency-domain far-field expressions of Section 3.2.5. As in Section 5.2.5 on the real time-domain far fields, we consider only the total far fields obtained by adding the contributions from propagating and evanescent plane waves. Thus, unlike Section 3.2.5, the present section does not include separate expressions for the far-field contributions from the propagating and evanescent plane waves.

Far Fields for $0 \le \theta < \pi/2$. Applying the analytic Fourier transform to the frequency-domain far-field formula (3.153) (which, according to (3.143), is valid also for $\theta = 0$) gives us immediately

$$\overset{+}{p}(\mathbf{r}, t) = \frac{\cos\theta}{cr}\frac{\partial}{\partial t}\overset{+}{T}(c^{-1}\cos\phi\sin\theta, c^{-1}\sin\phi\sin\theta, t - r/c) + O\left(\frac{1}{r^2}\right), \quad \mathrm{Im}(t) \le 0. \tag{5.271}$$

Applying (5.231) in taking the real part of (5.271) produces a far-field expression for the real time-domain acoustic field, which is identical to the real time-domain spectrum formula (5.186) and to the Green's function far-field formula (5.54). The formula (5.271) also shows that the analytic-signal far-field pattern can be expressed in terms of the analytic-signal spectrum as

$$\overset{+}{\mathcal{F}}(\theta, \phi, t) = \frac{\cos\theta}{c}\frac{\partial}{\partial t}\overset{+}{T}(c^{-1}\cos\phi\sin\theta, c^{-1}\sin\phi\sin\theta, t), \quad \mathrm{Im}(t) \le 0. \tag{5.272}$$

Similarly, the frequency-domain formula (3.156) (which, according to (3.146), is valid also for $\theta = 0$) shows that the analytic-signal electric far field is

$$\overset{+}{\mathbf{E}}(\mathbf{r}, t) = \frac{\cos\theta}{cr}\frac{\partial}{\partial t}\overset{+}{\mathbf{T}}(c^{-1}\cos\phi\sin\theta, c^{-1}\sin\phi\sin\theta, t - r/c) + O\left(\frac{1}{r^2}\right), \quad \mathrm{Im}(t) \le 0 \tag{5.273}$$

and its analytic-signal far-field pattern is

$$\overset{+}{\boldsymbol{\mathcal{F}}}(\theta, \phi, t) = \frac{\cos\theta}{c}\frac{\partial}{\partial t}\overset{+}{\mathbf{T}}(c^{-1}\cos\phi\sin\theta, c^{-1}\sin\phi\sin\theta, t), \quad \mathrm{Im}(t) \le 0. \tag{5.274}$$

Taking the real part of (5.273) and (5.274) produces, by use of (5.240), the spectrum formulas (5.188) and (5.189) for the real time-domain electric field. In addition, for the analytic-signal magnetic field we find that

$$\overset{+}{\mathbf{H}}(\mathbf{r}, t) = \sqrt{\frac{\epsilon}{\mu}}\frac{\cos\theta}{cr}\hat{\mathbf{r}} \times \frac{\partial}{\partial t}\overset{+}{\mathbf{T}}(c^{-1}\cos\phi\sin\theta, c^{-1}\sin\phi\sin\theta, t - r/c) + O\left(\frac{1}{r^2}\right), \quad \mathrm{Im}(t) \le 0 \tag{5.275}$$

and thus the analytic-signal electric and magnetic far-field patterns are related through

$$\overset{+}{\boldsymbol{\mathcal{F}}}{}^H(\theta, \phi, t) = \sqrt{\frac{\epsilon}{\mu}}\hat{\mathbf{r}} \times \overset{+}{\boldsymbol{\mathcal{F}}}(\theta, \phi, t), \quad \overset{+}{\boldsymbol{\mathcal{F}}}(\theta, \phi, t) = -\sqrt{\frac{\mu}{\epsilon}}\hat{\mathbf{r}} \times \overset{+}{\boldsymbol{\mathcal{F}}}{}^H(\theta, \phi, t), \quad \mathrm{Im}(t) \le 0. \tag{5.276}$$

Far Fields for $\theta = \pi/2$. Applying the analytic Fourier transform to the frequency-domain far-field formula (3.162) proves that the analytic-signal far-field and the corresponding far-field pattern are given in terms of the analytic-signal far-field function $\overset{+}{F}(\xi, \eta, t)$ as

$$\overset{+}{p}(\mathbf{r}, t) = \overset{+}{F}(c^{-1}\cos\phi, c^{-1}\sin\phi, t - r/c)\frac{1}{r} + O\left(\frac{1}{r^2}\right), \quad \text{Im}(t) \leq 0 \tag{5.277}$$

and

$$\overset{+}{\mathcal{F}}(\pi/2, \phi, t) = \overset{+}{F}(c^{-1}\cos\phi, c^{-1}\sin\phi, t), \quad \text{Im}(t) \leq 0. \tag{5.278}$$

For the analytic-signal electric far field, we find from the frequency-domain formula (3.163) that

$$\overset{+}{\mathbf{E}}(\mathbf{r}, t) = \overset{+}{\mathbf{F}}(c^{-1}\cos\phi, c^{-1}\sin\phi, t - r/c)\frac{1}{r} + O\left(\frac{1}{r^2}\right), \quad \text{Im}(t) \leq 0 \tag{5.279}$$

and

$$\overset{+}{\boldsymbol{\mathcal{F}}}(\pi/2, \phi, t) = \overset{+}{\mathbf{F}}(c^{-1}\cos\phi, c^{-1}\sin\phi, t), \quad \text{Im}(t) \leq 0. \tag{5.280}$$

Also, for these angles of observation, the analytic-signal magnetic far-field pattern can be determined from the analytic-signal electric far-field pattern by (5.276).

5.3.5 Near Fields in Terms of Far Fields in Spherical Coordinates

It was demonstrated in Section 5.3.3 that the analytic-signal spectrum, unlike the real time-domain spectrum, is determined completely for all real (ξ, η) from its corresponding far-field pattern evaluated at real and complex angles of observation. Therefore, the analytic-signal near fields can be expressed in terms of their analytic-signal far-field patterns. The formulas of this section are derived by applying the analytic Fourier transform to the frequency-domain formulas of Sections 3.2.6 and 3.2.7, and they are the analytic-signal analogs of the real time-domain formulas of Section 5.2.6. Throughout Section 5.3.5 it is assumed that the sources satisfy the conditions stated in Section 5.1.4, so that the analytic-signal far fields decay as $1/r$, or faster.

Let us start with the analytic-signal acoustic pressure $\overset{+}{p}(\mathbf{r}, t)$ and recall the formula (5.195), which states that for $\omega > 0$ and $z > 0$,

$$\begin{aligned} p_\omega(\mathbf{r}) = &-\frac{1}{2\pi c}\int_0^{2\pi}\int_0^{\pi/2}(-i\omega)\mathcal{F}_\omega(\sigma_\theta, \sigma_\phi)e^{i\omega\mathbf{r}\cdot\hat{\boldsymbol{\sigma}}_{\theta\phi}/c}\sin\sigma_\theta d\sigma_\theta d\sigma_\phi \\ &+\frac{i}{2\pi c}\int_0^{2\pi}\int_0^{+\infty}(-i\omega)\mathcal{F}_\omega(\pi/2 - i\alpha, \sigma_\phi)e^{i\omega(\cosh\alpha\mathbf{r}\cdot\hat{\boldsymbol{\sigma}}_\phi/c + iz\sinh\alpha/c)} \\ &\cdot\cosh\alpha d\alpha d\sigma_\phi \end{aligned} \tag{5.281}$$

where $\hat{\boldsymbol{\sigma}}_{\theta\phi} = \hat{\mathbf{x}}\cos\sigma_\phi\sin\sigma_\theta + \hat{\mathbf{y}}\sin\sigma_\phi\sin\sigma_\theta + \hat{\mathbf{z}}\cos\sigma_\theta$ and $\hat{\boldsymbol{\sigma}}_\phi = \hat{\mathbf{x}}\cos\sigma_\phi + \hat{\mathbf{y}}\sin\sigma_\phi$. The formula (5.281) expresses the frequency-domain acoustic pressure in terms of its far-field pattern evaluated at the real and complex angles of observation given by $0 \leq \phi \leq 2\pi$ and $\theta \in \Gamma$, where Γ is the contour in Figure 3.4. We can immediately apply the analytic Fourier transform (which involves only positive frequencies) to (5.281) to show that the analytic-signal acoustic near-field pressure in the region $z > 0$ can be expressed in terms of its far-field pattern as

$$\overset{+}{p}(\mathbf{r},t) = -\frac{1}{2\pi c}\int_0^{2\pi}\int_0^{\pi/2} \frac{\partial}{\partial t}\overset{+}{\mathcal{F}}(\sigma_\theta,\sigma_\phi,t-\mathbf{r}\cdot\hat{\boldsymbol{\sigma}}_{\theta\phi}/c)\sin\sigma_\theta d\sigma_\theta d\sigma_\phi$$
$$+\frac{i}{2\pi c}\int_0^{2\pi}\int_0^{+\infty}\frac{\partial}{\partial t}\overset{+}{\mathcal{F}}(\pi/2-i\alpha,\sigma_\phi,t-\cosh\alpha\mathbf{r}\cdot\hat{\boldsymbol{\sigma}}_\phi/c-iz\sinh\alpha/c)$$
$$\cdot\cosh\alpha d\alpha d\sigma_\phi,\quad \mathrm{Im}(t)\le 0. \tag{5.282}$$

Similarly, taking the inverse analytic Fourier transform of (3.186) shows that the analytic-signal velocity is

$$\overset{+}{\mathbf{v}}(\mathbf{r},t) = -\frac{1}{2\pi c^2\rho_{m0}}\int_0^{2\pi}\int_0^{\pi/2}\hat{\boldsymbol{\sigma}}_{\theta\phi}\frac{\partial}{\partial t}\overset{+}{\mathcal{F}}(\sigma_\theta,\sigma_\phi,t-\mathbf{r}\cdot\hat{\boldsymbol{\sigma}}_{\theta\phi}/c)\sin\sigma_\theta d\sigma_\theta d\sigma_\phi$$
$$+\frac{i}{2\pi c^2\rho_{m0}}\int_0^{2\pi}\int_0^{+\infty}\frac{\partial}{\partial t}\overset{+}{\mathcal{F}}(\pi/2-i\alpha,\sigma_\phi,t-\cosh\alpha\mathbf{r}\cdot\hat{\boldsymbol{\sigma}}_\phi/c-iz\sinh\alpha/c)$$
$$\cdot\left(\hat{\boldsymbol{\sigma}}_\phi\cosh\alpha+\hat{\mathbf{z}}i\sinh\alpha\right)\cosh\alpha\, d\alpha d\sigma_\phi,\quad \mathrm{Im}(t)\le 0. \tag{5.283}$$

The analytic-signal electric near field can similarly be expressed for $z>0$ in terms of the analytic-signal electric far-field pattern as

$$\overset{+}{\mathbf{E}}(\mathbf{r},t) = -\frac{1}{2\pi c}\int_0^{2\pi}\int_0^{\pi/2}\frac{\partial}{\partial t}\overset{+}{\boldsymbol{\mathcal{F}}}(\sigma_\theta,\sigma_\phi,t-\mathbf{r}\cdot\hat{\boldsymbol{\sigma}}_{\theta\phi}/c)\sin\sigma_\theta d\sigma_\theta d\sigma_\phi$$
$$+\frac{i}{2\pi c}\int_0^{2\pi}\int_0^{+\infty}\frac{\partial}{\partial t}\overset{+}{\boldsymbol{\mathcal{F}}}(\pi/2-i\alpha,\sigma_\phi,t-\cosh\alpha\mathbf{r}\cdot\hat{\boldsymbol{\sigma}}_\phi/c-iz\sinh\alpha/c)$$
$$\cdot\cosh\alpha d\alpha d\sigma_\phi,\quad \mathrm{Im}(t)\le 0 \tag{5.284}$$

which also holds for the magnetic field when $\overset{+}{\mathbf{E}}$ and $\overset{+}{\boldsymbol{\mathcal{F}}}$ are replaced by $\overset{+}{\mathbf{H}}$ and $\overset{+}{\boldsymbol{\mathcal{F}}}{}^H$, respectively. Moreover, applying the inverse analytic Fourier transform to the frequency-domain formula (3.188) proves that the analytic-signal magnetic near field can be expressed for $z>0$ in terms of the analytic-signal electric far-field pattern as

$$\overset{+}{\mathbf{H}}(\mathbf{r},t) = -\frac{\epsilon}{2\pi}\int_0^{2\pi}\int_0^{\pi/2}\hat{\boldsymbol{\sigma}}_{\theta\phi}\times\frac{\partial}{\partial t}\overset{+}{\boldsymbol{\mathcal{F}}}(\sigma_\theta,\sigma_\phi,t-\mathbf{r}\cdot\hat{\boldsymbol{\sigma}}_{\theta\phi}/c)\sin\sigma_\theta d\sigma_\theta d\sigma_\phi$$
$$+\frac{i\epsilon}{2\pi}\int_0^{2\pi}\int_0^{+\infty}(\hat{\boldsymbol{\sigma}}_\phi\cosh\alpha+\hat{\mathbf{z}}i\sinh\alpha) \tag{5.285}$$
$$\times\frac{\partial}{\partial t}\overset{+}{\boldsymbol{\mathcal{F}}}(\pi/2-i\alpha,\sigma_\phi,t-\cosh\alpha\mathbf{r}\cdot\hat{\boldsymbol{\sigma}}_\phi/c-iz\sinh\alpha/c)$$
$$\cdot\cosh\alpha d\alpha d\sigma_\phi,\quad \mathrm{Im}(t)\le 0.$$

A formula for the analytic-signal acoustic pressure, which holds throughout all space at all times, can be obtained by applying the inverse analytic Fourier transform to the formula

(3.195), expressing the frequency-domain acoustic pressure in terms of the advanced Green's function, to get

$$\overset{+}{p}(\mathbf{r},t) = \frac{1}{4\pi}\int_V \frac{\overset{+}{q}(\mathbf{r}',t+|\mathbf{r}-\mathbf{r}'|/c)}{|\mathbf{r}-\mathbf{r}'|}dV' - \frac{1}{2\pi c}\int_0^{2\pi}\int_0^{\pi}\frac{\partial}{\partial t}\overset{+}{\mathcal{F}}(\sigma_\theta,\sigma_\phi,t-\mathbf{r}\cdot\hat{\boldsymbol{\sigma}}_{\theta\phi}/c)\sin\sigma_\theta d\sigma_\theta d\sigma_\phi, \quad \mathrm{Im}(t)\leq 0 \tag{5.286}$$

which is valid throughout all space, including the source region. Note that the analytic-signal acoustic pressure is here expressed in terms of its far-field pattern over the entire far-field sphere and its source function evaluated at an advanced time.

Assume that the *real* source $q(\mathbf{r},t)$ turns off at $t = t_1$, so that $q(\mathbf{r},t)\equiv 0$ for $t > t_1$. Then the real part of the first integral in (5.286) is zero for $t > t_1$. However, $\mathrm{Im}[\overset{+}{q}(\mathbf{r},t)] = -\widetilde{\mathcal{H}}q(\mathbf{r},t)$ is not necessarily zero for $t > t_1$, so the imaginary part of the first integral in (5.286) may be nonzero for $t > t_1$. Hence, unlike the real time-domain acoustic pressure in (5.205), the analytic-signal acoustic pressure cannot be expressed solely in terms of its far-field pattern evaluated at real angles of observation at times after all real sources are turned off.

Taking the inverse analytic Fourier transform of (3.207) of Section 3.2.7 gives

$$\overset{+}{\mathbf{E}}(\mathbf{r},t) = \frac{1}{4\pi}\int_V \frac{\overset{+}{\mathbf{Q}}(\mathbf{r}',t+|\mathbf{r}-\mathbf{r}'|/c)}{|\mathbf{r}-\mathbf{r}'|}dV' - \frac{1}{2\pi c}\int_0^{2\pi}\int_0^{\pi}\frac{\partial}{\partial t}\overset{+}{\boldsymbol{\mathcal{F}}}(\sigma_\theta,\sigma_\phi,t-\mathbf{r}\cdot\hat{\boldsymbol{\sigma}}_{\theta\phi}/c)\sin\sigma_\theta d\sigma_\theta d\sigma_\phi, \quad \mathrm{Im}(t)\leq 0 \tag{5.287}$$

throughout all space. The analytic-signal source function

$$\overset{+}{\mathbf{Q}}(\mathbf{r},t) = -\mu\frac{\partial}{\partial t}\overset{+}{\mathbf{J}}(\mathbf{r},t) - \frac{1}{\epsilon}\nabla\overset{+}{\rho}(\mathbf{r},t) \tag{5.288}$$

is found by taking the inverse analytic Fourier transform of (3.205). We see from (5.287) that the analytic-signal electric field cannot be expressed solely in terms of its far-field pattern evaluated at real angles of observation at times after all real sources $\mathbf{Q}(\mathbf{r},t)$ are turned off.

When the sources are located in the half space $z > 0$, the fields in $z < 0$ are determined by (5.282), (5.283), (5.284), and (5.285) with $\int_0^{\pi/2}$ replaced by $\int_{\pi/2}^{\pi}$ and α replaced by $-\alpha$ ($d\alpha$ remains the same). The formulas (5.286) and (5.287) remain the same.

5.4 ELECTROMAGNETIC MISSILES

Wu [155] showed that the energy in an electromagnetic pulse radiated through a fixed-size finite area could decay more slowly than $1/r^2$ as r approaches infinity. He called these unusual far-field pulses "electromagnetic missiles" and proved their existence mathematically from the classical Maxwell equations for sources in a finite region of space that radiate a finite amount of energy. Let the fixed-size area (with fixed shape and orientation) be denoted by $A(r)$, where r denotes the distance from the source region to $A(r)$, and assume that the normal to $A(r)$ is $\hat{\mathbf{n}}$. Then the energy radiated through $A(r)$ is given from (2.191) as

$$\mathcal{E}(r) = \int_{A(r)}\int_{pulse}\hat{\mathbf{n}}\cdot\big[\mathbf{E}(\mathbf{r},t)\times\mathbf{H}(\mathbf{r},t)\big]dt\,dS \tag{5.289}$$

and the condition for the existence of an electromagnetic missile is that $r^2\mathcal{E}(r)\to\infty$ as

$r \to \infty$. Clearly, this condition cannot be satisfied unless the electromagnetic far fields decay slower than $1/r$ as $r \to \infty$. One example of $A(r)$ would be a circular disk, which has constant radius, is centered on the z axis, and has $\hat{\mathbf{n}} = \hat{\mathbf{z}}$. In that case the missile would propagate along the positive z axis, and r would be equal to z.

Because of the slow decay of the energy, electromagnetic missiles would be of great practical importance if, in fact, they could be generated by realizable antennas. For example, they could be used to transfer information with less energy and interference, or they could be used to concentrate large amounts of energy over large distances.

The main purpose of this section is to determine necessary conditions that must be satisfied by the sources if they are to produce electromagnetic missiles. We shall also determine the necessary conditions on the sources for the time-domain far fields to decay slower than $1/r$. (The necessary conditions for an electromagnetic missile are more restrictive than the necessary conditions for the time-domain far fields to decay slower than $1/r$, because the time duration of the far-field pulse may approach zero as r approaches infinity.) We shall derive these necessary conditions in both the time and frequency domains for three kinds of sources: (1) a finite number of moving point charges, (2) a continuum of current in a finite volume V, and (3) Huygens' sources on a surface S that surrounds V.

In addition, we determine expressions for the on-axis fields radiated by a uniform circular current disk with arbitrary time dependence, and for the fields everywhere outside the current disk with delta-function and rectangular-pulse time dependencies. For a step-function time dependence, the current disk generates an electromagnetic missile that travels in the z direction normal to the disk, and extends approximately to a distance from the z axis equal to the radius a of the current disk. Far from the current disk, this electromagnetic missile is essentially a pulse confined to a circular cylinder with radius approximately equal to a and with thickness (pulse width) that decays as $1/z$ (and thus the total energy in this electromagnetic missile decays as $1/z$).

One finding of this section is that the frequency spectrum of the source and fields of an electromagnetic missile must decay extremely slowly as $\omega \to \pm\infty$. To understand this slow far-field decay of the electromagnetic missile, recall that the distance to the boundary between frequency-domain near and far fields, along the boresight direction of a fixed-size aperture radiator, approaches infinity as $\omega \to \pm\infty$. If the frequency spectrum of the source decays slowly enough with frequency, the distance to the boundary between the frequency-domain near and far fields will not only approach infinity as $\omega \to \pm\infty$, but the time-domain fields (whose dominant frequencies become higher in magnitude as r gets larger) and the energy radiated through a fixed-size area can decay slower than $1/r$ and $1/r^2$, respectively, in the boresight direction.

5.4.1 EM Missiles for Accelerating Point Charges

For a finite number of separate point electric charges, each point electric charge q_e that moves with velocity $\mathbf{u}$ and acceleration $\dot{\mathbf{u}}$ radiates the fields $\mathbf{E}(\mathbf{r}, t)$ and $\mathbf{H}(\mathbf{r}, t)$ given in (2.99) and (2.100). These radiated fields satisfy

$$\lim_{r\to\infty} r\mathbf{E}(\mathbf{r}, t + r/c) = \mathcal{F}(\theta, \phi, t) = \frac{q_e \gamma_u^3}{4\pi\epsilon c^2}\hat{\mathbf{r}} \times \left[(\hat{\mathbf{r}} - \mathbf{u}(\tau)/c) \times \dot{\mathbf{u}}(\tau)\right]$$

$$= \frac{q_e \gamma_u^2}{4\pi\epsilon c^2}|\dot{\mathbf{u}}(\tau)| \left[\gamma_u(\hat{\mathbf{r}} - \mathbf{u}(\tau)/c)(\hat{\mathbf{r}} \cdot \hat{\dot{\mathbf{u}}}(\tau)) - \hat{\dot{\mathbf{u}}}(\tau)\right] \tag{5.290}$$

$$\mathbf{H}(\mathbf{r}, t) \sim \sqrt{\frac{\epsilon}{\mu}}\hat{\mathbf{r}} \times \mathbf{E}(\mathbf{r}, t), \quad r \to \infty \tag{5.291}$$

where $\gamma_u = 1/(1 - \hat{\mathbf{r}} \cdot \mathbf{u}(\tau)/c)$ and the velocity and acceleration are evaluated at the retarded time $\tau = t + \hat{\mathbf{r}} \cdot \mathbf{r}_q(\tau)/c$. The vector $\mathbf{r}_q(\tau)$ is the position of the charge at the retarded time τ. The origin of the coordinate system, and thus that of the vector $\mathbf{r}$, is chosen in the vicinity of the charge during the retarded times of interest.

Equations (5.290) and (5.291) reveal that the electric and magnetic far fields of the point charge decay as $1/r$ (or faster) unless the acceleration of the charge is infinite at some point (or points) in time. Thus, we conclude that *the far fields of a finite number of moving point charges decay more slowly than* $1/r$ *only if the acceleration of at least one of the point charges becomes infinite at some point in time.* (It is assumed that each point charge q_e is finite.) Also, the square brackets in (5.290) cannot be zero (for nonzero $\dot{\mathbf{u}}(\tau)$) in all directions of observation $\hat{\mathbf{r}}$. Thus, for a single point charge, the above necessary condition for slower than $1/r$ far-field decay is a sufficient condition as well.

For rectilinear motion, these points of infinite acceleration must be isolated points in time and must integrate over time to a value less than c, because the speed of each point charge must be less than the speed of light.

The magnitude of the time-domain acceleration can be written in terms of the Fourier transform of the velocity spectrum to show

$$|\dot{\mathbf{u}}(t)| = \left| \int_{-\infty}^{+\infty} -i\omega \mathbf{u}_\omega e^{-i\omega t} d\omega \right| \leq \int_{-\infty}^{+\infty} |\omega \mathbf{u}_\omega| d\omega \tag{5.292}$$

which is finite if $\mathbf{u}_\omega = O(|\omega|^{-(2+\alpha)})$, $\alpha > 0$, as $\omega \to \pm\infty$. Using the even and odd functional dependence on ω of the real and imaginary parts of the velocity spectrum $\mathbf{u}_\omega$, and assuming that $\dot{\mathbf{u}}(t)$ is zero before some initial time, it can also be shown that the condition $\mathbf{u}_\omega = O(|\omega|^{-2})$ as $\omega \to \pm\infty$ produces a finite acceleration. Thus, we conclude that *the far fields of a moving point charge can decay more slowly than* $1/r$ *only if the frequency spectrum of the velocity of the point charge decays slower than* $1/\omega^2$ *as* $\omega \to \pm\infty$. In addition, if $\mathbf{u}_\omega = O(|\omega|^{-(2-\alpha)})$, $\alpha > 0$, as $\omega \to \pm\infty$, examples of $\mathbf{u}_\omega$ can be found that produce infinite acceleration at some point in time.

The energy radiated per unit area in a far-field pulse can be expressed from (5.290) and (5.291) as

$$\begin{aligned} \lim_{r\to\infty} r^2 \int_{pulse} \hat{\mathbf{r}} \cdot \big[\mathbf{E}(\mathbf{r}, t + r/c) \times \mathbf{H}(\mathbf{r}, t + r/c)\big] dt &= \sqrt{\frac{\epsilon}{\mu}} \int_{pulse} |\boldsymbol{\mathcal{F}}(\theta, \phi, t)|^2 dt \\ &\propto \int_{pulse} |\dot{\mathbf{u}}(t)|^2 dt = 2\pi \int_{-\infty}^{+\infty} \omega^2 |\mathbf{u}_\omega|^2 d\omega. \end{aligned} \tag{5.293}$$

(Only $|\dot{\mathbf{u}}|^2$ is retained in (5.293) because the expression in the square brackets of the last part of (5.290) is never infinite but is nonzero for most far-field directions. Because $dt = dt'/\gamma_u$ and $1/\gamma_u$ is never infinite or zero, $1/\gamma_u$ has also been omitted from the time integration of $|\dot{\mathbf{u}}|^2$ in (5.293).) Thus, *the necessary condition for the finite number of point charges to radiate an electromagnetic missile (energy per unit area that decays slower than* $1/r^2$*) is that at least one point charge has infinite acceleration at some time, and in addition*

$$\int_{pulse} |\dot{\mathbf{u}}(t)|^2 dt = 2\pi \int_{-\infty}^{+\infty} \omega^2 |\mathbf{u}_\omega|^2 d\omega = \infty \tag{5.294}$$

which implies that the magnitude of the velocity spectrum $|\mathbf{u}_\omega|$ *decays as* $1/|\omega|^{3/2}$ *or slower as* $\omega \to \pm\infty$. Furthermore, examples of velocities (such as $t^\alpha \exp(-t)$, $t > 0$, and zero for

$t \leq 0$) are easily constructed that can produce electromagnetic missiles ($0 < \alpha \leq 1/2$), or no electromagnetic missile yet slower than $1/r$ far-field decay ($1/2 < \alpha < 1$).

We can also find conditions on the frequency-domain far-field pattern required to have slower than $1/r$ decay in the time-domain far field and slower than $1/r^2$ decay of the energy per unit area in the far-field pulse. The time-domain far-field pattern can be expressed in terms of the frequency-domain far-field pattern as

$$\mathcal{F}(\theta, \phi, t) = \int_{-\infty}^{+\infty} \mathcal{F}_\omega(\theta, \phi) e^{-i\omega t} d\omega. \tag{5.295}$$

If the time-domain far field is to decay slower than $1/r$ in some direction, the far-field pattern must be infinite in that direction. Thus, from (5.295) we have

$$|\mathcal{F}(\theta, \phi, t)| = \left| \int_{-\infty}^{+\infty} \mathcal{F}_\omega(\theta, \phi) e^{-i\omega t} d\omega \right| \leq \int_{-\infty}^{+\infty} |\mathcal{F}_\omega(\theta, \phi)| \, d\omega = \infty \tag{5.296}$$

or that the magnitude of the frequency-domain far-field pattern must decay slower than $1/|\omega|$ as $\omega \to \pm\infty$ for slower than $1/r$ far-field decay in the time domain. (By the reasoning explained after (5.292), when the magnitude of the frequency-domain far-field pattern decays as $1/|\omega|$, the time-domain far-field decays as $1/r$.)

Similarly, in order for the energy per unit area in the time-domain far-field pulse to decay slower than $1/r^2$, we must have

$$\int_{pulse} |\mathcal{F}(\theta, \phi, t)|^2 \, dt = 2\pi \int_{-\infty}^{+\infty} |\mathcal{F}_\omega(\theta, \phi)|^2 \, d\omega = \infty. \tag{5.297}$$

Equation (5.297) implies that the magnitude of the frequency-domain far-field pattern must decay as $1/|\omega|^{1/2}$ or slower as $\omega \to \pm\infty$ for the energy per unit area in the far-field pulse to decay slower than $1/r^2$, that is, for an electromagnetic missile to exist.

For rectilinear motion, either of the conditions (5.297) or (5.294) implies from (5.290) and (5.291) that the point charge radiates an infinite amount of energy because the energy radiated per unit area decays slower than $1/r^2$ over most of the far-field sphere.

5.4.2 EM Missiles for a Continuum of Current and for Huygens' Sources

Now assume that the source region consists of a continuum of current rather than a finite number of moving point charges. We proved in Section 5.1.4 that the electromagnetic far fields decay as $1/r$ (or faster) provided (i) the current $\mathbf{J}(\mathbf{r}, t)$ as well as its time derivative and time integral are integrable over the source volume V for all times, and (ii) there exists an integrable time-independent function $\alpha(\mathbf{r})$ such that $|\partial \mathbf{J}(\mathbf{r}, t)/\partial t| \leq \alpha(\mathbf{r})$. In the following we assume that the integrability condition (i) is always satisfied and that the first time derivative of the current (or Huygens' sources) exists.

Therefore, *only if the first time derivative of the current in a region of space is infinite at some point in time*[4] *can the far fields decay more slowly than* $1/r$. Similarly, the results of Section 5.1.4 prove the following statement involving the Huygens' sources $\hat{\mathbf{n}} \times \mathbf{E}(\mathbf{r}, t)$ and $\hat{\mathbf{n}} \times \mathbf{H}(\mathbf{r}, t)$ on the surface S that surrounds V. *Only if the first time derivative of* $\hat{\mathbf{n}} \times \mathbf{E}(\mathbf{r}, t)$

[4] Note that the first time derivative of the current can be infinite at points in space and not generate an electromagnetic missile. For example, the current parallel to a perfectly conducting sharp edge is infinite right at the edge. Yet this singularity is a function of the spatial coordinates and will not generate an electromagnetic missile because the singularity is integrable with respect to the spatial coordinates.

or $\hat{\mathbf{n}} \times \mathbf{H}(\mathbf{r}, t)$ *at some point* $\mathbf{r}$ *on* S *is infinite at some point in time can the far fields decay more slowly than* $1/r$. Thus, these are necessary (but not sufficient) conditions on the current and on the Huygens' sources for the existence of an electromagnetic missile.

Expressing the first time derivative of the current as a Fourier transform of the current spectrum shows that

$$\left|\frac{\partial}{\partial t}\mathbf{J}(\mathbf{r}, t)\right| = \left|\int_{-\infty}^{+\infty} -i\omega \mathbf{J}_\omega(\mathbf{r})e^{-i\omega t}\,d\omega\right| \leq \int_{-\infty}^{+\infty} |\omega \mathbf{J}_\omega(\mathbf{r})|\,d\omega \tag{5.298}$$

which implies that the necessary condition in the frequency domain for slower than $1/r$ time-domain far-field decay is that *the magnitude of the frequency spectrum of the current* $|\mathbf{J}_\omega(\mathbf{r})|$ *for some* $\mathbf{r}$ *decays more slowly than* $1/\omega^2$ *as* $\omega \to \pm\infty$. Similarly, we find that for the Huygens' sources to produce slower than $1/r$ time-domain far-field decay, *at some point* $\mathbf{r}$ *on* S *the magnitude of* $\hat{\mathbf{n}} \times \mathbf{E}_\omega(\mathbf{r})$ *or* $\hat{\mathbf{n}} \times \mathbf{H}_\omega(\mathbf{r})$ *must decay more slowly than* $1/\omega^2$ *as* $\omega \to \pm\infty$.

For an electromagnetic missile to exist, it is not sufficient for the far field to decay as $1/r$. In addition, the energy per unit area carried by the far-field pulse must decay slower than $1/r^2$. From equations (5.34) and (5.35), we find that the energy radiated per unit area in a far-field pulse behaves as

$$\begin{aligned}\lim_{r\to\infty} r^2 \int_{pulse} \hat{\mathbf{r}} \cdot \big[\mathbf{E}(\mathbf{r}, t + r/c) \times \mathbf{H}(\mathbf{r}, t + r/c)\big]dt &= \sqrt{\frac{\epsilon}{\mu}} \int_{pulse} |\mathcal{F}(\theta, \phi, t)|^2 dt \\ &\propto \int_V \int_V \left[\int_{pulse} \frac{\partial}{\partial t}\mathbf{J}(\mathbf{r}', \tau') \cdot \frac{\partial}{\partial t}\mathbf{J}(\mathbf{r}'', \tau'')dt\right] dV' dV''\end{aligned} \tag{5.299}$$

where $\tau' = t + \hat{\mathbf{r}} \cdot \mathbf{r}'/c$ and $\tau'' = t + \hat{\mathbf{r}} \cdot \mathbf{r}''/c$. (For surface current, the volume integrals in (5.299) are replaced by surface integrals.)

Equation (5.299) implies that the far-field energy per unit area in the pulse cannot decay more slowly than $1/r^2$ unless the time integral is infinite over some region of $(\mathbf{r}', \mathbf{r}'')$. Specifically

$$\int_{pulse} \frac{\partial}{\partial t}\mathbf{J}(\mathbf{r}', t') \cdot \frac{\partial}{\partial t}\mathbf{J}(\mathbf{r}'', t'')dt = \infty \tag{5.300}$$

for some $\mathbf{r}'$ and $\mathbf{r}''$. If we let $\mathbf{r}$ be the value of $\mathbf{r}'$ or $\mathbf{r}''$ in (5.300) that has the largest time derivative of the current, (5.300) implies that there must be some $\mathbf{r}$ for which

$$\int_{pulse} \left|\frac{\partial}{\partial t}\mathbf{J}(\mathbf{r}, t)\right|^2 dt = 2\pi \int_{-\infty}^{+\infty} \omega^2 |\mathbf{J}_\omega(\mathbf{r})|^2 d\omega = \infty. \tag{5.301}$$

Thus, in addition to the first time derivative of the current being infinite at some point in time, (5.301) is a necessary condition on the singularity of the current to produce an electromagnetic missile. Specifically, *the first time derivative of the current in a region of space must be infinite at some point in time, and the time integral of the magnitude squared of the first time derivative of the current must be infinite to produce an electromagnetic missile. In terms of the frequency spectrum of the current,* $|\mathbf{J}_\omega(\mathbf{r})|$ *for some* $\mathbf{r}$ *must decay as* $1/|\omega|^{3/2}$ *or slower as* $\omega \to \pm\infty$. For linearly polarized current, the singularities (infinities) of the first time derivative of the current must be isolated and integrate over time to a finite value if the current itself is to remain a finite function of time, even though the square of the first time derivative of current must integrate to an infinite value to produce an electromagnetic missile. The lower bound of $1/|\omega|^{3/2}$ as

$\omega \to \pm\infty$ for the frequency spectrum of the current producing an electromagnetic missile was also found by Myers [156] for current on a circular disk.

Similarly, we find from (5.37) and (5.38) that the Huygens' sources $\hat{\mathbf{n}} \times \mathbf{E}(\mathbf{r}, t)$ and $\hat{\mathbf{n}} \times \mathbf{H}(\mathbf{r}, t)$ on S must satisfy the following conditions. *At some point* $\mathbf{r}$ *on* S *the first time derivative of* $\hat{\mathbf{n}} \times \mathbf{E}(\mathbf{r}, t)$ *or* $\hat{\mathbf{n}} \times \mathbf{H}(\mathbf{r}, t)$ *must be infinite at some point in time, and the time integral of the magnitude squared of the first time derivative of* $\hat{\mathbf{n}} \times \mathbf{E}(\mathbf{r}, t)$ *or* $\hat{\mathbf{n}} \times \mathbf{H}(\mathbf{r}, t)$ *must be infinite to produce an electromagnetic missile. In terms of the frequency spectra of the Huygens' sources,* $|\hat{\mathbf{n}} \times \mathbf{E}_\omega(\mathbf{r})|$ *or* $|\hat{\mathbf{n}} \times \mathbf{H}_\omega(\mathbf{r})|$ *for some* $\mathbf{r}$ *on* S *must decay as* $1/|\omega|^{3/2}$ *or slower as* $\omega \to \pm\infty$.

For slower than $1/r$ decay of the time-domain far field in the direction (θ, ϕ), equation (5.296) must be satisfied, and $|\mathcal{F}_\omega(\theta, \phi)|$ must decay slower than $1/|\omega|$ as $\omega \to \pm\infty$. Also, for the existence of an electromagnetic missile in the direction (θ, ϕ), equation (5.297) must be satisfied, and $|\mathcal{F}_\omega(\theta, \phi)|$ must decay as $1/|\omega|^{1/2}$ or slower as $\omega \to \pm\infty$. Recall that the frequency-domain far-field pattern is defined as the limit as $r \to \infty$ in (3.31). Thus, when referring to the limit of the far-field pattern as $\omega \to \pm\infty$, it is assumed that the limit as $r \to \infty$ has already been taken.

The exact expressions for the time-domain fields radiated by a circular current disk, derived in Section 5.4.3 below, demonstrate that a spatial distribution of current can be found that will produce an electromagnetic missile with time or frequency dependence satisfying (5.301). It is emphasized that the condition (5.301) or (5.297) does not imply that the current source radiates an infinite amount of energy, because a continuum of current, unlike an accelerating point charge, can have infinite directivity. Therefore, the energy in the far-field pulse that decays slower than $1/r^2$ can be confined to a finite transverse area as $r \to \infty$. Conversely, if the total energy radiated by the sources is finite, electromagnetic missiles can occur only in isolated directions, that is, over a zero solid angle (namely, points and curves) on the far-field sphere (see Figure 5.8 below).

Also it is emphasized that (5.34)–(5.35) and (5.299) may no longer be valid expressions for evaluating the far fields and far-field energy per unit area when $\partial\mathbf{J}/\partial t$ becomes infinite at some point in time. To find the exact behavior of the far fields when $\partial\mathbf{J}/\partial t$ becomes infinite, we can determine exact closed-form expressions of the near fields, before letting $r \to \infty$, as is done in the next section for the circular current disk.

5.4.3 Fields from a Circular Disk of Uniform Surface Current

To illustrate the mathematical existence of electromagnetic missiles, exact closed-form expressions are derived for the on-axis electric and magnetic fields of a circular disk excited by a uniform surface current with arbitrary time dependence [157]. (Corresponding expressions hold for a uniform magnetic or electric field exciting the circular aperture.) We begin by deriving expressions for frequency-domain fields and then take the Fourier transform of the expressions to get the fields with arbitrary time dependence. Finally, we consider the special cases of delta-function, unit-step-function, and rectangular-pulse time dependencies for the current on the disk.

The frequency-domain magnetic field can be expressed in terms of the vector potential as

$$\mathbf{H}_\omega(\mathbf{r}) = \frac{1}{\mu}\nabla \times \mathbf{A}_\omega(\mathbf{r}). \tag{5.302}$$

By inserting this expression into $\mathbf{E}_\omega = -(1/(i\omega\epsilon))\nabla \times \mathbf{H}_\omega$, we find that the electric field outside the source region is given by

$$\mathbf{E}_\omega(\mathbf{r}) = -\frac{1}{i\omega\epsilon\mu}\left[k^2\mathbf{A}_\omega(\mathbf{r}) + \nabla\nabla \cdot \mathbf{A}_\omega(\mathbf{r})\right] \tag{5.303}$$

where we have used $\nabla \times \nabla \times = \nabla\nabla \cdot -\nabla^2$ and the fact that the vector potential satisfies the homogeneous Helmholtz equation $\nabla^2 \mathbf{A}_\omega + k^2 \mathbf{A}_\omega = 0$ outside the source region.

Let us now consider the circular disk of uniform, linearly polarized, surface electric current, $\mathbf{K}_\omega = K_\omega \hat{\mathbf{x}}$. This current is constant on the disk, which is located in the plane $z = 0$, is centered at the origin, and has radius a, as shown in Figure 5.6. The surface charge density of this uniform surface current is zero everywhere except at the rim of the disk where the surface charge density has a delta-function singularity. If the surface current is tapered to zero smoothly in a small distance from the rim, rather than abruptly at the rim, this rim charge can be made finite without appreciably changing the fields of the current disk away from the rim.

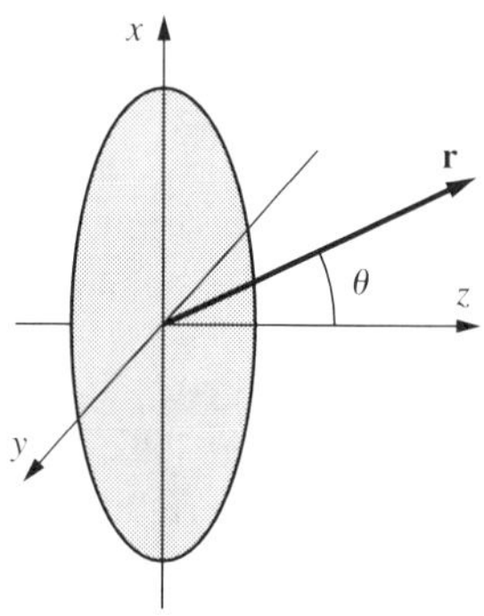

Figure 5.6 Circular disk of radius a with uniform surface current.

For the current on the disk, the vector potential at $\mathbf{r} = x\hat{\mathbf{x}} + y\hat{\mathbf{y}} + z\hat{\mathbf{z}}$ is given by

$$\mathbf{A}_\omega(\mathbf{r}) = \mu K_\omega \hat{\mathbf{x}} \mathcal{G}_\omega(x, y, z) \tag{5.304}$$

with

$$\mathcal{G}_\omega(x, y, z) = \frac{1}{4\pi} \int_{disk} \frac{e^{ik\sqrt{(x-x')^2+(y-y')^2+z^2}}}{\sqrt{(x-x')^2+(y-y')^2+z^2}} \, dS'. \tag{5.305}$$

(Although we have written $\mathcal{G}_\omega$ as a function of (x, y, z), in cylindrical coordinates (ρ, ϕ, z), $\mathcal{G}_\omega$ is a function only of $\rho = \sqrt{x^2 + y^2}$ and z, because of the circular symmetry.) Taking the curl of (5.304) produces the following expression for the magnetic field

$$\mathbf{H}_\omega(\mathbf{r}) = K_\omega \left(\hat{\mathbf{y}} \frac{\partial}{\partial z} - \hat{\mathbf{z}} \frac{\partial}{\partial y} \right) \mathcal{G}_\omega(x, y, z). \tag{5.306}$$

To evaluate this expression for $\mathbf{r} = z\hat{\mathbf{z}}$ with $z > 0$, we note that the function $\mathcal{G}_\omega(x, y, z)$ is infinitely differentiable with respect to x, y, and z for observation points on the positive z axis. Moreover, because of the symmetry of the region of integration, $\mathcal{G}_\omega(0, y, z)$ is an even function of y for $z > 0$, so $(\partial/\partial y)\mathcal{G}_\omega(0, y, z) = 0$ for $y = 0$, and (5.306) shows that the z component of the magnetic field is zero on the positive z axis. A simple integration in polar coordinates shows that

$$\mathcal{G}_\omega(0, 0, z) = \frac{1}{2ik} \left(e^{ik\sqrt{z^2+a^2}} - e^{ikz} \right) \tag{5.307}$$

and the remaining y component of the magnetic field $\mathbf{H}_\omega(z)$ on the positive z axis is therefore

$$\begin{aligned} \mathbf{H}_\omega(z) &= -\hat{\mathbf{y}} \frac{K_\omega}{2ik} \frac{\partial}{\partial z} \left(e^{ikz} - e^{ik\sqrt{z^2+a^2}} \right) \\ &= -\frac{1}{2} \hat{\mathbf{z}} \times \mathbf{K}_\omega \left(e^{ikz} - \frac{z}{\sqrt{z^2+a^2}} e^{ik\sqrt{z^2+a^2}} \right). \end{aligned} \tag{5.308}$$

The electric field for observation points outside the source region is found by inserting (5.304) into (5.303) to get

$$\mathbf{E}_\omega(\mathbf{r}) = \frac{iK_\omega}{\omega\epsilon}\left[\hat{\mathbf{x}}\left(k^2 + \frac{\partial^2}{\partial x^2}\right) + \hat{\mathbf{y}}\frac{\partial^2}{\partial x \partial y} + \hat{\mathbf{z}}\frac{\partial^2}{\partial x \partial z}\right]\mathcal{G}_\omega(x, y, z). \tag{5.309}$$

Because of the symmetry of the region of integration, we find that

$$\frac{\partial^2}{\partial x \partial y}\mathcal{G}_\omega(x, y, z) = \frac{\partial^2}{\partial x \partial z}\mathcal{G}_\omega(x, y, z) = 0, \quad \text{for } x = y = 0 \text{ and } z > 0. \tag{5.310}$$

Hence, the expression (5.309) shows that on the positive z axis the y and z components of the electric field are zero.

Also because of symmetry, the second derivatives of $\mathcal{G}_\omega(x, y, z)$ with respect to x and y are equal on the positive z axis, so that $\partial^2/\partial x^2$ in (5.309) can be replaced by $(\partial^2/\partial x^2 + \partial^2/\partial y^2)/2$ for observation points on the positive z axis. Moreover, $\mathcal{G}_\omega(x, y, z)$ satisfies the homogeneous Helmholtz equation everywhere outside the source region, so $\partial^2/\partial x^2 + \partial^2/\partial y^2$ can be replaced by $-\partial^2/\partial z^2 - k^2$ on the positive z axis, and the x component of the on-axis electric field is expressed in terms of $\mathcal{G}_\omega(x, y, z)$ and its second derivative with respect to z. Using the expression (5.307) finally proves that the on-axis electric field $\mathbf{E}_\omega(z)$ is given by

$$\begin{aligned}\mathbf{E}_\omega(z) &= \hat{\mathbf{x}}\frac{K_\omega}{4\omega\epsilon k}\left(k^2 - \frac{\partial^2}{\partial z^2}\right)\left(e^{ik\sqrt{z^2+a^2}} - e^{ikz}\right) \\ &= -\frac{1}{2}\sqrt{\frac{\mu}{\epsilon}}\mathbf{K}_\omega\left[e^{ikz} - \left(\frac{z^2 + a^2/2}{z^2 + a^2} + \frac{1}{2ik}\frac{a^2}{(z^2 + a^2)^{3/2}}\right)e^{ik\sqrt{z^2+a^2}}\right].\end{aligned} \tag{5.311}$$

For a *finite* frequency ω, the expressions on the right-hand sides of (5.308) and (5.311) can be expanded asymptotically to show that the electric and magnetic fields along the positive z axis behave as

$$\mathbf{H}_\omega(z) \sim \hat{\mathbf{z}} \times \mathbf{K}_\omega\frac{ika^2 e^{ikz}}{4z}, \quad z \to \infty \tag{5.312}$$

and

$$\mathbf{E}_\omega(z) \sim \mathbf{K}_\omega\sqrt{\frac{\mu}{\epsilon}}\frac{ika^2 e^{ikz}}{4z}, \quad z \to \infty. \tag{5.313}$$

The asymptotic expressions (5.312) and (5.313) become accurate when z is greater than the "Rayleigh distance," that is

$$z > 4a^2/\lambda. \tag{5.314}$$

Also, note from (5.312) and (5.313) that the electric and magnetic fields along the positive z axis obey the usual far-field relations

$$\mathbf{H}_\omega(z) \sim \sqrt{\frac{\epsilon}{\mu}}\hat{\mathbf{z}} \times \mathbf{E}_\omega(z), \quad \mathbf{E}_\omega(z) \sim -\sqrt{\frac{\mu}{\epsilon}}\hat{\mathbf{z}} \times \mathbf{H}_\omega(z), \quad z \to \infty. \tag{5.315}$$

The expression (5.308) for the magnetic field can be found in a number of references, for example, [158]. For the circular acoustic piston radiator, the on-axis expression corresponding to (5.308) can be found in numerous early references, [159] being the earliest of which we are aware. As far as we know, (5.311) was first derived by Wittmann [160] from a linear operator approach to calculating fields.

Having obtained the frequency-domain expressions (5.308) and (5.311) for the on-axis magnetic and electric fields, we take the Fourier transform of (5.308) and (5.311) to obtain the magnetic and electric fields for a circular disk of uniform surface current with arbitrary time dependence. Specifically, (5.308) and (5.311) transform to the time domain as

$$\mathbf{H}(z,t) = -\frac{1}{2}\hat{\mathbf{z}} \times \left[\mathbf{K}\left(t - \frac{z}{c}\right) - \frac{z}{\sqrt{z^2+a^2}}\mathbf{K}\left(t - \frac{\sqrt{z^2+a^2}}{c}\right)\right] \tag{5.316}$$

$$\mathbf{E}(z,t) = -\frac{1}{2}\sqrt{\frac{\mu}{\epsilon}}\left[\mathbf{K}\left(t - \frac{z}{c}\right) - \frac{z^2+a^2/2}{z^2+a^2}\mathbf{K}\left(t - \frac{\sqrt{z^2+a^2}}{c}\right) + \frac{1}{2}\frac{ca^2}{(z^2+a^2)^{3/2}}\int_{-\infty}^{t-\sqrt{z^2+a^2}/c}\mathbf{K}(t')dt'\right] \tag{5.317}$$

where, of course

$$\mathbf{K}(t) = \int_{-\infty}^{+\infty}\mathbf{K}_\omega e^{-i\omega t}d\omega. \tag{5.318}$$

Equation (5.316) was first derived, as far as we know, by Blejer [161]. The time-domain on-axis fields of a circular acoustic piston radiator could be obtained by Fourier transforming the frequency-domain expressions [159]. Alternatively, the time-domain on-axis fields of the piston radiator could be obtained from a convolution of its impulse response, which can be derived in closed form [162]. Similarly, the impulse response for the circular current disk (derived later in this section) can be used to find the on-axis expressions (5.316) and (5.317).

The first or leading pulse in (5.316) or (5.317) emanates from the center of the current disk. The second or trailing pulse in (5.316) or (5.317) emanates effectively from the rim of the current disk and trails the leading pulse. The time difference between the two pulses is equal to the difference in time that it takes light to travel to the observation point on the z axis from the rim and center of the disk. The trailing pulse travels faster than the speed of light and eventually approaches, but never quite catches, the leading pulse. The "superluminary" speed of the trailing pulse is a near-field phenomenon caused by the interference of the waves that radiate effectively from the rim of the current disk and cross the z axis at an angle. As one can show in general from Maxwell's equations, the speed of the pulse approaches the speed of light when the propagation distance becomes much greater than the effective size of the source distribution. And, of course, no signal reaches an observer in less time (after the sources are turned on) than the time interval equal to the minimum distance, from the source to the observer, divided by c.

Once the pulses have traveled a distance along the z axis much greater than the radius of the disk, (5.316) and (5.317) simplify to

$$\mathbf{H}(z,t) \sim -\frac{1}{2}\hat{\mathbf{z}} \times \left[\mathbf{K}\left(t - \frac{z}{c}\right) - \mathbf{K}\left(t - \frac{z}{c} - \frac{a^2}{2cz}\right)\right], \quad z \gg a \tag{5.319}$$

$$\mathbf{E}(z,t) \sim -\frac{1}{2}\sqrt{\frac{\mu}{\epsilon}}\left[\mathbf{K}\left(t - \frac{z}{c}\right) - \mathbf{K}\left(t - \frac{z}{c} - \frac{a^2}{2cz}\right)\right], \quad z \gg a. \tag{5.320}$$

If the first time derivative of the current exists and is finite, then by definition of the derivative, (5.319) and (5.320) reduce to

$$\mathbf{H}(z,t) \sim -\frac{a^2}{4cz}\hat{\mathbf{z}} \times \frac{\partial}{\partial t}\mathbf{K}\left(t^- - \frac{z}{c}\right), \quad z \gg a \tag{5.321}$$

$$\mathbf{E}(z,t) \sim -\sqrt{\frac{\mu}{\epsilon}}\frac{a^2}{4cz}\frac{\partial}{\partial t}\mathbf{K}\left(t^- - \frac{z}{c}\right), \quad z \gg a \tag{5.322}$$

provided z is large enough that the time derivative of the current does not change by a significant fraction during the time interval $a^2/(2cz)$. (The superscript "$-$" on t^- in (5.321) and (5.322) indicates that the time derivative is defined from the left in accordance with (5.319) and (5.320). This left-side definition of the time derivative allows $\partial\mathbf{K}/\partial t$ to have finite jump discontinuities [93, sec. 303].) This proviso can be expressed as

$$a^2/(2cz) \ll t_{min} \tag{5.323}$$

where t_{min} is the minimum time interval in which the time derivative of the current changes by a significant fraction in the neighborhood of t. When the current pulse has a finite frequency bandwidth (or at least an effectively finite frequency bandwidth) of maximum frequency ω_{max}, or minimum wavelength λ_{min}, t_{min} can be expressed as

$$t_{min} = 2\pi/\omega_{max} = \lambda_{min}/c \tag{5.324}$$

and (5.323) becomes

$$z \gg a^2/(2\lambda_{min}) \tag{5.325}$$

which is essentially the same as the Rayleigh criterion (5.314) for the minimum wavelength in the frequency spectrum of the finite bandwidth pulse.

Probably the most interesting results for the on-axis fields are obtained from (5.319) and (5.320) when the time derivative of the current pulse does not exist or is infinite at points in time, and thus (5.321) and (5.322) are no longer valid. In that case (5.319) and (5.320), for x-directed current $\mathbf{K}(t) = K(t)\hat{\mathbf{x}}$, can be written in the form

$$\mathbf{H}(z,t) \sim -\frac{a^2}{4cz}\hat{\mathbf{y}}S(t-\frac{z}{c},\frac{a^2}{2cz}), \quad z \gg a \tag{5.326}$$

$$\mathbf{E}(z,t) \sim -\sqrt{\frac{\mu}{\epsilon}}\frac{a^2}{4cz}\hat{\mathbf{x}}S(t-\frac{z}{c},\frac{a^2}{2cz}), \quad z \gg a \tag{5.327}$$

where $S(t,\Delta t)$ is the slope of the secant line [93, pp. 76–77] between the points on the current pulse at times t and $t-\Delta t$:

$$S(t,\Delta t) = \frac{K(t)-K(t-\Delta t)}{\Delta t}. \tag{5.328}$$

If the secant slope $S(t, a^2/(2cz))$ is bounded as $z\to\infty$, the far fields decay as $1/z$. If the secant slope is unbounded as $z\to\infty$, the far fields decay more slowly than $1/z$. When the secant slope has a limit, it equals the derivative of the current. That is,

$$\lim_{z\to\infty} S(t,\frac{a^2}{2cz}) = \frac{\partial K(t^-)}{\partial t} \tag{5.329}$$

when the limit in (5.329) exists. When the limit of the secant slope in (5.329) is $+\infty$ or $-\infty$, the derivative is said to be infinite [93, p. 68], and the fields decay more slowly than $1/z$ as $z\to\infty$. (The current pulse with time dependence given by $|t|^{1/2}$ has an infinite derivative at $t=0$. The current functions, $t\sin(1/t)$ and $|t|^{1/2}\sin(1/t)$, are examples of continuous current functions that have bounded and unbounded secant slopes, respectively, at the time $t=0$ where the limit in (5.329) and thus the derivative does not exist.)

Current with Delta-Function Time Dependence (Impulse Response). The impulse response, that is, the solution for the fields of the circular disk with surface current $\mathbf{K}(t) = \hat{\mathbf{x}}\delta(t)$, can be found in closed form throughout all space. To see this, take the inverse Fourier transform of (5.306) with $K_\omega = 1/(2\pi)$ to get for $z > 0$

$$\mathbf{H}^\delta(\mathbf{r}, t) = \frac{1}{4\pi}\left(\hat{\mathbf{y}}\frac{\partial}{\partial z} - \hat{\mathbf{z}}\frac{\partial}{\partial y}\right)\int\limits_{disk} \frac{\delta(t - |\mathbf{r} - \boldsymbol{\rho}'|/c)}{|\mathbf{r} - \boldsymbol{\rho}'|} dS' \tag{5.330}$$

where $\mathbf{r} = x\hat{\mathbf{x}} + y\hat{\mathbf{y}} + z\hat{\mathbf{z}}$ and $\boldsymbol{\rho}' = x'\hat{\mathbf{x}} + y'\hat{\mathbf{y}}$. This result can also be found from equations (2.53) and (2.55) with $\mathbf{J}(\mathbf{r}, t) = \delta(z)\delta(t)\hat{\mathbf{x}}$ on the disk. By shifting the origin of the primed coordinates to $\boldsymbol{\rho} = x\hat{\mathbf{x}} + y\hat{\mathbf{y}}$, the integral in (5.330) can be evaluated in closed form[5] to yield the following expression for the magnetic field

$$\mathbf{H}^\delta(\mathbf{r}, t) = \frac{1}{2\pi}\left(\hat{\mathbf{y}}\frac{\partial}{\partial z} - \hat{\mathbf{z}}\frac{\partial}{\partial y}\right)\mathcal{G}^\delta(\rho, z, t) \tag{5.332}$$

with

$$\mathcal{G}^\delta(\rho, z, t) = \int_{-\infty}^{+\infty} \mathcal{G}_\omega(x, y, z)e^{-i\omega t}d\omega = \frac{1}{2}\int\limits_{disk} \frac{\delta(t - |\mathbf{r} - \boldsymbol{\rho}'|/c)}{|\mathbf{r} - \boldsymbol{\rho}'|} dS' \tag{5.333}$$

given by

$$\mathcal{G}^\delta(\rho, z, t) = c\begin{cases} 0, & ct < z \\ \pi, & z < ct < D_1 \\ \arccos\left(\dfrac{\rho^2 - a^2 + c^2t^2 - z^2}{2\rho\sqrt{c^2t^2 - z^2}}\right), & D_1 < ct < D_2 \\ 0, & ct > D_2 \end{cases} \tag{5.334}$$

for $\rho < a$, and by

$$\mathcal{G}^\delta(\rho, z, t) = c\begin{cases} 0, & ct < D_1 \\ \arccos\left(\dfrac{\rho^2 - a^2 + c^2t^2 - z^2}{2\rho\sqrt{c^2t^2 - z^2}}\right), & D_1 < ct < D_2 \\ 0, & ct > D_2 \end{cases} \tag{5.335}$$

for $\rho > a$, where, of course, $\rho = \sqrt{x^2 + y^2}$. The distances

$$D_1 = \sqrt{(\rho - a)^2 + z^2} \tag{5.336}$$

and

$$D_2 = \sqrt{(\rho + a)^2 + z^2} \tag{5.337}$$

are the shortest and longest distance, respectively, from the field point (ρ, z) to a point on the rim of the current disk. The arccos functions in (5.334) and (5.335) range from 0 to π. For $\rho = 0$, $\mathcal{G}^\delta$ in (5.334) can be written simply as

$$\mathcal{G}^\delta(0, z, t) = -\pi c\left[h(t - z/c) - h\left(t - \sqrt{z^2 + a^2}/c\right)\right] \tag{5.338}$$

where $h(t)$ is the unit-step function. At each (ρ, z), $\mathcal{G}^\delta(\rho, z, t)$ is a continuous function of t except for the step discontinuity at $t = z/c$, and if $\rho = 0$, at $t = \sqrt{z^2 + a^2}/c$. At each (z, t), $\mathcal{G}^\delta(\rho, z, t)$ is a continuous function of ρ.

[5]The integral over the disk can also be computed from the following general result. Let the functions $f(x, y)$ and $g(x, y)$ be defined in the region D of the (x, y) plane. Assume that $f(x, y) = 0$ only on the curve C. If $\nabla f(x, y) \neq 0$ on C, we have

$$\int\limits_D \delta(f(x, y))\, g(x, y)dxdy = \int\limits_C \frac{g(x(s), y(s))}{|\nabla f(x(s), y(s))|}ds \tag{5.331}$$

where $(x(s), y(s))$ is a parametric representation for C with s denoting the arc length.

Similarly, the impulse electric field can be found by taking the inverse Fourier transform of (5.309) with $K_\omega = 1/(2\pi)$ to get

$$\mathbf{E}^\delta(\mathbf{r}, t) = \frac{1}{2\pi}\sqrt{\frac{\mu}{\epsilon}}\left[-\hat{\mathbf{x}}\frac{\partial}{c\partial t}\mathcal{G}^\delta(\rho, z, t) + c\nabla\frac{\partial}{\partial x}\int_{-\infty}^{t}\mathcal{G}^\delta(\rho, z, t')dt'\right]. \quad (5.339)$$

The function $\mathcal{G}^\delta(\rho, z, t)$ is proportional to the velocity potential of a rigid, circular acoustic piston radiator excited by an impulse (delta function) in velocity [162]–[165]. Plots of $\mathcal{G}^\delta(\rho, z, t)$ can be found in these references. Along the z axis, the expressions for the magnetic and electric fields in (5.332) and (5.339) reduce to (5.316) and (5.317) with $\mathbf{K}(t) = \hat{\mathbf{x}}\delta(t)$.

Current with Unit-Step Time Dependence. Consider the example of a surface current with time variation given by a unit-step function. The magnetic and electric fields generated by the step-function current can be found by integrating over time from $-\infty$ to t the impulse magnetic and electric fields $\mathbf{H}^\delta(\mathbf{r}, t)$ and $\mathbf{E}^\delta(\mathbf{r}, t)$ in (5.332) and (5.339). Here, however, we shall concentrate on the fields along the z axis and the far-field pattern. For a step-function current, the on-axis electric and magnetic fields in (5.319) and (5.320) are given by a rectangular pulse of constant amplitude independent of the large propagation distance z. The width of this constant-amplitude pulse is given by $a^2/(2z)$, which decreases as $1/z$ with the distance z to the pulse. In other words, a unit-step function in time in the surface current generates a far-field rectangular pulse along the z axis that does not decay in amplitude as $z \to \infty$, but decays in width like $1/z$; see Figure 5.7.

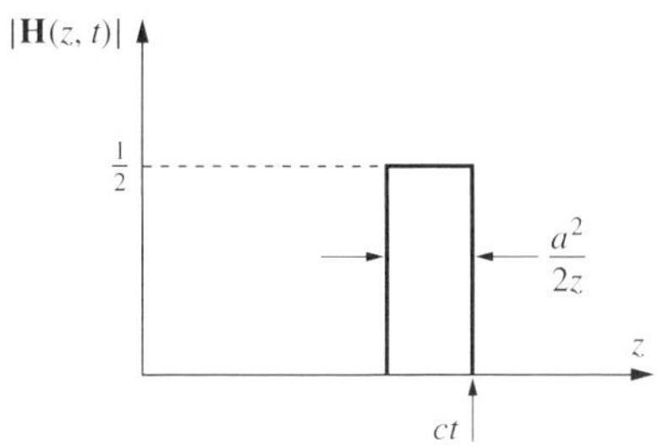

Figure 5.7 Magnetic field as a function of large z (at fixed time t) from a surface current with unit-step function time dependence.

The total energy in this step-function electromagnetic "missile" also decays as $1/z$. To show this, we need to determine how far from the z axis the pulse approximately maintains its amplitude and width. This can be done by evaluating the general radiation integral in (5.305) for fixed distances away from the z axis [155], [166]. However, a more direct and simpler way to determine the transverse dimensions of the pulse is to integrate the impulse fields $\mathbf{H}^\delta(\mathbf{r}, t)$ and $\mathbf{E}^\delta(\mathbf{r}, t)$ with respect to time to get (as mentioned in the previous paragraph) the step-function fields. Such an integration shows that the transverse dimensions of the pulse are on the order of the size of the circular current disk. In other words, the transverse area of the electromagnetic missile is roughly equal to the area of the current disk. Integrating the energy density $(\epsilon|\mathbf{E}|^2 + \mu|\mathbf{H}|^2)/2$ over this pulse with approximately constant amplitude and area, and with width equal to $a^2/(2z)$, produces a total energy in the "missile" that decays as $1/z$.

In brief, this electromagnetic "missile," excited by a step-function surface current, is a pulse, satisfying Maxwell's equations, that avoids spreading in the direction transverse to propagation by continuously decreasing its pulse width while increasing the average value of its frequency spectrum. Its total energy content vanishes as it propagates. Thus, its contribution to the total radiated energy in the far field is zero.

Note that the time derivative of the step-function surface current is a delta function, which is infinite and thus satisfies the necessary condition on the time dependence of the current derived in Section 5.4.2 for the presence of slower than $1/r$ far-field decay. Also, the integral of the square of the time derivative of the current, $\int_{-\infty}^{+\infty} \delta^2(t)dt$, is infinite. Thus, the step-function current also satisfies the necessary condition (5.301) for the existence of an electromagnetic missile (slower than $1/r^2$ decay of the energy per unit area in the far-field pulse). Other mathematical time dependencies for the surface current on the circular disk are easily constructed that do or do not produce electromagnetic missiles and yet exhibit slower than $1/r$ far-field behavior. (See, for example, the time functions given below (5.294).)

The magnitude of the time-domain far-field pattern of the unit-step far-field pulse shown in Figure 5.7 is infinite in the z direction (for times at which the far-field pattern is nonzero), since from (5.320)

$$|\mathcal{F}(\theta = 0, t)| = \lim_{z\to\infty} z\,|\mathbf{E}(z, t + z/c)| = \lim_{z\to\infty} \frac{z}{2}\sqrt{\frac{\mu}{\epsilon}} = \infty \tag{5.340}$$

which agrees with the general result (5.296) that the time-domain far-field pattern in the direction of the missile must be infinite.

The far-field pattern over the whole far-field sphere can be found in closed form in both the frequency and time domains for the circular current disk excited by the step function in time. The frequency-domain far-field pattern of the current disk can be found from (3.36), by integrating the surface current $K_\omega \hat{\mathbf{x}}$ over the disk instead of the volume current $\mathbf{J}_\omega$ in (3.36), to get

$$\begin{aligned} \mathcal{F}_\omega(\theta, \phi) &= -\sqrt{\frac{\mu}{\epsilon}} \frac{ia\,K_\omega\,J_1(ka\sin\theta)}{2\sin\theta} \hat{\mathbf{r}} \times (\hat{\mathbf{r}} \times \hat{\mathbf{x}}) \\ &= \sqrt{\frac{\mu}{\epsilon}} \frac{a\,J_1(ka\sin\theta)}{4\pi\omega\sin\theta}\, \hat{\mathbf{r}} \times (\hat{\mathbf{r}} \times \hat{\mathbf{x}}) \end{aligned} \tag{5.341}$$

where we have used the result

$$\int_{disk} e^{-ik\hat{\mathbf{r}}\cdot\mathbf{r}'} dS' = \frac{2\pi a J_1(ka\sin\theta)}{k\sin\theta} \tag{5.342}$$

and inserted the spectrum $K_\omega = -1/(2\pi i\omega)$ for the unit-step function time dependence. Notice that the frequency-domain far-field pattern (5.341) is well defined for all θ (in fact, it is an entire function of θ; see Section 3.2.4) and that in the direction of the missile its frequency dependence is

$$|\mathcal{F}_\omega(\theta = 0)| = \sqrt{\frac{\mu}{\epsilon}} \frac{a^2}{8\pi c}. \tag{5.343}$$

This behavior, which could also have been obtained from (5.313), agrees with the general result in Section 5.4.1 that the frequency-domain far-field pattern in the direction of the missile must decay slower than $1/|\omega|^{1/2}$ as $\omega \to \pm\infty$.

Taking the Fourier transform of (5.341) gives the time-domain far-field pattern

$$\mathcal{F}(\theta, \phi, t) = \sqrt{\frac{\mu}{\epsilon}} \frac{a\,\hat{\mathbf{r}} \times (\hat{\mathbf{r}} \times \hat{\mathbf{x}})}{2\pi\sin\theta} \int_0^\infty \frac{J_1(\omega(a/c)\sin\theta)}{\omega} \cos(\omega t)d\omega. \tag{5.344}$$

The integral in (5.344) can be evaluated in closed form [147, vol. 1, p. 43, eq. (4)], so that (5.344) reduces to

$$\mathcal{F}(\theta, \phi, t) = \sqrt{\frac{\mu}{\epsilon}} \frac{a\,\hat{\mathbf{r}} \times (\hat{\mathbf{r}} \times \hat{\mathbf{x}})}{2\pi\sin\theta} \begin{cases} \sqrt{1 - \left(\dfrac{ct}{a\sin\theta}\right)^2}, & 0 \le |t| \le (a/c)\sin\theta \\ 0, & |t| > (a/c)\sin\theta. \end{cases} \tag{5.345}$$

Equation (5.345) can also be derived directly from the integral in (5.49) that gives the time-domain far-field pattern in terms of the time-domain surface current [146]. We see from (5.345) that the time-domain far-field pattern has a $1/\theta$ singularity at $\theta = 0$ for this step-function time excitation in the surface current. This singularity along the z axis ($\theta = 0$) must be present because we have shown that an electromagnetic missile exists in the z direction for a surface current with a step-function time dependence. The time-domain far-field pattern in (5.345) is shown in Figure 5.8.

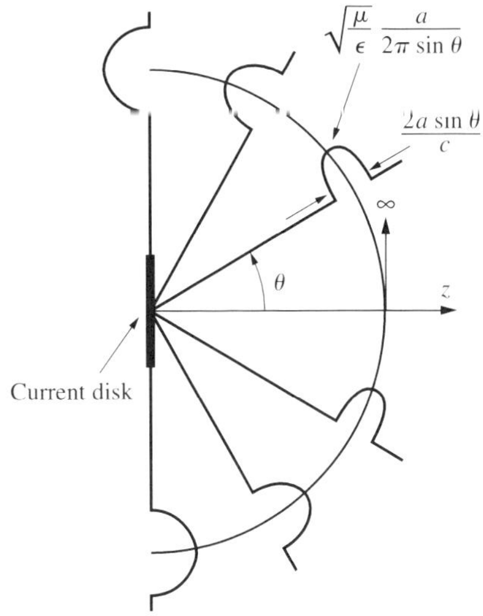

Figure 5.8 Time-domain far-field pattern of a current disk with unit-step time dependence. An electromagnetic missile is present in the direction $\theta = 0$.

An integration over all time and angles of the square of the time-domain far-field pattern in (5.345) shows that the total energy radiated by the current disk with step-function time dependence is finite. As a consequence of Parseval's theorem, the finiteness of the radiated energy can also be shown by integrating the frequency-domain far-field pattern in (5.341) over all angles and frequencies: Let's begin by noting that the far-field pattern in (5.341) satisfies

$$|\mathcal{F}_\omega(\theta, \phi)|^2 \le \frac{\mu a^2 |K_\omega|^2 [J_1(ka \sin\theta)]^2}{4\epsilon \sin^2\theta} \tag{5.346}$$

so that

$$\begin{aligned}
&\int_{-\infty}^{+\infty}\int_0^{2\pi}\int_0^{\pi} |\mathcal{F}_\omega(\theta, \phi)|^2 \sin\theta \, d\theta \, d\phi \, d\omega \\
&\le \frac{\pi \mu a^2}{2\epsilon} \int_{-\infty}^{+\infty} |K_\omega|^2 \int_0^{\pi} \frac{[J_1(ka \sin\theta)]^2}{\sin\theta} d\theta \, d\omega \\
&= \frac{\pi \mu a^2}{2\epsilon} \int_{-\infty}^{+\infty} |K_\omega|^2 \left(1 - \frac{J_1(2\omega a/c)}{\omega a/c}\right) d\omega
\end{aligned} \tag{5.347}$$

where we have used a formula in [167, p. 299, eq. (24)] to compute the integral over θ. The first-order Bessel function satisfies $(1 - J_1(2x)/x) = O(x^2)$ as $x \to 0$ and $(1 - J_1(2x)/x) \to 1$ as $x \to \pm\infty$. In addition, when the current has unit-step time dependence, its squared spectrum is given by $|K_\omega|^2 = 1/(2\pi\omega)^2$, and the last integral in (5.347) converges, so the energy radiated by the disk is indeed finite.

5.4.4 Concluding Remarks on Electromagnetic Missiles

In principle, a pulse can be sent to infinity with finite or even infinite energy through a finite area. For example, equations (5.319) and (5.320) show that a current source with delta-function time dependence on a circular disk radiates two delta-function pulses to an indefinitely large distance along the z axis normal to the center of the disk. Moreover, the general expressions for the impulse-response fields in (5.332) and (5.339) show that the transverse area of the leading delta-function pulse is equal to the area of the disk, and thus this electromagnetic missile carries an infinite amount of energy (since $\int_{-\infty}^{+\infty} \delta^2(t)dt = \infty$).

In general, however, we can argue that a current source of finite energy in a finite region of space cannot radiate an electromagnetic missile with nondecaying energy to an infinitely large distance from the source [155]. The argument goes as follows. Because a source pulse of finite bandwidth cannot generate an electromagnetic missile, any electromagnetic missile must maintain itself on the energy in the increasingly higher frequencies. If the total energy radiated by the sources is finite, the energy in the frequency spectrum at frequencies higher than any finite value must approach zero.[6]

Nonetheless, finite energy current sources are mathematically possible that will produce a pulse that travels without decay of the energy to a finite distance arbitrarily far from the source region. To see this, let the surface current on a circular disk of fixed radius a have a time dependence given by the rectangular pulse

$$\mathbf{K}_1(t) = \hat{\mathbf{x}} \begin{cases} \dfrac{1}{\sqrt{t_1}}, & 0 < t \le t_1 \\ 0, & t \le 0 \ \text{and} \ t > t_1 \end{cases} \tag{5.350}$$

with $t_1 > 0$. Equation (5.347) shows that this current radiates a finite energy since

$$2\pi \int_{-\infty}^{+\infty} |\mathbf{K}_{1\omega}|^2 d\omega = \int_0^{t_1} |\mathbf{K}_1(t)|^2 dt = 1. \tag{5.351}$$

(Note that $|\mathbf{K}_1(t)|^2$ behaves as a Dirac delta function as $t_1 \to 0$.)

Now equations (5.319) and (5.320) show that this current will radiate two separate nondecaying pulses to the far field along the z axis provided $a^2/(2cz) \ge t_1$. In other words, the pulses will travel without decay along the z axis to a distance z given by

$$z = \frac{a^2}{2ct_1}. \tag{5.352}$$

Moreover, the fields everywhere produced by the surface current $\mathbf{K}_1(t)$ can be found by con-

[6] In mathematical terms, given that

$$\lim_{U\to\infty} \int_0^U |\mathbf{K}_\omega|^2 d\omega = C < \infty \tag{5.348}$$

then

$$\begin{aligned} \lim_{L\to\infty} \lim_{U\to\infty} \int_L^U |\mathbf{K}_\omega|^2 d\omega &= \lim_{L\to\infty} \left[\lim_{U\to\infty} \int_0^U |\mathbf{K}_\omega|^2 d\omega - \int_0^L |\mathbf{K}_\omega|^2 d\omega \right] \\ &= C - \lim_{L\to\infty} \int_0^L |\mathbf{K}_\omega|^2 d\omega = 0. \end{aligned} \tag{5.349}$$

volving $|\mathbf{K}_1(t)|$ with the impulse-response fields given in (5.332) and (5.339). Specifically, we find that as $t_1 \to 0$, the magnetic field $\mathbf{H}_1(\mathbf{r}, t)$ produced by the surface current $\mathbf{K}_1(t)$ in (5.350) is given by

$$\mathbf{H}_1(\mathbf{r}, t) = -\frac{1}{2}\hat{\mathbf{z}} \times \begin{cases} \mathbf{K}_1\left(t - \frac{z}{c}\right) - \frac{z}{\sqrt{z^2 + a^2}}\mathbf{K}_1\left(t - \frac{\sqrt{z^2 + a^2}}{c}\right), & \rho = 0 \\ \mathbf{K}_1\left(t - \frac{z}{c}\right) + \mathbf{O}\left(\sqrt{t_1}\right), & 0 < \rho < a \\ \mathbf{O}\left(\sqrt{t_1}\right), & \rho > a \end{cases} \tag{5.353}$$

for $0 < z < a^2/(2ct_1)$. Similarly, the electric field produced by $\mathbf{K}_1(t)$ can be written as

$$\mathbf{E}_1(\mathbf{r}, t) = -\frac{1}{2}\sqrt{\frac{\mu}{\epsilon}} \begin{cases} \mathbf{K}_1\left(t - \frac{z}{c}\right) - \frac{z^2 + a^2/2}{z^2 + a^2}\mathbf{K}_1\left(t - \frac{\sqrt{z^2 + a^2}}{c}\right), & \rho = 0 \\ \mathbf{K}_1\left(t - \frac{z}{c}\right) + \mathbf{O}\left(\sqrt{t_1}\right), & 0 < \rho < a \\ \mathbf{O}\left(\sqrt{t_1}\right), & \rho > a\,. \end{cases} \tag{5.354}$$

The expressions (5.353) and (5.354) for the magnetic and electric fields reveal that the significant transverse area of the leading pulse is approximately equal to the area (πa^2) of the current disk, and thus this electromagnetic missile carries a finite nondecaying energy to a distance $z = a^2/(2ct_1)$. Since, in principle, t_1, the width of the current pulse, can be made arbitrarily small, the electromagnetic missile can be made to travel arbitrarily far in the z direction without any loss of energy. For a fixed nonzero value of t_1, the energy in the pulse will begin to decay beyond the distance z given in (5.352). Beyond the z in (5.352), the pulse remains an electromagnetic missile with energy decaying with z slower than $1/z^2$.

One can draw from high-frequency diffraction theory to explain why the fields in (5.353)–(5.354) are zero for $\rho > a$ as $t_1 \to 0$. In the frequency domain the fields diffracted from the edge of the current disk produce fields everywhere. However, for $\rho \neq 0$, the magnitude of the edge diffracted fields decreases with increasing frequency, because the effective length of the stationary phase integration along the circular edge of the disk decreases with increasing frequency. Since the magnitude of the frequency spectrum of $\mathbf{K}_1(t)$ behaves as $|\sin(\omega t_1/2)|/(\pi|\omega|\sqrt{t_1})$, as $t_1 \to 0$ the spectrum is practically zero except at very high frequencies where the edge diffracted fields are negligible for all $\rho \neq 0$. Thus, only the plane-wave pulse emanating from the disk and traveling in the z direction remains for $\rho \neq 0$. Along the z axis ($\rho = 0$), this leading plane-wave pulse is joined by a trailing pulse formed by the constructive interference of the fields diffracted from the entire circular edge of the disk.

In summary, classical sources of finite charge and current in a finite region of space can produce classical Maxwellian far-field pulses that decay slower than $1/r$ and carry energy radiated through a fixed-size area that decays slower than $1/r^2$. However, these anomalous pulses described by Maxwell's equations require either infinite acceleration of point charges, infinite first time derivatives in the macroscopic continuum of current, or infinite first time derivatives in the Huygens' sources, such that the frequency spectrum of the current or the Huygens' sources decays as $1/|\omega|^{3/2}$ or slower as $\omega \to \pm\infty$. The infinite acceleration of a point charge is presumably not possible, and all known sources have frequency spectra that decay much faster than $1/|\omega|^{3/2}$ as $\omega \to \pm\infty$. Thus, electromagnetic missiles described by

Maxwell's equations cannot be generated in the laboratory today or in the foreseeable future. Of course, the greater the maximum effective frequency ($\omega_{max} = 2\pi c/\lambda_{min}$) of the generator and the larger the maximum dimension (D) of the radiator, the further out one can push the distance to the boundary between the near and far fields; that is, the larger one can make the Rayleigh distance (D^2/λ_{min}).

6

Probe Correction in the Frequency Domain

Near-field antenna measurement techniques have been used extensively to determine antenna fields from measurements taken on scan surfaces located in the near field of the antenna [42], [55]. In particular, with the frequency-domain planar near-field measurement technique, the amplitude and phase[1] of the output of the measurement probe are recorded in the near field of the test antenna on a planar surface. The desired antenna fields are then computed by inserting the measured data into a set of planar near-field formulas [8]. This chapter presents the frequency-domain acoustic and electromagnetic probe-corrected planar near-field formulas that will subsequently be translated into the time domain in Chapter 7.

In Chapter 3, we considered acoustic and electromagnetic frequency-domain fields radiated by sources of finite extent that were located in the half space $z < z_0$. Formulas were derived that determine the fields in the source-free half space $z > z_0$ in terms of the fields in the scan plane $z = z_0$. Hence, if the field in the scan plane could be determined through measurements, the formulas of Chapter 3 could be used to determine the desired antenna field from planar near-field measurements. In practice, however, the field in the scan plane is usually measured with a nonideal probe whose output is not equal to the field at each point in the scan plane. Moreover, when the scan plane is in the near field of the source region, it is not possible, in general, to directly determine the field at a point in the scan plane from the output of a nonideal probe at that point.

This can be understood in the following way. When the scan plane is in the near field of the sources, the incident field in the scan plane consists of infinitely many propagating and evanescent plane waves. Each of these incident plane waves has an unknown amplitude. (In the expression (3.62) for the electric field, the plane-wave amplitudes are written in terms of the spectrum $\mathbf{T}_\omega(k_x, k_y)$.) The probe output at a point in the scan plane can be expressed as a superposition of the incident plane-wave amplitudes weighted by the plane-wave receiving characteristic of the probe. Assuming that the value of the probe output is known at a point

[1] Although progress has been made toward developing amplitude-only, frequency-domain, near-field measurements [168], in general, the accurate determination of the sidelobe fields of electrically large antennas requires the accurate measurement of both the phase and amplitude of the output of the probe. Direct time-domain near-field measurement techniques avoid the difficulty of accurately measuring the phase of the output of the probe but must deal with the corresponding difficulty of accurately synchronizing the output pulses of the probe with the input pulses of the test antenna.

in the scan plane, we then have one equation for determining the infinitely many unknown plane-wave amplitudes in the plane-wave expansion of the incident field at that point in the scan plane. From this single equation for a nonideal receiving probe, we cannot generally determine the unknown incident plane-wave amplitudes, and the field at a point in the scan plane cannot be determined from the probe output at that point. Thus, the formulas of Chapter 3 are not generally applicable when a nonideal probe is used to measure the field in a scan plane located in the near field of the source region.[2]

Therefore, it is of great practical interest to derive probe-corrected formulas, which express the field in the half space $z > z_0$ in terms of the output of a known nonideal probe in the scan plane $z = z_0$. Such probe-corrected formulas for planar scanning were first published in 1963 by Kerns [27] for frequency-domain fields and have been used extensively throughout the world to determine the fields of electromagnetic antennas [42]. The probe-corrected formulas can also be used to express the field at each point in the scan plane in terms of the probe output over the entire scan plane.

In this chapter, we shall rederive Kerns's formulas for both acoustic and electromagnetic fields using a straightforward approach, which will prove useful for the subsequent time-domain derivation in Chapter 7. In addition, alternative derivations of the probe-corrected formulas will be given based on the probe output satisfying the Helmholtz equation. Throughout this chapter, $e^{-i\omega t}$ time dependence is suppressed, and the frequency ω is assumed positive.

The schematic for the planar near-field measurement system is shown in Figure 6.1, where the source region that generates the field is called the test antenna. (To use the same terminology as for electromagnetic fields, the acoustic source will also be called the test antenna.) For both the acoustic and electromagnetic fields we shall use electromagnetic waveguide terminology to describe the probe that is used to measure the field generated by the test antenna. In addition, the feed system for the test antenna is also an electromagnetic waveguide (as indicated in Figure 6.1). That is, in acoustics the probe and test "antenna" are electroacoustic transducers. Therefore, we begin by describing the necessary waveguide quantities and their dependence on frequency.

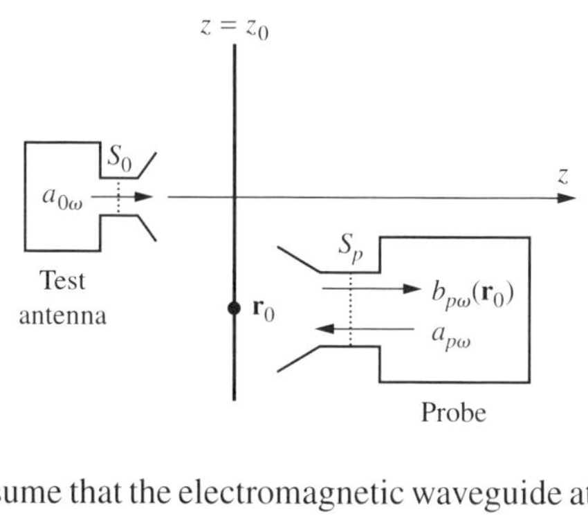

Figure 6.1 Planar scanning geometry for frequency-domain fields.

Assume that the electromagnetic waveguide attached to the probe supports just one propagating mode (or that only one propagating mode is excited), and let $\mathbf{E}^t_\omega(\boldsymbol{\rho})$ and $\mathbf{H}^t_\omega(\boldsymbol{\rho})$ denote the transverse electric and magnetic fields in the reference plane S_p of the probe waveguide. For the sake of simplicity, we assume that the waveguide is composed of perfect conductors separated by a homogeneous, isotropic medium, which is described by the constant permittivity ϵ_p and the constant permeability μ_p. Hence, the following discussion holds for coaxial

[2]If the probe is in the far field of the sources, the incident field at the probe consists locally of only one plane wave. Thus, when the receiving characteristic of the probe is known, the incident far field can be determined directly from the probe output, as demonstrated in Section 6.2.1. At the end of Section 6.1 and in Section 6.2.3, it is proven that an ideal acoustic monopole probe and an ideal electric dipole probe possess the proper receiving characteristics to measure the acoustic pressure and electric field, respectively.

transmission lines as well as rectangular and circular waveguides, and the transverse fields on S_p satisfy [8, p. 15]

$$\mathbf{H}_\omega^t(\boldsymbol{\rho}) = Y_{p\omega}^W \hat{\mathbf{n}} \times \mathbf{E}_\omega^t(\boldsymbol{\rho}). \tag{6.1}$$

Here $Y_{p\omega}^W$ is the wave admittance for the single propagating mode, and $\hat{\mathbf{n}}$ is the normal to S_p, which equals $-\hat{\mathbf{z}}$ in Figure 6.1. Silver [169, sec. 7.4] gives an excellent discussion of voltages and currents in waveguides, and refers to $1/Y_{p\omega}^W$ as the transverse wave impedance.

The field expressions in [169, p. 209, eq. (33a)] show that for the transverse electromagnetic (TEM) mode in a coaxial transmission line, the wave admittance is simply $Y_{p\omega}^W = (\epsilon_p/\mu_p)^{1/2}$, which is equal to the admittance of the medium that exists between the conductors. Similarly, the field expressions for the rectangular waveguide in [169, p. 228, eq. (77)] show that for the propagating TE_{10} mode, the wave admittance is $Y_{p\omega}^W = (\omega^2\epsilon_p\mu_p - (\pi/a)^2)^{1/2}/(\omega\mu_p)$, where a is the longest cross-sectional side of the waveguide. Note that the wave admittance is independent of frequency for the coaxial transmission line, whereas it depends on frequency for the rectangular waveguide.

The transverse electric and magnetic fields on the reference plane S_p can be expressed in terms of basis fields as [8, p. 51]

$$\mathbf{E}_\omega^t(\boldsymbol{\rho}) = (a_{p\omega} + b_{p\omega})\mathbf{e}_{b\omega}(\boldsymbol{\rho}) = V_{p\omega}\mathbf{e}_{b\omega}(\boldsymbol{\rho}) \tag{6.2}$$

and

$$\mathbf{H}_\omega^t(\boldsymbol{\rho}) = Y_p(a_{p\omega} - b_{p\omega})\mathbf{h}_{b\omega}(\boldsymbol{\rho}) = I_{p\omega}\mathbf{h}_{b\omega}(\boldsymbol{\rho}) \tag{6.3}$$

where Y_p is the characteristic admittance, $a_{p\omega}$ is the amplitude of the incident wave, and $b_{p\omega}$ is the amplitude of the emerging wave, as shown in Figure 6.1. The phases of $a_{p\omega}$ and $b_{p\omega}$ depend on the chosen position of the reference plane S_p in the probe waveguide. (If the surface S_p simply cuts two wires at quasi-static frequencies, $V_{p\omega}$ in (6.2) and $I_{p\omega}$ in (6.3) refer to conventional circuit voltages and currents which do not necessarily serve as modal coefficients. In that case, equations (6.2) and (6.3) become definitions of $a_{p\omega}$ and $b_{p\omega}$.)

From waveguide theory [8, p. 14], the basis fields can be chosen real and thus satisfy the relation

$$\mathbf{h}_{b\omega}(\boldsymbol{\rho}) = \frac{Y_{p\omega}^W}{Y_p} \hat{\mathbf{n}} \times \mathbf{e}_{b\omega}(\boldsymbol{\rho}). \tag{6.4}$$

We shall also choose the power normalization

$$\int_{S_p} \hat{\mathbf{n}} \cdot \left[\mathbf{e}_{b\omega}(\boldsymbol{\rho}) \times \mathbf{h}_{b\omega}(\boldsymbol{\rho})\right] dS = 1 \tag{6.5}$$

so that the characteristic admittance is given by

$$Y_p = Y_{p\omega}^W \int_{S_p} |\mathbf{e}_{b\omega}(\boldsymbol{\rho})|^2\, dS. \tag{6.6}$$

For waveguides composed of perfect conductors separated by a homogeneous, isotropic medium, the basis fields can always be written as $\mathbf{e}_{b\omega}(\boldsymbol{\rho}) = f_\omega \mathbf{e}_0(\boldsymbol{\rho})$, where $\mathbf{e}_0(\boldsymbol{\rho})$ is real and independent of the frequency and f_ω is a real frequency function. This follows from the fact that the separation constant determining the transverse dependence of the fields is independent of frequency [8, pp. 6–8]. Substituting $\mathbf{e}_{b\omega}(\boldsymbol{\rho}) = f_\omega \mathbf{e}_0(\boldsymbol{\rho})$ into (6.6) shows that the characteristic admittance Y_p is *independent* of frequency if we choose f_ω such that $Y_{p\omega}^W f_\omega^2$ is independent of frequency. In the following, we assume that Y_p is frequency independent in the frequency range where the probe operates, thereby simplifying the time-domain reciprocity relations of Chapter 7.

A consequence of choosing Y_p frequency independent is that the basis fields $\mathbf{e}_{b\omega}$ and $\mathbf{h}_{b\omega}$ become frequency dependent whenever the wave admittance $Y^W_{p\omega}$ is frequency dependent. Hence, for the TEM mode of a coaxial transmission line, the basis fields are independent of frequency, whereas for the rectangular waveguide the basis fields depend on frequency.

For broadband near-field measurements, it may be convenient to attach a coaxial transmission line to the probe because the wave admittance of its TEM mode is independent of frequency. Thus, the basis fields as well as the characteristic admittance can be chosen independent of frequency. Let a denote the radius of the inner conductor and b denote the radius of the outer conductor of the coaxial line. Then with $V_{p\omega}$ being the voltage between the conductors and $I_{p\omega}$ being the current flowing through each conductor, the characteristic admittance for the coaxial transmission line is simply

$$Y_p = \sqrt{\frac{\epsilon_p}{\mu_p}} \frac{2\pi}{\ln(b/a)} \tag{6.7}$$

where ϵ_p is the permittivity and μ_p the permeability of the medium that separates the conductors. The expression (6.7) is easily determined from (6.6) by inserting $Y^W_{p\omega} = (\epsilon_p/\mu_p)^{1/2}$ and $|\mathbf{e}_{b\omega}(\boldsymbol{\rho})| = 1/(\ln(b/a)|\boldsymbol{\rho}|)$; see [169, p. 218].

When the probe is used to measure the field of the test antenna, we assume that the probe waveguide is terminated in a perfectly matched load, so that the input to the probe $a_{p\omega}$ is zero. We refer to $b_{p\omega}$ as the probe output. The waveguide description given here for the probe applies also to the test antenna when it is attached to a waveguide feed.

6.1 ACOUSTIC FIELDS

The source that generates the acoustic field is located in a linear, quiescent, homogeneous, isotropic, and lossless fluid, as described in Section 2.4. Outside the source region, the acoustic pressure $p_\omega(\mathbf{r})$ satisfies the homogeneous Helmholtz equation $\nabla^2 p_\omega + k^2 p_\omega = 0$, where $k = \omega/c$ is the propagation constant and c is the constant wave speed.

Figure 6.1 shows the geometry for measuring the acoustic pressure $p_\omega(\mathbf{r})$ for the test antenna located in the half space $z < z_0$ and fed by a single propagating mode of amplitude $a_{0\omega}$ at the reference plane S_0. Alternatively, one can think of the test antenna (electroacoustic transducer) as a source region that includes the antenna, the feed system, and the generator, without specifying a waveguide mode with amplitude $a_{0\omega}$. This approach was taken in Chapter 3. The pressure generated by the test antenna is measured with the nonideal probe whose output $b_{p\omega}(\mathbf{r}_0)$ is recorded in the scan plane $z = z_0$. The phases of $a_{0\omega}$ and $b_{p\omega}$ depend on the chosen positions of the reference planes S_0 and S_p in the waveguide of the test antenna and probe, respectively. The goal of this section is to derive probe-corrected formulas that express the acoustic field in the half space $z > z_0$ (or more generally, to the "right" of the source region) in terms of the probe output in the scan plane $z = z_0$. The following plane-wave spectrum formulas, which were all proven in Chapter 3, will be used extensively throughout this section.

When the test antenna is the only object in all space, the formulas (3.56)–(3.57) show that the pressure can be expanded in terms of plane waves as

$$p_\omega(\mathbf{r}) = \frac{a_{0\omega}}{2\pi} \int_{-\infty}^{+\infty}\int_{-\infty}^{+\infty} T_{0\omega}(k_x, k_y) e^{i(k_x x + k_y y + \gamma z)} dk_x dk_y, \quad z \geq z_0 \tag{6.8}$$

where $T_{0\omega}(k_x, k_y)$ is the normalized transmitting spectrum of the test antenna given by

$$T_{0\omega}(k_x, k_y) = \frac{e^{-i\gamma z}}{2\pi a_{0\omega}} \int_{-\infty}^{+\infty}\int_{-\infty}^{+\infty} p_\omega(\mathbf{r}) e^{-i(k_x x + k_y y)} dx dy, \quad z \geq z_0. \tag{6.9}$$

The product $a_{0\omega}T_{0\omega}(k_x, k_y)$ equals the spectrum $T_\omega(k_x, k_y)$ defined in Chapter 3, the observation point is $\mathbf{r} = x\hat{\mathbf{x}} + y\hat{\mathbf{y}} + z\hat{\mathbf{z}}$, and the propagation vector for each of the plane waves $e^{i(k_x x + k_y y + \gamma z)}$ is

$$\mathbf{k} = k_x\hat{\mathbf{x}} + k_y\hat{\mathbf{y}} + \gamma\hat{\mathbf{z}} \tag{6.10}$$

with

$$\gamma = \begin{cases} \left|\sqrt{k^2 - k_x^2 - k_y^2}\right|, & k_x^2 + k_y^2 \le k^2 \\ i\left|\sqrt{k_x^2 + k_y^2 - k^2}\right|, & k_x^2 + k_y^2 > k^2. \end{cases} \tag{6.11}$$

When the test antenna is the only object in all space, its far-field pressure may be written in terms of the normalized far-field pattern as

$$p_\omega(\mathbf{r}) \sim a_{0\omega}\mathcal{F}_{0\omega}(\theta, \phi)\frac{e^{ikr}}{r}, \quad r \to \infty \tag{6.12}$$

and (3.147) together with (3.157) show that the normalized far-field pattern can be expressed in terms of the spectrum as

$$\mathcal{F}_{0\omega}(\theta, \phi) = -ik\cos\theta\; T_{0\omega}(k\cos\phi\sin\theta, k\sin\phi\sin\theta), \quad 0 \le \theta < \pi/2 \tag{6.13}$$

where (r, θ, ϕ) are the usual spherical coordinates defined by $x = r\cos\phi\sin\theta$, $y = r\sin\phi\sin\theta$, and $z = r\cos\theta$. In Chapter 3, formulas were also derived for the acoustic velocity $\mathbf{v}_\omega(\mathbf{r})$ in terms of the spectrum $T_\omega(k_x, k_y)$.

Let us begin the derivation of the probe-corrected formulas by assuming that the distance between the probe and the test antenna is always large enough that multiple interactions can be neglected. That is, the field scattered by the probe, rescattered by the test antenna, and returned to the probe, does not change the output of the probe. The effect of the multiple interactions can formally be taken into account by using the plane-wave scattering-matrix theory [8]. However, this theory does not provide quantitative information about these multiple interactions, and to obtain useful probe-corrected formulas they have to be neglected. Therefore, we will not use the plane-wave scattering-matrix formalism here, but instead we will use a more direct approach that clearly exhibits the physics of the transmit-receive process. Also, this direct approach can easily be extended to the time domain, as demonstrated in Chapter 7.

The location of a fixed reference point on the probe is denoted by $\mathbf{r}_0 = x_0\hat{\mathbf{x}} + y_0\hat{\mathbf{y}} + z_0\hat{\mathbf{z}}$, and the probe is simply said to be located at $\mathbf{r}_0$, when its reference point is located at $\mathbf{r}_0$. For planar scanning the probe translates without rotation in the z_0 plane. The output of the probe at the reference plane S_p is denoted by $b_{p\omega}(\mathbf{r}_0)$ when the probe is located at $\mathbf{r}_0$.

The receiving characteristic of the probe $R_{p\omega}(k_x, k_y)$ is defined to be the probe output when the probe is located at the origin with a fixed orientation and is illuminated by the incident plane wave $(2\pi)^{-1}e^{i(k_x x + k_y y + \gamma z)}$, that is, $b_{p\omega}(0) = R_{p\omega}(k_x, k_y)$. By linearity, this means that $b_{p\omega}(\mathbf{r}_0) = R_{p\omega}(k_x, k_y)e^{i(k_x x_0 + k_y y_0 + \gamma z_0)}$, when the incident field is the same plane wave $(2\pi)^{-1}e^{i(k_x x + k_y y + \gamma z)}$. Note that the plane-wave field $(2\pi)^{-1}e^{i(k_x x + k_y y + \gamma z)}$ propagates in the direction given by the spherical angles (θ, ϕ) determined from $k_x = k\cos\phi\sin\theta$ and $k_y = k\sin\phi\sin\theta$. When $k_x^2 + k_y^2 < k^2$, γ is real and the plane wave is propagating; when $k_x^2 + k_y^2 > k^2$, γ is imaginary and the plane wave is evanescent as $z \to \infty$.

Equation (6.8) shows that the plane-wave component of $p_\omega(\mathbf{r})$ for each (k_x, k_y) is given by

$$a_{0\omega}(2\pi)^{-1}T_{0\omega}(k_x, k_y)e^{i(k_x x + k_y y + \gamma z)}dk_x dk_y. \tag{6.14}$$

Thus, the contribution to the output of the linear probe from this plane-wave component is

$$db_{p\omega}(\mathbf{r}_0) = a_{0\omega}R_{p\omega}(k_x, k_y)T_{0\omega}(k_x, k_y)e^{i(k_x x_0 + k_y y_0 + \gamma z_0)}dk_x dk_y \tag{6.15}$$

provided multiple interactions between the probe and test antenna are negligible. Also, we are assuming throughout that the probe output line is perfectly matched, so that there is no reflection from its termination ($a_{p\omega} = 0$). The total output of the probe is thus seen from (6.8) to be the integral over all real (k_x, k_y) of the expression in (6.15), that is[3]

$$b_{p\omega}(\mathbf{r}_0) = a_{0\omega} \int_{-\infty}^{+\infty}\int_{-\infty}^{+\infty} R_{p\omega}(k_x, k_y) T_{0\omega}(k_x, k_y) e^{i(k_x x_0 + k_y y_0 + \gamma z_0)} dk_x dk_y \tag{6.16}$$

which is the final expression for the output of the probe located at $\mathbf{r}_0$. This acoustic transmission formula is valid from linearity provided the entire physical probe lies within the region $z \geq z_0$, so that the plane wave spectrum representation (6.8) is valid. More generally, (6.16) is valid for $z_0 > z_{00}$ where z_{00} is the value of z_0 for which the test and probe antennas would just touch during a scan. When the test and probe antennas are very close, multiple interactions, which (6.16) neglects, will not necessarily be negligible.

Taking the Fourier transform with respect to (x_0, y_0) of (6.16) shows that the spectrum $T_{0\omega}(k_x, k_y)$ for the acoustic pressure is given by [8], [29]

$$T_{0\omega}(k_x, k_y) = \frac{e^{-i\gamma z_0}}{(2\pi)^2 a_{0\omega} R_{p\omega}(k_x, k_y)} \int_{-\infty}^{+\infty}\int_{-\infty}^{+\infty} b_{p\omega}(\mathbf{r}_0) e^{-i(k_x x_0 + k_y y_0)} dx_0 dy_0 \tag{6.17}$$

where $\mathbf{r}_0 = x_0\hat{\mathbf{x}} + y_0\hat{\mathbf{y}} + z_0\hat{\mathbf{z}}$ and it has been assumed that $R_{p\omega}(k_x, k_y) \neq 0$. That is, the spectrum $T_{0\omega}(k_x, k_y)$ for the acoustic pressure $p_\omega(\mathbf{r})$ of the test antenna is simply the spatial Fourier transform of the probe output divided by the receiving characteristic of the probe.

If the probe is a receiving acoustic monopole with $R_{p\omega}(k_x, k_y)$ equal to a constant $R_{c\omega}$ for all (k_x, k_y), then (6.16) , with the help of (6.8), reduces simply to

$$b_{p\omega} = 2\pi R_{c\omega} p_\omega(\mathbf{r}_0). \tag{6.18}$$

That is, the output of a receiving monopole is proportional to the incident acoustic pressure at the reference point $\mathbf{r}_0$ of the monopole.

6.1.1 Derivation from the Helmholtz Equation

We shall show that for the probe outside the test antenna ("outside" being defined below), the probe output $b_{p\omega}(\mathbf{r})$ satisfies the homogeneous Helmholtz equation and the radiation condition. Based on these facts, we present an alternative derivation of the probe-corrected formula (6.17).

Let us begin by noting that the probe output $b_{p\omega}(\mathbf{r})$ can be expressed in terms of the plane-wave spectrum formula (6.16) in which each of the plane waves satisfies the homogeneous Helmholtz equation $(\nabla^2 + k^2)e^{i(k_x x + k_y y + \gamma z)} = 0$. Thus, the probe output satisfies the homogeneous Helmholtz equation $(\nabla^2 + k^2)b_{p\omega}(\mathbf{r}) = 0$ at points in space where the operator $\nabla^2 + k^2$ can be moved under the integral signs in (6.16). Moreover, the theorems of Kaplan [90, sec. 6.23] show that the operator $\nabla^2 + k^2$ can be brought under the integral signs in (6.16) if the integrand is exponentially decaying at infinity in the (k_x, k_y) plane. Hence, the probe output satisfies the homogeneous Helmholtz equation at points in space where the integrand in (6.16) decays exponentially as $|\gamma| \to \infty$. We shall therefore examine the asymptotic behavior of the integrand in (6.16).

Let z_s be the maximum (more precisely, the supremum) value of z for which the source function of the test antenna is nonzero; that is, the physical body of the test antenna is located

[3]Using adjoint reciprocity, we prove in Section 6.1.1 that the integral (6.16) indeed converges, provided the probe and test antennas do not touch during a scan.

in the region $z < z_s$. The formula (3.124) of Section 3.2.3 shows that the spectrum of the test antenna can be written as

$$T_{0\omega}(k_x, k_y) = C_{0\omega}(k_x, k_y)\, e^{|\gamma| z_s}, \quad k_x^2 + k_y^2 > k^2 \tag{6.19}$$

where $C_{0\omega}(k_x, k_y) \to 0$ as $|\gamma| \to \infty$.

A corresponding asymptotic formula for the receiving characteristic $R_{p\omega}(k_x, k_y)$ can also be found as follows. The reciprocity relations of [170, eq. (53c)] show that the receiving characteristic $R_{p\omega}(k_x, k_y)$ of the probe is proportional to $\gamma T^a_{p\omega}(-k_x, -k_y)$, which in turn is proportional to $F^a_{p\omega}(-k_x, -k_y)$; see Section 3.2.3. Here $T^a_{p\omega}(k_x, k_y)$ and $F^a_{p\omega}(k_x, k_y)$ are the transmitting spectrum and far-field function, respectively, of the adjoint probe when it is fed by the signal $a^a_{p\omega} = 1$, radiates into the region $z < 0$, and has its reference point located at the origin. This reciprocity relation holds when the probe is an electroacoustic transducer attached to an electromagnetic waveguide that supports just one propagating mode. Analogously to (3.113) we have

$$F^a_{p\omega}(k_x, k_y) = \frac{1}{4\pi} \int\limits_{V^a_p} q^a_{p\omega}(\mathbf{r}) e^{-i(k_x x + k_y y - \gamma z)} dV \tag{6.20}$$

where $q^a_{p\omega}(\mathbf{r})$ is the source function of the adjoint probe. Let z_p be the minimum (more precisely, the infimum) value of z for which the source function $q^a_{p\omega}(\mathbf{r})$ is nonzero; that is, the entire physical body of the adjoint probe is located in the region $z > z_p$. Repeating the derivation of Section 3.2.3 that led to (3.121) shows that

$$F^a_{p\omega}(k_x, k_y) = C^a_{p\omega}(k_x, k_y) e^{-|\gamma| z_p}, \quad k_x^2 + k_y^2 > k^2 \tag{6.21}$$

where $C^a_{p\omega}(k_x, k_y)$ is bounded as $|\gamma| \to \infty$. The receiving characteristic for the probe therefore satisfies

$$R_{p\omega}(k_x, k_y) = C_{p\omega}(k_x, k_y) e^{-|\gamma| z_p}, \quad k_x^2 + k_y^2 > k^2 \tag{6.22}$$

where $C_{p\omega}(k_x, k_y)$ is bounded as $|\gamma| \to \infty$. Since the physical dimensions of the adjoint probe equal the physical dimensions of the probe itself [170], the physical body of the probe is located in the region $z > z_p$, when its reference point is at the origin.

The asymptotic relations (6.19) and (6.22) show that the integrand of the integral representation (6.16) for the probe output satisfies

$$\begin{aligned} R_{p\omega}(k_x, k_y) T_{0\omega}(k_x, k_y) e^{i(k_x x_0 + k_y y_0 + \gamma z_0)} \\ = C_\omega(k_x, k_y) e^{|\gamma|(z_s - z_p - z_0)}, \quad k_x^2 + k_y^2 > k^2 \end{aligned} \tag{6.23}$$

where $C_\omega(k_x, k_y) \to 0$ as $|\gamma| \to \infty$. Consequently, when $z_s < z_0 + z_p$, the integrand of (6.16) decays exponentially as $|\gamma| \to \infty$, and the theorems of Kaplan [90, sec. 6.23] show that the operator $\nabla^2 + k^2$ can be brought inside the integral sign to prove that [43], [45]

$$\nabla^2 b_{p\omega}(\mathbf{r}) + k^2 b_{p\omega}(\mathbf{r}) = 0, \quad z > z_s - z_p. \tag{6.24}$$

When the reference point of the probe is at $\mathbf{r}_0 = x_0\hat{\mathbf{x}} + y_0\hat{\mathbf{y}} + z_0\hat{\mathbf{z}}$, the physical body of the probe is located in the region $z > z_0 + z_p$. Moreover, the physical body of the test antenna is in the region $z < z_s$, so the condition $z_0 > z_s - z_p$ (which is written as $z > z_s - z_p$ in (6.24), where $\mathbf{r}$ denotes the reference point of the probe) simply states that the probe cannot touch the test antenna during a scan. If, however, the probe gets too close to the test antenna during a scan, the formula (6.16) is no longer accurate because it neglects multiple interactions, and (6.24) no longer holds.

Since the far-field pressure of the test antenna

$$p_\omega(\mathbf{r}) \sim -ika_{0\omega} \cos\theta\, T_{0\omega}(k \cos\phi \sin\theta, k \sin\phi \sin\theta) \frac{e^{ikr}}{r} \tag{6.25}$$

is locally a plane wave, the definition of the receiving characteristic shows that the probe output is

$$\begin{aligned} b_{p\omega}(\mathbf{r}) \sim -2\pi a_{0\omega} ik \cos\theta\, R_{p\omega}(k\cos\phi\sin\theta, k\sin\phi\sin\theta) \\ \cdot\, T_{0\omega}(k\cos\phi\sin\theta, k\sin\phi\sin\theta)\frac{e^{ikr}}{r} \end{aligned} \tag{6.26}$$

when the reference point of the probe is at the far-field point (r, θ, ϕ). Equation (6.26) is the acoustic analog of the generalized Friis transmission formula [171, p. 6-8]. Hence, the probe output $b_{p\omega}(\mathbf{r})$ satisfies the homogeneous Helmholtz equation as well as the radiation condition in the region $z > z_s - z_p$. Consequently, it has the Green's function far-field representation (3.41)

$$b_{p\omega}(\mathbf{r}) \sim -\frac{e^{ikr}}{r}\frac{ik\cos\theta}{2\pi}\int_{-\infty}^{+\infty}\int_{-\infty}^{+\infty} b_{p\omega}(\mathbf{r}_0)\, e^{-ik\hat{\mathbf{r}}\cdot\mathbf{r}_0} dx_0 dy_0, \quad z_0 > z_s - z_p. \tag{6.27}$$

Comparing the far-field expressions (6.26) and (6.27) for the probe output shows that

$$\begin{aligned} a_{0\omega} T_{0\omega}(k_x, k_y) R_{p\omega}(k_x, k_y) = \frac{e^{-i\gamma z_0}}{(2\pi)^2}\int_{-\infty}^{+\infty}\int_{-\infty}^{+\infty} b_{p\omega}(\mathbf{r}_0)\, e^{-i(k_x x_0 + k_y y_0)} dx_0 dy_0, \\ k_x^2 + k_y^2 < k^2, \quad z_0 > z_s - z_p \end{aligned} \tag{6.28}$$

which is equivalent to (6.17) for propagating modes.

It was shown in Section 3.2.4 that the function $\gamma T_{0\omega}(k_x, k_y)$ is analytic around the real (k_x, k_y) axes. Moreover, the receiving characteristic $R_{p\omega}(k_x, k_y)$ is proportional to the far-field pattern of the adjoint probe, so the discussion in Section 3.2.4 also shows that the receiving characteristic is an analytic function of (k_x, k_y) around the real (k_x, k_y) axes. Therefore, $\gamma a_{0\omega} T_{0\omega}(k_x, k_y) R_{p\omega}(k_x, k_y)$ is an analytic function of (k_x, k_y) around the real (k_x, k_y) axes.

The right-hand side of (6.28) is equal to $T_{b\omega}(k_x, k_y)/(2\pi)$ for all (k_x, k_y), where $T_{b\omega}(k_x, k_y)$ is the spectrum of the probe output, which satisfies the homogeneous Helmholtz equation and the radiation condition. Therefore, the arguments in Section 3.2.4 can be applied to prove that $\gamma T_{b\omega}(k_x, k_y)$ is an analytic function around the real (k_x, k_y) axes.

We have now shown that after multiplying (6.28) by γ, both sides of the resulting equation are analytic functions of (k_x, k_y) around the real (k_x, k_y) axes. Since the two sides of the equation are identical for propagating modes, analytical continuation proves that the two sides are identical for all (k_x, k_y). Consequently, the spectrum $T_{b\omega}(k_x, k_y)$ given by the probe-corrected formula (6.28), which was obtained from the Helmholtz-equation derivation, is valid for all (k_x, k_y) and agrees with the probe-corrected formula (6.17) obtained from the inverse Fourier transform.

An alternative argument that does not require analytic continuation also shows that (6.28) is a valid expression for all real (k_x, k_y). Assume the probe was a point-source receiver that measured the pressure ($b_{p\omega} = p_\omega$) at that point. Such a probe would have a constant receiving characteristic, and (6.28) would reduce to simply the definition of the plane-wave spectrum for the acoustic pressure of the test antenna for all (k_x, k_y); see (6.9). Since an arbitrary probe can be represented by a superposition of point-source receivers [37], (6.28) is also valid for all (k_x, k_y) and an arbitrary linear probe.

The probe output $b_{p\omega}(\mathbf{r})$ (for a fixed orientation of the probe with respect to the test antenna) satisfies the homogeneous Helmholtz equation (6.24) not only in a half space in front of the test antenna, but in the entire region outside a closed surface (like the minimum convex surface described in Section 3.2.8) formed by the locus of points $(z_s - z_p)$ for all orientations of the z axis. This can be shown by repeating the analysis that led to (6.23) for different orientations of the z axis. The region outside the locus of points $(z_s - z_p)$, in which the probe output satisfies (6.24), is identical to the region in which the minimum convex surface enclosing the test antenna, and the minimum convex surface enclosing the probe, do not intersect. Of course, (6.24) holds only if multiple interactions between the probe and test antenna are negligible.

6.1.2 Formulas for Reciprocal Probes

If the probe is a reciprocal electroacoustic transducer [170], fed by an electromagnetic waveguide in which only a single mode is propagating, the receiving characteristic is given in terms of the normalized plane-wave transmitting spectrum $T_{p\omega}(k_x, k_y)$ for the probe by [170, eq. (10a)]

$$R_{p\omega}(k_x, k_y) = \frac{\gamma}{\omega Y_p \rho_{m0}} T_{p\omega}(-k_x, -k_y) \tag{6.29}$$

where Y_p is the characteristic admittance for the propagating mode in the waveguide feed, and ρ_{m0} is the mass density of the undisturbed fluid. For the TEM mode on a coaxial transmission line, Y_p is given in (6.7). The spectrum $T_{p\omega}(k_x, k_y)$ determines the field radiated by the probe into the half space $z < 0$ when the probe is fed by the signal $a_{p\omega} = 1$, the reference point of the probe is at the origin, and the test antenna is removed. The reciprocity relation (6.29) is satisfied when the materials used to build the probe satisfy the conditions stated in [170, p. 27]. In general, when the probe is not reciprocal, the receiving characteristic can be related to the transmitting characteristic of the adjoint probe [170, eq. (53c)].

In the introduction to the present chapter, we discussed the relation between $a_{p\omega}$ and the electromagnetic fields in the waveguide, and showed that Y_p can always be chosen to be independent of frequency.

Inserting the expression (6.29) for $R_{p\omega}(k_x, k_y)$ into the probe-corrected formula (6.17), we find that the spectrum for the test antenna is given by

$$T_{0\omega}(k_x, k_y) = \frac{Y_p \rho_{m0} \omega e^{-i\gamma z_0}}{(2\pi)^2 a_{0\omega} \gamma T_{p\omega}(-k_x, -k_y)} \int_{-\infty}^{+\infty}\int_{-\infty}^{+\infty} b_{p\omega}(\mathbf{r}_0) e^{-i(k_x x_0 + k_y y_0)} dx_0 dy_0. \tag{6.30}$$

Let us also express the far field pattern (6.13) of the test antenna in terms of the probe output and probe far-field pattern. The probe far-field pattern is defined when the probe is transmitting into the region $z < 0$ (with the test antenna removed) and its reference point is located at the origin. The probe's normalized far-field pattern $\mathcal{F}_{p\omega}(\theta, \phi)$ is defined such that the far-field of the probe is $p_{p\omega}(\mathbf{r}) \sim a_{p\omega}\mathcal{F}_{p\omega}(\theta, \phi)e^{ikr}/r$, where $a_{p\omega}$ is the amplitude of the input signal to the probe.

Since the probe radiates into the half space $z < 0$, the relation between its far-field pattern $\mathcal{F}_{p\omega}(\theta, \phi)$ and its spectrum $T_{p\omega}(k_x, k_y)$ is given by $\mathcal{F}_{p\omega}(\theta, \phi) = ik\cos\theta T_{p\omega}(k\cos\phi\sin\theta, k\sin\phi\sin\theta)$ for $\pi/2 < \theta \leq \pi$ (see Section 3.2.5). This shows that

$$T_{p\omega}(-k\cos\phi\sin\theta, -k\sin\phi\sin\theta) = \frac{\mathcal{F}_{p\omega}(\pi - \theta, \pi + \phi)}{ik\cos(\pi - \theta)}, \quad 0 \leq \theta < \pi/2. \tag{6.31}$$

Thus, from (6.30) and (6.13), the expression for the far-field pattern of the test antenna is

$$\mathcal{F}_{0\omega}(\theta, \phi) = -\frac{Y_p \rho_{m0} \omega k \cos\theta}{(2\pi)^2 a_{0\omega} \mathcal{F}_{p\omega}(\pi - \theta, \pi + \phi)} \int_{-\infty}^{+\infty}\int_{-\infty}^{+\infty} b_{p\omega}(\mathbf{r}_0) e^{-ik\hat{\mathbf{r}}\cdot\mathbf{r}_0} dx_0 dy_0 \tag{6.32}$$

where $\hat{\mathbf{r}} = \hat{\mathbf{x}}\cos\phi\sin\theta + \hat{\mathbf{y}}\sin\phi\sin\theta + \hat{\mathbf{z}}\cos\theta$ and it is assumed that $\mathcal{F}_{p\omega}(\pi - \theta, \pi + \phi) \neq 0$. The expression (6.32) gives the far-field pattern of the test antenna as the spatial Fourier transform of the output of the reciprocal probe divided by the far-field pattern of the probe.

6.2 ELECTROMAGNETIC FIELDS

Consider Figure 6.1 showing the geometry for measuring the electric field $\mathbf{E}_\omega(\mathbf{r})$ radiated into free space by the test antenna, which is located in the half space $z < z_0$ and is fed by a signal of amplitude $a_{0\omega}$. Alternatively, we can think of the test antenna, the feed system, and the generator as a source region without specifying any waveguide mode. The electric field is measured in the scan plane $z = z_0$ by a known probe, and the purpose of this section is to express the field of the test antenna in the half space $z \geq z_0$ (or more generally, to the "right" of the source region) in terms of the probe output. We begin by listing some useful plane-wave spectrum formulas that were derived in Chapter 3.

When the test antenna is the only object in all space, it radiates the field $\mathbf{E}_\omega(\mathbf{r})$, which, according to (3.62)–(3.63), can be expressed in terms of a spectrum of plane waves as

$$\mathbf{E}_\omega(\mathbf{r}) = \frac{a_{0\omega}}{2\pi} \int_{-\infty}^{+\infty}\int_{-\infty}^{+\infty} \mathbf{T}_{0\omega}(k_x, k_y) e^{i(k_x x + k_y y + \gamma z)} dk_x dk_y, \quad z \geq z_0 \tag{6.33}$$

where $\mathbf{T}_{0\omega}(k_x, k_y)$ is the transmitting spectrum of the test antenna given by

$$\mathbf{T}_{0\omega}(k_x, k_y) = \frac{e^{-i\gamma z}}{2\pi a_{0\omega}} \int_{-\infty}^{+\infty}\int_{-\infty}^{+\infty} \mathbf{E}_\omega(\mathbf{r}) e^{-i(k_x x + k_y y)} dx dy, \quad z \geq z_0. \tag{6.34}$$

The product $a_{0\omega}\mathbf{T}_{0\omega}(k_x, k_y)$ equals $\mathbf{T}_\omega(k_x, k_y)$ defined in Chapter 3, $\mathbf{r} = x\hat{\mathbf{x}} + y\hat{\mathbf{y}} + z\hat{\mathbf{z}}$, and γ is given in (6.11). The normalized far-field pattern of the test antenna is defined such that the electric far field is

$$\mathbf{E}_\omega(\mathbf{r}) \sim a_{0\omega}\boldsymbol{\mathcal{F}}_{0\omega}(\theta, \phi)\frac{e^{ikr}}{r}, \quad r \to \infty. \tag{6.35}$$

Equations (3.148) and (3.158) show that the normalized far-field pattern can be expressed in terms of the spectrum as

$$\boldsymbol{\mathcal{F}}_{0\omega}(\theta, \phi) = -ik\cos\theta\, \mathbf{T}_{0\omega}(k\cos\phi\sin\theta, k\sin\phi\sin\theta), \quad 0 \leq \theta < \pi/2. \tag{6.36}$$

The spectrum $\mathbf{T}_{0\omega}(k_x, k_y)$ of the field radiated by the test antenna will now be expressed in terms of the output in the scan plane $z = z_0$ of a known probe, as shown in Figure 6.1. We assume that the distance between the probe and the test antenna is always large enough for multiple interactions between the test antenna and the probe to be negligible. That is, the field scattered by the probe, rescattered by the test antenna, and returned to the probe does not change the output of the probe.

When the probe is located at $\mathbf{r}_0$, its scalar output is denoted by $b_{p\omega}(\mathbf{r}_0)$, and it proves convenient to define the vector output of the probe as the vector sum of the scalar output for two mutually perpendicular orientations [37], [43]:

$$\mathbf{b}_{p\omega}(\mathbf{r}_0) = b^x_{p\omega}(\mathbf{r}_0)\hat{\mathbf{x}} + b^y_{p\omega}(\mathbf{r}_0)\hat{\mathbf{y}} \tag{6.37}$$

where $b^x_{p\omega}(\mathbf{r}_0)$ and $b^y_{p\omega}(\mathbf{r}_0)$ are the scalar outputs of the probe when some reference line fixed to the probe is parallel to the x and y axis, respectively. For example, if the probe is an open-ended rectangular waveguide, the reference line could be chosen parallel to the longest cross-sectional side of the waveguide. The scalar output would be the amplitude of the propagating mode at the plane S_p in the output line of the probe, as discussed in the introduction to the present chapter.

The dyadic receiving characteristic of the probe $\bar{\mathbf{R}}_{p\omega}(k_x, k_y)$ is defined such that the vector output of the probe is $\mathbf{b}_{p\omega}(0) = \bar{\mathbf{R}}_{p\omega}(k_x, k_y) \cdot \mathbf{e}(k_x, k_y)$, when the probe is illuminated by an incident plane-wave electric field $(2\pi)^{-1}\mathbf{e}(k_x, k_y)e^{i(k_x x + k_y y + \gamma z)}$ with $\mathbf{e}(k_x, k_y) \cdot \mathbf{k} = 0$, and when the reference point of the probe is located at the origin. Since $\mathbf{e}(k_x, k_y) \cdot \mathbf{k}$ is zero, the dyadic receiving characteristic $\bar{\mathbf{R}}_{p\omega}(k_x, k_y)$ can be made unique by requiring

$$\bar{\mathbf{R}}_{p\omega}(k_x, k_y) \cdot \mathbf{k} = 0. \tag{6.38}$$

By linearity, $\mathbf{b}_{p\omega}(\mathbf{r}_0) = \bar{\mathbf{R}}_{p\omega}(k_x, k_y) \cdot \mathbf{e}(k_x, k_y)e^{i(k_x x_0 + k_y y_0 + \gamma z_0)}$, when the plane wave $(2\pi)^{-1}\mathbf{e}(k_x, k_y)e^{i(k_x x + k_y y + \gamma z)}$ is the illuminating field and the probe is located at $\mathbf{r}_0 = x_0\hat{\mathbf{x}} + y_0\hat{\mathbf{y}} + z_0\hat{\mathbf{z}}$. The plane-wave expansion (6.33) for the electric field radiated by the test antenna shows that the plane-wave component for each (k_x, k_y) is

$$a_{0\omega}(2\pi)^{-1}\mathbf{T}_{0\omega}(k_x, k_y)e^{i(k_x x + k_y y + \gamma z)}dk_x dk_y \tag{6.39}$$

so the vector output of the probe due to this plane-wave component is given by

$$d\mathbf{b}_{p\omega}(\mathbf{r}_0) = a_{0\omega}\bar{\mathbf{R}}_{p\omega}(k_x, k_y) \cdot \mathbf{T}_{0\omega}(k_x, k_y)e^{i(k_x x_0 + k_y y_0 + \gamma z_0)}dk_x dk_y. \tag{6.40}$$

The total vector output of the linear probe is seen from (6.33) to be the integral over all real (k_x, k_y) of the expression in (6.40), that is

$$\mathbf{b}_{p\omega}(\mathbf{r}_0) = a_{0\omega} \int_{-\infty}^{+\infty}\int_{-\infty}^{+\infty} \bar{\mathbf{R}}_{p\omega}(k_x, k_y) \cdot \mathbf{T}_{0\omega}(k_x, k_y)e^{i(k_x x_0 + k_y y_0 + \gamma z_0)}dk_x dk_y \tag{6.41}$$

which is the final expression or plane-wave "transmission formula" for the vector output of the probe located at $\mathbf{r}_0$. Recall that we are assuming that multiple interactions between the probe and test antenna are negligible, and that the output line of the probe is perfectly matched ($a_{p\omega} = 0$). Taking the inverse Fourier transform with respect to (x_0, y_0) of this transmission formula shows that

$$\bar{\mathbf{R}}_{p\omega}(k_x, k_y) \cdot \mathbf{T}_{0\omega}(k_x, k_y) = \frac{e^{-i\gamma z_0}}{(2\pi)^2 a_{0\omega}} \int_{-\infty}^{+\infty}\int_{-\infty}^{+\infty} \mathbf{b}_{p\omega}(\mathbf{r}_0)e^{-i(k_x x_0 + k_y y_0)}dx_0 dy_0 \tag{6.42}$$

from which $\mathbf{T}_{0\omega}(k_x, k_y)$ can be determined. Since the vector output of the probe $\mathbf{b}_{p\omega}(\mathbf{r}_0)$ is perpendicular to $\hat{\mathbf{z}}$, (6.42) gives only two equations for determining the three rectangular components of the spectrum $\mathbf{T}_{0\omega}(k_x, k_y)$. However, $\mathbf{T}_{0\omega}(k_x, k_y)$ satisfies the well-known orthogonality relation $\mathbf{k} \cdot \mathbf{T}_{0\omega}(k_x, k_y) = 0$, where $\mathbf{k}$ is given in (6.10), so only two equations are necessary for determining $\mathbf{T}_{0\omega}(k_x, k_y)$. Because the vector output of the probe (6.37) is perpendicular to $\hat{\mathbf{z}}$, the receiving characteristic can be written as

$$\bar{\mathbf{R}}_{p\omega}(k_x, k_y) = \hat{\mathbf{x}}\mathbf{R}^x_{p\omega}(k_x, k_y) + \hat{\mathbf{y}}\mathbf{R}^y_{p\omega}(k_x, k_y) \tag{6.43}$$

with $\mathbf{R}^x_{p\omega}(k_x, k_y) \cdot \mathbf{k} = 0$ and $\mathbf{R}^y_{p\omega}(k_x, k_y) \cdot \mathbf{k} = 0$ (in accordance with (6.38)). To solve (6.42) for the spectrum $\mathbf{T}_{0\omega}(k_x, k_y)$ of the electric field, we can define the inverse of the dyadic $\bar{\mathbf{R}}_{p\omega}(k_x, k_y)$ by [43]

$$\bar{\mathbf{R}}^{-1}_{p\omega}(k_x, k_y) \cdot \bar{\mathbf{R}}_{p\omega}(k_x, k_y) = \hat{\boldsymbol{\theta}}_\omega(k_x, k_y)\hat{\boldsymbol{\theta}}_\omega(k_x, k_y) + \hat{\boldsymbol{\phi}}_\omega(k_x, k_y)\hat{\boldsymbol{\phi}}_\omega(k_x, k_y) \tag{6.44}$$

where $\hat{\boldsymbol{\theta}}_\omega$ and $\hat{\boldsymbol{\phi}}_\omega$ are unit vectors orthogonal to $\mathbf{k}$ that satisfy the relations $\hat{\boldsymbol{\theta}}_\omega \times \hat{\boldsymbol{\phi}}_\omega = \hat{\mathbf{k}}$, $\hat{\mathbf{k}} \times \hat{\boldsymbol{\theta}}_\omega = \hat{\boldsymbol{\phi}}_\omega$, and $\hat{\boldsymbol{\phi}}_\omega \times \hat{\mathbf{k}} = \hat{\boldsymbol{\theta}}_\omega$, with $\hat{\mathbf{k}} = k^{-1}\mathbf{k} = (k_x\hat{\mathbf{x}} + k_y\hat{\mathbf{y}} + \gamma\hat{\mathbf{z}})/k$. Specifically, $\hat{\boldsymbol{\theta}}_\omega$ and $\hat{\boldsymbol{\phi}}_\omega$ can be expressed in terms of k_x and k_y as

$$\hat{\boldsymbol{\theta}}_\omega(k_x, k_y) = \frac{\gamma k_x}{k\sqrt{k_x^2 + k_y^2}}\hat{\mathbf{x}} + \frac{\gamma k_y}{k\sqrt{k_x^2 + k_y^2}}\hat{\mathbf{y}} - \frac{\sqrt{k_x^2 + k_y^2}}{k}\hat{\mathbf{z}} \tag{6.45}$$

and

$$\hat{\boldsymbol{\phi}}_\omega(k_x, k_y) = -\frac{k_y}{\sqrt{k_x^2 + k_y^2}}\hat{\mathbf{x}} + \frac{k_x}{\sqrt{k_x^2 + k_y^2}}\hat{\mathbf{y}}. \tag{6.46}$$

If $k_x = k\cos\phi\sin\theta$ and $k_y = k\sin\phi\sin\theta$, then $\gamma = k\cos\theta$ and the unit vectors $\hat{\boldsymbol{\theta}}_\omega(k_x, k_y)$ and $\hat{\boldsymbol{\phi}}_\omega(k_x, k_y)$ are simply the usual angular unit vectors of the spherical coordinate system (k, θ, ϕ) [83, app. 1, eqs. 115–116].

Writing $\bar{\mathbf{R}}^{-1}_{p\omega}(k_x, k_y)$ as $\boldsymbol{\alpha}\hat{\mathbf{x}} + \boldsymbol{\beta}\hat{\mathbf{y}}$ and assuming that $\hat{\mathbf{k}} \cdot \left[\mathbf{R}^y_{p\omega} \times \mathbf{R}^x_{p\omega}\right] \neq 0$, we find that the inverse receiving characteristic is given by

$$\bar{\mathbf{R}}^{-1}_{p\omega}(k_x, k_y) = \frac{\hat{\mathbf{k}} \times \left[\mathbf{R}^y_{p\omega}(k_x, k_y)\hat{\mathbf{x}} - \mathbf{R}^x_{p\omega}(k_x, k_y)\hat{\mathbf{y}}\right]}{\hat{\mathbf{k}} \cdot \left[\mathbf{R}^y_{p\omega}(k_x, k_y) \times \mathbf{R}^x_{p\omega}(k_x, k_y)\right]}. \tag{6.47}$$

Multiplying (6.42) with $\bar{\mathbf{R}}^{-1}_{p\omega}(k_x, k_y)$, and using the definition of the inverse receiving characteristic (6.44), shows that the spectrum $\mathbf{T}_{0\omega}(k_x, k_y)$ is given by

$$\mathbf{T}_{0\omega}(k_x, k_y) = \frac{e^{-i\gamma z_0}}{(2\pi)^2 a_{0\omega}} \bar{\mathbf{R}}^{-1}_{p\omega}(k_x, k_y) \cdot \int_{-\infty}^{+\infty}\int_{-\infty}^{+\infty} \mathbf{b}_{p\omega}(\mathbf{r}_0) e^{-i(k_x x_0 + k_y y_0)} dx_0 dy_0 \tag{6.48}$$

provided $\hat{\mathbf{k}} \cdot \left[\mathbf{R}^y_{p\omega}(k_x, k_y) \times \mathbf{R}^x_{p\omega}(k_x, k_y)\right] \neq 0$. This is the final probe-corrected formula giving the spectrum $\mathbf{T}_{0\omega}(k_x, k_y)$ for the electric field of the test antenna as the product of the inverse receiving characteristic of the probe and the spatial Fourier transform of the vector output of the probe in the scan plane $z = z_0$.

In principle, once the spectrum is computed from the measured data in the plane $z = z_0$, (6.33) can be used to compute the electric field in any z plane (not just $z \geq z_0$) up to the test antenna. In practice, however, the plane $z = z_0$ is usually chosen a few wavelengths away from the test antenna to keep multiple reflections between the probe and test antenna at an acceptably low level. This means that most of the evanescent fields have decayed to a value too small to be detected by the measurement system. Thus, most of the evanescent part of the spectrum cannot be computed with any accuracy from (6.48), and the evanescent fields, which may dominate in the region a few wavelengths from the test antenna, cannot be accurately computed from (6.33).

6.2.1 Derivation from the Helmholtz Equation

With the probe outside the test antenna ("outside" being defined below), we show that the rectangular components of the vector output of the probe $\mathbf{b}_{p\omega}(\mathbf{r})$ satisfy the homogeneous Helmholtz equation and the scalar radiation condition. These properties of the vector output will subsequently be used to rederive the probe-corrected formula (6.48). The derivation will follow the lines of Section 6.1.1, where the probe-corrected formula for acoustic fields was derived using the Helmholtz equation.

The vector output of the probe has only x and y components and is given by the plane-wave expression (6.41) in which each plane wave satisfies the homogeneous Helmholtz equation. As in the acoustic Section 6.1.1, we find from the theorems of Kaplan [90, sec. 6.23] that the vector output satisfies the homogeneous Helmholtz equation $(\nabla^2 + k^2)\mathbf{b}_{p\omega}(\mathbf{r}) = 0$ at spatial points where the integrand of (6.41) decays exponentially as $|\gamma| \to \infty$.

With the vector formulas (3.130)–(3.131), and the adjoint reciprocity relations in [8, p. 106], we can perform a derivation similar to the one that led to (6.23) in order to show that the integrand of (6.41) satisfies

$$\bar{\mathbf{R}}_{p\omega}(k_x, k_y) \cdot \mathbf{T}_{0\omega}(k_x, k_y) e^{i(k_x x_0 + k_y y_0 + \gamma z_0)} = \mathbf{C}_\omega(k_x, k_y) e^{|\gamma|(z_s - z_p - z_0)}, \quad k_x^2 + k_y^2 > k^2 \tag{6.49}$$

where $\mathbf{C}_\omega(k_x, k_y) \to 0$ as $|\gamma| \to \infty$. As in the acoustic case, z_s is the maximum (more precisely, the supremum) value of z for which the source function of the test antenna is nonzero;

that is, the physical body of the test antenna is located in the region $z < z_s$. Similarly, z_p is the minimum (more precisely, the infimum) value of z for which the source function of the adjoint probe is nonzero, when it is radiating into the region $z < 0$ with its reference point located at the origin. Since the physical dimensions of the adjoint probe equal the physical dimensions of the probe itself, the physical body of the probe is located in the region $z > z_p$ when its reference point is at the origin.

Thus, when the reference point of the probe is at $\mathbf{r} = x\hat{\mathbf{x}} + y\hat{\mathbf{y}} + z\hat{\mathbf{z}}$, the integrand of the plane-wave representation (6.41) is exponentially decaying as $|\gamma| \to \infty$ when $z > z_s - z_p$, so the vector output of the probe satisfies the homogeneous Helmholtz equation [43], [45]

$$\nabla^2 \mathbf{b}_{p\omega}(\mathbf{r}) + k^2 \mathbf{b}_{p\omega}(\mathbf{r}) = 0, \quad z > z_s - z_p. \tag{6.50}$$

The condition $z > z_s - z_p$ in (6.50) simply states that the probe cannot touch the test antenna during a scan. As in the acoustics case, if multiple interactions are not negligible, (6.41), and thus (6.50), no longer hold.

The frequency-domain vector Helmholtz equation (6.50) has been used in the past to efficiently compute the near-field coupling between co-sited antennas [45], to suggest a novel method of spherical near-field scanning with probe correction [172], and to derive probe-corrected transmission formulas for plane-polar near-field antenna measurements [43].

Since the electric far field of the test antenna

$$\mathbf{E}_\omega(\mathbf{r}) \sim -ika_{0\omega} \cos\theta \; \mathbf{T}_{0\omega}(k\cos\phi \sin\theta, k\sin\phi\sin\theta)\frac{e^{ikr}}{r} \tag{6.51}$$

is locally a plane wave, the definition of the receiving characteristic shows that the vector probe output is

$$\begin{aligned} \mathbf{b}_{p\omega}(\mathbf{r}) \sim -2\pi a_{0\omega} ik\cos\theta \; \bar{\mathbf{R}}_{p\omega}(k\cos\phi\sin\theta, k\sin\phi\sin\theta) \\ \cdot \mathbf{T}_{0\omega}(k\cos\phi\sin\theta, k\sin\phi\sin\theta)\frac{e^{ikr}}{r} \end{aligned} \tag{6.52}$$

when the probe is located at the far-field point (r, θ, ϕ). Equation (6.52) is the generalized Friis transmission formula [171, p. 6-8]. Equations (6.50) and (6.52) show that the x and y components of the probe output $\mathbf{b}_{p\omega}(\mathbf{r})$ satisfy the homogeneous Helmholtz equation as well as the scalar radiation condition in the region $z > z_s - z_p$. Hence, by applying (3.41) to the x and y components of $\mathbf{b}_{p\omega}(\mathbf{r})$, we obtain the Green's function far-field representation

$$\mathbf{b}_{p\omega}(\mathbf{r}) \sim -\frac{e^{ikr}}{r}\frac{ik\cos\theta}{2\pi}\int_{-\infty}^{+\infty}\int_{-\infty}^{+\infty} \mathbf{b}_{p\omega}(\mathbf{r}_0)\, e^{-ik\hat{\mathbf{r}}\cdot\mathbf{r}_0} dx_0 dy_0, \quad z_0 > z_s - z_p. \tag{6.53}$$

Comparing the far-field expressions (6.52) and (6.53) reveals that

$$\begin{aligned} a_{0\omega}\bar{\mathbf{R}}_{p\omega}(k_x, k_y)\cdot\mathbf{T}_{0\omega}(k_x, k_y) = \frac{e^{-i\gamma z_0}}{(2\pi)^2}\int_{-\infty}^{+\infty}\int_{-\infty}^{+\infty} \mathbf{b}_{p\omega}(\mathbf{r}_0)\, e^{-i(k_x x_0 + k_y y_0)} dx_0 dy_0, \\ k_x^2 + k_y^2 < k^2, \quad z_0 > z_s - z_p. \end{aligned} \tag{6.54}$$

The right-hand side of (6.54) is equal to $\mathbf{T}_{b\omega}(k_x, k_y)/(2\pi)$ for all (k_x, k_y), where $\mathbf{T}_{b\omega}(k_x, k_y)$ is the plane-wave spectrum for the vector output of the probe. The formula (6.54) is equivalent to (6.42) for propagating modes and can be made valid for all (k_x, k_y) by analytic continuation; see Section 3.2.4 and the discussion near the end of Section 6.1.1. We can use the expression (6.47) for the inverse receiving characteristic to obtain the probe-corrected formula (6.48). Thus, the probe-corrected formulas for electromagnetic fields have been derived from the homogeneous Helmholtz equation for the vector output of the probe.

As in the acoustic case, the vector output of the probe $\mathbf{b}_{p\omega}(\mathbf{r})$ (for a fixed orientation of the probe with respect to the test antenna) satisfies the homogeneous Helmholtz equation in the entire region outside a minimum convex closed surface (like the convex surface in Section 3.2.8) formed by the locus of points $(z_s - z_p)$ for all orientations of the z axis. This is shown by repeating the analysis of this section for different orientations of the z axis. The exterior region in which the probe can be located and have its output obey (6.50) is given by the region in which the minimum convex surface enclosing the test antenna, and the minimum convex surface enclosing the probe antenna, do not intersect (provided multiple interactions are neglected).

6.2.2 Formulas for Reciprocal Probes

Consider a reciprocal probe attached to an electromagnetic waveguide that supports just one propagating mode. It follows from [8, ch. 2, eq. (1.6-20a)] that the normalized transmitting spectrum of the probe $\mathbf{T}_{p\omega}(k_x, k_y)$ is related to the vector receiving characteristic $\mathbf{R}_{p\omega}(k_x, k_y)$ by the reciprocity relation

$$\mathbf{R}_{p\omega}(k_x, k_y) = \sqrt{\frac{\epsilon}{\mu}}\,\frac{\gamma}{Y_p k}\,\mathbf{T}_{p\omega}(-k_x, -k_y) \tag{6.55}$$

where ϵ, μ, and Y_p are the permittivity of free space, the permeability of free space, and the characteristic admittance for the propagating mode in the waveguide feed, respectively. For the TEM mode on a coaxial transmission line, Y_p is given in (6.7). The spectrum $\mathbf{T}_{p\omega}(k_x, k_y)$ determines the field radiated into the half space $z < 0$ (with the test antenna removed) by the probe when it is fed with the input signal $a_{p\omega} = 1$ and has its reference point located at the origin.[4]

The reciprocity relation (6.55) is satisfied when the conditions in [8, p. 71, eq. (1.5-6)] are satisfied by the materials used to build the probe. In particular, (6.55) is satisfied if these materials can be described by a scalar permittivity, a scalar permeability, and a scalar conductivity. When the probe is not reciprocal, the receiving characteristic can be related to the transmitting characteristic of the adjoint probe [8, p. 106]. In the introduction to the present chapter, we showed that the characteristic admittance Y_p can always be chosen to be independent of frequency.

In Kerns's work [8], $\mathbf{s}'_{20}$ and $\mathbf{s}'_{02}$ correspond to our $\mathbf{T}_{p\omega}(k_x, k_y)$ and $\mathbf{R}_{p\omega}(k_x, k_y)$, and $\mathbf{s}_{10}$ corresponds to $\mathbf{T}_{0\omega}(k_x, k_y)$. The vector receiving characteristic $\mathbf{R}_{p\omega}(k_x, k_y)$ of the probe is defined according to [8, ch. 2, eqs. (1.3-8), (1.6-14a)], such that the scalar output of the probe is given by $b_{p\omega} = \mathbf{R}_{p\omega}(k_x, k_y) \cdot \mathbf{e}(k_x, k_y)$ when the reference point of the probe is located at the origin, and the incident field is the plane wave $(2\pi)^{-1}\mathbf{e}(k_x, k_y)e^{i(k_x x + k_y y + \gamma z)}$

[4] A very general reciprocity relation also holds between the outputs of any two reciprocal antennas [173, part 1, sec. 4.2]. Suppose the test antenna in Figure 6.1 is fed by an input modal coefficient $a_{0\omega}$, and the output of the probe antenna is $b_{p\omega}$ when the probe antenna is terminated in a load with zero reflection coefficient so that $a_{p\omega} = 0$ (and thus $I_{p\omega} = -Y_p b_{p\omega}$, where Y_p is the characteristic admittance of the probe's waveguide). Next feed the probe antenna with an input modal coefficient $a'_{p\omega}$, and denote the output of the test antenna by $b'_{0\omega}$ when the test antenna is terminated in a load with zero reflection coefficient so that $a'_{0\omega} = 0$ (and thus $I'_{0\omega} = -Y_0 b'_{0\omega}$, where Y_0 is the characteristic admittance of the test antenna's waveguide). Then the Lorentz reciprocity theorem [173, part 1, sec. 4.2], applied to the two closed surfaces formed by the waveguide reference plane of each antenna and the surface enclosing the shielded power supply or detector (rectangular boxes in Figure 6.1) of each antenna, proves for reciprocal electromagnetic antennas (or for reciprocal electroacoustic transducers) that

$$\frac{I_{p\omega}}{a_{0\omega}} = \frac{I'_{0\omega}}{a'_{p\omega}}. \tag{6.56}$$

Moreover, this "system 2-port" reciprocity relation (6.56) holds even if multiple interactions between the two reciprocal antennas are not negligible, and if passive reciprocal scatterers are present.

with $\mathbf{e}(k_x, k_y) \cdot \mathbf{k} = 0$.[5] Letting $\mathbf{T}^x_{p\omega}(k_x, k_y)$ and $\mathbf{T}^y_{p\omega}(k_x, k_y)$ be the transmitting spectra of the probe when it is located at the origin and its reference line is parallel to the x and y axis, respectively, we see from (6.43) that $\mathbf{R}^x_{p\omega}(k_x, k_y)$ and $\mathbf{R}^y_{p\omega}(k_x, k_y)$ are the corresponding vector receiving characteristics of the probe. Applying the reciprocity relation (6.55) to these x and y receiving-transmitting characteristics gives

$$\mathbf{R}^x_{p\omega}(k_x, k_y) = \sqrt{\frac{\epsilon}{\mu}}\,\frac{\gamma}{Y_p k}\,\mathbf{T}^x_{p\omega}(-k_x, -k_y) \tag{6.57}$$

and

$$\mathbf{R}^y_{p\omega}(k_x, k_y) = \sqrt{\frac{\epsilon}{\mu}}\,\frac{\gamma}{Y_p k}\,\mathbf{T}^y_{p\omega}(-k_x, -k_y). \tag{6.58}$$

The equations (6.57) and (6.58) allow the dyadic receiving characteristic $\bar{\mathbf{R}}_{p\omega}(k_x, k_y)$ in (6.43) to be written in terms of the transmitting spectrum of the probe in two different orientations. We may also insert the values of $\mathbf{R}^x_{p\omega}(k_x, k_y)$ and $\mathbf{R}^y_{p\omega}(k_x, k_y)$ from (6.57) and (6.58) into the expression (6.47) for the inverse of $\bar{\mathbf{R}}_{p\omega}(k_x, k_y)$, and then use (6.48) to calculate the transmitting spectrum of the test antenna.

Let us finally express the far-field pattern (6.36) of the test antenna in terms of the vector output of the reciprocal probe and the normalized far-field patterns $\mathcal{F}^x_{p\omega}(\theta, \phi)$ and $\mathcal{F}^y_{p\omega}(\theta, \phi)$ of the probe in the two orientations, when the probe is located at the origin and fed by the input signal $a_{p\omega} = 1$. To do this, use the far-field relation (6.36) and the expression (6.48) for the transmitting spectrum to show that the far-field pattern for the test antenna is given by

$$\mathcal{F}_{0\omega}(\theta, \phi) = -\frac{ik\cos\theta}{(2\pi)^2 a_{0\omega}}\,\bar{\mathbf{R}}^{-1}_{p\omega}(k\cos\phi\sin\theta, k\sin\phi\sin\theta) \cdot \int_{-\infty}^{+\infty}\int_{-\infty}^{+\infty} \mathbf{b}_{p\omega}(\mathbf{r}_0) e^{-ik\hat{\mathbf{r}}\cdot\mathbf{r}_0} dx_0 dy_0. \tag{6.59}$$

Since the far field of the probe is to be calculated in the hemisphere with $z < 0$, it is found that $\mathcal{F}^x_{p\omega}(\theta, \phi) = ik\cos\theta\mathbf{T}^x_{p\omega}(k\cos\phi\sin\theta, k\sin\phi\sin\theta)$ for $\pi/2 < \theta \leq \pi$ (see Section 3.1.1). The reciprocity relations (6.57) and (6.58) then show that

$$\mathbf{R}^x_{p\omega}(k\cos\phi\sin\theta, k\sin\phi\sin\theta) = \sqrt{\frac{\epsilon}{\mu}}\,\frac{i}{Y_p k}\,\mathcal{F}^x_{p\omega}(\pi - \theta, \pi + \phi) \tag{6.60}$$

and similarly for $\mathbf{R}^y_{p\omega}(k_x, k_y)$. Inserting these results into the expression for the inverse receiving characteristic (6.47) yields

$$\bar{\mathbf{R}}^{-1}_{p\omega}(k\cos\phi\sin\theta, k\sin\phi\sin\theta) = \sqrt{\frac{\mu}{\epsilon}}\,\frac{iY_p k\,\hat{\mathbf{r}} \times \left[\mathcal{F}^y_{p\omega}(\pi-\theta, \pi+\phi)\hat{\mathbf{x}} - \mathcal{F}^x_{p\omega}(\pi-\theta, \pi+\phi)\hat{\mathbf{y}}\right]}{\hat{\mathbf{r}} \cdot \left[\mathcal{F}^x_{p\omega}(\pi-\theta, \pi+\phi) \times \mathcal{F}^y_{p\omega}(\pi-\theta, \pi+\phi)\right]} \tag{6.61}$$

which is the final expression for the inverse dyadic receiving characteristic in terms of the far-field patterns of the probe in two orientations. Inserting (6.61) into (6.59) gives the far-field pattern of the test antenna in terms of the far-field patterns of the reciprocal probe and the

[5] One can also see that $\mathbf{R}^x_{p\omega}(k_x, k_y)$ and $\mathbf{R}^y_{p\omega}(k_x, k_y)$ are vector receiving characteristics corresponding to $\mathbf{s}'_{02}$ in Kerns's work [8] by comparing the x and y components of the vector transmission formula (6.41) with Kerns's scalar transmission formula [8, ch. 2, eq. (3.1–1); and ch. 3, eq. (1–12)].

spatial Fourier transform of the vector output of the probe

$$
\begin{aligned}
&\boldsymbol{\mathcal{F}}_{0\omega}(\theta,\phi) \\
&= \sqrt{\frac{\mu}{\epsilon}}\,\frac{Y_p k^2\cos\theta}{(2\pi)^2 a_{0\omega}}\,\frac{\hat{\mathbf{r}}\times\left[\mathcal{F}^y_{p\omega}(\pi-\theta,\pi+\phi)\hat{\mathbf{x}}-\mathcal{F}^x_{p\omega}(\pi-\theta,\pi+\phi)\hat{\mathbf{y}}\right]}{\hat{\mathbf{r}}\cdot\left[\boldsymbol{\mathcal{F}}^x_{p\omega}(\pi-\theta,\pi+\phi)\times\boldsymbol{\mathcal{F}}^y_{p\omega}(\pi-\theta,\pi+\phi)\right]} \\
&\quad\cdot\int_{-\infty}^{+\infty}\int_{-\infty}^{+\infty}\mathbf{b}_{p\omega}(\mathbf{r}_0)e^{-ik\hat{\mathbf{r}}\cdot\mathbf{r}_0}dx_0dy_0
\end{aligned}
\tag{6.62}
$$

provided $\hat{\mathbf{r}}\cdot\left[\boldsymbol{\mathcal{F}}^x_{p\omega}(\pi-\theta,\pi+\phi)\times\boldsymbol{\mathcal{F}}^y_{p\omega}(\pi-\theta,\pi+\phi)\right]\neq 0$.

6.2.3 The Reciprocal Elementary Electric Dipole Probe

In this section we consider the reciprocal elementary electric dipole probe and show that the probe-corrected formulas for this probe reduce to the non-probe-corrected formulas of Chapter 3. The reciprocal elementary electric dipole probe is attached to a coaxial transmission line that supports only the TEM mode. When this dipole is used as a transmitting antenna and fed with an input signal $a_{p\omega}$ in the plane S_p of the coaxial transmission line, its current distribution is given by $\mathbf{J}_\omega(\mathbf{r}) = -i\omega p^d_\omega\delta(\mathbf{r})\hat{\mathbf{p}}$, where p^d_ω and $\hat{\mathbf{p}}$ are the amplitude and direction of the dipole moment, respectively. The transmitting spectrum that describes the dipole radiation into the half space $z<0$ is found from (3.109) with $\mathbf{k}$ replaced by $\mathbf{k}^-$

$$
\begin{aligned}
\mathbf{T}_{p\omega}(k_x,k_y) &= -\frac{i\omega p^d_\omega}{4\pi a_{p\omega}k\gamma}\sqrt{\frac{\mu}{\epsilon}}\,\mathbf{k}^-\times\left(\mathbf{k}^-\times\hat{\mathbf{p}}\right) \\
&= -\frac{i\omega C_p}{8\pi^2 k\gamma}\sqrt{\frac{\mu}{\epsilon}}\,\mathbf{k}^-\times\left(\mathbf{k}^-\times\hat{\mathbf{p}}\right)
\end{aligned}
\tag{6.63}
$$

where $\mathbf{k}^- = k_x\hat{\mathbf{x}}+k_y\hat{\mathbf{y}}-\gamma\hat{\mathbf{z}}$ is the propagation vector for plane waves propagating into the half space $z<0$, and $C_p = 2\pi p^d_\omega/a_{p\omega}$. (In Section 7.2.6 we show that the constant C_p is independent of frequency if the reference plane S_p is chosen at the base of the dipole.) The reciprocity relation (6.55) shows that the vector receiving characteristic for this probe is [8, p. 131]

$$
\mathbf{R}_{p\omega}(k_x,k_y) = -\frac{i\omega C_p}{8\pi^2 Y_p}\hat{\mathbf{k}}\times\left(\hat{\mathbf{k}}\times\hat{\mathbf{p}}\right)
\tag{6.64}
$$

where Y_p is given in (6.7). When the dipole probe is located at $\mathbf{r}_0$ and is illuminated by the field (6.33) of the test antenna, the output $b_{p\omega}(\mathbf{r}_0)$ in the plane S_p of the matched coaxial transmission line is

$$
\begin{aligned}
b_{p\omega}(\mathbf{r}_0) &= a_{0\omega}\int_{-\infty}^{+\infty}\int_{-\infty}^{+\infty}\mathbf{R}_{p\omega}(k_x,k_y)\cdot\mathbf{T}_{0\omega}(k_x,k_y)e^{i(k_xx_0+k_yy_0+\gamma z_0)}dk_xdk_y \\
&= \frac{i\omega C_p}{4\pi Y_p}\,\hat{\mathbf{p}}\cdot\mathbf{E}_\omega(\mathbf{r}_0)
\end{aligned}
\tag{6.65}
$$

where $\mathbf{E}_\omega(\mathbf{r}_0)$ is the field of the test antenna at $\mathbf{r}_0$. The expression (6.65) agrees with Kerns's result [8, p. 131], and is obtained by use of the identity $[\hat{\mathbf{k}}\times(\hat{\mathbf{k}}\times\hat{\mathbf{p}})]\cdot\mathbf{T}_{0\omega}(k_x,k_y) = -\hat{\mathbf{p}}\cdot\mathbf{T}_{0\omega}(k_x,k_y)$, which follows from (3.64).

When the dipole is used to measure the field of the test antenna, we direct it so that $\hat{\mathbf{p}}$ equals $\hat{\mathbf{x}}$ or $\hat{\mathbf{y}}$. We find from (6.64) that the vector receiving characteristics for these two orientations are

$$\mathbf{R}^{x}_{p\omega}(k_x, k_y) = -\frac{i\omega C_p}{8\pi^2 Y_p}\hat{\mathbf{k}} \times \left(\hat{\mathbf{k}} \times \hat{\mathbf{x}}\right) \tag{6.66}$$

and

$$\mathbf{R}^{y}_{p\omega}(k_x, k_y) = -\frac{i\omega C_p}{8\pi^2 Y_p}\hat{\mathbf{k}} \times \left(\hat{\mathbf{k}} \times \hat{\mathbf{y}}\right). \tag{6.67}$$

Thus, the dyadic receiving characteristic is

$$\bar{\mathbf{R}}_{p\omega}(k_x, k_y) = -\frac{i\omega C_p}{8\pi^2 Y_p}\left[\hat{\mathbf{x}}\,\hat{\mathbf{k}} \times \left(\hat{\mathbf{k}} \times \hat{\mathbf{x}}\right) + \hat{\mathbf{y}}\,\hat{\mathbf{k}} \times \left(\hat{\mathbf{k}} \times \hat{\mathbf{y}}\right)\right]. \tag{6.68}$$

Moreover, the vector output of the dipole probe is found from (6.65) to be

$$\mathbf{b}_{p\omega}(\mathbf{r}_0) = b^{x}_{p\omega}(\mathbf{r}_0)\hat{\mathbf{x}} + b^{y}_{p\omega}(\mathbf{r}_0)\hat{\mathbf{y}} = \frac{i\omega C_p}{4\pi Y_p}\mathbf{E}^{tan}_{\omega}(\mathbf{r}_0) \tag{6.69}$$

where $\mathbf{E}^{tan}_{\omega}(\mathbf{r}_0) = \hat{\mathbf{x}} \cdot \mathbf{E}_{\omega}(\mathbf{r}_0)\hat{\mathbf{x}} + \hat{\mathbf{y}} \cdot \mathbf{E}_{\omega}(\mathbf{r}_0)\hat{\mathbf{y}}$ is the transverse electric field of the test antenna in the scan plane $z = z_0$. Substituting $\mathbf{R}^{x}_{p\omega}(k_x, k_y)$ and $\mathbf{R}^{y}_{p\omega}(k_x, k_y)$ from (6.66) and (6.67) into (6.47), and simplifying the resulting dyadic expression, shows that $\bar{\mathbf{R}}^{-1}_{p\omega}(k_x, k_y)$ for the dipole probe can be written as

$$\bar{\mathbf{R}}^{-1}_{p\omega}(k_x, k_y) = -\frac{8\pi^2 Y_p}{i\omega\gamma C_p}\mathbf{k} \times \left(\hat{\mathbf{y}}\hat{\mathbf{x}} - \hat{\mathbf{x}}\hat{\mathbf{y}}\right) = -\frac{8\pi^2 Y_p}{i\omega\gamma C_p}\mathbf{k} \times \left[\hat{\mathbf{z}} \times (\hat{\mathbf{x}}\hat{\mathbf{x}} + \hat{\mathbf{y}}\hat{\mathbf{y}})\right]. \tag{6.70}$$

With $\mathbf{b}_{p\omega}(\mathbf{r}_0)$ and $\bar{\mathbf{R}}^{-1}_{p\omega}(k_x, k_y)$ from (6.69) and (6.70) inserted into (6.48), the spectrum of the test antenna takes the form

$$\begin{aligned}\mathbf{T}_{\omega}(k_x, k_y) &= a_{0\omega}\mathbf{T}_{0\omega}(k_x, k_y)\\ &= -\frac{e^{-i\gamma z_0}}{2\pi\gamma}\mathbf{k} \times \int_{-\infty}^{+\infty}\int_{-\infty}^{+\infty} \hat{\mathbf{z}} \times \mathbf{E}^{tan}_{\omega}(\mathbf{r}_0)e^{-i(k_x x_0 + k_y y_0)}dx_0 dy_0\end{aligned} \tag{6.71}$$

or

$$\mathbf{T}_{\omega}(k_x, k_y) = \frac{e^{-i\gamma z_0}}{2\pi}\int_{-\infty}^{+\infty}\int_{-\infty}^{+\infty}\left[\mathbf{E}^{tan}_{\omega}(\mathbf{r}_0) - \frac{1}{\gamma}\hat{\mathbf{z}}\,\mathbf{k} \cdot \mathbf{E}^{tan}_{\omega}(\mathbf{r}_0)\right]e^{-i(k_x x_0 + k_y y_0)}dx_0 dy_0. \tag{6.72}$$

Since $\mathbf{k} \cdot \mathbf{T}_{\omega}(k_x, k_y) = 0$, we have $\hat{\mathbf{z}} \cdot \mathbf{T}_{\omega}(k_x, k_y) = -\gamma^{-1}\mathbf{k} \cdot \mathbf{T}^{tan}_{\omega}(k_x, k_y)$, and the square bracket in (6.72) can be replaced by $\mathbf{E}_{\omega}(\mathbf{r}_0)$. Hence, the probe-corrected formula (6.48) does indeed reduce to the non-probe-corrected formula (6.34) when the probe is a reciprocal electric dipole.

7

Probe Correction in the Time Domain

In the present chapter we derive the probe-corrected formulas for planar near-field measurements of time-domain acoustic and electromagnetic fields [23]. These formulas are the time-domain analogs of the frequency-domain formulas derived in Chapter 6 and can be used to determine the time-domain fields of pulsed antennas from time-domain measurements taken in a plane with an arbitrary linear probe having known time-domain receiving characteristics. Based on the formulas in this chapter, sampling theorems and computation schemes, which express the antenna fields in terms of space-time sampled values of the probe output, are presented in Chapter 8. Here in Chapter 7, as in Chapter 6, we refer to both acoustic and electromagnetic radiators as antennas.

Time-domain planar near-field measurements have been performed in Russia since 1984, as described by Skulkin et al. [174]–[177]. These near-field measurements were carried out with stroboscopic transformers, and probe correction was not included. The far fields were calculated with a simple and efficient far-field formula that holds when the time dependence of the field is the same throughout the scan plane [177, eqs. (1) and (2)]. That is, the field in the scan plane was assumed to have the form $\mathbf{E}(\mathbf{r}_0, t) = f(t)\mathbf{E}_0(\mathbf{r}_0)$, where $f(t)$ is a time function and $\mathbf{E}_0(\mathbf{r})$ depends only on the spatial variables [177]. The far-field formula based on this assumption is accurate when the test antenna has a large and uniformly excited aperture, but it is inaccurate when the time dependence of the field varies significantly in the scan plane. In [175], Skulkin presents solutions to a number of practical problems and explains how time gating can be used to enhance the accuracy of the calculated far fields. More recently, Dominek and Shamansky [56] used a measurement system with a bandwidth of 10 GHz to measure the near field of a pulsed aperture antenna. They subsequently determined its far field using the probe-corrected formulas that will be derived in the present chapter.

To determine the behavior of antennas in the frequency domain over wide bandwidths, de Jongh, Hajian, and Ligthart [57] employed a time-domain measurement technique in which the test antenna is fed with a pulse. In [57] they report that this time-domain technique has the following advantages over conventional frequency-domain measurements. Unwanted reflections due to antenna mismatching, surrounding walls, and nearby objects can be eliminated by time gating. The measurement time can be reduced because only a single time-domain scan is needed to determine the wideband frequency-domain response of the test antenna. Out-of-band properties of the test antenna can be easily determined in the time domain. Errors resulting from the truncation of the scan plane can be removed by time gating. The measurement system for the time-domain technique is less complex than the system required if frequency-domain measurements are to be carried out at discrete frequencies. The main disadvantage that they reported was a lower accuracy at higher frequencies because the spectrum of the measurement pulse decays with frequency, while the noise level remains constant.

The probe-corrected formulas of the present chapter will be derived without making any restrictive assumptions on the behavior of the field in the scan plane. The measurement geometry is shown in Figure 7.1. The test antenna is located in the half space $z < 0$ and radiates an acoustic or electromagnetic field that we want to determine in the half space $z > 0$. The output of the probe, whose reference point $\mathbf{r}_0$ is located in the scan plane $z = 0$, is recorded as it translates without rotation throughout the scan plane in the field of the test antenna. The probe-corrected formulas for acoustic and electromagnetic fields are presented in Sections 7.1 and 7.2, respectively. General probe-corrected formulas are first derived that hold for arbitrary linear probes. The general formulas are then simplified through the use of "time-derivative" probes, and references are given that discuss the practical construction of such probes. Time-domain reciprocity relations are also derived that determine the receiving characteristics of reciprocal probes from their transmitting characteristics.

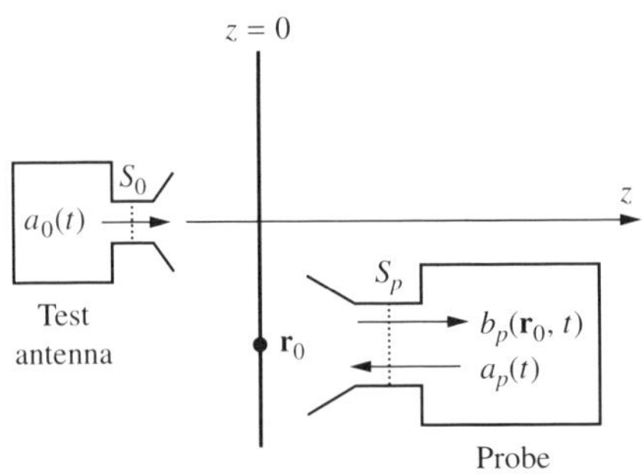

Figure 7.1 Planar scanning geometry for time-domain fields.

As in Chapter 6 on frequency-domain scanning, we begin by describing the electromagnetic waveguide that is attached to the measurement probe. Assume that this waveguide supports just one propagating mode (or that only one propagating mode is excited) and is composed of perfect conductors separated by a homogeneous, isotropic medium. (Such waveguides include coaxial transmission lines as well as rectangular and circular waveguides.) According to the discussion in the introduction to Chapter 6, we can choose the characteristic admittance Y_p of the propagating mode so that it is independent of frequency. The Fourier transform applied to (6.2) and (6.3) shows that the transverse time-domain electric and magnetic fields in a reference plane S_p of the probe waveguide are given by

$$\mathbf{E}^t(\boldsymbol{\rho}, t) = \frac{1}{2\pi} \int_{-\infty}^{+\infty} \left[a_p(t') + b_p(t')\right] \mathbf{e}_b(\boldsymbol{\rho}, t - t')dt' \tag{7.1}$$

and

$$\mathbf{H}^t(\boldsymbol{\rho}, t) = \frac{Y_p}{2\pi} \int_{-\infty}^{+\infty} \left[a_p(t') - b_p(t')\right] \mathbf{h}_b(\boldsymbol{\rho}, t - t')dt' \tag{7.2}$$

where $a_p(t)$ is the amplitude of the incident wave, and $b_p(t)$ is the amplitude of the emerging wave at the reference plane S_p of the probe waveguide. The basis fields $\mathbf{e}_b(\boldsymbol{\rho}, t)$ and $\mathbf{h}_b(\boldsymbol{\rho}, t)$ depend on time whenever the wave admittance $Y^W_{p\omega}$ (defined by (6.1)) for the propagating mode depends on frequency. As illustrated by (7.1) and (7.2), a convolution is in general required to compute the transverse fields from the values of $a_p(t)$ and $b_p(t)$. However, when the wave admittance $Y^W_{p\omega}$ for the propagating mode is independent of frequency, the basis fields are also independent of frequency, and (7.1) and (7.2) simplify to

$$\mathbf{E}^t(\boldsymbol{\rho}, t) = \left[a_p(t) + b_p(t)\right]\mathbf{e}_b(\boldsymbol{\rho}), \quad \mathbf{H}^t(\boldsymbol{\rho}, t) = Y_p\left[a_p(t) - b_p(t)\right]\mathbf{h}_b(\boldsymbol{\rho}). \tag{7.3}$$

The transverse electromagnetic fields in a coaxial transmission line satisfy (7.3) with Y_p given in (6.7). The sum $a_p(t) + b_p(t)$ is the voltage between the inner and outer conductors at the reference plane S_p, and the difference $Y_p\left[a_p(t) - b_p(t)\right]$ is the current flowing in each conductor.

When the probe is used to measure the field of the test antenna, we assume that the probe's waveguide is perfectly matched so that the incident wave amplitude $a_p(t)$ is zero, and we refer to $b_p(t)$ as the probe output. As in Chapter 6, the waveguide description given here for the probe also applies to the test antenna when it is attached to a waveguide feed, as indicated in Figure 7.1.

7.1 ACOUSTIC FIELDS

Figure 7.1 shows the acoustic source region (called the test antenna), which is located in the half space $z < 0$ and generates the time-domain acoustic pressure $p(\mathbf{r}, t)$ that satisfies the scalar wave equation (5.1). The test antenna lies in a linear, quiescent, homogeneous, isotropic, and lossless fluid. The probe output $b_p(\mathbf{r}_0, t)$ depends on the time t and the position $\mathbf{r}_0$ of the reference point of the probe (and, of course, on the chosen position of the reference plane S_p in the probe's waveguide). Time-domain probe-corrected formulas, which determine the acoustic field of the test antenna in the half space $z > 0$ in terms of the probe output in the scan plane $z = 0$, will be derived in three different ways. First, these formulas are obtained in Section 7.1.1 by applying the Fourier inversion formula

$$p(\mathbf{r}, t) = \int_{-\infty}^{+\infty} p_\omega(\mathbf{r}) e^{-i\omega t} d\omega \tag{7.4}$$

with $p_\omega(\mathbf{r})$ being the frequency-domain acoustic pressure, to the frequency-domain formulas of Section 6.1. Second, the probe-corrected formulas are derived in Section 7.1.2 by working directly in the time domain using time-domain plane waves and a time-domain receiving characteristic for the probe. Third, the probe-corrected formulas are obtained in Section 7.1.3 from the result that the probe output satisfies the homogeneous wave equation.

As shown in Section 7.1.4, the probe-corrected formulas simplify greatly when the output of the probe illuminated by a plane wave is proportional to the time derivative of the field of that plane wave. Finally, in Section 7.1.5 we present reciprocity relations that determine the receiving characteristics of reciprocal electroacoustic transducer probes from their transmitting characteristics.

Before deriving the time-domain probe-corrected formulas, we summarize the necessary time-domain plane-wave expansions from Section 5.2. According to (5.86) and (5.87), the acoustic pressure $p(\mathbf{r}, t)$ radiated into the half space $z > 0$ by the test antenna can be expressed as

$$p(\mathbf{r},t) = -\frac{1}{2\pi} \iint_{\xi^2+\eta^2<c^{-2}} \frac{\partial^2}{\partial t^2} T(\xi,\eta,t-\xi x-\eta y-\zeta z)d\xi d\eta$$

$$-\frac{1}{2\pi^2} \iint_{\xi^2+\eta^2>c^{-2}} \int_{-\infty}^{+\infty} \frac{\partial^2}{\partial t^2} \frac{|\zeta| z\, T(\xi,\eta,t-t'-\xi x-\eta y)}{|\zeta|^2 z^2+t'^2} dt' d\xi d\eta, \quad z>0 \tag{7.5}$$

with

$$\zeta = \begin{cases} \left|\sqrt{c^{-2}-\xi^2-\eta^2}\right|, & \xi^2+\eta^2 \le c^{-2} \\ i\left|\sqrt{\xi^2+\eta^2-c^{-2}}\right|, & \xi^2+\eta^2 > c^{-2} \end{cases}. \tag{7.6}$$

The time-domain spectrum $T(\xi,\eta,t)$ can be determined from

$$T(\xi,\eta,t) = \frac{1}{2\pi} \int_{-\infty}^{+\infty}\int_{-\infty}^{+\infty} p(\mathbf{r}_0, t+\xi x_0+\eta y_0) dx_0 dy_0, \quad \mathbf{r}_0 = x_0\hat{\mathbf{x}} + y_0\hat{\mathbf{y}}. \tag{7.7}$$

Thus, the time-domain spectrum is a space-time Radon transform of the pressure in the scan plane $z=0$. In the frequency-domain derivations of Section 6.1, it was convenient to employ the spectrum $T_{0\omega}(k_x,k_y)$, normalized with respect to the input signal $a_{0\omega}$ that feeds the test antenna, so that $T_\omega(k_x,k_y) = a_{0\omega}T_{0\omega}(k_x,k_y)$. In the time domain, however, we shall employ the spectrum $T(\xi,\eta,t)$ for the actual acoustic pressure radiated by the test antenna. The time-domain spectra $T(\xi,\eta,t) = \int_{-\infty}^{+\infty} T_\omega(\omega\xi,\omega\eta)e^{-i\omega t}d\omega$ and $T_0(\xi,\eta,t) = \int_{-\infty}^{+\infty} T_{0\omega}(\omega\xi,\omega\eta)e^{-i\omega t}d\omega$, with $a_0(t) = \int_{-\infty}^{+\infty} a_{0\omega}e^{-i\omega t}d\omega$, are related through the convolution

$$T(\xi,\eta,t) = \frac{1}{2\pi}\int_{-\infty}^{+\infty} a_0(t-t')T_0(\xi,\eta,t')dt'. \tag{7.8}$$

The normalized time-domain plane-wave spectrum $T_0(\xi,\eta,t)$ is the spectrum for the acoustic pressure when the input signal feeding the test antenna at the reference plane S_0 of the feed waveguide is a delta function $a_0(t) = 2\pi\delta(t)$. In the following time-domain derivations, we treat the test antenna as a source region without specifying any particular input signal.

The time-domain far-field pattern, defined so that the acoustic far-field pressure is

$$p(\mathbf{r},t) \sim \frac{\mathcal{F}(\theta,\phi,t-r/c)}{r}, \quad r\to\infty \tag{7.9}$$

can be expressed from (5.187) in terms of the spectrum $T(\xi,\eta,t)$ as

$$\mathcal{F}(\theta,\phi,t) = \frac{\cos\theta}{c}\frac{\partial}{\partial t}T(c^{-1}\cos\phi\sin\theta, c^{-1}\sin\phi\sin\theta, t), \quad 0\le\theta<\pi/2. \tag{7.10}$$

The formulas (7.5)–(7.10) constitute the time-domain field representation that gives the pressure in the half space $z>0$ in terms of the spectrum $T(\xi,\eta,t)$, which is determined in (7.7) in terms of the pressure in the plane $z=0$. In Chapter 5, formulas were also derived for the acoustic velocity $\mathbf{v}(\mathbf{r},t)$ in terms of the time-domain spectrum $T(\xi,\eta,t)$. In the following we shall derive expressions that determine the time-domain spectrum $T(\xi,\eta,t)$, or the time-domain far-field pattern $\mathcal{F}(\theta,\phi,t)$, in terms of the output and receiving characteristic of the probe.

7.1.1 Formulas Obtained from the Fourier Transform

We first derive the time-domain probe-corrected formulas by applying the Fourier transform to the frequency-domain formulas of Section 6.1. The frequency-domain probe-corrected formulas (6.16) and (6.17) show that with the reference point $\mathbf{r}_0 = x_0\hat{\mathbf{x}} + y_0\hat{\mathbf{y}}$ of the receiving probe kept in the scan plane, the frequency-domain plane-wave transmission formula and spectrum for the pressure radiated by the test antenna are given by

$$b_{p\omega}(\mathbf{r}_0) = \omega^2 \int_{-\infty}^{+\infty}\int_{-\infty}^{+\infty} R_{p\omega}(\omega\xi, \omega\eta) T_\omega(\omega\xi, \omega\eta)\, e^{i\omega(\xi x_0 + \eta y_0)} d\xi d\eta \tag{7.11}$$

and

$$T_\omega(\omega\xi, \omega\eta) = \frac{1}{(2\pi)^2 R_{p\omega}(\omega\xi, \omega\eta)} \int_{-\infty}^{+\infty}\int_{-\infty}^{+\infty} b_{p\omega}(\mathbf{r}_0) e^{-i\omega(\xi x_0 + \eta y_0)} dx_0 dy_0 \tag{7.12}$$

where $R_{p\omega}(\omega\xi, \omega\eta)$ and $b_{p\omega}(\mathbf{r}_0)$ are the frequency-domain receiving characteristic and the output of the probe, respectively. Defining the direct and inverse time-domain receiving characteristics of the probe as[1]

$$R_p(\xi, \eta, t) = \int_{-\infty}^{+\infty} R_{p\omega}(\omega\xi, \omega\eta)\, e^{-i\omega t} d\omega \tag{7.13}$$

and

$$R_p^{-1}(\xi, \eta, t) = \int_{-\infty}^{+\infty} \frac{e^{-i\omega t}}{R_{p\omega}(\omega\xi, \omega\eta)} d\omega \tag{7.14}$$

then using the convolution theorem

$$\int_{-\infty}^{+\infty} f_\omega g_\omega e^{-i\omega t} d\omega = \frac{1}{2\pi} \int_{-\infty}^{+\infty} f(t') g(t - t') dt' \tag{7.15}$$

in taking the inverse Fourier transform of (7.11) and (7.12), produces the time domain plane wave transmission formula

$$b_p(\mathbf{r}_0, t) = -\frac{1}{2\pi}\frac{\partial^2}{\partial t^2} \int_{-\infty}^{+\infty}\int_{-\infty}^{+\infty}\int_{-\infty}^{+\infty} R_p(\xi, \eta, t - t' - \xi x_0 - \eta y_0) T(\xi, \eta, t') dt' d\xi d\eta \tag{7.16}$$

and the following three alternative formulas for the time-domain spectrum

$$T(\xi, \eta, t) = \frac{1}{(2\pi)^3} \int_{-\infty}^{+\infty} R_p^{-1}(\xi, \eta, t') \int_{-\infty}^{+\infty}\int_{-\infty}^{+\infty} b_p(\mathbf{r}_0, t - t' + \xi x_0 + \eta y_0) dx_0 dy_0 dt' \tag{7.17}$$

$$T(\xi, \eta, t) = \frac{1}{(2\pi)^3} \int_{-\infty}^{+\infty}\int_{-\infty}^{+\infty}\int_{-\infty}^{+\infty} R_p^{-1}(\xi, \eta, t - t' + \xi x_0 + \eta y_0) b_p(\mathbf{r}_0, t') dx_0 dy_0 dt' \tag{7.18}$$

[1] Both the time-domain receiving characteristic and its inverse defined in (7.13) and (7.14) may involve delta functions and their derivatives with respect to time. This will be seen in Section 7.1.4 for time-derivative probes whose receiving characteristics are proportional to $d\delta(t)/dt$.

and

$$T(\xi, \eta, t) = \frac{1}{(2\pi)^3} \int_{-\infty}^{+\infty} \int_{-\infty}^{+\infty} \int_{-\infty}^{+\infty} R_p^{-1}(\xi, \eta, t - t' + \xi x_0 + \eta y_0) b_p(\mathbf{r}_0, t') dt' dx_0 dy_0 \tag{7.19}$$

where

$$b_p(\mathbf{r}_0, t) = \int_{-\infty}^{+\infty} b_{p\omega}(\mathbf{r}_0) e^{-i\omega t} d\omega \tag{7.20}$$

is the time-domain output of the probe, when it is located at $\mathbf{r}_0$. The formulas (7.17), (7.18), and (7.19) are the general time-domain probe-corrected formulas for the time-domain spectrum of the acoustic field in terms of the output and receiving characteristic of the probe. It is assumed that the distance between the probe and the test antenna is large enough that multiple interactions can be neglected. That is, the test antenna fields scattered by the probe, rescattered by the test antenna, and returned to the probe, do not affect the output of the probe.

Assume that the test antenna is turned on at $t = t_0$, so that all fields are zero for $t \leq t_0$. Then the output of the probe $b_p(\mathbf{r}_0, t)$ is zero for $\mathbf{r}_0 \notin \mathcal{R}(t)$, where $\mathcal{R}(t)$ is the finite region of the scan plane in which the field is nonzero at time t. (For late times, $\mathcal{R}(t)$ is approximately a circular disk with radius $c(t - t_0)$.) Without making any assumption about the time dependence of the inverse receiving characteristic $R_p^{-1}(\xi, \eta, t)$, the probe-corrected formulas all involve at least one integral over an infinite domain. For example, the (x_0, y_0) integral in (7.18) becomes an integral over the finite region $\mathcal{R}(t')$, but the t' integration goes from $t' = t_0$ to $t' = +\infty$ and thus involves an integration over an infinite domain. Similar statements hold for the probe-corrected formulas (7.17) and (7.19). However, as we shall see in Section 7.1.4 below, only integrals over finite domains need to be calculated to get the time-domain far-field pattern when a "time-derivative" probe is used.

Equation (7.19) may be the most attractive for measurement purposes since the inner time integral can be calculated at each scan point $\mathbf{r}_0$ for the desired values of (ξ, η, t). Thus, one would not have to store the values of the output of the probe $b_p(\mathbf{r}_0, t')$ over time to calculate the time-domain spectrum of the field radiated by the test antenna.

Having determined the spectrum $T(\xi, \eta, t)$, we can then use (7.5) to calculate the time-domain pressure everywhere in the half space $z > 0$, or use (7.10) to calculate the time-domain far-field pattern. Instead of the real time-domain formulas (7.7)–(7.10), we could have used the analytic-signal formulas of Section 5.3 involving the analytic fields. The expressions for the field in the half space $z > 0$ would then have been given by the simpler expression (5.233) which, however, involves the analytic-signal fields that cannot be measured directly. Hence, for measurement purposes, the real formulas used in this chapter seem advantageous compared to the analytic-signal formulas. This issue was discussed more fully in Section 5.3.

7.1.2 Formulas Obtained Directly in the Time Domain

The probe-corrected acoustic formulas can also be derived directly in the time domain. The derivation goes as follows. First, a time-domain receiving characteristic is used along with a time-domain plane-wave expansion for the pressure radiated by the test antenna to obtain an expression for the probe output. This expression involves the receiving characteristic of the probe and the spectrum of the acoustic pressure radiated by the test antenna. Employing identities from the theory of Radon transforms, we then solve for the spectrum, which determines the pressure everywhere in the half space $z > 0$ through the formula (7.5).

To define a useful time-domain receiving characteristic of the probe, we have to specify the time-domain basis fields that will be used to expand the field radiated by the test antenna. We choose the time-domain basis fields to be simply the inverse Fourier transform of the propagating and evanescent plane waves used to expand the frequency-domain fields in Chapter 6. These time-domain basis fields are called time-domain plane waves and were discussed in Section 5.2. The time-domain plane waves are denoted by $\Phi^B(\mathbf{r}, t, \xi, \eta)$ and are given in (5.98). They are functions of space, time, and the two spectral variables ξ and η. When $\xi^2 + \eta^2 < c^{-2}$, the time-domain plane wave is propagating (see Figure 5.2), and when $\xi^2 + \eta^2 > c^{-2}$, the time-domain plane wave is evanescent as $z \to \infty$ (see Figure 5.3).

The time-domain receiving characteristic of the probe is denoted by $R_p(\xi, \eta, t)$, and is defined to be the output of the probe when the reference point of the probe is located at the origin, and the incoming pressure field is the plane wave $\Phi^B(\mathbf{r}, t, \xi, \eta)$, that is $b_p(0, t) = R_p(\xi, \eta, t)$. The expressions (5.98) and (5.99) for the time-domain plane waves, and the definition of $R_p(\xi, \eta, t)$, proves that $b_p(\mathbf{r}_0, t) = R_p(\xi, \eta, t - \xi x_0 - \eta y_0)$ when the reference point of the probe is at $\mathbf{r}_0 = x_0\hat{\mathbf{x}} + y_0\hat{\mathbf{y}}$ and the incident field is $\Phi^B(\mathbf{r}, t, \xi, \eta)$.[2]

To calculate the probe output when the incident field is that of the test antenna, we recall from (5.100) that the time-domain pressure radiated by the test antenna can be written in terms of the time-domain plane waves as

$$p(\mathbf{r}, t) = -\frac{1}{2\pi} \int_{-\infty}^{+\infty} \int_{-\infty}^{+\infty} \int_{-\infty}^{+\infty} \Phi^B(\mathbf{r}, t - t', \xi, \eta) \frac{\partial^2}{\partial t'^2} T(\xi, \eta, t') dt' d\xi d\eta, \quad z > 0. \tag{7.21}$$

Inserting the expression (5.98) for the time-domain plane waves into (7.21), we obtain the formula (7.5).

From linearity and the definition of the time-domain receiving characteristic, the probe output due to the plane-wave component

$$dp(\mathbf{r}_0, t) = -\frac{1}{2\pi} \Phi^B(\mathbf{r}, t - t', \xi, \eta) \frac{\partial^2}{\partial t'^2} T(\xi, \eta, t') dt' d\xi d\eta \tag{7.22}$$

becomes

$$db_p(\mathbf{r}_0, t) = -\frac{1}{2\pi} R_p(\xi, \eta, t - t' - \xi x_0 - \eta y_0) \frac{\partial^2}{\partial t'^2} T(\xi, \eta, t') dt' d\xi d\eta. \tag{7.23}$$

Hence, by superposition the total output of the linear probe due to the entire field radiated by the test antenna is found from (7.21) as

$$b_p(\mathbf{r}_0, t) = -\frac{1}{2\pi} \int_{-\infty}^{+\infty} \int_{-\infty}^{+\infty} \int_{-\infty}^{+\infty} R_p(\xi, \eta, t - t' - \xi x_0 - \eta y_0) \frac{\partial^2}{\partial t'^2} T(\xi, \eta, t') dt' d\xi d\eta. \tag{7.24}$$

A simple manipulation of the time-convolution integral in (7.24) shows that the probe output can also be expressed as

$$b_p(\mathbf{r}_0, t) = -\frac{1}{2\pi} \frac{\partial^2}{\partial t^2} \int_{-\infty}^{+\infty} \int_{-\infty}^{+\infty} \int_{-\infty}^{+\infty} R_p(\xi, \eta, t - t' - \xi x_0 - \eta y_0) T(\xi, \eta, t') dt' d\xi d\eta \tag{7.25}$$

[2]From the definition of the frequency-domain receiving characteristic $R_{p\omega}(k_x, k_y)$ in Section 6.1, it follows that the time-domain output of the probe at $\mathbf{r}_0 = x_0\hat{\mathbf{x}} + y_0\hat{\mathbf{y}}$, due to the incoming plane-wave field $\Phi^B(\mathbf{r}, t, \xi, \eta)$, should be $b_p(\mathbf{r}_0, t) = (2\pi)^{-1} \int_{-\infty}^{+\infty} (2\pi)\Phi^B(\mathbf{r}_0, t - t', \xi, \eta) R_p(\xi, \eta, t') dt' = R_p(\xi, \eta, t - \xi x_0 - \eta y_0)$. The last part of this equation is obtained from (5.98) and (5.99), which shows that in the scan plane $\Phi^B(\mathbf{r}_0, t - t', \xi, \eta) = \delta(t - t' - \xi x_0 - \eta y_0)$ for all (ξ, η). Thus, the time-domain definition of the receiving characteristic concurs with the frequency-domain definition used in Section 6.1.

which agrees with the time-domain transmission formula (7.16) obtained by inverse Fourier transforming (7.11). As previously stated, we assume that the distance between the probe and the test antenna is large enough that multiple interactions can be neglected.

The probe output in (7.25) can also be written as

$$b_p(\mathbf{r}_0, t) = -\frac{1}{2\pi}\frac{\partial^2}{\partial t^2}\int_{-\infty}^{+\infty}\int_{-\infty}^{+\infty} T_b(\xi, \eta, t - \xi x_0 - \eta y_0)d\xi d\eta, \quad \mathbf{r}_0 = x_0\hat{\mathbf{x}} + y_0\hat{\mathbf{y}} \tag{7.26}$$

where the spectrum of the probe output $T_b(\xi, \eta, t)$ is given by

$$T_b(\xi, \eta, t) = \int_{-\infty}^{+\infty} R_p(\xi, \eta, t - t')T(\xi, \eta, t')dt'. \tag{7.27}$$

The Radon transform identities (5.101) and (5.102) reveal that $T_b(\xi, \eta, t)$ is the space-time Radon transform of the probe output in the scan plane; that is

$$T_b(\xi, \eta, t) = \frac{1}{2\pi}\int_{-\infty}^{+\infty}\int_{-\infty}^{+\infty} b_p(\mathbf{r}_0, t + \xi x_0 + \eta y_0)dx_0 dy_0, \quad \mathbf{r}_0 = x_0\hat{\mathbf{x}} + y_0\hat{\mathbf{y}}. \tag{7.28}$$

Probably the simplest way to verify that (7.28) implies (7.26) (and vice versa) is to Fourier transform (7.28) into the frequency domain, use the two-dimensional spatial Fourier transform to get an expression for $b_{p\omega}(\mathbf{r}_0)$ in terms of $T_{b\omega}(k_x, k_y)$, and finally Fourier transform back into the time domain to get (7.26).

The inverse time-domain receiving characteristic $R_p^{-1}(\xi, \eta, t)$ for the probe, in accordance with (7.13) and (7.14), can be defined as

$$\int_{-\infty}^{+\infty} R_p^{-1}(\xi, \eta, t')R_p(\xi, \eta, t - t')dt' = 4\pi^2\delta(t). \tag{7.29}$$

Then we can multiply the expression (7.27) by $R_p^{-1}(\xi, \eta, t'' - t)$ and integrate with respect to t to get

$$T(\xi, \eta, t) = \frac{1}{(2\pi)^2}\int_{-\infty}^{+\infty} R_p^{-1}(\xi, \eta, t')T_b(\xi, \eta, t - t')dt'. \tag{7.30}$$

The final step in deriving the probe-corrected formula directly in the time domain is to insert the expression (7.28) for $T_b(\xi, \eta, t)$ into (7.30) to show that the spectrum of the field radiated by the test antenna is given by

$$T(\xi, \eta, t) = \frac{1}{(2\pi)^3}\int_{-\infty}^{+\infty} R_p^{-1}(\xi, \eta, t')\int_{-\infty}^{+\infty}\int_{-\infty}^{+\infty} b_p(\mathbf{r}_0, t - t' + \xi x_0 + \eta y_0)dx_0 dy_0 dt'. \tag{7.31}$$

This expression is seen to agree with the result (7.17). Interchanging the orders of integration in (7.31) produces the probe-corrected formulas (7.18) and (7.19).

7.1.3 Derivation from the Wave Equation

The probe-corrected formulas (7.17)–(7.19) can also be derived from the result that the probe output satisfies the wave equation. In Section 6.1.1, it was shown that the frequency-domain output of the probe satisfies the Helmholtz equation. Taking the Fourier transform of the Helmholtz equation (6.24), we obtain the homogeneous time-domain wave equation for the time-domain output of the probe

$$\nabla^2 b_p(\mathbf{r},t) - \frac{1}{c^2}\frac{\partial^2}{\partial t^2} b_p(\mathbf{r},t) = 0, \quad z > z_s - z_p \tag{7.32}$$

where z_s is the maximum (more precisely, the supremum) value of z for which the source function of the test antenna is nonzero. That is, the physical body of the test antenna is located in the region $z < z_s$. Similarly, z_p is the minimum (more precisely, the infimum) value of z for which the source function $q_p^a(\mathbf{r},t)$ of the adjoint probe is nonzero, when the adjoint probe is radiating into the region $z < 0$ with its reference point located at the origin. Since the physical dimensions of the adjoint probe equal the physical dimensions of the probe itself, the condition $z > z_s - z_p$ simply states that the test and probe antennas cannot touch during a scan. As in the preceding section, we assume that $z_s - z_p \le 0$, so that $z = 0$ can be used as the scan plane and the homogeneous wave equation (7.32) is satisfied in the half space $z > 0$. Moreover, as explained at the end of Section 6.1.1, for a probe that translates without rotation with respect to the test antenna, the probe output $b_p(\mathbf{r},t)$ satisfies the homogeneous wave equation (7.32) in the entire region outside a closed surface formed by the locus of points $(z_s - z_p)$ for all orientations of the z axis; that is, in the region in which the minimum convex surface enclosing the test antenna, and the minimum convex surface enclosing the probe, do not intersect.

Like the corresponding frequency-domain equation (6.24), the time-domain wave equation (7.32) holds only if multiple interactions between the probe and test antennas are negligible. However, in the time domain, unlike the frequency domain, the multiple interactions can often be eliminated by time gating the output of the probe.

The expressions (7.9) and (7.10) show that the time-domain acoustic far-field pressure can be expressed in terms of its time-domain spectrum as

$$p(\mathbf{r},t) \sim \frac{\cos\theta}{cr}\frac{\partial}{\partial t} T(c^{-1}\cos\phi\sin\theta, c^{-1}\sin\phi\sin\theta, t - r/c). \tag{7.33}$$

Since the far-field is locally a plane wave, the output of the probe, when the probe is located at the far-field point (r,θ,ϕ) and is illuminated by the pressure field (7.33), is determined from linearity and the time-domain receiving characteristic $R_p(\xi,\eta,t)$ as

$$\begin{aligned} b_p(\mathbf{r},t) &\sim \frac{\cos\theta}{cr}\int_{-\infty}^{+\infty} R_p(c^{-1}\cos\phi\sin\theta, c^{-1}\sin\phi\sin\theta, t - t' - r/c) \\ &\qquad \cdot \frac{\partial}{\partial t'} T(c^{-1}\cos\phi\sin\theta, c^{-1}\sin\phi\sin\theta, t')dt' \\ &= \frac{\cos\theta}{cr}\frac{\partial}{\partial t}\int_{-\infty}^{+\infty} R_p(c^{-1}\cos\phi\sin\theta, c^{-1}\sin\phi\sin\theta, t - t' - r/c) \\ &\qquad \cdot T(c^{-1}\cos\phi\sin\theta, c^{-1}\sin\phi\sin\theta, t')dt'. \end{aligned} \tag{7.34}$$

The far-field output of the probe in (7.34), which is the time-domain acoustic analog of the generalized Friis transmission formula [171, p. 6-8], has the form $\mathcal{F}_b(\theta,\phi,t - r/c)/r$, so that the probe output satisfies the outgoing radiation condition at infinity. Since the probe output satisfies the wave equation (7.32) for $z > 0$ and the outgoing radiation condition, it obeys the Green's function far-field representation (5.54)

$$b_p(\mathbf{r},t) \sim \frac{\cos\theta}{2\pi\, cr}\frac{\partial}{\partial t}\int_{-\infty}^{+\infty}\int_{-\infty}^{+\infty} b_p(\mathbf{r}_0, t - r/c + \mathbf{r}_0\cdot\hat{\mathbf{r}})\, dx_0 dy_0, \quad z_s - z_p < 0. \tag{7.35}$$

Comparing the two far-field expressions (7.34) and (7.35) yields the relation

$$\int_{-\infty}^{+\infty} R_p(\xi,\eta,t-t')T(\xi,\eta,t')dt' = \frac{1}{2\pi}\int_{-\infty}^{+\infty}\int_{-\infty}^{+\infty} b_p(\mathbf{r}_0, t+\xi x_0+\eta y_0)\, dx_0 dy_0, \qquad \xi^2+\eta^2<c^{-2}, \quad z_s - z_p < 0. \tag{7.36}$$

To isolate $T(\xi,\eta,t)$ from this equation, we multiply both sides by the inverse receiving characteristic $R_p^{-1}(\xi,\eta,t''-t)$ and integrate the resulting equation with respect to t from $-\infty$ to $+\infty$. We then use the relation (7.29) between $R_p^{-1}(\xi,\eta,t)$ and $R_p(\xi,\eta,t)$ to obtain

$$T(\xi,\eta,t'') = \frac{1}{(2\pi)^3}\int_{-\infty}^{+\infty} R_p^{-1}(\xi,\eta,t''-t)\int_{-\infty}^{+\infty}\int_{-\infty}^{+\infty} b_p(\mathbf{r}_0, t+\xi x_0+\eta y_0)\, dx_0 dy_0 dt, \qquad \xi^2+\eta^2<c^{-2}, \quad z_s - z_p < 0 \tag{7.37}$$

which is equivalent to the probe-corrected formulas (7.17)–(7.19) for $\xi^2+\eta^2<c^{-2}$.

The wave-equation derivation has thus led to an expression for the spectrum $T(\xi,\eta,t)$ that is valid for propagating modes. We know from (7.17) that (7.37) holds for all (ξ,η). However, as discussed in Section 5.2.4, the function $T(\xi,\eta,t)$ is not, in general, analytic with respect to (ξ,η), and we cannot, by analytic continuation in the time domain, determine its values for all real (ξ,η) from its values in the propagating region $\xi^2+\eta^2<c^{-2}$.

Nonetheless, we can argue the validity of (7.37) for all (ξ,η) by returning to (7.36). Assume that the probe is a point-source receiver at $\mathbf{r}_0$ that measures the pressure ($b_p(\mathbf{r}_0,t) = p(\mathbf{r}_0,t)$). Such a probe would have $R_p(\xi,\eta,t)=\delta(t)$, and (7.36) would reduce to the definition of the time-domain plane-wave spectrum for the acoustic pressure of the test antenna for all (ξ,η); see (7.7). Since an arbitrary probe can be reproduced by a superposition of point-source receivers [37], [58], equations (7.37) and (7.36) are valid for all (ξ,η) and an arbitrary linear probe.

7.1.4 Formulas for Time-Derivative Probes

The general time-domain probe-corrected formulas (7.17)–(7.19) require the calculation of a double spatial integral over the scan plane and a time integral. We will now show that for a special type of probe, the time integral can be evaluated analytically and the resulting probe-corrected formula is greatly simplified. The output of such a probe, due to an incoming plane wave Φ^B, is proportional to the time derivative of Φ^B. That is, the frequency-domain receiving characteristic is proportional to ω and can be written as

$$R_{p\omega}(\omega\xi,\omega\eta) = -\frac{i\omega Q_p(\xi,\eta)}{2\pi}, \qquad (\xi,\eta)\in D \tag{7.38}$$

where the angularly dependent function $Q_p(\xi,\eta)$ is real and independent of frequency for $(\xi,\eta)\in D$, with D being part of the (ξ,η) plane. Some probes, such as the one that has $Q_p(\xi,\eta)=1$ for all (ξ,η), act as time-derivative probes for all time-domain plane waves, so that the region D is the entire (ξ,η) plane. Other probes, like the one considered in Section 8.3.2, act as time-derivative probes only for certain time-domain plane waves, and the region D is only part of the (ξ,η) plane. Moreover, physically realizable probes only approximate the linear frequency dependence in (7.38) over a finite bandwidth.

With the receiving characteristic (7.38) inserted into the probe-corrected formula (7.12), the inverse Fourier transform applied to (7.12) gives

$$\frac{\partial}{\partial t}T(\xi,\eta,t)=\frac{1}{2\pi\, Q_p(\xi,\eta)}\int\limits_{-\infty}^{+\infty}\int\limits_{-\infty}^{+\infty} b_p(\mathbf{r}_0, t+\xi x_0+\eta y_0)dx_0dy_0, \quad (\xi,\eta)\in D \tag{7.39}$$

with $\mathbf{r}_0=\hat{\mathbf{x}}x_0+\hat{\mathbf{y}}y_0$.

Multiplying (7.39) by $Q_p(\xi,\eta)$ and taking the inverse Radon transform yields

$$b_p(\mathbf{r}_0,t)=-\frac{1}{2\pi}\frac{\partial}{\partial t}\int\limits_{-\infty}^{+\infty}\int\limits_{-\infty}^{+\infty} Q_p(\xi,\eta)\frac{\partial^2}{\partial t^2}T(\xi,\eta,t-\xi x_0-\eta y_0)d\xi\, d\eta. \tag{7.40}$$

If the time-derivative probe is an omnidirectional receiving acoustic monopole with $Q_p(\xi,\eta)$ equal to a constant Q_c for all (ξ,η), then (7.40) reduces, with the help of (7.5) and (5.99), to simply

$$b_p(\mathbf{r}_0,t)=Q_c\frac{\partial}{\partial t}p(\mathbf{r}_0,t). \tag{7.41}$$

That is, the output of the receiving monopole is proportional to the time derivative of the incident acoustic pressure at the location of the monopole.

The far-field expression (7.10) shows that when D covers the propagation region $\xi^2+\eta^2<c^{-2}$, the time-domain far-field pattern is given from (7.39) as

$$\mathcal{F}(\theta,\phi,t)=\frac{\cos\theta}{2\pi c Q_p(c^{-1}\cos\phi\sin\theta, c^{-1}\sin\phi\sin\theta)} \cdot\int\limits_{-\infty}^{+\infty}\int\limits_{-\infty}^{+\infty} b_p(\mathbf{r}_0, t+\hat{\mathbf{r}}\cdot\mathbf{r}_0/c)dx_0dy_0 \tag{7.42}$$

where $0\le\theta<\pi/2$, $\hat{\mathbf{r}}=\hat{\mathbf{x}}\cos\phi\sin\theta+\hat{\mathbf{y}}\sin\phi\sin\theta+\hat{\mathbf{z}}\cos\theta$, and $\mathbf{r}_0=\hat{\mathbf{x}}x_0+\hat{\mathbf{y}}y_0$. Hence, when the probe has the simple receiving characteristic (7.38), the far-field pattern of the test antenna is simply the integral over the scan plane of the probe output divided by the angular dependence of the receiving characteristic. The probe-corrected formula (7.42) for the far field of the test antenna possesses the simplicity of the formula (7.7) for the time-domain spectrum (or far-field pattern) in terms of the pressure measured by an ideal probe.

The time-derivative probe may also be characterized in the time domain by the time-domain receiving characteristic

$$R_p(\xi,\eta,t)=Q_p(\xi,\eta)\frac{\partial}{\partial t}\Phi^B(0,t,\xi,\eta)=Q_p(\xi,\eta)\delta'(t), \quad (\xi,\eta)\in D \tag{7.43}$$

where Φ^B is the time-domain plane wave defined by (5.98)–(5.99). Since $R_p(\xi,\eta,t)$ is defined as the output of the probe in a plane wave, (7.43) confirms that the output of a time-derivative probe in a plane wave is proportional to the time derivative of the plane-wave pressure field. The time-domain expression (7.43) can be found by taking the inverse Fourier transform of the frequency-domain result (7.38). Combining the definition (7.29) of the time-domain inverse receiving characteristic with the expression (7.43) of the receiving characteristic shows that

$$\frac{\partial}{\partial t}R_p^{-1}(\xi,\eta,t)=\frac{(2\pi)^2\delta(t)}{Q_p(\xi,\eta)}, \quad (\xi,\eta)\in D. \tag{7.44}$$

By taking the time derivative of the probe-corrected formula (7.31) and inserting this value for the time derivative of $R_p^{-1}(\xi,\eta,t)$, we recover the probe-corrected formula (7.39).

Let us again assume that the test antenna is turned on at $t = t_0$, so that the output of the probe $b_p(\mathbf{r}_0, t)$ is zero when $\mathbf{r}_0$ is outside the region $\mathcal{R}(t)$ in the scan plane. Then the integrand of (7.39) is zero when $\mathbf{r}_0$ is outside the region $\mathcal{R}(t + \xi x_0 + \eta y_0)$. For late times, when the finite region $\mathcal{R}(t)$ is approximately a circle of radius $c(t - t_0)$, this translates into: $b_p(\mathbf{r}_0, t) = 0$ for $|\mathbf{r}_0| > c(t - t_0) + c\xi x_0 + c\eta y_0$. When $\xi^2 + \eta^2 < c^{-2}$, this condition is satisfied for all $|\mathbf{r}_0|$ larger than some finite number, and the region of integration in (7.39) (and (7.42)) is finite. Consequently, when the receiving characteristic of the probe satisfies (7.38), with D covering the propagating region $\xi^2 + \eta^2 < c^{-2}$, the far-field pattern of the test antenna can be calculated from measured data taken in a *finite* scan plane without introducing any error into the calculation. This property of the special time-domain probe-corrected formula (7.42) is not possessed by any of the frequency-domain probe-corrected formulas and, as will be discussed in Chapter 8, makes it possible to use planar time-domain scanning for broadbeam antennas.

7.1.5 Formulas for Reciprocal Probes

Assume that the probe is a reciprocal electroacoustic transducer [170], whose electromagnetic waveguide supports just one propagating mode. We show that in this case there is a simple relation between the time-domain receiving characteristic $R_p(\xi, \eta, t)$ and the time-domain transmitting characteristic $T_p^\delta(\xi, \eta, t)$ of the probe. Here $T_p^\delta(\xi, \eta, t)$ is simply the spectrum (7.7) of the field radiated by the probe into the half space $z < 0$, when the reference point of the probe is at the origin and the probe is fed by the delta-function input signal $a_p(t) = 2\pi\delta(t)$; see Figure 7.1.

The field radiated by the probe into the half space $z < 0$ in this situation is denoted by $p_p^\delta(\mathbf{r}, t)$, and is given by the time-convolution integral (7.21) with $T(\xi, \eta, t)$ replaced by $T_p^\delta(\xi, \eta, t)$, and $\Phi^B(\mathbf{r}, t, \xi, \eta)$ replaced by the time-domain plane waves for the half space $z < 0$, given in (5.98) with z replaced by $-z$. The delta-function far-field pattern of the probe is given by the negative of (7.10); that is,

$$\mathcal{F}_p^\delta(\theta, \phi, t) = -\frac{\cos\theta}{c}\frac{\partial}{\partial t}T_p^\delta(c^{-1}\cos\phi\sin\theta, c^{-1}\sin\phi\sin\theta, t), \quad \pi/2 < \theta \le \pi. \tag{7.45}$$

The frequency-domain reciprocity relation (6.29) and the formula $\gamma = \omega\zeta$, which both hold for positive frequencies, imply that the receiving characteristic $R_{p\omega}(k_x, k_y)$ and the spectrum $T_{p\omega}^\delta(k_x, k_y)$ for all real (ξ, η) are related through

$$R_{p\omega}(\xi\omega, \eta\omega) = \frac{\zeta}{Y_p\rho_{m0}}T_{p\omega}^\delta(-\xi\omega, -\eta\omega), \quad \omega > 0 \tag{7.46}$$

where ρ_{m0} is the mass density of the undisturbed fluid and Y_p is the characteristic admittance for the propagating mode in the electromagnetic waveguide. For the TEM mode on a coaxial transmission line, Y_p is given in (6.7). The reciprocity relation (7.46) holds when the materials used to build the probe satisfy the conditions stated in [170, p. 27]. If the probe is not reciprocal, the receiving characteristic can be related to the transmitting characteristic of the adjoint probe [170, eq. (53c)].

Since the frequency spectrum f_ω of any real time function $f(t)$ must satisfy $f_\omega = f_{-\omega}^*$, the formula (7.46) proves that for all real frequencies we have

$$R_{p\omega}(\xi\omega, \eta\omega) = \begin{cases} \dfrac{\zeta}{Y_p\rho_{m0}}T_{p\omega}^\delta(-\xi\omega, -\eta\omega), & \xi^2 + \eta^2 < c^{-2} \\ i\dfrac{\omega}{|\omega|}\dfrac{|\zeta|}{Y_p\rho_{m0}}T_{p\omega}^\delta(-\xi\omega, -\eta\omega), & \xi^2 + \eta^2 > c^{-2} \end{cases}. \tag{7.47}$$

Taking the inverse Fourier transform of (7.47), and using the Hilbert transform formula (5.94), reveals the following time-domain reciprocity relation

$$R_p(\xi, \eta, t) = \begin{cases} \dfrac{\zeta}{Y_p \rho_{m0}} T_p^{\delta}(-\xi, -\eta, t), & \xi^2 + \eta^2 < c^{-2} \\ \dfrac{|\zeta|}{Y_p \rho_{m0}} \widetilde{\mathcal{H}} T_p^{\delta}(-\xi, -\eta, t), & \xi^2 + \eta^2 > c^{-2} \end{cases} \tag{7.48}$$

where $\widetilde{\mathcal{H}}$ denotes the Hilbert transform defined in (5.93). Note that the time-domain reciprocity relation (7.48) takes the simple form of the corresponding frequency-domain reciprocity relation (6.29) for $\xi^2 + \eta^2 < c^{-2}$ (propagating plane waves); but it has the more complicated form involving the Hilbert transform for $\xi^2 + \eta^2 > c^{-2}$ (evanescent plane waves).

The time-domain reciprocity relation (7.48) with $\xi = c^{-1} \cos\phi \sin\theta$ and $\eta = c^{-1} \sin\phi \sin\theta$ gives the following relation between the delta-function far-field pattern in (7.45) and the receiving characteristic of the probe

$$\frac{\partial}{\partial t} R_p(c^{-1} \cos\phi \sin\theta, c^{-1} \sin\phi \sin\theta, t) = \frac{\mathcal{F}_p^{\delta}(\pi - \theta, \phi + \pi, t)}{Y_p \rho_{m0}}, \quad 0 \le \theta < \pi/2. \tag{7.49}$$

Consequently, for reciprocal electroacoustic transducers, the delta-function far-field pattern is proportional to the time derivative of the time-domain receiving characteristic. Since the delta-function far-field pattern is equal to the time derivative of the step-function far-field pattern, it is seen that the step-function far-field pattern of a reciprocal probe is proportional to its time-domain receiving characteristic. Because the time-domain receiving characteristic $R_p(c^{-1} \cos\phi \sin\theta, c^{-1} \sin\phi \sin\theta, t)$ of the probe is defined as the output of the probe $b_p(\mathbf{r}, t)$ at $\mathbf{r} = 0$ in an incident plane wave $\delta(t - xc^{-1} \cos\phi \sin\theta - yc^{-1} \sin\phi \sin\theta - zc^{-1} \cos\theta)$, equation (7.49) implies that the far-field pattern in a given direction from a reciprocal transducer fed by an input pulse $2\pi a_p(t)$ is equal to the time derivative of the output of the transducer when it is illuminated by a plane wave incident from the opposite direction and with the time dependence $a_p(t)$.

Reciprocal Time-Derivative Probes. Finally let us consider a reciprocal electroacoustic time-derivative transducer whose receiving characteristic satisfies (7.43), with D covering the propagating region $\xi^2 + \eta^2 < c^{-2}$. Then (7.43) combines with (7.49) to produce

$$Q_p(c^{-1} \cos\phi \sin\theta, c^{-1} \sin\phi \sin\theta)\delta''(t) = \frac{\mathcal{F}_p^{\delta}(\pi - \theta, \phi + \pi, t)}{Y_p \rho_{m0}}, \quad 0 \le \theta < \pi/2. \tag{7.50}$$

Thus, the angular dependence of the delta-function far-field pattern is that of the receiving characteristic, and the time dependence of the delta-function far-field pattern is the second derivative of a delta function. Consequently, the far-field of a reciprocal, time-derivative probe is proportional to the second time derivative of the signal feeding the probe.

The formulas derived in this section allow us to calculate the receiving characteristic from the spectrum or the far-field pattern of the probe. If the probe is not reciprocal, the receiving characteristic is determined by the spectrum or the far-field pattern of the adjoint probe [170].

From its definition in Section 7.1.4, the time-domain receiving characteristic $R_p(\xi, \eta, t)$ of the probe for $\xi^2 + \eta^2 < c^{-2}$ can be determined by measuring its output at all orientations in an incident plane-wave pulse that has a constant frequency spectrum over the required usable bandwidth (so as to approximate a plane wave with a delta-function time dependence). The measurements could also be taken in the frequency domain to determine $R_{p\omega}(k_x, k_y)$ over the required usable bandwidth, and the time-domain receiving characteristic could be computed from the Fourier transform relationship (7.13).

7.2 ELECTROMAGNETIC FIELDS

The derivations of the probe-corrected formulas for electromagnetic fields are similar to the ones presented in Section 7.1 for acoustic fields. In Section 7.2.1 we obtain the time-domain probe-corrected formulas by Fourier transforming the frequency-domain formulas of Section 6.2. Then, in Section 7.2.2 we derive the probe-corrected formulas directly in the time domain. In Section 7.2.3, we present an alternative derivation that relies on the rectangular components of the vector output of the probe satisfying the wave equation. Formulas for time-derivative probes (so-called electromagnetic D-dot probes) are presented in Section 7.2.4, and it is shown that the probe-corrected formulas simplify greatly when such probes are used. In Section 7.2.5, the receiving characteristics for reciprocal probes are expressed in terms of spectra and far-field functions, and in Section 7.2.6 we consider the reciprocal elementary electric dipole probe whose output is proportional to the time derivative of the incident field. Finally, the probe-corrected formula for a reciprocal probe with a frequency-independent far-field pattern is derived in Section 7.2.7.

We begin by summarizing the relevant time-domain plane-wave spectrum formulas for the electric field, which allow us to calculate the electric field everywhere in the free-space region $z > 0$ in terms of the field in the plane $z = 0$. From (5.105) and (5.107) it follows that the time-domain electric field radiated by the test antenna in Figure 7.1 is given by

$$\mathbf{E}(\mathbf{r}, t) = -\frac{1}{2\pi} \iint_{\xi^2+\eta^2<c^{-2}} \frac{\partial^2}{\partial t^2} \mathbf{T}(\xi, \eta, t - \xi x - \eta y - \zeta z) d\xi d\eta$$

$$-\frac{1}{2\pi^2} \iint_{\xi^2+\eta^2>c^{-2}} \int_{-\infty}^{+\infty} \frac{\partial^2}{\partial t^2} \frac{z|\zeta|\,\mathbf{T}(\xi, \eta, t - t' - \xi x - \eta y)}{|\zeta|^2 z^2 + t'^2} dt' d\xi d\eta, \quad z > 0 \tag{7.51}$$

with ζ defined in (7.6) and

$$\mathbf{T}(\xi, \eta, t) = \frac{1}{2\pi} \int_{-\infty}^{+\infty} \int_{-\infty}^{+\infty} \mathbf{E}(\mathbf{r}_0, t + \xi x_0 + \eta y_0) dx_0 dy_0, \quad \mathbf{r}_0 = x_0 \hat{\mathbf{x}} + y_0 \hat{\mathbf{y}}. \tag{7.52}$$

Because $\nabla \cdot \mathbf{E} = 0$, the spectrum of the electric field satisfies the orthogonality relations (5.111) and (5.112), and since $\nabla \times \mathbf{E} = -\mu \partial \mathbf{H}/\partial t$, the spectrum for the magnetic field $\mathbf{T}^H(\xi, \eta, t)$ is determined by the spectrum for the electric field by (5.116) and (5.117). The time-domain far-field pattern for the electric field is defined such that

$$\mathbf{E}(\mathbf{r}, t) \sim \frac{\mathcal{F}(\theta, \phi, t - r/c)}{r}, \quad r \to \infty \tag{7.53}$$

and is given by the simple expression (5.188)

$$\mathcal{F}(\theta, \phi, t) = \frac{\cos\theta}{c} \frac{\partial}{\partial t} \mathbf{T}(c^{-1} \cos\phi \sin\theta, c^{-1} \sin\phi \sin\theta, t), \quad 0 \le \theta < \pi/2. \tag{7.54}$$

The equations (7.51) and (7.54) give the electric near and far fields in the half space $z > 0$ in terms of the plane-wave time-domain spectrum $\mathbf{T}(\xi, \eta, t)$. Corresponding expressions for the magnetic field and its spectrum $\mathbf{T}^H(\xi, \eta, t)$ hold, as explained in Section 5.2.

7.2.1 Formulas Obtained from the Fourier Transform

The probe-corrected plane-wave transmission formula for the near-field scanning of electromagnetic fields can be found by taking the inverse Fourier transform of (6.41) to get

$$\mathbf{b}_p(\mathbf{r}_0, t) = -\frac{1}{2\pi}\int_{-\infty}^{+\infty}\int_{-\infty}^{+\infty}\int_{-\infty}^{+\infty} \bar{\mathbf{R}}_p(\xi, \eta, t - t' - \xi x_0 - \eta y_0) \cdot \frac{\partial^2}{\partial t'^2}\mathbf{T}(\xi, \eta, t')dt'd\xi d\eta$$

$$= -\frac{1}{2\pi}\frac{\partial^2}{\partial t^2}\int_{-\infty}^{+\infty}\int_{-\infty}^{+\infty}\int_{-\infty}^{+\infty} \bar{\mathbf{R}}_p(\xi, \eta, t - t' - \xi x_0 - \eta y_0) \cdot \mathbf{T}(\xi, \eta, t')dt'd\xi d\eta. \tag{7.55}$$

Since $\mathbf{T}_\omega(\omega\xi, \omega\eta) = a_{0\omega}\mathbf{T}_{0\omega}(\omega\xi, \omega\eta)$, the time-domain spectra, defined through $\mathbf{T}(\xi, \eta, t) = \int_{-\infty}^{+\infty} \mathbf{T}_\omega(\omega\xi, \omega\eta)e^{-i\omega t}d\omega$ and $\mathbf{T}_0(\xi, \eta, t) = \int_{-\infty}^{+\infty} \mathbf{T}_{0\omega}(\omega\xi, \omega\eta)e^{-i\omega t}d\omega$, are related by the convolution integral

$$\mathbf{T}(\xi, \eta, t) = \frac{1}{2\pi}\int_{-\infty}^{+\infty} a_0(t - t')\mathbf{T}_0(\xi, \eta, t')dt'. \tag{7.56}$$

The time-domain dyadic receiving characteristic of the probe is defined from the inverse Fourier transform of (6.43) as

$$\bar{\mathbf{R}}_p(\xi, \eta, t) = \hat{\mathbf{x}}\mathbf{R}_p^x(\xi, \eta, t) + \hat{\mathbf{y}}\mathbf{R}_p^y(\xi, \eta, t) = \int_{-\infty}^{+\infty} \bar{\mathbf{R}}_{p\omega}(\omega\xi, \omega\eta)\, e^{-i\omega t}d\omega \tag{7.57}$$

and the time-domain vector output of the probe is found from (6.37) as

$$\mathbf{b}_p(\mathbf{r}_0, t) = b_p^x(\mathbf{r}_0, t)\hat{\mathbf{x}} + b_p^y(\mathbf{r}_0, t)\hat{\mathbf{y}} = \int_{-\infty}^{+\infty} \mathbf{b}_{p\omega}(\mathbf{r}_0)\, e^{-i\omega t}d\omega. \tag{7.58}$$

The inverse Fourier transform of (6.38) along with the definition of $\widetilde{\boldsymbol{\chi}}(\xi, \eta)$ in (5.92)

$$\widetilde{\boldsymbol{\chi}}(\xi, \eta) = \begin{cases} \xi\hat{\mathbf{x}} + \eta\hat{\mathbf{y}} + \zeta\hat{\mathbf{z}}, & \xi^2 + \eta^2 \le c^{-2} \\ \xi\hat{\mathbf{x}} + \eta\hat{\mathbf{y}} + |\zeta|\widetilde{\mathcal{H}}\hat{\mathbf{z}}, & \xi^2 + \eta^2 > c^{-2} \end{cases} \tag{7.59}$$

reveals that $\mathbf{R}_p^x(\xi, \eta, t)$ and $\mathbf{R}_p^y(\xi, \eta, t)$ satisfy $\widetilde{\boldsymbol{\chi}} \cdot \mathbf{R}_p^x(\xi, \eta, t) = 0$ and $\widetilde{\boldsymbol{\chi}} \cdot \mathbf{R}_p^y(\xi, \eta, t) = 0$.

Next take the inverse Fourier transform of $\mathbf{T}_\omega(\omega\xi, \omega\eta) = a_{0\omega}\mathbf{T}_{0\omega}(\omega\xi, \omega\eta)$ in (6.48) with $z_0 = 0$ to get the following three alternative time-domain probe-corrected formulas for the time-domain spectrum of the test antenna

$$\mathbf{T}(\xi, \eta, t) = \frac{1}{(2\pi)^3}\int_{-\infty}^{+\infty} \bar{\mathbf{R}}_p^{-1}(\xi, \eta, t') \cdot \int_{-\infty}^{+\infty}\int_{-\infty}^{+\infty} \mathbf{b}_p(\mathbf{r}_0, t - t' + \xi x_0 + \eta y_0)dx_0dy_0dt' \tag{7.60}$$

$$\mathbf{T}(\xi, \eta, t) = \frac{1}{(2\pi)^3}\int_{-\infty}^{+\infty}\int_{-\infty}^{+\infty}\int_{-\infty}^{+\infty} \bar{\mathbf{R}}_p^{-1}(\xi, \eta, t - t' + \xi x_0 + \eta y_0) \cdot \mathbf{b}_p(\mathbf{r}_0, t')dx_0dy_0dt' \tag{7.61}$$

$$\mathbf{T}(\xi, \eta, t) = \frac{1}{(2\pi)^3}\int_{-\infty}^{+\infty}\int_{-\infty}^{+\infty}\int_{-\infty}^{+\infty} \bar{\mathbf{R}}_p^{-1}(\xi, \eta, t - t' + \xi x_0 + \eta y_0) \cdot \mathbf{b}_p(\mathbf{r}_0, t')dt'dx_0dy_0 \tag{7.62}$$

where $\mathbf{r}_0 = x_0\hat{\mathbf{x}} + y_0\hat{\mathbf{y}}$ is a point in the scan plane. The inverse time-domain receiving characteristic of the probe is defined from (6.47) and (5.91) as

$$\bar{\mathbf{R}}_p^{-1}(\xi,\eta,t) = \int_{-\infty}^{+\infty} \frac{\boldsymbol{\chi}_\omega(\xi,\eta) \times \left[\mathbf{R}^y_{p\omega}(\omega\xi,\omega\eta)\hat{\mathbf{x}} - \mathbf{R}^x_{p\omega}(\omega\xi,\omega\eta)\hat{\mathbf{y}}\right]}{\boldsymbol{\chi}_\omega(\xi,\eta) \cdot \left[\mathbf{R}^y_{p\omega}(\omega\xi,\omega\eta) \times \mathbf{R}^x_{p\omega}(\omega\xi,\omega\eta)\right]} e^{-i\omega t} d\omega. \tag{7.63}$$

We can show from (6.44) that for all (ξ, η) the dyadic receiving characteristic and its inverse satisfy the relation

$$\int_{-\infty}^{+\infty} \bar{\mathbf{R}}_p^{-1}(\xi,\eta,t-t') \cdot \bar{\mathbf{R}}_p(\xi,\eta,t')dt' = (2\pi)^2 \left[\hat{\boldsymbol{\theta}}(\xi,\eta)\hat{\boldsymbol{\theta}}(\xi,\eta) + \hat{\boldsymbol{\phi}}(\xi,\eta)\hat{\boldsymbol{\phi}}(\xi,\eta)\right] \delta(t) \tag{7.64}$$

where $\hat{\boldsymbol{\theta}}(\xi,\eta)$ and $\hat{\boldsymbol{\phi}}(\xi,\eta)$ are time-independent unit vectors found from (6.45) and (6.46)

$$\hat{\boldsymbol{\theta}}(\xi,\eta) = \begin{cases} \dfrac{c\zeta\xi}{\sqrt{\xi^2+\eta^2}}\hat{\mathbf{x}} + \dfrac{c\zeta\eta}{\sqrt{\xi^2+\eta^2}}\hat{\mathbf{y}} - c\sqrt{\xi^2+\eta^2}\,\hat{\mathbf{z}}, & \xi^2+\eta^2 \le c^{-2} \\[2ex] \dfrac{c|\zeta|\xi\widetilde{\mathcal{H}}}{\sqrt{\xi^2+\eta^2}}\hat{\mathbf{x}} + \dfrac{c|\zeta|\eta\widetilde{\mathcal{H}}}{\sqrt{\xi^2+\eta^2}}\hat{\mathbf{y}} - c\sqrt{\xi^2+\eta^2}\,\hat{\mathbf{z}}, & \xi^2+\eta^2 > c^{-2} \end{cases} \tag{7.65}$$

and

$$\hat{\boldsymbol{\phi}}(\xi,\eta) = -\frac{\eta}{\sqrt{\xi^2+\eta^2}}\hat{\mathbf{x}} + \frac{\xi}{\sqrt{\xi^2+\eta^2}}\hat{\mathbf{y}} \tag{7.66}$$

with $\widetilde{\mathcal{H}}$ being the Hilbert transform (5.93). The expression for $\hat{\boldsymbol{\theta}}(\xi,\eta)$ is obtained in the following way. First note that (6.45) gives $\hat{\boldsymbol{\theta}}_\omega(\omega\xi,\omega\eta)$ for $\omega > 0$ and then use the fact that $\hat{\boldsymbol{\theta}}_{-\omega} = \hat{\boldsymbol{\theta}}^*_\omega$ to determine an expression valid for all real ω. This expression can then be Fourier transformed with the help of the Hilbert transform formula (5.94) to get (7.65). The expression for $\hat{\boldsymbol{\phi}}$ can be obtained directly from (6.46), which satisfies $\hat{\boldsymbol{\phi}}_{-\omega} = \hat{\boldsymbol{\phi}}^*_\omega$ and is therefore valid for all real ω. Note that $\widetilde{\mathcal{H}}^2 = -1$ so that $\hat{\boldsymbol{\theta}}(\xi,\eta)$ is indeed a unit vector for all real (ξ,η). The relation (7.64) will be used in Section 7.2.4 to determine the inverse receiving characteristics of time-derivative probes.

The formulas (7.60)–(7.62) express the time-domain spectrum for the field of the test antenna in terms of the inverse dyadic receiving characteristic and the vector output of the probe in the scan plane. They were derived under the assumption that multiple interactions between the test antenna and the probe can be neglected. That is, the fields scattered by the probe, rescattered by the test antenna, and returned to the probe do not affect the output of the probe. Equation (7.62) may be the most attractive for measurement purposes since the inner time integral can be calculated at each scan point $\mathbf{r}_0$ for the desired values of (ξ,η,t). Thus, one would not have to store the values of the vector output of the probe over time to calculate the spectrum of the electric field radiated by the test antenna.

The general time-domain probe-corrected formulas (7.60)–(7.62) require the evaluation of triple integrals and the determination of the inverse receiving characteristic of the probe. In Sections 7.2.4 and 7.2.7 we present results for certain types of probes for which the inverse receiving characteristic can be determined explicitly, and the probe-corrected formula (7.62) reduces to a double spatial integral over the scan plane.

7.2.2 Formulas Obtained Directly in the Time Domain

The probe-corrected time-domain transmission formula (7.55) and the formulas (7.60)–(7.62) for the time-domain plane-wave spectrum can also be derived directly starting from the definition of the time-domain receiving characteristic of the probe. The time-domain dyadic receiving characteristic $\bar{\mathbf{R}}_p(\xi, \eta, t)$ of the probe can be defined such that $\bar{\mathbf{R}}_p(\xi, \eta, t-\xi x_0-\eta y_0) \cdot \mathbf{e}(\xi, \eta)$ gives the vector output of the probe in (7.58) for an incident time-domain plane-wave electric field $\Phi^B(\mathbf{r}, t, \xi, \eta)\mathbf{e}(\xi, \eta)$ with $\widetilde{\boldsymbol{\chi}}(\xi, \eta) \cdot \mathbf{e}(\xi, \eta)\Phi^B(\mathbf{r}, t, \xi, \eta) = 0$, that is

$$\mathbf{b}_p(\mathbf{r}_0, t) = \bar{\mathbf{R}}_p(\xi, \eta, t - \xi x_0 - \eta y_0) \cdot \mathbf{e}(\xi, \eta). \tag{7.67}$$

The reference point of the probe is located at $\mathbf{r}_0 = x_0\hat{\mathbf{x}}+y_0\hat{\mathbf{y}}$, $\Phi^B(\mathbf{r}, t, \xi, \eta)$ is defined in (5.98)–(5.99), and $\widetilde{\boldsymbol{\chi}}(\xi, \eta)$ is defined in (7.59). Since $\widetilde{\boldsymbol{\chi}}(\xi, \eta) \cdot \mathbf{e}(\xi, \eta)\Phi^B(\mathbf{r}, t, \xi, \eta) = 0$, the dyadic receiving characteristic $\bar{\mathbf{R}}_p(\xi, \eta, t)$ can be made unique by requiring that $\widetilde{\boldsymbol{\chi}} \cdot \mathbf{R}_p^x(\xi, \eta, t) = 0$ and $\widetilde{\boldsymbol{\chi}} \cdot \mathbf{R}_p^y(\xi, \eta, t) = 0$. The expression (5.106) for the electric field radiated by the test antenna shows that when a linear probe located at $\mathbf{r}_0$ is illuminated by the test antenna, the total probe output determined by plane-wave superposition is given by

$$\mathbf{b}_p(\mathbf{r}_0, t) = -\frac{1}{2\pi}\int_{-\infty}^{+\infty}\int_{-\infty}^{+\infty}\int_{-\infty}^{+\infty} \bar{\mathbf{R}}_p(\xi, \eta, t - t' - \xi x_0 - \eta y_0) \cdot \frac{\partial^2}{\partial t'^2}\mathbf{T}(\xi, \eta, t')dt'd\xi d\eta \tag{7.68}$$

which agrees with (7.55) obtained from the inverse Fourier transform of (6.41).

Since (7.68) is in the form of an inverse space-time Radon transform of the time-domain plane-wave spectrum of the vector output of the probe

$$\mathbf{T}_b(\xi, \eta, t) = \int_{-\infty}^{+\infty} \bar{\mathbf{R}}_p(\xi, \eta, t - t') \cdot \mathbf{T}(\xi, \eta, t')dt' \tag{7.69}$$

we can take the Radon transform of (7.68) to obtain the relation

$$\int_{-\infty}^{+\infty} \bar{\mathbf{R}}_p(\xi, \eta, t - t') \cdot \mathbf{T}(\xi, \eta, t')dt' = \frac{1}{2\pi}\int_{-\infty}^{+\infty}\int_{-\infty}^{+\infty} \mathbf{b}_p(\mathbf{r}_0, t + \xi x_0 + \eta y_0)dx_0dy_0 \tag{7.70}$$

with $\mathbf{r}_0 = x_0\hat{\mathbf{x}} + y_0\hat{\mathbf{y}}$. To solve for $\mathbf{T}(\xi, \eta, t)$, the spectrum of the test antenna, take the dot product of (7.70) with the time-domain inverse dyadic receiving characteristic $\bar{\mathbf{R}}_p^{-1}(\xi, \eta, t''-t)$ and integrate the resulting equation over t from $-\infty$ to $+\infty$. The relationship (7.64) between $\bar{\mathbf{R}}_p^{-1}(\xi, \eta, t)$ and $\bar{\mathbf{R}}_p(\xi, \eta, t)$ then transforms (7.70) to

$$\mathbf{T}(\xi, \eta, t) = \frac{1}{(2\pi)^3}\int_{-\infty}^{+\infty} \bar{\mathbf{R}}_p^{-1}(\xi, \eta, t') \cdot \int_{-\infty}^{+\infty}\int_{-\infty}^{+\infty} \mathbf{b}_p(\mathbf{r}_0, t - t' + \xi x_0 + \eta y_0)dx_0dy_0dt' \tag{7.71}$$

which is equivalent to (7.60)–(7.62).

7.2.3 Derivation from the Wave Equation

In this section we present a wave-equation derivation, corresponding to the acoustic derivation in Section 7.1.3, of the electromagnetic probe-corrected formula (7.60). It was shown in Section 6.2.1 that when multiple interactions can be neglected, the frequency-domain vector output of the probe $\mathbf{b}_{p\omega}(\mathbf{r})$ in (6.37) satisfies the homogeneous Helmholtz equation (6.50) in the region $z > z_s - z_p$. Fourier transforming this frequency-domain result (6.50) produces the time-domain wave equation

$$\nabla^2 \mathbf{b}_p(\mathbf{r}, t) - \frac{1}{c^2}\frac{\partial^2}{\partial t^2}\mathbf{b}_p(\mathbf{r}, t) = 0, \quad z > z_s - z_p \tag{7.72}$$

where z_s is the maximum (more precisely, the supremum) value of z for which the source function of the test antenna is nonzero. That is, the physical body of the test antenna is located in the region $z < z_s$. Moreover, z_p is the minimum (more precisely, the infimum) value of z for which the source function of the adjoint probe is nonzero, when the adjoint probe is radiating into the region $z < 0$ with its reference point located at the origin. Since the physical dimensions of the probe and its adjoint are the same, the condition $z > z_s - z_p$ simply states that the test and probe antennas cannot touch during a scan. We assume that $z_s - z_p \leq 0$, so that $z = 0$ can be used as the scan plane and the homogeneous vector wave equation (7.72) is satisfied in the half space $z > 0$. Moreover, as explained in Section 6.2.1, for a probe that translates without rotation with respect to the test antenna, the probe output $\mathbf{b}_p(\mathbf{r}_0, t)$ satisfies the homogeneous wave equation (7.72) in the entire region outside a minimum convex closed surface formed by the locus of points $(z_s - z_p)$ for all orientations of the z axis; that is, in the region in which the minimum convex surfaces enclosing the test and probe antennas do not intersect. Also, (7.72) holds even if multiple interactions between the probe and test antennas are significant but are eliminated by time gating.

Through the equations (7.53) and (7.54), the time-domain electric far field can be expressed in terms of its time-domain spectrum as

$$\mathbf{E}(\mathbf{r}, t) \sim \frac{\cos\theta}{cr}\frac{\partial}{\partial t}\mathbf{T}(c^{-1}\cos\phi\sin\theta, c^{-1}\sin\phi\sin\theta, t - r/c). \tag{7.73}$$

Because the far field is locally a plane wave, the vector output of the probe, when the probe is located at the far-field point (r, θ, ϕ) and is illuminated by the electric field (7.73), is determined from linearity and the definition of the dyadic time-domain receiving characteristic $\bar{\mathbf{R}}_p(\xi, \eta, t)$ as

$$\begin{aligned}\mathbf{b}_p(\mathbf{r}, t) &\sim \frac{\cos\theta}{cr}\int_{-\infty}^{+\infty} \bar{\mathbf{R}}_p(c^{-1}\cos\phi\sin\theta, c^{-1}\sin\phi\sin\theta, t - t' - r/c) \\ &\qquad \cdot\frac{\partial}{\partial t'}\mathbf{T}(c^{-1}\cos\phi\sin\theta, c^{-1}\sin\phi\sin\theta, t')dt' \\ &= \frac{\cos\theta}{cr}\frac{\partial}{\partial t}\int_{-\infty}^{+\infty} \bar{\mathbf{R}}_p(c^{-1}\cos\phi\sin\theta, c^{-1}\sin\phi\sin\theta, t - t' - r/c) \\ &\qquad \cdot\mathbf{T}(c^{-1}\cos\phi\sin\theta, c^{-1}\sin\phi\sin\theta, t')dt'\end{aligned} \tag{7.74}$$

which is the time-domain version of the generalized Friis transmission formula [171, p. 6-8]. Equation (7.74) reveals that the x and y components of $\mathbf{b}_p(\mathbf{r}, t)$ have far-field expressions of the form $\mathcal{F}(\theta, \phi, t - r/c)/r$. Hence, the x and y components $\mathbf{b}_p(\mathbf{r}, t)$ satisfy the outgoing radiation condition at infinity, as well as the homogeneous wave equation, so $\mathbf{b}_p(\mathbf{r}, t)$ obeys the Green's function far-field relation (5.54)

$$\mathbf{b}_p(\mathbf{r}, t) \sim \frac{\cos\theta}{2\pi\, cr}\frac{\partial}{\partial t}\int_{-\infty}^{+\infty}\int_{-\infty}^{+\infty} \mathbf{b}_p(\mathbf{r}_0, t - r/c + \mathbf{r}_0\cdot\hat{\mathbf{r}})\, dx_0 dy_0, \quad z_s - z_p < 0. \tag{7.75}$$

Comparing the two far-field expressions (7.74) and (7.75) gives the relation

$$\int_{-\infty}^{+\infty} \bar{\mathbf{R}}_p(\xi, \eta, t - t') \cdot \mathbf{T}(\xi, \eta, t')dt' = \frac{1}{2\pi} \int_{-\infty}^{+\infty}\int_{-\infty}^{+\infty} \mathbf{b}_p(\mathbf{r}_0, t + \xi x_0 + \eta y_0)\, dx_0 dy_0, \quad \xi^2 + \eta^2 < c^{-2}, \quad z_s - z_p < 0. \tag{7.76}$$

To derive a formula for the time derivative of $\mathbf{T}(\xi, \eta, t)$, we multiply both sides of (7.76) by the inverse receiving characteristic $\bar{\mathbf{R}}_p^{-1}(\xi, \eta, t'' - t)$ and integrate the resulting equation with respect to t from $-\infty$ to $+\infty$. The relation (7.64) between $\bar{\mathbf{R}}_p^{-1}(\xi, \eta, t)$ and $\bar{\mathbf{R}}_p(\xi, \eta, t)$ then shows that

$$\mathbf{T}(\xi, \eta, t) = \frac{1}{(2\pi)^3} \int_{-\infty}^{+\infty} \bar{\mathbf{R}}_p^{-1}(\xi, \eta, t') \cdot \int_{-\infty}^{+\infty}\int_{-\infty}^{+\infty} \mathbf{b}_p(\mathbf{r}_0, t - t' + \xi x_0 + \eta y_0) dx_0 dy_0 dt' \tag{7.77}$$

which is equivalent to the probe-corrected formulas (7.60)–(7.62) for $\xi^2 + \eta^2 < c^{-2}$. Thus, we have derived an expression for the spectrum $\mathbf{T}(\xi, \eta, t)$ that is valid for propagating modes. From the discussion at the end of Section 7.1.3, it follows that the function $\mathbf{T}(\xi, \eta, t)$ is determined from (7.77) for $\xi^2 + \eta^2 > c^{-2}$ as well as for $\xi^2 + \eta^2 < c^{-2}$.

7.2.4 Formulas for D-Dot Probes

The general probe-corrected formulas (7.60)–(7.62) are rather cumbersome to use because they involve a time integration in addition to the double integration over the scan plane. Moreover, the general inverse dyadic receiving characteristic in (7.63) is laborious to determine from the receiving characteristic. It is therefore important to consider special probes for which the probe-corrected formulas simplify. In this section, we present results for so-called D-dot probes, which lead to considerably simplified probe-corrected formulas.

If the electromagnetic probe is a so-called D-dot probe [146], [178], [179], [180], its output from a plane wave is proportional to the time derivative of the plane-wave field, and its frequency-domain receiving characteristic is

$$\bar{\mathbf{R}}_{p\omega}(\omega\xi, \omega\eta) = -\frac{i\omega}{2\pi}\, \bar{\mathbf{Q}}_p(\xi, \eta) = -\frac{i\omega}{2\pi} \left[\hat{\mathbf{x}}\mathbf{Q}_p^x(\xi, \eta) + \hat{\mathbf{y}}\mathbf{Q}_p^y(\xi, \eta)\right], \quad (\xi, \eta) \in D \tag{7.78}$$

where the angularly dependent function $\bar{\mathbf{Q}}_p(\xi, \eta)$ is real and independent of frequency for $(\xi, \eta) \in D$, with D being a part of the (ξ, η) plane. Some probes, such as the reciprocal elementary electric dipole considered in Section 7.2.6 below, satisfy (7.78) for propagating plane waves $\xi^2 + \eta^2 < c^{-2}$, and satisfy (7.79) and (7.81) for all (ξ, η) if $\mathbf{Q}_p^x(\xi, \eta)$ and $\mathbf{Q}_p^y(\xi, \eta)$ are simple operators for the evanescent plane waves $\xi^2 + \eta^2 > c^{-2}$. Other probes may act as D-dot probes only for certain time-domain plane waves, and D is then only part of the (ξ, η) plane. Moreover, physically realizable probes only approximate the linear frequency dependence in (7.78) over a finite bandwidth.

The time-domain receiving characteristic corresponding to (7.78) is

$$\begin{aligned}\bar{\mathbf{R}}_p(\xi, \eta, t) &= \bar{\mathbf{Q}}_p(\xi, \eta)\, \frac{\partial}{\partial t}\Phi^B(0, t, \xi, \eta) = \bar{\mathbf{Q}}_p(\xi, \eta)\delta'(t) \\ &= \left[\hat{\mathbf{x}}\mathbf{Q}_p^x(\xi, \eta) + \hat{\mathbf{y}}\mathbf{Q}_p^y(\xi, \eta)\right]\delta'(t), \quad (\xi, \eta) \in D\end{aligned} \tag{7.79}$$

where the angularly dependent functions satisfy $\widetilde{\boldsymbol{\chi}}(\xi, \eta)\cdot\mathbf{Q}_p^x(\xi, \eta) = 0$ and $\widetilde{\boldsymbol{\chi}}(\xi, \eta)\cdot\mathbf{Q}_p^y(\xi, \eta) = 0$. Because $\bar{\mathbf{R}}_p(\xi, \eta, t)\cdot\mathbf{e}(\xi, \eta)$ is defined as the vector output of the probe when it is illuminated by the plane wave $\mathbf{e}(\xi, \eta)\Phi^B(0, t, \xi, \eta)$, (7.79) confirms that the output of a D-dot probe illuminated by a plane wave is a linear combination of the time derivative of the plane-wave electric field components.

By inserting the formula (7.79) into the relation (7.64), we find that the time derivative of the inverse dyadic receiving characteristic must satisfy

$$\frac{\partial}{\partial t}\bar{\mathbf{R}}_p^{-1}(\xi, \eta, t) \cdot \left[\hat{\mathbf{x}}\mathbf{Q}_p^x(\xi, \eta) + \hat{\mathbf{y}}\mathbf{Q}_p^y(\xi, \eta)\right]$$
$$= (2\pi)^2 \left[\hat{\boldsymbol{\theta}}(\xi, \eta)\hat{\boldsymbol{\theta}}(\xi, \eta) + \hat{\boldsymbol{\phi}}(\xi, \eta)\hat{\boldsymbol{\phi}}(\xi, \eta)\right] \delta(t), \quad (\xi, \eta) \in D. \tag{7.80}$$

Writing $\bar{\mathbf{R}}_p^{-1}(\xi, \eta, t)$ as $\alpha\hat{\mathbf{x}} + \beta\hat{\mathbf{y}}$ and using the facts that $\hat{\boldsymbol{\theta}}(\xi, \eta) \times \hat{\boldsymbol{\phi}}(\xi, \eta) = c\widetilde{\boldsymbol{\chi}}(\xi, \eta)$, $c\widetilde{\boldsymbol{\chi}}(\xi, \eta) \times \hat{\boldsymbol{\theta}}(\xi, \eta) = \hat{\boldsymbol{\phi}}(\xi, \eta)$, and $c\hat{\boldsymbol{\phi}}(\xi, \eta) \times \widetilde{\boldsymbol{\chi}}(\xi, \eta) = \hat{\boldsymbol{\theta}}(\xi, \eta)$, we can show that

$$\widetilde{\boldsymbol{\chi}}(\xi, \eta) \cdot \left[\mathbf{Q}_p^y(\xi, \eta) \times \mathbf{Q}_p^x(\xi, \eta)\right] \frac{\partial}{\partial t}\bar{\mathbf{R}}_p^{-1}(\xi, \eta, t)$$
$$= (2\pi)^2\, \widetilde{\boldsymbol{\chi}}(\xi, \eta) \times \left[\mathbf{Q}_p^y(\xi, \eta)\hat{\mathbf{x}} - \mathbf{Q}_p^x(\xi, \eta)\hat{\mathbf{y}}\right] \delta(t), \quad (\xi, \eta) \in D. \tag{7.81}$$

This relationship can also be derived by multiplying (6.47) by its denominator and taking the inverse Fourier transform. The formula (7.59) shows that in the evanescent region $\xi^2+\eta^2 > c^{-2}$ the vector $\widetilde{\boldsymbol{\chi}}$ involves the Hilbert transform. Therefore, in the evanescent region we cannot just divide (7.81) by $\widetilde{\boldsymbol{\chi}}(\xi, \eta) \cdot \left[\mathbf{Q}_p^y(\xi, \eta) \times \mathbf{Q}_p^x(\xi, \eta)\right]$ to get an expression for the inverse receiving characteristic. To solve (7.81) for the inverse receiving characteristic in the evanescent region, we must have additional information about the functions $\mathbf{Q}_p^x$ and $\mathbf{Q}_p^y$. (Equation (7.81) is used in Section 7.2.6 to determine the inverse receiving characteristic for all (ξ, η) of an elementary dipole probe.)

In the propagating region $\xi^2+\eta^2 < c^{-2}$, on the other hand, the vector $\widetilde{\boldsymbol{\chi}}$ does not involve any operators, and when $\widetilde{\boldsymbol{\chi}}(\xi, \eta) \cdot \left[\mathbf{Q}_p^y(\xi, \eta) \times \mathbf{Q}_p^x(\xi, \eta)\right] \neq 0$ equation (7.81) proves that

$$\frac{\partial}{\partial t}\bar{\mathbf{R}}_p^{-1}(\xi, \eta, t) = (2\pi)^2\, \bar{\mathbf{Q}}_p^{-1}(\xi, \eta)\, \delta(t), \quad (\xi, \eta) \in D_p \tag{7.82}$$

where

$$\bar{\mathbf{Q}}_p^{-1}(\xi, \eta) = \frac{\widetilde{\boldsymbol{\chi}}(\xi, \eta) \times \left[\mathbf{Q}_p^y(\xi, \eta)\hat{\mathbf{x}} - \mathbf{Q}_p^x(\xi, \eta)\hat{\mathbf{y}}\right]}{\widetilde{\boldsymbol{\chi}}(\xi, \eta) \cdot \left[\mathbf{Q}_p^y(\xi, \eta) \times \mathbf{Q}_p^x(\xi, \eta)\right]}, \quad (\xi, \eta) \in D_p \tag{7.83}$$

and D_p is that part of D in which $\xi^2 + \eta^2 < c^{-2}$. Inserting the inverse receiving characteristic (7.82) into the probe-corrected formula (7.62) proves that the time derivative of the spectrum of the test antenna is given by

$$\frac{\partial}{\partial t}\mathbf{T}(\xi, \eta, t) = \frac{1}{2\pi}\bar{\mathbf{Q}}_p^{-1}(\xi, \eta) \cdot \int_{-\infty}^{+\infty}\int_{-\infty}^{+\infty} \mathbf{b}_p(\mathbf{r}_0, t + \xi x_0 + \eta y_0) dx_0 dy_0, \quad (\xi, \eta) \in D_p \tag{7.84}$$

with $\mathbf{r}_0 = x_0\hat{\mathbf{x}} + y_0\hat{\mathbf{y}}$. Combining the expression (7.84) for the time derivative of the spectrum with the expression (7.54) for the far-field pattern, we see that when D covers the entire propagating region $\xi^2 + \eta^2 < c^{-2}$, the far-field pattern is given by

$$\mathcal{F}(\theta, \phi, t) = \frac{\cos\theta}{2\pi c}\, \bar{\mathbf{Q}}_p^{-1}(c^{-1}\cos\phi\sin\theta, c^{-1}\sin\phi\sin\theta)$$
$$\cdot \int_{-\infty}^{+\infty}\int_{-\infty}^{+\infty} \mathbf{b}_p(\mathbf{r}_0, t + \hat{\mathbf{r}} \cdot \mathbf{r}_0/c) dx_0 dy_0 \tag{7.85}$$

where $0 \leq \theta < \pi/2$, $\hat{\mathbf{r}} = \hat{\mathbf{x}}\cos\phi\sin\theta + \hat{\mathbf{y}}\sin\phi\sin\theta + \hat{\mathbf{z}}\cos\theta$, and $\mathbf{r}_0 = x_0\hat{\mathbf{x}} + y_0\hat{\mathbf{y}}$.

Thus, for D-dot probes, the probe-corrected formulas simplify such that only the calculation of a double integral over the scan plane is needed to find the spectrum $\mathbf{T}(\xi, \eta, t)$ in D_p or the far-field pattern $\mathcal{F}(\theta, \phi, t)$. This makes it very attractive to use D-dot probes for practical measurements. In Section 7.2.5 we show how the functions $\mathbf{Q}_p^x(\xi, \eta)$ and $\mathbf{Q}_p^y(\xi, \eta)$ can be determined from the far-field patterns of reciprocal D-dot probes. Practical aspects and examples of different types of probes, which can be useful for near-field measurements, have been discussed in numerous articles [178]–[191]. These references facilitate choosing probes for the practical implementation of planar near-field scanning in the time-domain.

A number of D-dot probes and their practical constructions have been described in the informative article by Baum et al. [178, sec. II-A]. In addition to detailed descriptions of the probes, the paper [178] lists numerous valuable references. Practical aspects as well as measurement results for D-dot probes have also been discussed by Shen, King, and Wu [181]. Theoretical considerations, which show that a thin cylindrical dipole of height h is a D-dot probe in the frequency range where $kh \leq 1$, have been given by King [182].

Examples of D-dot probes listed in the above-mentioned literature are

- The hollow spherical dipole
- The asymptotic conical dipole
- The flush plate dipole
- The conforming flush plate dipole
- The electrically thin long wire
- The V-conical monopole
- The asymptotic conical monopole.

7.2.5 Formulas for Reciprocal Probes

If the probe is reciprocal and is attached to a waveguide that supports just one propagating mode, we can derive time-domain reciprocity relations between the transmitting and receiving characteristics of the probe using the Fourier transform along with the frequency-domain reciprocity relations of Section 6.2.2. These reciprocity relations hold when the conditions in [8, p. 71, eq. (1.5-6)] are satisfied by the materials used to build the probe. In particular, they hold when these materials can be described by a scalar permittivity, a scalar permeability, and a scalar conductivity. For a nonreciprocal probe, the receiving characteristic can be related to the transmitting characteristic of the adjoint probe [8, p. 106].[3]

The time-domain transmitting characteristic $\mathbf{T}_p^{\delta x}(\xi, \eta, t)$ of the probe is defined to be

[3]As in the frequency domain (see Footnote 4 of Chapter 6), a very general time-domain reciprocity relation holds between the outputs of any two reciprocal antennas fed by the same input pulse. Let the test antenna in Figure 7.1 be fed by an input pulse $a_0(t)$, and denote by $b_p(t)$ the output pulse of the probe antenna terminated in a load with zero reflection coefficient so that $a_p(t) = 0$. (The relative position and orientation of the two antennas are fixed.) Next feed the probe antenna with an input pulse $a'_p(t)$ *that is identical to* $a_0(t)$, and denote by $b'_0(t)$ the output of the test antenna terminated in a load with zero reflection coefficient so that $a'_0(t) = 0$. Then in the frequency domain, $a_{0\omega} = a'_{p\omega}$, and (6.56) from Footnote 4 of Chapter 6 shows that for reciprocal electromagnetic antennas (or for reciprocal electroacoustic transducers)

$$I_{p\omega} = I'_{0\omega}. \tag{7.86}$$

Taking the inverse Fourier transform of (7.86) reveals the following general time-domain reciprocity relation between the output of each antenna when the input to the other antenna has the same waveform

$$I_p(t) = I'_0(t). \tag{7.87}$$

Like the frequency-domain "system 2-port" reciprocity relations (6.56) and (7.86), the time-domain "system 2-port" reciprocity relation (7.87) holds even if multiple interactions between the two reciprocal antennas are not negligible and if passive reciprocal scatterers are present.

the space-time Radon transform (7.52) of the field radiated by the probe into the half space $z < 0$ when: (i) the reference line of the probe is parallel to the x axis, (ii) the reference point of the probe is at the origin, and (iii) the probe is fed by the delta-function input signal $a_p(t) = 2\pi\delta(t)$. In the introduction to the present chapter, the time-domain fields in the waveguide were expressed in terms of $a_p(t)$. The frequency-domain reciprocity relation (6.57) shows that $\mathbf{R}^x_{p\omega}(\xi\omega, \eta\omega)$ for all real ω can be written like the acoustic formula (7.47)

$$\mathbf{R}^x_{p\omega}(\xi\omega, \eta\omega) = \begin{cases} \dfrac{\zeta}{\mu Y_p}\mathbf{T}^{\delta x}_{p\omega}(-\xi\omega, -\eta\omega), & \xi^2+\eta^2 < c^{-2} \\ i\dfrac{\omega}{|\omega|}\dfrac{|\zeta|}{\mu Y_p}\mathbf{T}^{\delta x}_{p\omega}(-\xi\omega, -\eta\omega), & \xi^2+\eta^2 > c^{-2}. \end{cases} \tag{7.88}$$

Taking the inverse Fourier transform of (7.88) and using the Hilbert transform formula (5.94) reveals the time-domain reciprocity relation

$$\mathbf{R}^x_p(\xi, \eta, t) = \begin{cases} \dfrac{\zeta}{\mu Y_p}\mathbf{T}^{\delta x}_p(-\xi, -\eta, t), & \xi^2+\eta^2 < c^{-2} \\ \dfrac{|\zeta|}{\mu Y_p}\widetilde{\mathcal{H}}\mathbf{T}^{\delta x}_p(-\xi, -\eta, t), & \xi^2+\eta^2 > c^{-2} \end{cases} \tag{7.89}$$

where $\widetilde{\mathcal{H}}$ is the Hilbert transform given in (5.93), Y_p is the frequency-independent characteristic admittance for the propagating mode in the waveguide feed, and μ is the permeability of free space. The characteristic admittance for the TEM mode on a coaxial transmission line is given in (6.7).

The delta-function far-field pattern of the probe is given by the negative of (7.54)

$$\boldsymbol{\mathcal{F}}^{\delta x}_p(\theta, \phi, t) = -\frac{\cos\theta}{c}\frac{\partial}{\partial t}\mathbf{T}^{\delta x}_p(c^{-1}\cos\phi\sin\theta, c^{-1}\sin\phi\sin\theta, t), \quad \pi/2 < \theta \le \pi \tag{7.90}$$

or from (7.89)

$$\frac{\partial}{\partial t}\mathbf{R}^x_p(c^{-1}\cos\phi\sin\theta, c^{-1}\sin\phi\sin\theta, t) = \frac{\boldsymbol{\mathcal{F}}^{\delta x}_p(\pi-\theta, \phi+\pi, t)}{\mu Y_p}, \quad 0 \le \theta < \pi/2. \tag{7.91}$$

Similarly, when the reference line of the probe is parallel to the y axis, we have

$$\mathbf{R}^y_p(\xi, \eta, t) = \begin{cases} \dfrac{\zeta}{\mu Y_p}\mathbf{T}^{\delta y}_p(-\xi, -\eta, t), & \xi^2+\eta^2 < c^{-2} \\ \dfrac{|\zeta|}{\mu Y_p}\widetilde{\mathcal{H}}\mathbf{T}^{\delta y}_p(-\xi, -\eta, t), & \xi^2+\eta^2 > c^{-2} \end{cases} \tag{7.92}$$

$$\boldsymbol{\mathcal{F}}^{\delta y}_p(\theta, \phi, t) = -\frac{\cos\theta}{c}\frac{\partial}{\partial t}\mathbf{T}^{\delta y}_p(c^{-1}\cos\phi\sin\theta, c^{-1}\sin\phi\sin\theta, t), \quad \pi/2 < \theta \le \pi \tag{7.93}$$

and

$$\frac{\partial}{\partial t}\mathbf{R}^y_p(c^{-1}\cos\phi\sin\theta, c^{-1}\sin\phi\sin\theta, t) = \frac{\boldsymbol{\mathcal{F}}^{\delta y}_p(\pi-\theta, \phi+\pi, t)}{\mu Y_p}, \quad 0 \le \theta < \pi/2. \tag{7.94}$$

Reciprocal D-Dot Probes. Let us assume that the probe is a reciprocal D-dot probe, with D covering the entire propagating region $\xi^2+\eta^2 < c^{-2}$. Then we find from the expression for the receiving characteristic (7.78), and the frequency-domain reciprocity relation (6.60),

that

$$\mathbf{Q}_p^x(c^{-1}\cos\phi\sin\theta, c^{-1}\sin\phi\sin\theta) = -\frac{2\pi}{\mu Y_p\omega^2}\mathcal{F}_{p\omega}^x(\pi-\theta,\pi+\phi), \quad 0\le\theta<\pi/2 \tag{7.95}$$

where $\mathcal{F}_{p\omega}^x(\theta,\phi)$ is the frequency-domain far-field pattern of the probe when it is located at the origin, has its reference line parallel to the x axis, and is fed by the unit-amplitude frequency-domain input signal $a_{p\omega}=1$. Note that in this situation the probe radiates into the half space $z<0$, and because of the special form of the receiving characteristic for the D-dot probe, the frequency-domain far-field pattern $\mathcal{F}_{p\omega}^x(\theta,\phi)$ is real and proportional to ω^2. That is, multiplying (7.95) by ω^2 and taking the inverse Fourier transform yields

$$\mathbf{Q}_p^x(c^{-1}\cos\phi\sin\theta, c^{-1}\sin\phi\sin\theta)\delta''(t) = \frac{\mathcal{F}_p^{\delta x}(\pi-\theta,\phi+\pi,t)}{\mu Y_p}, \quad 0\le\theta<\pi/2 \tag{7.96}$$

which can also be found by combining (7.91) with (7.79).

Similarly, we find that

$$\mathbf{Q}_p^y(c^{-1}\cos\phi\sin\theta, c^{-1}\sin\phi\sin\theta) = -\frac{2\pi}{\mu Y_p\omega^2}\mathcal{F}_{p\omega}^y(\pi-\theta,\pi+\phi), 0\le\theta<\pi/2 \tag{7.97}$$

and

$$\mathbf{Q}_p^y(c^{-1}\cos\phi\sin\theta, c^{-1}\sin\phi\sin\theta)\delta''(t) = \frac{\mathcal{F}_p^{\delta y}(\pi-\theta,\phi+\pi,t)}{\mu Y_p}, \quad 0\le\theta<\pi/2. \tag{7.98}$$

The formulas (7.95) and (7.97) express the functions $\mathbf{Q}_p^x(\xi,\eta)$ and $\mathbf{Q}_p^y(\xi,\eta)$ directly in terms of the frequency-domain far-field patterns of the probe in the two orientations. They enable us through (7.83) to determine the function $\bar{\mathbf{Q}}_p^{-1}(\xi,\eta)$ required in the time-domain probe-corrected formula (7.85) for the time-domain far-field pattern in terms of the time-domain measured output of the probe in two mutually perpendicular directions.

7.2.6 The Reciprocal Elementary Electric Dipole Probe

In this section we determine the time-domain receiving characteristic of the reciprocal elementary electric dipole probe, whose frequency-domain receiving characteristic was found in Section 6.2.3. It is shown for this dipole probe that the probe-corrected formulas reduce to the non-probe-corrected formulas of Chapter 5.

As in Section 6.2.3, the dipole probe is attached to a coaxial transmission line that supports only the TEM mode whose characteristic admittance Y_p is given in (6.7). We require that when the dipole is used as a transmitting antenna and is fed with an input signal $a_p(t)$, the magnitude of its dipole moment $p^d(t)$ has the same time dependence as $a_p(t)$. (Any time delay between the reference plane S_p in the coaxial line and the dipole can be eliminated by shifting the reference plane S_p to the base of the dipole.) We will show that this requirement is met by any dipole whose characteristic length is much shorter than the minimum wavelength occurring in the input signal $a_p(t)$. The minimum wavelength is $\lambda_{min}=2\pi c/\omega_{max}$, where ω_{max} is the maximum effective frequency occurring in the input signal $a_p(t)$. When the dipole is short, $V_{p\omega}$ and $I_{p\omega}$ occurring in (6.2) and (6.3) are the uniquely defined quasi-static circuit voltage and current, respectively. The voltage is given by $V_{p\omega}=q_{e\omega}/C$, where $q_{e\omega}$ is the separated electric charge on the dipole and C is the capacitance.

The dipole moment p_ω^d equals $\ell q_{e\omega}$, where ℓ is a characteristic length of the dipole, assumed to be much smaller than λ_{min}. In the limit $\ell\to 0$ the quasi-static analysis presented here is valid for all input signals $a_p(t)$, regardless of their bandwidths. The expression for the dipole moment in terms of $q_{e\omega}$ proves that $V_{p\omega} = p^d_\omega/(\ell C)$, so that the dipole moment has the same frequency dependence as $V_{p\omega}$. Since $I_{p\omega} = -i\omega q_{e\omega}$, we have the relation $I_{p\omega} = -i\omega C V_{p\omega} \approx 0$ for low frequencies. When $I_{p\omega} \approx 0$, (6.3) shows that $a_{p\omega} \approx b_{p\omega}$ so that $V_{p\omega} = a_{p\omega} + b_{p\omega} \approx 2a_{p\omega}$. Thus, we have the desired result that in the quasi-static approximation (short dipole), the ratio $p^d_\omega/a_{p\omega} \approx 2\ell C$ is independent of frequency.[4]

We begin by obtaining an expression for the probe output in terms of the electric field in the scan plane. When $a_p(t) = 2\pi\delta(t)$, the dipole current distribution is given by $\mathbf{J}^\delta(\mathbf{r}, t) = C_p\delta'(t)\delta(\mathbf{r})\hat{\mathbf{p}}$, where $C_p = 2\pi p^d_\omega/a_{p\omega} \approx 4\pi\ell C$ is a constant and $\hat{\mathbf{p}}$ is the direction of the dipole moment. The transmitting spectrum $\mathbf{T}^\delta_p(\xi, \eta, t)$ that describes the radiation of the dipole into the half space $z < 0$ can be determined by Fourier transforming (6.63) to get

$$\mathbf{T}^\delta_p(\xi, \eta, t) = \begin{cases} \dfrac{C_p}{4\pi\epsilon\zeta}\widetilde{\boldsymbol{\chi}}^-(\xi, \eta) \times \left(\widetilde{\boldsymbol{\chi}}^-(\xi, \eta) \times \hat{\mathbf{p}}\right)\delta'(t), & \xi^2 + \eta^2 < c^{-2} \\ -\dfrac{C_p\widetilde{\mathcal{H}}}{4\pi\epsilon|\zeta|}\widetilde{\boldsymbol{\chi}}^-(\xi, \eta) \times \left(\widetilde{\boldsymbol{\chi}}^-(\xi, \eta) \times \hat{\mathbf{p}}\right)\delta'(t), & \xi^2 + \eta^2 > c^{-2} \end{cases} \tag{7.99}$$

where $\widetilde{\boldsymbol{\chi}}^-(\xi, \eta)$ is the propagation vector for plane waves propagating into the half space $z < 0$, given by (7.59) with $\hat{\mathbf{z}}$ replaced by $-\hat{\mathbf{z}}$. Alternatively, we could derive (7.99) directly in the time domain by inserting the time-domain current distribution for the dipole into the analogs of (5.150), (5.152), and (5.153) that hold for radiation into the region $z < 0$.

With the fact that $\widetilde{\mathcal{H}}^2 = -1$, the reciprocity relation (7.89) shows that the vector receiving characteristic for this probe is

$$\mathbf{R}_p(\xi, \eta, t) = \frac{C_p}{4\pi\mu\epsilon Y_p}\widetilde{\boldsymbol{\chi}}(\xi, \eta) \times \left(\widetilde{\boldsymbol{\chi}}(\xi, \eta) \times \hat{\mathbf{p}}\right)\delta'(t) \tag{7.100}$$

which holds for all (ξ, η) and concurs with the frequency-domain formula (6.64). When the probe is located at $\mathbf{r}_0 = x_0\hat{\mathbf{x}} + y_0\hat{\mathbf{y}}$ and is illuminated by the general field (7.51) of the test antenna, the scalar output $b_p(\mathbf{r}_0, t)$ at the reference plane S_p of the coaxial transmission line is found from (7.55) as

$$\begin{aligned} b_p(\mathbf{r}_0, t) &= -\frac{1}{2\pi}\int_{-\infty}^{+\infty}\int_{-\infty}^{+\infty}\int_{-\infty}^{+\infty} \mathbf{R}_p(\xi, \eta, t - t' - \xi x_0 - \eta y_0) \cdot \frac{\partial^2}{\partial t'^2}\mathbf{T}(\xi, \eta, t')dt'd\xi d\eta \\ &= \frac{C_p}{8\pi^2c^2\epsilon\mu Y_p}\hat{\mathbf{p}} \cdot \frac{\partial}{\partial t}\int_{-\infty}^{+\infty}\int_{-\infty}^{+\infty}\frac{\partial^2}{\partial t^2}\mathbf{T}(\xi, \eta, t - \xi x_0 - \eta y_0)d\xi d\eta \\ &= -\frac{C_p}{4\pi Y_p}\hat{\mathbf{p}} \cdot \frac{\partial}{\partial t}\mathbf{E}(\mathbf{r}_0, t) \end{aligned} \tag{7.101}$$

where $\mathbf{E}(\mathbf{r}_0, t)$ is the field of the test antenna at $\mathbf{r}_0$. In deriving (7.101), we have used (5.108) to show that

$$\mathbf{E}(\mathbf{r}_0, t) = -\frac{1}{2\pi}\int_{-\infty}^{+\infty}\int_{-\infty}^{+\infty}\frac{\partial^2}{\partial t^2}\mathbf{T}(\xi, \eta, t - \xi x_0 - \eta y_0)d\xi d\eta, \quad \mathbf{r}_0 = x_0\hat{\mathbf{x}} + y_0\hat{\mathbf{y}} \tag{7.102}$$

[4] For a small magnetic dipole (current loop) the magnitude of its magnetic moment m_ω equals $I_{p\omega}S_A$, where S_A is the area of the loop. Thus, $I_{p\omega}$ has the same frequency dependence as m_ω, and in the quasi-static limit, $V_{p\omega} = -i\omega L I_{p\omega} \approx 0$ so that from (6.2) we have $a_{p\omega} \approx -b_{p\omega}$. This in turn proves that $I_{p\omega} = (a_{p\omega} - b_{p\omega})Y_p \approx 2Y_p a_{p\omega}$, and thus for a small current loop the ratio $m_\omega/a_{p\omega}$ is independent of frequency.

and the facts that $\widetilde{\chi} \cdot \mathbf{T} = 0$ and $\widetilde{\chi} \cdot \widetilde{\chi} = c^{-2}$. The time-domain expression (7.101), which can also be obtained by Fourier transforming the frequency-domain formula (6.65), shows that the elementary electric dipole probe measures the time derivative of the component of the incident electric field in the direction of the dipole moment at the point in space where the dipole is located.

If the dipole probe is oriented so that $\hat{\mathbf{p}} = \hat{\mathbf{x}}$ or $\hat{\mathbf{p}} = \hat{\mathbf{y}}$, we see from (7.100) that for all (ξ, η) the dyadic receiving characteristic is given by

$$\bar{\mathbf{R}}_p(\xi, \eta, t) = \frac{C_p}{4\pi\epsilon\mu Y_p}\left[\hat{\mathbf{x}}\widetilde{\chi}(\xi, \eta) \times \left(\widetilde{\chi}(\xi, \eta) \times \hat{\mathbf{x}}\right) + \hat{\mathbf{y}}\widetilde{\chi}(\xi, \eta) \times \left(\widetilde{\chi}(\xi, \eta) \times \hat{\mathbf{y}}\right)\right]\delta'(t) \tag{7.103}$$

and from (7.101) that the vector output of the probe is

$$\mathbf{b}_p(\mathbf{r}_0, t) = -\frac{C_p}{4\pi Y_p}\frac{\partial}{\partial t}\mathbf{E}^{tan}(\mathbf{r}_0, t) \tag{7.104}$$

where $\mathbf{E}^{tan}(\mathbf{r}_0, t) = \hat{\mathbf{x}} \cdot \mathbf{E}(\mathbf{r}_0, t)\hat{\mathbf{x}} + \hat{\mathbf{y}} \cdot \mathbf{E}(\mathbf{r}_0, t)\hat{\mathbf{y}}$ is the transverse electric field of the test antenna in the scan plane $z = 0$. The formula (7.104) reveals that the probe output is proportional to the time derivative of the tangential electric field of an arbitrary incident field. Hence, this elementary electric dipole is a D-dot probe for all time-domain plane waves.

Having determined the probe output, we can also derive an expression for the inverse dyadic receiving characteristic of the probe. Comparing (7.79) and (7.103) shows that[5]

$$\mathbf{Q}_p^x(\xi, \eta) = \frac{C_p}{4\pi\epsilon\mu Y_p}\widetilde{\chi}(\xi, \eta) \times \left(\widetilde{\chi}(\xi, \eta) \times \hat{\mathbf{x}}\right) \tag{7.105}$$

and

$$\mathbf{Q}_p^y(\xi, \eta) = \frac{C_p}{4\pi\epsilon\mu Y_p}\widetilde{\chi}(\xi, \eta) \times \left(\widetilde{\chi}(\xi, \eta) \times \hat{\mathbf{y}}\right). \tag{7.106}$$

Inserting these expressions for $\mathbf{Q}_p^x(\xi, \eta)$ and $\mathbf{Q}_p^y(\xi, \eta)$ into (7.81) proves that the inverse dyadic receiving characteristic satisfies the relation

$$\widetilde{\chi}(\xi, \eta) \cdot \hat{\mathbf{z}}\frac{\partial}{\partial t}\bar{\mathbf{R}}_p^{-1}(\xi, \eta, t) = \frac{16\pi^3 Y_p}{C_p}\widetilde{\chi}(\xi, \eta) \times \left[\hat{\mathbf{y}}\hat{\mathbf{x}} - \hat{\mathbf{x}}\hat{\mathbf{y}}\right]\delta(t). \tag{7.107}$$

The formula (7.59) for $\widetilde{\chi}$ and the fact that $\widetilde{\mathcal{H}}^2 = -1$ show that the time derivative of the inverse dyadic receiving characteristic is given by

$$\frac{\partial}{\partial t}\bar{\mathbf{R}}_p^{-1}(\xi, \eta, t) = \begin{cases} \dfrac{16\pi^3 Y_p}{\zeta C_p}\widetilde{\chi}(\xi, \eta) \times \left[\hat{\mathbf{y}}\hat{\mathbf{x}} - \hat{\mathbf{x}}\hat{\mathbf{y}}\right]\delta(t), & \xi^2 + \eta^2 < c^{-2} \\ -\dfrac{16\pi^3 Y_p \widetilde{\mathcal{H}}}{|\zeta| C_p}\widetilde{\chi}(\xi, \eta) \times \left[\hat{\mathbf{y}}\hat{\mathbf{x}} - \hat{\mathbf{x}}\hat{\mathbf{y}}\right]\delta(t), & \xi^2 + \eta^2 > c^{-2} \end{cases}. \tag{7.108}$$

Now that both the probe output and the inverse dyadic receiving characteristic are determined, we take the time derivative of the expression (7.62) for the spectrum of the test antenna

[5]For the evanescent spectrum $(\xi^2 + \eta^2 > c^{-2})$, $\mathbf{Q}_p^x(\xi, \eta)$ and $\mathbf{Q}_p^y(\xi, \eta)$ in (7.105) and (7.106) are operators involving the Hilbert transform. Thus, $\mathbf{Q}_p^x(\xi, \eta)$ and $\mathbf{Q}_p^y(\xi, \eta)$ for the reciprocal elementary electric dipole probe do not satisfy the frequency-domain relation (7.78) for $\xi^2 + \eta^2 > c^{-2}$ with $\mathbf{Q}_p^x(\xi, \eta)$ and $\mathbf{Q}_p^y(\xi, \eta)$ independent of frequency for all real ω. Still, it is a D-dot probe in that (i) it satisfies (7.79) and (7.81) for all (ξ, η) with $\mathbf{Q}_p^x(\xi, \eta)$ and $\mathbf{Q}_p^y(\xi, \eta)$ given by the operators in (7.105) and (7.106), and (ii) its output is proportional to the time derivative of the electric-field components of an arbitrary incident field.

and insert (7.104) and (7.108) to get

$$\frac{\partial}{\partial t}\mathbf{T}(\xi,\eta,t) = \frac{1}{2\pi}\int_{-\infty}^{+\infty}\int_{-\infty}^{+\infty}\frac{\partial}{\partial t}\Big[\mathbf{E}^{tan}(\mathbf{r}_0, t+\xi x_0+\eta y_0)$$

$$-\hat{\mathbf{z}}\,\zeta^{-1}(\xi\hat{\mathbf{x}}+\eta\hat{\mathbf{y}})\cdot\mathbf{E}^{tan}(\mathbf{r}_0, t+\xi x_0+\eta y_0)\Big]dx_0dy_0, \quad \xi^2+\eta^2 < c^{-2} \tag{7.109}$$

and

$$\frac{\partial}{\partial t}\mathbf{T}(\xi,\eta,t) = \frac{1}{2\pi}\int_{-\infty}^{+\infty}\int_{-\infty}^{+\infty}\frac{\partial}{\partial t}\Big[\mathbf{E}^{tan}(\mathbf{r}_0, t+\xi x_0+\eta y_0)$$

$$+\hat{\mathbf{z}}\,|\zeta|^{-1}(\xi\hat{\mathbf{x}}+\eta\hat{\mathbf{y}})\cdot\widetilde{\mathcal{H}}\mathbf{E}^{tan}(\mathbf{r}_0, t+\xi x_0+\eta y_0)\Big]dx_0dy_0, \quad \xi^2+\eta^2 > c^{-2}. \tag{7.110}$$

Since $\widetilde{\boldsymbol{\chi}}(\xi,\eta)\cdot\mathbf{T}(\xi,\eta,t) = 0$ for all (ξ,η), we find that the expressions (7.109) and (7.110) for the time derivative of the spectrum for the test antenna both equal the result obtained from the non-probe-corrected formula (7.52). Thus, we have demonstrated that the probe-corrected formulas reduce to the non-probe-corrected formulas when an elementary electric dipole probe is used to measure the field of the test antenna.

7.2.7 A Reciprocal Probe with a Frequency-Independent Far Field

We finally consider a reciprocal probe whose frequency-domain far-field pattern is real and independent of frequency when the probe is used as a transmitting antenna and fed with a frequency-domain signal of unit amplitude, that is $a_{p\omega} = 1$. As in the previous sections, $a_{p\omega}$ denotes the amplitude of a single propagating waveguide mode at the reference plane S_p in the electromagnetic waveguide attached to the probe. This means that for an arbitrary input pulse $a_p(t)$, the time dependence of the time-domain far-field pattern is the same as the time dependence of $a_p(t)$. A probe that approximates this behavior in the frequency range from 2 to 18 GHz has been reported in [192].

We denote the frequency-independent (over the measurement bandwidth) far-field pattern of the probe by $\mathcal{F}^{x}_{pB}(\theta,\phi)$ and $\mathcal{F}^{y}_{pB}(\theta,\phi)$ when a reference line attached to the probe is parallel to $\hat{\mathbf{x}}$ and $\hat{\mathbf{y}}$, respectively. Taking the Fourier transform of the formula (6.62), and noting that the frequency-domain far-field patterns $\mathcal{F}^{x}_{pB}(\theta,\phi)$ and $\mathcal{F}^{y}_{pB}(\theta,\phi)$ for the probe are real and independent of ω, shows that the time-domain far-field pattern of the test antenna is given by

$$\mathcal{F}(\theta,\phi,t) = -\sqrt{\frac{\mu}{\epsilon}}\,\frac{Y_p\cos\theta}{(2\pi c)^2}\,\frac{\hat{\mathbf{r}}\times\Big[\mathcal{F}^{y}_{pB}(\pi-\theta,\pi+\phi)\hat{\mathbf{x}} - \mathcal{F}^{x}_{pB}(\pi-\theta,\pi+\phi)\hat{\mathbf{y}}\Big]}{\hat{\mathbf{r}}\cdot\Big[\mathcal{F}^{x}_{pB}(\pi-\theta,\pi+\phi)\times\mathcal{F}^{y}_{pB}(\pi-\theta,\pi+\phi)\Big]}$$

$$\cdot\int_{-\infty}^{+\infty}\int_{-\infty}^{+\infty}\frac{\partial^2}{\partial t^2}\mathbf{b}_p(\mathbf{r}_0, t+\hat{\mathbf{r}}\cdot\mathbf{r}_0/c)dx_0dy_0 \tag{7.111}$$

where Y_p is the characteristic admittance of the propagating mode in the probe waveguide (given by (6.7) for the TEM mode on a coaxial transmission line), and $\hat{\mathbf{r}} = \hat{\mathbf{x}}\cos\phi\sin\theta + \hat{\mathbf{y}}\sin\phi\sin\theta + \hat{\mathbf{z}}\cos\theta$. Equation (7.111) shows that to calculate the far-field pattern of the test antenna from measurements taken with a probe whose far-field pattern is frequency independent over the measurement bandwidth, we have to take the second time derivative of the probe output.

8

Sampling Theorems and Computation Schemes

In Chapters 5 and 7, we derived formulas that express the acoustic and electromagnetic time-domain far fields as integrations of data over an infinite scan plane. The data consists of either the time derivative of the fields (Chapter 5) or the output of an arbitrary, known measurement probe (Chapter 7).

In the present chapter, we show that for bandlimited time-domain fields the infinite integrations over the scan plane can be converted into summations, which require knowledge only of the data in the scan plane at discrete points and at discrete times. Moreover, we investigate what errors are introduced into the far-field calculation when the scan plane is truncated to a finite size and the data in the scan plane is known only in a finite time interval. The results will be cast in the form of sampling theorems and computation schemes that determine time-domain far fields from space-time sampled values of the data in the scan plane.

Two different types of computation schemes will be presented: frequency-domain computation schemes and time-domain computation schemes. The frequency-domain computation schemes are presented in Section 8.1 and involve the following three steps. (1) Calculate the frequency-domain data in the scan plane from the corresponding time-domain (measured) data. (2) Calculate the frequency-domain far field from the frequency-domain data in the scan plane. (3) Calculate the time-domain far field from the frequency-domain far field. These schemes make use of the frequency-domain far-field formulas in Chapters 3 and 6, known frequency-domain sampling theorems, and the fast Fourier transform (FFT).

The time-domain computation schemes, which are presented in Section 8.2, use the formulas in Chapters 5 and 7 that determine the time-domain far fields directly from the time-domain data in the scan plane. Time-domain sampling theorems and reconstruction formulas are derived to determine the sample spacing between points in the scan plane, and the time-sampling intervals, required to calculate the far fields accurately.

In Section 8.3 the frequency-domain and time-domain computation schemes are used to calculate the far-field patterns of simple acoustic sources from data in a finite scan plane, and the two types of computation schemes are compared. Formulas are given for the number of operations required by each of the computation schemes to compute the far-field patterns.

The sources (test antenna), whose far fields we want to determine, are located in the half space $z < 0$ behind the finite scan plane at $z = 0$, as shown in Figure 8.1. (Here in Chapter 8,

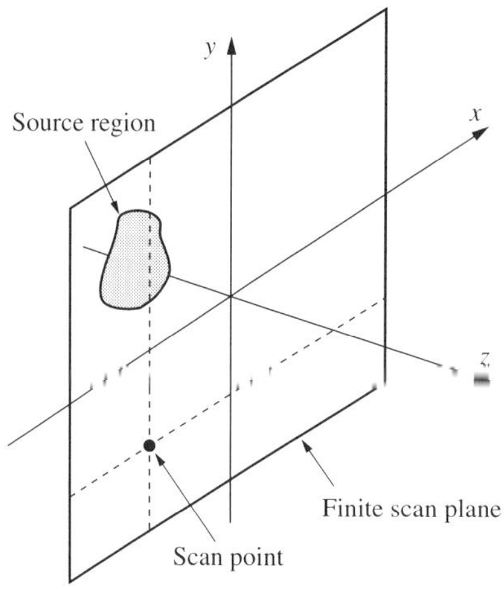

Figure 8.1 The source region (test antenna) located in the half space $z < 0$ behind a finite scan plane at $z = 0$.

as in Chapters 6 and 7, we shall refer to an acoustic radiator, as well as an electromagnetic radiator, as a test antenna.) The source-free region is lossless and has the same properties as in the previous chapters. Specifically, for the acoustic field the source-free region is a linear, quiescent, homogeneous, isotropic, lossless fluid with constant wave speed c and undisturbed mass density ρ_{m0}. For the electromagnetic field, the source-free region is described by the constant permeability μ and the constant permittivity ϵ, so that the constant "speed of light" is given by $c = 1/\sqrt{\epsilon\mu}$.

To ensure that the probe-corrected formulas are valid, it is assumed, as explained in Chapters 6 and 7, that multiple interactions between the test antenna and the probe are negligible. One advantage of making near-field measurements in the time domain is that multiple interactions can often be eliminated simply by time gating the output of the probe. We shall show in Section 8.3, that time gating also can often be used to eliminate finite-scan errors, that is, errors introduced by truncating the infinite scan plane to a finite area.

In the application of the probe-corrected formulas to the evaluation of far fields radiated by pulsed antennas, we assume that the near-field data in the scan plane of Figure 8.1 is obtained through measurements in the following way. With the probe antenna located at a scan point in the scan plane, the test antenna is fed with a pulse, and the resulting probe output is sampled in time. The probe is then moved to the next scan point in the scan plane. The test antenna is fed with another pulse (identical to the first one), and the resulting probe output is again sampled in time. This procedure is continued until the probe output at all scan points has been sampled. Thus, in principle, the number of pulses fed to the test antenna equals the number of scan points in the scan plane. In practice, of course, repetitive pulses may be used to reduce the measurement errors.

8.1 FREQUENCY-DOMAIN SAMPLING THEOREMS AND COMPUTATION SCHEMES

Before deriving the discrete formulas for the frequency-domain computation schemes, we present some general representations and sampling theorems for bandlimited time functions. A time function $f(t)$ is bandlimited with bandlimit ω_{max} when its Fourier transform

$$f_\omega = \frac{1}{2\pi}\int_{-\infty}^{+\infty} f(t)\, e^{i\omega t} dt \tag{8.1}$$

is zero for $|\omega| > \omega_{max}$. For $|\omega| < \omega_{max}$ we can expand f_ω in a Fourier series and find that the Fourier coefficients are determined from $f(t)$ evaluated at discrete times. Specifically, the Fourier expansion is

$$f_\omega = \frac{\Delta t}{2\pi} \sum_{l_t=-\infty}^{+\infty} f(l_t \Delta t)\, e^{il_t\omega\Delta t}, \quad |\omega| < \omega_{max} \tag{8.2}$$

where $\Delta t = \pi/\omega_{max}$. The assumption that $f_\omega = 0$ for $|\omega| > \omega_{max}$ implies that $f_\omega = 0$ for $|\omega| > \omega_1$ for any ω_1 greater than ω_{max}. Therefore, (8.2) is valid for any positive Δt which is less than or equal to π/ω_{max}. The representation (8.2) shows that f_ω for $|\omega| < \omega_{max}$ is determined from $f(t)$ sampled at the discrete points $t = l_t \Delta t$, where l_t can take any integer value.

The Fourier inversion formula

$$f(t) = \int_{-\omega_{max}}^{\omega_{max}} f_\omega\, e^{-i\omega t} d\omega \tag{8.3}$$

together with the theorem of Whittaker and Watson [88, sec. 5.31, p. 92] show that bandlimited functions with continuous spectra are entire functions of complex time. Therefore, such time functions cannot be zero in any time interval with nonzero length, unless they are zero for all times. However, in practice many bandlimited functions are significantly nonzero only in a time interval of the form $t_f < t < t_f + T$. Writing the approximate duration T as $(N_t - 1)\Delta t$, so that N_t is the number of time samples required to record the entire approximate duration of $f(t)$, and shifting the time-sample points from $l_t \Delta t$ to $t_f + l_t \Delta t$, the formula (8.2) becomes

$$f_\omega = \frac{\Delta t}{2\pi} \sum_{l_t=0}^{N_t-1} f(t_f + l_t \Delta t)\, e^{i\omega(t_f + l_t\Delta t)}, \quad |\omega| < \omega_{max}. \tag{8.4}$$

Equation (8.4) will be used in the frequency-domain computation schemes to compute the frequency-domain data in the scan plane from the corresponding time-domain data.

Now introduce a time function $g(t)$, which has bandlimit ω_{max} and is significantly nonzero only in the interval $t_g < t < t_g + T$. Then $g(t)$ can be expanded in a Fourier series for $t_g < t < t_g + T$, and we find that the Fourier coefficients are determined from g_ω evaluated at the discrete frequencies $l_\omega \Delta\omega$, where $\Delta\omega$ must be chosen less than or equal to $2\pi/T$. In order to use the FFT in the computation schemes, we choose $\Delta\omega = 2\pi/(N_t \Delta t) = 2\pi/(T + \Delta t)$, where $\Delta t = \pi/\omega_{max}$ is the usual time-sampling interval. Since $g_\omega = 0$ for $|\omega| > \omega_{max}$, the Fourier expansion for $g(t)$ contains only a finite number of terms so that

$$g(t) = \Delta\omega \sum_{l_\omega=-N_t/2}^{N_t/2} g_{l_\omega \Delta\omega}\, e^{-il_\omega \Delta\omega\, t}, \quad t_g < t < t_g + T \tag{8.5}$$

where N_t is assumed even. Note that $\Delta\omega = 2\omega_{max}/N_t$ and that the right-hand side of (8.5) is a periodic function with period $2\pi/\Delta\omega = T + \Delta t = N_t \Delta t$. The formula (8.5) will be used in the frequency-domain computation schemes to compute the time-domain far-field patterns from the frequency-domain far-field patterns.

8.1.1 Acoustic Fields Computed from the Pressure in the Scan Plane

Assume that the time-domain pressure is effectively bandlimited with bandlimit ω_{max}, and that the time derivative of the pressure is known in the scan plane. This data in the scan plane could be obtained by measuring the acoustic field in the scan plane with a probe (such as the acoustic monopole discussed in Section 7.1.4) whose output is the time derivative of the pressure. The first step in the frequency-domain computation scheme is to use the Fourier transform to calculate the frequency-domain pressure.

Let the pressure at a point in the scan plane begin at some time t_0 and end approximately at a later time $t_0 + T_0$. The initial time t_0 may depend on the position in the scan plane, whereas the duration T_0 is assumed to be the same for all scan points. By expressing the duration T_0 as $(N_t - 1)\Delta t$, with the number N_t of time samples chosen large enough for T_0 to cover the entire time duration of the signal at any point in the scan plane, the formula (8.4) gives

$$-i\omega p_\omega(\mathbf{r}_0) = \frac{\Delta t}{2\pi} \sum_{l_t=0}^{N_t-1} \frac{\partial p}{\partial t}(\mathbf{r}_0, t_0 + l_t \Delta t)\, e^{i\omega(t_0 + l_t \Delta t)}, \quad |\omega| < \omega_{max} \tag{8.6}$$

where $\Delta t = \pi/\omega_{max}$ is the time-sampling interval, and we have introduced the convention

$$\frac{\partial p}{\partial t}(\mathbf{r}, T) = \frac{\partial}{\partial t} p(\mathbf{r}, t)\bigg|_{t=T} \tag{8.7}$$

which will be used throughout this chapter. If the pressure itself, rather than its time derivative, is known in the scan plane, (8.6) is merely replaced by

$$p_\omega(\mathbf{r}_0) = \frac{\Delta t}{2\pi} \sum_{l_t=0}^{N_t-1} p(\mathbf{r}_0, t_0 + l_t \Delta t)\, e^{i\omega(t_0 + l_t \Delta t)}, \quad |\omega| < \omega_{max}. \tag{8.8}$$

The frequency-domain far-field pattern, defined such that the frequency-domain far-field pressure is

$$p_\omega(\mathbf{r}) \sim \mathcal{F}_\omega(\theta, \phi)\, \frac{e^{ikr}}{r}, \quad r \to \infty \tag{8.9}$$

can be determined from the frequency-domain pressure in the scan plane from (3.42) as

$$\mathcal{F}_\omega(\theta, \phi) = \frac{\cos\theta}{2\pi c} \int_{-\infty}^{+\infty}\int_{-\infty}^{+\infty} (-i\omega) p_\omega(\mathbf{r}_0) e^{-i\omega \hat{\mathbf{r}}\cdot\mathbf{r}_0/c} dx_0 dy_0 \tag{8.10}$$

where $\hat{\mathbf{r}} = \hat{\mathbf{x}} \cos\phi \sin\theta + \hat{\mathbf{y}} \sin\phi \sin\theta + \hat{\mathbf{z}} \cos\theta$ is the direction to the far-field observation point, and $\mathbf{r}_0 = x_0\hat{\mathbf{x}} + y_0\hat{\mathbf{y}}$ is a point in the scan plane. If the scan plane is located outside the reactive zone of the radiator, so that the evanescent plane-wave contributions are negligible, the pressure in the scan plane can be sampled with a maximum uniform sample spacing of approximately $\Delta x_0 = \Delta y_0 = \lambda_{min}/2$, where $\lambda_{min} = 2\pi c/\omega_{max}$ is the minimum effective wavelength occurring in the time-domain field. If the scan plane is located close enough to the radiator that evanescent plane-wave contributions cannot be neglected, the maximum uniform sampling spacing must be decreased accordingly [41], [42], [44].

Even if the scan plane does not lie outside the reactive zone of the fields at the lower frequencies, the sample spacing of $\lambda_{min}/2$ is less than $\lambda/2$ for $|\omega| < \omega_{max}$. Thus, usually the $\lambda_{min}/2$ sample spacing can adequately detect the spatial variation in the evanescent fields at the lower frequencies. This is confirmed in the numerical examples of Section 8.3. If there is an extraordinary amount of energy at a low frequency, for example, at a resonant frequency of the antenna structure, the spatial sample spacing may have to be decreased to accommodate the high spatial variation of the evanescent fields at this low frequency. Alternatively, the distance between the antenna and the scan plane can be increased to reduce the evanescent fields that illuminate the probe [44].

The sampling theorem allows the double integral in (8.10) to be replaced by an infinite double summation. Moreover, because actual measurements are taken on a finite scan area, the infinite double summation becomes finite, and we have for $0 \le \theta < \pi/2$ and $0 \le \phi \le 2\pi$

$$\begin{aligned}\mathcal{F}_\omega(\theta, \phi) &= \frac{\Delta x_0 \Delta y_0 \cos\theta}{2\pi c} \sum_{m=-N_x}^{N_x} \sum_{n=-N_y}^{N_y} (-i\omega) p_\omega(\mathbf{r}_{0mn})\, e^{-i\omega \hat{\mathbf{r}}\cdot\mathbf{r}_{0mn}/c} \\ &= \frac{\Delta x_0 \Delta y_0 \cos\theta}{2\pi c} \sum_{m=-N_x}^{N_x} e^{-i(\omega/c)m\Delta x_0 \cos\phi \sin\theta} \\ &\quad \cdot \sum_{n=-N_y}^{N_y} (-i\omega) p_\omega(\mathbf{r}_{0mn})\, e^{-i(\omega/c)n\Delta y_0 \sin\phi \sin\theta}\end{aligned} \tag{8.11}$$

where $\mathbf{r}_{0mn} = m\Delta x_0 \hat{\mathbf{x}} + n\Delta y_0 \hat{\mathbf{y}}$ is a sampling point in the scan plane and the integers N_x and N_y are determined by the size of the scan plane. Note that (8.11) holds for all ω because $p_\omega(\mathbf{r}) = 0$ for $|\omega| > \omega_{max}$.

We also note that by singling out the far-field phase factor $e^{i\omega r/c}$ from the radiated fields, Bucci and Franceschetti [50] have derived a method for reducing the effective spatial bandwidth of the radiated fields, thereby increasing the required uniform sampling increments beyond $\lambda/2$ if the scan plane is further from the test antenna's surface than half the maximum dimension of the test antenna. Thus, for scan planes further than this distance from the test antenna, the method could be applied to our frequency-domain computation scheme to increase the spatial sampling increments Δx_0 and Δy_0 to a value larger than $\lambda_{min}/2$.

The time-domain far-field pattern is defined such that the time-domain far-field pressure is

$$p(\mathbf{r}, t) \sim \frac{\mathcal{F}(\theta, \phi, t - r/c)}{r}, \qquad r \to \infty. \tag{8.12}$$

We assume that the far-field pattern is significantly nonzero only in the interval $t_\mathcal{F} < t < t_\mathcal{F} + T_\mathcal{F}$, where the initial time $t_\mathcal{F}$ may depend on (θ, ϕ), whereas the duration $T_\mathcal{F}$ is chosen large enough to be independent of (θ, ϕ). If the duration $T_\mathcal{F}$ of the far-field pattern is less than or equal to the duration T_0 of the field at any point in the scan plane, (8.5) shows that for $t_\mathcal{F} < t < t_\mathcal{F} + T_0$

$$\begin{aligned}\mathcal{F}(\theta, \phi, t) - \Delta\omega \sum_{l_\omega=-N_t/2}^{N_t/2} \mathcal{F}_{l_\omega \Delta\omega}(\theta, \phi)\, e^{-il_\omega \Delta\omega\, t} &= \Delta\omega\, e^{-i\omega_{max} t}\, \mathcal{F}_{\omega_{max}}(\theta, \phi) \\ &\quad + \Delta\omega\, e^{i\omega_{max} t} \sum_{l_\omega=0}^{N_t-1} \mathcal{F}_{(l_\omega - N_t/2)\Delta\omega}(\theta, \phi)\, e^{-il_\omega \Delta\omega\, t}\end{aligned} \tag{8.13}$$

where $\Delta\omega = 2\omega_{max}/N_t$, and N_t is assumed even. In (8.13) the far-field pattern is written in a form that is suitable for the FFT, and the values of $\mathcal{F}_\omega$ for negative ω can be obtained from the reality condition $\mathcal{F}_\omega = \mathcal{F}^*_{-\omega}$, where * indicates complex conjugation. Note that the far-field pattern computed from (8.13) is a periodic function with period $2\pi/\Delta\omega = T_0 + \Delta t$.

If $T_\mathcal{F} > T_0$, we have to decrease the frequency sample spacing $\Delta\omega$ by increasing N_t in (8.13) to avoid significant time-domain aliasing. If it is necessary to increase N_t in (8.13), we will also increase N_t in (8.6) by the same amount in order to compute the summation in (8.6) with the FFT. Specifically, the frequency sample spacing $\Delta\omega$ should be chosen less than or equal to $2\pi/T_\mathcal{F}$, so that the number of time samples becomes $N_t = \omega_{max}(T_\mathcal{F} + \Delta t)/\pi$; see (8.5). In Section 8.3 it will be shown that $T_\mathcal{F}$, in general, depends not only on the duration T_0 of the field in the scan plane, but also on the size of the scan plane. This is because the artificial edges of the finite scan plane produce a diffracted field that will make the time duration of the far-field pattern calculated from (8.11) longer than the duration of the exact far-field pattern (see also Section 8.3.5). The formulas (8.6), (8.11), and (8.13) constitute the frequency-domain computation scheme without probe correction for acoustic fields.

We shall now present this frequency-domain computation scheme in a more compact form that is easy to implement. Begin by defining A_{m,n,l_t} to be the time derivative of the acoustic pressure in the scan plane at the space-time point $(x, y, t) = (m\Delta x_0, n\Delta y_0, t_0 + l_t\Delta t)$, that is

$$A_{m,n,l_t} = \frac{\partial p}{\partial t}(\mathbf{r}_{0mn}, t_0 + l_t\Delta t) \tag{8.14}$$

where $\mathbf{r}_{0mn} = m\Delta x_0\hat{\mathbf{x}} + n\Delta y_0\hat{\mathbf{y}}$, $\Delta x_0 = \Delta y_0 = \lambda_{min}/2$, and $\Delta t = \pi/\omega_{max} = \Delta x_0/c = \Delta y_0/c$. The integers m, n, and l_t can have the following values: $m = -N_x, \ldots, 0, \ldots, N_x$, $n = -N_y, \ldots, 0, \ldots, N_y$, and $l_t = 0, 1, 2, \ldots, N_t - 1$. With this definition we find from (8.6), (8.11), and (8.13) that for $t_{\mathcal{F}} < t < t_{\mathcal{F}} + T_0$ the time-domain far-field pattern is given by

$$\mathcal{F}(\theta, \phi, t) = \frac{\Delta x_0 \Delta y_0 \cos\theta}{2\pi N_t c} \cdot \sum_{l_\omega=-N_t/2}^{N_t/2} \sum_{m=-N_x}^{N_x} \sum_{n=-N_y}^{N_y} \sum_{l_t=0}^{N_t-1} A_{m,n,l_t}\, e^{il_\omega\Delta\omega(t_0+l_t\Delta t-\hat{\mathbf{r}}\cdot\mathbf{r}_{0mn}/c-t)} \tag{8.15}$$

where $\Delta\omega = 2\omega_{max}/N_t$, and $\hat{\mathbf{r}} = \cos\phi\sin\theta\hat{\mathbf{x}} + \sin\phi\sin\theta\hat{\mathbf{y}} + \cos\theta\hat{\mathbf{z}}$. The formula (8.15) holds when the duration $T_{\mathcal{F}}$ of the calculated far-field pattern is less than or equal to the duration T_0 of the field at a point in the scan plane. If $T_{\mathcal{F}} > T_0$, the frequency spacing $\Delta\omega$ must be decreased, as discussed below (8.13). Using the identity

$$\sum_{l_\omega=-N_t/2}^{N_t/2} e^{il_\omega C} = \frac{\sin[C(N_t+1)/2]}{\sin[C/2]} \tag{8.16}$$

the summation over l_ω in (8.15) can be performed in closed form, so that

$$\mathcal{F}(\theta, \phi, t) = \frac{\Delta x_0 \Delta y_0 \cos\theta}{2\pi c} \cdot \sum_{m=-N_x}^{N_x} \sum_{n=-N_y}^{N_y} \sum_{l_t=0}^{N_t-1} \frac{\sin\left[\pi\left(\frac{t+\hat{\mathbf{r}}\cdot\mathbf{r}_{0mn}/c-t_0}{\Delta t} - l_t\right)\frac{N_t+1}{N_t}\right]}{N_t\sin\left[\pi\left(\frac{t+\hat{\mathbf{r}}\cdot\mathbf{r}_{0mn}/c-t_0}{\Delta t} - l_t\right)\frac{1}{N_t}\right]} A_{m,n,l_t}. \tag{8.17}$$

The alternative expression (8.17) for the time-domain far-field pattern will prove convenient when the fields computed by the frequency-domain and time-domain computation schemes are compared. However, equation (8.17) is not computationally efficient since it does not allow use of the FFT.

Number of Operations. We shall estimate the number of operations required to compute the time-domain far-field pattern with the frequency-domain computation scheme consisting of the three separate calculations given by (8.6), (8.11), and (8.13). According to Nussbaumer [193, p. 88], if N_t is chosen such that $N_t = 2^I$ where I is an integer, the FFT requires $O(N_t \log_2(N_t))$ operations to evaluate (8.6) for the following N_t values of ω: $\omega = l_\omega\Delta\omega$, $l_\omega = 0, 1, 2, \ldots, N_t - 1$ with $\Delta\omega = 2\omega_{max}/N_t$. (The values of the right-hand side of (8.6) with $\omega = l_\omega\Delta\omega$, $l_\omega = N_t/2+1, N_t/2+2, \ldots, N_t - 1$ can be used to determine p_ω for negative values of ω. Although the reality condition $p_{-\omega} = p^*_\omega$ can also be used to find p_ω for negative ω, there appears to be no significant reduction in the number of operations if p_ω is calculated only for $\omega = l_\omega \Delta\omega$, with $l_\omega = 0, 1, 2, \ldots, N_t/2$. In *Numerical Recipes* [194, p. 499], to minimize computation time it is recommended that only FFT algorithms be used that require N_t to be an integer power of 2. We can always obtain such a value of N_t by adding zeros to the summation.) The frequency-domain pressure in (8.6), at $(2N_x + 1)(2N_y + 1)$ scan points and N_t frequencies, can thus be computed by the FFT in $O(N_x N_y N_t \log_2(N_t))$ operations.

Next consider the calculation of the frequency-domain far-field pattern (8.11). We first show that the two-dimensional FFT is not well suited for computing the frequency-domain far-field pattern, if subsequently the time-domain far-field pattern is to be determined from (8.13). However, the two-dimensional FFT is well suited if only the frequency-domain far-field pattern is sought. To use the two-dimensional FFT in the far-field calculation, we shift the limits of the summations in (8.11) to range from 0 to $2(N_x + 1) - 1$ and $2(N_y + 1) - 1$, and add zeros to the summations to make $N_x + 1$ and $N_y + 1$ equal 2 raised to integers. Then the two-dimensional FFT requires $O(N_x N_y \log_2(N_x N_y))$ operations to calculate the frequency-domain far-field pattern (8.11) for a fixed ω at $4(N_x + 1)(N_y + 1)$ different angles of observation [193, p. 103]. These angles of observation $(\theta_{m'n'}, \phi_{m'n'})$ are determined by the FFT and satisfy the equations

$$\cos\phi_{m'n'} \sin\theta_{m'n'} = \frac{\pi c\, m'}{(N_x + 1)\omega\, \Delta x_0}, \qquad \sin\phi_{m'n'} \sin\theta_{m'n'} = \frac{\pi c\, n'}{(N_y + 1)\omega\, \Delta y_0} \tag{8.18}$$

where $m' = 0, 1, \ldots, 2N_x + 1$ and $n' = 0, 1, \ldots, 2N_y + 1$. The values of $\cos\phi_{m'n'} \sin\theta_{m'n'}$ and $\sin\phi_{m'n'} \sin\theta_{m'n'}$ for $m' = N_x + 2, N_x + 3, \ldots, 2N_x + 1$ and $n' = N_y + 2, N_y + 3, \ldots, 2N_y + 1$ determine $\mathcal{F}_\omega(\theta, \phi)$ in (8.11) for negative values of $\cos\phi$ and negative values of $\sin\phi$, respectively. The angles $(\theta_{m'n'}, \phi_{m'n'})$ determined by (8.18) are equal to real angles of observation (θ, ϕ) only for values of m' and n' that keep the right-hand side of each equation in (8.18) less than or equal to unity. We can solve the equations in (8.18) to find that

$$\sin\theta_{m'n'} = \frac{\pi c}{\omega} \sqrt{\frac{m'^2}{(N_x + 1)^2 (\Delta x_0)^2} + \frac{n'^2}{(N_y + 1)^2 (\Delta y_0)^2}}, \qquad \omega > 0 \tag{8.19}$$

$$\cos\phi_{m'n'} = \frac{m'}{\sqrt{m'^2 + n'^2 \dfrac{(N_x + 1)^2 (\Delta x_0)^2}{(N_y + 1)^2 (\Delta y_0)^2}}} \tag{8.20}$$

and

$$\sin\phi_{m'n'} = \frac{n'}{\sqrt{n'^2 + m'^2 \dfrac{(N_y + 1)^2 (\Delta y_0)^2}{(N_x + 1)^2 (\Delta x_0)^2}}}. \tag{8.21}$$

(Formulas similar to (8.19)–(8.21) are derived in [39].) To calculate the time-domain far-field pattern from (8.13), the frequency-domain far-field pattern must be determined for $\omega = l_\omega \Delta\omega$, with $l_\omega = 0, 1, 2, \ldots, N_t/2$ (since $\mathcal{F}_\omega = \mathcal{F}^*_{-\omega}$). The equations (8.19)–(8.21) show that the angles of observation $(\theta_{m'n'}, \phi_{m'n'})$ depend on ω, so the FFT determines the frequency-domain far-field patterns at angles of observation that depend on frequency. Hence, we cannot immediately use the formula (8.13) to calculate the time-domain far-field pattern from frequency-domain far-field patterns obtained from the FFT, because the frequency-domain far-field patterns are not determined for the same angles of observation.

By letting N_x, N_y, Δx_0, and Δy_0 be functions of ω, we could possibly make the angles in (8.19)–(8.21) independent of ω. However, with $N_x + 1$ and $N_y + 1$ equal to 2 raised to integer powers, we would have to let the sample spacings Δx_0 and Δy_0 be of the form $u\lambda_{min}/2$, where u is a noninteger. Thus, the frequency-domain pressure must be known at points in the scan plane that lie between the sample points $\mathbf{r}_{0mn}$. Several excellent interpolation schemes exist that could determine the pressure at these intermediate points [51], [195], but for our purposes they would significantly increase the complexity of the computation scheme. Therefore, we will not discuss the issue of interpolation further, and when the final goal is to determine the time-domain far-field pattern, we shall not use the FFT to compute the frequency-domain far-field pattern. If we only want to characterize the radiator in the frequency domain, it is not necessary to know the frequency-domain far-field patterns at the same angles of observation for different frequencies, and the FFT can then be used directly without performing any interpolation.

Without using the two-dimensional FFT, we will now estimate the number of operations required to compute the frequency-domain far-field pattern (8.11). In general, to calculate the double summation (8.11) for $N_{\theta\phi}$ different angles of observation and $N_t/2$ different frequencies requires $O(N_x N_y N_{\theta\phi} N_t)$ operations. However, we can significantly reduce the number of operations by considering angles of observation $(\theta_{m'n'}, \phi_{m'n'})$ that satisfy

$$\cos\phi_{m'n'} \sin\theta_{m'n'} = L_1(m'), \quad \sin\phi_{m'n'} \sin\theta_{m'n'} = L_2(n') \tag{8.22}$$

where $L_1(m')$ is a function of m' only and $L_2(n')$ is a function of n' only. Moreover, $m' = 1, 2, \ldots, (N_{\theta\phi})^{1/2}$ and $n' = 1, 2, \ldots, (N_{\theta\phi})^{1/2}$, so that $N_{\theta\phi}$ again is the total number of observation angles for which the far-field pattern is calculated. When the angles of observation satisfy (8.22), the summation over n in (8.11) is independent of m', and its calculation for all m and n' and for $N_t/2$ different frequencies takes $O(N_x N_y (N_{\theta\phi})^{1/2} N_t)$ operations. It takes $O(N_x N_{\theta\phi} N_t)$ operations to evaluate remaining summation over m for $N_t/2$ different frequencies, so the total number of operations it takes to evaluate the frequency-domain far-field pattern (8.11) for $N_{\theta\phi}$ different observation angles (θ, ϕ) satisfying (8.22) at $N_t/2$ different frequencies is $O(N_x N_y (N_{\theta\phi})^{1/2} N_t) + O(N_x N_{\theta\phi} N_t)$.

We can solve (8.22) to get the following explicit expressions for the angles of observation

$$\sin\theta_{m'n'} = \sqrt{[L_1(m')]^2 + [L_2(n')]^2} \tag{8.23}$$

and

$$\cos\phi_{m'n'} = \frac{L_1(m')}{\sqrt{[L_1(m')]^2 + [L_2(n')]^2}}, \quad \sin\phi_{m'n'} = \frac{L_2(n')}{\sqrt{[L_1(m')]^2 + [L_2(n')]^2}}. \tag{8.24}$$

Finally, to calculate the time-domain far-field pattern in (8.13) with the FFT for $N_{\theta\phi}$ different observation angles at N_t different times (with spacing Δt) requires $O(N_{\theta\phi} N_t \log_2(N_t))$ operations. Consequently, the time-domain far-field pattern can be computed at $N_{\mathcal{F}} = N_{\theta\phi} N_t$ different space-time points (θ, ϕ, t) by the frequency-domain computation scheme in a total number of operations given by

$$\begin{aligned} &O(N_x N_y N_t \log_2(N_t)) + O(N_x N_y (N_{\theta\phi})^{1/2} N_t) \\ &\qquad + O(N_x N_{\theta\phi} N_t) + O(N_{\theta\phi} N_t \log_2(N_t)). \end{aligned} \tag{8.25}$$

8.1.2 Electromagnetic Fields Computed from the Electric Field in the Scan Plane

Assume now that the time derivative of the tangential electric field is known in the scan plane. As discussed in Section 7.2.6, this information could be obtained by measuring the field in the scan plane with a reciprocal elementary dipole probe. Using the same notation as in Section 8.1.1, and assuming the same duration and bandlimit for the fields, we find that the formulas for the electromagnetic frequency-domain computation scheme are given as follows.

The frequency-domain electric field in the scan plane is determined from the sampling theorem (8.4) as

$$-i\omega\,\hat{\mathbf{z}}\times\mathbf{E}_\omega(\mathbf{r}_0)=\frac{\Delta t}{2\pi}\sum_{l_t=0}^{N_t-1}\hat{\mathbf{z}}\times\frac{\partial\mathbf{E}}{\partial t}(\mathbf{r}_0,t_0+l_t\Delta t)\,e^{i\omega(t_0+l_t\Delta t)},\qquad|\omega|<\omega_{max}\tag{8.26}$$

where

$$\frac{\partial\mathbf{E}}{\partial t}(\mathbf{r},T)=\left.\frac{\partial}{\partial t}\mathbf{E}(\mathbf{r},t)\right|_{t=T}.\tag{8.27}$$

The frequency-domain far-field pattern is found by applying the sampling theorem to (3.44) to get

$$\begin{aligned}\mathcal{F}_\omega(\theta,\phi)=&-\frac{\Delta x_0\Delta y_0}{2\pi c}\,\hat{\mathbf{r}}\times\sum_{m=-N_x}^{N_x}e^{-i(\omega/c)m\Delta x_0\cos\phi\sin\theta}\\&\cdot\sum_{n=-N_y}^{N_y}(-i\omega)\,\hat{\mathbf{z}}\times\mathbf{E}_\omega(\mathbf{r}_{0mn})\,e^{-i(\omega/c)n\Delta y_0\sin\phi\sin\theta}\end{aligned}\tag{8.28}$$

which is valid for $0\le\theta<\pi/2$, $0\le\phi\le2\pi$, and all ω. The time-domain far-field pattern for $t_{\mathcal{F}}<t<t_{\mathcal{F}}+T_0$ is computed from the sampling theorem (8.5) as

$$\begin{aligned}\mathcal{F}(\theta,\phi,t)&=\Delta\omega\sum_{l_\omega=-N_t/2}^{N_t/2}\mathcal{F}_{l_\omega\Delta\omega}(\theta,\phi)\,e^{-il_\omega\Delta\omega t}\\&=\Delta\omega\,e^{-i\omega_{max}t}\,\mathcal{F}_{\omega_{max}}(\theta,\phi)\\&\quad+\Delta\omega\,e^{i\omega_{max}t}\sum_{l_\omega=0}^{N_t-1}\mathcal{F}_{(l_\omega-N_t/2)\Delta\omega}(\theta,\phi)\,e^{-il_\omega\Delta\omega t}\end{aligned}\tag{8.29}$$

when the duration $T_{\mathcal{F}}$ of the calculated far-field pattern is less than or equal to the duration T_0 of the field at a point in the scan plane. If $T_{\mathcal{F}}>T_0$, we must increase N_t as discussed below (8.13).

With

$$\mathbf{A}_{m,n,l_t}=\hat{\mathbf{z}}\times\frac{\partial\mathbf{E}}{\partial t}(\mathbf{r}_{0mn},t_0+l_t\Delta t)\tag{8.30}$$

the far-field pattern can be expressed as

$$\begin{aligned}\mathcal{F}(\theta,\phi,t)=&-\frac{\Delta x_0\Delta y_0}{2\pi N_t c}\\&\cdot\sum_{l_\omega=-N_t/2}^{N_t/2}\sum_{m=-N_x}^{N_x}\sum_{n=-N_y}^{N_y}\sum_{l_t=0}^{N_t-1}\hat{\mathbf{r}}\times\mathbf{A}_{m,n,l_t}\,e^{il_\omega\Delta\omega(t_0+l_t\Delta t-\hat{\mathbf{r}}\cdot\mathbf{r}_{0mn}/c-t)}.\end{aligned}\tag{8.31}$$

The number of operations required by the frequency-domain computation scheme to compute the time-domain far-field pattern at $N_{\mathcal{F}}=N_{\theta\phi}N_t$ different space-time points (θ,ϕ,t) is given in (8.25) when (θ,ϕ) satisfies (8.22).

8.1.3 Acoustic Fields Computed from the Probe Output in the Scan Plane

Assume that the acoustic field in the scan plane is measured with a known time-derivative probe, as described in Section 7.1.4. The receiving characteristic of such a probe is in the form of (7.38), and we assume that the acoustic field has the same bandlimits and time duration as in Section 8.1.1. The scan plane is far enough from the acoustic radiator (test antenna) that the probe and test antennas lie outside each other's reactive fields.

Then the frequency-domain output of the probe is calculated from the sampling theorem (8.4) as

$$b_{p\omega}(\mathbf{r}_0) = \frac{\Delta t}{2\pi} \sum_{l_t=0}^{N_t-1} b_p(\mathbf{r}_0, t_0 + l_t \Delta t)\, e^{i\omega(t_0 + l_t \Delta t)}, \quad |\omega| < \omega_{max}. \tag{8.32}$$

By combining the probe-corrected formula (6.17) with the far-field result (6.13), and inserting the expression (7.38) for the receiving characteristic, we find that the frequency-domain far-field pattern $\mathcal{F}_\omega(\theta, \phi) = a_{0\omega}\mathcal{F}_{0\omega}(\theta, \phi)$ is given in terms of the frequency-domain probe output as

$$\mathcal{F}_\omega(\theta, \phi) = \frac{\cos\theta}{2\pi c Q_p(c^{-1}\cos\phi\sin\theta, c^{-1}\sin\phi\sin\theta)} \cdot \int_{-\infty}^{+\infty}\int_{-\infty}^{+\infty} b_{p\omega}(\mathbf{r}_0)\, e^{-i\omega\hat{\mathbf{r}}\cdot\mathbf{r}_0/c} dx_0 dy_0 \tag{8.33}$$

with $\mathbf{r}_0 = x_0\hat{\mathbf{x}} + y_0\hat{\mathbf{y}}$. Repeating the derivation that led from (8.10) to (8.11) yields

$$\mathcal{F}_\omega(\theta, \phi) = \frac{\Delta x_0 \Delta y_0 \cos\theta}{2\pi c Q_p(c^{-1}\cos\phi\sin\theta, c^{-1}\sin\phi\sin\theta)} \cdot \sum_{m=-N_x}^{N_x} e^{-i(\omega/c)m\Delta x_0\cos\phi\sin\theta} \sum_{n=-N_y}^{N_y} b_{p\omega}(\mathbf{r}_{0mn})\, e^{-i(\omega/c)n\Delta y_0\sin\phi\sin\theta} \tag{8.34}$$

which is valid for $0 \le \theta < \pi/2$, $0 \le \phi \le 2\pi$, and all ω. The time-domain far-field pattern is then found from (8.13) when the duration of the calculated far-field pattern is equal to or shorter than the duration of the field at a point in the scan plane. If the duration of the calculated far-field pattern is longer, we must decrease the frequency spacing $\Delta\omega$ as explained below (8.13) in Section 8.1.1.

By defining b_{m,n,l_t} to be the output of the probe in the scan plane at the space-time point $(x, y, t) = (m\Delta x_0, n\Delta y_0, t_0 + l_t\Delta t)$, that is,

$$b_{m,n,l_t} = b_p(\mathbf{r}_{0mn}, t_0 + l_t\Delta t) \tag{8.35}$$

we find from (8.32) and (8.34) that the time-domain far-field pattern is given by

$$\mathcal{F}(\theta, \phi, t) = \frac{\Delta x_0 \Delta y_0 \cos\theta}{2\pi c N_t Q_p(c^{-1}\cos\phi\sin\theta, c^{-1}\sin\phi\sin\theta)} \cdot \sum_{l_\omega=-N_t/2}^{N_t/2} \sum_{m=-N_x}^{N_x} \sum_{n=-N_y}^{N_y} \sum_{l_t=0}^{N_t-1} b_{m,n,l_t}\, e^{il_\omega\Delta\omega(t_0 + l_t\Delta t - \hat{\mathbf{r}}\cdot\mathbf{r}_{0mn}/c - t)}. \tag{8.36}$$

The number of operations required by the frequency-domain computation scheme to compute the time-domain far-field pattern at $N_{\mathcal{F}} = N_{\theta\phi}N_t$ different space-time points (θ, ϕ, t) is given in (8.25) when (θ, ϕ) satisfies (8.22).

8.1.4 Electromagnetic Fields Computed from the Probe Output in the Scan Plane

Let the electric field in the scan plane be measured with a D-dot probe as described in Section 7.2.4. The receiving characteristic of such a probe is in the form (7.78), so that the frequency-domain inverse dyadic receiving characteristic is

$$\bar{\mathbf{R}}_{p\omega}^{-1}(k_x, k_y) = \frac{2\pi i}{\omega} \bar{\mathbf{Q}}_p^{-1}(k_x/\omega, k_y/\omega) \tag{8.37}$$

where $\bar{\mathbf{Q}}_p^{-1}(\xi, \eta)$ is given by (7.83).

When the electric field has the same bandlimit and duration as the acoustic field in Section 8.1.1, the frequency-domain probe output can be computed from

$$\mathbf{b}_{p\omega}(\mathbf{r}_0) = \frac{\Delta t}{2\pi} \sum_{l_t=0}^{N_t-1} \mathbf{b}_p(\mathbf{r}_0, t_0 + l_t \Delta t)\, e^{i\omega(t_0+l_t\Delta t)}, \quad |\omega| < \omega_{max}. \tag{8.38}$$

Again, we are assuming that the probe and test antennas are far enough apart that they lie outside each other's reactive fields. Using the expression (8.37) for the inverse receiving characteristic in the formula (6.59) and recalling that $\mathcal{F}_\omega(\theta, \phi) = a_{0\omega}\mathcal{F}_{0\omega}(\theta, \phi)$, we find that the frequency-domain far-field pattern for $0 \le \theta < \pi/2$, $0 \le \phi \le 2\pi$, and all ω is

$$\begin{aligned}\mathcal{F}_\omega(\theta, \phi) = {} & \frac{\Delta x_0 \Delta y_0 \cos\theta}{2\pi c} \bar{\mathbf{Q}}_p^{-1}(c^{-1}\cos\phi\sin\theta, c^{-1}\sin\phi\sin\theta) \\ & \cdot \sum_{m=-N_x}^{N_x} e^{-i(\omega/c)m\Delta x_0 \cos\phi\sin\theta} \sum_{n=-N_y}^{N_y} \mathbf{b}_{p\omega}(\mathbf{r}_{0mn})\, e^{-i(\omega/c)n\Delta y_0 \sin\phi\sin\theta}.\end{aligned} \tag{8.39}$$

The time-domain far-field pattern is determined from (8.29) if the far-field pattern has the same or shorter duration as the field at a point in the scan plane. If the time duration of the calculated far-field pattern is longer, the frequency spacing $\Delta\omega$ must be decreased as described below (8.13) in Section 8.1.1.

The discrete formulas (8.38)–(8.39) can be written in a more compact form by first defining

$$\mathbf{b}_{m,n,l_t} = \mathbf{b}_p(\mathbf{r}_{0mn}, t_0 + l_t \Delta t) \tag{8.40}$$

and then combining (8.38) and (8.39) to get

$$\begin{aligned}\mathcal{F}(\theta, \phi, t) = {} & \frac{\Delta x_0 \Delta y_0 \cos\theta}{2\pi c N_t} \bar{\mathbf{Q}}_p^{-1}(c^{-1}\cos\phi\sin\theta, c^{-1}\sin\phi\sin\theta) \\ & \cdot \sum_{l_\omega=-N_t/2}^{N_t/2} \sum_{m=-N_x}^{N_x} \sum_{n=-N_y}^{N_y} \sum_{l_t=0}^{N_t-1} \mathbf{b}_{m,n,l_t}\, e^{il_\omega\Delta\omega(t_0+l_t\Delta t - \hat{\mathbf{r}}\cdot\mathbf{r}_{0mn}/c - t)}.\end{aligned} \tag{8.41}$$

The number of operations required by the frequency-domain computation scheme to compute the time-domain far-field pattern at $N_\mathcal{F} = N_{\theta\phi}N_t$ different space-time points (θ, ϕ, t) is given in (8.25) when (θ, ϕ) satisfies (8.22).

8.2 TIME-DOMAIN SAMPLING THEOREMS AND COMPUTATION SCHEMES

Before presenting the formulas for the time-domain computation schemes, we derive the standard sampling theorem and reconstruction formula for bandlimited time functions. Let the time function $f(t)$ be bandlimited with bandlimit ω_{max} so that $f_\omega = 0$ for $|\omega| > \omega_{max}$. Inserting the expression (8.2) for f_ω into the Fourier integral (8.3) proves the reconstruction formula

$$f(t) = \sum_{l_t=-\infty}^{+\infty} \frac{\sin\left[\pi\left(\frac{t}{\Delta t} - l_t\right)\right]}{\pi\left(\frac{t}{\Delta t} - l_t\right)} f(l_t\Delta t) \tag{8.42}$$

where $\Delta t = \pi/\omega_{max}$ is the time-sample spacing. As discussed below equation (8.3), in practice many bandlimited functions are significantly nonzero only in a time interval of the form $t_f < t < t_f + T$. By expressing the approximate duration T as $(N_t - 1)\Delta t$ and shifting the sample points from $l_t \Delta t$ to $t_f + l_t \Delta t$, the formula (8.42) becomes

$$f(t) = \sum_{l_t=0}^{N_t-1} \frac{\sin\left[\pi\left(\dfrac{t-t_f}{\Delta t} - l_t\right)\right]}{\pi\left(\dfrac{t-t_f}{\Delta t} - l_t\right)} f(t_f + l_t \Delta t) \tag{8.43}$$

which could also have been obtained by inserting (8.4) into (8.3). The reconstruction formula (8.43) will be used in the time-domain computation schemes to compute the time-domain data in the scan plane at an arbitrary time from the time-domain data at a set of discrete times.

8.2.1 Acoustic Fields Computed from the Pressure in the Scan Plane

The time-domain computation scheme for acoustic fields consists simply of using the time-domain formula (5.55), which expresses the far-field pattern in terms of the time derivative of the pressure in the scan plane as

$$\mathcal{F}(\theta, \phi, t) = \frac{\cos\theta}{2\pi c} \int_{-\infty}^{+\infty}\int_{-\infty}^{+\infty} \frac{\partial}{\partial t} p(\mathbf{r}_0, t + \hat{\mathbf{r}} \cdot \mathbf{r}_0/c) dx_0 dy_0, \quad 0 \le \theta < \pi/2. \tag{8.44}$$

We assume that the time derivative of the pressure is known in the scan plane and that the pressure is bandlimited with bandlimit ω_{max}, so that the minimum wavelength is $\lambda_{min} = 2\pi c/\omega_{max}$.

Equation (8.44) can be obtained by Fourier transforming the frequency-domain formula (8.10), which was converted to the double summation in (8.11) by means of the sampling theorem. Since (8.11) holds for all frequencies, (8.44) can be converted to a summation by Fourier transforming (8.11) to get

$$\mathcal{F}(\theta, \phi, t) = \frac{\Delta x_0 \Delta y_0 \cos\theta}{2\pi c} \sum_{m=-N_x}^{N_x} \sum_{n=-N_y}^{N_y} \frac{\partial}{\partial t} p(\mathbf{r}_{0mn}, t + \hat{\mathbf{r}} \cdot \mathbf{r}_{0mn}/c) \tag{8.45}$$

where $\mathbf{r}_{0mn} = m\Delta x_0 \hat{\mathbf{x}} + n\Delta y_0 \hat{\mathbf{y}}$ is a sampling point in the scan plane and $\Delta x_0 = \Delta y_0 = \lambda_{min}/2$. This formula represents a time-domain sampling theorem that requires one to sample the time-domain pressure in the scan plane at a spatial sample spacing of $\lambda_{min}/2$, the same spacing required by the frequency-domain computation scheme.

Valuable time-domain sampling theorems and reconstruction formulas for plane-polar samples have been obtained by Bucci, D'Elia, and Migliore [196]. In certain cases, these plane-polar sampling theorems allow use of sample spacing that is larger than $\lambda_{min}/2$, thereby reducing the number of scan points in the scan plane. However, in general, the sample spacing can be larger than $\lambda_{min}/2$ only if $2d/D > 1$, where d is the minimum distance between the scan plane and the test antenna and D is the maximum dimension of the test antenna. Hence, if the test antenna is large and located close to the scan plane so that $2d/D < 1$, no increase in the sample spacing is obtained by using the plane-polar sampling theorems in [196]. When the measurements are performed with a plane-polar scanner, the plane-polar reconstruction formulas are useful for calculating the field values at rectangular grid points in the scan plane. These field values can then be used to efficiently compute the frequency-domain far fields using the FFT.

The direct time-domain formula (8.45) reveals a useful property of the time-domain computation scheme. If the source is turned on at $t = t_0$, we can calculate the far-field pattern for times $t < t_1$ by measuring the pressure in the scan plane only for a limited time $t < t_2$. (This is not true for the frequency-domain computation scheme because the calculation of the frequency-domain pressure in the scan plane requires the values of the time-domain pressure over its entire duration.) To determine t_2 as a function of t_1 and the angle of observation θ, we assume that the scan plane is large enough that the field has not reached its edges at $t = t_2$. Then the truncation of the scan plane does not introduce any error into the far-field calculation, so for $t < t_1$ the far-field pattern (8.45) equals the exact far-field pattern (8.44).

Because the source is turned on at $t = t_0$, the pressure $p(\mathbf{r}_0, t)$ is zero for $|\mathbf{r}_0| > c(t - t_0)$ when $\mathbf{r}_0$ is far from the source region, so that the minimum distance from the source region to $\mathbf{r}_0$ is approximately $|\mathbf{r}_0|$. From the time-domain formula (8.45) it is then seen that the latest time t_2 required to calculate the far-field pattern at the angle of observation θ and time t_1 is approximately given by $t_2 \approx t_1 + |\mathbf{r}_{0max}|c^{-1} \sin\theta$, where $|\mathbf{r}_{0max}|$ is the radius of the circle in the scan plane circumscribing the nonzero field in the scan plane at $t = t_2$. Consequently, $|\mathbf{r}_{0max}| = c(t_2 - t_0)$, and we have

$$t_2 \approx \frac{t_1 - t_0}{1 - \sin\theta} + t_0. \tag{8.46}$$

Therefore, to calculate the far-field pattern at an angle θ for $t_0 < t < t_1$, we only need to measure the pressure in the scan plane for $t_0 < t < t_2$, with t_2 given in (8.46). In particular, $t_2 = t_1$ if the far-field pattern is calculated on the z axis ($\theta = 0$), as can be seen immediately from the time-domain formula (8.45) since $\hat{\mathbf{r}} \cdot \mathbf{r}_{0mn} \equiv 0$ when $\theta = 0$. In contrast to the frequency-domain computation scheme, the time-domain computation scheme therefore can calculate the far-field pattern for early times from measurements taken at early times.

To calculate the far field at the angles of observation (θ, ϕ) and time t from the formula (8.45), we have to know the time derivative of the pressure at the point $\mathbf{r}_{0mn}$ in the scan plane at the times $\tau_{mn}(\theta, \phi) = t + m\Delta t \cos\phi \sin\theta + n\Delta t \sin\phi \sin\theta$, where $\Delta t = \Delta x_0/c = \Delta y_0/c$. Similarly, to calculate the far field at the angles of observation (θ', ϕ') and time t, we have to know the time derivative of the pressure at the point $\mathbf{r}_{0mn}$ at the times $\tau_{mn}(\theta', \phi')$. The time difference $|\tau_{mn}(\theta, \phi) - \tau_{mn}(\theta', \phi')|$ for the two different angles of observation can be smaller than Δt. Therefore, in general, we need the value of the time derivative of the pressure at each $\mathbf{r}_{0mn}$ at times that lie between time samples with spacing Δt.

One way of obtaining the values of the time derivative of the pressure in the scan plane at times that lie between samples is to use the reconstruction formula (8.43). To use this formula, we assume that the time-domain field at $\mathbf{r}_{0mn}$ is significant only in the time interval $t_0 < t < t_0 + T_0$, where the initial time t_0 may depend on the position $\mathbf{r}_{0mn}$ and the duration T_0 is independent of $\mathbf{r}_{0mn}$. With the duration written as $T_0 = (N_t - 1)\Delta t$, the reconstruction formula (8.43) shows that

$$\frac{\partial}{\partial t} p(\mathbf{r}_{0mn}, t) = \sum_{l_t=0}^{N_t-1} \frac{\sin\left[\pi\left(\dfrac{t - t_0}{\Delta t} - l_t\right)\right]}{\pi\left(\dfrac{t - t_0}{\Delta t} - l_t\right)} \frac{\partial p}{\partial t}(\mathbf{r}_{0mn}, t_0 + l_t\Delta t) \tag{8.47}$$

where N_t is the integer occurring in the frequency-domain formula (8.6), and we have used the convention (8.7).

For each $\mathbf{r}_{0mn}$, this reconstruction formula requires the calculation of N_t values of the $\sin(x)/x$ function, N_t real multiplications, and N_t real additions, so when N_t is large it may require a considerable amount of computer time. Moreover, because the pressure at *all* the time sample points $t_0 + l_t\Delta t$ must be known to compute the pressure at any time t, we cannot compute the early-time far field from the early-time pressure in the scan plane when

(8.47) is employed. Hence, definite disadvantages are associated with the reconstruction formula (8.47).

For practical applications it may not be necessary to use the exact reconstruction formula (8.47). Instead, we can calculate the time derivative of the pressure at a time t between the two time-sample points $t_0 + l_t\Delta t$ and $t_0 + (l_t + 1)\Delta t$ by using the linear interpolation formula

$$
\begin{aligned}
\frac{\partial}{\partial t} p(\mathbf{r}_{0mn}, t) = {} & \frac{t_0 + (l_t + 1)\Delta t - t}{\Delta t} \frac{\partial p}{\partial t}(\mathbf{r}_{0mn}, t_0 + l_t\Delta t) \\
& + \frac{t - t_0 - l_t\Delta t}{\Delta t} \frac{\partial p}{\partial t}(\mathbf{r}_{0mn}, t_0 + (l_t + 1)\Delta t).
\end{aligned}
\tag{8.48}
$$

By using a value of Δt which is smaller than $\pi/\omega_{max} = \Delta x_0/c = \Delta y_0/c$, the accuracy of this linear formula can be increased, as will be demonstrated in Section 8.3.1.

The computation scheme involving the reconstruction formula (8.47) can be recast in a more compact form that is easy to implement. Combining the time-domain formula (8.45) and the reconstruction formula (8.47) shows that the time-domain far-field pattern is given by

$$
\begin{aligned}
\mathcal{F}(\theta, \phi, t) = {} & \frac{\Delta x_0 \Delta y_0 \cos\theta}{2\pi c} \\
& \cdot \sum_{m=-N_x}^{N_x} \sum_{n=-N_y}^{N_y} \sum_{l_t=0}^{N_t-1} \frac{\sin\left[\pi\left(\dfrac{t + \hat{\mathbf{r}}\cdot\mathbf{r}_{0mn}/c - t_0}{\Delta t} - l_t\right)\right]}{\pi\left(\dfrac{t + \hat{\mathbf{r}}\cdot\mathbf{r}_{0mn}/c - t_0}{\Delta t} - l_t\right)} A_{m,n,l_t}
\end{aligned}
\tag{8.49}
$$

where $\hat{\mathbf{r}} = \cos\phi\sin\theta\hat{\mathbf{x}} + \sin\phi\sin\theta\hat{\mathbf{y}} + \cos\theta\hat{\mathbf{z}}$, and A_{m,n,l_t} is the time derivative of the acoustic pressure in the scan plane at the space-time points with coordinates $(m\Delta x_0, n\Delta y_0, t_0 + l_t\Delta t)$ given in (8.14). The formula (8.49) expresses the time-domain far-field pattern directly in terms of the space-time sampled values (8.14) of the time derivative of the acoustic pressure in the scan plane. Alternatively, we could use the linear interpolation formula (8.48) in (8.45) to derive an expression for the far-field pattern in terms of the sampled data in (8.14).

At this point, it is worth comparing the expression (8.49) obtained from the time-domain computation scheme with the analogous expression (8.17) obtained from the frequency-domain computation scheme. The two expressions are different. In particular, the far-field pattern (8.17) is a periodic function of time with period $N_t\Delta t$, whereas the far-field pattern in (8.49) is not periodic. Moreover, the expression (8.17) approaches (8.49) as $N_t \to \infty$. These considerations indicate that the frequency-domain computation scheme can have aliasing problems if the value of N_t is not large enough, whereas the time-domain computation scheme has no aliasing problems in time. The numerical example in Section 8.3.1 shows that the two computation schemes indeed agree when the aliasing in (8.17) is avoided by increasing N_t.

Number of Operations. We shall estimate the number of operations required to calculate the time-domain far-field pattern with the time-domain computation scheme. Let $N_{\mathcal{F}}$ denote the number of space-time far-field points (θ, ϕ, t) at which the far-field pattern is calculated. (For the time-domain computation scheme, $N_{\mathcal{F}}$ need not equal $N_{\theta\phi}N_t$, where $N_{\theta\phi}$ is the number of (θ, ϕ) at which the far-field pattern is calculated, because the far-field pattern may not be required at N_t times for each (θ, ϕ). In the frequency-domain computation scheme, the FFT applied to (8.13) automatically computed the far-field pattern at N_t times.)

If the formula (8.49) is used to calculate the far-field pattern at one far-field point (θ, ϕ, t), the number of operations is $O(N_x N_y N_t)$. To calculate the far-field pattern at $N_{\mathcal{F}}$ different

(θ, ϕ, t), the time-domain scheme in (8.49) therefore requires

$$O(N_x N_y N_t N_{\mathcal{F}}) \tag{8.50}$$

operations.

The number of operations at each time t required by the linear interpolation formula (8.48) is independent of N_t, so to calculate the far-field pattern from (8.45) using the linear formula (8.48) at one far-field point (θ, ϕ, t) requires $O(N_x N_y)$ operations. Hence, if (8.45) and (8.48) are used to calculate the far-field pattern at $N_{\mathcal{F}}$ different space-time points, the number of operations is

$$O(N_x N_y N_{\mathcal{F}}). \tag{8.51}$$

We see from (8.50) and (8.51) that for large N_t, we can reduce the number of operations significantly by using the linear interpolation (8.48) instead of the exact reconstruction formula (8.47), which is used in (8.45) to get (8.49).

With a computer consisting of many parallel processors, we can perform many operations simultaneously, and for certain computations we can thereby significantly reduce the computer time. The formula (8.49) for the time-domain far-field pattern is well suited for implementation on a parallel computer. To see this, define the array

$$S_{m,n,l_t}(\theta, \phi, t) = \frac{\sin\left[\pi\left(\dfrac{t + \hat{\mathbf{r}} \cdot \mathbf{r}_{0mn}/c - t_0}{\Delta t} - l_t\right)\right]}{\pi\left(\dfrac{t + \hat{\mathbf{r}} \cdot \mathbf{r}_{0mn}/c - t_0}{\Delta t} - l_t\right)} \tag{8.52}$$

so that (8.49) can be written as

$$\mathcal{F}(\theta, \phi, t) = \frac{\Delta x_0 \Delta y_0 \cos\theta}{2\pi c} \sum_{m=-N_x}^{N_x} \sum_{n=-N_y}^{N_y} \sum_{l_t=0}^{N_t-1} S_{m,n,l_t}(\theta, \phi, t)\, A_{m,n,l_t} \tag{8.53}$$

and note that the terms $S_{m,n,l_t}(\theta, \phi, t)\, A_{m,n,l_t}$ in the triple summation can be computed in parallel. That is, $S_{m,n,l_t}(\theta, \phi, t)\, A_{m,n,l_t}$ and $S_{m',n',l'_t}(\theta, \phi, t)\, A_{m',n',l'_t}$, with $(m, n, l_t) \neq (m', n', l'_t)$, can be computed simultaneously because they are independent.

Once the products $S_{m,n,l_t}(\theta, \phi, t)\, A_{m,n,l_t}$ are calculated for all (m, n, l_t), the far-field pattern is obtained by the triple summation, which can also be computed efficiently on a parallel computer. This was demonstrated in [26], where it is shown how the formula (8.53) was implemented on a parallel computer called the Connection Machine.

8.2.2 Electromagnetic Fields Computed from the Electric Field in the Scan Plane

Let the time derivative of the tangential electric field be known in the scan plane, and assume that the electromagnetic fields have the same bandlimit and duration as the acoustic field in Section 8.2.1. Using the same notation as in Section 8.2.1, we find that the electromagnetic formulas for the time-domain computation scheme are

$$\mathcal{F}(\theta, \phi, t) = -\frac{\Delta x_0 \Delta y_0}{2\pi c}\, \hat{\mathbf{r}} \times \sum_{m=-N_x}^{N_x} \sum_{n=-N_y}^{N_y} \hat{\mathbf{z}} \times \frac{\partial}{\partial t} \mathbf{E}(\mathbf{r}_{0mn}, t + \hat{\mathbf{r}} \cdot \mathbf{r}_{0mn}/c) \tag{8.54}$$

and

$$\mathcal{F}(\theta,\phi,t) = -\frac{\Delta x_0 \Delta y_0}{2\pi c}$$

$$\cdot \sum_{m=-N_x}^{N_x} \sum_{n=-N_y}^{N_y} \sum_{l_t=0}^{N_t-1} \frac{\sin\left[\pi\left(\dfrac{t+\hat{\mathbf{r}}\cdot\mathbf{r}_{0mn}/c - t_0}{\Delta t} - l_t\right)\right]}{\pi\left(\dfrac{t+\hat{\mathbf{r}}\cdot\mathbf{r}_{0mn}/c - t_0}{\Delta t} - l_t\right)} \hat{\mathbf{r}} \times \mathbf{A}_{m,n,l_t} \tag{8.55}$$

where $\mathbf{A}_{m,n,l_t}$ is given by (8.50). This computation scheme has the same properties as its acoustic analog described in Section 8.2.1. The number of operations required in (8.55) to calculate the far-field pattern at $N_{\mathcal{F}}$ different space-time points (θ,ϕ,t) is $O(N_x N_y N_t N_{\mathcal{F}})$. When the linear interpolation formula

$$\begin{aligned} \frac{\partial}{\partial t}\mathbf{E}(\mathbf{r}_{0mn},t) = {} & \frac{t_0+(l_t+1)\Delta t - t}{\Delta t}\frac{\partial \mathbf{E}}{\partial t}(\mathbf{r}_{0mn}, t_0 + l_t\Delta t) \\ & + \frac{t - t_0 - l_t\Delta t}{\Delta t}\frac{\partial \mathbf{E}}{\partial t}(\mathbf{r}_{0mn}, t_0 + (l_t+1)\Delta t) \end{aligned} \tag{8.56}$$

can be used to accurately calculate the field between the time samples $t_0 + l_t\Delta t$ and $t_0 + (l_t+1)\Delta t$, the number of operations required to calculate the far-field pattern at $N_{\mathcal{F}}$ different space-time points (θ,ϕ,t) is $O(N_x N_y N_{\mathcal{F}})$.

8.2.3 Acoustic Fields Computed from the Probe Output in the Scan Plane

Now let the acoustic field in the scan plane be measured with a time-derivative probe as described in Section 7.1.4, and assume that the acoustic field has the same bandlimit and time duration as in Section 8.2.1. The time-domain far-field pattern is then given by (7.42), whose discrete analog is found by Fourier transforming (8.34) to get

$$\mathcal{F}(\theta,\phi,t) = \frac{\Delta x_0 \Delta y_0 \cos\theta}{2\pi c Q_p(c^{-1}\cos\phi\sin\theta, c^{-1}\sin\phi\sin\theta)} \cdot \sum_{m=-N_x}^{N_x}\sum_{n=-N_y}^{N_y} b_p(\mathbf{r}_{0mn}, t + \hat{\mathbf{r}}\cdot\mathbf{r}_{0mn}/c). \tag{8.57}$$

By applying the reconstruction formula (8.43) to the probe output, this formula for the time-domain far-field pattern may also be written directly in terms of the sampled data b_{m,n,l_t} in (8.35) as

$$\mathcal{F}(\theta,\phi,t) = \frac{\Delta x_0 \Delta y_0 \cos\theta}{2\pi c Q_p(c^{-1}\cos\phi\sin\theta, c^{-1}\sin\phi\sin\theta)}$$

$$\cdot \sum_{m=-N_x}^{N_x} \sum_{n=-N_y}^{N_y} \sum_{l_t=0}^{N_t-1} \frac{\sin\left[\pi\left(\dfrac{t+\hat{\mathbf{r}}\cdot\mathbf{r}_{0mn}/c - t_0}{\Delta t} - l_t\right)\right]}{\pi\left(\dfrac{t+\hat{\mathbf{r}}\cdot\mathbf{r}_{0mn}/c - t_0}{\Delta t} - l_t\right)} b_{m,n,l_t}. \tag{8.58}$$

The number of operations required by (8.58) to calculate the far-field pattern at $N_{\mathcal{F}}$ different space-time points (θ,ϕ,t) is $O(N_x N_y N_t N_{\mathcal{F}})$.

When the linear interpolation formula

$$b_p(\mathbf{r}_{0mn}, t) = \frac{t_0 + (l_t + 1)\Delta t - t}{\Delta t} b_p(\mathbf{r}_{0mn}, t_0 + l_t \Delta t) + \frac{t - t_0 - l_t \Delta t}{\Delta t} b_p(\mathbf{r}_{0mn}, t_0 + (l_t + 1)\Delta t) \tag{8.59}$$

can be used to accurately calculate the probe output between the time samples $t_0 + l_t \Delta t$ and $t_0 + (l_t + 1)\Delta t$, the number of operations required to calculate the far-field pattern at $N_{\mathcal{F}}$ different space-time points (θ, ϕ, t) is $O(N_x N_y N_{\mathcal{F}})$.

8.2.4 Electromagnetic Fields Computed from the Probe Output in the Scan Plane

Assume that the electric field in the scan plane is measured with a D-dot probe as described in Section 7.2.4, so that the time-domain far-field pattern is given by (7.85). When the electric field has the same bandlimit and duration as the acoustic field in Section 8.2.1, it is found by Fourier transforming (8.39) that

$$\mathcal{F}(\theta, \phi, t) = \frac{\Delta x_0 \Delta y_0 \cos\theta}{2\pi c} \bar{\mathbf{Q}}_p^{-1}(c^{-1}\cos\phi\sin\theta, c^{-1}\sin\phi\sin\theta) \cdot \sum_{m=-N_x}^{N_x} \sum_{n=-N_y}^{N_y} \mathbf{b}_p(\mathbf{r}_{0mn}, t + \hat{\mathbf{r}} \cdot \mathbf{r}_{0mn}/c). \tag{8.60}$$

Applying the reconstruction formula (8.43) to the vector output of the probe shows that the far-field pattern may also be expressed as

$$\mathcal{F}(\theta, \phi, t) = \frac{\Delta x_0 \Delta y_0 \cos\theta}{2\pi c} \bar{\mathbf{Q}}_p^{-1}(c^{-1}\cos\phi\sin\theta, c^{-1}\sin\phi\sin\theta) \cdot \sum_{m=-N_x}^{N_x} \sum_{n=-N_y}^{N_y} \sum_{l_t=0}^{N_t-1} \frac{\sin\left[\pi\left(\dfrac{t + \hat{\mathbf{r}} \cdot \mathbf{r}_{0mn}/c - t_0}{\Delta t} - l_t\right)\right]}{\pi\left(\dfrac{t + \hat{\mathbf{r}} \cdot \mathbf{r}_{0mn}/c - t_0}{\Delta t} - l_t\right)} \mathbf{b}_{m,n,l_t} \tag{8.61}$$

where $\mathbf{b}_{m,n,l_t}$ is given by (8.40). The number of operations required by (8.61) to calculate the far-field pattern at $N_{\mathcal{F}}$ different space-time points (θ, ϕ, t) is equal to $O(N_x N_y N_t N_{\mathcal{F}})$.

When the linear interpolation formula

$$\mathbf{b}_p(\mathbf{r}_{0mn}, t) = \frac{t_0 + (l_t + 1)\Delta t - t}{\Delta t} \mathbf{b}_p(\mathbf{r}_{0mn}, t_0 + l_t \Delta t) + \frac{t - t_0 - l_t \Delta t}{\Delta t} \mathbf{b}_p(\mathbf{r}_{0mn}, t_0 + (l_t + 1)\Delta t) \tag{8.62}$$

can be used to accurately calculate the vector output of the probe between the time samples $t_0 + l_t \Delta t$ and $t_0 + (l_t + 1)\Delta t$, the number of operations required to calculate the far-field pattern at $N_{\mathcal{F}}$ different space-time points (θ, ϕ, t) is $O(N_x N_y N_{\mathcal{F}})$.

8.3 NUMERICAL EXAMPLES

The foregoing frequency-domain and time-domain computation schemes will now be used in numerical examples involving acoustic radiators and an open-ended electromagnetic waveguide. These examples illustrate general phenomena encountered in time-domain scanning and a number of important issues that arise when planar near-field scanning is used to evaluate the time-domain far fields of typical acoustic and electromagnetic radiators. Finally, we summarize the advantages and disadvantages of the frequency-domain and time-domain computation schemes.

8.3.1 The Acoustic Point Source

We begin by considering a Gaussian acoustic point source located on the z axis at $z = -d$ with $d > 0$, as shown in Figure 8.2. The acoustic pressure of this point source is given by

$$p(\mathbf{r}, t; d) = \frac{f^G(t - |\mathbf{r} + d\hat{\mathbf{z}}|/c)}{4\pi |\mathbf{r} + d\hat{\mathbf{z}}|} \tag{8.63}$$

where $f^G(t)$ is the Gaussian time function

$$f^G(t) = e^{-4t^2/\tau^2}, \qquad f^G_\omega = \frac{\tau}{4\sqrt{\pi}} e^{-\omega^2\tau^2/16} \tag{8.64}$$

and τ equals half the signalwidth defined such that $|f^G(t)| < 0.02|f^G(0)|$ for $|t| > \tau$. The Gaussian pulse $f^G(t)$ is not perfectly bandlimited, but $f^G_\omega / f^G_0 < 1.3 \cdot 10^{-4}$ for $\omega > 12/\tau$, and $f^G(t)$ can therefore be approximated well by a pulse with bandlimit $\omega_{max} = 12/\tau$. With $N_t = 10$ and $\Delta t = \pi\tau/12$, the reconstruction formula (8.43) reproduces the Gaussian pulse $f^G(t)$ to within an absolute error of $3 \cdot 10^{-4}$. The sampling theorem with $\omega_{max} = 12/\tau$ shows that the sample spacing is $\Delta x_0 = \Delta y_0 = \lambda_{min}/2 = \pi c/\omega_{max} \approx c\tau/4$. We choose $d = 2\lambda_{min} \approx c\tau$, and the scan plane is taken to be a square of sidelength $10d$ (corresponding to $N_x = N_y = 20$), as shown in Figure 8.2.

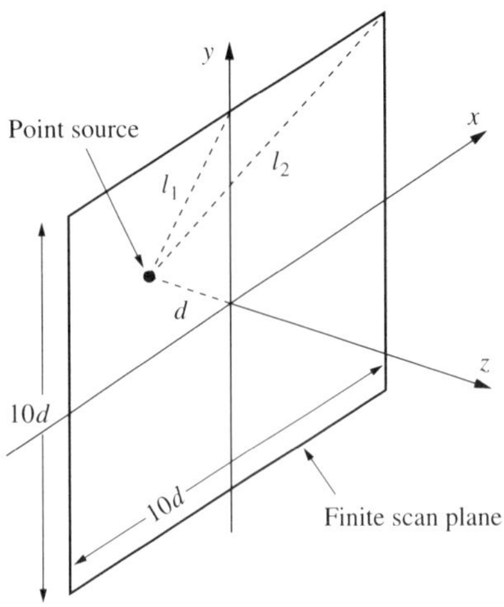

Figure 8.2 The acoustic point source located at a distance d behind a finite scan plane.

We start by showing results for the on-axis far-field pattern $\mathcal{F}(\theta = 0, t)$ obtained from the time-domain formula (8.45) and the linear interpolation formula (8.48). Figure 8.3 shows (a) the exact on-axis far-field pattern, (b) the on-axis far-field pattern obtained from the formulas (8.45) and (8.48) with $\Delta t = \pi/\omega_{max}$ (as prescribed by the standard sampling theorem), and (c) the on-axis far-field pattern obtained from (8.45) and (8.48) with $\Delta t = \pi/(3\omega_{max})$ (that is, a value obtained by over sampling in time).

The exact curve (a) has the Gaussian waveform and is significantly nonzero only in the interval $-0.1\tau < t < 2.1\tau$. The curve (b) has some visible discontinuities in its slope and approximates the exact curve well in the interval $-\tau < t < 4\tau$. In the interval $4\tau < t < 8\tau$ it is negative and erroneous. The curve (c) cannot be distinguished from the exact curve in the interval $-\tau < t < 4\tau$ and is erroneous in the interval $4\tau < t < 8\tau$. Hence, by oversampling

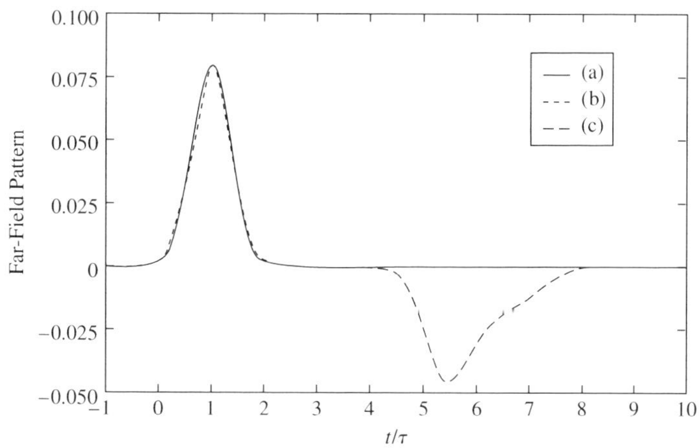

Figure 8.3 The on-axis far-field pattern for a Gaussian point source: (a) the exact far-field pattern; (b) the far-field pattern obtained from the time-domain computation scheme with linear interpolation and $\Delta t = \pi/\omega_{max}$; (c) the far-field pattern obtained from the time-domain computation scheme with linear interpolation and $\Delta t = \pi/(3\omega_{max})$.

in time with $\Delta t = \pi/(3\omega_{max})$ we can use the simple linear interpolation formula (8.48) to obtain an excellent approximation to the exact curve in the interval $-\tau < t < 4\tau$.

It will now be shown that the erroneous behavior of the two approximate curves (b) and (c) in the interval $4\tau < t < 8\tau$ is due to the fact that the scan plane is finite. To show this, note that the time between the arrival of the direct signal ($t \approx -0.1\tau$) and the arrival of the erroneous signal ($t \approx 4.1\tau$) is equal to the time it takes the signal to travel the distance $l_1 - d$, where $l_1 = \sqrt{26}d$ is the distance from the source to the midpoint of the edges of the scan plane shown in Figure 8.2 (recall that $\tau \approx d/c$).[1] Similarly, the time difference between the end of the direct signal ($t \approx 2.1\tau$) and the end of the erroneous signal ($t \approx 8\tau$) is the time it takes the signal to travel the distance $l_2 - d$, where $l_2 = \sqrt{51}d$ is the distance from the source to the corners of the scan plane.[2]

Thus, the erroneous parts of the approximate far-field patterns represent diffraction from the artificial edges of the truncated scan plane. By enlarging the scan plane, the erroneous parts of the approximate far-field patterns will move to later times, but the erroneous parts will never disappear when the scan plane is finite. Furthermore, from the exact integral (5.54) it is seen that for $\theta = 0$ and $t < 4\tau$ there is no contribution from the region outside the square scan plane with sidelength $10d$. Hence, because the signal has not yet reached the edges of the scan plane for $t < 4\tau$, the truncated scan plane does not introduce any error into the calculation of the far-field pattern for $\theta = 0$ and $t < 4\tau$. However, as time goes on, the signal eventually reaches the edges, and the finite scan plane introduces an error that can be separated in time from the exact far-field pattern.

Having shown and discussed the results obtained from the time-domain computation scheme, we now examine the results obtained from the frequency-domain computation scheme. As explained in Section 8.1, the frequency sample spacing $\Delta\omega = 2\omega_{max}/N_t$ depends on the bandwidth ω_{max} of the field and on the duration of the far-field pattern. Moreover, we have just shown that the duration of the far-field pattern is erroneously extended beyond that of the

[1]The acoustic far-field pressure is given by $p(\mathbf{r}, t) \sim \mathcal{F}(\theta, \phi, t - r/c)/r$, so the time of arrival at the actual far-field point (r, θ, ϕ) is $t + r/c$ for a signal that arrives at the time t in the far-field pattern.

[2]Because the observation point is at $\theta = 0$, all corners of the scan plane contribute to the erroneous signal at the same time.

field at a point in the scan plane by the truncation of the scan plane. Therefore, in practice, the required frequency sample spacing $\Delta\omega$ depends not only on the bandwidth ω_{max} (and the duration of the field at a point in the scan plane), but also on the size of the scan plane.

From Figure 8.3 showing the far-field pattern calculated from the time-domain computation scheme, it is seen that with our chosen scan-plane size the duration of the calculated time-domain on-axis far-field pattern is $T_f = 8.2\tau$. Therefore, to avoid significant time-domain aliasing, one must choose $\Delta\omega = 2\pi/T_f = 0.77\tau^{-1}$. This means that the number of time samples needed to avoid significant time-domain aliasing is $N_t = 2\omega_{max}/\Delta\omega \approx 32$.

Figure 8.4 shows the following three plots of the on-axis time-domain far-field pattern $\Gamma(\theta = 0, t)$: (a) the exact far-field pattern, (b) the far-field pattern obtained from the frequency-domain formulas (8.6), (8.11), and (8.13) with $N_t = 16$ (which is sufficient for reconstructing the time derivative of the pressure at a point in the scan plane), and (c) the far-field pattern obtained from the frequency-domain formulas (8.6), (8.11), and (8.13) with $N_t = 32$ (the number prescribed by the sampling theorem to compute the far field extended in time by the truncated scan plane). The exact far-field pattern (a) is significantly nonzero only in the interval $-0.1\tau < t < 2.1\tau$. The curve (b) is periodic with period $T = 2\pi/\Delta\omega = \pi N_t/\omega_{max} = 4.2\tau$ and is clearly erroneous due to time-domain aliasing. The curve (c) is periodic with period $T = 8.4\tau$ and cannot be distinguished from the exact curve (a) in the interval $-0.5\tau < t < 4.1\tau$. No significant time-domain aliasing occurs for the curve (c), and from Figure 8.3 it is seen that the curve (c) is simply a periodic repetition of the far-field pattern calculated from the time-domain computation scheme. Thus, when $\Delta\omega$ is chosen small enough to avoid time-domain aliasing, the frequency-domain computation scheme produces the same time-domain far-field pattern as the time-domain computation scheme.

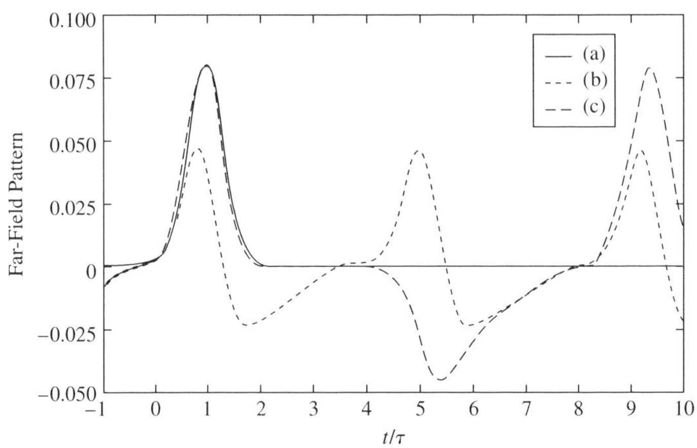

Figure 8.4 The on-axis far-field pattern for a Gaussian point source: (a) the exact far-field pattern; (b) the far-field pattern obtained from the frequency-domain computation scheme with $N_t = 16$; (c) the far-field pattern obtained from the frequency-domain computation scheme with $N_t = 32$.

Let us finally show how the location in time of the erroneous signal, which is caused by the edges of the finite scan plane, depends on the angles of observation (θ, ϕ). Figures 8.5, 8.6, and 8.7 show the far-field pattern of the acoustic point source at the far-field angles $(\theta, \phi) = (30^0, 0)$, $(\theta, \phi) = (45^0, 0)$, and $(\theta, \phi) = (60^0, 0)$, respectively. The computations were made with the time-domain formula (8.49), which employs $\sin(x)/x$ interpolation. It is seen that the erroneous signal caused by the finite scan plane moves to earlier times as the angle of observation θ increases. In fact, for $\theta = 45^0$ and $\theta = 60^0$ the erroneous signal interferes

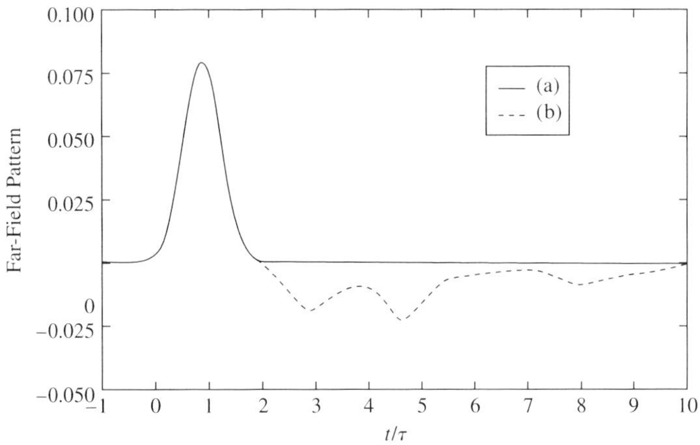

Figure 8.5 The far-field pattern for a Gaussian point source evaluated at $(\theta, \phi) = (30^0, 0)$: (a) the exact far-field pattern; (b) the far-field pattern obtained from the time-domain computation scheme.

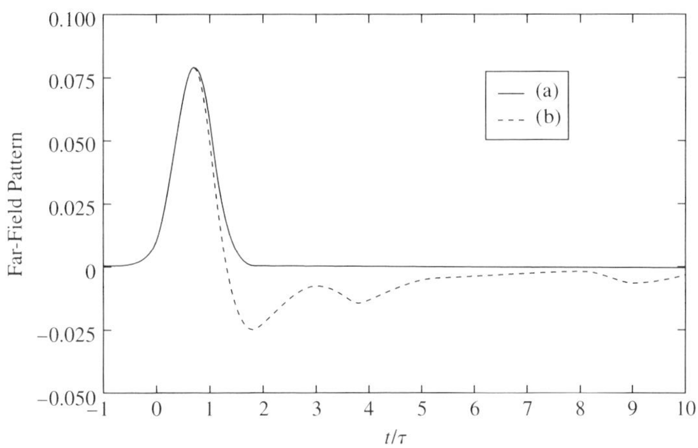

Figure 8.6 The far field pattern for a Gaussian point source evaluated at $(\theta, \phi) = (45^0, 0)$: (a) the exact far-field pattern; (b) the far-field pattern obtained from the time-domain computation scheme.

with the exact signal and makes the computed far-field patterns very inaccurate. By enlarging the scan plane, we can move the erroneous signal to later times, so that the correct far-field pattern again can be obtained by time gating.

To explain the movements of the erroneous signal, consider Figure 8.8 showing the point source, the scan plane, and the direction to the observation point. The erroneous signal caused by the finite scan plane shows up in the far-field pattern at time $t_{er}(\theta) = t_0 + l_1/c - \Delta r/c$, where t_0 is the time at which the point source is turned on, and Δr and l_1 are distances shown in Figure 8.8. It is seen that $\Delta r = \frac{1}{2} s \sin\theta$ where $s = 10d$ is the sidelength of the square scan plane. The exact signal shows up in the far-field pattern at time $t_{ex}(\theta) = t_0 + (d/c)\cos\theta$, where d is the distance between the point source and the scan plane. Consequently, the time difference between the exact signal (Gaussian curve) and the erroneous signal (caused by the edges of the finite scan plane) is $\Delta T(\theta) = t_{er}(\theta) - t_{ex}(\theta) = (\sqrt{26} - 5\sin\theta - \cos\theta)d/c$, where we have used the facts that $l_1 = \sqrt{26}d$ and $s = 10d$. For $\theta = 30^0$, $\theta = 45^0$, and $\theta = 60^0$ it is found that $\Delta T = 1.8\tau$, $\Delta T = 0.9\tau$, and $\Delta T = 0.3\tau$, respectively. Hence for $\theta = 45^0$

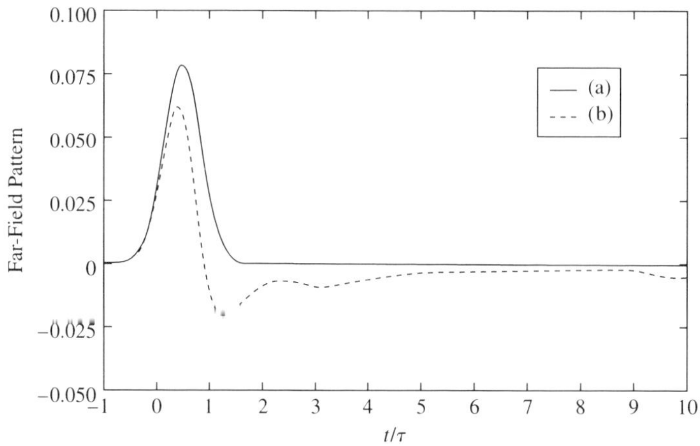

Figure 8.7 The far-field pattern for a Gaussian point source evaluated at $(\theta, \phi) = (60^0, 0)$: (a) the exact far-field pattern; (b) the far-field pattern obtained from the time-domain computation scheme.

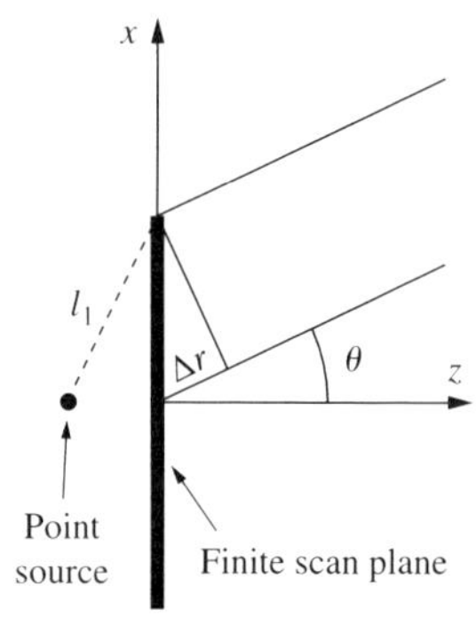

Figure 8.8 The point source, the square scan plane, and the direction θ to the observation point.

and $\theta = 60^0$, the duration of the exact signal (which is approximately 2τ) is significantly longer than ΔT, and the erroneous signal is no longer separated in time from the exact signal. This example illustrates that when the source has a broad beam, one should make sure that the scan plane is large enough for all angles of observation of interest to avoid interference of the erroneous signal caused by the finite scan plane.

Let us now compare the number of operations required by the two computation schemes to evaluate the time-domain far-field pattern of the acoustic point source with Gaussian time dependence. We have $N_x = N_y = 20$ and assume that the far-field pattern is calculated at $N_{\theta\phi} = 100$ different angles of observation satisfying (8.22). (This number of far-field points would adequately resolve the far-field angular variation of a radiator whose maximum dimension is small compared to the minimum wavelength.) To avoid time-domain aliasing caused by the edges of the scan plane, the number of time samples is $N_t = 32$ for the frequency-domain computation scheme. For these values of N_x, N_y, $N_{\theta\phi}$, and N_t, the arguments of the $O(N)$ functions in (8.25) add up to approximately $3 \cdot 10^5$.

When using the time-domain computation scheme, we need only calculate the far-field pattern at 10 different times, so with $N_{\theta\phi} = 100$ we have $N_{\mathcal{F}} = 1000$. With these values of N_x, N_y, and $N_{\mathcal{F}}$, the argument of the $O(N)$ function in (8.51), which is derived for linear interpolation, is approximately $4 \cdot 10^5$. (Similarly, if the formula (8.49) that employs $\sin(x)/x$ interpolation is used, $N_t = 10$ and $N_{\mathcal{F}} = 1000$, and the argument in the $O(N)$ function in (8.50) is approximately $4 \cdot 10^6$.)

Hence, the number of operations required by the frequency-domain and time-domain computation schemes are of the order $3 \cdot 10^5$ and $4 \cdot 10^5$, respectively. Thus, for this Gaussian point source the number of operations for the two schemes appears to be roughly the same.

8.3.2 An Acoustic Example Involving Probe Correction

The probe-corrected formula (8.57), obtained from the time-domain computation scheme, will now be verified numerically by calculating the far-field pattern of an acoustic point source with Gaussian time dependence from measurements taken with a known nonideal probe.

The acoustic point source with Gaussian time dependence is located on the z axis at $z = -d$ and radiates the pressure $p(\mathbf{r}, t; d)$ given in (8.63). The corresponding frequency-domain pressure is $p_\omega(\mathbf{r}; d)$. The pressure in the scan plane is measured with a probe whose frequency-domain receiving characteristic is

$$R_{p\omega}(k_x, k_y) = -\frac{ic\gamma}{2\pi}, \quad \omega > 0 \tag{8.65}$$

with γ given by (6.11). Inserting $(k_x, k_y) = (k\cos\phi\sin\theta, k\sin\phi\sin\theta)$ into (8.65) shows that

$$R_{p\omega}(k\cos\phi\sin\theta, k\sin\phi\sin\theta) = -\frac{i\omega\cos\theta}{2\pi}, \quad 0 \le \theta < \pi/2 \tag{8.66}$$

which is identical to (7.38) with $Q_p = \cos\theta$. The probe under consideration is therefore a time-derivative probe (see Section 7.1.4) for propagating plane waves and (8.57) applies. Since γ is complex for evanescent plane waves, the expression (8.65) shows that this probe is not a time-derivative probe for evanescent plane waves.

To calculate the double summation (8.57) for the far-field pattern, the output of the probe must be known at each scan point in the scan plane. Since the receiving characteristic of the probe is defined in terms of plane waves, we will derive an expression for the probe output by expanding the acoustic point-source pressure in terms of plane waves. We start by using the Weyl identity (3.61) to express the frequency-domain acoustic point-source pressure in terms of plane waves as

$$\begin{aligned} p_\omega(\mathbf{r}; d) &= \frac{f_\omega^G\, e^{ik|\mathbf{r}+d\hat{\mathbf{z}}|}}{4\pi|\mathbf{r}+d\hat{\mathbf{z}}|} \\ &= \frac{if_\omega^G}{8\pi^2}\int_{-\infty}^{+\infty}\int_{-\infty}^{+\infty}\frac{1}{\gamma}e^{i[k_x x + k_y y + \gamma(z+d)]}dk_x dk_y, \quad \omega > 0,\ z > -d \end{aligned} \tag{8.67}$$

where f_ω^G is the Gaussian spectrum in (8.63). The plane-wave spectrum $T_\omega(k_x, k_y)$ for the point-source pressure is found by combining (6.8) and (8.67) to get

$$T_\omega(k_x, k_y) = a_{0\omega}T_{0\omega}(k_x, k_y) = \frac{if_\omega^G}{4\pi\gamma}e^{i\gamma d}, \quad \omega > 0. \tag{8.68}$$

This expression for the spectrum can also be obtained directly from the far-field pressure by means of (6.13). Inserting the expression (8.65) for the receiving characteristic and (8.68) for the spectrum into the formula (6.16) for the probe output gives

$$b_{p\omega}(\mathbf{r}_0) = \frac{cf_\omega^G}{8\pi^2}\int_{-\infty}^{+\infty}\int_{-\infty}^{+\infty} e^{i(k_x x_0 + k_y y_0 + \gamma d)}dk_x dk_y, \quad \omega > 0 \tag{8.69}$$

which combined with the formula (8.67) shows that the probe output at $\mathbf{r}_0 = x_0\hat{\mathbf{x}} + y_0\hat{\mathbf{y}}$ can be expressed in terms of the point-source pressure at $\mathbf{r}_0$ as

$$b_{p\omega}(\mathbf{r}_0) = -c\frac{\partial}{\partial d}p_\omega(\mathbf{r}_0; d) \tag{8.70}$$

for all real ω. This expression could also have been given in terms of a partial derivative with respect to z, but since $z = 0$ in the scan plane it is more convenient to write it in terms of the distance d between the point source and the scan plane. Taking the inverse Fourier transform of (8.70) shows that the time-domain output of the probe is simply

$$b_p(\mathbf{r}_0, t) = -c\frac{\partial}{\partial d}p(\mathbf{r}_0, t; d). \tag{8.71}$$

We have thus shown that the output of the probe equals minus c times the partial derivative with respect to d (or z) of the point-source pressure. This result could also have been obtained by using the formulas in [37], where receiving antennas are described as linear differential operators. When the distance between the probe and the point source is large, (8.71) can be used to show that the probe output is $b_p(\mathbf{r}_0, t) \approx \cos\theta(\partial/\partial t)p(\mathbf{r}_0, t; d)$, where θ is the angle between the z axis and the vector $\mathbf{r}_0 + d\hat{\mathbf{z}}$. This result is explained by noting that the point-source pressure at a position $\mathbf{r}_0$ in the far field consists of one locally plane wave propagating in the direction given by θ, and the output of the special probe equals $\cos\theta$ times the time derivative of an incident plane-wave field.

Inserting the expression (8.71) for the probe output and $Q_p = \cos\theta$ into (8.57) gives the following expression from which the far-field pattern can be computed

$$\mathcal{F}(\theta, \phi, t) = -\frac{\Delta x_0 \Delta y_0}{2\pi}\sum_{m=-N_x}^{N_x}\sum_{n=-N_y}^{N_y}\frac{\partial}{\partial d}p(\mathbf{r}_{0mn}, t + \hat{\mathbf{r}}\cdot\mathbf{r}_{0mn}/c; d). \tag{8.72}$$

As in Section 8.3.1, $\omega_{max} \approx 12/\tau$, $\lambda_{min} = 2\pi c/\omega_{max} \approx c\tau/2$, $d = 2\lambda_{min} \approx c\tau$, and $\Delta x_0 = \Delta y_0 = \lambda_{min}/2 \approx c\tau/4$. The far-field pattern will be calculated from (8.72) for $(\theta, \phi) = (45^0, 0)$, and the scan plane is chosen to be a square of sidelength $20d$. With this size of the scan plane and this observation direction, the erroneous signal (caused by the edges of the finite scan plane) does not interfere with the correct signal. The scanning geometry is shown in Figure 8.1. In Figure 8.9 the curve (a) is the exact Gaussian far-field pattern, and the curve (b) is the far-field pattern calculated by the probe-corrected formula (8.72).

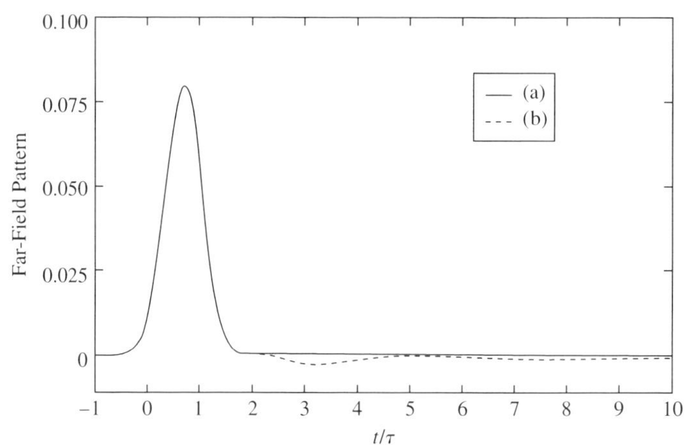

Figure 8.9 The far-field pattern of the Gaussian acoustic point source evaluated at $(\theta, \phi) = (45^0, 0)$: (a) the exact far-field pattern; (b) the far-field pattern obtained from the probe-corrected formula.

The exact curve (a) has the Gaussian wave form and is significantly nonzero only in the interval $-0.3\tau < t < 1.8\tau$. The curve (b) cannot be distinguished from the exact curve in the interval $-\tau < t < 2.4\tau$. Because the scan plane is finite, the curve (b) is erroneous (negative) in parts of the interval from $t = 2.4\tau$ to $t = 23\tau$, where the end of the erroneous signal at $t = 23\tau$ is found by determining the time after which the field is identically zero in the finite scan plane; see Sections 8.2.1 and 8.3.1. However, the erroneous part of the curve (b) is zero in the interval where the exact far-field pattern is nonzero, so the exact far-field pattern can be obtained from the curve (b) by time gating. Furthermore, it is demonstrated again that time-domain planar near-field measurements can be used to accurately determine the far-field pattern of broadbeam antennas that cannot be accurately determined in the frequency domain because of the errors introduced by the finite scan plane.

8.3.3 A Directive Acoustic Source

Having presented results for the acoustic point source that produces an isotropic far field, we shall consider the source in Figure 8.10, which produces a field with a narrow beam. The time-domain pressure of this source can be obtained by complex extension of the space-time source coordinates of an acoustic point source with delta-function time dependence [197]. Specifically, the pressure is

$$p(\mathbf{r}, t) = \mathrm{Re}\left(\frac{\tau\overset{+}{\delta}(t - it_1 - s/c)}{4\pi s}\right) \tag{8.73}$$

where $\overset{+}{\delta}(t)$ is the analytic delta function found by inserting $\delta(t')$ into (5.227). Equation (8.73) requires $\overset{+}{\delta}(t)$ only at $\mathrm{Im}(t) < 0$, for which the lower equation in (5.227) gives

$$\overset{+}{\delta}(t) = \frac{1}{\pi i t}, \quad \mathrm{Im}(t) < 0. \tag{8.74}$$

In (8.73), τ is a positive constant, and

$$s = \sqrt{x^2 + y^2 + (z - ib + d)^2}, \quad \mathrm{Re}(s) \geq 0 \tag{8.75}$$

is the complex distance to the source point. Moreover, it_1 is the imaginary initial time with real t_1 satisfying $0 < b < ct_1$. The source region is the disk of radius b given by $z = -d$, $\sqrt{x^2 + y^2} < b$, as shown in Figure 8.10. We choose the parameters for the field (8.73) so that $d = c\tau$, $b = 0.95d$, and $t_1 = \tau$.

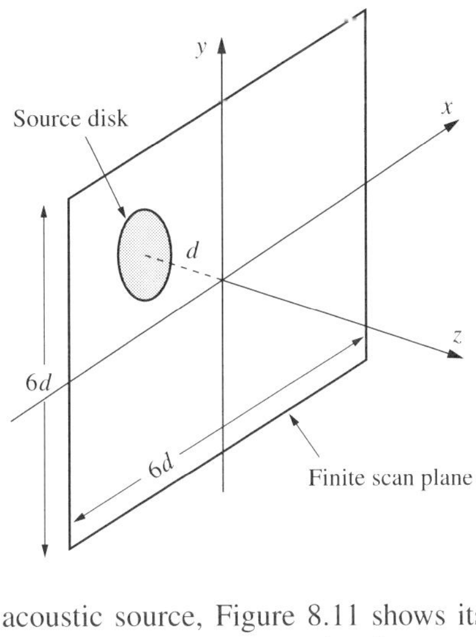

Figure 8.10 A directive acoustic source located in the disk of radius b at $z = -d$.

To illustrate the properties of this acoustic source, Figure 8.11 shows its pressure at different points in the square scan plane with sidelength $6d$. We see that the pressure decays as we move away from the center of the scan plane, so the strongest radiation occurs around the

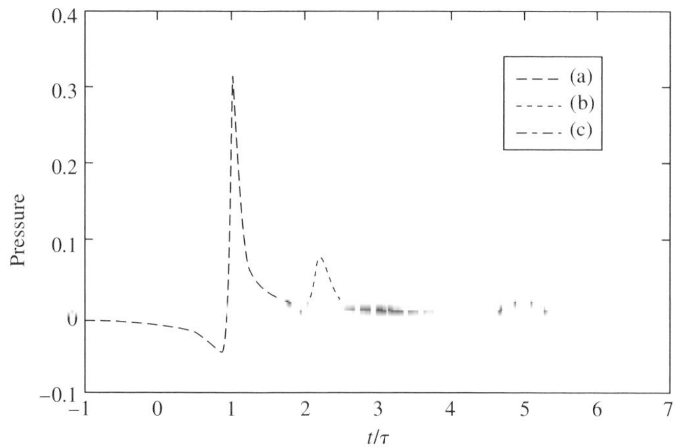

Figure 8.11 The pressure generated by the source disk at three different points in the scan plane: (a) $x = y = 0$; (b) $x = y = d$; (c) $x = y = 3d$.

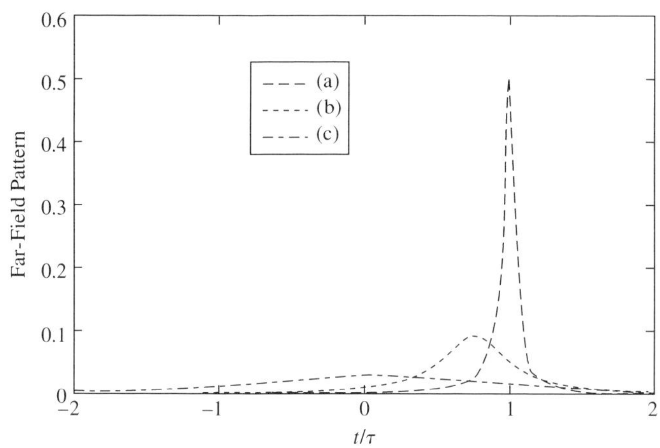

Figure 8.12 The exact far-field pattern of the source disk evaluated at three different angles of observation: (a) $(\theta, \phi) = (0, 0)$; (b) $(\theta, \phi) = (40^0, 0)$; (c) $(\theta, \phi) = (80^0, 0)$.

direction $\theta = 0$. Figure 8.12 shows the exact far-field pattern computed from (8.73) for three different angles of observation in the plane $\phi = 0$, and demonstrates the directive properties of the field radiated by the source disk.

A numerical calculation of the frequency-domain spectrum of the pressure (8.73) in the scan plane shows that this pressure is approximately bandlimited with bandlimit $\omega_{max} = 80/\tau$, so that we can choose $\Delta x_0 = \Delta y_0 = 0.04c\tau$, $N_x = N_y = 75$, and $\Delta t = 0.04\tau$. The time-domain computation scheme (8.49) was used to compute the on-axis far-field pattern of the source disk, and the results are shown in Figure 8.13. It is seen that the erroneous signal caused by the edges of the scan plane is not nearly as strong as in the case of the acoustic point source. The reason for this reduction in finite-scan error is that the directive radiation from the source disk produces a much weaker field at the edges of the scan plane than at the center of the scan plane.

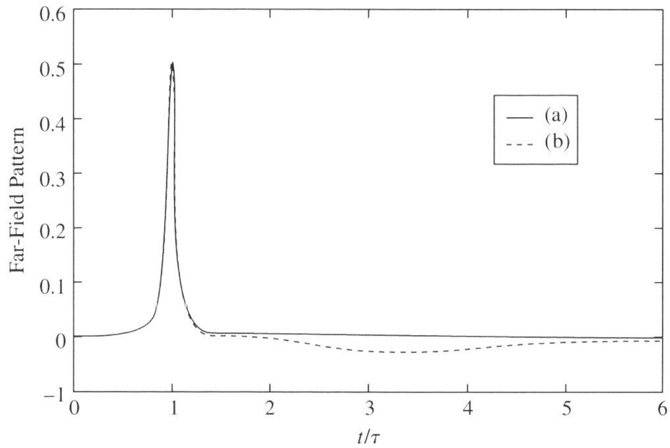

Figure 8.13 The on-axis far-field pattern of the source disk: (a) the exact far-field pattern; (b) the far-field pattern obtained from the time-domain computation scheme.

8.3.4 The Open-Ended Electromagnetic Waveguide Antenna

Finally, consider the open-ended rectangular-waveguide electromagnetic antenna shown in Figure 8.14. For a special Gaussian excitation of this antenna, we will use approximate formulas to determine the time dependence of the field in a scan plane and of the on-axis far-field pattern. Based on these results, we then discuss the advantages and disadvantages of using the frequency-domain and time-domain computation schemes to compute the time-domain far-field pattern. (We shall not use any of the computation schemes to perform actual far-field calculations.) The radiated near and far fields of this open-ended waveguide have properties that are typical of electromagnetic antennas fed by short pulses, so the discussions and conclusions of this section apply to a large class of antennas.

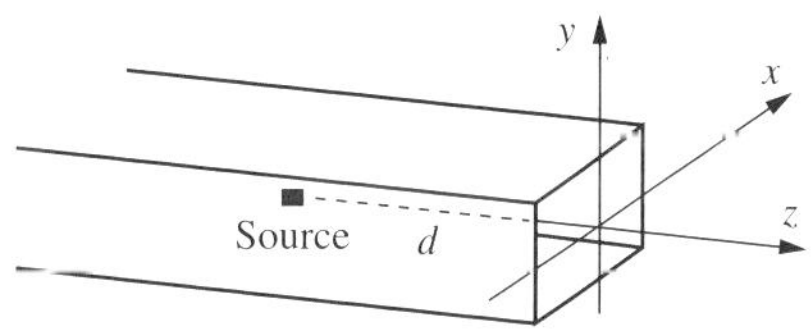

Figure 8.14 The open-ended rectangular-waveguide electromagnetic antenna.

The open-ended waveguide is fed by a source that is located at a distance d from the aperture of the waveguide and that has the Gaussian time dependence $f^G(t)$ given in (8.64). Assuming that only the TE_{10} mode is excited, we find that the electric waveguide field, which propagates in the direction of the positive z axis, is given by

$$\mathbf{E}_{10}(\mathbf{r}) = \hat{\mathbf{y}} f^G_\omega \cos\left(\frac{\pi x}{a}\right) e^{i\beta z} \tag{8.76}$$

where a is the length of the longest cross-sectional side and β is the longitudinal propagation constant given by

$$\beta = \begin{cases} ik\sqrt{(\omega_c/\omega)^2 - 1}, & \omega \le \omega_c \\ k\sqrt{1 - (\omega_c/\omega)^2}, & \omega > \omega_c \end{cases} . \tag{8.77}$$

Furthermore, $\omega_c = \pi c/a$ is the cutoff frequency for the TE_{10} mode, and f^G_ω is the Gaussian spectrum of the source given in (8.64). The spectrum of the electric field is exponentially attenuated below cutoff, and its value at the center of the aperture is approximately given by

$$\mathbf{E}_\omega^A \approx \hat{\mathbf{y}} E_\omega^A \approx \hat{\mathbf{y}} \begin{cases} e^{-kd\sqrt{(\omega_c/\omega)^2-1}}\, f_\omega^G, & 0 < \omega \le \omega_c \\ e^{ikd\sqrt{1-(\omega_c/\omega)^2}}\, f_\omega^G, & \omega > \omega_c \end{cases} . \tag{8.78}$$

We assume that the reflection at the aperture does not significantly change the qualitative frequency dependence of the aperture field. We also assume that the frequency dependence of the field in a scan plane not far from the aperture is approximately the same as the frequency dependence (8.78) of the field in the aperture.

In [198, eqs. (1a), (2) and (9)] it is shown that the on-axis far-field pattern $\mathcal{F}_\omega(\theta = 0)$ for the open-ended waveguide is given approximately by

$$\boldsymbol{\mathcal{F}}_\omega(\theta = 0) = \mathcal{F}_\omega(\theta = 0)\hat{\mathbf{y}} = K_0 \omega E_\omega^A \left[1 + \Gamma_W + \frac{\beta}{k}(1 - \Gamma_W) \right] \hat{\mathbf{y}} \tag{8.79}$$

where we have set C_0 in equation (9) of [198] equal to zero. (This simplification reduces the far-field pattern to the less accurate expression obtained with the Stratton-Chu formulas [169, ch. 10].) The constant K_0 is frequency independent, and Γ_W is the reflection coefficient of the TE_{10} mode at the open end of the waveguide. This reflection coefficient is approximately frequency independent over the recommended usable bandwidth of the waveguide.

Assuming that the waveguide is designed for X-band, we see that the TE_{10} cutoff frequency is $\omega_c = \pi c/a = 4.1 \cdot 10^{10}\ \mathrm{s}^{-1}$, $\Gamma_W \approx 0.27 e^{i1.4}$, $a = 2.286$ cm, and $b = 1.016$ cm (b is the length of the shortest cross-sectional side). The Gaussian source is chosen so that its spectrum, at the midpoint of the interval from ω_c to the next cutoff frequency $2\omega_c = 2\pi c/a = 8.2 \cdot 10^{10}\ \mathrm{s}^{-1}$, equals half its value at $\omega = 0$. The input pulse and its spectrum are then given by (8.64) with $\tau = 5.4 \cdot 10^{-11}$ s, and we assume that $d = 5$ cm.

Figure 8.15 shows the amplitudes of the frequency spectra of the Gaussian input pulse in (8.64), the approximate electric field in the aperture given by (8.78), and the approximate on-axis far-field pattern given by (8.79). It is seen that the frequency spectrum of the aperture electric field (b) is significantly attenuated below cutoff and that its slope is large around the cutoff frequency. This spectrum is very different from that of the Gaussian input pulse (a) because of the dispersion of the TE_{10} mode over the 5 centimeters of travel in the waveguide. Also, the spectrum of the on-axis far-field pattern (c) is similar to that of the aperture electric field. Thus, it is very different from the spectrum of the Gaussian input pulse (a). Near the cutoff frequency, the spectrum of the far-field pattern (c) is much smoother than the spectrum of the aperture electric field (b), because of the frequency dependence of β/k in the spectrum of the far-field pattern in (8.79).

Figure 8.16 shows the time dependence of the aperture electric field obtained by Fourier transforming (8.78). The time dependence of the electric field in the aperture is very different from the time dependence of the Gaussian input pulse (8.64). In particular, the aperture electric field has a pulsewidth of approximately 80τ, while the Gaussian input pulse has a pulsewidth of approximately 2τ. Hence, the number of time samples required for the aperture electric field is 40 times the number of time samples required for the Gaussian input pulse. As mentioned above, the Gaussian input pulse required approximately 10 time samples, so the aperture electric field requires approximately 400 time samples. In other words, to reconstruct the aperture electric field by use of the reconstruction theorem (8.43), we have to sample this field at 400 different times. Therefore, when a far-field calculation is performed for this antenna, we have to use 400 time samples in the scan plane to make sure that the far field can be calculated accurately for all times. (We assume that the time dependence of the field in the scan plane is approximately the same as the time dependence of the aperture field.)

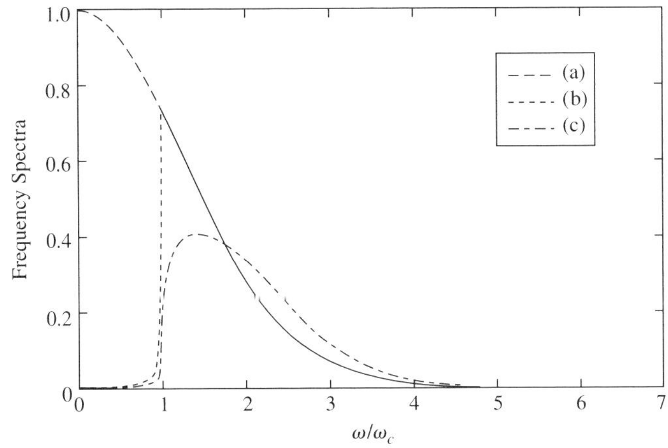

Figure 8.15 The amplitudes of the spectra for the open-ended rectangular waveguide: (a) the Gaussian input pulse; (b) the approximate aperture electric field; (c) the approximate on-axis far-field pattern. Different normalizations are used for the three curves.

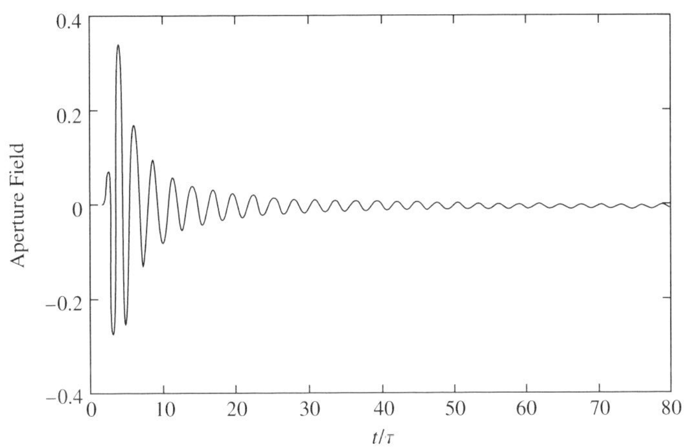

Figure 8.16 The approximate time dependence of the aperture electric field for the open-ended rectangular-waveguide antenna.

The time dependence of the approximate on-axis far-field pattern is determined by Fourier transforming (8.79) and is shown in Figure 8.17. This time dependence is very similar to the time dependence of the aperture electric field. However, because the far-field frequency spectrum has less low-frequency content than the aperture-field frequency spectrum, the time-domain on-axis far-field pattern dies off faster with time than the aperture electric field, and the pulsewidth of the far-field pattern is approximately 40τ, which is 20 times that of the Gaussian input pulse.

In summary, for the open-ended waveguide antenna the time dependence of the aperture electric field and of the far-field pattern are very different from the time dependence of the input pulse because of the dispersive effects of the waveguide on the TE_{10} mode. In particular, for a Gaussian input pulse the durations of the aperture electric field and far-field pattern are approximately 40 and 20 times larger than the duration of the input pulse, respectively.

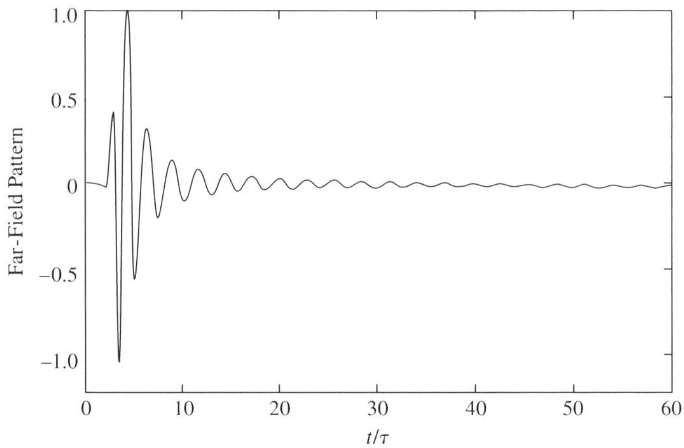

Figure 8.17 The approximate time dependence of the on-axis far-field pattern for the open-ended rectangular-waveguide antenna.

Assuming that the time dependence of the field in the scan plane is similar to that of the aperture field shown in Figure 8.16, we find that 400 time samples for each point in the scan plane are required to calculate the far-field pattern for all times. (We also assume that the open-ended waveguide antenna is directive enough that the diffraction due to the edges of the finite scan plane can be neglected.) The on-axis far-field pattern of the open-ended waveguide antenna, shown in Figure 8.17, consists of a main (early-time) part ($0 < t < 10\tau$), which contains most of the power and an oscillatory part ($t > 10\tau$). For some applications one may be interested only in the main part of the far-field pattern. In Section 8.2.1 it was shown that the time-domain computation scheme can determine the main part of the on-axis far-field pattern from measured data taken in the time interval $0 < t < 10\tau$. Since the duration of the field at a point in the scan plane is approximately 80τ (see Figure 8.16), this means that only $\frac{1}{8} \cdot 400 = 50$ time samples are needed to calculate the main part of the on-axis far-field pattern using the time-domain computation scheme.

To calculate the frequency-domain field in the scan plane, as required by the frequency-domain computation scheme, one has to sample the field at each point in the scan plane over its entire duration. Hence, with the frequency-domain computation scheme, one cannot reduce the number of time samples of the field in the scan plane, even if only the early-time far-field pattern is sought. This example shows that if one is interested only in the time-domain far-field pattern for early times, the number of time samples of the field in the scan plane required by the time-domain computation scheme can be much smaller than the number required by the frequency-domain computation scheme.

Number of Operations. Let us estimate the number of operations required by the computation schemes to compute the far-field pattern when $\omega_{max} \approx 4\omega_c = 1.6 \cdot 10^{11}\text{s}^{-1}$. The scan plane is taken to be a square of sidelength 20 cm, and since $\lambda_{min} = 2\pi c/\omega_{max} = 1.1$ cm, we find that $N_x = N_y = 18$, resulting in $(2N_x + 1)(2N_y + 1) = 1369$ scan points. As discussed above, the number of time samples needed to record the entire duration of the field at a point in the scan plane is $N_t = 400$. For the frequency-domain computation scheme, N_t must be an integer power of 2, so for the frequency-domain scheme we let $N_t = 512$. Moreover, assume that the far-field pattern is computed over the entire forward hemisphere at $N_{\theta\phi} = 16(\omega_{max} b/c)^2 \approx 500$ different angles of observation that satisfy (8.22). With this number of observation points, the angular variation of the far-field pattern will be adequately resolved. Adding up the arguments of the $O(N)$ functions in (8.25) shows that the number of operations required by the frequency-domain scheme is of the order 10^7.

If a time-domain computation scheme is used, we need only $N_t = 400$ time samples to record the entire duration (80τ) of the field at a point in the scan plane and 200 time samples to record the entire duration (40τ) of the far-field pattern. Hence, the total number of space-time far-field points (θ, ϕ, t) is $N_{\mathcal{F}} = 200 \cdot N_{\theta\phi} = 10^5$. The number of operations required by the formulas (8.55) and (8.61), which employ $\sin(x)/x$ interpolation, is found from the argument of the $O(N)$ function in (8.50) to be of the order 10^{10}. If instead the linear interpolation formulas (8.56) and (8.62) are used with (8.54) and (8.60), the number of operations is found from the argument of the $O(N)$ function in (8.51) to be of the order $3 \cdot 10^7$.

We see that the number of operations required by the frequency-domain scheme is of the same order as the number of operations required by the time-domain scheme that uses linear interpolations. Moreover, the number of operations required by the time-domain scheme that uses $\sin(x)/x$ interpolations is a factor of 10^3 higher.

As discussed above, if one is interested in calculating the far field only for early times, the number of operations required by the time-domain computation schemes decreases significantly, whereas the number of operations required by the frequency-domain computation schemes does not. For example, if the early-time far-field pattern is calculated, only 50 far-field time samples are needed, so that $N_{\mathcal{F}} = 50 \cdot N_{\theta\phi} = 2.5 \cdot 10^4$. In that case, the number of operations required by the time-domain schemes that use $\sin(x)/x$ interpolation and linear interpolations are of the order $3 \cdot 10^9$ and $8 \cdot 10^6$, respectively.

8.3.5 Comparison of the Computation Schemes

To compare the frequency-domain and time-domain computation schemes and discuss their relative merits, begin by considering the effect of truncating the infinite scan plane to a finite scan plane.

Aliasing Problems Caused by the Finite Scan Plane. All the computation schemes determine the far-field patterns from time-domain data in a finite (truncated) scan plane. For example, the two computation schemes presented in Sections 8.1.1 and 8.2.1 both calculate the finite-scan far-field pattern

$$\mathcal{F}^{fs}(\theta, \phi, t) = \frac{\cos\theta}{2\pi c} \int_{-Y_m}^{Y_m} \int_{-X_m}^{X_m} \frac{\partial}{\partial t} p(\mathbf{r}_0, t + \hat{\mathbf{r}} \cdot \mathbf{r}_0/c) dx_0 dy_0 \tag{8.80}$$

where X_m and Y_m determine the size of the scan plane. When the source is highly directive and the edges of the finite scan plane are illuminated by a very weak field, the effect of the truncation is negligible, and the finite-scan far-field pattern is practically equal to the exact far-field pattern.

When the source has a broad beam, however, and the edges of the finite scan plane are illuminated by a strong field, the finite-scan far-field pattern is significantly different from the exact far-field pattern. We can visualize the effect of the finite scan area by imagining the source (test antenna) radiating through an aperture in an infinite, perfectly absorbing screen placed in the scan plane. The aperture in the screen coincides with the finite scan area, and thus the rim (edges) of the aperture is excited by the test antenna and re-radiates to the observation point. The total field at an observation point consists of the field radiated directly by the test antenna as well as the field radiated by the rim of the aperture. The time delay between the direct pulse and the rim pulse can be determined, for each observation point, from the geometry of the test antenna and the rim of the scan area, as illustrated in Figure 8.8 and explained in Section 8.3.1.

We can express the finite-scan far-field pattern as $\mathcal{F}^{fs} = \mathcal{F} + \mathcal{F}^e$, where $\mathcal{F}$ is the exact far-field pattern and $\mathcal{F}^e$ is referred to as the erroneous (rim-induced) far-field pattern. Then it follows from Section 8.3.1 that if the scan plane is large enough, $\mathcal{F}^e$ can be separated in time from $\mathcal{F}$, so that $\mathcal{F}$ can be determined from $\mathcal{F}^{fs}$ by time gating. Moreover, the numerical examples in Section 8.3.1 confirm that the duration and location in time of $\mathcal{F}^e$ depend not only on the size of the test antenna and scan plane, but also on the angles of observation.

The time-domain computation scheme computes the exact value of the finite-scan far-field pattern (8.80), provided the spatial sample spacing and the time sample spacing (interval) are chosen in accordance with the sampling theorems in Section 8.2.1. Hence, if the scan plane is large enough, by time gating we can effectively eliminate $\mathcal{F}^e$ and determine the exact far-field pattern $\mathcal{F}$ from the time-domain computation scheme without any prior information about the duration of the finite-scan far-field pattern.

To use the frequency-domain scheme, the spatial sample spacing and the time sample spacing must be chosen in accordance with the sampling theorems in Section 8.1.1. (The spacing between sample points in the scan plane is the same for the frequency-domain and time-domain schemes.) In addition, to avoid time-domain aliasing, the frequency sample spacing $\Delta\omega$ (which occurs only in the frequency-domain scheme) must be chosen less than or equal to $2\pi/T^{fs}$, where T^{fs} is the duration of the finite-scan far-field pattern (8.80). Thus, to use the frequency-domain computation scheme, the duration of the finite-scan far-field pattern, which depends on the angles of observation as well as on the size of the test antenna and finite scan plane, must be known *a priori*. In practical situations, this information may be difficult to estimate when the test antenna is a complicated radiator. Hence, the time-domain computation scheme is easier to use when the edges of the scan plane are illuminated by a strong field that makes the finite-scan far-field pattern significantly different from the exact far-field pattern.

The preceding discussion applies to all the computation schemes of this chapter because they all determine finite-scan far-field patterns that can be written in the form (8.80).

Early-Time Far-Field Calculation. In Section 8.3.4 we analyzed the open-ended electromagnetic waveguide fed with a Gaussian source. The far-field pattern of this radiator has an early-time part that contains most of the energy and a late-time part that contains little energy. This is a typical far-field pattern for large electromagnetic antennas fed by short pulses, and for some applications we may be interested only in the far-field pattern at early times. Section 8.2.1 showed that the time-domain computation schemes determine early-time far fields from early-time data in the scan plane. Thus, when a time-domain computation scheme is used, it is not necessary to use (or even measure) the data in the scan plane for its entire duration if we are interested only in the far-field pattern for early times. This is unlike the frequency-domain computation schemes, which require measured time-domain data in the scan plane for its entire duration in order to calculate the frequency-domain data in the scan plane. (The frequency-domain data in the scan plane is obtained from the corresponding time-domain data by the Fourier-transform formulas (8.6), (8.26), (8.32), or (8.38).)

Hence, if only the early-time far-field pattern is needed, the amount of data required by the time-domain computation schemes can be significantly smaller than the amount required by the frequency-domain computation schemes. Moreover, the measurement time may also be reduced when the data in the scan plane is needed only in a shortened time interval.

Number of Operations. The formula (8.25) determines the number of operations required by the frequency-domain computation schemes, and the formulas (8.50)–(8.51) determine the number of operations for the time-domain computation schemes. To compare the number of operations when N_x, N_y, N_t, $N_{\theta\phi}$, and $N_{\mathcal{F}}$, are large, we assume that

$$N_x = O(N), \quad N_y = O(N), \quad N_t = O(N), \quad N_{\theta\phi} = O(N^2), \quad N_{\mathcal{F}} = O(N^3) \qquad (8.81)$$

where N is large. The number of scan points in the scan plane is thus of the same order as the number of far-field angles of observations, and the number of time samples used in the scan plane is the same as the number of times for which the far field is calculated at each observation point.

The formula (8.25) shows that in this case the frequency-domain scheme requires $O(N^4)$ operations. Similarly, the formulas (8.50) and (8.51) show that the time-domain scheme that uses $\sin(x)/x$ interpolation requires $O(N^6)$ operations, whereas the time-domain scheme that uses linear interpolation requires $O(N^5)$ operations. These considerations show that for large radiators, for which N_x, N_y, N_t, $N_{\theta\phi}$, and $N_{\mathcal{F}}$ can be chosen in accordance with (8.81), the frequency-domain scheme is in general the fastest. Before using a frequency-domain scheme, however, we must estimate the duration of the finite-scan far-field pattern to avoid time-domain aliasing, as described above. When only part of the far field is calculated, the difference in number of operations becomes smaller, and the time-domain computation schemes therefore become more advantageous because of their simplicity and their lack of aliasing problems.

In practice, it would be ideal to compute the fields using both the time-domain and frequency-domain computation schemes to help verify that the computer programs are working properly. Regardless of what computation scheme is used, planar time-domain near-field measurements, unlike single-frequency near-field measurements [125], have the capability of eliminating the errors caused by the finite scan plane (through the use of time gating). For some radiators, such as antennas with broad beams, this may be ample reason to consider time-domain measurements even when the final characterization of the radiator is required in the frequency domain.

A

Uniqueness of Solution to Laplace's Equation

We want to prove that the only solution to Laplace's equation

$$\nabla^2 \psi(\mathbf{r}) = 0 \tag{A.1}$$

is the trivial solution $\psi(\mathbf{r}) = 0$, if $\psi(\mathbf{r})$ is a twice continuously differentiable solution to (A.1) throughout all space, and

$$\lim_{r \to \infty} \psi(\mathbf{r}) = 0 \tag{A.2}$$

for all directions of $\mathbf{r}$ (and for any chosen origin of the coordinate system). In spherical coordinates (r, θ, ϕ), (A.2) can be expressed as

$$\lim_{r \to \infty} \psi(r, \theta, \phi) = 0 \tag{A.3}$$

for all (θ, ϕ) (and for any chosen origin of the coordinate system).

Begin the proof by integrating (A.1) over an arbitrary regular volume V_a enclosed by the surface S_a, and apply the divergence theorem to get

$$\int_{V_a} \nabla^2 \psi(\mathbf{r}) dV = \int_{S_a} \frac{\partial \psi}{\partial n} dS = 0. \tag{A.4}$$

The divergence theorem can be applied rigorously in (A.4) because the first and second derivatives of $\psi(\mathbf{r})$ are continuous.

Next apply Green's second identity to $\psi(\mathbf{r}')$ and the function $1/|\mathbf{r} - \mathbf{r}'|$ in the volume $V_a - V_\delta^s$ bounded by $S_a - S_\delta^s$, where V_δ^s is an infinitesimal spherical volume with surface S_δ^s surrounding the singularity of $1/|\mathbf{r} - \mathbf{r}'|$, to get the familiar result [1, p. 167, eq. (14)]

$$\psi(\mathbf{r}) = \frac{1}{4\pi} \int_{S_a} \frac{1}{|\mathbf{r} - \mathbf{r}'|} \frac{\partial}{\partial n'} \psi(\mathbf{r}') dS' - \frac{1}{4\pi} \int_{S_a} \psi(\mathbf{r}') \frac{\partial}{\partial n'} \left(\frac{1}{|\mathbf{r} - \mathbf{r}'|} \right) dS', \quad \mathbf{r} \in V_a. \tag{A.5}$$

By letting S_a be the surface S_R of a sphere of radius R centered at the origin ($\mathbf{r}$=0), (A.5) becomes for $\mathbf{r} = 0$

$$\psi(\mathbf{r} = 0) = \frac{1}{4\pi R} \int_{S_R} \frac{\partial \psi}{\partial n'} dS' + \frac{1}{4\pi} \int_{S_R} \frac{\psi}{R^2} dS'. \tag{A.6}$$

According to (A.4), the first integral in (A.6) is zero, and (A.6) reduces to

$$\psi(\mathbf{r}=0) = \frac{1}{4\pi}\int_{4\pi} \psi(R,\theta',\phi')d\Omega'. \tag{A.7}$$

The surface integral in (A.7) has been written in spherical coordinates with $dS'/R^2 = \sin\theta' d\theta' d\phi' = d\Omega'$.

Take the absolute value of (A.7) to get

$$|\psi(\mathbf{r}=0)| = \frac{1}{4\pi}\left|\int_{4\pi} \psi(R,\theta',\phi')d\Omega'\right| \leq \frac{1}{4\pi}\int_{4\pi} |\psi(R,\theta',\phi')|d\Omega' \leq M(R) \tag{A.8}$$

where $M(R)$ is the maximum value of the continuous function $|\psi(R,\theta,\phi)|$ at a fixed radius R. Since (A.3) implies that $\lim_{R\to\infty} M(R) = 0$, taking the limit of (A.8) as $R \to \infty$ gives

$$|\psi(\mathbf{r}=0)| \leq \lim_{R\to\infty} M(R) = 0. \tag{A.9}$$

That is, $\psi(\mathbf{r}=0) = 0$. Since the origin ($\mathbf{r}=0$) of the coordinate system can be chosen at any point in space, $\psi(\mathbf{r}) = 0$ for all $\mathbf{r}$.

Thus, the uniqueness of solution to Laplace's equation is proved under the far-field condition (A.2). It is reasonable that this uniqueness theorem should hold because the separable spherical-coordinate solutions that satisfy Laplace's equation outside the source region and approach zero as $r \to \infty$ have $1/r^n$ radial dependence with integer $n \geq 1$ [1, p. 202]. However, these solutions do not satisfy Laplace's equation (A.1) at $r = 0$, unless their coefficients are zero and thus $\psi(\mathbf{r}) = 0$.

B

Proofs of Theorems 2-I and 2-II

The sufficient conditions on the current density $\mathbf{J}(\mathbf{r}, t)$ for the existence of solution to the time-domain Maxwell equations, and sufficient conditions on the mass-flow and external force densities, $s(\mathbf{r}, t)$ and $\mathbf{f}(\mathbf{r}, t)$, for the existence of solution to the time-domain linear acoustic equations, are derived in Sections 2.1.6 and 2.2.5 using Theorems 2-I and 2-II stated in Section 2.1.6. These two theorems are proven in this appendix.

B.1 PROOF OF THEOREM 2-I

To prove Theorem 2-I of Section 2.1.6, we want to show that the first partial derivatives with respect to x, y, and z of

$$\lim_{\delta\to 0} \int\limits_{V-V_\delta} \frac{f(\mathbf{r}', t - |\mathbf{r}-\mathbf{r}'|/c)}{|\mathbf{r}-\mathbf{r}'|} dV' \tag{B.1}$$

can be brought inside the integral sign, if $f(\mathbf{r}, t)$ and $\partial f(\mathbf{r}, t)/\partial t$ are continuous functions of $\mathbf{r}$ and t, and $f(\mathbf{r}, t)$ is zero outside a bounded volume V. The partial derivative of (B.1) with respect to x , if it exists, is defined as

$$\frac{\partial}{\partial x}\lim_{\delta\to 0} \int\limits_{V-V_\delta} \frac{f(\mathbf{r}', t - |\mathbf{r}-\mathbf{r}'|/c)}{|\mathbf{r}-\mathbf{r}'|} dV'$$

$$= \lim_{\Delta x\to 0}\lim_{\delta\to 0} \int\limits_{V-V^s_{\delta,1}-V^s_{\delta,2}} \frac{1}{\Delta x}\left[\frac{f(\mathbf{r}', t - |\Delta x\hat{\mathbf{x}}+\mathbf{r}-\mathbf{r}'|/c)}{|\Delta x\hat{\mathbf{x}}+\mathbf{r}-\mathbf{r}'|} - \frac{f(\mathbf{r}', t - |\mathbf{r}-\mathbf{r}'|/c)}{|\mathbf{r}-\mathbf{r}'|}\right] dV' \tag{B.2}$$

where the volumes $V^s_{\delta,1}$ and $V^s_{\delta,2}$ are initially chosen as spherical principal volumes that exclude the singularities within the source region at $|\mathbf{r}-\mathbf{r}'| = 0$ and $|\Delta x\hat{\mathbf{x}}+\mathbf{r}-\mathbf{r}'| = 0$, respectively. The limit as δ approaches zero in (B.2) is taken before the limit of Δx approaches zero. Thus, $|\Delta x|$ can be chosen as much larger than the diameter δ of $V^s_{\delta,1}$ and $V^s_{\delta,2}$.

Since $||\Delta x\hat{\mathbf{x}} + \mathbf{r} - \mathbf{r}'| - |\mathbf{r} - \mathbf{r}'|| \le |\Delta x|$, and $\partial f(\mathbf{r}, t)/\partial t$ is a continuous function of $\mathbf{r}$ and t, $f(\mathbf{r}', t - |\Delta x\hat{\mathbf{x}} + \mathbf{r} - \mathbf{r}'|/c)$ can be expanded in the following two terms of a Taylor series about $t - |\mathbf{r} - \mathbf{r}'|/c$ as $|\Delta x| \to 0$

$$\begin{aligned} f(\mathbf{r}', t - |\Delta x\hat{\mathbf{x}} + \mathbf{r} - \mathbf{r}'|/c) = \\ f(\mathbf{r}', t - |\mathbf{r} - \mathbf{r}'|/c) - \frac{1}{c}\frac{\partial f}{\partial t}(\mathbf{r}', t - |\mathbf{r} - \mathbf{r}'|/c)(|\Delta x\hat{\mathbf{x}} + \mathbf{r} - \mathbf{r}'| - |\mathbf{r} - \mathbf{r}'|) \\ + o(|\Delta x\hat{\mathbf{x}} + \mathbf{r} - \mathbf{r}'| - |\mathbf{r} - \mathbf{r}'|) \end{aligned} \tag{B.3}$$

where $o(|\Delta x\hat{\mathbf{x}} + \mathbf{r} - \mathbf{r}'| - |\mathbf{r} - \mathbf{r}'|)$ signifies a function that approaches zero faster than $(|\Delta x\mathbf{x} + \mathbf{r} - \mathbf{r}'| - |\mathbf{r} - \mathbf{r}'|)$, and that is also continuous with respect to $\mathbf{r}'$ and Δx (since $f(\mathbf{r}', t - |\Delta x\hat{\mathbf{x}} + \mathbf{r} - \mathbf{r}'|/c)$ on the left-hand side of (B.3), and the first two terms on the right-hand side of (B.3), are continuous functions of $\mathbf{r}'$ and Δx). Inserting (B.3) into the right-hand side of (B.2) gives

$$\begin{aligned} &\lim_{\Delta x\to 0}\lim_{\delta\to 0}\int\limits_{V-V^s_{\delta,1}-V^s_{\delta,2}} \frac{f(\mathbf{r}', t - |\mathbf{r} - \mathbf{r}'|/c)}{\Delta x}\left(\frac{1}{|\Delta x\hat{\mathbf{x}} + \mathbf{r} - \mathbf{r}'|} - \frac{1}{|\mathbf{r} - \mathbf{r}'|}\right) dV' \\ &- \lim_{\Delta x\to 0}\lim_{\delta\to 0}\int\limits_{V-V^s_{\delta,2}} \left(\frac{1}{c}\frac{\partial f}{\partial t}(\mathbf{r}', t - |\mathbf{r} - \mathbf{r}'|/c)\frac{(|\Delta x\hat{\mathbf{x}} + \mathbf{r} - \mathbf{r}'| - |\mathbf{r} - \mathbf{r}'|)}{\Delta x|\Delta x\hat{\mathbf{x}} + \mathbf{r} - \mathbf{r}'|}\right. \\ &\qquad\left. + \frac{o(|\Delta x\hat{\mathbf{x}} + \mathbf{r} - \mathbf{r}'| - |\mathbf{r} - \mathbf{r}'|)}{\Delta x|\Delta x\hat{\mathbf{x}} + \mathbf{r} - \mathbf{r}'|}\right) dV'. \end{aligned} \tag{B.4}$$

The first integral in (B.4) can be treated in the same way that was used by Kellogg in the corresponding proof for static potentials in the source region. Specifically, with $f(\mathbf{r}', t - |\mathbf{r} - \mathbf{r}'|/c)$ a continuous function of $\mathbf{r}'$, [76, pp. 151–152] shows that the first term in (B.4) exists and reduces to

$$\begin{aligned} &\lim_{\Delta x\to 0}\lim_{\delta\to 0}\int\limits_{V-V^s_{\delta,1}-V^s_{\delta,2}} \frac{f(\mathbf{r}', t - |\mathbf{r} - \mathbf{r}'|/c)}{\Delta x}\left(\frac{1}{|\Delta x\hat{\mathbf{x}} + \mathbf{r} - \mathbf{r}'|} - \frac{1}{|\mathbf{r} - \mathbf{r}'|}\right) dV' \\ &= \lim_{\delta\to 0}\int\limits_{V-V_\delta} f(\mathbf{r}', t - |\mathbf{r} - \mathbf{r}'|/c)\frac{\partial}{\partial x}\left(\frac{1}{|\mathbf{r} - \mathbf{r}'|}\right) dV' \end{aligned} \tag{B.5}$$

where V_δ is an arbitrarily shaped principal volume that excludes the singularity at $\mathbf{r}' = \mathbf{r}$.

The need for $V^s_{\delta,2}$ in the second integral in (B.4) can be eliminated by making the change of integration variable $\mathbf{r}'' = \mathbf{r}' - \mathbf{r} - \Delta x\hat{\mathbf{x}}$ and writing dV'' in spherical coordinates as $r''^2 \sin\theta'' d\theta'' d\phi'' dr''$. Then the second integral term in (B.4) becomes

$$\begin{aligned} \lim_{\Delta x\to 0}\int \left(\frac{1}{c}\frac{\partial f}{\partial t}(\mathbf{r}'' + \mathbf{r} + \Delta x\hat{\mathbf{x}}, t - |\mathbf{r}'' + \Delta x\hat{\mathbf{x}}|/c)\frac{(r'' - |\mathbf{r}'' + \Delta x\hat{\mathbf{x}}|)}{\Delta x}\right. \\ \left. + \frac{o(r'' - |\mathbf{r}'' + \Delta x\hat{\mathbf{x}}|)}{\Delta x}\right) r'' \sin\theta'' d\theta'' d\phi'' dr''. \end{aligned} \tag{B.6}$$

Since $\partial f(\mathbf{r}, t)/\partial t$ is a continuous function of $\mathbf{r}$ and t, $\partial f(\mathbf{r}'' + \mathbf{r} + \Delta x\hat{\mathbf{x}}, t - |\mathbf{r}'' + \Delta x\hat{\mathbf{x}}|/c)/\partial t$ is a continuous function of $\mathbf{r}''$ and Δx, and

$$\lim_{\Delta x\to 0}\frac{\partial f}{\partial t}(\mathbf{r}'' + \mathbf{r} + \Delta x\hat{\mathbf{x}}, t - |\mathbf{r}'' + \Delta x\hat{\mathbf{x}}|/c) = \frac{\partial f}{\partial t}(\mathbf{r}'' + \mathbf{r}, t - r''/c). \tag{B.7}$$

Also, $(r'' - |\mathbf{r}'' + \Delta x\hat{\mathbf{x}}|)/\Delta x$ and $o(r'' - |\mathbf{r}'' + \Delta x\hat{\mathbf{x}}|)/\Delta x$ are continuous functions of $\mathbf{r}''$ and

Δx, such that

$$\lim_{\Delta x \to 0} \frac{(r'' - |\mathbf{r}'' + \Delta x \hat{\mathbf{x}}|)}{\Delta x} = -\frac{x''}{r''} = -\sin\theta'' \cos\phi'' \tag{B.8}$$

and

$$\lim_{\Delta x \to 0} \frac{o(r'' - |\mathbf{r}'' + \Delta x \hat{\mathbf{x}}|)}{\Delta x} = 0. \tag{B.9}$$

Therefore, the limit in (B.6) exists and can be brought inside the integral sign [71, vol. 2, sec. 225] to reduce (B.6) to

$$\begin{aligned}
&-\int_V \frac{1}{c}\frac{\partial f}{\partial t}(\mathbf{r}'' + \mathbf{r}, t - r''/c) x'' \sin\theta'' d\theta'' d\phi'' dr'' \\
&\quad = -\lim_{\delta \to 0} \int_{V - V_\delta} \frac{1}{c}\frac{\partial f}{\partial t}(\mathbf{r}'' + \mathbf{r}, t - r''/c) x'' \sin\theta'' d\theta'' d\phi'' dr'' \\
&\quad = -\lim_{\delta \to 0} \int_{V - V_\delta} \frac{1}{c}\frac{\partial f}{\partial t}(\mathbf{r}'' + \mathbf{r}, t - r''/c) \frac{x''}{r''^2} dV''.
\end{aligned} \tag{B.10}$$

With a change of integration variable given by $\mathbf{r}' = \mathbf{r}'' + \mathbf{r}$, the last integral term in (B.10) becomes

$$\begin{aligned}
&\lim_{\delta \to 0} \int_{V - V_\delta} \frac{1}{c}\frac{\partial f}{\partial t}(\mathbf{r}', t - |\mathbf{r} - \mathbf{r}'|/c) \frac{(x - x')}{|\mathbf{r} - \mathbf{r}'|^2} dV' \\
&\quad = -\lim_{\delta \to 0} \int_{V - V_\delta} \frac{1}{|\mathbf{r} - \mathbf{r}'|} \frac{\partial f}{\partial x}(\mathbf{r}', t - |\mathbf{r} - \mathbf{r}'|/c) dV'.
\end{aligned} \tag{B.11}$$

Combining (B.5) and (B.11) with (B.4) and (B.2) gives

$$\frac{\partial}{\partial x} \lim_{\delta \to 0} \int_{V - V_\delta} \frac{f(\mathbf{r}', t - |\mathbf{r} - \mathbf{r}'|/c)}{|\mathbf{r} - \mathbf{r}'|} dV' = \lim_{\delta \to 0} \int_{V - V_\delta} \frac{\partial}{\partial x} \left[\frac{f(\mathbf{r}', t - |\mathbf{r} - \mathbf{r}'|/c)}{|\mathbf{r} - \mathbf{r}'|} \right] dV'. \tag{B.12}$$

Moreover, the left- and right-hand sides of (B.12) can be proven to be continuous functions of $\mathbf{r}$ and t as follows. From (B.10) and (B.11), we see that

$$\lim_{\delta \to 0} \int_{V - V_\delta} \frac{1}{|\mathbf{r} - \mathbf{r}'|} \frac{\partial f}{\partial x}(\mathbf{r}', t - |\mathbf{r} - \mathbf{r}'|/c) dV' \tag{B.13}$$

is a continuous function of $\mathbf{r}$ and t, because $\partial f(\mathbf{r}, t)/\partial t$ is a continuous function of $\mathbf{r}$ and t [90, p. 273]. Thus, we are left with proving the continuity with respect to $\mathbf{r}$ and t of

$$\begin{aligned}
&\lim_{\delta \to 0} \int_{V - V_\delta} f(\mathbf{r}', t - |\mathbf{r} - \mathbf{r}'|/c) \frac{\partial}{\partial x} \left(\frac{1}{|\mathbf{r} - \mathbf{r}'|} \right) dV' \\
&\quad = -\lim_{\delta \to 0} \int_{V - V_\delta} f(\mathbf{r}', t - |\mathbf{r} - \mathbf{r}'|/c) \frac{(x - x')}{|\mathbf{r} - \mathbf{r}'|^3} dV' \\
&\quad = \int_V f(\mathbf{r}'' + \mathbf{r}, t - r''/c) \sin^2\theta'' \cos\phi'' d\theta'' d\phi'' dr''
\end{aligned} \tag{B.14}$$

where the change of integration variable $\mathbf{r}'' = \mathbf{r}' - \mathbf{r}$ has been made to obtain the last integral, and x''/r'' has been written as $\sin\theta''\cos\phi''$. The last integral in (B.14) is continuous with respect to $\mathbf{r}$ and t because $f(\mathbf{r}'' + \mathbf{r}, t - r''/c)$ is a continuous function of $\mathbf{r}$, $\mathbf{r}''$ and t (given that $f(\mathbf{r}, t)$ is a continuous function of $\mathbf{r}$ and t) [90, p. 273]. Consequently, the right-hand side (and thus the left-hand side) of (B.12) is a continuous function of $\mathbf{r}$ and t.

Since the derivation of the equality (B.12) and the continuity of the left- and right-hand sides of (B.12) can be repeated for the partial derivatives with respect to y and z of (B.1), instead of the partial derivative with respect to x, Theorem 2-I of Section 2.1.6 is proven.

B.2 PROOF OF THEOREM 2-II

To prove Theorem 2-II of Section 2.1.6, we begin with the result from Theorem 2-I that

$$\begin{aligned}
\frac{\partial}{\partial x}\lim_{\delta\to 0}\int\limits_{V-V_\delta}\frac{f(\mathbf{r}', t-|\mathbf{r}-\mathbf{r}'|/c)}{|\mathbf{r}-\mathbf{r}'|}dV'
&= \lim_{\delta\to 0}\int\limits_{V-V_\delta} f(\mathbf{r}', t-|\mathbf{r}-\mathbf{r}'|/c)\frac{\partial}{\partial x}\left(\frac{1}{|\mathbf{r}-\mathbf{r}'|}\right)dV' \\
&\quad - \lim_{\delta\to 0}\int\limits_{V-V_\delta}\frac{1}{c}\frac{\partial f}{\partial t}(\mathbf{r}', t-|\mathbf{r}-\mathbf{r}'|/c)\frac{(x-x')}{|\mathbf{r}-\mathbf{r}'|^2}dV'
\end{aligned} \tag{B.15}$$

for $f(\mathbf{r}, t)$ a continuous function of $\mathbf{r}$ and t; we then take a second partial derivative, $\partial/\partial x$, for example, to get

$$\begin{aligned}
\frac{\partial^2}{\partial x^2}\lim_{\delta\to 0}\int\limits_{V-V_\delta}\frac{f(\mathbf{r}', t-|\mathbf{r}-\mathbf{r}'|/c)}{|\mathbf{r}-\mathbf{r}'|}dV'
&= \frac{\partial}{\partial x}\lim_{\delta\to 0}\int\limits_{V-V_\delta} f(\mathbf{r}', t-|\mathbf{r}-\mathbf{r}'|/c)\frac{\partial}{\partial x}\left(\frac{1}{|\mathbf{r}-\mathbf{r}'|}\right)dV' \\
&\quad - \frac{\partial}{\partial x}\lim_{\delta\to 0}\int\limits_{V-V_\delta}\frac{1}{c}\frac{\partial f}{\partial t}(\mathbf{r}', t-|\mathbf{r}-\mathbf{r}'|/c)\frac{(x-x')}{|\mathbf{r}-\mathbf{r}'|^2}dV'.
\end{aligned} \tag{B.16}$$

The definition of the partial derivative allows the last integral in (B.16) to be written in the form of (B.4), namely

$$\begin{aligned}
&\lim_{\Delta x\to 0}\lim_{\delta\to 0}\int\limits_{V-V^s_{\delta,1}-V^s_{\delta,2}}\frac{1}{c}\frac{\partial f}{\partial t}(\mathbf{r}', t-|\mathbf{r}-\mathbf{r}'|/c)\frac{1}{\Delta x}\left(\frac{(\Delta x\hat{\mathbf{x}}+x-x')}{|\Delta x\hat{\mathbf{x}}+\mathbf{r}-\mathbf{r}'|^2}-\frac{(x-x')}{|\mathbf{r}-\mathbf{r}'|^2}\right)dV' \\
&- \lim_{\Delta x\to 0}\lim_{\delta\to 0}\int\limits_{V-V^s_{\delta,2}}\left(\frac{1}{c^2}\frac{\partial^2 f}{\partial t^2}(\mathbf{r}', t-|\mathbf{r}-\mathbf{r}'|/c)\frac{(\Delta x\hat{\mathbf{x}}+x-x')}{|\Delta x\hat{\mathbf{x}}+\mathbf{r}-\mathbf{r}'|}\right. \\
&\qquad\qquad \cdot\frac{(|\Delta x\hat{\mathbf{x}}+\mathbf{r}-\mathbf{r}'|-|\mathbf{r}-\mathbf{r}'|)}{\Delta x|\Delta x\hat{\mathbf{x}}+\mathbf{r}-\mathbf{r}'|} \\
&\qquad\qquad \left. + \frac{(\Delta x\hat{\mathbf{x}}+x-x')}{|\Delta x\hat{\mathbf{x}}+\mathbf{r}-\mathbf{r}'|}\frac{o(|\Delta x\hat{\mathbf{x}}+\mathbf{r}-\mathbf{r}'|-|\mathbf{r}-\mathbf{r}'|)}{\Delta x|\Delta x\hat{\mathbf{x}}+\mathbf{r}-\mathbf{r}'|}\right)dV'.
\end{aligned} \tag{B.17}$$

In the same way that we proved the second term in (B.4) exists and is equal to (B.11), we find the second term in (B.17) exists and is given by

$$
\begin{aligned}
\lim_{\Delta x\to 0}\lim_{\delta\to 0}\int\limits_{V-V^s_{\delta,2}} \Bigg(&\frac{1}{c^2}\frac{\partial^2 f}{\partial t^2}(\mathbf{r}',t-|\mathbf{r}-\mathbf{r}'|/c)\frac{(\Delta x\hat{\mathbf{x}}+x-x')}{|\Delta x\hat{\mathbf{x}}+\mathbf{r}-\mathbf{r}'|}\\
&\cdot\frac{(|\Delta x\hat{\mathbf{x}}+\mathbf{r}-\mathbf{r}'|-|\mathbf{r}-\mathbf{r}'|)}{\Delta x|\Delta x\hat{\mathbf{x}}+\mathbf{r}-\mathbf{r}'|}\\
&+\frac{(\Delta x\hat{\mathbf{x}}+x-x')}{|\Delta x\hat{\mathbf{x}}+\mathbf{r}-\mathbf{r}'|}\frac{o(|\Delta x\hat{\mathbf{x}}+\mathbf{r}-\mathbf{r}'|-|\mathbf{r}-\mathbf{r}'|)}{\Delta x|\Delta x\hat{\mathbf{x}}+\mathbf{r}-\mathbf{r}'|}\Bigg)dV'\\
=\lim_{\delta\to 0}\int\limits_{V-V_\delta}&\frac{1}{c^2}\frac{\partial^2 f}{\partial t^2}(\mathbf{r}',t-|\mathbf{r}-\mathbf{r}'|/c)\frac{(x-x')^2}{|\mathbf{r}-\mathbf{r}'|^3}dV'\\
=-\lim_{\delta\to 0}\int\limits_{V-V_\delta}&\frac{(x-x')}{c|\mathbf{r}-\mathbf{r}'|^2}\frac{\partial^2 f}{\partial x\partial t}(\mathbf{r}',t-|\mathbf{r}-\mathbf{r}'|/c)dV'
\end{aligned}
\tag{B.18}
$$

if $\partial f(\mathbf{r},t)/\partial t$ and $\partial^2 f(\mathbf{r},t)/\partial t^2$ are continuous functions of $\mathbf{r}$ and t.

The first integral in (B.17) can be evaluated much like Kellogg [76, pp. 151–152] evaluated the first integral in (B.4), if one uses the identity

$$
\begin{aligned}
\frac{1}{\Delta x}\left(\frac{(\Delta x\hat{\mathbf{x}}+x-x')}{|\Delta x\hat{\mathbf{x}}+\mathbf{r}-\mathbf{r}'|^2}-\frac{(x-x')}{|\mathbf{r}-\mathbf{r}'|^2}\right)&=\frac{\partial}{\partial x}\left(\frac{x-x'}{|\mathbf{r}-\mathbf{r}'|^2}\right)\\
&+\frac{\left[\left(\frac{4(x-x')^2}{|\mathbf{r}-\mathbf{r}'|^2}-3\right)(x-x')+\left(\frac{2(x-x')^2}{|\mathbf{r}-\mathbf{r}'|^2}-1\right)\Delta x\right]\Delta x}{|\mathbf{r}-\mathbf{r}'|^2(|\mathbf{r}-\mathbf{r}'|^2+(\Delta x)^2+2(x-x')\Delta x)}
\end{aligned}
\tag{B.19}
$$

to get

$$
\begin{aligned}
\lim_{\Delta x\to 0}\lim_{\delta\to 0}\int\limits_{V-V^s_{\delta,1}-V^s_{\delta,2}}&\frac{1}{c}\frac{\partial f}{\partial t}(\mathbf{r}',t-|\mathbf{r}-\mathbf{r}'|/c)\frac{1}{\Delta x}\\
&\cdot\left(\frac{(\Delta x\hat{\mathbf{x}}+x-x')}{|\Delta x\hat{\mathbf{x}}+\mathbf{r}-\mathbf{r}'|^2}-\frac{(x-x')}{|\mathbf{r}-\mathbf{r}'|^2}\right)dV'\\
=\lim_{\delta\to 0}\int\limits_{V-V_\delta}&\frac{1}{c}\frac{\partial f}{\partial t}(\mathbf{r}',t-|\mathbf{r}-\mathbf{r}'|/c)\frac{\partial}{\partial x}\left(\frac{x-x'}{|\mathbf{r}-\mathbf{r}'|^2}\right)dV'\\
-\lim_{\Delta x\to 0}\lim_{\delta\to 0}\int\limits_{V-V^s_{\delta,1}-V^s_{\delta,2}}&\frac{1}{c}\frac{\partial f}{\partial t}(\mathbf{r}''+\mathbf{r},t-r''/c)\\
&\cdot\frac{\left[\left(\frac{4(x''^2)}{(r''^2)}-3\right)x''-\left(\frac{2(x''^2)}{(r''^2)}-1\right)\Delta x\right]\Delta x}{r''^2(r''^2+(\Delta x)^2+2x''\Delta x)}dV''
\end{aligned}
\tag{B.20}
$$

where the first term on the right-hand side of (B.20) exists and the change of integration variable $\mathbf{r}''=\mathbf{r}'-\mathbf{r}$ has been made in the last integral of (B.20). To show that this last integral is zero when the limits are taken, note that $\partial f/\partial t$ is bounded, and express x'' and dV'' in spherical coordinates as $r'' \sin\theta'' \cos\phi''$ and $r''^2 \sin\theta'' d\theta'' d\phi'' dr''$, respectively. Then we have

$$\lim_{\delta\to 0} \int\limits_{V-V^s_{\delta,1}-V^s_{\delta,2}} \frac{1}{c}\frac{\partial f}{\partial t}(\mathbf{r}''+\mathbf{r}, t - r''/c) \frac{\left[\left(\frac{4(x''^2)}{(r''^2)} - 3\right)x'' - \left(\frac{2(x''^2)}{(r''^2)} - 1\right)\Delta x\right]\Delta x}{r''^2(r''^2 + (\Delta x)^2 + 2x''\Delta x)} dV''$$

$$\le c_1 \Delta x \int_0^{c_0}\int_0^{2\pi}\int_0^{\pi} \frac{r''}{(r''^2 + (\Delta x)^2 - 2\Delta x r'' \sin\theta'' \cos\phi'')} d\theta'' d\phi'' dr''$$

$$+ c_2(\Delta x)^2 \int_0^{c_0}\int_0^{2\pi}\int_0^{\pi} \frac{1}{(r''^2 + (\Delta x)^2 - 2\Delta x r'' \sin\theta'' \cos\phi'')} d\theta'' d\phi'' dr'' \tag{B.21}$$

where c_0, c_1, and c_2 are finite positive constants. The two triple integrals on the right-hand side of (B.21) can be evaluated to show that the first triple integral approaches a finite value as $\Delta x \to 0$, and the second triple integral is of order $1/\Delta x$ as $\Delta x \to 0$. Thus, the right-hand side of (B.21) is zero in the limit as $\Delta x \to 0$ and (B.20) becomes

$$\lim_{\Delta x\to 0}\lim_{\delta\to 0} \int\limits_{V-V^s_{\delta,1}-V^s_{\delta,2}} \frac{1}{c}\frac{\partial f}{\partial t}(\mathbf{r}', t - |\mathbf{r}-\mathbf{r}'|/c) \frac{1}{\Delta x}\left(\frac{(\Delta x\hat{\mathbf{x}} + x - x')}{|\Delta x\hat{\mathbf{x}} + \mathbf{r} - \mathbf{r}'|^2} - \frac{(x-x')}{|\mathbf{r}-\mathbf{r}'|^2}\right) dV'$$

$$= \lim_{\delta\to 0} \int\limits_{V-V_\delta} \frac{1}{c}\frac{\partial f}{\partial t}(\mathbf{r}', t - |\mathbf{r}-\mathbf{r}'|/c) \frac{\partial}{\partial x}\left(\frac{x-x'}{|\mathbf{r}-\mathbf{r}'|^2}\right) dV'. \tag{B.22}$$

Substitution of (B.22) and (B.18) into (B.17) shows the desired result that the second term on the right-hand side of (B.16) exists and the $\partial/\partial x$ partial derivative can be brought inside the integral sign, that is,

$$\frac{\partial}{\partial x}\lim_{\delta\to 0} \int\limits_{V-V_\delta} \frac{1}{c}\frac{\partial f}{\partial t}(\mathbf{r}', t - |\mathbf{r}-\mathbf{r}'|/c) \frac{(x-x')}{|\mathbf{r}-\mathbf{r}'|^2} dV'$$

$$= \lim_{\delta\to 0} \int\limits_{V-V_\delta} \frac{\partial}{\partial x}\left[\frac{1}{c}\frac{\partial f}{\partial t}(\mathbf{r}', t - |\mathbf{r}-\mathbf{r}'|/c) \frac{(x-x')}{|\mathbf{r}-\mathbf{r}'|^2}\right] dV' \tag{B.23}$$

given that $\partial f(\mathbf{r}, t)/\partial t$ and $\partial^2 f(\mathbf{r}, t)/\partial t^2$ are continuous functions of $\mathbf{r}$ and t.

It remains to prove that the first term on the right-hand side of (B.16) also exists and that the partial derivative $\partial/\partial x$ can be brought inside the integral sign. To do this, divide the volume V into $V - V^s$ and V^s, where V^s is a sphere centered on a chosen observation point $\mathbf{r}$. The sphere V^s is fixed in size and space, so that $\lim_{\delta\to 0}(V_\delta/V^s) = 0$. Nonetheless, V^s can be chosen arbitrarily small. The integrals in (B.23) over $V - V^s$ contain no singularities, and Leibnitz's rule [90, sec. 4.9] can be applied to get

$$\frac{\partial}{\partial x} \int\limits_{V-V^s} f(\mathbf{r}', t - |\mathbf{r}-\mathbf{r}'|/c) \frac{\partial}{\partial x}\left(\frac{1}{|\mathbf{r}-\mathbf{r}'|}\right) dV'$$

$$= \int\limits_{V-V^s} \frac{\partial}{\partial x}\left[f(\mathbf{r}', t - |\mathbf{r}-\mathbf{r}'|/c) \frac{\partial}{\partial x}\left(\frac{1}{|\mathbf{r}-\mathbf{r}'|}\right)\right] dV' \tag{B.24}$$

provided $\partial f(\mathbf{r}, t)/\partial t$, in addition to $f(\mathbf{r}, t)$, is a continuous function of $\mathbf{r}$ and t. Thus, we need only prove the existence of

$$\frac{\partial}{\partial x}\lim_{\delta\to 0}\int\limits_{V^s-V_\delta} f(\mathbf{r}',t-|\mathbf{r}-\mathbf{r}'|/c)\frac{\partial}{\partial x}\left(\frac{1}{|\mathbf{r}-\mathbf{r}'|}\right)dV' \tag{B.25}$$

and evaluate (B.25) as V^s becomes arbitrarily small.

Since V^s can be chosen arbitrarily small, even though it is fixed as $\delta\to 0$, $|\mathbf{r}-\mathbf{r}'|$ can be made arbitrarily small, and $f(\mathbf{r}',t-|\mathbf{r}-\mathbf{r}'|/c)$ can be expanded in a Taylor series about t as

$$f(\mathbf{r}',t-|\mathbf{r}-\mathbf{r}'|/c) = f(\mathbf{r}',t) - \frac{1}{c}\frac{\partial f}{\partial t}(\mathbf{r}',t)|\mathbf{r}-\mathbf{r}'| + o(|\mathbf{r}-\mathbf{r}'|) \tag{B.26}$$

which allows (B.25) to be written as

$$\frac{\partial}{\partial x}\lim_{\delta\to 0}\int\limits_{V^s-V_\delta} f(\mathbf{r}',t)\frac{\partial}{\partial x}\left(\frac{1}{|\mathbf{r}-\mathbf{r}'|}\right)dV'$$
$$-\frac{\partial}{\partial x}\lim_{\delta\to 0}\int\limits_{V^s-V_\delta}\left[\frac{1}{c}\frac{\partial f}{\partial t}(\mathbf{r}',t)|\mathbf{r}-\mathbf{r}'| + o(|\mathbf{r}-\mathbf{r}'|)\right]\frac{\partial}{\partial x}\left(\frac{1}{|\mathbf{r}-\mathbf{r}'|}\right)dV'. \tag{B.27}$$

Kellogg [76, pp. 153 and 19] has shown that the first term in (B.27) exists and can be evaluated in closed form as $\delta\to 0$, and the diameter D^s of V^s approaches 0, to give

$$\lim_{D^s\to 0}\frac{\partial}{\partial x}\lim_{\delta\to 0}\int\limits_{V^s-V_\delta} f(\mathbf{r}',t)\frac{\partial}{\partial x}\left(\frac{1}{|\mathbf{r}-\mathbf{r}'|}\right)dV' = -\frac{4}{3}\pi f(\mathbf{r},t) \tag{B.28}$$

provided $f(\mathbf{r},t)$ is Hölder continuous with respect to $\mathbf{r}$ (for each t); that is, if for each $\mathbf{r}$ and t there are three positive constants c, A and α, such that

$$|f(\mathbf{r}',t) - f(\mathbf{r},t)| \le A|\mathbf{r}-\mathbf{r}'|^\alpha \tag{B.29}$$

for all $\mathbf{r}'$ for which $|\mathbf{r}-\mathbf{r}'|\le c$. The Hölder condition (B.29) is stronger than continuity for all positive α and weaker than differentiability if $\alpha<1$. The analysis of [76, pp. 153–155] also shows that the second integral term in (B.27) vanishes as the diameter of V^s approaches zero, because the bracketed function in its integrand equals zero at $\mathbf{r}'=\mathbf{r}$ and satisfies the Hölder condition

$$\left|\frac{1}{c}\frac{\partial f}{\partial t}(\mathbf{r}',t)|\mathbf{r}-\mathbf{r}'| + o(|\mathbf{r}-\mathbf{r}'|)\right| \le A|\mathbf{r}-\mathbf{r}'| \tag{B.30}$$

for all $|\mathbf{r}-\mathbf{r}'|\le c$. Therefore, we have shown that the first term on the right-hand side of (B.16) exists and can be expressed as

$$\frac{\partial}{\partial x}\lim_{\delta\to 0}\int\limits_{V-V_\delta} f(\mathbf{r}',t-|\mathbf{r}-\mathbf{r}'|/c)\frac{\partial}{\partial x}\left(\frac{1}{|\mathbf{r}-\mathbf{r}'|}\right)dV'$$
$$= -\frac{4}{3}\pi f(\mathbf{r},t) + \lim_{D^s\to 0}\int\limits_{V-V^s}\frac{\partial}{\partial x}\left[f(\mathbf{r}',t-|\mathbf{r}-\mathbf{r}'|/c)\frac{\partial}{\partial x}\left(\frac{1}{|\mathbf{r}-\mathbf{r}'|}\right)\right]dV'$$
$$= -\frac{4}{3}\pi f(\mathbf{r},t) + \lim_{\delta\to 0}\int\limits_{V-V_\delta^s}\frac{\partial}{\partial x}\left[f(\mathbf{r}',t-|\mathbf{r}-\mathbf{r}'|/c)\frac{\partial}{\partial x}\left(\frac{1}{|\mathbf{r}-\mathbf{r}'|}\right)\right]dV' \tag{B.31}$$

where δ in the last integral of (B.31) is the diameter of the spherical volume $V_\delta^s = V^s$. To arrive at (B.31), it has been assumed that $f(\mathbf{r}, t)$ and $\partial f(\mathbf{r}, t)/\partial t$ are continuous functions of $(\mathbf{r}, t)$ and that $f(\mathbf{r}, t)$ is Hölder continuous in $\mathbf{r}$ at each t.

Substitution from (B.31) and (B.23) into the right-hand side of (B.16) yields

$$\begin{aligned}
&\frac{\partial^2}{\partial x^2} \lim_{\delta\to 0} \int_{V-V_\delta} \frac{f(\mathbf{r}', t - |\mathbf{r}-\mathbf{r}'|/c)}{|\mathbf{r}-\mathbf{r}'|} dV' \\
&= -\frac{4}{3}\pi f(\mathbf{r}, t) + \lim_{\delta\to 0} \int_{V-V_\delta^s} \frac{\partial}{\partial x}\left[f(\mathbf{r}', t - |\mathbf{r}-\mathbf{r}'|/c) \frac{\partial}{\partial x}\left(\frac{1}{|\mathbf{r}-\mathbf{r}'|}\right)\right. \\
&\qquad \left. - \frac{1}{c}\frac{\partial f}{\partial t}(\mathbf{r}', t - |\mathbf{r}-\mathbf{r}'|/c) \frac{(x - x')}{|\mathbf{r}-\mathbf{r}'|^2}\right] dV' \\
&= -\frac{4}{3}\pi f(\mathbf{r}, t) + \lim_{\delta\to 0} \int_{V-V_\delta^s} \frac{\partial^2}{\partial x^2}\left[\frac{f(\mathbf{r}', t - |\mathbf{r}-\mathbf{r}'|/c)}{|\mathbf{r}-\mathbf{r}'|}\right] dV'
\end{aligned} \tag{B.32}$$

where, in all, we have assumed that $f(\mathbf{r}, t)$, $\partial f(\mathbf{r}, t)/\partial t$ and $\partial^2 f(\mathbf{r}, t)/\partial t^2$ are continuous functions of $(\mathbf{r}, t)$, and $f(\mathbf{r}, t)$ is Hölder continuous in $\mathbf{r}$ at each t. When the partial derivative $\partial/\partial y$ or $\partial/\partial z$ of (B.15) is taken instead of $\partial/\partial x$, the only change that occurs in the analysis leading to (B.32) is that the right-hand side of (B.28) equals zero instead of $-4\pi f(\mathbf{r}, t)/3$. Specifically

$$\lim_{D^s\to 0} \frac{\partial}{\partial y}\left[f(\mathbf{r}, t) \lim_{\delta\to 0} \int_{V^s - V_\delta} \frac{\partial}{\partial x}\left(\frac{1}{|\mathbf{r}-\mathbf{r}'|}\right) dV' \right] = 0 \tag{B.33}$$

and instead of (B.32), we obtain the result

$$\begin{aligned}
&\frac{\partial^2}{\partial y \partial x} \lim_{\delta\to 0} \int_{V-V_\delta} \frac{f(\mathbf{r}', t - |\mathbf{r}-\mathbf{r}'|/c)}{|\mathbf{r}-\mathbf{r}'|} dV' \\
&\qquad = \lim_{\delta\to 0} \int_{V-V_\delta^s} \frac{\partial^2}{\partial y \partial x}\left[\frac{f(\mathbf{r}', t - |\mathbf{r}-\mathbf{r}'|/c)}{|\mathbf{r}-\mathbf{r}'|}\right] dV'.
\end{aligned} \tag{B.34}$$

Because the derivations of (B.32) and (B.34) can be repeated with x replaced by y or z, and with y replaced by x or z, Theorem 2-II of Section 2.1.6 is proven.

C

Uniqueness of Solution to the Scalar Helmholtz Equation

We want to prove that the only solution to the homogeneous scalar Helmholtz equation

$$\nabla^2\psi(\mathbf{r}) + k^2\psi(\mathbf{r}) = 0 \tag{C.1}$$

is the trivial solution $\psi(\mathbf{r}) = 0$, if $\psi(\mathbf{r})$ is a twice continuously differentiable solution to (C.1) throughout all space that satisfies the following outgoing wave condition (for $e^{-i\omega t}$ time dependence)

$$\frac{\partial\psi}{\partial r} - ik\psi = o(1/r) \tag{C.2}$$

for all directions of $\mathbf{r}$ (and for any chosen origin of the coordinate system). The $o(1/r)$ on the right-hand side of (C.2) denotes a function of $\mathbf{r}$ that approaches zero faster than $1/r$ as $r \to \infty$; that is, $\lim_{r\to\infty}[ro(1/r)] = 0$. The propagation constant $k = \omega/c$ can be any positive or negative real number, but we assume that $k \neq 0$. For $k = 0$, (C.1) reduces to Laplace's equation, for which uniqueness of solution was proven in Appendix A. The outgoing wave condition in (C.2) is weaker than the Sommerfeld radiation conditions [1, sec. 9.1] because ψ can satisfy (C.2), and yet both ψ and $\partial\psi/\partial r$ can decay slower than $1/r$ as $r \to \infty$. However, it is assumed that $\psi(\mathbf{r})$ and its first and second partial derivatives are continuous functions of $\mathbf{r}$. Thus, (C.2) implies that $o(1/r)$ on the right-hand side of (C.2) is a continuous function of $\mathbf{r}$. In spherical coordinates (r, θ, ϕ), (C.2) can be expressed as

$$\frac{\partial}{\partial r}\psi(r,\theta,\phi) - ik\psi(r,\theta,\phi) = o(1/r) \tag{C.3}$$

for all (θ, ϕ) (and for any chosen origin of the coordinate system). Rellich [199] and Wilcox [200] proved the uniqueness of solution to the homogeneous Helmholtz equation in the free space outside scatterers under the same outgoing wave condition.

We begin the proof by substituting $\psi(\mathbf{r})$ and its complex conjugate $\psi^*(\mathbf{r})$ into Green's first identity [1, sec. 3.3] to obtain

$$\int_{V_R} (|\nabla\psi|^2 + \psi\nabla^2\psi^*)dV = \int_{S_R} \psi\frac{\partial\psi^*}{\partial r}dS \tag{C.4}$$

where V_R is a sphere with surface S_R of radius R. Substituting $\nabla^2\psi^*$ from (C.1) into the volume integral of (C.4), and $\partial\psi^*/\partial r$ from (C.3) into the surface integral of (C.4), then taking the imaginary part of (C.4) and squaring the result produces

$$k^2\left(\int_{S_R}|\psi(R,\theta,\phi)|^2 dS\right)^2 = \left(\int_{S_R}\mathrm{Im}[\psi(R,\theta,\phi)o(1/R)]dS\right)^2$$

$$\leq \left(\int_{S_R}|\psi(R,\theta,\phi)||o(1/R)|dS\right)^2 \leq \int_{S_R}|\psi(R,\theta,\phi)|^2 dS \int_{S_R}|o(1/R)|^2 dS \tag{C.5}$$

where (R,θ,ϕ) in (C.5) are the spherical coordinates of $\mathbf{r}$. The last inequality in (C.5) is obtained from the Schwarz inequality for integrals. From (C.5) we can write

$$\left(k^2\int_{S_R}|\psi|^2 dS - \int_{S_R}|o(1/R)|^2 dS\right)\int_{S_R}|\psi|^2 dS \leq 0 \tag{C.6}$$

which implies that

$$k^2\int_{S_R}|\psi(R,\theta,\phi)|^2 dS \leq \int_{S_R}|o(1/R)|^2 dS. \tag{C.7}$$

Taking the limit of (C.7) as $R\to\infty$ shows that

$$\lim_{R\to\infty}\int_{4\pi} R^2|\psi(R,\theta,\phi)|^2 d\Omega = 0 \tag{C.8}$$

where dS has been written in spherical coordinates as $dS = R^2\sin\theta d\theta d\phi = R^2 d\Omega$.

Substituting ψ (instead of $\partial\psi^*/\partial r$) from (C.3) into the surface integral of (C.4) and repeating the above steps produces

$$\lim_{R\to\infty}\int_{4\pi} R^2\left|\frac{\partial}{\partial R}\psi(R,\theta,\phi)\right|^2 d\Omega = 0. \tag{C.9}$$

Next we apply Green's second identity to $\psi(\mathbf{r}')$ and the free-space Green's function $\exp(ik|\mathbf{r}-\mathbf{r}'|)/|\mathbf{r}-\mathbf{r}'|$ in the volume $V_a - V_\delta^s$ bounded by $S_a - S_\delta^s$, where V_δ^s is an infinitesimal spherical volume with surface S_δ^s surrounding the singularity of $1/|\mathbf{r}-\mathbf{r}'|$, to obtain the familiar result [1, secs. 3.4 and 8.13]

$$\psi(\mathbf{r}) = \frac{1}{4\pi}\int_{S_a}\left[\frac{e^{ik|\mathbf{r}-\mathbf{r}'|}}{|\mathbf{r}-\mathbf{r}'|}\frac{\partial}{\partial n'}\psi(\mathbf{r}') - \psi(\mathbf{r}')\frac{\partial}{\partial n'}\left(\frac{e^{ik|\mathbf{r}-\mathbf{r}'|}}{|\mathbf{r}-\mathbf{r}'|}\right)\right]dS', \quad \mathbf{r}\in V_a. \tag{C.10}$$

Equation (C.10) is an identity for any twice continuously differentiable solution to the homogeneous scalar Helmholtz equation (C.1).

By letting S_a be the surface of the sphere S_R centered at the origin ($\mathbf{r}=0$) and taking the absolute value, (C.10) becomes for $\mathbf{r}=0$

$$|\psi(\mathbf{r}=0)| = \frac{1}{4\pi}\left|\int_{4\pi}\left[\frac{\partial\psi}{\partial R}(R,\theta',\phi') + \left(\frac{1}{R}-ik\right)\psi(R,\theta',\phi')\right]Re^{ikR}d\Omega'\right|$$

$$\leq \frac{1}{4\pi}\int_{4\pi} R\left|\frac{\partial\psi}{\partial R}\right|d\Omega' + \frac{(1/R^2+k^2)^{\frac{1}{2}}}{4\pi}\int_{4\pi} R|\psi|d\Omega'. \tag{C.11}$$

The Schwarz integral inequality applied to $|\psi|$ and R reveals that

$$\int_{4\pi} R|\psi|d\Omega' \le \left(4\pi R^2 \int_{4\pi} |\psi|^2 d\Omega'\right)^{\frac{1}{2}}. \tag{C.12}$$

Similarly, the Schwarz inequality applied to $|\partial\psi/\partial R|$ and R gives

$$\int_{4\pi} R\left|\frac{\partial\psi}{\partial R}\right| d\Omega' \le \left(4\pi R^2 \int_{4\pi} \left|\frac{\partial\psi}{\partial R}\right|^2 d\Omega'\right)^{\frac{1}{2}}. \tag{C.13}$$

Utilizing (C.12) and (C.13) in (C.11) produces the inequality

$$|\psi(\mathbf{r}=0)| \le \left(\frac{1}{4\pi}\int_{4\pi} R^2 \left|\frac{\partial\psi}{\partial R}\right|^2 d\Omega'\right)^{\frac{1}{2}} + \left(\frac{(1/R^2+k^2)}{4\pi}\int_{4\pi} R^2|\psi|^2 d\Omega'\right)^{\frac{1}{2}}. \tag{C.14}$$

Taking the limit of (C.14) as $R \to \infty$ and using (C.8) and (C.9) shows that $\psi(\mathbf{r}=0)=0$. Since the origin ($\mathbf{r}=0$) of the coordinate system can be chosen arbitrarily, $\psi(\mathbf{r})=0$ for all $\mathbf{r}$.

Thus, the uniqueness of solution to the scalar Helmholtz equation is proved under the outward wave condition (C.2). It is not surprising that this uniqueness theorem should hold because the separable spherical-wave solutions that satisfy the Helmholtz equation outside the source region and the outgoing wave condition (C.2) have a radial dependence given by spherical Hankel functions of the first kind. However, the spherical Hankel functions of the first kind, or any linear combination of them, do not satisfy the homogeneous Helmholtz equation (C.1) at $r=0$, unless their coefficients are zero and thus $\psi(\mathbf{r})=0$.

D

Validation of the Plane-Wave Spectrum Representation

In this appendix we prove the validity of the frequency-domain plane-wave spectrum formulas derived in Section 3.2. Since the proofs for the acoustic and electromagnetic formulas are very similar, we shall consider only the acoustic formulas, which are given in (3.56)–(3.57) and state that

$$p_\omega(\mathbf{r}) = \frac{1}{2\pi}\int_{-\infty}^{+\infty}\int_{-\infty}^{+\infty} T_\omega(k_x, k_y)e^{i(k_x x+k_y y+\gamma z)}dk_x dk_y, \quad z \geq z_0 \tag{D.1}$$

where the plane-wave spectrum is given by

$$T_\omega(k_x, k_y) = \frac{e^{-i\gamma z_0}}{2\pi}\int_{-\infty}^{+\infty}\int_{-\infty}^{+\infty} p_\omega(\mathbf{r}_0)e^{-i(k_x x_0+k_y y_0)}dx_0 dy_0 \tag{D.2}$$

and $\mathbf{r}_0 = x_0\hat{\mathbf{x}} + y_0\hat{\mathbf{y}} + z_0\hat{\mathbf{z}}$ is a point in the scan plane $z = z_0$.

We note that the acoustic pressure is neither absolutely integrable nor square integrable in any plane $z = z_0$. This follows from the fact that the acoustic pressure behaves as $p_\omega(\mathbf{r}_0) = O[(x_0^2 + y_0^2)^{-1/2}]$ for $(x_0^2 + y_0^2)^{1/2} \to \infty$, so that both the integrals $\int_{-\infty}^{+\infty}\int_{-\infty}^{+\infty} |p_\omega(\mathbf{r}_0)|dx_0 dy_0$ and $\int_{-\infty}^{+\infty}\int_{-\infty}^{+\infty} |p_\omega(\mathbf{r}_0)|^2 dx_0 dy_0$ are infinite. Consequently, the theory of the Fourier transform in the Lebesgue spaces L_1 or L_2 [140], [141] cannot be used to prove the validity of (D.1) and (D.2) because the acoustic pressure in the plane $z = z_0$ belongs neither to L_1 nor to L_2. (Equation (3.243) shows that the power radiated through the plane $z = z_0$ is proportional to the integral of $p_\omega(\partial p_\omega^*/\partial z)$ over this plane. This energy, which is calculated in Section 3.2.9, is finite, even though the integral of $|p_\omega|^2$ over the plane $z = z_0$ is infinite.)

In Section 3.2.2 we proved rigorously from the Weyl identity (3.61) that the expression (D.1) produces the correct pressure when the plane-wave spectrum is given by the Huygens' source expression (3.105). Using the facts that

$$\frac{\partial}{\partial n}e^{-i(k_x x+k_y y+\gamma z)} = -i\mathbf{k}\cdot\hat{\mathbf{n}}e^{-i(k_x x+k_y y+\gamma z)}, \quad \frac{\partial}{\partial n}p_\omega = i\omega\rho_{m0}\,\hat{\mathbf{n}}\cdot\mathbf{v}_\omega \tag{D.3}$$

with $\hat{\mathbf{n}}$ being the outward normal to S, the formula (3.105) becomes

$$T_\omega(k_x, k_y) = \frac{i}{4\pi\gamma}\int_S \left[p_\omega(\mathbf{r})\frac{\partial}{\partial n}e^{-i(k_x x+k_y y+\gamma z)} - e^{-i(k_x x+k_y y+\gamma z)}\frac{\partial}{\partial n}p_\omega(\mathbf{r})\right]dS \tag{D.4}$$

where S is a finite smooth surface located in the source-free part of the half space $z < z_0$ and enclosing the finite source region V, as shown in Figure 3.1. Consequently, if we can prove that the spectrum given by (D.2) equals the spectrum given by (D.4), the validity of the representation (D.1)–(D.2) has been established. Below we show that the expressions (D.2) and (D.4) indeed agree, provided the integral over the scan plane in (D.2) is calculated in a prescribed manner.

Since the integrand in (D.2) is not absolutely integrable, the value of the integral in (D.2) depends on the order of integration. To ensure convergence and to produce the correct spectrum, we specify that the double integral in (D.2) be calculated in the following way. First, introduce a convex scan area D_R that is part of the plane $z = z_0$ and lies within the circle of radius R. Moreover, require that the scan area D_R has a fixed shape and, as $R \to \infty$, it covers the entire plane $z = z_0$. For example, D_R can be the circular scan area given by $x_0^2 + y_0^2 \le R^2$ or the rectangular scan area given by $|x_0| \le R/2$, $|y_0| \le R/2$. These two scan areas are the ones used for most practical planar near-field measurements. The integration in (D.2) is then performed by first integrating over D_R and subsequently letting $R \to \infty$. Hence

$$T_\omega(k_x, k_y) = \frac{e^{-i\gamma z_0}}{2\pi} \lim_{R\to\infty} \iint_{D_R} p_\omega(\mathbf{r}_0) e^{-i(k_x x_0 + k_y y_0)} dx_0 dy_0. \tag{D.5}$$

To show that (D.5) converges, first write the integration (D.5) in polar coordinates (ρ_0, ϕ_0) given by $x_0 = \rho_0 \cos\phi_0$ and $y_0 = \rho_0 \sin\phi_0$, as

$$T_\omega(k_x, k_y) = \frac{e^{-i\gamma z_0}}{2\pi} \lim_{R\to\infty} \int_0^R \rho_0 \int_0^{2\pi} I_R(\rho_0, \phi_0) p_\omega(\mathbf{r}_0) e^{-i\rho_0(k_x \cos\phi_0 + k_y \sin\phi_0)} d\phi_0 d\rho_0 \tag{D.6}$$

where $I_R(\rho_0, \phi_0) = 1$ when the point (ρ_0, ϕ_0) is in D_R, and $I_R(\rho_0, \phi_0) = 0$ otherwise. In the plane $z = z_0$, the acoustic far-field pressure can be written as

$$p_\omega(\mathbf{r}_0) = \mathcal{F}_\omega(\pi/2, \phi_0) \frac{e^{ik\rho_0}}{\rho_0} + O\left(\frac{e^{ik\rho_0}}{\rho_0^2}\right) \tag{D.7}$$

where $\mathcal{F}_\omega(\theta, \phi)$ is the far-field pattern of p_ω. Inserting the far-field expression (D.7) into (D.6) and performing an asymptotic evaluation shows that for $(k_x, k_y) \ne (0, 0)$ the ϕ_0 integral is asymptotically of order $\rho_0^{-1/2}(k_x^2 + k_y^2)^{-1/2} e^{-i\rho_0(\pm(k_x^2+k_y^2)^{1/2} - k)}$, as $\rho_0 \to \infty$. Integrating the resulting ρ_0 integral by parts shows that the integral (D.6) indeed converges when $(k_x, k_y) \ne (0, 0)$ and $\gamma \ne 0$. Moreover, this asymptotic analysis shows that the integral in (D.6) has a possible γ^{-1} singularity at $\gamma = 0$, which was discussed in detail in Section 3.2.3. The possible divergence of the integral (D.6) for $(k_x, k_y) = (0, 0)$ will be discussed below. Let us first prove that for $(k_x, k_y) \ne (0, 0)$ the formula (D.6) produces the correct spectrum given in (D.4).

D.1 EVALUATION OF THE SPECTRUM FOR $(k_x, k_y) \ne (0, 0)$

The asymptotic analysis above shows that the value of the integral in (D.6) is independent of the shape of D_R for $(k_x, k_y) \ne (0, 0)$. In the following derivation, we let D_R be a circular disk centered at $(0, 0, z_0)$ so that

$$T_\omega(k_x, k_y) = \frac{e^{-i\gamma z_0}}{2\pi} \int_0^{+\infty} \rho_0 \int_0^{2\pi} p_\omega(\mathbf{r}_0) e^{-i\rho_0(k_x \cos\phi_0 + k_y \sin\phi_0)} d\phi_0 d\rho_0, \quad (k_x, k_y) \ne (0, 0). \tag{D.8}$$

It should be emphasized that the integration in (D.8) would not converge if the ρ_0 integration was performed before the ϕ_0 integration. This can be seen directly from the far-field expression (D.7) for the pressure in the scan plane.

The Helmholtz representation (2.382) shows that the pressure in the scan plane $z = z_0$ can be expressed in terms of the Huygens' sources on S as

$$p_\omega(\mathbf{r}_0) = \frac{1}{4\pi} \int_S \left[p_\omega(\mathbf{r}) \frac{\partial}{\partial n} \frac{e^{ik|\mathbf{r}_0 - \mathbf{r}|}}{|\mathbf{r}_0 - \mathbf{r}|} - \frac{e^{ik|\mathbf{r}_0 - \mathbf{r}|}}{|\mathbf{r}_0 - \mathbf{r}|} \frac{\partial}{\partial n} p_\omega(\mathbf{r}) \right] dS. \tag{D.9}$$

Inserting this expression into the formula (D.8) gives

$$\begin{aligned} T_\omega(k_x, k_y) &= \frac{e^{-i\gamma z_0}}{8\pi^2} \int_0^{+\infty} \rho_0 \int_S \int_0^{2\pi} p_\omega(\mathbf{r}) \frac{\partial}{\partial n} \frac{e^{ik|\mathbf{r}_0 - \mathbf{r}|}}{|\mathbf{r}_0 - \mathbf{r}|} e^{-i\rho_0(k_x \cos\phi_0 + k_y \sin\phi_0)} d\phi_0 dS d\rho_0 \\ &- \frac{e^{-i\gamma z_0}}{8\pi^2} \int_0^{+\infty} \rho_0 \int_S \int_0^{2\pi} \frac{e^{ik|\mathbf{r}_0 - \mathbf{r}|}}{|\mathbf{r}_0 - \mathbf{r}|} \frac{\partial}{\partial n} p_\omega(\mathbf{r}) e^{-i\rho_0(k_x \cos\phi_0 + k_y \sin\phi_0)} d\phi_0 dS d\rho_0. \end{aligned} \tag{D.10}$$

We have used the facts that the pressure and its normal derivative are continuous functions on the bounded closed surface S to interchange the integration over S with the integration over ϕ_0 [71, vol. 2, p. 339]. Since the Green's function $e^{ik|\mathbf{r}_0 - \mathbf{r}|}/|\mathbf{r}_0 - \mathbf{r}|$ and its normal derivative satisfy the same asymptotic far-field conditions as the acoustic pressure, the asymptotic analysis below (D.6) shows that the integrals

$$I_1(k_x, k_y, \mathbf{r}, z_0) = \int_0^{+\infty} \rho_0 \int_0^{2\pi} \frac{e^{ik|\mathbf{r}_0 - \mathbf{r}|}}{|\mathbf{r}_0 - \mathbf{r}|} e^{-i\rho_0(k_x \cos\phi_0 + k_y \sin\phi_0)} d\phi_0 d\rho_0 \tag{D.11}$$

and

$$I_2(k_x, k_y, \mathbf{r}, z_0) = \int_0^{+\infty} \rho_0 \int_0^{2\pi} \frac{\partial}{\partial n} \frac{e^{ik|\mathbf{r}_0 - \mathbf{r}|}}{|\mathbf{r}_0 - \mathbf{r}|} e^{-i\rho_0(k_x \cos\phi_0 + k_y \sin\phi_0)} d\phi_0 d\rho_0 \tag{D.12}$$

both converge for all $\mathbf{r} \in S$. Consequently, the conditions are met for the theorem in Hobson [71, vol. 2, p. 348] to hold, and the integration over S can be interchanged with the integration over ρ_0 in (D.10). Hence

$$T_\omega(k_x, k_y) = \frac{e^{-i\gamma z_0}}{8\pi^2} \int_S \left[p_\omega(\mathbf{r}) I_2(k_x, k_y, \mathbf{r}, z_0) - I_1(k_x, k_y, \mathbf{r}, z_0) \frac{\partial}{\partial n} p_\omega(\mathbf{r}) \right] dS. \tag{D.13}$$

We have already shown that the value of the integral (D.5) is independent of the shape of D_R. Therefore, without changing the value of the integral (D.11), we can change the polar integration variables such that they are centered at $x\hat{\mathbf{x}} + y\hat{\mathbf{y}} + z_0\hat{\mathbf{z}}$ to get[1]

$$\begin{aligned} I_1(k_x, k_y, \mathbf{r}, z_0) &= e^{-i(k_x x + k_y y)} \int_0^{+\infty} \rho_0 \int_0^{2\pi} \frac{e^{ik\sqrt{\rho_0^2 + (z_0 - z)^2}}}{\sqrt{\rho_0^2 + (z_0 - z)^2}} e^{-i\rho_0(k_x \cos\phi_0 + k_y \sin\phi_0)} d\phi_0 d\rho_0 \\ &= 2e^{-i(k_x x + k_y y)} \int_0^{+\infty} \rho_0 \frac{e^{ik\sqrt{\rho_0^2 + (z_0 - z)^2}}}{\sqrt{\rho_0^2 + (z_0 - z)^2}} \int_0^{\pi} \cos(\rho_0 \sqrt{k_x^2 + k_y^2} \sin\phi_0) d\phi_0 d\rho_0. \end{aligned} \tag{D.14}$$

From the integral expression for the Bessel function of zero order it is found that the integral

[1] By considering the integration over two nonconcentric circular disks with equal radii, one can prove directly from (D.11) that the center of the polar integration variables can be shifted without changing the value of the integral. This approach is used in the evaluation of the integral (D.22) below.

over ϕ_0 is simply $\pi J_0(\rho_0\sqrt{k_x^2+k_y^2})$. For $(k_x, k_y) \neq (0, 0)$ the remaining integral over ρ_0 can then be calculated from the formula in [147, vol. 2, eq. (25), p. 9], and we find that

$$I_1(k_x, k_y, \mathbf{r}, z_0) = \frac{2\pi i}{\gamma} e^{-i(k_x x + k_y y + \gamma(z - z_0))}, \quad (k_x, k_y) \neq (0, 0). \tag{D.15}$$

Since $I_2(k_x, k_y, \mathbf{r}, z_0)$ is a continuous function on S, the theorem in [71, vol. 2, p. 357] shows that the normal derivative can be brought outside the integral in (D.12) to give

$$\begin{aligned} I_2(k_x, k_y, \mathbf{r}, z_0) &= \frac{\partial}{\partial n} I_1(k_x, k_y, \mathbf{r}, z_0) \\ &= \frac{2\pi i}{\gamma} \frac{\partial}{\partial n} e^{-i(k_x x + k_y y + \gamma(z - z_0))}, \quad (k_x, k_y) \neq (0, 0). \end{aligned} \tag{D.16}$$

Substituting (D.15) and (D.16) into (D.13) reveals that, for $(k_x, k_y) \neq (0, 0)$, the spectrum given by the integration over the scan plane in (D.8) equals the spectrum given in terms of the Huygens' sources in (D.4). Note that although the integral (D.8) diverges by oscillation for $(k_x, k_y) = (0, 0)$, the expression for the spectrum obtained by substituting (D.15) and (D.16) into (D.13) is identical to (D.4) and therefore remains valid for $(k_x, k_y) = (0, 0)$. This shows that the integral (D.8) has a removable singularity at $(k_x, k_y) = (0, 0)$.

We have now shown that if we let $T_\omega(k_x, k_y)$ be defined by (D.5) for $(k_x, k_y) \neq (0, 0)$ and define $T_\omega(0, 0) = \lim_{(k_x,k_y)\to(0,0)} T_\omega(k_x, k_y)$, then the plane-wave spectrum formula (D.1) produces the correct acoustic pressure. (Arbitrarily, we can let $T_\omega(0, 0) = 0$ since the value of the double integral (D.1) is unaffected by the value of $T_\omega(k_x, k_y)$ at a single point.) *Under the condition that the spectrum in (D.2) be calculated as specified by (D.5), we have now proven that the plane-wave spectrum formulas (D.1) and (D.2) indeed are valid.*

D.2 EVALUATION OF THE SPECTRUM FOR $(k_x, k_y) = (0, 0)$

Although the single point $(k_x, k_y) = (0, 0)$ is unimportant when the pressure is calculated by the plane-wave spectrum integral (D.1), it is very important for far-field calculations because the far-field pattern is given directly in terms of the spectrum as

$$\mathcal{F}_\omega(\theta, \phi) = -ik\cos\theta\, T_\omega(k\cos\phi\sin\theta, k\sin\phi\sin\theta) \tag{D.17}$$

which shows that the on-axis far-field pressure is determined by $T_\omega(0, 0)$. Moreover, the Huygens' source expression (D.4) shows that

$$T_\omega(0, 0) = \frac{i}{4\pi k} \int_S \left[p_\omega(\mathbf{r}) \frac{\partial}{\partial n} e^{-ikz} - e^{-ikz} \frac{\partial}{\partial n} p_\omega(\mathbf{r}) \right] dS \tag{D.18}$$

which, through (D.17), is in agreement with the Huygens' source expression (3.38) for the far-field pattern.

The far-field expression (D.7) for the acoustic pressure in the scan plane shows that for $(k_x, k_y) = (0, 0)$, the integral over the scan plane (D.8) diverges by oscillation. Consequently, for a circular scan area, the on-axis far-field pattern cannot be determined directly from the integration (D.8) over the scan plane. We shall now show that the correct value of $T_\omega(0, 0)$ given in (D.18) is related to the oscillating result obtained from (D.8) through the formula

$$T_\omega(0, 0) = \frac{e^{-ikz_0}}{2\pi} \lim_{R\to\infty} \left[\int_0^R \rho_0 \int_0^{2\pi} p_\omega(\mathbf{r}_0) d\phi_0 d\rho_0 - \frac{e^{ikR}}{ik} \int_0^{2\pi} \mathcal{F}_\omega(\pi/2, \phi_0) d\phi_0 \right]. \tag{D.19}$$

To prove (D.19), insert the Huygens' source expressions (D.9) and

$$\mathcal{F}_\omega(\pi/2, \phi_0) = \frac{1}{4\pi} \int_S \left[p_\omega(\mathbf{r}) \frac{\partial}{\partial n} e^{-ik(x\cos\phi_0 + y\sin\phi_0)} - e^{-ik(x\cos\phi_0 + y\sin\phi_0)} \frac{\partial}{\partial n} p_\omega(\mathbf{r}) \right] dS \tag{D.20}$$

into the integrals in (D.19) to get

$$\int_0^R \rho_0 \int_0^{2\pi} p_\omega(\mathbf{r}_0) d\phi_0 d\rho_0 - \frac{e^{ikR}}{ik} \int_0^{2\pi} \mathcal{F}_\omega(\pi/2, \phi_0) d\phi_0$$

$$= \frac{1}{4\pi} \int_S \left[p_\omega(\mathbf{r}) \frac{\partial}{\partial n} \Lambda_R(\mathbf{r}, z_0) - \Lambda_R(\mathbf{r}, z_0) \frac{\partial}{\partial n} p_\omega(\mathbf{r}) \right] dS \tag{D.21}$$

where

$$\Lambda_R(\mathbf{r}, z_0) = \int_0^{2\pi} \left[\int_0^R \rho_0 \frac{e^{ik|\mathbf{r}_0 - \mathbf{r}|}}{|\mathbf{r}_0 - \mathbf{r}|} d\rho_0 - \frac{e^{ikR}}{ik} e^{-ik(x\cos\phi_0 + y\sin\phi_0)} \right] d\phi_0. \tag{D.22}$$

We have used the facts that R is finite and that the pressure and its normal derivative are continuous on the surface S to interchange the integrations and derivatives required to obtain (D.21). To evaluate (D.22), introduce new polar integration variables (ρ_1, ϕ_1) that are centered at $x\hat{\mathbf{x}} + y\hat{\mathbf{y}} + z_0\hat{\mathbf{z}}$. For $R > \sqrt{x^2 + y^2}$, the circular disk given by $0 \le \phi_0 < 2\pi$ and $0 \le \rho_0 \le R$ can be described by the new variables as $0 \le \phi_1 < 2\pi$ and $0 \le \rho_1 \le R_1(\phi_1)$ with

$$R_1(\phi_1) = \sqrt{x^2 + y^2} \cos\phi_1 + \sqrt{R^2 - (x^2 + y^2)\sin^2\phi_1} \tag{D.23}$$

and (D.22) can be expressed as

$$\Lambda_R(\mathbf{r}, z_0) = -\frac{2\pi}{ik} e^{ik(z_0 - z)} + \frac{1}{ik} \int_0^{2\pi} \left[e^{ik\sqrt{R_1^2(\phi_1) + (z_0 - z)^2}} - e^{ik\left(R + \sqrt{x^2 + y^2}\cos\phi_1\right)} \right] d\phi_1. \tag{D.24}$$

The asymptotic relation $\sqrt{R_1^2(\phi_1) + (z_0 - z)^2} = R + \sqrt{x^2 + y^2}\cos\phi_1 + O(R^{-1})$ as $R \to \infty$ shows that the integral in (D.24) approaches zero as $R \to \infty$. Hence

$$\Lambda_R(\mathbf{r}, z_0) \to -\frac{2\pi}{ik} e^{ik(z_0 - z)}, \quad R \to \infty \tag{D.25}$$

and (D.21) along with the Huygens' source expression (D.18) reveal that the formula (D.19) for the spectrum is indeed correct. We have now shown that for a circular scan area, the value of the spectrum at $(k_x, k_y) = (0, 0)$ can be determined through (D.19).

Let us finally consider the square scan area D_R given by $|x_0| \leq R/2$, $|y_0| \leq R/2$. For this scan area the expression (D.5) with $(k_x, k_y) = (0, 0)$ becomes

$$\frac{e^{-ikz_0}}{2\pi} \lim_{R\to\infty} \iint_{D_R} p_\omega(\mathbf{r}_0) dx_0 dy_0$$

$$= \frac{e^{-ikz_0}}{2\pi} \lim_{R\to\infty} \left[\frac{1}{ik} \int_0^{2\pi} \mathcal{F}_\omega(\pi/2, \phi_0)\, e^{ik(R/2)/f(\phi_0)} d\phi_0 \right.$$

$$\left. - \frac{e^{ikR/2}}{ik} \int_0^{2\pi} \mathcal{F}_\omega(\pi/2, \phi_0) d\phi_0 + \int_0^{R/2} \rho_0 \int_0^{2\pi} p_\omega(\mathbf{r}_0) d\phi_0 d\rho_0 \right] \tag{D.26}$$

where

$$f(\phi_0) = \begin{cases} \cos\phi_0, & 0 \leq \phi_0 < \pi/4 \\ \sin\phi_0, & \pi/4 \leq \phi_0 < 3\pi/4 \\ -\cos\phi_0, & 3\pi/4 \leq \phi_0 < 5\pi/4 \\ -\sin\phi_0, & 5\pi/4 \leq \phi_0 < 7\pi/4 \\ \cos\phi_0, & 7\pi/4 \leq \phi_0 < 2\pi \end{cases} \tag{D.27}$$

so that $(R/2)/f(\phi_0)$ is the radial distance to the edge of the square scan area. A stationary-phase evaluation shows that the first term in the square brackets of (D.26) approaches zero as $R \to \infty$. From (D.19) we see that the expression in (D.26) equals $T_\omega(0, 0)$. We have thus shown that for a square scan area, the expression (D.5) for the spectrum holds for all (k_x, k_y), including $(k_x, k_y) = (0, 0)$. This result is easily generalized to a rectangular scan area.

References

[1] J. A. Stratton, *Electromagnetic Theory*, New York: McGraw-Hill, 1941.

[2] L. B. Felsen and N. Marcuvitz, *Radiation and Scattering of Waves*, Englewood Cliffs, NJ: Prentice-Hall, 1973.

[3] M. Born and E. Wolf, *Principles of Optics*, 6th Edition, Oxford: Pergamon, (reprint with corrections)1993.

[4] L. Mandel and E. Wolf, *Optical Coherence and Quantum Optics*, New York: Cambridge University Press, 1995.

[5] G. S. Smith, *An Introduction to Classical Electromagnetic Radiation*, Cambridge, UK: Cambridge University Press, 1997.

[6] L. M. Brekhovskikh, *Waves in Layered Media*, New York: Academic Press, 1980.

[7] P. C. Clemmow, *The Plane Wave Spectrum Representation of Electromagnetic Fields*, Oxford: Pergamon, 1966.

[8] D. M. Kerns, *Plane-Wave Scattering-Matrix Theory of Antennas and Antenna-Antenna Interactions*, NBS Monograph 162, Washington, DC: U.S. Government Printing Office, 1981.

[9] M. Nieto-Vesperinas, *Scattering and Diffraction in Physical Optics*, New York: John Wiley, 1991.

[10] K. E. Oughstun and G. C. Sherman, *Electromagnetic Pulse Propagation in Causal Dielectrics*, New York: Springer-Verlag, 1994.

[11] P. S. Schultz and J. F. Claerbout, "Velocity estimation and downward continuation by wavefront synthesis," *Geophysics*, vol. 43, pp. 691–714, June 1978.

[12] C. H. Chapman, "A new method for computing synthetic seismograms," *Geophys. J. Royal Astr. Soc.*, vol. 54, pp. 481–518, September 1978.

[13] C. H. Chapman, "Generalized Radon transforms and slant stacks," *Geophys. J. Royal Astr. Soc.*, vol. 66, pp. 445–453, August 1981.

[14] R. A. Phinney, K. R. Chowdhury, and L. N. Frazer, "Transformation and analysis of record sections," *J. Geophys. Res.*, vol. 86, pp. 359–377, January 1981.

[15] P. L. Stoffa, P. Buhl, J. B. Diebold, and F. Wenzel, "Direct mapping of seismic data to the domain of intercept time and ray parameter–A plane-wave decomposition," *Geophysics*, vol. 46, pp. 255–267, March 1981.

[16] M. Tygel and P. Hubral, *Transient Waves in Layered Media*, New York: Elsevier, 1987.

[17] E. Heyman, "Time-dependent plane-wave spectrum representations for radiation from volume source distributions," *J. Math. Phys.*, vol. 37, pp. 658–681, February 1996.

[18] B. Z. Steinberg, E. Heyman, and L. B. Felsen, "Phase-space beam summation for time-dependent radiation from large apertures: Continuous parameterization," *J. Opt. Soc. Am. A*, vol. 8, pp. 943–958, June 1991.

[19] T. Melamed and E. Heyman "Phase-space beam summation for time-harmonic and time-dependent radiation from extended apertures: 3D formulation," *SPIE*, vol. 1629, pp. 336–352, 1992.

[20] E. A. Marengo and A. J. Devaney, "Time-dependent plane wave and multipole expansions of the electromagnetic field," *J. Math. Phys.*, vol. 39, pp. 3643–3660, July 1998.

[21] T. B. Hansen and A. D. Yaghjian, "Planar near-field scanning in the time domain, Part 1: Formulation," *IEEE Trans. Antenna Propagat.*, vol. AP-42, pp. 1280–1291, September 1994.

[22] T. B. Hansen and A. D. Yaghjian, "Planar near-field scanning in the time domain, Part 2: Sampling theorems and computation schemes," *IEEE Trans. Antenna Propagat.*, vol. AP-42, pp. 1292–1300, September 1994.

[23] T. B. Hansen and A. D. Yaghjian, "Probe-corrected planar near-field scanning in the time domain," *IEEE Trans. Antenna Propagat.*, vol. AP-43, pp. 569–584, June 1995.

[24] A. D. Yaghjian and T. B. Hansen, "Time-domain far fields," *J. Appl. Phys.*, vol. 79, pp. 2822–2830, March 1996; also "Theorems on time-domain far fields," pp. 165–176 in *Ultra-Wideband Short-Pulse Electromagnetics 3*, Eds. C. E. Baum, L. Carin, and A. P. Stone, New York: Plenum, 1997.

[25] A. D. Yaghjian, T. B. Hansen, and A. J. Devaney, "Minimum source region for a given far-field pattern," *IEEE Trans. Antenna Propagat.*, vol. 45, pp. 911–912, May 1997.

[26] T. B. Hansen, *Time-Domain Planar Near-Field Scanning Using a Connection Machine*, Technical Report, Department of Signal Theory and Communications, Polytechnic University of Catalonia, Barcelona, Spain, December 1995.

[27] D. M. Kerns, "Analytical techniques for the correction of near-field antenna measurements made with an arbitrary but known measuring antenna," *Abstracts of URSI-IRE Meeting*, Washington, DC, pp. 6–7, April–May 1963.

[28] D. M. Kerns, "Correction of near-field antenna measurements made with an arbitrary but known measuring antenna," *Electron. Lett.*, vol. 6, pp. 346–347, May 1970.

[29] D. M. Kerns, "Scattering-matrix description and near-field measurements of electroacoustic transducers," *J. Acoust. Soc. Am.*, vol. 57, pp. 497–507, February 1975.

[30] D. T. Paris, W. M. Leach, Jr., and E. B. Joy, "Basic Theory of probe-compensated near-field measurements," *IEEE Trans. Antenna Propagat.*, vol. AP-26, pp. 373–379, May 1978.

[31] Y. Rahmat-Samii, V. Galindo-Israel, and R. Mittra, "A plane-polar approach for far-field construction from near-field measurements," *IEEE Trans. Antennas Propagat.*, vol. AP-28, pp. 216–230, March 1980.

[32] W. M. Leach and D. T. Paris, "Probe compensated near-field measurements on a cylinder," *IEEE Trans. Antenna Propagat.*, vol. AP-21, pp. 435–445, July 1973.

[33] A. D. Yaghjian, *Near-Field Antenna Measurements on a Cylindrical Surface: A Source Scattering-Matrix Formulation*, NBS Technical Note 696 (revised), National Bureau of Standards, Boulder, CO, 1977.

[34] F. Jensen, "On the probe compensation for near-field measurements on a sphere," *AEU*, vol. 29, pp. 305–308, July–August 1975.

[35] P. F. Wacker, "Near-field antenna measurements using a spherical scan: Efficient data reduction with probe correction," *IEE Conf. Publ. 113*, Conf. Precision Electromagn. Measurements, London, pp. 286–288, July 1974.

[36] F. H. Larsen, "Probe correction of spherical near-field measurements," *Electron. Lett.*, vol. 13, pp. 393–395, July 1977.

[37] A. D. Yaghjian and R. C. Wittmann, "The receiving antenna as a linear differential operator: Application to spherical near-field measurements," *IEEE Trans. Antenna Propagat.*, vol. AP-33, pp. 1175–1185, November 1985.

[38] R. C. Baird, A. C. Newell, P. F. Wacker, and D. M. Kerns, "Recent experimental results in near-field antenna measurements," *Electron. Lett.*, vol. 6, pp. 349–351, May 1970.

[39] E. B. Joy, W. M. Leach, Jr., G. P. Rodrigue, and D. T. Paris, "Applications of probe-compensated near-field measurements," *IEEE Trans. Antenna Propagat.*, vol. AP-26, pp. 379–389, May 1978.

[40] J. E. Hansen, Ed., J. Hald, F. Jensen, and F. H. Larsen, *Spherical Near-Field Antenna Measurements*, London: Peter Peregrinus, 1988.

[41] E. B. Joy and D. T. Paris, "Spatial sampling and filtering in near-field measurements," *IEEE Trans. Antenna Propagat.*, vol. AP-20, pp. 253–261, May 1972.

[42] A.D. Yaghjian, "An overview of near-field antenna measurements," *IEEE Trans. Antenna Propagat.*, vol. AP-34, pp. 30–45, January 1986.

[43] A. D. Yaghjian, "Antenna coupling and near-field sampling in plane-polar coordinates," and "Sampling in plane-polar coordinates," *IEEE Trans. Antenna Propagat.*, vols. 40 and 44, pp. 304–312 and pp. 696–700, March 1992 and May 1996.

[44] A. D. Yaghjian, "Sampling criteria for resonant antennas and scatterers," *J. Appl. Phys.*, vol. 79, pp. 7474–7482, May 1996; correction, vol. 79, p. 2547, August 1996.

[45] A. D. Yaghjian, "Efficient computation of antenna coupling and fields within the near-field region," *IEEE Trans. Antenna Propagat.*, vol. AP-30, pp. 113–128, January 1982.

[46] A. D. Yaghjian, *Upper-Bound Errors in Far-Field Antenna Parameters Determined from Planar Near-Field Measurements*, NBS Technical Note 667, National Bureau of Standards, Boulder, CO, 1975.

[47] A. C. Newell, "Error analysis techniques for planar near-field measurements," *IEEE Trans. Antenna Propagat.*, vol. 36, pp. 754–768, June 1988.

[48] E. B. Joy and A. D. Dingsor, "Computer simulation of cylindrical surface near-field measurement system errors," *IEEE Antenna Propagat. Soc. International Symposium Digest*, Seattle WA, pp. 565–568, June 1979.

[49] F. Jensen, "Computer simulations as a design tool in near-field testing," *IEE Conf. Publ. 169, Pt. 1*, Seattle WA, pp. 111–114, November 1978.

[50] O. M. Bucci and G. Franceschetti, "On the spatial bandwidth of scattered fields," *IEEE Trans. Antenna Propagat.*, vol. AP-35, pp. 1445–1455, December 1987.

[51] R. C. Wittmann, B. K. Alpert, and M. H. Frances, "Near-field antenna measurements using nonideal measurement locations," *IEEE Trans. Antenna Propagat.*, vol. 46, pp. 716–722, May 1998.

[52] D. Slater, *Near-Field Antenna Measurements*, Boston: Artech House, 1991.

[53] R. C. Johnson, H. A. Ecker, and J. S. Hollis, "Determination of far-field antenna patterns from near-field measurements," *Proc. IEEE*, vol. 61, pp. 1668–1694, December 1973.

[54] W. H. Kummer and E. S. Gillespie, "Antenna measurements-1978," *Proc. IEEE*, vol. 66, pp. 483–507, April 1978.

[55] J. Appel-Hansen, J. D. Dyson, E. S. Gillespie, and T. G. Hickman, "Antenna measurements," chapter 8 in *The Handbook of Antenna Design*, Eds. A. W. Rudge, K. Milne, A. D. Olver, and P. Knight, London: Peter Peregrinus, 1982.

[56] A. Dominek and H. Shamansky, "Planar, time domain, near-field measurements," *Proceedings of Antenna Measurement Techniques Association (18th Meeting and Symposium)*, Seattle, WA, pp. 92–96, September–October 1996.

[57] R. V. de Jongh, M. Hajian, and L. P. Ligthart, "Implementation and operation of the time domain antenna measurement (ATDM) technique," *Proceedings of Antenna Measurement Techniques Association (19th Meeting and Symposium)*, Boston, MA, pp. 400–404, November 1997.

[58] T. B. Hansen, "Spherical expansions of time-domain acoustic fields: Application to near-field scanning," *J. Acoust. Soc. Am.*, vol. 98, pp. 1204–1215, August 1995.

[59] T. B. Hansen, "Formulation of spherical near-field scanning for electromagnetic fields in the time domain," *IEEE Trans. Antenna Propagat.*, vol. 45, pp. 620–630, April 1997.

[60] T. B. Hansen, "Formulation of spherical near-field scanning in the time domain," in *Computational Wave Propagation*, Eds. G. Kriegsman and B. Enquist, New York: Springer-Verlag, 1997, pp. 45–74.

[61] I. Christiansen, "Time-domain spherical near-field formulas in the case where the radial components of the electromagnetic field are measured," *IEEE Trans. Antenna Propagat.*, vol. 46, pp. 595–596, April 1998.

[62] I. Christiansen, J. Appel-Hansen, and T. B. Hansen, "Time-domain cylindrical scanning of scalar fields," *Proceedings of the Fourth International Symposium on Antennas & EM Theory (ISAE'97)*, Xi'an, China, pp. 40–43, August 1997.

[63] J.D. Jackson, *Classical Electrodynamics*, 3rd Edition, New York: John Wiley, 1998.

[64] W. K. H. Panofsky and M. Phillips, *Classical Electricity and Magnetism*, 2nd Edition, Reading, MA: Addison-Wesley, 1962.

[65] L. D. Landau, E. M. Lifshitz, and L. P. Pitaevskii, *Electrodynamics of Continuous Media*, 2nd Edition, New York: Pergamon Press, 1984.

[66] D. S. Jones, *The Theory of Electromagnetism*, New York: Macmillan, 1964.

[67] D. S. Jones, *Acoustic and Electromagnetic Waves*, New York: Oxford University Press, 1986.

[68] P. M. Morse and K. U. Ingard, *Theoretical Acoustics*, New York: McGraw-Hill, 1968.

[69] L. D. Landau and E. M. Lifshitz, *Fluid Mechanics*, 2nd Edition, Oxford: Butterworth-Heinemann, 1995.

[70] A. D. Pierce, *Acoustics*, New York: McGraw-Hill, 1981.

[71] E. W. Hobson, *The Theory of Functions of a Real Variable*, New York: Dover, Vols. 1 and 2, 1957.

[72] A. D. Yaghjian, "Electric dyadic Green's functions in the source region," *Proc. IEEE*, vol. 68, pp. 248–263, February 1980; see also *Proc. IEEE*, vol. 69, pp. 282–285, February 1981.

[73] A.D. Yaghjian, "Maxwellian and cavity electromagnetic fields within continuous sources," *Am. J. Phys.*, vol. 53, pp. 859–863, September 1985.

[74] M. Silberstein, "Application of a generalized Leibniz rule for calculating electromagnetic fields within continuous source regions," *Radio Science*, vol. 26, pp. 183–190, January–February 1991.

[75] A. D. Yaghjian, "A delta-distribution derivation of the electric field in the source region," *Electromagnetics*, vol. 2, pp. 161–167, 1982.

[76] O. D. Kellogg, *Foundations of Potential Theory*, New York: Springer-Verlag, 1967.

[77] P. Penfield, Jr. and H. A. Haus, *Electrodynamics of Moving Media*, Cambridge, MA: M.I.T. Press, 1967.

[78] S. R. de Groot and L. G. Suttorp, *Foundations of Electrodynamics*, Amsterdam: North-Holland, 1972.

[79] A. M. Portis, *Electromagnetic Fields: Sources and Media*, New York: John Wiley, 1978.

[80] R. M. Fano, L. J. Chu, and R. B. Adler, *Electromagnetic Fields, Energy, and Forces*, New York: John Wiley, 1960.

[81] J. C. Maxwell, *A Treatise on Electricity and Magnetism*, 3rd revised Edition, New York: Dover, Vols. 1 and 2, 1954.

[82] E. Whittaker, *A History of the Theories of Aether & Electricity*, New York: Dover, Vols. 1 and 2, 1989.

[83] J. van Bladel, *Electromagnetic Fields*, New York: McGraw-Hill, 1964.

[84] H. B. Phillips, *Vector Analysis*, New York: John Wiley, 1933.

[85] J. A. Kong, *Electromagnetic Wave Theory*, New York: John Wiley, 1986.

[86] A. D. Yaghjian, "Equivalence of surface current and aperture field integrations for reflector antennas," *IEEE Trans. Antennas Propagat.*, vol. AP-32, pp. 1355–1358, December 1984.

[87] B. Jiang, J. Wu, and L. A. Povinelli, "The origin of spurious solutions in computational electromagnetics," *J. Computational Physics*, vol. 125, pp. 104–123, 1996.

[88] E. T. Whittaker and G. N. Watson, *A Course of Modern Analysis*, London: Cambridge University Press, 1952.

[89] W. Rudin, *Principles of Mathematical Analysis*, New York: McGraw-Hill, 1953.

[90] W. Kaplan, *Advanced Calculus*, 4th Edition, Reading, MA: Addison-Wesley, 1991.

[91] C. Müller, *Foundations of the Mathematical Theory of Electromagnetic Waves*, New York: Springer-Verlag, 1969.

[92] C. Wilcox, "The mathematical foundations of diffraction theory," pp. 65–97 in *Electromagnetic Waves*, Ed. R. E. Langer, Madison WI: University of Wisconsin Press, 1962.

[93] J. M. H. Olmsted, *Advanced Calculus*, Englewood Cliffs, NJ: Prentice-Hall, 1961.

[94] F. Kottler, "Elektromagnetische Theorie der Beugung an schwarzen Schirmen," *Ann. Phys.*, vol. 71, pp. 457–508, 1923.

[95] W. Franz, "Zur Formulierung des Huygensschen Prinzips," *Z. Naturforschg.*, vol. 3a, pp. 500–506, 1948; also section I.10 in *Theorie der Beugung Elektromagnetischer Wellen*, Berlin: Springer-Verlag, 1957.

[96] B. Baker and E. T. Copson, *The Mathematical Theory of Huygens' Principle*, 2nd Edition, London: Oxford University Press, 1950.

[97] H. A. Haus and J. R. Melcher, *Electromagnetic Fields and Energy*, Englewood Cliffs, NJ: Prentice-Hall, 1989.

[98] J. J. Bowman, T. B. A. Senior, and P. L. E. Uslenghi, *Electromagnetic and Acoustic Scattering by Simple Shapes* (revised printing), New York: Hemisphere, 1987.

[99] J. S. Asvestas and R. E. Kleinman, "Low-frequency scattering by perfectly conducting obstacles," *J. Math. Phys.*, vol. 12, pp. 795–811, May 1971.

[100] H. A. Haus and P. Penfield, Jr., "Force on a current loop," *Physics Letters*, vol. 26A, pp. 412–413, 25 March 1968.

[101] B. D. H. Tellegen, "Magnetic-dipole models," *Am. J. Phys.*, vol. 30, pp. 650–652, September 1962.

[102] T. H. Boyer, "The force on a magnetic dipole," *Am. J. Phys.*, vol. 56, pp. 688–692, August 1988.

[103] A. D. Yaghjian, *Relativistic Dynamics of a Charged Sphere: Updating the Lorentz-Abraham Model*, New York: Springer-Verlag, 1992.

[104] W. Kaplan, *Operational Methods for Linear Systems*, Reading, MA: Addison-Wesley, 1962.

[105] R. Courant, *Differential and Integral Calculus*, 2nd Edition (revised), New York: Interscience, 1937.

[106] W. K. Saunders, "On solutions of Maxwell's equations in an exterior region," *Proc. Natl. Acad. Sci.*, vol. 38, pp. 342–348, 1952.

[107] N. M. Günter, *Potential Theory and Its Applications to Basic Problems of Mathematical Physics*, New York: Frederick Ungar, 1967.

[108] L. J. Chu, "Physical limitations of omni-directional antennas," *J. Appl. Phys.*, vol. 19, pp. 1163–1175, December 1948.

[109] R. F. Harrington, "Effect of antenna size on gain, bandwidth, and efficiency," *J. Res. Nat. Bur. Stands.*, vol. 64D, pp. 1–12, January–February 1960.

[110] R. E. Collin and S. R. Rothschild, "Evaluation of antenna Q," *IEEE Trans. Antennas Propagat.*, vol. AP-12, pp. 23–27, January 1964.

[111] D. R. Rhodes, "On the stored energy of planar apertures," *IEEE Trans. Antennas Propagat.*, vol. AP-14, pp. 676–683, November 1966.

[112] G. V. Borgiotti, "On the reactive energy of an aperture," *IEEE Trans. Antennas Propagat.*, vol. AP-15, pp. 565–566, July 1967.

[113] E. J. Post, *Formal Structures of Electromagnetics*, Amsterdam: North-Holland, 1962.

[114] N. I. Muskhelishvili, *Singular Integral Equations*, Groningen-Holland: Noordhoff, 1953.

[115] P. M. Morse and H. Feshbach, *Methods of Theoretical Physics*, New York: McGraw-Hill, 1953.

[116] M. O. O'Donnell, E. T. Jaynes, and J. G. Miller, "General relationships between ultrasonic attenuation and dispersion," *J. Acoust. Soc. Am.*, vol. 63, pp. 1935–1937, June 1978.

[117] J. J. Maciel and L. B. Felsen, "Systematic study of fields due to extended apertures by Gaussian beam discretization," *IEEE Trans. Antenna Propagat.*, vol. AP-37, pp. 884–892, July 1989.

[118] B. Z. Steinberg, E. Heyman, and L. B. Felsen, "Phase-space beam summation for time-harmonic radiation from large apertures," *J. Opt. Soc. Am. A*, vol. 8, pp. 41–59, January 1991.

[119] A. N. Norris and T. B. Hansen, "Wavefield representation using compact and directionally localized beams," pp. 179–196 in *Computational Wave Propagation*, Eds. B. Enquist and G. Kriegsman, *IMA Volumes in Mathematics and Its Applications*, New York: Springer-Verlag, 1997.

[120] A. N. Norris and T. B. Hansen, "Exact complex source representations of time-harmonic radiation," *Wave Motion*, vol. 25, pp. 127–141, 1997.

[121] Lord Rayleigh, "On the passage of waves through apertures in plane screens and allied problems," *Phil. Mag.*, vol. 43, ser. 5, pp. 259–272, January–June 1897.

[122] C. T. Tai, *Dyadic Green's Functions in Electromagnetic Theory*, 2nd Edition, New York: IEEE Press, 1994.

[123] W. R. Smythe, "The double current sheet in diffraction," *Phys. Rev.*, vol. 72, pp. 1066–1070, December 1947.

[124] H. Weyl, "Ausbreitung elektromagnetischer Wellen über einem ebenen Leiter," *Ann. Physik*, vol. 60, pp. 481–500, 1919.

[125] M. L. Crawford, "Calibration of broadbeam antennas using planar near-field measurements," *Digest Conf. Precision Electromagnetic Measurements*, Boulder, CO, pp. 53–56, June–July 1976.

[126] N. Bleistein and R. A. Handelsman, *Asymptotic Expansions of Integrals*, New York: Dover, 1986.

[127] I. S. Gradshteyn and I. M. Ryzhik, *Tables of Integrals, Series, and Products*, 5th Edition, Ed. A. Jeffrey, Boston: Academic Press, 1994.

[128] E. Wolf and M. Nieto-Vesperinas, "Analyticity of the angular spectrum amplitude of scattered fields and some of its consequences," *J. Opt. Soc. Am. A*, vol. 2, pp. 886–889, June 1985.

[129] G. C. Sherman, J. J. Stamnes, and E. Lalor, "Asymptotic approximations to angular-spectrum representations," *J. Math. Phys.*, vol. 17, pp. 760–776, May 1976.

[130] E. Wolf and J. T. Foley, "Do evanescent waves contribute to the far field?," *Optics Letters*, vol. 23, pp. 16–18, January 1998.

[131] A. J. Devaney and E. Wolf, "Multipole expansions and plane wave representations of the electromagnetic field," *J. Math. Phys.*, vol. 15, pp. 234–244, February 1974.

[132] N. N. Bojarski, "A survey of the near-field far-field inverse scattering inverse source integral equation," *IEEE Trans. Antenna Propagat.*, vol. AP-30, pp. 975–979, September 1982.

[133] R. P. Porter and A. J. Devaney, "Holography and the inverse source problem," *J. Opt. Soc. Am*, vol. 72, pp. 327–330, March 1982.

[134] D. S. Jones, *Methods in Electromagnetic Wave Propagation*, 2nd Edition, New York: IEEE/OUP, 1994.

[135] R. C. Wittmann,"Spherical-wave operators and the translation formulas," *IEEE Trans. Antenna Propagat.*, vol. 36, pp. 1078–1087, August 1988.

[136] A. D. Yaghjian, "The planar near-field reconstruction (PNFR) approach to high-resolution remote sensing of subsurface anomalies," chapter 4 in *High Resolution Sensing Techniques for Slope Stability Studies*, R. L. Jesch et al., Federal Highway Administration Report No. FHWA-RD-79-32 (prepared by NIST-formerly NBS, Boulder, CO, and available from National Technical Information Service, 5285 Port Royal Road, Springfield, Virginia 22161, USA), January 1979.

[137] L. Chommeloux, C. Pichot, and J.-C. Bolomey, "Electromagnetic modeling for microwave imaging of cylindrical buried inhomogeneities," *IEEE Trans. Microwave Theory and Techniques*, vol. MTT-34, pp. 1064–1076, October 1986.

[138] R. P. Porter,"Generalized holography with application to inverse scattering and inverse source problems," pp. 315–397 in *Progress in Optics*, Volume 27, Ed. E. Wolf, Amsterdam: North-Holland, 1989.

[139] R. J. Blakely, *Potential Theory in Gravity and Magnetic Applications*, Cambridge, UK: Cambridge University Press, 1995.

[140] K. Chandrasekharan, *Classical Fourier Transforms*, Berlin: Springer-Verlag, 1989.

[141] S. Bochner and K. Chandrasekharan, *Fourier Transforms*, Princeton, NJ: Princeton University Press, 1949.

[142] C. E. Baum, *Focused Aperture Antennas*, Kirtland AFB, NM, Sensor and Simulation Notes, Note 306, 1987.

[143] C. E. Baum, *A Simplified Two-Dimensional Model for the Fields above the Distributed-Source Surface Transmission Line*, Kirtland AFB, NM, Sensor and Simulation Notes, Note 66, 1968.

[144] J. W. Dettman, *Applied Complex Variables*, New York: Macmillan, 1965 (Dover, 1984).

[145] R. J. Luebbers, K. S. Kunz, M. Schneider, and F. Hunsberger, "A finite-difference time-domain near zone to far zone transformation," *IEEE Trans. Antenna Propagat.*, vol. AP-39, pp. 429–433, April 1991.

[146] D. A. Hill, *Far-Field Transient Response of an Antenna from Near-Field Data*, NBS Internal Report 86-3063, National Bureau of Standards, Boulder, CO, 1986.

[147] A. Erdélyi, W. Magnus, F. Oberhettinger, and G. Francesco, *Tables of Integral Transforms*, New York: McGraw-Hill, Vols. 1 and 2, 1954.

[148] S. R. Deans, *The Radon Transform and Some of Its Applications*, New York: John Wiley, 1983.

[149] H. E. Moses, R. J. Nagem, and G. V. H. Sandri, "The general solution of the three-dimensional acoustic equation and of Maxwell's equations in the infinite domain in terms of the asymptotic solution in the wave zone," *J. Math. Phys.*, vol. 33, pp. 86–101, January 1992.

[150] H. E. Moses and R. T. Prosser, "Acoustic and electromagnetic bullets: Derivation of new exact solutions of the acoustic and Maxwell's equations," *SIAM J. Appl. Math.*, vol. 50, pp. 1325–1340, October 1990.

[151] C. A. Balanis, *Antenna Theory: Analysis and Design*, 2nd Edition, New York: John Wiley, 1997.

[152] O. E. Allen, D. A. Hill, and A. R. Ondrejka, "Time-domain antenna characterizations," *IEEE Trans. Electromag. Compat.*, vol. 35, pp. 339–346, August 1993.

[153] F. Jay, Ed., *IEEE Standard Dictionary of Electrical and Electronics Terms*, 3rd Edition, New York: John Wiley, 1984.

[154] D. Gabor, "Theory of Communication," *J. Institution of Electrical Engineers.*, vol. 93, pt. III, pp. 429–457, 1946

[155] T. T. Wu, Electromagnetic Missiles, *J. Appl. Phys.*, vol. 57, pp. 2370–2373, April 1985.

[156] J. M. Myers, *Developmental Study of Electromagnetic Missiles*, Gordon McKay Laboratory Annual Progress Report No. 3, Harvard University, Cambridge, MA, sec. 5, 1989.

[157] D. J. Blejer, R. C. Wittmann, and A. D. Yaghjian, "On-axis fields from a circular uniform surface current," pp. 285–292 in *Ultra-Wideband, Short-Pulse Electromagnetics*, New York: Plenum Press, 1993.

[158] R. C. Rudduck and C. J. Chen, "New plane wave spectrum formulations for the near fields of circular and strip apertures," *IEEE Trans. Antenna Propagat.*, vol. AP-24, pp. 438–447, July 1976.

[159] H. Backhaus and F. Trendelenburg, "Über die Richtwirkung von Kolbenmembranen," *Z. Tech. Phys.*, vol. 7, pp. 630–635, December 1926.

[160] R. C. Wittmann, "Spherical-wave expansions of the electromagnetic fields of a uniformly excited circular aperture," *Abstracts of URSI Meeting*, Boulder, CO, January 1992 p. 53; see also R. C. Wittmann and A. D. Yaghjian, "Spherical-wave expansions of piston-radiator fields," *J. Acoust. Society Am.*, vol. 90, pp. 1647–1655, September 1991.

[161] D. J. Blejer, "On-axis transient fields from a uniform circular distribution of electric or magnetic current," p. 363 in *Proc. Progress in Electromagnetic Research Symp. (PIERS)*, Boston MA, June 1989.

[162] M. Greenspan, "Piston radiator: Some extensions of the theory," *J. Acoust. Society Am.*, vol. 65, pp. 608–621, March 1979.

[163] J. W. Miles, "Transient loading of a baffled piston," *J. Acoust. Society Am.*, vol. 25, pp. 200–203, March 1953.

[164] F. Oberhettinger, "On transient solutions of the "baffled piston" problem," *J. Res. Nat. Bur. Stands.*, vol. 65B, pp. 1–6, March 1961.

[165] D. E. Robinson, S. Lees, and L. Bess, "Near field transient radiation patterns for circular pistons," *IEEE Trans. Acoust. Speech and Signal Processing*, vol. ASSP-22, pp. 395–403, December 1974.

[166] H. M. Shen and T. T. Wu, "The transverse energy pattern of an electromagnetic missile from a circular current disk," *SPIE*, vol. 1061, pp. 352–360, 1989.

[167] Y.L. Luke, *Integrals of Bessel Functions*, New York: McGraw-Hill, 1962.

[168] [illegible], "[illegible] pattern evaluation from near field intensities on planes," *IEEE Trans. Antenna Propagat.*, vol. 46, pp. 701–710, May 1996.

[169] S. Silver, *Microwave Antenna Theory and Design*, New York: Dover, 1965.

[170] A. D. Yaghjian, "Generalized or adjoint reciprocity relations for electroacoustic transducers," *J. Res. Nat. Bur. Stands.*, vol. 79B, pp. 17–39, January–June 1975.

[171] P. K. Park and C. T. Tai, "Receiving antennas," chapter 6 in *Antenna Handbook*, Eds. Y. T. Lo and S. W. Lee, New York: Van Nostrand Reinhold, 1988.

[172] F. H. Larsen, *Probe-Corrected Spherical Near-Field Antenna Measurements*, Technical University of Denmark Report LD 36, Lyngby, Denmark, pp. 63–65, December 1980.

[173] R. E. Collin and F. J. Zucker, *Antenna Theory*, New York: McGraw-Hill, Parts 1 and 2, 1969.

[174] S. P. Skulkin and V. I. Turchin, "The time-pulse method of measuring antenna characteristics in the near zone," *Radiophysics and Quantum Electronics*, vol. 32, pp. 61–70, July 1989.

[175] S. P. Skulkin, "Time domain near field antenna measurements," in *Electromagnetic Environments and Consequences*, Part 2, Eds. D. J. Serafin et al., France: EUROEM, pp. 1492–1497, 1994.

[176] S. P. Skulkin, "Radiation of nonsinusoidal waves by aperture antennas," in *Electromagnetic Environments and Consequences*, Part 2, Eds. D.J. Serafin et al., France: EUROEM, pp. 1498–1504, 1994.

[177] S. P. Skulkin and Yu. Sorpov, "An analysis of range distance requirements for large antenna measurements by the use of the transient characteristics," *IEEE Antenna Propagat. Soc. International Symposium Digest*, Newport Beach, CA, pp. 1758–1761, June 1995.

[178] C. E. Baum, E. L. Breen, J. C. Giles, J. O'Neill, and G. D. Sower, "Sensors for electromagnetic pulse measurements both inside and away from nuclear source regions," *IEEE Trans. Antenna Propagat.*, vol. AP-26, pp. 22–35, January 1978; see also C. E. Baum, "Electromagnetic sensors and measurement techniques," pp. 77–144 in *Fast Electrical and Optical Measurements*, Eds. J. E. Thompson and L. H. Luessen, Holland: Martinus-Nijhoff, 1986.

[179] M. Kanda, "Time domain sensors for radiated impulsive measurements," *IEEE Trans. Antenna Propagat.*, vol. AP-31, pp. 438–444, May 1983.

[180] R. W. Ziolkowski, "Properties of electromagnetic beams generated by ultra-wide bandwidth pulse-driven arrays," *IEEE Trans. Antenna Propagat.*, vol. AP-40, pp. 888–905, August 1992.

[181] H. M. Shen, R. W. P. King, and T. T. Wu, "New sensors for measuring very short electromagnetic pulses," *IEEE Trans. Antenna Propagat.*, vol. AP-38, pp. 838–846, June 1990.

[182] R. W. P. King, "The cylindrical dipole as a sensor or probe," *IEEE Trans. Electromagn. Compat.*, vol. EMC-24, pp. 364–367, August 1982.

[183] H. M. Shen, R. W. P. King, and T. T. Wu, "V-conical antenna," *IEEE Trans. Antenna Propagat.*, vol. AP-36, pp. 1519–1525, November 1988.

[184] T. T. Wu and R. W. P. King, "The cylindrical antenna and nonreflecting resistive loading," *IEEE Trans. Antenna Propagat.*, vol. AP-13, pp. 369–374, May 1965.

[185] M. Kanda, "An electromagnetic near-field sensor for simultaneous electric and magnetic-field measurements," *IEEE Trans. Electromagn. Compat.*, vol. EMC-26, pp. 102–110, August 1994.

[186] M. Kanda and L. D. Driver, "An isotropic electric-field probe with tapered resistive dipoles for broad-band use, 100 kHz to 18 GHz," *IEEE Trans. Microwave Theory Tech.*, vol. MTT-35, pp. 124–130, February 1987.

[187] M. Kanda, "A relatively short cylindrical broadband antenna with tapered loading for picosecond pulse measurements," *IEEE Trans. Antenna Propagat.*, vol. AP-26, pp. 439–447, May 1978.

[188] J. Randa and M. Kanda, "Multiple-source, multiple-frequency error of an electric field meter," *IEEE Trans. Antenna Propagat.*, vol. AP-33, pp. 2–9, January 1985.

[189] M. Kanda, "Standard probes for electromagnetic field measurements," *IEEE Trans. Antenna Propagat.*, vol. AP-41, pp. 1349–1364, October 1993.

[190] G. S. Smith, "Limitations on the size of miniature electric-field probes," *IEEE Trans. Microwave Theory Tech.*, vol. MTT-32, pp. 594–599, June 1984.

[191] N. S. Nahman, "Picosecond-domain waveform measurements," *Proc. IEEE*, vol. 66, pp. 441–454, April 1978.

[192] W. H. Darden and W. D. Burnside, *The Design of Ultra-Wide-Band Antennas with Narrow Beamwidths*, Technical Report 721929-62, ElectroScience Laboratory, The Ohio State University, Columbus, OH, October 1993.

[193] H. J. Nussbaumer, *Fast Fourier Transform and Convolution Algorithms*, New York: Springer-Verlag, 1982.

[194] W. H. Press, S. A. Teukolsky, W. T. Vetterling, and B. P. Flannery, *Numerical Recipes in Fortran*, 2nd Edition, Cambridge, UK: Cambridge University Press, 1992.

[195] O. M. Bucci, C. Gennarelli, and C. Savarse, "Fast and accurate near-field far-field transformation by sampling interpolation of plane polar measurements," *IEEE Trans. Antenna Propagat.*, vol. AP-39, pp. 48–55, January 1991.

[196] O. M. Bucci, G. D'Elia, and M. D. Migliore, "Optimal time-domain field interpolation from plane-polar samples," *IEEE Trans. Antenna Propagat.*, vol. AP-45, pp. 989–994, June 1997.

[197] E. Heyman and L. B. Felsen, "Complex-source pulsed-beam fields," *J. Opt. Soc. Am. A*, vol. 6, pp. 806–817, June 1989.

[198] A. D. Yaghjian, "Approximate formulas for the far field and gain of open-ended rectangular waveguide," *IEEE Trans. Antenna Propagat.*, vol. AP-32, pp. 378–384, April 1984.

[199] F. Rellich, "Über das asymptotische Verhalten der Lösungen von $\Delta u + ku = 0$ in unendlichen Gebieten," *Jber. Deutschen Math. Verein.*, vol. 53, pp. 57–64, 1943.

[200] C. H. Wilcox, "A generalization of theorems of Rellich and Atkinson," *Proc. Am. Math. Soc.*, vol. 7, pp. 271–276, 1956.

Glossary of Symbols

The principal symbols are listed with page numbers for the pages on which the symbols are defined or explained.

ROMAN LETTERS

$a_0(t)$	time-domain amplitude of incident wave in test-antenna waveguide 274
$a_p(t)$	time-domain amplitude of incident wave in probe waveguide 272
$a_{0\omega}$	frequency-domain amplitude of incident wave in test-antenna waveguide 256
$a_{p\omega}$	frequency-domain amplitude of incident wave in probe waveguide 255
A_{m,n,l_t}	array containing time derivative of pressure 302
$\mathbf{A}_{m,n,l_t}$	array containing time derivative of electric field 305
$\mathbf{A}(\mathbf{r}, t)$	time-domain vector potential 19
$\mathbf{A}_\omega(\mathbf{r})$	frequency-domain vector potential 79
b_{m,n,l_t}	array containing output of acoustic probe 306
$\mathbf{b}_{m,n,l_t}$	array containing vector output of electromagnetic probe 307
$b_p(t), b_p(\mathbf{r}, t)$	time-domain amplitude of emerging wave in probe waveguide (scalar output of probe) 272, 275
$b_{p\omega}, b_{p\omega}(\mathbf{r})$	frequency-domain amplitude of emerging wave in probe waveguide (scalar output of probe) 255, 257
$\mathbf{b}_p(\mathbf{r}, t)$	time-domain vector output of electromagnetic probe 287
$\mathbf{b}_{p\omega}(\mathbf{r})$	frequency-domain vector output of electromagnetic probe 262
$\mathbf{B}(\mathbf{r}, t)$	time-domain magnetic induction 10
$\mathbf{B}_c(\mathbf{r}, t)$	time-domain cavity-defined magnetic induction 27
$\mathbf{B}_\omega(\mathbf{r})$	frequency-domain magnetic induction 10
c	constant acoustic or electromagnetic wave speed 12, 55
$\mathbf{C}_{el}(t)$	electromagnetic torque (couple) 48
$\mathbf{D}(\mathbf{r}, t)$	time-domain electric displacement 10
$\mathbf{e}_b(\boldsymbol{\rho})$	time-independent electric basis field in waveguide 273
$\mathbf{e}_b(\boldsymbol{\rho}, t)$	time-domain electric basis field in waveguide 272

Symbol	Description	Page
$\mathbf{e}_{b\omega}(\boldsymbol{\rho})$	frequency-domain electric basis field in waveguide	255
$\mathbf{E}(\mathbf{r})$	static electric field	12
$\mathbf{E}(\mathbf{r}, t)$	time-domain electric field	9
$\overset{+}{\mathbf{E}}(\mathbf{r}, t)$	analytic-signal electric field	224
$\mathbf{E}_c(\mathbf{r}, t)$	time-domain cavity-defined electric field	27
$\mathbf{E}^t(\boldsymbol{\rho}, t)$	transverse time-domain electric field in waveguide	272
$\mathbf{E}_\omega(\mathbf{r})$	frequency-domain electric field	72
$\mathbf{E}_\omega^a(\mathbf{r})$	advanced frequency-domain electric field	140
$\mathbf{E}_\omega^e(\mathbf{r})$	evanescent frequency-domain electric field	129
$\mathbf{E}_\omega^h(\mathbf{r})$	homogeneous frequency-domain electric field	141
$\mathbf{E}_\omega^p(\mathbf{r})$	propagating frequency-domain electric field	129
$\mathbf{E}_\omega^{tan}(\mathbf{r})$	transverse frequency-domain electric field in scan plane	269
$\mathbf{E}_\omega^t(\boldsymbol{\rho})$	transverse frequency-domain electric field in waveguide	254
$f^G(t)$	Gaussian time function	314
f_ω^G	Gaussian spectrum	314
$\mathbf{f}(\mathbf{r})$	time-independent acoustic external force density	54
$\mathbf{f}(\mathbf{r}, t)$	time-domain acoustic external force density	52
$\mathbf{f}_\omega(\mathbf{r})$	frequency-domain acoustic external force density	87
$F(\xi, \eta, t)$	far-field function for time-domain pressure	205
$\overset{+}{F}(\xi, \eta, t)$	far-field function for analytic-signal pressure	228
$F_\omega(k_x, k_y)$	far-field function for frequency-domain pressure	124
$\mathbf{F}_{ac}(t)$	acoustic wave force	69
$\mathbf{F}_{el}(t)$	electromagnetic force	37
$\mathbf{F}_{ac}^{ave}$	time-average acoustic wave force	92
$\mathbf{F}_{el}^{ave}$	time-average electromagnetic force	84
$\mathbf{F}(\xi, \eta, t)$	far-field function for time-domain electric field	206
$\overset{+}{\mathbf{F}}(\xi, \eta, t)$	far-field function for analytic-signal electric field	229
$\mathbf{F}^H(\xi, \eta, t)$	far-field function for time-domain magnetic field	207
$\overset{+}{\mathbf{F}}{}^H(\xi, \eta, t)$	far-field function for analytic-signal magnetic field	229
$\mathbf{F}_\omega(k_x, k_y)$	far-field function for frequency-domain electric field	124
$\mathbf{F}_\omega^H(k_x, k_y)$	far-field function for frequency-domain magnetic field	124
$\mathcal{F}(\theta, \phi, t)$	far-field pattern for time-domain pressure	185
$\overset{+}{\mathcal{F}}(\theta, \phi, t)$	far-field pattern for analytic-signal pressure	230
$\mathcal{F}_p^\delta(\theta, \phi, t)$	far-field pressure pattern of probe fed by delta function in time	282
$\mathcal{F}_\omega(\theta, \phi)$	far-field pattern for frequency-domain pressure	109
$\mathcal{F}_{0\omega}(\theta, \phi)$	normalized frequency-domain far-field pressure pattern of acoustic radiator	257
$\mathcal{F}_{p\omega}(\theta, \phi)$	normalized frequency-domain far-field pressure pattern of probe (when radiating)	261
$\boldsymbol{\mathcal{F}}(\theta, \phi, t)$	far-field pattern for time-domain electric field	185
$\overset{+}{\boldsymbol{\mathcal{F}}}(\theta, \phi, t)$	far-field pattern for analytic-signal electric field	233
$\boldsymbol{\mathcal{F}}^H(\theta, \phi, t)$	far-field pattern for time-domain magnetic field	185
$\overset{+}{\boldsymbol{\mathcal{F}}}{}^H(\theta, \phi, t)$	far-field pattern for analytic-signal magnetic field	233
$\boldsymbol{\mathcal{F}}_\omega(\theta, \phi)$	far-field pattern for frequency-domain electric field	110
$\boldsymbol{\mathcal{F}}_\omega^H(\theta, \phi)$	far-field pattern for frequency-domain magnetic field	110
$\boldsymbol{\mathcal{F}}_{0\omega}(\theta, \phi)$	normalized frequency-domain electric far-field pattern of test antenna	262

Symbol	Description	Page
$\mathcal{F}^x_{p\omega}(\theta, \phi)$	normalized frequency-domain electric far-field pattern of x-directed probe (when radiating)	267
$\mathcal{F}^y_{p\omega}(\theta, \phi)$	normalized frequency-domain electric far-field pattern of y-directed probe (when radiating)	267
$G(\mathbf{r}, \mathbf{r}')$	static Green's function	157
$G^D(\mathbf{r}, \mathbf{r}')$	Dirichlet static Green's function	158
$G^N(\mathbf{r}, \mathbf{r}')$	Neumann static Green's function	158
$G(\mathbf{r}, \mathbf{r}', t, t')$	free-space time-domain Green's function	31
$G^D(\mathbf{r}, \mathbf{r}', t, t')$	Dirichlet time-domain Green's function	176
$G^e(\mathbf{r}, \mathbf{r}', t, t')$	evanescent time-domain Green's function	198
$G^p(\mathbf{r}, \mathbf{r}', t, t')$	propagating time-domain Green's function	198
$G_\omega(\mathbf{r}, \mathbf{r}')$	free-space frequency-domain Green's function	77
$G^a_\omega(\mathbf{r}, \mathbf{r}')$	advanced frequency-domain Green's function	138
$G^D_\omega(\mathbf{r}, \mathbf{r}')$	Dirichlet frequency-domain Green's function	105
$G^e_\omega(\mathbf{r}, \mathbf{r}')$	evanescent frequency-domain Green's function	117
$G^p_\omega(\mathbf{r}, \mathbf{r}')$	propagating frequency-domain Green's function	117
$\bar{\mathbf{G}}^D(\mathbf{r}, \mathbf{r}', t, t')$	Dirichlet dyadic time-domain Green's function	178
$\bar{\mathbf{G}}_\omega(\mathbf{r}, \mathbf{r}')$	free-space dyadic frequency-domain Green's function	78
$\bar{\mathbf{G}}^D_\omega(\mathbf{r}, \mathbf{r}')$	Dirichlet dyadic frequency-domain Green's function	107
$\mathbf{h}_b(\boldsymbol{\rho})$	time-independent magnetic basis field in waveguide	273
$\mathbf{h}_b(\boldsymbol{\rho}, t)$	time-domain magnetic basis field in waveguide	272
$\mathbf{h}_{b\omega}(\boldsymbol{\rho})$	frequency-domain magnetic basis field in waveguide	255
$\mathbf{H}(\mathbf{r})$	static magnetic field	12
$\mathbf{H}(\mathbf{r}, t)$	time-domain magnetic field	9
$\overset{+}{\mathbf{H}}(\mathbf{r}, t)$	analytic-signal magnetic field	225
$\mathbf{H}_c(\mathbf{r}, t)$	time-domain cavity-defined magnetic field	27
$\mathbf{H}^t(\boldsymbol{\rho}, t)$	transverse time-domain magnetic field in waveguide	272
$\mathbf{H}_\omega(\mathbf{r})$	frequency-domain magnetic field	72
$\mathbf{H}^h_\omega(\mathbf{r})$	homogeneous frequency-domain magnetic field	141
$\mathbf{H}^t_\omega(\boldsymbol{\rho})$	transverse frequency-domain magnetic field in waveguide	254
$\widetilde{\mathcal{H}}$	Hilbert transform	193
$I_{p\omega}$	frequency-domain current in probe waveguide	255
$I(t; a, b)$	retangular-pulse time function	198
$\bar{\mathbf{I}}$	identity dyadic	25
$j(\mathbf{r}, t)$	time-domain velocity-potential source function	59
$\mathbf{J}(\mathbf{r})$	static current density	12
$\mathbf{J}(\mathbf{r}, t)$	time-domain current density	9
$\overset{+}{\mathbf{J}}(\mathbf{r}, t)$	analytic-signal current density	227
$\mathbf{J}_\omega(\mathbf{r})$	frequency-domain current density	72
k	propagation constant equal to $\omega/c = \sqrt{\mathbf{k} \cdot \mathbf{k}}$	74, 87
k_x	x component of propagation vector $\mathbf{k}$	113, 167
k_y	y component of propagation vector $\mathbf{k}$	113, 167
$\mathbf{k}$	frequency-domain and statics propagation vector	114, 168
$\hat{\mathbf{k}}$	unit frequency-domain propagation vector ($\mathbf{k}/k$)	115
$L(k_x, k_y)$	plane-wave spectrum of electrostatic potential	168
$L^H(k_x, k_y)$	plane-wave spectrum of magnetostatic potential	171

Symbol	Description	Page
$\bar{\mathbf{L}}$	source or depolarization dyadic	26
$\mathbf{m}$	magnetic dipole moment	42,165
$\mathbf{M}(\mathbf{r}, t)$	time-domain magnetic polarization density (magnetization)	10
$\mathbf{M}_\omega(\mathbf{r})$	frequency-domain magnetic polarization density (magnetization)	72
$\hat{\mathbf{n}}$	unit normal	11
N_t	number of time samples	299
N_x	number of spatial samples in x direction	301
N_y	number of spatial samples in y direction	301
$N_{\theta\phi}$	number of spatial far-field points	304
$N_{\not{t}}$	number of space-time far-field points	304
$p(\mathbf{r})$	time-independent acoustic pressure	54
$p(\mathbf{r}, t)$	time-domain acoustic pressure	53
$\overset{+}{p}(\mathbf{r}, t)$	analytic-signal acoustic pressure	222
$p_\omega(\mathbf{r})$	frequency-domain acoustic pressure	87
$p_\omega^a(\mathbf{r})$	advanced frequency-domain acoustic pressure	139
$p_\omega^e(\mathbf{r})$	evanescent frequency-domain acoustic pressure	129
$p_\omega^h(\mathbf{r})$	homogeneous frequency-domain acoustic pressure	139
$p_\omega^p(\mathbf{r})$	propagating frequency-domain acoustic pressure	129
$\mathbf{p}$	electric dipole moment	42, 157
$P_{ac}(t)$	acoustic power	69
P_{ac}^{ave}	time-average acoustic power	92
$P_{el}(t)$	electromagnetic power	37
P_{el}^{ave}	time-average electromagnetic power	84
$\mathbf{P}(\mathbf{r}, t)$	time-domain electric polarization density	10
$\mathbf{P}_\omega(\mathbf{r})$	frequency-domain electric polarization density	72
$\mathcal{P}_{ac}^c(z_1)$	complex acoustic power transmitted across z_1 plane	149
$\mathcal{P}_{ac}^h$	time-average acoustic power radiated into hemisphere	133, 149
$\mathcal{P}_{el}^c(z_1)$	complex electromagnetic power transmitted across z_1 plane	147
$\mathcal{P}_{el}^h$	time-average electromagnetic power radiated into hemisphere	134, 148
q^{tot}	static electric charge	157
$q(\mathbf{r}, t)$	time-domain pressure source function	55
$\overset{+}{q}(\mathbf{r}, t)$	analytic-signal pressure source function	226
$q_\omega(\mathbf{r})$	frequency-domain pressure source function	87
$\mathbf{Q}(\mathbf{r}, t)$	time-domain electromagnetic source function	217
$\overset{+}{\mathbf{Q}}(\mathbf{r}, t)$	analytic-signal electromagnetic source function	236
$\mathbf{Q}_\omega(\mathbf{r})$	frequency-domain electromagnetic source function	140
$Q_p(\xi, \eta)$	angularly-dependent function for time-derivative probe	280
$\mathbf{Q}_p^x(\xi, \eta)$	vector angularly-dependent function for x-directed D-dot probe	289
$\mathbf{Q}_p^y(\xi, \eta)$	vector angularly-dependent function for y-directed D-dot probe	289
$\bar{\mathbf{Q}}_p(\xi, \eta)$	dyadic angularly-dependent function for D-dot probe	289
$\bar{\mathbf{Q}}_p^{-1}(\xi, \eta)$	inverse dyadic angularly-dependent function for D-dot probe	290
r	radial spherical coordinate ($\|\mathbf{r}\|$)	9
$\mathbf{r}$	point in space given by $x\hat{\mathbf{x}} + y\hat{\mathbf{y}} + z\hat{\mathbf{z}}$	9
$\hat{\mathbf{r}}$	radial unit vector in spherical coordinates ($\mathbf{r}/r$)	9
$\mathbf{r}_i$	image of $\mathbf{r}$ with respect to plane	105
$\mathbf{r}_{0mn}$	scan point given by $m\Delta x_0\hat{\mathbf{x}} + n\Delta y_0\hat{\mathbf{y}}$	301

$\mathbf{T}_p^{\delta y}(\xi, \eta, t)$ plane-wave spectrum of y-directed electromagnetic probe fed by delta function in time 292
$\mathbf{T}(k_x, k_y)$ plane-wave spectrum of static electric field 168
$\mathbf{T}^H(k_x, k_y)$ plane-wave spectrum of static magnetic field 171
$\mathbf{T}_\omega(k_x, k_y)$ frequency-domain plane-wave spectrum of electric field 114
$\mathbf{T}_\omega^H(k_x, k_y)$ frequency-domain plane-wave spectrum of magnetic field 115
$\mathbf{T}_{b\omega}(k_x, k_y)$ frequency-domain plane-wave spectrum of vector output of electromagnetic probe 265
$\mathbf{T}_{0\omega}(k_x, k_y)$ normalized frequency-domain plane-wave spectrum of test antenna 262
$\mathbf{T}_{p\omega}(k_x, k_y)$ normalized frequency-domain plane-wave spectrum of electromagnetic probe (when radiating) 266
$\mathbf{T}_{p\omega}^x(k_x, k_y)$ normalized frequency-domain plane-wave spectrum of x-directed electromagnetic probe (when radiating) 267
$\mathbf{T}_{p\omega}^y(k_x, k_y)$ normalized frequency-domain plane-wave spectrum of y-directed electromagnetic probe (when radiating) 267
$\bar{\mathbf{T}}_{ac}(\mathbf{r}, t)$ time-domain acoustic wave-stress dyadic 69
$\bar{\mathbf{T}}_{ac}^c(\mathbf{r})$ complex acoustic wave-stress dyadic 92
$\bar{\mathbf{T}}_{el}(\mathbf{r}, t)$ time-domain electromagnetic stress dyadic 37
$\bar{\mathbf{T}}_{el}^c(\mathbf{r})$ complex electromagnetic stress dyadic 84
$\mathbf{v}(\mathbf{r})$ time-independent acoustic velocity 54
$\mathbf{v}(\mathbf{r}, t)$ time-domain acoustic velocity 52
$\overset{+}{\mathbf{v}}(\mathbf{r}, t)$ analytic-signal acoustic velocity 224
$\mathbf{v}_\omega(\mathbf{r})$ frequency-domain acoustic velocity 87
V volume containing the sources 11
V_0 infinite volume outside S 30
V_a arbitrary volume 17
V_δ arbitrarily shaped principal volume 19
V_δ^s spherical principal volume 21
$V_{p\omega}$ frequency-domain voltage in probe waveguide 255
$\mathbf{w}(\mathbf{r})$ time-independent vorticity 53
$\hat{\mathbf{x}}$ unit vector in the x direction 9
$\hat{\mathbf{y}}$ unit vector in the y direction 9
Y_p characteristic admittance in probe waveguide 255
$Y_{p\omega}^W$ wave admittance in probe waveguide 255
$\hat{\mathbf{z}}$ unit vector in the z direction 9

GREEK LETTERS

γ z component of frequency-domain propagation vector $\mathbf{k}$ 113
Γ integration contour in complex plane 136
$\delta(t)$ Dirac delta function 7
$\overset{+}{\delta}(t)$ analytic delta function 321
Δt time sampling interval equal to π/ω_{max} 299
Δx_0 sample spacing in x direction 300
Δy_0 sample spacing in y direction 300
$\Delta\omega$ frequency sampling increment 299
ϵ permittivity of free space 9

Index

D

E

R

W

About the Authors

Thorkild B. Hansen received the M.S. and Ph.D. degrees from the Technical University of Denmark, Lyngby, Denmark, in 1989 and 1991. From 1992 to 1997 he worked as a research scientist in the Electromagnetics Directorate of the Air Force Research Laboratory, Hanscom Air Force Base, MA. In 1997 he joined the staff of Schlumberger-Doll Research, Ridgefield, CT. Dr. Hansen has worked for the past 10 years on research problems in electromagnetics and acoustics. Within high-frequency diffraction theory, he has developed methods for calculating the scattering from corners, narrow grooves, and shadow boundaries. He has formulated planar and spherical near-field scanning techniques in the time domain and derived exact beam representations that are alternatives to the classical multipole expansions. Most recently, he has worked on forward and inverse scattering problems for ground penetrating radar.

Dr. Hansen won the 1992 R.W.P. King Prize Paper Award for his paper on corner diffraction published in the *IEEE Transactions on Antennas and Propagation*. In 1995 Dr. Hansen, together with Dr. Yaghjian, won the IEEE S.A. Schelkunoff Prize Paper Award for their papers on time-domain near-field scanning. Dr. Hansen is a member of the IEEE, the Optical Society of America, the American Association for the Advancement of Science, and URSI Commission B.

Arthur D. Yaghjian received the B.S., M.S., and Ph.D. degrees in electrical engineering from Brown University, Providence, RI, in 1964, 1966, and 1969. During the spring semester of 1967, he taught mathematics at Tougaloo College, MS. After receiving the Ph.D. degree he taught mathematics and physics for a year at Hampton University, VA, and in 1971 he joined the research staff of the Electromagnetics Division of the National Institute of Standards and Technology (NIST), Boulder, CO. In 1983 he transferred to the Electromagnetics Directorate of the Air Force Research Laboratory (AFRL), Hanscom Air Force Base, MA, where he was employed as a research scientist until 1996. In 1989 he took an eight-month leave of absence to accept a visiting professorship in the Electromagnetics Institute of the Technical University of Denmark. He presently works as an independent consultant in electromagnetics.

Dr. Yaghjian's research in electromagnetics has led to the development of exact, numerical, and high-frequency methods for predicting and measuring the near and far fields of antennas and scatterers. He has contributed to the determination of electromagnetic fields in continuous media, and to the formulation of classical equations of motion of charged particles.

Dr. Yaghjian is a Fellow of the IEEE, has served as an associate editor for the IEEE and URSI, and is a member of Sigma Xi. He has received best paper awards from the IEEE, NIST, and AFRL.